Cellular Ceramics

Edited by

M. Scheffler, P. Colombo

Cellular Ceramics

Structure, Manufacturing,
Properties and Applications

Edited by
Michael Scheffler, Paolo Colombo

WILEY-VCH Verlag GmbH & Co. KGaA

Editors

Dr. Michael Scheffler
Department of Materials Science & Engineering,
University of Washington
418 Roberts Hall, Box 352120
Seattle, WA 98195-2120, USA
e-mail: mscheff@gmx.de
Prof. Ing. Paolo Colombo
Dipartimento di Chimica Applicata
e Scienza dei Materiali,
Universita' di Bologna
V. le Risorgimento 2, 40136 Bologna, Italy
e-mail: paolo.colombo@unipd.it

Cover Picture
Top left: Periodic cellular structure. Colloidal inks were extruded by robotic deposition. Sub-millimeter filaments of extruded colloidal gel are deposited layer-by-layer to assemble the structure in the *z* stacking direction followed by drying and sintering. The white-colored *x*-*y*-*z* axes are 400 μm in length (Image courtesy of Prof. J. Lewis, University of Illinois; see also Chapter 2.3).

Bottom left: Hierarchically built porous material. Rattan palm wood was transformed into char and infiltrated at high temperature with liquid silicon retaining its cellular channel structure. The Si/SiC porous material was then used for hydrothermal zeolite crystallisation under partial transformation of the excess silicon. MFI type zeolite was formed in the longitudinal channels of the material. The open channel diameter is 300–320 μm and the zeolite layer is 40–60 μm (Image courtesy of Dr. F. Scheffler, University of Erlangen-Nuremberg, Germany; see Chapter 2.5 and Ref. [29] in Chapter 5.4).

Right: Prototype of a silicon carbide foam heater element. The electrical conductive ceramic foam heats up when electrical power is applied to top and bottom end. Here a power of 750 W was applied. The ceramic foam is 30 mm in diameter (Photo taken by Friedrich Weimer, Dresden. Image courtesy of J. Adler, Fraunhofer-IKTS, Dresden, Germany).

Library of Congress Card No.:
applied for

British Library Cataloguing-in-Publication Data
A catalogue record for this book is available from the British Library.

Bibliographic information published by Die Deutsche Bibliothek
Die Deutsche Bibliothek lists this publication in the Deutsche Nationalbibliografie; detailed bibliographic data is available in the Internet at <http://dnb.ddb.de>.

Printed in the Federal Republic of Germany.

Printed on acid-free paper.

Typesetting Kühn & Weyh, Satz und Medien, Freiburg
Printing Strauss GmbH, Mörlenbach
Bookbinding J. Schäffer GmbH i. G., Grünstadt

ISBN-13: 978-3-527-31320-4
ISBN-10: 3-527-31320-6

Foreword

For many years, the presence of porosity in ceramics was often seen to be problematic and a significant scientific effort was made to devise processing routes that produced ceramics with zero porosity. An exception to this philosophy was the refractory industry, in which it was understood that the presence of porosity is critical in controlling thermal conductivity. A sophisticated example of this concept was the development of refractory tiles for the thermal protection system of the Space Shuttle. In other branches of materials science, similar ideas were recognized. For example, rigid and flexible foams had been developed in polymer science and engineering. In these materials, porosity is controlled to optimize the elastic behavior and weight. In more recent times, scientific developments have touched on new areas such as biomimetics, in which scientists aim to duplicate natural structures. There has also been the push (and pull) to design materials and devices at smaller scale levels. Materials are becoming multifunctional with designed hierarchical structures, and porous ceramics can be seen in this light. The challenge now is for materials scientists to produce ceramics with porosity of any fraction, shape, and size. This also leads to new directions in the scientific understanding of porous structures and their properties. For the above reasons and my personal involvement in this field, I am pleased to see this new book on porous ceramics. This book takes a broad view of the field, while still allowing some detailed scientific aspects to be addressed. The book considers novel processing approaches, structure characterization, advances in understanding structure–property relationships and the challenges in all these areas. It is interesting to see the structural variety that forms the "pallette" for the materials scientist and the wide range of properties that are controlled by porosity and therefore require careful optimization. Finally, the book gives examples of technologies in which porous ceramics are being exploited and the demands that arise as products move to commercial use. I applaud the editors for their vision and the authors for sharing their insight. I wish you a successful outcome for your efforts.

David J. Green
State College, Pennsylvania, USA
October 29, 2004

Cellular Ceramics: Structure, Manufacturing, Properties and Applications.
Michael Scheffler, Paolo Colombo (Eds.)

ISBN: 3-527-31320-6

Contents

Cellular Ceramics: Structure, Manufacturing, Properties and Applications.
Michael Scheffler, Paolo Colombo (Eds.)

ISBN: 3-527-31320-6

Preface

Porosity in materials can be arranged in a well-defined and homogeneous manner or heterogeneously. It can be oriented, separated, or interconnected. From these possibilities pores of different shape, size, and interconnectivity arise. The three-dimensional assemblage of a large number of pores possessing a specific shape leads to a solid monolith displaying what can be termed a cellular structure.

A close analysis of materials found in nature reveals that most of them have a cellular structure and thus contain a significant amount of porosity, which plays a key role in optimizing their properties for a specific function. Indeed, Robert Hooke (1635–1703), a natural philosopher, experimental scientist, inventor, and architect, realized this in his investigations of the natural world and coined the term "cell" for describing the basic unit of the structure of cork, which reminded him of the cells of a monastery. In "Observation XVIII" of his book "Micrographia: or Some Physiological Descriptions of Minute Bodies Made by Magnifying Glasses with Observations and Inquiries Thereupon" (London: J. Martyn and J. Allestry, 1665), he wrote:

"... I could exceedingly plainly perceive it to be all perforated and porous, much like a Honey-comb, but that the pores of it were not regular ... these pores, or cells, ... were indeed the first microscopical pores I ever saw, and perhaps, that were ever seen, for I had not met with any Writer or Person, that had made any mention of them before this ..."

Similarly, with an updated pool of knowledge and equipped with higher resolution analytical instruments, 300 years or so later researchers around the globe are interested in investigating and exploiting the advantages and peculiarities of cellular materials. Indicators of the increasing importance of this field are the numerous international conferences devoted to all three classes of cellular materials (metals, plastics, and ceramics), special issues of various scientific journals, and a rising number of specific books discussing either cellular structures in general or, more specifically, cellular metals and cellular plastics, among them:

L.J. Gibson, M.F. Ashby, *Cellular Solids: Structure and Properties,* Cambridge University Press, 1999;
D.L. Weaire, *The Physics of Foams,* Oxford University Press, 2001;
S. Perkowitz, *Universal Foam: From Cappuccino to the Cosmos,* Walker & Co., New York, 2000;

Cellular Ceramics: Structure, Manufacturing, Properties and Applications.
Michael Scheffler, Paolo Colombo (Eds.)

ISBN: 3-527-31320-6

H.-P. Degischer, B. Kriszt (eds.), *Handbook of Cellular Metals: Production, Processing, Applications,* Wiley-VCH, Weinheim, 2002;
M.F. Ashby, A. Evans, N.A. Fleck, L.J. Gibson, J.W. Hutchinson, H.N.G. Wadley, *Metal Foams: A Design Guide,* Butterworth-Heinemann, Oxford, 2000;
S.-T Lee, N.S. Ramesh, *Polymeric Foams: Mechanisms and Materials,* CRC Press, Boca Raton, FL, 2004;
A.H. Landrock, *Handbook of Plastic Foams,* Noyes Publications, Park Ridge, NJ, 1995.

The reason for this considerable interest in cellular materials derives from the recognition that porosity affords further functionalities to a material, ranging from an increased surface area, to permeability, to the control of heat transport within the structure, to the maximization of the strength/density ratio.

An analysis of the published literature by searching just the terms "ceramic" and "foam" revealed an exponential increase in scientific papers and patents with a total of 26 publications in 1977, 64 in 1992, 133 in 1998, and 167 in 2004.

Books dealing with porous ceramics have also been published (e.g., R.W. Rice, *Porosity of Ceramics,* Marcel Dekker, New York, 1998), but no publication specifically concerning cellular ceramics was available yet. Thus, the idea was born to fill this gap with a focused book and to provide students, researchers, manufacturers, and users with a comprehensive discussion of the most relevant aspects of this topic, covering manufacturing processes, structure characterization, analysis of the properties/structure relationship, and examples of applications. As such, this book does not deal, on purpose, with all classes of porous ceramic materials, disregarding, for instance, membranes, zeolites, and low-porosity solids, for which excellent reviews and books are already available. It is also not a collection of publications deriving from a conference, but rather represents the contribution of specialists from academia and industry who are at the forefront of this innovative field. This book contains an updated set of references allowing the reader to gain further insight into specific issues of this fascinating class of advanced materials.

We are deeply grateful to the authors for their enthusiasm and willingness to contribute to this project and to the referees for their critical involvement in the peer-reviewing process. Dr. Jörn Ritterbusch (Wiley-VCH) deserves special recognition for displaying the necessary foresight for embracing this endeavor, and we are indebted to Heike Höpcke (Wiley-VCH) for her graceful and helpful assistance throughout the editorial process.

Finally, we appreciate the patience, support, and encouragement of our families, to which this book is dedicated.

Michael Scheffler
Seattle, WA, USA

Paolo Colombo
Bologna, Italy

19/11/2004

List of Contributors

Jörg Adler
Fraunhofer Institut Keramische
Technologien und Sinterwerkstoffe
Winterbergstr. 28
01277 Dresden
Germany

Michael F. Ashby
Cambridge University
Department of Engineering
Trumpington Street,
Cambridge, CB2 1PZ
United Kingdom

Enrico Bernardo
Dipartimento di Ingegneria
Meccanica – Settore Materiali
University of Padova
via Marzolo, 9
35131 Padova
Italy

John Binner
IPTME
Loughborough University,
Loughborough
LE11 3TU
UK

Aldo R. Boccaccini
Imperial College London
Department of Materials
South Kensington Campus
Prince Consort Road
London SW7 2BP
UK

Srinivasa Rao Boddapati
Dept. of Materials Science and
Engineering
Box No. 352120
University of Washington
Seattle, WA 98195
USA

Rajendra K. Bordia
Dept. of Materials Science and
Engineering
Box No. 352120
University of Washington
Seattle, WA 98195
USA

Ken Brakke
Mathematics Department
Susquehanna University
Selinsgrove, Pennsylvania
USA

Cellular Ceramics: Structure, Manufacturing, Properties and Applications.
Michael Scheffler, Paolo Colombo (Eds.)

ISBN: 3-527-31320-6

Giovanna Brusatin
Dipartimento di Ingegneria
Meccanica – Settore Materiali
University of Padova
via Marzolo, 9
35131 Padova
Italy

Peter Claus
Darmstadt University of Technology
Department of Chemistry
Ernst-Berl-Institute, Technical
Chemistry II
Petersenstrasse 20
64287 Darmstadt
Germany

Paolo Colombo
Department of Applied Chemistry and
Material Science
Universita' di Bologna
V.le Risorgimento, 2
40136 Bologna
Italy

Simon Cox
Department of Physics
Trinity College
Dublin
Ireland

Janet B. Davis
Rockwell Scientific
1049 Camino Dos Rios
Thousand Oaks, CA 91360
USA

Philip de Goey
Eindhoven University of Technology
Department of Mechanical
Engineering
PO Box 513
5600 MB Eindhoven
The Netherlands

Ann P. Dowling
Cambridge University
Engineering Department
Trumpington Street
Cambridge CB2 1PZ
UK

Iain D.J. Dupère
Cambridge University
Engineering Department
Trumpington Street
Cambridge CB2 1PZ
UK

Thomas Fend
German Aerospace Center
Institute of Technical Thermodynamics
Linder Höhe
51147 Köln
Germany

Debora Fino
Department of Materials Science and
Chemical Engineering
Politecnico di Torino
Corso Duca degli Abruzzi 24
10129 Torino
Italy

Michael Grutzeck
The Pennylvania State University 104
Materials Research Laboratory
University Park, PA 16802
USA

Bernhard Hoffschmidt
University of Applied Scienes Aachen
Solar-Institute Jülich
Aachen
Germany

Murilo Daniel de Mello Innocentini
Universidade de Ribeirão Preto
Rua Santa Cruz, 974, Centro
13560-680 São Carlos – SP
Brazil

Julian R. Jones
Imperial College London
Department of Materials
South Kensington Campus
Prince Consort Road
London SW7 2BP
UK

James Klett
Oak Ridge National Laboratory
P.O. Box 2008
Oak Ridge
TN 37931-6087
USA

Jennifer A. Lewis
University of Illinois
Department of Materials Science and Engineering
1304 W. Green Street
Urbana IL 61801
USA

Tian J. Lu
Cambridge University
Engineering Department
Trumpington Street
Cambridge CB2 1PZ
UK

Martin Lucas
Darmstadt University of Technology
Department of Chemistry
Ernst-Berl-Institute, Technical Chemistry II
Petersenstrasse 20
64287 Darmstadt
Germany

Jan Luyten
Flemish Institute for Technological Research
Materials Technology Boeretang 200
2400 Mol
Belgium

David B. Marshall
Rockwell Scientific
1049 Camino Dos Rios
Thousand Oaks
CA 91360
USA

Hans-Peter Martin
Fraunhofer Institut Keramische Technologien und Sinterwerkstoffe
Winterbergstr. 28
01277 Dresden
Germany

Luiz C. B. Martins
SELEE Corporation,
700 Shepherd St.
Hendersonville, NC 28792
USA

Steven Mullens
Flemish Institute for Technological Research
Materials Technology Boeretang 200
2400 Mol
Belgium

Andrew Norris
Vesuvius Hi-Tech Ceramics
PO Box 788
Alfred, NY 14802
USA

Rudolph A. Olson III
SELEE Corporation
700 Shepherd St.
Hendersonville, NC 28792
USA

Fernando dos Santos Ortega
Universidade do Vale do Paraíba
São José dos Campos, SP
Brazil

Olaf Pickenäcker
Röhm GmbH & Co. KG
Abt. VT-P
Kirschenallee
64293 Darmstadt
Germany

Robert Pitz-Paal
German Aerospace Center
Institute of Technical Thermodynamics
Linder Höhe
51147 Köln
Germany

Oliver Reutter
German Aerospace Center
Institute of Technical Thermodynamics
Linder Höhe
51147 Köln
Germany

Roy W. Rice
5411 Hopark Dr.
Alexandria, VA 22310-1109
USA

Anthony P. Roberts
Queensland University of Technology
2 George St.
GPO Box 2434
Brisbane, Qld 4001
Australia

Guido Saracco
Department of Materials Science and Chemical Engineering
Politecnico di Torino
Corso Duca degli Abruzzi 24
10129 Torino
Italy

Giovanni Scarinci
Dipartimento di Ingegneria Meccanica – Settore Materiali
University of Padova
via Marzolo, 9
35131 Padova
Italy

Franziska Scheffler
University of British Columbia
Department of Chemistry
2036 Main Wall
Vancouver, BC, V6T1Z1
Canada

Michael Scheffler
University of Washington
Department of Materials Science & Engineering
Box 352120
Seattle, WA 98195-2120
USA

Sabine Schimpf
Darmstadt University of Technology
Department of Chemistry
Ernst-Berl-Institute, Technical Chemistry II
Petersenstrasse 20
64287 Darmstadt
Germany

Koen Schreel
Eindhoven University of Technology
Department of Mechanical Engineering
PO Box 513
5600 MB Eindhoven
The Netherlands

Pilar Sepulveda
Engineering and Physical Sciences Research Council
Swindon
UK

Heino Sieber
University of Erlangen-Nürnberg
Materials Science (III), QSS Group
Martensstrasse 5
91058 Erlangen
Germany

Mrityunjay Singh
NASA Glenn Research Centre
Cleveland
Ohio
USA

James E. Smay
Oklahoma State University
School of Chemical Engineering
Stillwater, Oklahoma
USA

Edwin P. Stankiewicz
Ultramet
12173 Montague Street
Pacoima, CA 91331
USA

Dimosthenis Trimis
Institute of Fluid Mechanics and Combustion
University of Erlangen-Nuremberg
Cauerstr. 4
91058 Erlangen
Germany

Klemens Wawrzinek
enginion AG
Gustav-Meyer-Allee 25
13355 Berlin
Germany

Denis Weaire
Trinity College Dublin
Physics Department
Dublin 2
Ireland

Miroslaw Weclas
University of Applied Sciences
Institute of Vehicle Technology
Nürnberg
Germany

John Wight
Corning Inc.
SP-DV-01-09 Corning Inc.
Corning, NY 14831
USA

Juergen Zeschky
University of Erlangen-Nuremberg
Department of Materials Science
91058 Erlangen
Germany

Part 1
Introduction

Cellular Ceramics: Structure, Manufacturing, Properties and Applications.
Michael Scheffler, Paolo Colombo (Eds.)

ISBN: 3-527-31320-6

1.1
Cellular Solids – Scaling of Properties

Michael F. Ashby

1.1.1
Introduction

Cellular solids – ceramics, polymers, metals – have properties that depend on both topology and material. Of the three classes, polymer foams are the most widely investigated, and it is from these studies that much of the current understanding derives. Recent advances in techniques for foaming metals has led to their intense study, extending the understanding. Of the three classes, ceramic foams are the least well characterized. Their rapidly growing importance as filters, catalyst supports, membranes, and scaffolds for cell growth has stimulated much recent work, making this book both relevant and timely.

The underlying principles that influence cellular properties are common to all three classes. Three factors dominate (Fig. 1):

- The properties of the solid of which the foam is made.
- The topology (connectivity) and shape of the cells.
- The relative density $\tilde{\rho}/\rho_s$ of the foam, where $\tilde{\rho}$ is the density of the foam and ρ_s that of the solid of which it is made.

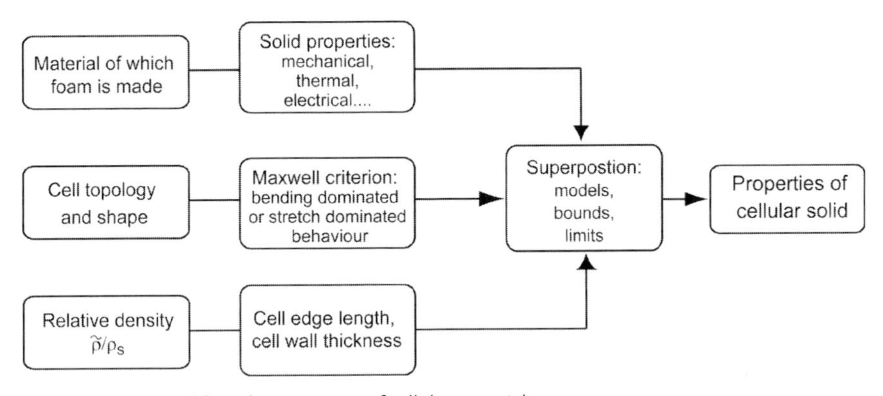

Fig. 1 Design variables. The properties of cellular materials depend on the material of the cell walls, the cell topology, and the relative density $\tilde{\rho}/\rho_s$. The contents of the boxes are explained in the text.

Cellular Ceramics: Structure, Manufacturing, Properties and Applications.
Michael Scheffler, Paolo Colombo (Eds.)

ISBN: 3-527-31320-6

This chapter summarizes these principles, providing an introduction to the more specialized chapters that follow.

1.1.2
Cellular or "Lattice" Materials

A *lattice* is a connected network of struts. In the language of structural engineering, a lattice truss or space frame means an array of struts, pin-jointed or rigidly bonded at their connections, usually made of one of the conventional materials of construction: wood, steel, or aluminum. Their purpose is to create stiff, strong, load-bearing structures using as little material as possible, or, where this is useful, to be as light as possible. The word "lattice" is used in other contexts: in the language of crystallography, for example, a lattice is a hypothetical grid of connected lines with three-dimensional translational symmetry. The intersections of the lines define the atom sites in the crystal; the unit cell and symmetry elements of the lattice characterize the crystal class.

Here we are concerned with lattice or *cellular materials*. Like the trusses and frames of the engineer, these are made up of a connected array of struts or plates, and like the crystal lattice, they are characterized by a typical cell with certain symmetry elements; some, but not all, have translational symmetry. But lattice materials differ from the lattices of the engineer in one important regard: that of scale. That of the unit cell of lattice materials is one of millimeters or micrometers, and it is this that allows them to be viewed both as structures and as materials. At one level, they can be analyzed by using classical methods of mechanics, just as any space frame is analyzed. But at another we must think of the lattice not only as a set of connected struts, but as a "material" in its own right, with its own set of effective properties, allowing direct comparison with those of monolithic materials.

Historically, *foams*, a particular subset of lattice-structured materials, were studied long before attention focused on lattices of other types. Early studies assumed that foam properties depended linearly on relative density $\tilde{\rho}/\rho_s$ (i.e., the volume fraction of solid in the material), but for most foams this is not so. A sound understanding of their mechanical properties began to emerge in the 1960s and 1970 with the work of Gent and Thomas [1] and Patel and Finnie [2]. Work since then has built a comprehensive understanding of mechanical, thermal, and electrical properties of foams, summarized in the texts "Cellular Solids" [3], "Metal Foams, a Design Guide" [4], and a number of conference proceedings [5–9]. The ideas have been applied with success to ceramic foams, notably by Green et al. [10–13], Gibson et al. [14–17], and Vedula et al. [18, 19].

The central findings of this body of research are summarized briefly in Section 1.1.3. One key finding is that the deformation of most foams, whether open- or closed-cell, is *bending-dominated* – a term that is explained more fully below. A consequence of this is that their stiffnesses and strength (at a given relative density) fall far below the levels that would be expected of *stretch-dominated* structures, typified by a fully triangulated lattice. To give an idea of the difference: a low-connectivity lattice, typified by a foam, with a relative density of 0.1 (meaning that the solid cell walls occupy 10 % of the volume) is less stiff by a factor of 10 than a stretch-dominated, triangulated lattice of the same relative density.

In this section we explore the significant features of both bending- and stretch-dominated structures, using dimensional methods to arrive at simple, approximate scaling laws for mechanical, thermal, and electrical properties. Later chapters deal with these in more detail; the merit of the method used here is that of retaining physical clarity and mathematical simplicity. The aim is to provide an overview, setting the scene for what is to come.

1.1.3 Bending-Dominated Structures

Figure 2 is an image of an open-cell foam. It typifies one class of lattice-structured material. It is made up of struts connected at joints, and the characteristic of this class is the low connectivity of the joints (the number of struts that meet there).

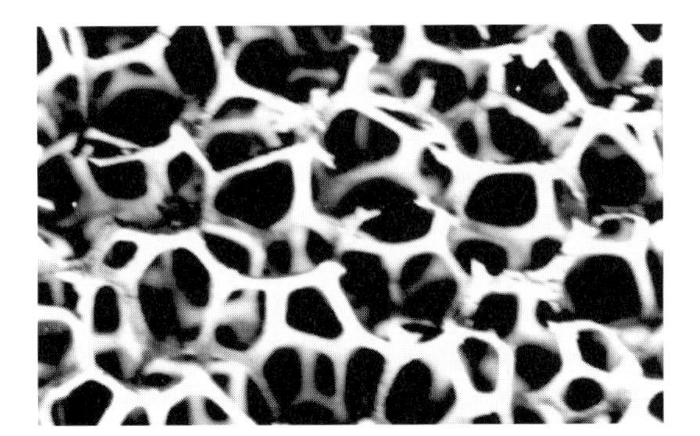

Fig. 2 A typical cellular structure. The topology of the cells causes the cell edges to bend when the structure is loaded. Even when the cells are closed, the deformation is predominantly bending because the thin cell faces buckle easily.

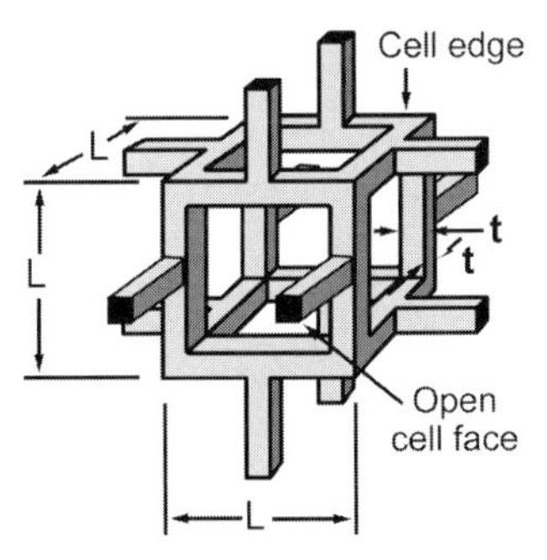

Fig. 3 An idealized cell in an open-cell foam.

Figure 3 is an idealization of a unit cell of the structure. It consists of solid struts surrounding a void space containing a gas or fluid. Lattice-structured materials (often called cellular solids) are characterized by their relative density, which for the structure shown here (with $t \ll L$) is

$$\frac{\tilde{\rho}}{\rho_s} \propto \left(\frac{t}{L}\right)^2 \tag{1}$$

where $\tilde{\rho}$ is the density of the foam, ρ_s is the density of the solid of which it is made, L is the cell size, and t is the thickness of the cell edges.

1.1.3.1
Mechanical Properties

Figure 4 shows the compressive stress–strain curve of bending-dominated lattice. The material is linear-elastic, with modulus $\tilde{E}$ up to its elastic limit, at which point the cell edges yield, buckle, or fracture. The structure continues to collapse at a

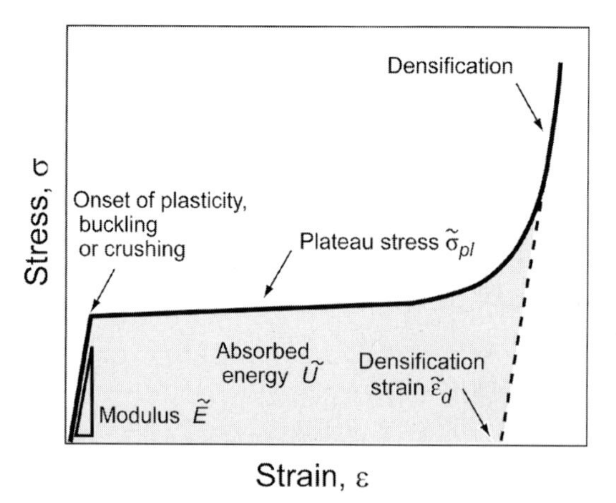

Fig. 4 Stress–strain curve of a cellular solid, showing the important parameters.

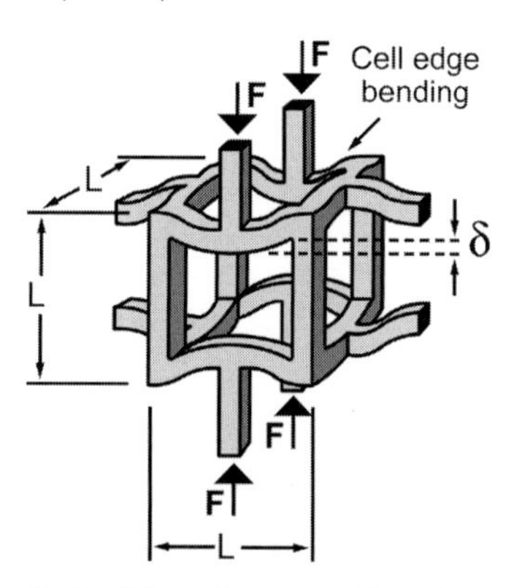

Fig. 5 When a low-connectivity structure is loaded, the cell edges bend, giving a low modulus.

nearly constant stress (plateau stress $\tilde{\sigma}_{pl}$) until opposite sides of the cells impinge (densification strain $\tilde{\varepsilon}_d$), when the stress rises steeply. The three possible collapse mechanisms compete; the one that requires the lowest stress wins. The mechanical properties are calculated in the ways developed below, details of which can be found in Ref. [3].

A remote compressive stress σ exerts a force $F \propto \sigma L^2$ on the cell edges, causing them to bend, as shown in Fig. 5, with bending deflection δ. A strut of length L, loaded at its midpoint by a force F, deflects by a distance δ

$$\delta \propto \frac{FL^3}{E_s I} \tag{2}$$

where E_s is the modulus of the solid of which the strut is made, and $I = t^4/12$ the second moment of area of the cell edge of square cross section $t \times t$. The compressive strain suffered by the cell as a whole is then $\varepsilon \propto 2\delta/L$. Assembling these results gives the modulus $\tilde{E} = \sigma/\varepsilon$ of the foam as

$$\frac{\tilde{E}}{E_s} \propto \left(\frac{\tilde{\rho}}{\rho_s}\right)^2 \quad \text{(bending-dominated behavior).} \tag{3}$$

Since $\tilde{E} = E_s$ when $\tilde{\rho} = \rho_s$, we expect the constant of proportionality to be close to unity – a speculation confirmed both by experiment and by numerical simulation.

A similar approach can be used to model the collapse load, and thus the plateau stress of the structure. The cell walls yield, as shown in Fig. 6, when the force exerted on them exceeds their fully plastic moment

$$M_f = \frac{\sigma_{y,s}\, t^3}{4} \tag{4}$$

where $\sigma_{y,s}$ is the yield strength of the solid of which the foam is made. This moment is related to the remote stress by $M \propto FL \propto \sigma L^3$. Assembling these results gives the failure strength $\tilde{\sigma}_{pl}$:

$$\frac{\tilde{\sigma}_{pl}}{\sigma_{y,s}} \propto \left(\frac{\tilde{\rho}}{\rho_s}\right)^{3/2} \quad \text{(bending-dominated behavior).} \tag{5}$$

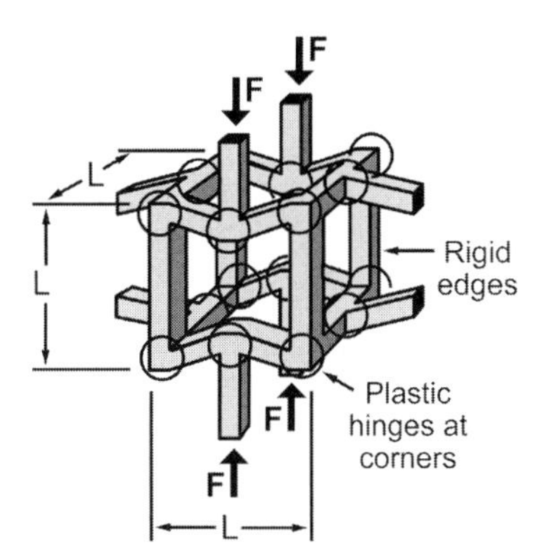

Fig. 6 Foams made of ductile materials collapse by plastic bending of the cell edges.

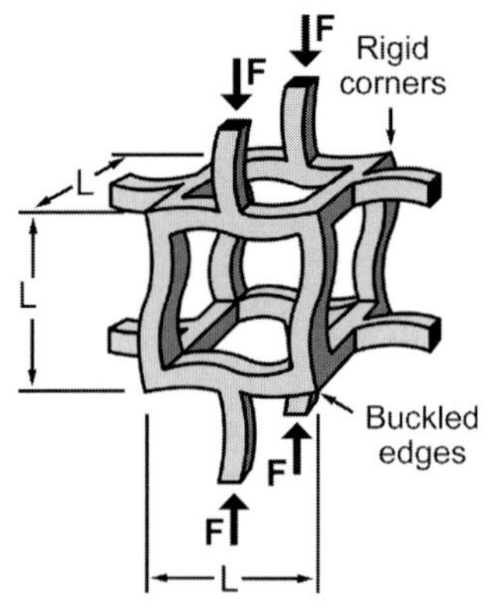

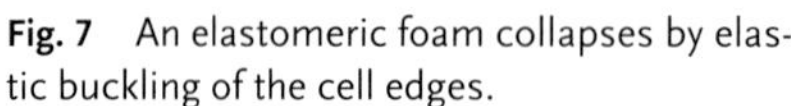

Fig. 7 An elastomeric foam collapses by elastic buckling of the cell edges.

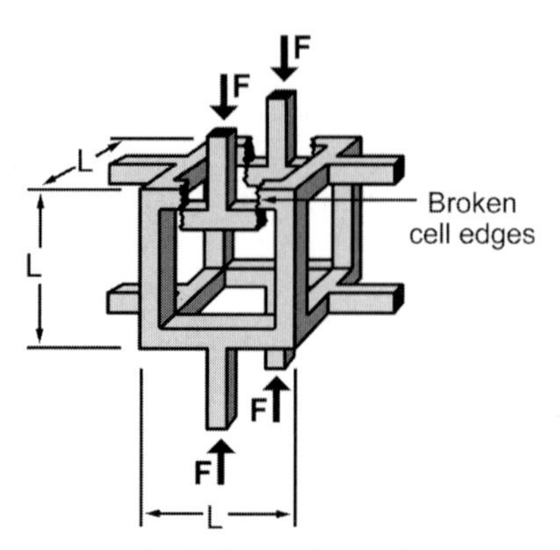

Fig. 8 A brittle foam collapses by successive fracturing of the cell edges. Ceramic foams generally show this collapse mechanism.

The constant of proportionality has been established both by experiment and by numerical computation; its value is approximately 0.3.

Elastomeric foams collapse not by yielding but by elastic buckling, and brittle foams by cell-wall fracture (Figs. 7 and 8). As with plastic collapse, simple scaling laws describe this behavior well. A strut of length L buckles under a compressive load F_b, the Euler buckling load, where

$$F_b \propto \frac{E_s I}{L^2} \propto \frac{E_s t^4}{L^2}. \tag{6}$$

Since $F = \sigma F^2$, the stress that causes the foam to collapse by elastic buckling $\tilde{\sigma}_{el}$ scales as

$$\frac{\tilde{\sigma}_{el}}{E_s} \propto \left(\frac{\tilde{\rho}}{\rho_s}\right)^2 \quad \text{(buckling-dominated behavior).} \tag{7}$$

More sophisticated modeling gives the constant of proportionality as 0.05. Cell walls fracture when the bending moment exceeds that given by Eq. (4) with $\sigma_{y,s}$ replaced by $\sigma_{cr,s}$, the modulus of rupture of a strut. The crushing stress therefore scales in the same way as the plastic collapse stress, giving

$$\frac{\tilde{\sigma}_{cr}}{\sigma_{cr,s}} \propto \left(\frac{\tilde{\rho}}{\rho_s}\right)^{3/2} \quad \text{(fracturing-dominated behavior)} \tag{8}$$

with a constant of proportionality of about 0.2.

Densification when the stress rises steeply is a purely geometric effect: the opposite sides of the cells are forced into contact and further bending or buckling is not possible. If we think of compression as a strain-induced increase in relative density, then simple geometry gives the densification strain $\tilde{\varepsilon}_d$ as

$$\tilde{\varepsilon}_d = 1 - \left(\frac{\tilde{\rho}}{\rho_s}\right) \Big/ \left(\frac{\rho_{crit}}{\rho_s}\right) \tag{9}$$

where ρ_{crit}/ρ_s is the relative density at which the structure locks up. Experiments broadly support this estimate, and indicate a value for the lock up density as $\rho_{crit}/\rho_s \approx 0.6$.

Foamlike lattices are often used for cushioning, packaging, or to protect against impact, by utilizing the long, flat plateau of their stress–strain curves. The useful energy that they can absorb per unit volume $\tilde{U}$ (Fig. 4) is approximated by

$$\tilde{U} \approx \tilde{\sigma}_{pl}\,\tilde{\varepsilon}_d \tag{10}$$

where $\tilde{\sigma}_{pl}$ is the plateau stress – the yield, buckling, or fracturing strength of Equations (6), (7), or (8), whichever is least.

This bending-dominated behavior is not limited to open-cell foams with structures like that of Fig. 2. Most closed-cell foams also follow these scaling laws, at first sight an unexpected result because the cell faces must carry membrane stresses when the foam is loaded, and these should lead to a linear dependence of both stiffness and strength on relative density. The explanation lies in the fact that the cell faces are very thin; they buckle or rupture at stresses so low that their contribution to stiffness and strength is small, and the cell edges carry most of the load.

1.1.3.2 Thermal Properties

Cellular solids have useful heat-transfer properties. The cells are sufficiently small that convection of the gas within them is usually suppressed. Heat transfer through the lattice is then the sum of that conducted though the struts and that through the still air (or other fluid) contained in the cells. On average one-third of the struts lie parallel to each axis, and this suggests that the conductivity might be described by

$$\tilde{\lambda} = \frac{1}{3}\left(\frac{\tilde{\rho}}{\rho_s}\right)\lambda_s + \left(1 - \left(\frac{\tilde{\rho}}{\rho_s}\right)\right)\lambda_g. \tag{11}$$

Here the first term on the right describes conduction through the solid cell walls and edges (conductivity λ_s) and the second that through the gas contained in the cells (conductivity λ_g; for dry air $\lambda_g = 0.025\ \mathrm{W\ m^{-1}\ K^{-1}}$). This is an adequate approximation for very low density foams, but it obviously breaks down as $\tilde{\rho}/\rho_s$ approaches unity. This is because joints are shared by the struts, and as $\tilde{\rho}/\rho_s$ rises, the joints occupy a larger and larger fraction of the volume. This volume scales as t^3/L^3 or, via Eq. (1), as $(\tilde{\rho}/\rho_s)^{3/2}$, so we need an additional term to allow for this:

$$\tilde{\lambda} = \frac{1}{3}\left(\left(\frac{\tilde{\rho}}{\rho_s}\right) + 2\left(\frac{\tilde{\rho}}{\rho_s}\right)^{3/2}\right)\lambda_s + \left(1 - \left(\frac{\tilde{\rho}}{\rho_s}\right)\right)\lambda_g \tag{12}$$

which now correctly reduces to $\tilde{\lambda} = \lambda_s$ at $\tilde{\rho} = \rho_s$. The term associated with the gas, often negligible, becomes important in foams intended for thermal insulation, since these have a low relative density and a conductivity approaching λ_g.

The thermal diffusivities of lattice structures scale in a different way. Thermal diffusivity is defined as

$$a = \frac{\lambda}{\rho C_p} \tag{13}$$

where C_p is the specific heat in J kg^{-1} K^{-1}. The specific heat $\tilde{C}_p$ of a cellular structure is the same as that of the solid of which it is made (because of its units). Thus, neglecting for simplicity any conductivity through the gas, we find the thermal diffusivity $\tilde{a}$ to be

$$\tilde{a} = \frac{\tilde{\lambda}}{\tilde{\rho}\,\tilde{C}_p} \approx \frac{1}{3}\left(1 + 2\left(\frac{\tilde{\rho}}{\rho_s}\right)^{1/2}\right)\frac{\lambda_s}{\rho_s\, C_{p,s}}. \tag{14}$$

a surprising result, since it is almost independent of relative density.

The thermal expansion coefficient of a cellular material is less interesting: it is the same as that of the solid from which it is made.

1.1.3.3
Electrical Properties

Insulating lattices are attractive as structural materials with low dielectric constant, which falls towards 1 (the value for air or vacuum) as the relative density decreases:

$$\tilde{\varepsilon} = 1 + (\varepsilon_s - 1)\left(\frac{\tilde{\rho}}{\rho_s}\right) \tag{15}$$

where ε_s is the dielectric constant of the solid of which the cell walls are made. Those that conduct have electrical conductivities that follow the same scaling law as the thermal conductivity (Eq. (11) with thermal conductivities replaced by electrical conductivities); here the conductivity of the gas can usually be ignored.

1.1.4
Maxwell's Stability Criterion

If lattice-structure materials with low strut connectivity, like those of Figs. 2 and 3, have low stiffness because the configuration of their cell edges allows them to bend, might it not be possible to devise other configurations in which the cell edges are made to stretch instead? This thinking leads to the idea of *microtruss lattice structures.* To understand these we need Maxwell's stability criterion, a deceptively simple yet profoundly fundamental rule [20].

The condition for a pin-jointed frame (i.e., one that is hinged at its joints) made up of b struts and j frictionless joints, like those in Fig. 9, to be both statically and kinematically determinate (i.e., it is rigid and does not fold up when loaded) in two dimensions, is:

$$M = b - 2j + 3 = 0. \tag{16}$$

In three dimensions the equivalent equation is

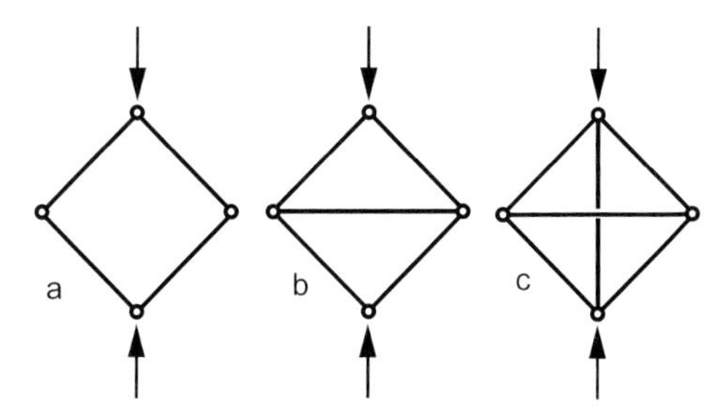

Fig. 9 The pin-jointed frame in a) folds up when loaded – it is a mechanism. If its joints are welded together the struts bend (as in Fig. 5) – it becomes a *bending-dominated structure*. The triangulated frame in b) is stiff when loaded because the transverse strut carries tension – it is a *stretch-dominated structure*. The frame in c) is over-constrained; if the horizontal bar is shortened, the vertical one is put into tension even when no external loads are applied (giving a state of self-stress).

$$M = b - 3j + 6 = 0. \tag{17}$$

If $M < 0$, as in Fig. 9a, the frame is a *mechanism*; it has one or more degrees of freedom, and – in the directions that these allow displacements – it has no stiffness or strength. If its joints are locked (as they are in the lattice structures that concern us here) the bars of the frame *bend* when the structure is loaded, as in Fig. 5. If, instead, $M = 0$, as in Fig. 9b, the frame ceases to be a mechanism. If it is loaded, its members carry tension or compression (even when pin-jointed), and it becomes a *stretch-dominated* structure. Locking the joints now makes little difference because slender structures are much stiffer when stretched than when bent. There is an underlying principle here: the structural efficiency of stretch-dominated structures is high; that of bending-dominated structures is low.

Fig. 9c introduces a further concept, that of self-stress. It is a structure with $M > 0$. If the vertical strut is shortened, it pulls the other struts into compression, which is balanced by the tension it carries. The struts carry stress even though there are no external loads on the structure. The criteria of Eqs. (14) and (15) are necessary conditions for rigidity, but are not in general sufficient conditions, as they do not account for the possibility of states of self-stress and mechanisms. A generalization of the Maxwell rule in three dimensions is given by Calladine [21]:

$$M = b - 3j + 6 = s - m \tag{18}$$

where s and m are the number of states of self-stress and of mechanisms, respectively. Each can be determined by finding the rank of the equilibrium matrix that describes the frame in a full structural analysis [22]. A *just-rigid* framework (a lattice that is both statically and kinematically determinate) has $s = m = 0$. The nature of Maxwell's rule as a necessary rather than sufficient condition is made clear by examination of Eq. (16): vanishing of the left-hand side only implies that the number of mechanisms and states of self-stress are equal, not that each equals zero.

Maxwell's criterion gives insight into the design of lattice materials, and reveals why foams are almost always bending-dominated [23–25]. Examples of some idealized cell shapes are shown in Fig. 10. Isolated cells that satisfy Maxwell's criterion and are rigid are labeled "YES", while "NO" means the Maxwell condition is not satisfied and that the cell is a mechanism. It is generally assumed that the best model for a cell in a foam approximates a space-filling shape. However, none of the space-filling shapes (indicated by numbers 2, 3, 4, 6, and 8) are rigid. In fact no single space-filling polyhedral cell has $M \geq 0$. Space-filling combinations of cell shapes, by contrast, exist that have $M \geq 0$; for example, the tetrahedron and octahedron in combination fill space to form a rigid framework.

Maxwell's criterion gives a prescription for designing stretch-dominated lattices, which we now examine.

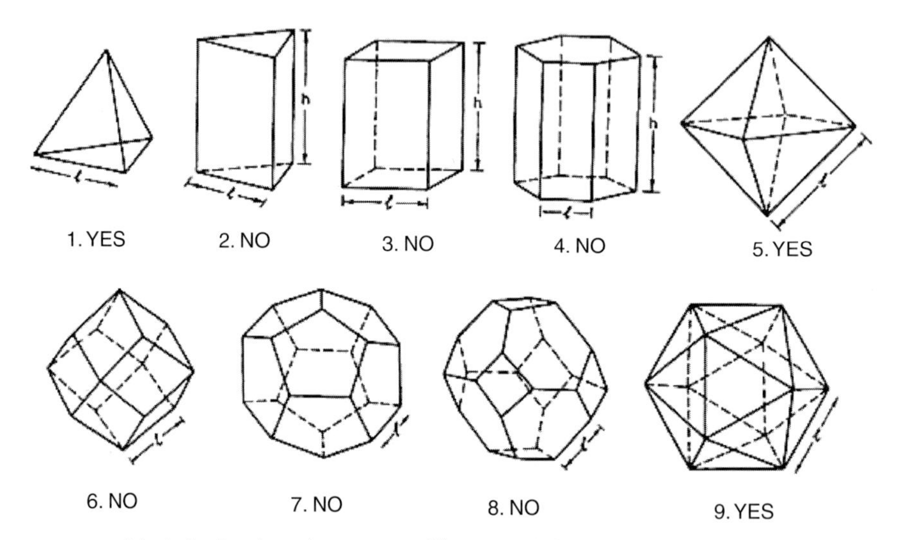

Fig. 10 Polyhedral cells. Those that are space filling (2–4, 6, 8) all have $M < 0$, that is, they are bending-dominated structures.

1.1.5
Stretch-Dominated Structures

Figure 11 shows an example of a microtruss lattice structure. For this structure $M = 18$; it has no mechanism and many possible states of self-stress. It is one of many structures for which $M \geq 0$, and its mechanical response is stretch-dominated. In this section we review briefly the properties of stretch-dominated microtruss lattice materials, using the same approach as that of Section 1.1.3.

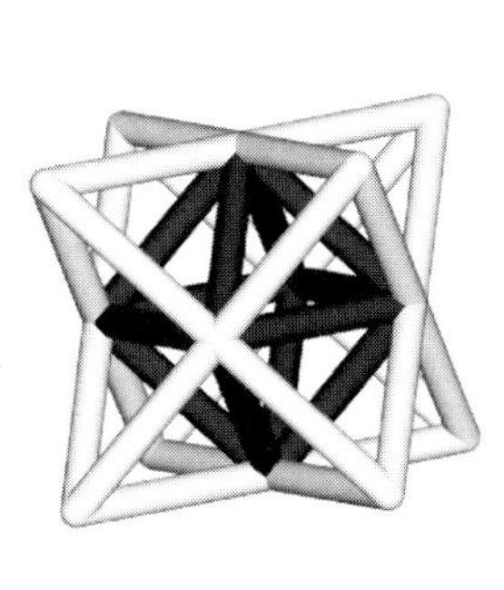
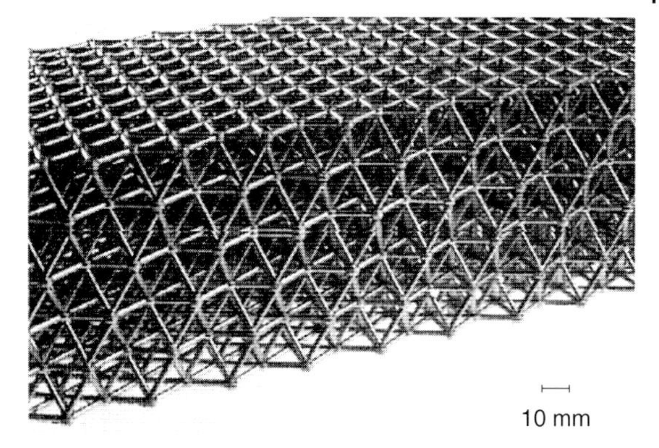

Fig. 11 A microtruss structure with $M > 0$, together with its unit cell.

Consider the tensile loading of the material. Since it has no mechanisms, the structure first responds by elastic stretching of the struts. On average one-third of its struts carry tension when the structure is loaded in simple tension, regardless of the loading direction. Thus

$$\frac{\tilde{E}}{E_s} \approx \frac{1}{3}\left(\frac{\tilde{\rho}}{\rho_s}\right) \quad \text{(stretch-dominated behavior).} \tag{19}$$

The elastic limit is reached when one or more sets of struts yields plastically, buckles, or fractures; the mechanism with the lowest collapse load determines the strength of the structure as a whole. If the struts are plastic, the collapse stress – by the same argument as before – is

$$\frac{\tilde{\sigma}_{pl}}{\sigma_{y,s}} \approx \frac{1}{3}\left(\frac{\tilde{\rho}}{\rho_s}\right) \quad \text{(plastic stretch-dominated behavior).} \tag{20}$$

This is an upper bound since it assumes that the struts yield in tension or compression when the structure is loaded. If the struts are slender, they may buckle before they yield. Then, following the same reasoning that led to Eq. (7), the "buckling strength" scales as

$$\frac{\tilde{\sigma}_{el}}{E_s} \propto \left(\frac{\tilde{\rho}}{\rho_s}\right)^2 \quad \text{(buckling-dominated behavior).} \tag{21}$$

The only difference is the magnitude of the constant of proportionality, which depends on the details of the connectivity of the strut. But remembering that buckling of a strut depends most importantly on its slenderness t/L, and that this is directly related to relative density, we do not expect the dependence on configuration to be strong. In practice elastomeric foams always fail by buckling; rigid polymer and metallic foams buckle before they yield when $\tilde{\rho}/\rho_s \leq 0.05$ and $\tilde{\rho}/\rho_s \leq 0.01$, respectively.

Failure can also occur by strut fracture. A lattice structure made from a ceramic or other brittle solid will collapse when the struts start to break. Stretch domination means that the struts carrying tension will fail first. Following the argument that led to Eq. (18) we anticipate a collapse stress $\tilde{\sigma}_{cr}$ that scales as

$$\frac{\tilde{\sigma}_{cr}}{\sigma_{cr,s}} \propto \left(\frac{\tilde{\rho}}{\rho_s}\right) \text{ (stretch-fracture-dominated behavior)} \quad (22)$$

where $\sigma_{cr,s}$ is now the tensile fracture strength of the material of a strut. Here the constant of proportionality is less certain. Brittle fracture is a stochastic process, dependent on the presence and distribution of defects in the struts. Depending on the width of this distribution, the failure of the first strut may or may not trigger the failure of the whole.

The main thing to be learnt from these results is that both the modulus and initial collapse strength of a stretching-dominated lattice are much greater than those of a bending-dominated cellular material of the same relative density. This makes stretch-dominated cellular solids the best choice for lightweight structural applications. But because the mechanisms of deformation now involve "hard" modes (tension, compression) rather than "soft" ones (bending), initial yield is followed by plastic buckling or brittle collapse of the struts, which leads to post-yield softening (Fig. 12). This makes them less good for energy-absorbing applications, which require a stress–strain curve with a long, flat plateau. This post-yield regime ends and the stress rises steeply at the densification strain, given, as before, by Eq. (9).

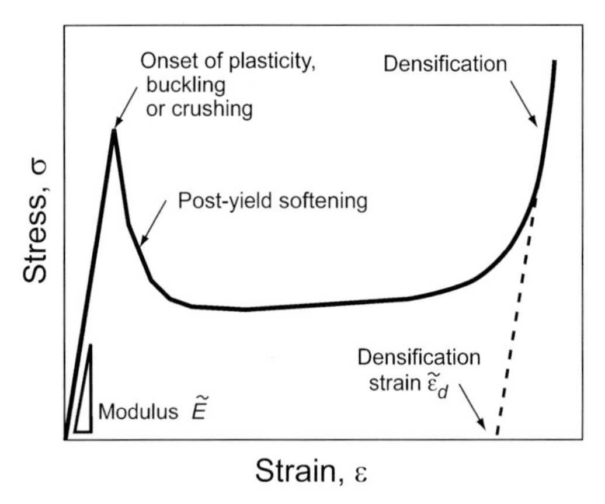

Fig. 12 Schematic stress–strain curve for a stretch-dominated structure. It has high stiffness and high initial strength, but can show post-yield softening.

These results are summarized in Figs. 13 and 14, in which the relative modulus $\tilde{E}/E_s$ and strength $\tilde{\sigma}/\sigma_s$ are plotted against relative density $\tilde{\rho}/\rho_s$. They show the envelopes within which the currently-researched cellular structures lie. In Fig. 13, the two broken lines show the locus of relative stiffness as the relative density changes for ideal stretch- and bending-dominated lattices made of the material lying at the point (1,1). Stretch-dominated, prismatic microstructures have moduli that scale as $\tilde{\rho}/\rho_s$ (slope 1); bending-dominated, cellular microstructures have moduli that scale as $(\tilde{\rho}/\rho_s)^2$ (slope 2). *Honeycombs*, a prime choice as cores for sandwich panels and as supports for exhaust catalysts, are extraordinarily efficient; if loaded precisely parallel to the axis of the hexagons, they lie on the "ideal-stretch" line. In directions normal to this they are exceptionally compliant. *Foams*, available in a wide range of densities, epitomize bending-dominated behavior. If ideal, their relative moduli would lie along the lower broken line. Many do, but some fall below. This is because of the way they are made [4]; their structure is often heterogeneous, strong in some places, weak in others; the weak regions drag down both stiffness and strength. *Woven structures* are lattices made by three-dimensional weaving of wires; at present these are synthesized by brazing stacks of two-dimensional wire meshes, giving configurations that are relatively dense and have essentially ideal bending-dominated properties. There is potential for efficient low-density lattices here; it requires the ability to weave three-dimensional meshes. *Pyramidal lattices* have struts configured as if along the edges and base of a pyramid – Fig. 11 is an example. They are fully triangulated and show stretch-dominated properties but lie by a factor of 3 below the ideal line. *Kagome lattices* – the name derives from that of Japanese weaves – are more efficient; they offer the lowest mass-to-stiffness ratio.

Strength (Fig. 14) has much in common with stiffness, but there are some differences. The "ideals" are again shown as broken lines. Stretch-dominated, prismatic microstructures have strengths that scale as $\tilde{\rho}/\rho_s$ (slope 1); bending dominated scale as $(\tilde{\rho}/\rho_s)^{3/2}$ (slope 1.5). Honeycombs, even when compressed parallel to the hexagon

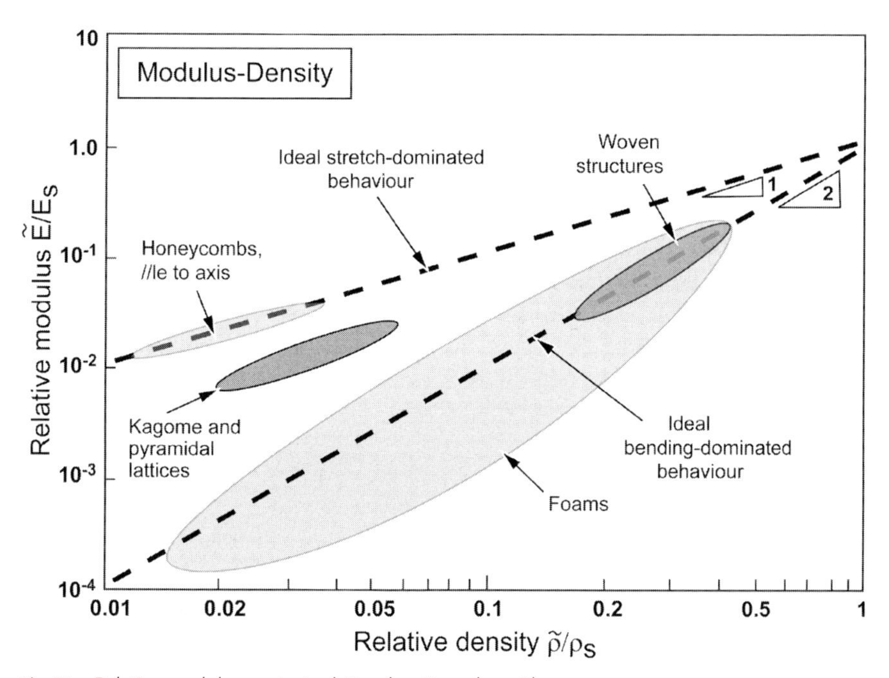

Fig. 13 Relative modulus against relative density on logarithmic scales for cellular structures with different topologies. Bending-dominated structures lie along a trajectory of slope 2, and stretch-dominated structures along a line of slope 1.

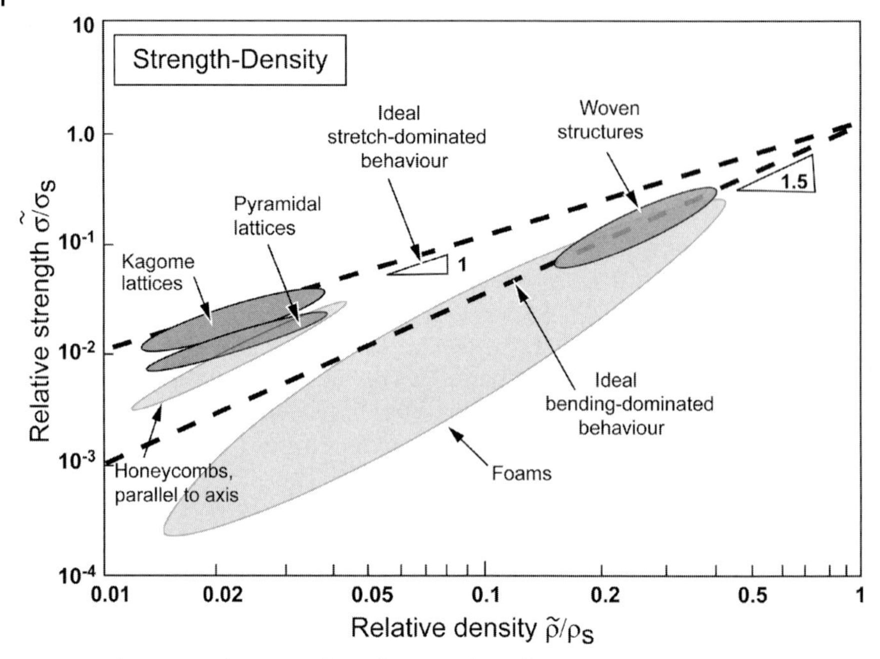

Fig. 14 Relative strength against relative density on logarithmic scales for cellular structures with different topologies. Bending-dominated structures lie along a trajectory of slope 1.5, and stretch-dominated structures along a line of slope 1.

axis, fall below the ideal because the thin cell walls buckle easily. Metallic foams, similarly underperform – none reach the ideal bending-dominated performance line, a consequence of their imperfections. The current generation of woven structures lies on the bending-dominated ideal. As with stiffness, pyramidal and Kagome lattices offer near-ideal stretch-dominated performance.

The bending/stretching distinction influences mechanical properties profoundly, but has no effect on *thermal* or *electrical properties*. At the approximate level we seek in this overview, they are adequately described by Eqs. (11)–(13).

1.1.6
Summary

Structural engineers have known and used latticelike structures for generations, but it is only in the last 20 years that an understanding of *materials* with a lattice-like structure has emerged. Many of these respond to stress in precisely the way engineers seek to avoid – by bending deformation of the struts that make up the structure. As materials, these are interesting for their low stiffness and strength, and the large strains they can accommodate – properties that are attractive in cushioning, packaging, and energy absorption and in accommodating thermal shock. But if stiffness and strength at low weight are sought, the lattice must be configured in such a way that bending is prevented, leaving strut-stretching as the dominant mode of deformation. This suggests the possibility of a family of microtruss structured materials, many as yet unexplored.

With this introduction, we are ready to explore the design and characterisation of ceramic cellular materials in more detail in the chapters that follow.

Acknowledgements

Many people have contributed to the ideas reported in this chapter. I particularly wish to recognize the contributions of Profs. L.J. Gibson, N.A. Fleck, A.G. Evans, J.W. Hutchinson, and H.N.G. Wadley, the fruits of long collaborations.

References

1 Gent, A.N., Thomas, A.G. *J. Appl. Polym. Sci.* **1959**, *1*, 107.
2 Patel, M.R., Finnie, I. *J. Mater.* **1970**, *5*, 909
3 Gibson, L.J., Ashby, M.F. *Cellular Solids, Structure and Properties*, 2nd ed., Cambridge University Press, Cambridge, UK, 1997.
4 Ashby, M.F., Evans, A.G., Fleck, N.A., Gibson, L.J., Hutchinson, J.W., Wadley, H.N.G. *Metal Foams: A Design Guide*, Butterworth Heinemann, Oxford, UK., 2000.
5 Banhart, J. (Ed.) *Metallschäume*, MIT Verlag, Bremen, Germany, 1997.
6 Banhart, J., Ashby, M.F., Fleck, N.A. (Eds.) *Metal Foams and Foam Metal Structures*, Proc. Int. Conf. Metfoam '99, MIT Verlag, Bremen, Germany, 1999.
7 Banhart, J., Ashby, M.F., Fleck, N.A. (Eds.) *Metal Foams and Foam Metal Structures*, Proc. Int. Conf. Metfoam '01, MIT Verlag, Bremen, Germany, 2001.
8 Banhart J., Fleck, N.A. (Eds), *Metal Foams and Foam Metal Structures*, Proc. Int. Conf. Metfoam '03, MIT Verlag, Bremen, Germany, 2003.
9 Shwartz, D.S., Shih, D.S., Evans, A.G., Wadley, H.N.G. (Eds.) *Porous and Cellular Materials for Structural Applications*, Materials Research Society Proceedings Vol. 521, MRS, Warrendale, PA, USA, 1998.
10 Brezny, R., Green, D.J. *J. Am. Ceram. Soc.* **1989**, *72*, 1145–1152.
11 Brezny, R., Green, D.J. *Acta Metall. Mater.* **1990**, *38*, 2517–2526.
12 Brezny, R., Green, D.J. *J. Am. Ceram. Soc.* **1991**, *74*, 1061–1065.
13 Nanjangud. S.C., Brezny, R., Green, D.J. *J. Am. Ceram. Soc.* **1995**, *78*, 266–268.
14 Huang, J.S., Gibson, L.J. *Acta Metall. Mater.* **1991**, *39*, 1617–1626.
15 Huang, J.S., Gibson, L.J. *Acta Metall. Mater.* **1991**, *39*, 1627–1636.
16 Huang, J.S., Gibson, L.J. *J. Mater. Sci. Lett.* **1993**, *12*, 602–604.
17 Triantafillou, T.C., Gibson, L.J. *Int. J. Mech. Sci.* **1990**, *32*, 479–496.
18 Vedula, V.R., Green, D.J., Hellman, J.R. *J. Eur. Ceram. Soc.* **1998**, *18*, 2073–2080.
19 Vedula, V.R., Green, D.J., Hellman, J.R., Segall, A.E. *J. Mater. Sci.* **1998**, *33*, 5427–5432.
20 Maxwell, J.C. *Philos. Mag.* **1864**, *27*, 294.
21 Calladine, C.R. *Theory of Shell Structures*, Cambridge University Press, Cambridge, UK, 1983.
22 Pellegrino, S., Calladine, C.R. *Int. J. Solids Struct.* **1986**, *22*(4), 409.
23 Deshpande, V.S., Ashby, M.F., Fleck, N.A. *Acta Metall. Mater.* **2001**, *49*, 1035–1040.
24 Deshpande, V.S., Fleck, N.A., Ashby, M.F. *J. Mech. Phys. Solids* **2001**.
25 Guest, S.D. *Philos. Trans. R. Soc. Lond. A*, **2000**, *358*, 229–243.

1.2
Liquid Foams – Precursors for Solid Foams

Denis Weaire, Simon Cox, and Ken Brakke

1.2.1
The Structure of a Liquid Foam

In 1873 the blind Belgian physicist Joseph Plateau (Fig. 1), Professor at the University of Ghent, published his life's work, "Experimental and Theoretical Statics of Liquids Subject to Molecular Forces Only" [1]. Translated in this way the title is clumsy in English, and it is elusive in French. The book is mainly based upon a series of papers in the Memoirs of the Belgian Academy which referred to "figures of equilibrium of a liquid mass without weight". In both cases a direct reference to effects of surface tension as the main object of enquiry might have been more helpful.

Fig. 1 Joseph Plateau, the blind Belgian physicist who established the basic geometrical and topological laws of foam equilibrium.

The book deserves its frequent citation as the standard historical reference on the geometric structure of a liquid foam. Plateau's name is often invoked for the attribution of basic geometrical and topological laws of equilibrium, which he established experimentally (with theoretical underpinning from his mathematical colleague Ernest Lamarle). But the scope of Plateau's work is really much wider. It included topics considered "advanced" today, such as rupture mechanisms, surface viscosity,

Cellular Ceramics: Structure, Manufacturing, Properties and Applications.
Michael Scheffler, Paolo Colombo (Eds.)

ISBN: 3-527-31320-6

and the Marangoni effect. His work is not matched in its thoroughness by any modern text.

The books of Lawrence [2] and Mysels et al. [3], from the 1900s, are largely concerned with individual films or bubbles. The chemical industry sponsored extensive research, usually aimed at understanding the role of surfactants. The book of Bikerman [4] gives the flavor of that work. Generally speaking, the essential simplicity of Plateau's vision was lost in this application-driven research. Inspired by C.S. Smith [5], various physicists turned their attention to the subject in the 1980s. Drawing on the best of the work in chemical engineering, including that of Princen (see his review article [6]) and Lemlich (e.g., [7]), they began to build a systematic theory of foam properties. The recent *Physics of Foams* [8] is a broad introduction, mostly concentrating on liquid foams, based on two decades of progress towards that goal. It concludes with a figure (Fig. 2) which has proven useful as a conceptual map of our present stage of knowledge. This is intended to convey the existence of successful theory and experiment for dry, static foams, while a host of interesting and awkward problems are posed by wet and/or deforming foams. By *wet* we imply a relatively high value of the liquid fraction Φ_l. Other more recent works include those of Isenburg [9], which describes the mathematics of soap films and bubbles, and of Exerowa and Kruglyakov [10], which provides in particular a detailed discussion of the physical chemistry of soap films.

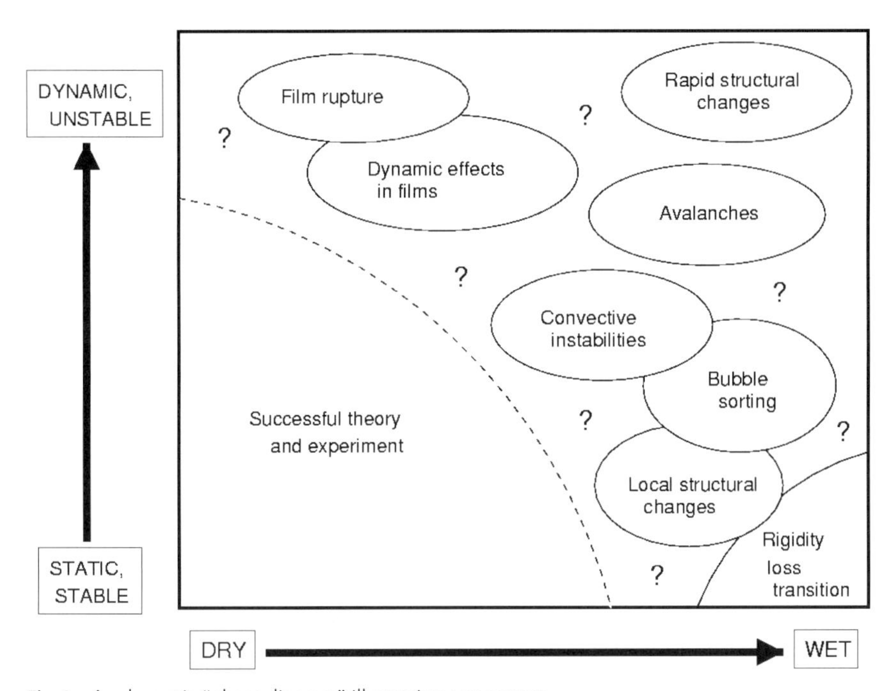

Fig. 2 A schematic "phase-diagram" illustrating our current understanding of liquid foams, extending from dry to wet structures and from static to dynamic processes.

A well-developed science of dry, static (or quasistatic) foams is a valuable first step. Most common foams are quite dry when in equilibrium under gravity, and we can understand most of their properties when they are in this condition. These include the essential structure, as Plateau described it long ago, the statistics associated with that structure, its coarsening with time due to gas diffusion, its elastic and plastic properties, and the drainage of liquid through it when equilibrium is disturbed.

The completion of this picture by the pursuit of dynamic effects and high liquid fractions is underway. Experiments are even performed in the microgravity environment of space, where homogeneous wet foams can exist in equilibrium [11].

Part of the motivation of the subject lies in its relation to solid foams. Most of these have a liquid foam as a precursor, so that a solid foam is usually a frozen (and perhaps processed) liquid foam. As such, it retains all or some of the characteristics of the pre-existing liquid structure. That elegant form has advantages for some purposes. It is, for example, close to the optimum for electrical or thermal conductivity [12]. For other properties, such as mechanical strength, it is more of a mixed blessing.

Many examples of such solid foams are to be found in this book. Others include the metallic foams that have begun to attract serious commercial interest, and biological foams such as bone and cork. At this point we must also draw attention to the distinction between open- and closed-cell foams. The former have two continuous phases, while in the latter one of the phases is separated into discrete units. In both cases, the structure is similar: open-cell foams can be thought of as solid closed-cell foams with the thin films removed. This is precisely how they are made in many cases.

In the case of ceramics, such open-cell foams are commonly made by replication of a polymer foam. On the other hand, closed-cell foams may be formed by direct foaming of a ceramic slurry, followed by solidification. These processes are explained in Chapter 2.1.

So a good starting point for understanding such foams and striving for innovation in their fabrication is a study of the liquid foam structure from which they are derived.

The favored tool for approaching the subject by simulation is Surface Evolver [13]. This consists of software expressly designed for modeling soap bubbles, foams, and other liquid surfaces shaped by minimizing energy (such as surface tension and gravity) and subject to various constraints (such as bubble volumes and fixed frames). The surface is represented as a collection of triangular tiles. The complicated topologies found in foams are routinely handled. In particular, Surface Evolver can deal with the topological changes encountered during foam coarsening and quasistatic flow. It provides interactive 3D graphics and an extensive command language. Surface Evolver is freely available from http://www.susqu.edu/brakke/evolver/ and is regularly updated. With the help of this software, realistic structures representing foams in equilibrium can be created and used for a variety of purposes in the physics of both liquid foams and their solidified counterparts. We used it to create many of the illustrations that follow.

1.2.2
The Elements of Liquid Foam Structure

Foam can be analyzed in terms of distinct interrelated structural elements (bubbles, films, Plateau borders, and junctions) with precise and simple geometry. This elegant structure is illustrated beautifully in the images of the photographer-artist Michael Boran, one of which is shown in Fig. 3.

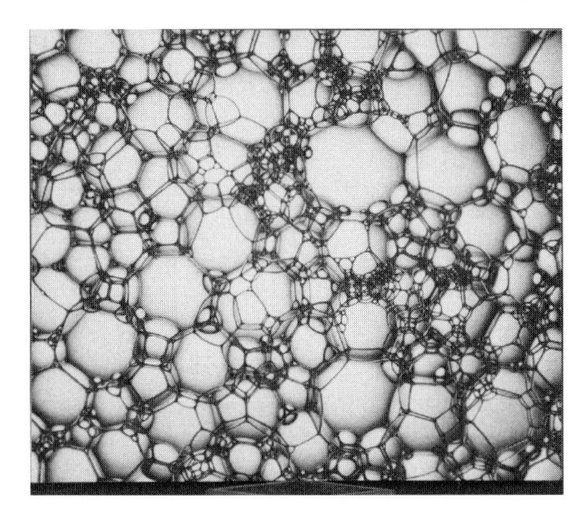

Fig. 3 The structure of a liquid foam as seen by the photographer-artist Michael Boran.

The bubbles which are pressed together to form the foam are separated by thin *films*. Although these are the most evident feature of the foam structure, they play only an incidental role in many of its properties. (One exception is stability, since foams collapse because of film rupture.) Where films meet along a line or curve, there is a liquid-filled interstitial channel called a *Plateau border*. Where several Plateau borders meet to form an interconnected network, they do so at a *junction*.

Hence, each bubble may be represented as a polyhedral shape, such that faces are identified with films, edges with Plateau borders, and junctions with vertices. Plateau's equilibrium rules apply to a dry foam (a liquid fraction of $\Phi_l = 0$ or, in practice, much less than 0.01) at equilibrium. They are:

- Only *three* films may meet at a Plateau border, and they meet at equal angles (Fig. 4a).
- Only *four* Plateau borders may meet at a junction, and they meet at equal angles (Fig. 4b).

These have been proved most rigorously in ref. [14].

To these rules we can add the Laplace–Young law, which states that the pressure difference Δp across a film is proportional to its mean curvature κ

$$\Delta p = 4\gamma\kappa \quad (1)$$

where γ is the surface tension of a single air–liquid interface: the surface tension of the film is 2γ. The mean curvature κ is the mean of the reciprocals of the radii of curvature in two orthogonal directions

$$\kappa = \frac{1}{2}\left(\frac{1}{r_1} + \frac{1}{r_2}\right). \tag{2}$$

This law dictates, in particular, that the film must have curvatures consistent with the pressures in adjacent bubbles. If we look closely at a single surface of a Plateau border the same law applies, with an extra factor of one-half, and it implies that they have negative pressure with respect to the adjacent bubbles, as in Fig. 4c and d.

A foam can have any liquid fraction in a range from zero to about 36 %, the *wet limit*, at which the bubbles come apart. This is the limit appropriate to a typical disordered foam, being related to the dense random packing of hard spheres with a relative density of 64 %. At both the wet and dry limits the structure is relatively simple. Figure 5 shows simulated foams with $\Phi_l \approx 0.5$ and 5 %. In the middle of this range of liquid fraction, however, the structure is much more difficult to analyze and its properties remain somewhat obscure. Moreover, real foams are disordered, as indicated in Fig. 6. Beyond a liquid fraction of a few percent, multiple junctions may form, instead of the tetrahedral ones which are idealized in the confluence of four lines in the dry limit. At that limit such multiple junctions are explicitly forbidden by Plateau's rules, but they can become stable for finite liquid fraction, as illustrated in Fig. 7.

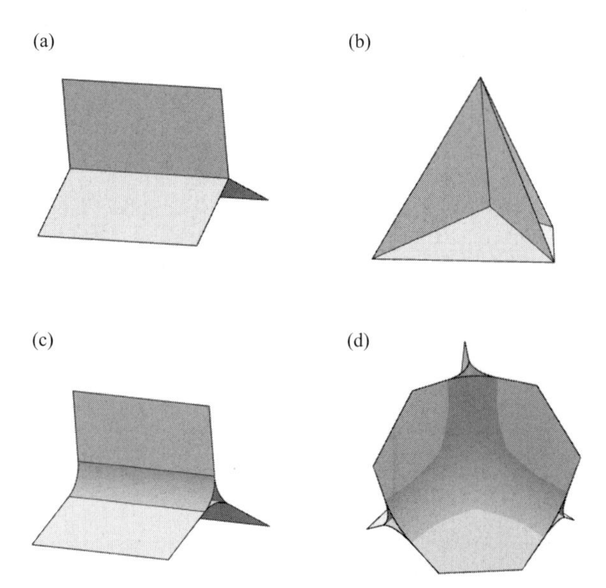

Fig. 4 The elements of a foam, illustrating Plateau's laws. In the dry limit, three films meet in a single Plateau border (a) and four Plateau borders meet in a single junction (b). This junction is formed at the meeting point of six films in a tetrahedral-shaped wire frame, such as Plateau used to discover the laws which bear his name. When liquid is added to a foam, the Plateau borders (c) and junctions (d) become swollen, meeting the films with zero contact angle. These images were generated with Surface Evolver.

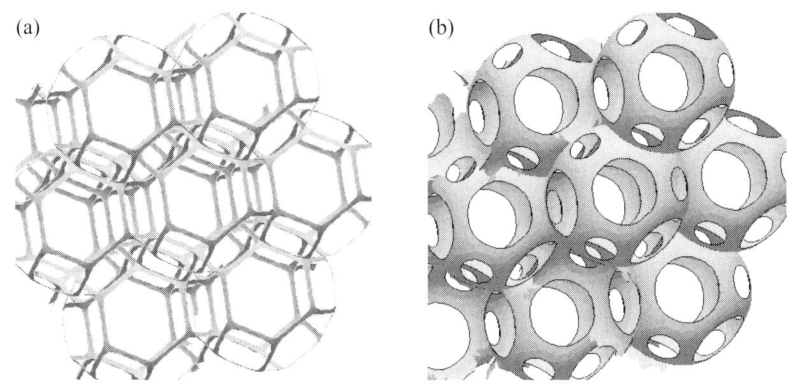

Fig. 5 A simulated wet foam with liquid fraction of a) 0.5 and b) 5 % for the case of the tetrakaidecahedral structure proposed by Lord Kelvin for the structure of the ether. The films have been removed from this Surface Evolver calculation, giving the impression of an open-celled rather than a closed-cell foam.

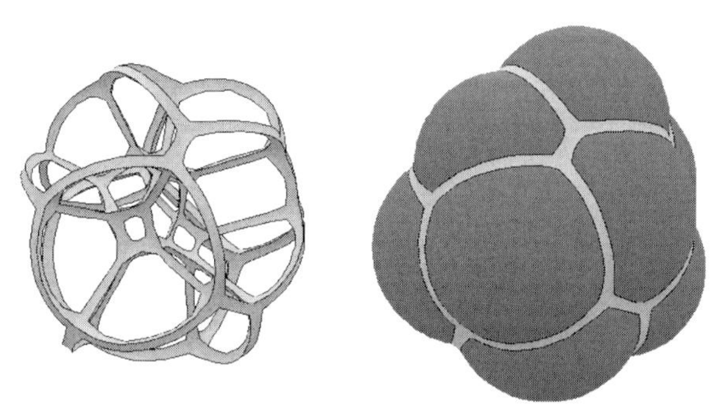

Fig. 6 A finite, wet, polydisperse, closed-cell foam, such as this simulated cluster of bubbles, reveals a complex liquid network when the films are removed.

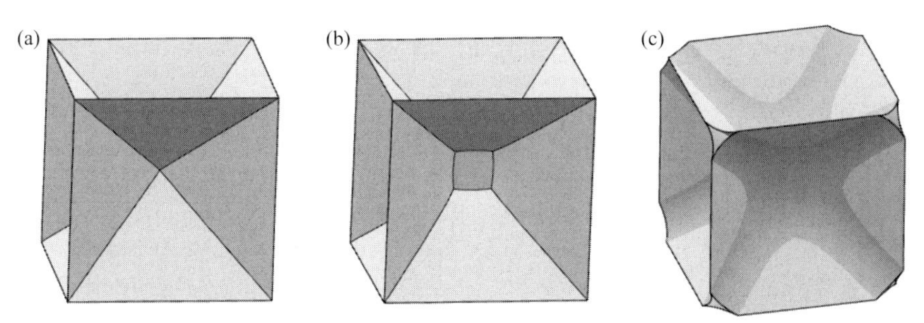

Fig. 7 Higher order junctions are only possible when the structure is sufficiently wet, as seen in these Surface Evolver simulations. In a cubic wire frame, such as that used by Plateau, the dry eightfold symmetric vertex (a) is unstable: the vertex dissociates into four fourfold junctions (b). When sufficient liquid is added, however, the eightfold vertex (c) can be stable.

1.2.3
Real Liquid Foams

Real liquid foams are generally only stable if they contain surfactants. Their most important effect is not simply that of lowering the surface tension, although it is closely related. Rather, it is the dynamical properties of the surface tension which serve to stabilize the films. The surface tension γ increases whenever a film is suddenly stretched locally, and this results in increased forces which oppose that tendency.

Also of importance for the stability of the films is the so-called *disjoining pressure*, which expresses the mutually repulsive force between the two faces of a film. It opposes further thinning once the film thickness has been reduced to a value at which this repulsive force is significant. As the liquid pressure decreases to a large, negative, value (relative to atmospheric pressure) the Plateau borders shrink and increase in curvature in accordance with the Laplace–Young law, but the films do not thin indefinitely. The film thickness at which equilibrium is achieved is therefore determined by the balance between the disjoining pressure and the bulk pressure of the liquid.

Up to a point all this can be ignored and the films assumed to be stable and of negligible thickness in accounting for the total liquid content or such properties as conductivity. Some of the simplicity of the various formulae which express properties in terms of liquid fraction [8] arises from such an assumption.

At the leading edge of the subject, the films are coming back into play. And there can be no doubting their primary importance in relation to stability.

1.2.4
Quasistatic Processes

A quasistatic process in a foam is one in which the relaxation of the structure back to equilibrium is much faster than the timescale on which the foam is perturbed. In this limit, we can use Surface Evolver to probe the slow dynamics of foams. Topics of interest in this area fall roughly into three categories: gas diffusion, liquid motion, and bubble motion.

Coarsening is the gradual change of the foam structure due to gas diffusion through the films. This diffusion is driven by the pressure differences between bubbles and can therefore be related to the curvature of the films by the Laplace–Young law. Small bubbles have high pressure, so they lose gas and disappear. Thus, the average bubble size increases with time, that is, the foam coarsens, typically on a timescale of about 30 min [15]. In two dimensions an exact law [16] states that the rate of change of area of a bubble depends only upon its number of sides.

In three dimensions there is no such exact result. It is known that the growth of a bubble does not depend only upon its number of faces, although this is a reasonable approximation for most cases, as evidenced by theory, simulation, and experiment [17–19].

Whenever there is a finite amount of liquid in the foam, it is subject to gravity-driven *drainage*, unless it has already come into equilibrium under gravity [20]. In this state, there is a vertical profile of liquid fraction, related to a variation of the local pressure in the liquid according to the hydrostatic pressure law, necessary for equilibrium under gravity. More generally, the liquid moves through the Plateau borders and junctions of the static foam structure (and to a lesser extent the films) in response to gravity and local pressure variations. An understanding of this process, particularly at the limit of low liquid fraction, has been aided by the development of nonlinear foam drainage equations [21–23].

The most striking manifestation of drainage is the solitary wave front which propagates down through the foam when liquid is added to the top at constant flow rate (Fig. 8). This is known as *forced drainage*. The velocity of the front is constant to a good approximation, and easily observed. This simple measurement is a good starting point in exploring drainage and its dependence on the nature of surfactants [24].

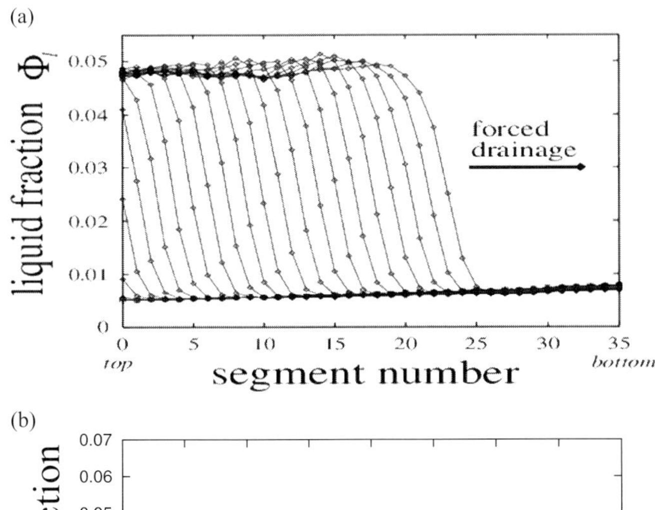

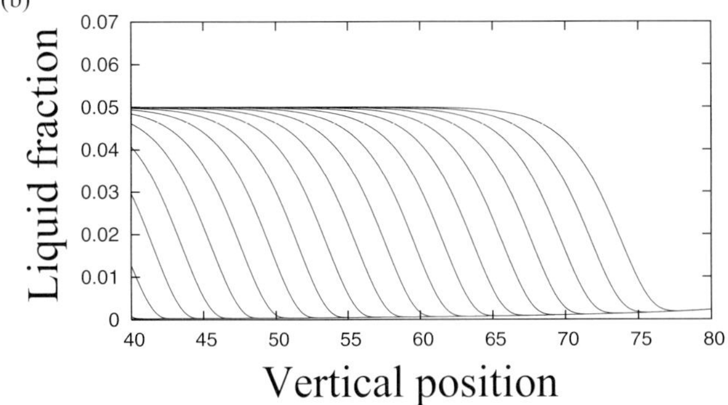

Fig. 8 a) The forced-drainage experiment, in which a solitary wave is observed to descend through the foam, has greatly improved our understanding of liquid drainage through foams. In this case the position of the wave front was measured by means of conductivity. b) The numerical solution of a foam drainage equation allows us to verify the agreement between theory and experiment.

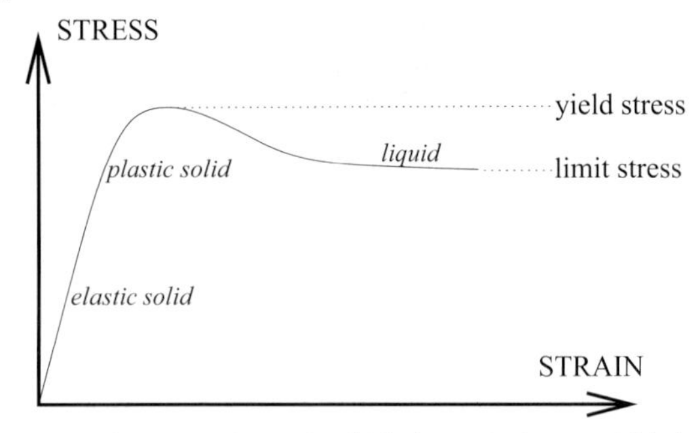

Fig. 9 In an experiment in which the strain is slowly increased, the stress first increases linearly. The slope is the elastic shear modulus. This increase continues until the plastic limit, at which the first topological changes occur and the foam yields. The system shows hysteresis in the sense that once the foam has yielded, the stress drops to a lower limit stress characterized by further intermittent topological changes and liquidlike behavior (flow). In practice, the difference between the limit stress and the yield stress is small, and both depend significantly upon the strain rate.

In all of the above, the bubbles themselves essentially remained stationary. But recently the deformation and flow (*rheology*) of the foam, in which the bubbles are rearranged, has been a focus of experiment and theoretical debate [25]. It has long been known that a liquid foam has a well-defined shear modulus, proportional to surface tension and many orders of magnitude less than its bulk modulus (which is essentially that of the enclosed gas). Moreover, large shear strains may be imposed before substantial rearrangements (topological changes) are incurred, leading to plastic behavior, a yield stress, and so on. This is summarized in Figure 9, which schematically illustrates the results of an experiment in which a foam is sheared. In the elastic regime, the structure deforms without topological changes. At the yield stress, topological changes occur and the plastic regime is entered. Finally, the topological changes lead to flow.

Beyond this simple and well-understood picture, much remains to be explored as regards strain-rate-dependent effects, localization of shear (shear banding), and so on. Further details can be found in a review [26] and recent references [27–30].

1.2.5 Beyond Quasistatics

To understand the role of viscosity, for example, or the effects of combinations of the above processes, we must progress beyond the quasistatic picture.

In the latter category, one area of interest is *convective instability,* which occurs when liquid is added to a dry foam at a high flow rate. Drainage causes the foam to become wet enough that it overcomes its yield stress and begins to move, as illustrated in Fig. 10. There are at least two possible modes of instability in which the foam circulates [31–33]. To predict parameters such as the critical flow rate for the onset of motion as a function of bubble size, we must combine theories of drainage and rheology, a challenge yet to be met.

For the rheology of real foams, we must in some sense understand the effects of viscous drag on the structure. Our attempts to model this in two dimensions [34] require a modification of the Laplace–Young law (Eq. 1) to the following form

$$\Delta p = 2\gamma\kappa + \lambda v \tag{3}$$

where v is the normal velocity of a film, and all viscous effects are included in the parameter λ. (The factor of 2 rather than 4 arises because we are in two dimensions.) This *viscous froth model* shows good qualitative agreement with experiment, but has yet to be extended to three dimensions. It is no more than a tentative start to a difficult problem.

Three-dimensional modeling of viscous effects is essential to a full understanding of the formation of solid foam, whether consisting of ceramics, metals, or plastics. Although we have a good understanding of the static properties of foams [35], the effects of rheology on a solidifying foam remain a challenge, despite being of great interest. There has undoubtedly been a good deal of work done on this in the polyurethane industry in particular, but much of it seems to have been proprietary. There is also extensive literature in the domain of chemical engineering for viscous effects in ordinary liquid foams (e.g., [6, 37, 38]). It is largely empirically based, and this makes it difficult to assess its validity. Most of the subject of foam rheology remains in the terra incognita depicted in Fig. 2.

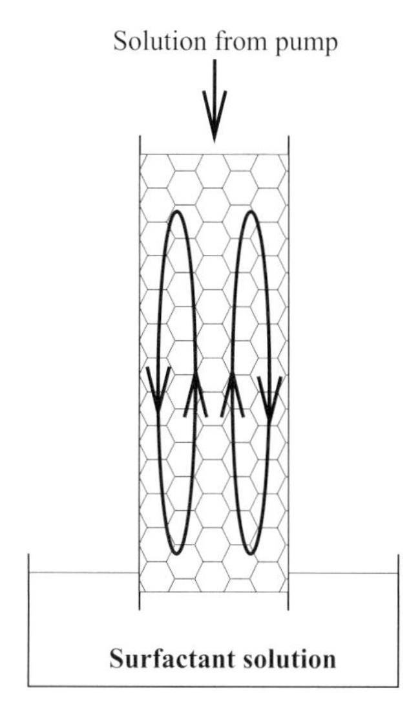

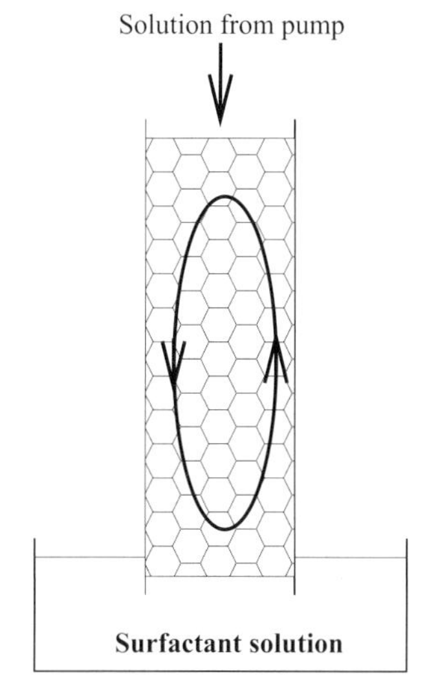

Fig. 10 A forced drainage experiment at high flow-rate leads to one of a number of convective instabilities, in which the foam itself is continuously deformed. Two possibilities are illustrated here: the first is a cylindrically symmetric motion in which bubbles in the center of the foam rise and those on the outside descend, while the second is a solid-body rotation.

1.2.6 Summary

The equilibrium structure of a foam is an elegant and well-defined arrangement of films, Plateau borders, and junctions. This structure is largely independent of the material from which the foam is constructed and its method of production. As ever more diverse materials are foamed with many different applications in mind, it should be remembered that the 100-year-old laws of Plateau are still essential in describing the structure of the foam.

In the static state, our knowledge of the behavior of both solid and liquid foams is good, particularly at the dry limit. Moreover, this understanding extends to many properties of a liquid foam's quasistatic motion. We now seek to extend theories to cope with wet foams in dynamic situations. A promising step in this direction is the use of *microgravity* experiments aboard parabolic flights, sounding rockets, and potentially the International Space Station. The current generation of these experiments aims to study the drainage and rheology of wet foams and the formation and collapse of metallic foams in the absence of gravity.

Acknowledgements

D.W. and S.J.C. acknowledge financial assistance from the European Space Agency. K.B. thanks Trinity College Dublin for hospitality during a sabbatical period.

References

1 Plateau, J.A.F. *Statique Expérimentale et Théorique des Liquides Soumis aux Seules Forces Moléculaires.* Gauthier-Villars, Paris, **1873**.
2 Lawrence, A.S.C. *Soap Films.* Bell, London, **1929**.
3 Mysels, K.J., Shinoda, K., Frankel, S. *Soap Films: Studies of their Thinning.* Pergamon, New York, **1959**.
4 Bikerman, J.J. *Foams: Theory and Industrial Applications.* Reinhold, New York, **1953**.
5 Smith, C.S. *Sci. Am.* **1954**, *190*, 58–64.
6 Princen, H.M. in Sjöblom, J. (ed.) Encyclopedic Handbook of Emulsion Technology, Marcel Dekker, New York, **2000**, pp. 243–278.
7 Leonard, R.A., Lemlich, R. *A.I.Ch.E. J.* **1965**, *11*, 18–29.
8 Weaire, D., Hutzler, S. *The Physics of Foams.* Clarendon Press, Oxford, **1999**.
9 Isenberg, C. *The Science of Soap Films and Soap Bubbles.* Dover, New York, **1992**.
10 Exerowa, D., Kruglyakov, P.M. *Foam and Foam Films.* Elsevier, Amsterdam, **1998**.
11 Passerone, A., Weaire, D. In *A World Without Gravity (SP-1251)*, ESA, The Netherlands, **2001**, pp. 241–253.
12 Durand, M., Sadoc, J.-F., Weaire, D. *Proc. R. Soc. London Ser. A* **2004**, *460*, 1269–1285.
13 Brakke, K. *Exp. Math.* **1992**, *1*, 141–165.
14 Taylor, J.E. *Ann. Math.* **1976**, *103*, 489–539.
15 Glazier, J.A., Gross, S.P., Stavans, *J. Phys. Rev. A* **1987**, *36*, 306–312.
16 von Neumann, J. in *Metal Interfaces*, American Society for Metals, Cleveland, **1952**, pp 108–110.
17 Hilgenfeldt, S., Kraynik, A.M., Koehler, S.A., Stone, H.A. *Phys. Rev. Lett.* **2001**, *86*, 2685–2689.
18 Cox, S.J., Graner, F. *Phys. Rev. E* **2004**, *69*, 031409.
19 Monnereau, C., Vignes-Adler, M. *Phys. Rev. Lett.* **1998**, *80*, 5228–5231.

20 Weaire, D., Hutzler, S., Verbist, G., Peters, E. *Adv. Chem. Phys.* **1997**, *102*, 315–374.
21 Goldfarb, I.I., Kann, K.B., Schreiber, I.R. *Fluid Dynamics* **1988**, *23*, 244–249.
22 Verbist, G., Weaire, D., Kraynik, A.M. *J. Phys. Condensed Matter* **1996**, *8*, 3715–3731.
23 Koehler, S.A., Hilgenfeldt, S., Stone, H.A. *Phys. Rev. Lett.* **1999**, *82*, 4232–4235.
24 Saint-Jalmes, A., Langevin, D. *J. Phys. Condens. Matter* **2002**, *14*, 9397–9412.
25 Kraynik, A.M. *Ann. Rev. Fluid Mech.* **1988**, *20*, 325–357.
26 Weaire, D., Fortes, M.A. *Adv. Phys.* **1994**, *43*, 685–738.
27 Debregeas, G., Tabuteau, T., di Meglio, J.M. *Phys. Rev. Lett.* **2001**, *87*, 178305.
28 Lauridsen, J., Twardos, M., Dennin, M. *Phys. Rev. Lett.* **2002**, *89*, 098303.
29 Gopal, A.D., Durian, D.J. *Phys. Rev. Lett.* **2003**, *91*, 188303.
30 Cohen-Addad, S., Höhler, R., Khidas, Y. *Phys. Rev. Lett.* **2004**, *93*, 028302.
31 Hutzler, S., Weaire, D., Crawford, R. *Europhysics Lett.* **1998**, *41*, 461–465.
32 Vera, M.U., Saint-Jalmes, A., Durian, D.J. *Phys. Rev. Lett.* **2000**, *84*, 3001–3004.
33 Weaire, D., Hutzler, S., Cox, S., Kern, N., Alonso, M.D., Drenckhan, W. *J. Phys. Condens. Matter* **2003**, *15*, S65–S73.
34 Kern, N., Weaire, D., Martin, A., Hutzler, S., Cox, S.J. *Phys. Rev. E* **2004**, *70*, 041411.
35 Kraynik, A., Reinelt, D., van Swol, F. *Phys. Rev. E* **2003**, *67*, 031403.
36 Reinelt, D.A., Kraynik, A.M. *J. Colloid Interf. Sci.* **1989**, *132*, 491–503.
37 Neethling, S.J., Cilliers, J.J. *Minerals Eng.* **1998**, *11*, 1035–1046.
38 Gardiner, B.S., Dlugogorski, B.Z., Jameson, G.J. *Ind. Eng. Chem. Res.* **1999**, *38*, 1099–1106.

Part 2
Manufacturing

Cellular Ceramics: Structure, Manufacturing, Properties and Applications.
Michael Scheffler, Paolo Colombo (Eds.)

ISBN: 3-527-31320-6

2.1
Ceramic Foams

Jon Binner

2.1.1
Introduction

Over the past few years there has been a significant increase in interest in the production and use of highly porous ceramic materials. This is associated mainly with the properties such materials offer, such as high surface area, high permeability, low density, low specific heat, and high thermal insulation. These characteristics are essential for technological applications such as catalyst supports, filters for molten metals, hot gases, and ion exchange, refractory linings for furnaces, thermal protection systems, heat exchangers, and as porous implants in the area of biomaterials. Cell size, morphology, and degree of interconnectedness are also important factors that influence potential applications for these materials. Predominantly closed-cell materials are needed for thermal insulation, while open-cell, interconnected materials are required for uses involving fluid transport such as filters and catalysts.

The different properties required of cellular ceramics mean that a range of processing routes is needed to manufacture them; no one route is sufficiently flexible to yield all the necessary structures [1]. This has led to a wide variety of routes being developed and patented in many countries around the world; however, they can be crudely divided into three categories with a series of variations on the basic themes. One of the oldest approaches is based on the replication of polymer foams by applying a ceramic slurry that is dried in place prior to the polymer template's being burnt out and the ceramic sintered. While this leads to very open, reticulated foams, burning out of the polymer leaves hollow and damaged struts that can reduce the mechanical properties of the final foam significantly. Despite this, these foams are manufactured in large quantities and used extensively in industry, often as filters for molten metals. The second basic approach relies on foaming a ceramic slurry by mechanical agitation or in situ evolution of gases. These approaches probably yield the widest range of cellular structures and hence properties, but they are generally less open than the replicated foams. While a very wide range of applications are now being considered it is probably true to say that most of them are still in the development stage. The final approach relies on the incorporation of sacrificial additives in the form of beads or related materials. Depending on the quantity added, the foams can be predominantly open or closed in nature. In the sections below, each of the

Cellular Ceramics: Structure, Manufacturing, Properties and Applications.
Michael Scheffler, Paolo Colombo (Eds.)

ISBN: 3-527-31320-6

processes outlined above is examined in more depth. An examination of the typical properties obtainable and the applications for which the ceramics are being used or considered for use can be found in other chapters of this book.

2.1.2
Replication Techniques

2.1.2.1
Slurry Coating and Combustion of Polymer Foams

The replication of polymer foams was one of the first manufacturing techniques developed for producing ceramics with controlled macroporosity, the first patent being taken out in 1963 [2]. However, despite its age it is still the most common and widely used technique in industry. The process involves coating a flexible, open-cell polymer foam with ceramic slurries [3–5]. After removal of the excess slip by squeezing and subsequent drying, the polymer is burned out and the ceramic sintered in a single step. A flowchart is shown in Fig. 1 [6].

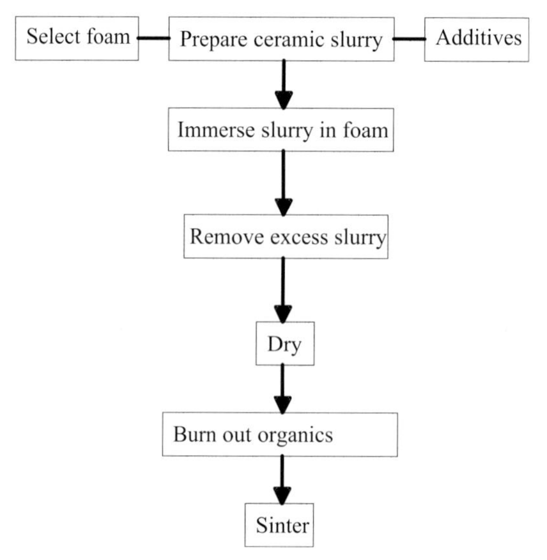

Fig. 1 Flowchart of the production of cellular ceramics by the replica process, adapted from Ref. [6].

Many different polymers can be used for the precursor foam; these include polyurethane (PU), poly(vinyl chloride) (PVC), polystyrene (PS), and cellulose. Reproducibility of the properties of the organic foam is extremely important. It must be able to spring back after being squeezed out and have controlled tolerances to ensure consistency in the ceramic product. Finally, the foam must burn out cleanly and completely during sintering without damaging the ceramic replica. As an example, Selee Corp. in the USA uses an interconnected, open-cell polyurethane foam with 97% void volume as organic precursor, the structure of which consists of a complex pattern of dodecahedra repeated in three dimensions [4].

The production of a reticulated ceramic component begins with the machining of the original polymer foam to the desired shape. While theoretically there are relatively few limitations on the shape, in practice the brittle nature of the final ceramic replica means that failure can occur during burn out, sintering, or usage if the shape is too complex. Since the final ceramic foam is a direct replica of the original foam (Fig. 2), the polymer foam structure and pore size are critical in determining the properties of the final component, for example, its density and permeability.

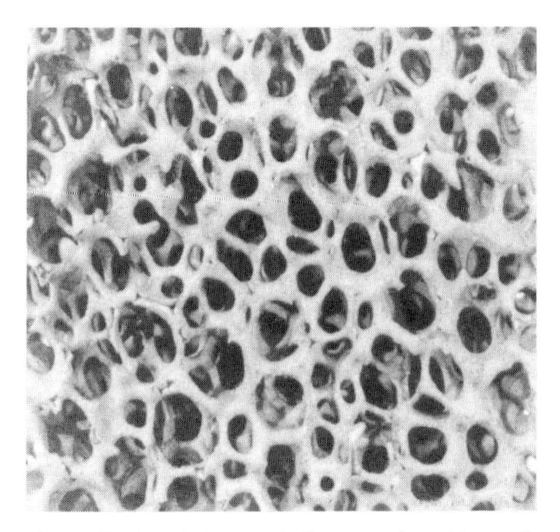

Fig. 2 Reticulated ceramic foam produced by replication of a polyurethane foam.

Generally, any fine ceramic powder can be used that can be made into a suitable suspension; the rheological characteristics are similar to those required for other slurry-based production routes for ceramic components. The solids content usually lies in the range 50–70 wt%; a higher value can lead to an excessively high viscosity that can cause difficulties during infiltration. Typically, the suspension is thixotropic, that is, the viscosity decreases with time at a fixed rate of shear, and on removal of the stress the material regains its original structure with a consequent increase in viscosity. This type of slurry allows coating of the organic foam without excessive drainage; in some cases it is desirable for the slurry to bridge the polymer foam struts. This creates smaller window openings between the cells and thus increases the tortuosity of the flow path for fluids and therefore increases efficiency when the ceramic is used as a filter. Flocculating agents can also be added to improve the adherence of the slurry to the polymer material and thus reduce the chances of poor coating.

Once the polymer foam and ceramic slurry are ready the coating process is carried out. This involves immersing the foam in the slurry and compressing it to remove air. While still in the slurry the foam is allowed to expand again, causing the slurry to be sucked into the open cells of the foam. This step can be repeated several times to achieve the desired coating density, after which pores that have become blocked with excess slip are cleared, typically by a defined compression of the foam.

When the foam has been coated appropriately, it is dried in an oven to solidify the ceramic structure, after which it is exposed to an initial thermal treatment that is designed to burn out the polymer from inside the ceramic struts as well as remove any organic additives from the ceramic slurry. Typically this involves heating the dried foam to a temperature in the range of 350–800 °C, depending on the composition of the precursor polymer. The process can be performed in air or any other type of atmosphere as appropriate, and the heating rate must be carefully controlled, and generally it is very slow to prevent formation of residual stresses or cracking of the ceramic network due to volatilization of the polymer. This results in the formation of a green ceramic replica foam in which the struts are hollow and which requires a normal sintering schedule to achieve the required density and strength of the ceramic in the struts. The conditions required depend on the composition of the ceramic material used.

While this approach generally yields a very open reticulated structure with high permeability, the main disadvantage is associated with the hollow struts and large number of flaws that result from burning out the polymer foam substrate (Fig. 3); low mechanical properties are therefore a typical characteristic of this route. Another complication is related to the quantity and toxicity of the gases that are released during polymer burn out, for example, hydrogen cyanide in the case of polyurethane, which can make expensive scrubbing of the waste gases necessary.

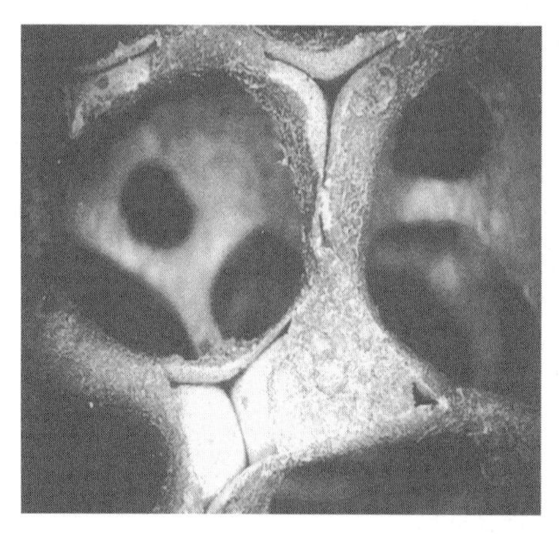

Fig. 3 Ceramic struts resulting from the polymer foam reticulation process. Note how they are hollow and have triangular holes caused by escape of the polymer during combustion.

In an attempt to overcome the low strengths that arise from the replication technique, Luyten et al. [7] replaced the conventional (alumina) ceramic slurry with one that contains a passivated mixture of aluminum and alumina. During the sintering stage, the aluminum is oxidized to yield a reaction-bonded alumina foam. While the grain structure was finer and thus higher strength foams were achieved compared to conventionally produced foams by this route, the approach could not overcome the intrinsic disadvantage of hollow and cracked struts arising from the burnout of the polymer.

However, this has been achieved by Edirisinghe et al. [8], who electrosprayed alumina slurry onto a polyurethane foam prior to removing the polymer by pyrolysis after drying. The electrospray technique involved making the slurry flow through a nozzle kept at a high voltage relative to a ground electrode. This electrostatically atomized the suspension, creating very fine droplets that were used to coat a commercial polymer packaging foam (Fig. 4). To produce about 30 mm cubic samples capable of withstanding significant handling required 5 h of electrospraying and sintering at 1200 °C. The most interesting feature of the process is that the strut cross

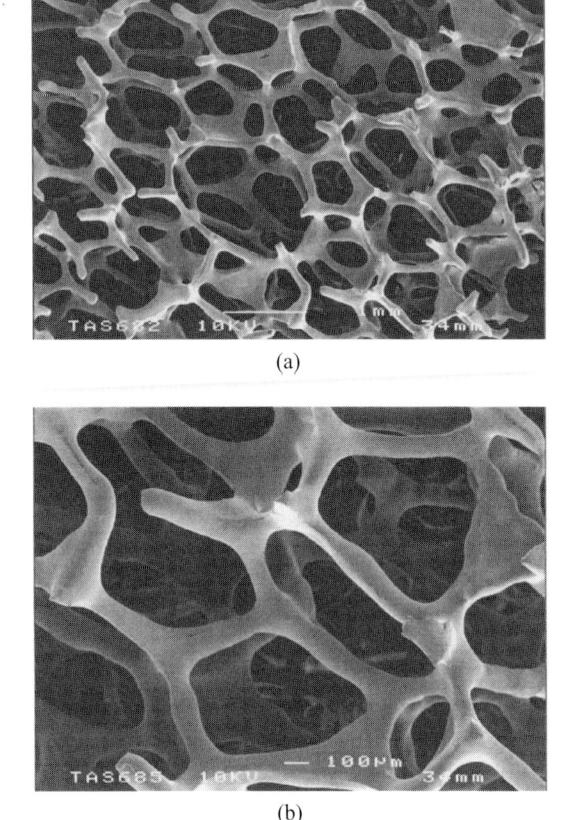

(a)

(b)

Fig. 4 Alumina foam produced by pyrolyzing electrosprayed polyurethane foam showing a) the open cell structure and b) the struts; after Ref. [8]. (S.N. Jayasinghe and M.J. Edrisinghe, J. Por. Mat. **2002** 9 265–273). Reprinted with kind permission of Springer Science and Business Media.

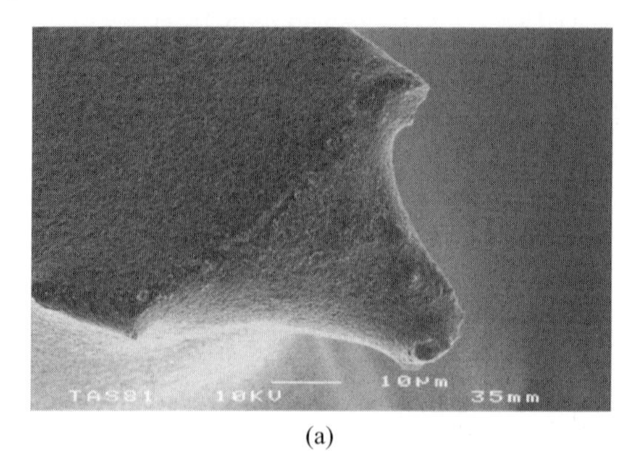

(a)

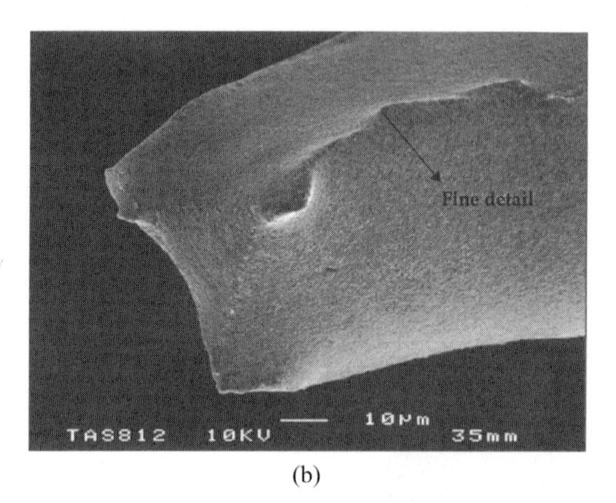

(b)

Fig. 5 Struts of the alumina foam prepared by pyrolyzing electrosprayed polyurethane foam showing a) a solid cross section and b) a homogeneous crack-free microstructure incorporating fine details of the template; after Ref [8]. (S.N. Jayasinghe and M.J. Edrisinghe, J. Por. Mat. **2002** *9* 265–273). Reprinted with kind permission of Springer Science and Business Media.

sections show an homogeneous, crack-free microstructure, intriguingly without the central void typical of ceramic foams produced by immersion in a slurry (Fig. 5). Although the authors claim that the latter can be explained by the very fine size of the droplets, this argument is not convincing.

2.1.2.2
Pyrolysis and CVD Coating of Polymer Foams

A variation of the replication process described above is the pyrolysis process, in which the polymer material is not burned out but pyrolyzed to yield a carbonaceous skeleton that can then be coated by the appropriate ceramic. The process begins with the pyrolysis of a resin-impregnated thermosetting foam to obtain a reticulated carbon skeleton. This can then be coated with a ceramic slurry as described above, but it is more usual to use chemical vapor deposition (CVD) [9, 10] to coat the individual ligaments. Any material that can be deposited by the CVD process can be used for the coating, including oxides, nitrides, carbides, borides, silicides, and metals.

The reticulated carbon network is heated to the required deposition temperature, and then a gaseous precursor compound is passed through the hot body. The gas is reduced or decomposed on the carbon surfaces throughout the internal structure of the foam according to any one of a number of different chemical reactions (Tab. 1), to form a uniform coating that is typically 10–1000 μm thick and has up to 50 % of the theoretical density. The process utilizes the high rates of deposition available with CVD, typically 100–400 μm h^{-1}.

Table 1. Chemical vapor deposition; examples of precursors and reaction temperatures; after Ref. [65].

Coating	Reaction	Temperature/ °C
AlB_2	$AlCl_3 + BCl_3$	1000
Al_2O_3	$AlCl_3 + CO_2 + H_2$	800–1300
B_4C	$BCl_3 + CO + H_2$	1200–1800
	$B_2H_6 + CH_4$	1200
	$(CH_3)_3B$	560
BN	$BCl_3 + NH_3$	1000–2000
	$B_3N_3H_3Cl_3$	1000–1500
SiC	$CH_3SiCl_3 + H_2$	1000
	$SiCl_4 + C_6H_5CH_3$	1500–1800
Si_3N_4	$SiH_4 + NH_3$	950–1050
	$SiCl_4 + NH_3$	1000–1500
TiB_2	$TiCl_4 + BBr_3$	300–850
TiC	$TiCl_4 + H_2 + CH_4$	980–1400
ZrB_2	$ZrCl_4 + BBr_3$	1700–2500

To increase the flexural and tensile properties of the foam, a dense face sheet can be applied to one or more surfaces of the reticulated material by changing the gas flow patterns during the CVD process. Other unusual structures developed include a material that is thermally insulating at one end and thermally conducting at the other by deposition of different materials on each half of the preform and fabrication of insulators with a density gradient throughout the material. The latter method makes it possible to blend a high-density surface into an area of low thermal conductivity.

2.1.2.3 Structure of Reticulated Ceramics

An oft-quoted characteristic of reticulated ceramics is the number of pores per linear inch (ppi value), even in countries where SI units are used. Ceramic foams produced by the replication methods described above usually have pore sizes between 5 and

65 ppi (2–25 pores per cm) and densities ranging between 5 and 30 % of theoretical. The foams prepared are typically 10–100 cm wide and 1–10 cm thick.

Reticulated ceramics produced by replication of a polyurethane foam exhibit large cracks along the struts, which are also hollow (Fig. 3). Both these features occur as a result of the elimination of the polymeric precursor. Green et al. [11–14] have examined these materials in some depth and concluded that while the strength could be improved if the strut cracks and other large flaws were eliminated, the large triangular holes in the struts do not necessarily reduce the strut strength unless the apex is near the outside of the strut or is associated with a strut crack. In addition, they found that most of the theoretical relationships developed by Gibson and Ashby [15] for the mechanical behavior of cellular materials hold quite well for these reticulated ceramics. While some differences were apparent, these could be attributed to microstructural defects.

Gauckler et al. [16] used stereological methods to characterize the microstructure of reticulated ceramics produced by replication of a polyurethane foam to understand their flow characteristics since they were being used for metal infiltration. The basic stereological parameters used are shown in Fig. 6. The structure of the reticulated ceramic consisted of rounded polyhedra with a nominal diameter P, connected by openings or windows with diameter ϕ. The total porosity of the reticulated body was designated f, the total internal cell surface area per unit volume S_v and the distance between two pore centers S. The structure revealed anisotropy; in the plane of the filter plate the pores were spherical, whereas in the perpendicular direction the pores were both larger and elongated. With respect to the permeability of the ceramics, as expected the most important features were the pore size P, and window size ϕ. The relationship between pore and window sizes (Fig. 7a), was linear, and that between the internal surface area per unit volume and window size is shown in Fig. 7b.

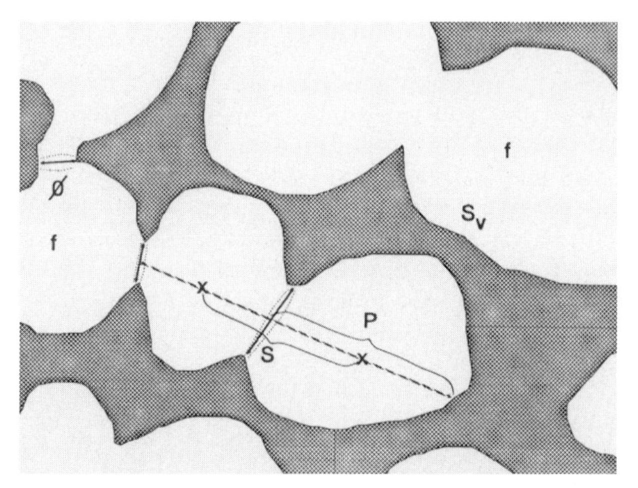

Fig. 6 Stereological parameters of reticulated ceramics; after Ref. [16]. (L.J. Gauckler and M.M. Waeber, in Light Metals 1985, Proc. 114th Ann. Meet. Metal. Soc. AIME, 1985, pp. 1261–1283.) Reprinted with permission of The Minerals, Metals and Materials Society.

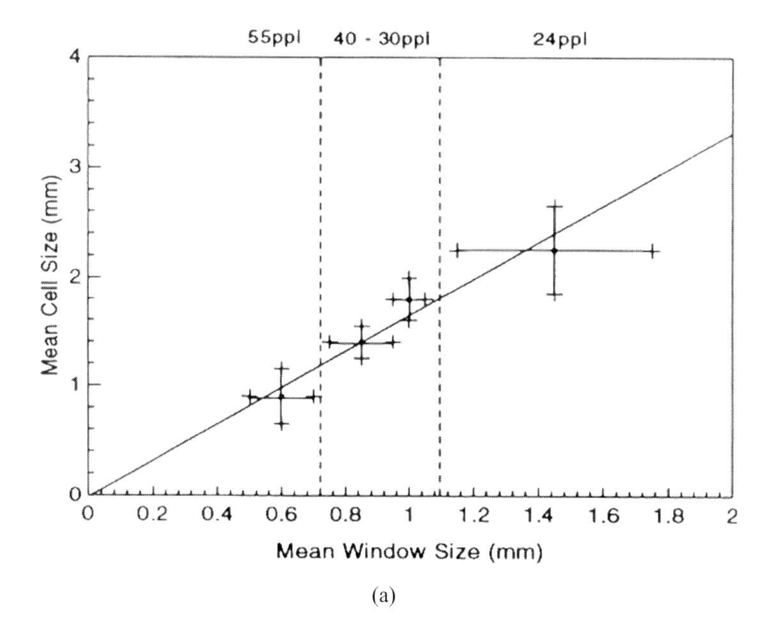

(a)

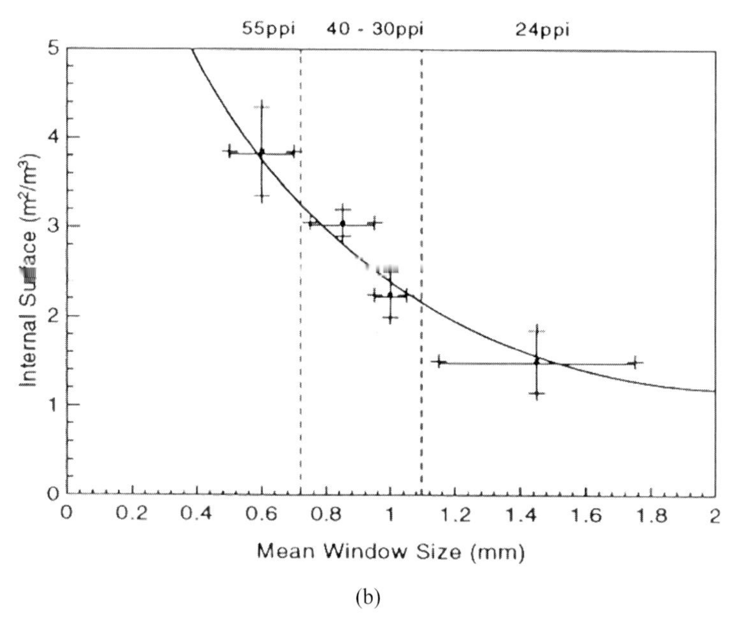

(b)

Fig. 7 Relation between window size and a) mean cell size and b) internal surface area per unit volume for reticulated ceramics; after Ref. [16]. (L.J. Gauckler and M.M. Waeber, in Light Metals 1985, Proc. 114th Ann. Meet. Metal. Soc. AIME, 1985, pp. 1261–1283.) Reprinted with permission of The Minerals, Metals and Materials Society.

2.1.3 Foaming Techniques

The foaming of ceramic slurries involves dispersing a gas in the form of bubbles into a ceramic suspension. There are two basic approaches for achieving this: 1) incorporating an external gas by mechanical frothing, injection of a gas stream, or application of an aerosol propellant, and 2) evolution of a gas in situ. In most cases the addition of a surfactant is required to reduce the surface tension of the gas–liquid interfaces and thus stabilize the gas bubbles developed within the slurry. Such stabilization only works for a limited period of time, however, since several transformations in the bubble structure may subsequently occur due to thinning of their surrounding lamellae. This necessitates a further mechanism to provide a longer term form of stabilization, since any changes in the foam structure prior to solidification influence the final cell size distribution, wall thickness, and microstructure of the solid foams. These in turn have a major role in determining properties such as permeability and strength. For example, when the films surrounding bubbles remain intact until solidification, a closed-cell foam is formed. Open-celled foams are produced when the films partially rupture. In extreme cases, excessive film rupture can lead to foam collapse.

The sections below examine each of the options associated with the different foaming techniques in turn. The process of foaming itself is a well-documented science and although the majority of the literature available relates to the production of foams from surfactant solutions, the same principles apply when foaming ceramic slurries. Therefore, a discussion of the important factors necessary in producing a stable foam, such as the effect of surfactants on surface tension, film elasticity, and foam persistence, will not be reviewed here. Interested readers are referred to treatises such as that by Rosen [17].

2.1.3.1 Incorporation of an External Gas Phase

Various processes have been patented for the manufacture of foam ceramics by entraining an external gas phase [18–20]; all of them incorporate a material that orientates itself at the gas–water interface, thereby stabilizing the gas phase, and one or more additives that cause the structure to set prior to drying and firing.

One of the earliest, patented in 1967 by Du Pont [18], focused on the production of foams from aqueous suspensions of very fine (< 200 nm), negatively charged colloidal silica particles and a cationic surfactant. Other additives included materials that served as binding agents to stabilize the foams, and the wide range listed in the patent includes water-soluble polymers, sugars, starches, resins, and gums. Once formed, the slurries were converted to foams by mechanical agitation by a wide range of approaches, including beater- or blender-type mixers, injection of a stream of gas, and introduction of an aerosol propellant. Once formed into the desired shape of product, the final step involved drying the foam prior to sintering.

A slightly more sophisticated approach was patented by Mitsubishi Chemical Industries [19] for the production of alumina-based foams. The process consisted of mixing gibbsite powder and at least one other powder selected from a group consisting of pseudoboehmite, amorphous aluminum hydroxide, and/or alumina cement. These powders were ground and then mixed with water to form a slurry. A thickener and, if necessary, a surfactant or a binder, were also added if required, depending on the desired structure of the foam and how likely it was to collapse before drying could be completed. While various types of binders were listed, those preferred by Mitsubishi included Portland cement, magnesia cement, and gypsum. A slight variation of the process involved reversing the order by producing the aqueous suspension of the water, thickener, and surfactant first and then adding the ceramic materials. Either way, the actual foaming process was achieved by mechanical agitation to produce the desired quantity of foam, which depended on the required bulk density and the shape of the product. After the foamed slurry had solidified in the mold, it was subjected to hydrothermal treatment to precipitate boehmite crystals from the boehmite-forming compounds and thus develop increased strength.

Four years later, in 1989, a patent application by BASF [20] outlined yet another similar approach to the production of foam ceramics from an aqueous slurry, though the patent was subsequently withdrawn in 1992. This approach involved the desired ceramic powder, water, polymer binder, and, optionally, a surfactant, gelling agents, and a rheology-control agent. An extensive list of polymeric binders was discussed in the patent; those used were all film-forming below room temperature, capable of deformation under pressure, and able to cross-link or react with a cross-linking agent on input of energy from a range of sources such as heating, an electron beam, ultraviolet light, or X-ray irradiation. Foaming was carried out continuously in a high-shear foaming head such as an Oakes mixer or batchwise with a conventional rotary beater such as a Hobart mixer. The patent indicated that for stable foams the slurry should be foamed to between three and five times its original volume, although experiments were conducted with a wider range of volume increases. The foam produced was shaped by placing it in a mold or by spreading it to form a sheet of material that could be cut into slabs. Setting could be achieved with or without the addition of gelling agents, the latter giving greater green strength and allowing thicker foams to be achieved; the patent talks of increasing the thickness possible from 3.75 to 7.5 cm by using a gelling agent. It also indicated that it was possible to produce thicker foams, but these led to problems with sintering, since foams are extremely effective insulators and thus large thermal gradients resulted. It was also noted that foams which were gelled tended to shrink, whereas systems without gelling agents tended to expand. Drying was accomplished by heating; for foams less than 3.75 cm thick this typically involved 100–120 °C for 15–35 min. After drying, the green body could be sintered at the appropriate temperature.

In the mid-1990s a new process was developed and patented that utilized the in situ polymerization of an organic monomer to stabilize foams produced from aqueous ceramic powder suspensions and prevent them from draining and ultimately collapsing [21, 22]. The monomer had the added advantage of significantly improving the green strength of the foam to such an extent that it was readily machineable after drying.

The process for fabricating the ceramic foams, which has now been commercialized by a new company, Hi-Por Ceramics (http://www.hi-por.com/), is given by the flowchart in Fig. 8. A stable, well-dispersed, high-solids, aqueous ceramic suspension is prepared that also incorporates an acrylate monomer together with an initiator and catalyst to provide in situ polymerization. After the addition of a foaming agent, a high-shear mixer is used to provide mechanical agitation that results in the formation of a wet ceramic foam. Process variables include the degree of mechanical agitation and the pressure in the system. The latter affects the degree to which the bubbles formed can expand and hence can be used to control pore size if required. One of the advantages of the in situ polymerization method is that it is common to observe a period of inactivity between the addition of reagents and the actual beginning of the polymerization reaction. This is known as the induction period or idle time t_i and is beneficial since it allows the casting of the fluid foam into a mold prior to polymerization and stabilization of the foam structure. However, on the negative side, the induction period also allows time for bubble enlargement and lamella thinning. These can result in the presence of flaws in the cell walls if excessive disruption of the films occurs before polymerization takes place. Good control of the induction time is primarily achieved by altering the concentration of initiator and catalyst, although other parameters are also very important in determining the induction time, such as temperature and pH.

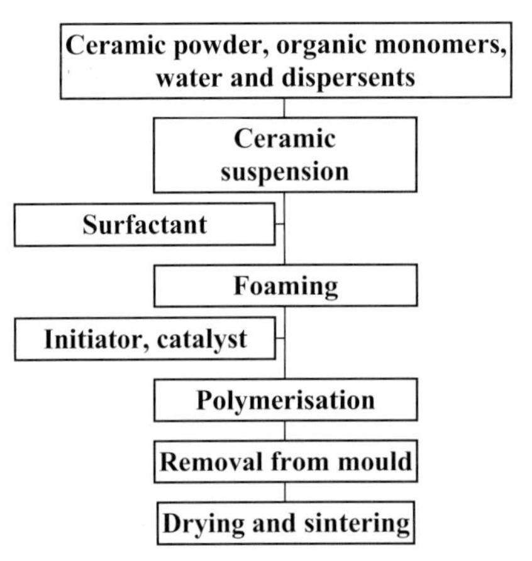

Fig. 8 Process flowchart for the production of ceramic foams by mechanical agitation and the in situ polymerization of organic monomers; after Ref. [22].

The manufacture of a ceramic foam begins with selection of the appropriate ceramic powder. This depends on the end use of the foam and the physical and chemical properties required. However the heart of the process is the creation of a stable ceramic slurry, which necessitates the use of fine ceramic powders, typically $D_{50} < 3\,\mu m$. Materials have been processed with larger particle sizes, but the quality of the dispersion is inferior and the ability of the powder to densify within the foam struts during sintering is reduced. Once selected, the powder is mixed with deionized water containing dispersants, initiator, catalyst, and a premix solution containing about 6 wt% organic monomers. High-shear mixing is used to destroy any agglomerates. The organic monomer solution provides a low-viscosity liquid medium that can be readily polymerized to form a strong cross-linked polymer–water gel. The polymerization reaction traps both the ceramic powder and the water in the foamed structure. A foaming agent (surfactant) is then added to the slurry, the amount of which depends on the viscosity of the slurry and the final density of the foam desired, prior to transfer to a continuous foaming unit. This is then configured to produce a foam with the required density and cell size consistently and repeatably. Usually the slurry is foamed to between 2 and 7 times its original volume. The onset of polymerization is controlled to allow enough time for the foam to exit the foaming unit and be cast into the appropriate mold. After polymerization is complete the structure is strong enough to be demolded and transferred to an oven for drying. The polymeric binder is burnt out as a controlled step in the sintering cycle to leave a mechanically strong, highly porous ceramic. Foam densities in the range of 7–50 % of theory can be produced with cell sizes varying from about 30 to 2000 μm. In general, the lower the density, the larger the cell size, but there has been substantial work in recent years to obtain foams with densities of 30 % of theory or more and cell sizes greater than 1000 μm to provide a combination of high permeability and strength.

The gel-foaming method is able to produce parts with a high degree of complexity and excellent mechanical properties. The complexity can be achieved in part due to the casting process, which makes it possible to shape forms without additional machining. If further complexity is required, such as holes or slots, the dried green foams are strong enough to be clamped and easy to machine. The good mechanical properties result from the foam structure (Fig. 9); the struts between the cells can be fully sintered and typically contain minimal defects. However, it is also possible, at the sacrifice of some of the strength, to only partially sinter the foams and thus obtain porosity in the foam struts. It has been shown [23] that when used for the

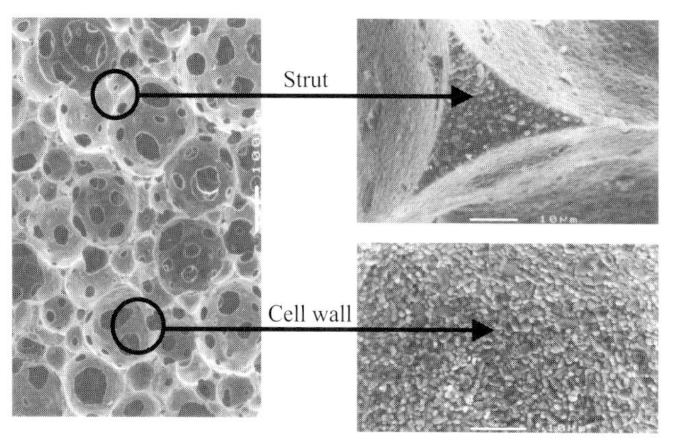

Fig. 9 Structure of a 30 % dense alumina ceramic foam produced by gel casting; after Ref. [64]. (J.G.P. Binner, Brit. Ceram. Trans. **1997** *96* [6] 247–249.) Reprinted with permission from the Institue of Materials, Minerals and Mining.

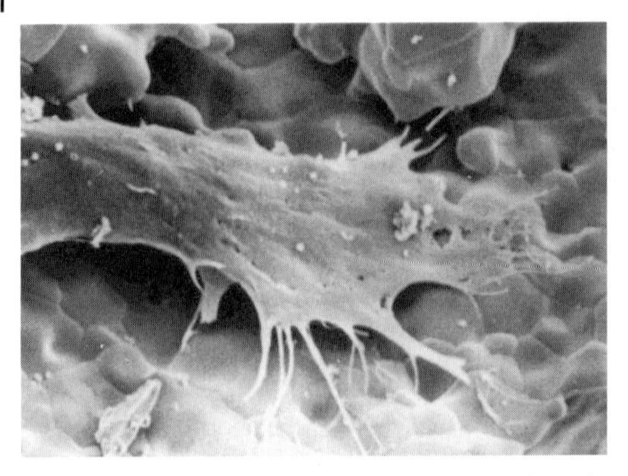

Fig. 10 A human long bone cell growing on the surface of an hydroxyapatite ceramic foam. The presence of microporosity in the foam walls allows the cell to attach easily.

bioceramic hydroxyapatite this results in foams that are extremely good for encouraging bone ingrowth (Fig. 10).

2.1.3.2 In Situ Gas Evolution

The common theme in foaming ceramics by this approach is the presence of a foaming agent that decomposes due to heat or a chemical reaction to generate a gas within a ceramic slurry.

An early patent by A.C.I. Operations in 1973 described a method of producing a foam ceramic of predominantly open-celled porosity as a result of internal reactions to liberate hydrogen [24]. In the process, an amphoteric metal powder, which could be aluminum, zinc, or tin, reacted with sodium silicate to liberate hydrogen and hence foam a ceramic slurry. The mix was then cured by heating at a temperature of at least 80 °C to produce a lightweight foamed product.

A patent for a similar process was obtained by the Duriron company [25] in 1989. The initial slurry was a mixture of the appropriate ceramic powder, a water-soluble source of silicate and aluminate, a particulate metal, a surfactant system, and a gel strengthening agent. On mixing, an aluminosilicate hydrogel was formed that served to bind together the components in a generally self supporting structure over a brief period of time. During this in situ setting reaction, an additional reaction took place in which the particulate metal reacted with the alkali compounds present, which were typically sodium-based, to produce hydrogen gas in situ. As the internal evolution of the gas expanded the slurry, the hydrogel reaction was timed to coincide with the end of hydrogen gas evolution and thus set the porous structure formed. The surfactant present in the composition served to break up the bubbles of the evolving gas into suitably small bubbles as well as stabilizing them and thus ensuring that the porosity developed in the structure was predominantly of an open-celled nature.

Around the same time, Fujiu et al. [26] used the rapid increase in viscosity that can be induced in sol–gels to stabilize ceramic foams. Commercially available 40 wt% silica sols were combined with Freon (CCl_3F) as foaming agent, which was used because of its low boiling point of 23.8 °C and limited solubility in water. As a result of these properties, Freon could be dispersed in water as an emulsion and then heated to induce foaming by its vaporization. The precursor solution was first stored in a bath at 30 °C until the sol viscosity had increased to the desired level for foaming, typically 20–40 min. Foaming was then promoted by stirring the gel, the pH being adjusted to increase the gelation rate. The foam was then stored for 12 h at 30 °C followed by 6 d at room temperature before being dried at 70 °C for a further 7 d, heated to 400 °C for a further 2 d, and finally sintered in air at 1000–1100 °C. Production was therefore not fast by any standard, but porous bodies were produced with a minimum cell size of 90 μm at 31 % density and 400 μm at 17 % density.

Just two years later, Minnear [5] produced foamed ceramics by a much faster process combining a prepolymer and acetone solution with a suspension of deionized water, alumina, hydrochloric acid, and a surfactant. The two systems were stirred vigorously until a creamy consistency was achieved, and then the mixture allowed to rise as a result of the carbon dioxide produced. Mixing typically took only about 15 s, the rise time was 2–3 min, curing took about 5 min, and the drying time was 24 h in a laboratory hood. Examples of the ceramics produced are shown in Fig. 11.

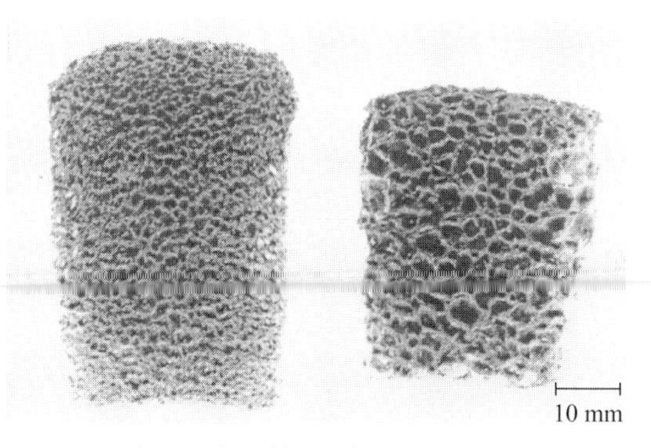

Fig. 11 Typical examples of foamed ceramics produced by in situ gas evolution; after Ref. [5]. (W.P. Minnear, Ceram. Trans., **1992** *26* 149–156.) Reprinted with permission of The American Ceramic Society, www.ceramics.org. Copyright [1992]. All rights reserved.

More recently, Colombo and Bernardo [27] produced macrocellular silicon oxycarbide (SiOC) open-cell ceramic foams by in situ gas evolution in solutions of preceramic polymers. They mixed a methyl polysiloxane with precursors for polyurethane (polyols and isocyanates) in dichloromethane, together with appropriate surfactants and catalysts. Blowing was started by vigorously stirring the mixture and inserting the sample into an oven at a controlled temperature in the range 25–40 °C. The expansion was caused by a combination of evaporation of the solvent as a result of the exothermic reactions occurring in the solution (physical blowing) and chemical blowing arising from the reaction between the water generated by condensation of the SiOH groups in the silicone resin with the isocyanate to form carbon dioxide gas [28].

The choice of polyurethane precursors and the type and amount of surfactants influenced the viscosity of the mixture and hence the final foam architecture and characteristics, while the type and amount of catalysts controlled the rising profile of the foam. Due to immiscibility with the preceramic polymer, during the blowing stage the polyurethane phase separated into small islands, typically measuring 50–300 μm in diameter, embedded within the preceramic polymer matrix. This controlled the resulting green foam morphology and characteristics [28]. Specifically, by using polyurethane precursors as both blowing aids and structural templates for the preparation of the ceramic foams, a wide range of bulk densities and cell sizes could be fabricated. Depending whether the precursors were the basis for so-called flexible, semirigid, or rigid polyurethane foams, the foams produced ranged from completely open-celled, owing to the presence of windows in the cell walls, to completely closed-celled.

Work by Greil, Scheffler and co-workers [29–32] has shown the possibility of foaming poly(silsesquioxane) without an additional foaming agent. A specific poly (phenyl methyl silsesquioxane) containing small amounts (ca. 7 mol%) of ethoxyl and hydroxyl groups was foamed by an in situ blowing technique. When heated above 200 °C, condensation reactions involving the functional groups released water and ethanol, which triggered pore formation in the polymer melt. Simultaneously, an increase in the number of ≡Si–O–Si≡ intermolecular cross-links led to an increase in the viscosity of the polymer melt and thus prevented collapse of the nascent foam structure. The resulting stabilized, thermoset preceramic polymer foam could be easily machined, and subsequent heat treatment in an inert atmosphere above 1000 °C led to the formation of amorphous or partly crystalline Si–C–O ceramic. The properties of the open-celled ceramic foams could be controlled by mixing the preceramic polymer with inert or reactive fillers prior to the foaming step and by pyrolysis parameters such as atmosphere and temperature [33].

A single-stage process has recently been developed and commercialized for the production of ultralight cellular ceramics by Grader et al. [34–37]. The foams are generated by the simple heat treatment of a single-crystalline precursor that contains all the necessary foaming functions. These are crystals of an aluminum chloride isopropyl ether complex [$AlCl_3(iPr_2O)$], which are obtained by mixing concentrated solutions of $AlCl_3$, iPr_2O, and CH_2Cl_2. The foaming mechanism is based on the decomposition of the precursor crystals, which yield polymerizing species dissolved in liquid isopropyl chloride. As long as the solvent and growing AlO_x–$Cl_y(OiPr)_z$ species are mixed homogeneously, the boiling point of the solution is above that of pure isopropyl chloride and its vapor pressure is low. The inventors believe that polymerization takes place in the liquid until a critical polymer size is attained, whereupon phase separation into polymer-rich and solvent-rich regions occurs. Since the expelled solvent is suddenly above its boiling point, bubbles start forming instantly and rise to the surface. The foam is stabilized as a result of gelation in the polymer-rich regions, creating the cell walls. The resultant foams have porosities that are typically in the range 94–99% of theoretical and consist of an arrangement of closed cells, 50–300 μm in diameter, having cell walls about 1–2 μm thick. The very high surface area of the foams, 200 m^2 g^{-1} at 650 °C, suggests that the cell walls contain nanometer-sized pores. While the cellular structure is retained during heating to 1500 °C, the surface area of the foams decreases rapidly with increasing sintering temperature [31]. Products made by this process are now available from the company Cellaris in Israel (http://www.cellaris.com).

A different approach by Matthews et al. has yielded a process that is based on the use of supercritical carbon dioxide ($scCO_2$) and which allows the extrusion or injection molding of both open- and closed-cell porous ceramics from ceramic–polymer mixes [38, 39]. A supercritical fluid is a substance that exists above its critical temperature and pressure, that is, the liquid and gaseous phases are in equilibrium and the fluid has the properties of both phases. In this process, specific functional materials are dissolved in $scCO_2$ and this mixture is then incorporated into the bulk polymer–ceramic mix during processing where the solution of functional material dissolves in the bulk material. Altering the solution conditions of temperature and pressure then facilitates the creation of a cellular structure from the system due to its inherent thermal instability. The polymer can be burnt out to leave behind a porous green ceramic that can subsequently be sintered.

There are a number of advantages to using a supercritical fluid, for example, it has a plasticizing effect on polymers, reducing the viscosity by about 50 % on average, which in turn allows lower processing temperatures and injection pressures to be used. This lowers clamping forces and decreases cycle times. Thus, the addition of $scCO_2$ to ceramic–polymer mixes significantly alters the rheological behavior, and this allows loadings of ceramic powder in the mix to be increased and more complex shaped parts to be molded. In addition, the supercritical fluid can be recycled and used again. Preliminary patent applications have been made [34] and the process is now being commercialized by SCF Processing Ltd in Ireland (http://www.scf.ie).

2.1.3.2
Gelation

Foam stability can be increased by various factors. For foams with thick lamellae, the major influence on foam stability is the bulk viscosity of the solution or suspension. This is the main reason why thickeners or gelling agents such as high molecular weight polymers were often reported as being used in the patents discussed above. They have been well reviewed by Moreno [40] and hence will not be discussed further here.

A number of approaches have been adopted for the setting of fluid suspensions, regardless of whether they have been previously foamed. They include the incorporation of gelling substances such as cellulose derivatives [40] and alginates [41], the use of compositions that can intrinsically form a gel [42], and in situ polymerization of monomers [43].

The most commonly used gelling additives are cellulose and its derivatives. These include methyl cellulose, hydroxycellulose and carboxymethyl cellulose. Usually binders gel on cooling, but methyl cellulose derivatives gel on heating due to hydrophobic self-aggregation processes to form a three-dimensional network with an interstitial liquid. Although cellulose-derived materials and agar are members of the same general family, their gelation behavior is quite different. Agar and its purified derivative agarose are soluble in hot water and gel on cooling. Agar is a naturally occurring polysaccharide, derived from the red algae class of seaweed by a series of extraction and bleaching operations, and has been used in the production of hydroxyapatite foams for wastewater filters [44]. Foams with densities of less than 20% of theoretical and cell sizes from 50 to 1000 μm could be produced; the presence of agar allowed lower solids suspensions to be used and reduced the risk of formation of drying cracks. Agarose is composed of alternating units of two forms of the sugar molecule galactose. The polymer chains aggregate to form rigid bundles of twisted helical chains, yielding a cage network that immobilizes water in the cages by strong interaction between the hydroxyl groups along the polymer backbone and the water molecules. Agarose has found use in ceramic processing as a gelling agent in the production of injection-molded ceramic pieces [45]. Other polymers that can fulfil this role include poly(vinyl alcohol), poly(acrylic acid), polyacrylamide, and polyvinylpyrrolidone [46].

The use of polymerizable materials was proposed by Landham et al. [47] in the context of tape casting, and the idea was further extended for bulk ceramic bodies in a series of papers by Omatete et al. during the development of the process of gel casting [43, 48–51]. The concept involves using a monomeric solvent that is polymerized after casting to form a rigid body. The benefits compared to using a gelling agent that is already polymerized are the ability to formulate slurries with a lower viscosity, since the size of the organic molecule is smaller, so that higher solids loadings can be achieved while still achieving a good packing density, and the excellent green strengths that can be achieved. The organic monomers used must be soluble in water and retain high reactivity. Commonly used materials include methyl methacrylate, butyl acrylate, acrylamide, and other acrylates. The polymerization of these vinyl monomers can be brought about by a variety of initiating systems that have a marked influence on the rate at which the reaction proceeds. For example, chemical initiation with systems such as peroxide–amine, as initiator and catalyst, respectively, can be very effective in producing fast polymerization [43].

The primary disadvantage of all systems that are based on the addition of gelling agents, whether they be polymeric when used or polymerize in situ during processing, is that they need burning out at the end of the process, either in a separate stage or by including a medium-temperature (typically 400–500 °C) hold into the sintering cycle. While this can be a significant problem for the manufacture of dense ceramics due to the relatively long time periods required for the potentially large volumes of gas to escape, for ceramic foams it is usually little more than an inconvenience, especially for the more open structures.

2.1.3.4
Ceramic Foam Structure

Ceramic foams produced by the foaming processes described above differ from those produced by the replication processes in a number of ways. Perhaps the two most important are that, firstly, they are generally less open and, secondly, they do not possess hollow struts created by the loss of the original precursor material. These characteristics can lead to their having lower permeability and greater strength.

To understand the factors associated with the manufacture of foam ceramics from a slurry, Minnear [5] examined foam ceramics produced by in situ evolution of gas. The samples produced were characterized with respect to cell size, degree of cell-wall openness, and what he called the "rise height", which equates to the degree of expansion of the slurry as it became the foam. The latter was estimated to the nearest millimeter by averaging through the normally dome-shaped top (see Fig. 11). The pore size was determined on a cross section under low-power magnification by the linear intercept method. To determine the degree of openness of the structure, a reticulation scale was devised ranging from 1 to 5. A rating of "1" was defined as a structure having a few small holes in the cell walls (< 10 % of the wall area) and only an occasional cell wall completely broken. A rating of "5" indicated very good reticulation with more than 90 % of the cells having many holes and numerous cell walls completely missing. A rating of "3" meant that the walls between adjacent cells contained approximately equal amounts of open and closed area.

The foams produced by Minnear had appreciable porosity and a foam volume of 3–6 times the original slurry volume. The average pore size varied from about 1 to 5 mm, except for the pores adjacent to walls and free surfaces. A strong inverse correlation was noted between pore size and reticulation factor (Fig. 12); foams with finer pores tended to have more open structures. Minear explained this by noting that for a given foam volume, smaller pores have more wall area and therefore thinner walls on average. These were more likely to break and so would yield a more open struc-

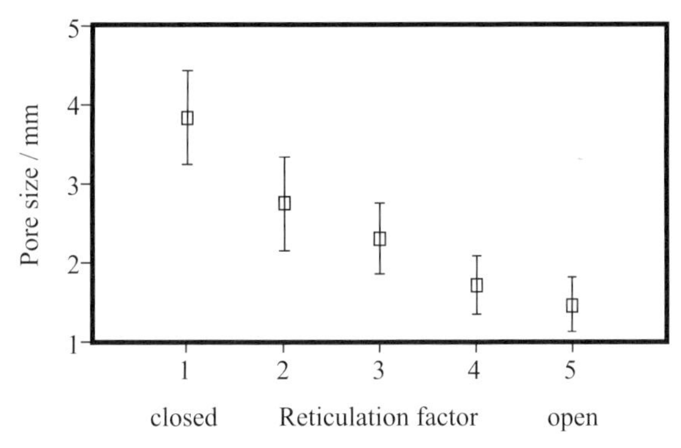

Fig. 12 Pore size versus reticulation factor for cellular ceramics produced by in situ gas evolution; redrawn from Ref. [5].

ture. A detailed mathematical argument, based on his foam constant concept, can be found in his paper on the subject [5]. However, if the foam volume, (i.e., foam density), is allowed to change, then the opposite behavior is observed. Characterization of foams produced by the gel casting technique [52] has shown that more open structures are associated with lower density foams and these, in turn, tended to have larger cell sizes. This again can be explained quite simply using the same arguments put forward by Minnear: the larger the foam volume the thinner the walls will be on average and hence the weaker they will be.

When a foam is produced, the two most important factors are the foam cell size and the wet foam density. These were systematically investigated by Gido et al. [53] with respect to machine variables for an Oakes foaming unit, albeit for a liquid surfactant solution rather than a ceramic suspension. The foam was produced by injecting pressurized nitrogen gas into a stream of surfactant solution, which was then mixed in a rotary-head mixer from which it exited as a foam. Characteristic shear rates were between 2000 s^{-1} for a motor speed of 600 rpm and 4000 s^{-1} for a speed of 1200 rpm. The variables investigated were the liquid and gas flow rates, the speed of the rotary-head mixer and the viscosity of the liquid phase. From the results obtained it was observed that the average mean bubble diameter decreased as the rotational speed increased, and at comparable foam quantities and rotation speeds, the mean bubble diameter was significantly smaller for a foam generated from a higher viscosity liquid. However, one of the most interesting aspects of the work is the lack of correlation between bubble size and wet foam density, and the range of wet foam densities and bubble sizes achievable.

2.1.4
Other Techniques

Volatile or combustible additives that are lost during firing can be incorporated into the ceramic, and the volume, size, shape, and distribution of the resulting porosity is determined by the amount and nature of the fugitive phase [54]. One particular approach for forming porous ceramics used starch as both binder and pore former [55, 56]. It is well known that starches can act as binders due to their gelling ability in water [57], but during firing they result in residual porosity. While this has always been seen as a disadvantage for dense ceramics, it was used to advantage to generate sintered porous ceramics with porosities in the range 23–70 % [55]. The overall pore structure was dominated by the 10–80 μm spherical pores left by the starch particles, and the average size of the small pores connecting the large pores was controlled by the total solids loading and starch content in the precursor slips to between about 0.5 and 10 μm. Chemically modified starch was found to give better dimensional control and uniformity than native starch with regard to the average size of the connecting pores owing to its more stable properties during water processing.

Another organic material that has been used to promote the creation of foams is egg white; since albumin has amphiphilic properties its solutions are prone to foaming. Hence it is possible to generate foams by the simple process of vigorously mixing or ball milling the ceramic suspension with the addition of albumin [58, 59]. Once again, the organic additive acts as both a binder and pore-forming agent.

While the majority of such applications lead to porous, as distinct from cellular, ceramics, if the additives comprise a large enough fraction of the whole then they can result in the material's developing a cellular nature. For example, poly(methyl methacrylate), PMMA, microbeads have been added to heat-treated methylsilicone resin powder in a weight ratio of 80:20 prior to uniaxial pressing [60]. The PMMA microbeads were then burnt out and the porous green bodies pyrolyzed in an inert atmosphere to yield silicon oxycarbide (SiOC) ceramic microcellular foams. Further information on these foams can be found in Chapter 5.11.

Inorganic pore formers can also be used; these include alkaline earth metals and their oxides, as well as a range of other oxides, nitrides, and carbides. The mineral perlite is also commonly used [61]. After forming and drying, the green bodies are fired, and the pore formers partially or completely melt and/or react with the ceramic powder. As with the organic additives, the structure and porosity of the final porous ceramic can be adjusted by means of the shape, size, and amount of the pore formers. The final pore size of these structures can be in the range of a few micrometers up to several centimeters, the latter usually involving the addition of hollow additives, for example, glass spheres. The use of inorganic pore formers that do not vaporize can have the advantage that they do not pollute the environment, but since the material is retained in the final body the chemistry of the pore former and its effect on the host material increase in importance. An example is the formation of porous silicon carbide ceramics by the addition of graphite powders, the pore diameter being controlled by the size of the graphite particles [62]. In this work, the SiC particles were bonded together by taking advantage of an oxidation-bonding process in which the powder compacts were heated in air so that SiC particles were bonded to each other by oxidation-derived SiO_2 glass.

Closely related in conceptual terms is the chemical leaching of one phase in systems that contain two or more phases to yield connected porosity. For example, in 1996 Japanese researchers investigated the consequences of 1 M hydrochloric acid on cordierite [63]. They found that both magnesium and aluminum ions were leached from the structure in the same molar ratio, leaving a porous structure with a large surface area, proportional to the amount of cations removed. The pore radius distribution ranged from 0.4 to 0.9 nm, which, being comparable to those of typical molecular sieves, encouraged them to try simple adsorption tests. They found that the porous substrate was capable of adsorbing some amine compounds.

A final approach, which is slightly different to either of the above since it does not rely on the removal of a discrete phase, is the sintering together of hollow spheres (or other shapes) to yield closed-cell structures with excellent control of cell size. For example, in recent work by Luyten et al. [7], spheres of both polymeric and biological origin were coated with an alumina slurry, packed into a mold and then slurry-coated a second time. After drying, the spheres were burnt out, and the porous green body sintered, to yield a relatively strong and lightweight material that was built up of hollow alumina spheres; the density ranged from 15 to 20% of theoretical. A wide diversity of materials could be produced by varying the diameters and shapes of the hollow precursor material. Further information on the concept can be found in Chapter 2.8.

2.1.6 Summary

It is clear that the field of cellular ceramics is currently very active with new process routes constantly being developed and reported in the scientific literature and at conferences. Relatively few of them, however, are inherently novel; rather most are modifications or adaptations of existing processes, so that the concept of having three basic categories of manufacturing routes is still fundamentally true. What has changed over the past decade is that more processes are being commercialized. Whereas the polymer replication process was one of the few routes to be in commercial use as little as 15 years ago, now there are several routes available, each of which yields cellular ceramics with different pore structures and degrees of porosity, and hence different properties and potential applications. This provides much greater choice for the end user and far greater potential for the tailoring of structures to meet specific end-use requirements.

Acknowledgements

The author would like to acknowledge the work of some of his former research students whose work on the processing of ceramic foams has helped to contribute to this chapter. In chronological order: Dr. Rob Smith, Dr. Jutta Reichert, Dr. Pilar Sepulveda, Dr. Yongheng Zhang, and Mr. Lars Monson.

References

1 Sepulveda, P., *Am. Ceram. Bull.* **1997** *76* [10] 61–65.
2 Schwartzwalder, K. and Somers, A.V., US Pat. No. 3090094, 1963.
3 Lange, F.F. and Miller, K.T., *J. Am. Ceram. Soc.* **1987** *2* [4] 827–831.
4 Brockmeyer, J. and Pizzirusso, J.F., *Mater. Eng.* **1988** *105* [7] 39–41.
5 Minnear, W.P., *Ceram. Trans.* **1992** *26* 149–156.
6 Saggio-Woyansky, J., Scott, C.E. and Minnear, W.P., *Am. Ceram. Bull.* **1992** *71* [11] 1674–1682.
7 Luyten, J., Mullens, S., Cooymans, J., De Wilde, A. and Thijs, I. in *Shaping II*, Proc. 2nd Int. Conf. Shaping Adv. Ceram., Gent, Belgium, 2002 (eds. J. Luyten and J.-P. Erauw), 43–48.
8 Jayasinghe, S.N. and Edrisinghe, M.J., *J. Porous Mater.* **2002** 9 265–273.
9 Sherman, A.J., Tuffias, R.H. and Kaplan, R.B., *Am. Ceram. Bull.* **1991** *70* [6] 1025–1029.
10 Lin, Y.S. and Burggraaf, A.J., *AIChE J.* **1992** *38* [3] 445–454.
11 Hagiwara, H. and Green, D.J., *J. Am. Ceram. Soc.* **1987** *70* [11] 811–815.
12 Brezny, R., Green, D.J. and Dam, C.Q., *J. Am. Ceram. Soc.* **1989** *72* [6] 885–889.
13 Brezny, R. and Green, D.J., *J. Am. Ceram. Soc.* **1989** *72* [7] 1145–1152.
14 Van Voorhees, E.J. and Green, D.J., *J. Am. Ceram. Soc.* 1991 *74* [11] 2747–2752.

15 Gibson, L.J. and Ashby, M.F., *Cellular Solids: Structure and Properties*, Pergamon, Oxford, 1988.

16 Gauckler, L.J. and Waeber, M.M. in *Light Metals 1985*, Proc. 114th Ann. Meet. Metal. Soc. AIME, 1985, 1261–1283.

17 Rosen, M.J., *Surfactants and Interfacial Phenomena*, 2nd ed., John Wiley & Sons, New York, 1989, chap. 7.

18 E.I. Du Pont de Nemours, GB Pat. No. 1 175 760, 1967.

19 Mitsubishi Chemical Industries Ltd, US Pat. No. 4505866, 1985.

20 BASF Aktiengesellschaft, Europ. Pat. Appln. No. EP 0330963, 1989.

21 Sambrook, R.M., Binner, J.G.P., Smith, R.T. and Reichert, J., World Pat. No. WO 9304013, 1993.

22 Sepulveda, P. and Binner, J.G.P., *J. Eur. Ceram. Soc.* **1999** *19* 2059–2066.

23 Sepulveda, P., Binner, J.G.P., Rogero, S.O., Higa, O.Z. and Bressiani, J.C., *J. Biomed. Mater. Res.* **2000** *50* 27-34.

24 A.C.I. Operations pty, British Pat. No. GB 1321093, 1973.

25 The Duriron Co. Inc., World Pat. No. WO 8905285, 1989.

26 Fujiu, T., Messing, G.L. and Huebner, W., *J. Am. Ceram. Soc.* **1990** *73* [1] 85–90.

27 Colombo, P. and Bernardo, E., *Compos. Sci. Technol.* **2003** *63* 2353–2359.

28 Takahashi, T., Münsted, H., Colombo, P. and Modesti, M., *J. Mater. Sci.* **2001** *36* 1627–1639.

29 Gambaryan-Roisman, T., Scheffler, M., Buhler, P. and Greil, P., *Ceram. Trans.* **2000** *108* 121–130.

30 Gambaryan-Roisman, T., Scheffler, M., Takahashi, T., Buhler P. and Greil, P. in *Euromat 99*, vol. 12, Ceramics Processing, Reliability, Tribology and Wear (ed. G. Müller), DGM, Frankfurt, 2000, pp. 247–251.

31 Zeschky, J., Neunhoeffer, F.G., Neubauer, J., Lo, S.H.J., Kummer, B., Scheffler, M. and Greil, P., *Compos. Sci. Technol.* **2003** *63* 2361–2370.

32 Zeschky, J., Scheffler, M., Colombo, P. and Greil, P., *Ceram. Eng. Sci. Proc.* **2002** *23* [4] 285–290.

33 Greil, P., *J. Am. Ceram. Soc.* **1995** *78* 835–848.

34 Grader, G.S., Shter, G.E. and de Hazan, Y., *J. Mater. Res.* **1999** *14* [4] 1485–1494.

35 Grader, G.S., de Hazan, Y., Natali, G., Dadosh, T. and Shter, G.E., *J. Mater. Res.* **1999** *14* [10] 4020–4024.

36 Mann, M., Shter, G.E. and Grader, G.S., *J. Mater. Res.* **2002** *17* [4] 831–837.

37 Grader, G.S., Shter, G.E. and Hazan, Y., US Patent No. 6 602 449, 2003.

38 Matthews, S. and Matthews, J., presented at the 27th International Cocoa Beach Conference on Advanced Ceramics and Composites, January 2003.

39 Hornsby, P.R. and Matthews, S., US Patent Application No. 0084795, 2004.

40 Moreno, R., *Am. Ceram. Bull.* **1992** *71* [11] 1647–1657.

41 Katsuki, H., Kawahara, A. and Ichinose, H., *J. Mater. Sci.* **1992** *27* 6067–6070.

42 Bagwell, R.B. and Messing, G.L., *Key Eng. Mater.* **1996** *115* 45–64.

43 Young, A.C., Omatete, O.O., Janney, M.A. and Menchhofer, P.A., *J. Am. Ceram. Soc.* **1991** *74* [3] 612–618.

44 Binner, J.G.P. and Reichert, J., *J. Mater. Sci.* **1996** *31* 5717–5723.

45 Fanelli, A.J., Silvers, R.D., Frei, W.S., Burlew, J.V. and Marsh, G.B., *J. Am. Ceram. Soc.* **1989** *72* [10] 1833–1836.

46 Asahi Kogaku Kogyo Kabushiki Kaisha, Eur. Pat. No. EP 0360244, 1994.

47 Landham, R.R., Nahass, P., Leung, D.C., Ungureit, M., Rhine, W.E., Bowen, H.K. and Calvert, P.D., *Am. Ceram. Bull.* **1987** 66 [10] 1513 1516.

48 Omatete, O.O., Janney, M.A. and Strehlow, R.A., *Am. Ceram. Bull.* **1991** *70* [10] 1641–1649.

49 Omatete, O.O., Tiegs, T.N. and Young, A.C., *Ceram. Eng. Sci. Proc.* **1991** *12* 1257–1264.

50 Omatete, O.O., Bleier, A., Westmoreland, C.G. and Young, A.C., *Ceram. Eng. Sci. Proc.* **1991** *12* 2084–2094.

51 Omatete, O.O., Strehlow, R.A. and Walls, C.A., *Ceram. Trans.* **1992** *26* 101–107.

52 Binner, J.G.P., *Int. Ceram.* **1998** *2* 69–71.

53 Gido, S.P., Hirt, G.E., Mongomery, S.M., Prud'homme, R.K. and Rebenfeld, L., *J. Disp. Sci. Technol.* **1989** *10* 785–793.

54 Komarneni, S., Pach, L. and Pidugu, R., *Mater. Res. Soc. Symp. Proc.* **1995** *371* 285–290.

55 Lyckfeldt, O. and Ferreira, J.M.F., *J. Euro. Ceram. Soc.* **1998** *18* 131–140.

56 Lemos, A.F. and Ferreira, J.M.F., *Mater. Sci. Eng.* **2000** *C 11* 35–40.

57 Rutenberg, M.W. in *Handbook of Water-Soluble Gums and Resins* (ed. R.L. Davidsson), McGraw-Hill, New York, **1980**, pp. 22.1–22.83.

58 Dhara, S. and Bhargava, P., *J. Am. Ceram. Soc.* **2003** *86* [10] 1645–50.

59 Garrna, I., Reetza, C., Brandesb, N., Krohb, L.W. and Schuberta, H., *J. Eur. Ceram. Soc.* **2004** *24* 579–587.

60 Colombo, P., Bernardo, E. and Biasetto, L., *J. Am. Ceram. Soc.* **2004** *87* [1] 152–154.

61 H.R. Maier, C.E. Scott and W.P. Minnear, German Pat. No. DE 19605149, 1997.

62 She, J.H., Ohji, T. and Kanzaki, S., *J. Eur. Ceram. Soc.* **2003** *24* 331–334.

63 Abe, H., Tsuzuki, H., Fukunaga, A., Tateyama, H. and Egashira, M., *Key Eng. Mater.* **1996** *115* 159–166.

64 Binner, J.G.P., *Brit. Ceram. Trans.* **1997** 96 [6] 247–249.

65 Binner, J., *Met. Mater.* **1992** 8 [10] 534–537.

2.2 Honeycombs

John Wight

2.2.1 Introduction

A paste is a viscoplastic formable body of dispersed particles in a polymer solution. This chapter describes the paste process for manufacturing ceramic and glass honeycombs with an emphasis on direct extrusion of the honeycomb shape. This paste process creates a porous, particulate honeycomb, the porous microstructure of which can be processed to change its composition, porosity, and connectivity. The honeycomb shape can also be changed by reduction extrusion and hot draw, enabling a wide range of honeycomb channel diameters from bigger than 1 cm to smaller than 1 µm. Glass honeycombs can be hot drawn down to fibers with channels and specific surface area comparable to the pore structure of the assembled particles from which they were first made – except that the channels are linear instead of random. The honeycomb extrusion process is a mechanical way to create linear porosity.

This chapter starts with the highest volume product, automotive catalyst supports, and ends with the lowest volume, photonic crystal fibers. The progression from one to the other is also a progression from large channels to small channels, and it demonstrates something fundamental to the honeycomb paste extrusion process: the interaction of shape and microstructure.

2.2.2 Forming the Honeycomb Geometry

2.2.2.1 Background

A porous ceramic is a composite of nonsolid (gas and/or liquid) and ceramic. It can have the surface area of ceramic powder in the shape of an object. If the nonsolid phase is open and continuous, the entire volume of the object is readily accessible for reaction. Ranges of specific strength (Chapter 4.1), thermal conductivity (Chapter 4.3), and permeability (Chapter 4.2) are possible, and depend upon the microstructure, that is, the connectivity of the solid and nonsolid phases. A disconnected

Cellular Ceramics: Structure, Manufacturing, Properties and Applications.
Michael Scheffler, Paolo Colombo (Eds.)

ISBN: 3-527-31320-6

solid phase (e.g., packed beads) is immune to thermal shock since it is already "broken", but is prone to self attrition [1]. A reticulated foam gives beneficial flow bifurcation for mixing, but at the cost of backpressure. Each arrangement has its own advantages and disadvantages.

There are many ways to distribute a nonsolid in a solid: random (dispersed bubbles and percolating paths) and nonrandom (honeycombs and woven structures). For some objects that are much more nonsolid than solid, the nonsolid and the solid percolate in such a way as to achieve optimum strength and permeability for a given porosity for a given application. The honeycomb (linear cellular) geometry is a popular design solution in nature and in industry (Fig. 1) [2].

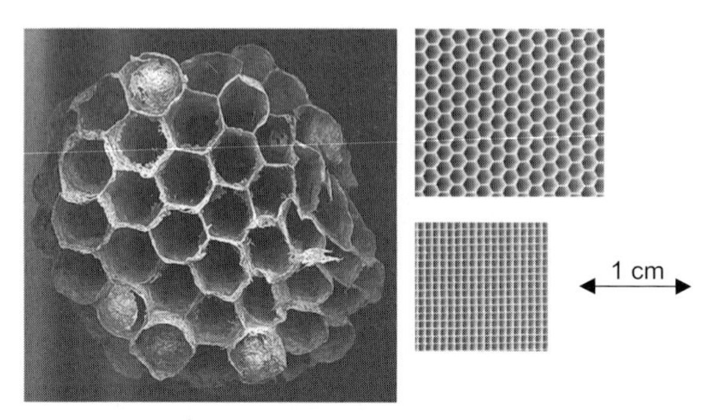

Fig. 1 Popular design solutions: Wasp nest (left) and ceramic honeycombs (right).

The largest volume honeycomb application is the automotive catalytic support: A typical automobile has about 800 m of honeycomb channels in its catalytic converter, which has 4 m^2 of geometric surface in 1.2 L of substrate. It has two types of porosity: geometric (channels) and microstructural (pores and microcracks). These honeycombs have about 20 000 km of pore paths in the microstructure of its webs. The honeycomb market is about one hundred million substrates (two per car) per year for automotive catalytic converter substrates alone, and this is divided between Corning, NGK, Denso, Emitec, and others. Other markets include diesel particulate filter (Chapter 5.2), molten-metal filters (Chapter 5.1), membrane supports, and composite core material.

The cross-sectional geometry of the honeycomb is a tessellating pattern; the channels can be triangular, square, hexagonal, round, or of other shapes [3]. There can be gradients in pitch and web thickness in the radial [4, 5] and axial [6] directions, if this offers an advantage. Industrial honeycombs come in a variety of sizes: anodized alumina membranes (Anapore®: 0.02–0.2 μm diameter channels [7]), photonic crystal fibers (0.5 μm [8]), glass capillary arrays (Burle: 2–50 μm [9]), multicapillary bundles (Alltech: 40 μm), "microcombs" (Philips: 100 μm [10]), automotive catalytic converter substrates (Corning: 0.8–1.3 mm [11]), lightweight mirror cores (10–150 mm [12], Fig. 2). And they are available in a variety of compositions: alumina, carbon, cordierite, fused silica, mullite, potassium borosilicate glass, silicon carbide, and so on. The combination of cell geometry and composition is designed to meet the application.

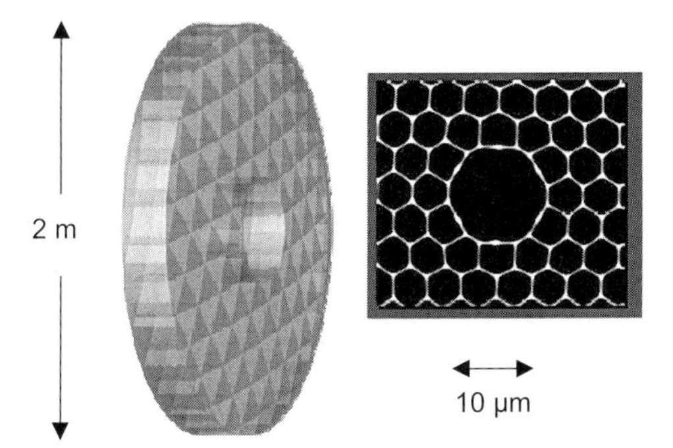

Fig. 2 Computer drawing of the lightweight primary mirror substrate (left) ([13], used with permission, http://snap.lbl.gov/pubdocs/Lampton.PDF) and a micrograph of a photonic crystal fiber (right) ([14], used with permission, http://www.nature.com). Both are honeycomb structures, but their channel size differs by almost five orders of magnitude. For different reasons they have a very similar geometry (honeycomb annulus). The photonic crystal fiber is one of the finest structures ever made; it can have a web thickness of less than 200 nm and lengths of more than a hundred meters. This one was made by a stack and hot draw down technique.

The honeycomb is a logical way of scaling up the benefits of the tube geometry. This is called "numbering up" in microreactor terminology. Numbering up from the laboratory to an actual industrial application may entail going from one tube to more than 50 000 tubes. The honeycomb is an orderly arrangement/bundle of a large number of tubes into a unified whole. The disadvantage of tube bundling is the task of managing the size and shape distributions/variabilities of the tube population, slip planes, and other contributions to packing defects. The shape or the consolidation of interstices of the tubes must be managed. The advantage of tube bundling is that individual tubes can be formed, inspected, and culled before being assembled into an array.

"Massive parallel processing" describes the direct extrusion of a honeycomb. The extrusion of 50 000 channels simultaneously is both a humane and an economical alternative to the bundling of 50 000 tubes to obtain the same geometry. In extrusion, a honeycomb monolith results from the assembly of flow streams within the die. There is flow communication within the die (adjacent flow momentum exchange, shear, and knit) and flow compensation external to the die (web thinning, thickening, and buckling).

2.2.2.2 Honeycomb Extrusion Die

Tubes can be extruded from paste (plastically formable mixture of dispersed particles in a binder) [15] and in some cases from hot melt (viscously and/or plastically formable liquid or solid). The tube-forming die is just a rod extrusion die with the addition of a spider, a tool that defines/creates the inner diameter and is held in place in the flow stream by a set of vanes. During tube extrusion, the batch diverges, splits across the vanes and then converges and knits (welds/melds) to form a monolithic tube with knit lines that run the length of the tube. To build a honeycomb extrusion die with an array of holes and spiders would be a Herculean task due to all of the precision machined and mated surfaces, especially for high cell densities (number of channels per square centimeter) and large honeycomb diameters.

In 1971, R. Bagley invented a practical honeycomb extrusion die (Celcor® die) [16]. He had the die machined from a single plate of metal: an array of feed holes was drilled partway into the back of the plate, criss-crossing slits were milled partway into the front of the plate, and everything was registered so that the feedholes fed the slits, and the slits (array of pins) defined the honeycomb, the exiting extrudate (Fig. 3) [17].

This honeycomb forming process entails an inversion (Fig. 3): the entrance of the die is designed for die strength in that the metal is the continuous phase and the paste is the dispersed phase, and the exit of the die is designed for honeycomb extrudate strength in that the metal is the dispersed phase and the paste is the continuous phase. The distribution of metal from the entrance to the exit is for strength to resist the extrusion pressure and for shape to reduce the pressure from the drag of shaping. Inbetween the entrance and the exit of the die, an inversion occurs in the extrusion direction, and here is where the process "miracle" occurs: the array of paste filaments transforms into an array of crosses whose legs elongate until they impinge and knit with the adjacent crosses to form a monolithic honeycomb before

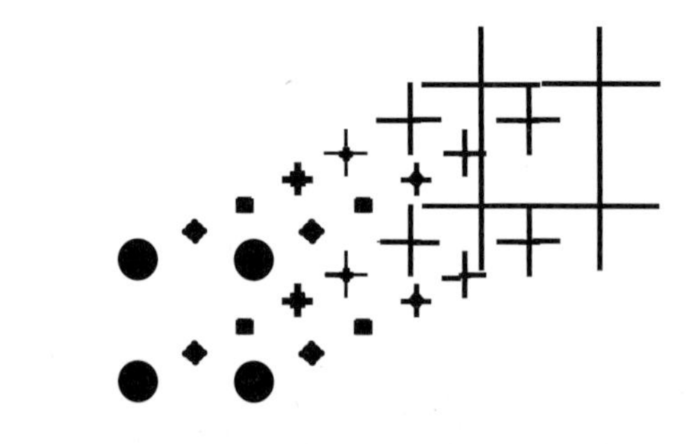

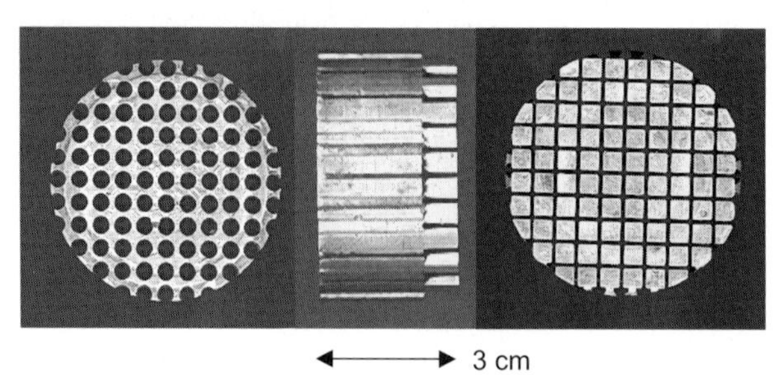

Fig. 3 Honeycomb die. A schematic (top) of the paste at different stages of going through the inversion from the entrance (left) to the exit (right) of the honeycomb die. At the entrance of the die (bottom left), the metal is the continuous phase and the paste is the disperse phase, and at the exit of the die (bottom right), the metal is dispersed and the paste is continuous.

exiting the die. The array of metal filaments (pins) in a matrix of paste becomes an array of air filaments in a matrix of paste as the honeycomb exits the die (it actually inhales). Buried in this matrix of paste is an array of knit lines whose properties are not detrimental.

This process creates a huge amount of well-defined porosity by purely mechanical means on the order of a 1600 km of channel per hour. A solid billet of paste or melt is pressurized to flow through a die to be plastically formed into a honeycomb with 90 % open frontal area (OFA, the channel face area divided by the total face area multiplied by 100) which has one tenth the bulk density of the billet (paste). Push 1 m of billet into the die and 10 m of honeycomb is extruded with the same diameter (Fig. 4). No chemistry is necessary: no pore formers [18], no etchants [12, 19–22], no burnout and intrude procedures [23, 24], and no washcoat and burnout procedures [25, 26] (Chapter 2.1).

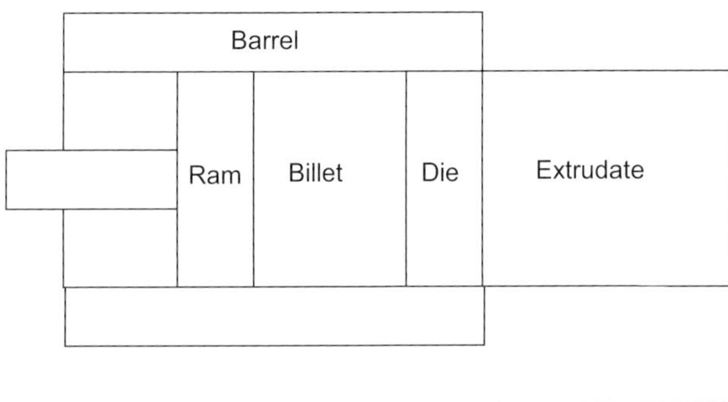

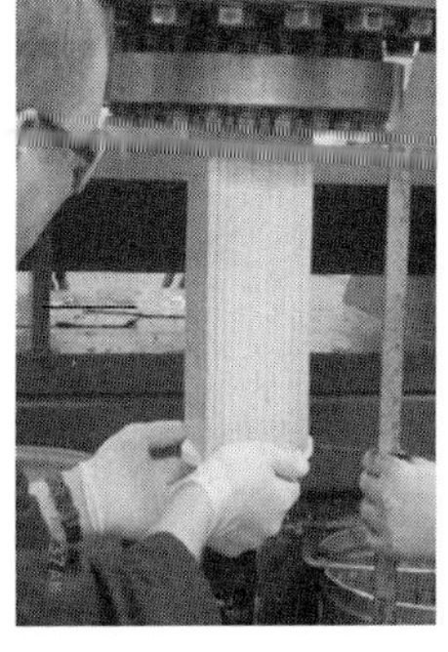

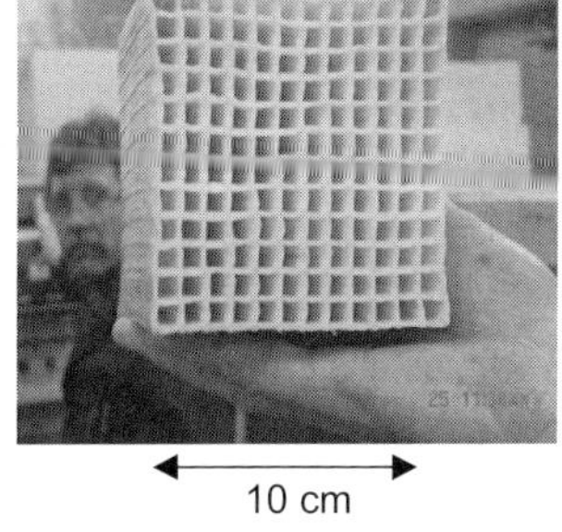

Fig. 4 Paste extrusion of honeycomb. Different views of ram extrusion of a ULE® paste: the initial push through the die (bottom left), the honeycomb extruding/hanging down (bottom center), and the extrudate horizontally supporting itself (bottom-right). A schematic of a ram extruder is shown in the top figure.

High OFA is greatly valued in that the higher the OFA, the lower the backpressure in a flow-through application (Chapter 4.2). However, in extrusion die design, the higher the OFA, the higher the extrusion pressure, for a given cell density and paste. A substrate which is designed for low backpressure makes for a die that has high backpressure; in some ways the extrusion and the application are the flow complements of one another. Ceramic foam forming processes (Chapter 2.1) do not have this particular high-OFA challenge.

2.2.2.3
Nonextrusion Fabrication Processes

Honeycombs can be constructed from lower dimensional parts (points, lines, and planes), which are much lower force forming operations than direct honeycomb extrusion, because the cross-sectional area of these extruded components is smaller than that of their assembled honeycomb. Another advantage of rods and sheets is that they are easier to coextrude (there is more room/dimension to manifold a rod or tape coextrusion die) and to decorate than directly extruded honeycombs. Once assembled these "two-color" composite honeycombs can have increased functionality [27–29]. Here are some examples of lower dimensional builds of honeycombs:

- A "point" can draw a honeycomb by a variety of rapid prototyping schemes: MIT's 3DP [30], Rutgers' FDC [31], Sandia's Robocast (Chapter 2.3) [32], and others. These methods offer the opportunity to fabricate high aspect ratio walls [33] and periodic structures in the axial direction of the honeycomb, too [34, 35]. Here axial periodicity can be used to create a mixing exchange between adjacent channels by having a rod stacking sequence that bifurcates the flow (like reticulated foam), so as to prevent boundary layer formation on the channel wall, which otherwise would reduce the reaction rate [36–38]. Axial periodicity can also be achieved in direct honeycomb extrusion by employing moving dielets [39, 40], but this is not practical for substrates with high cell density.
- "Lines" can be bundled to make a honeycomb, as has already been described in the bundling of tubes to make a honeycomb. This is the quickest path to making a honeycomb from a tooling point of view. The technique has ancient roots in millefiori [41] and has modern practitioners. Different colored rods and tubes can be assembled to form a "dot-matrix" cross-section pattern: a simple build is a single core/clad design, and a more complicated build is a complete array/matrix honeycomb. The resulting composite billet can be reduction-extruded (MFCX: microfabrication by coextrusion) [27, 42–44] or hot drawn down (as is commercially done with core/clad rods by Galilieo Electro-Optics, Burle Technologies [9], and Schott Fiber Optics [45]) to miniaturize the dot-matrix pattern. The process is iterative in that the reduced parts can be bundled and reduced again and again to achieve high cell densities and large cross-sectional areas. The resulting composites have a range of useful properties:
  - A refractive index contrast between the matrix and the array [46] results in image transfer fibers and taper optics.
  - A leachable matrix creates flexible fiber bundles for fiberscopes [45, 47].
  - A crackable matrix creates tough composites [48, 49].
  - A leachable array results in thin channel array plates for photomultipliers [9, 50, 51].

Note that honeycomb plates can be made by etching out the array of filaments due to their relatively small aspect ratio and thickness, but hollow honeycomb fibers cannot be made by etching due to their extremely high channel aspect ratio and length and insufficient etching contrast between the array and matrix. Hot drawing a honeycomb with an array of glass filaments is less prone to distortion (radial pitch gradient and loss of channel volume) than an array of air filaments, which does not have a constant-volume condition. Air filaments can be inflated and deflated as required during sintering and hot drawing so as to oppose surface tension (loss of channel volume by viscous sintering) and to modify OFA [21, 22, 52].

- "Planes" can be stacked to make a honeycomb. Sheets can be made by a papermaking process (like the Cercor® substrate [53], which was the predecessor of the more cost effective Celcor® substrate [1]), by tape casting and by extrusion and/or calendering with or without ribs [54]. Sheets can be embossed and corrugated to create troughs that become channels upon stacking. These can be rolled up or stacked to assemble honeycomb structures. Sheets can be selectively laminated into a stack, and then the stack can be expanded into a honeycomb [10, 55]. The sheets can be decorated before assembly so as to have periodic chemistry, windows, and so forth on the sheet [56]. The advantage of sheets is that there is no limit to the face area of the honeycomb made by stacking or rolling, because it occurs outside of the die. The face area of a honeycomb made by direct extrusion is mechanically and practically limited by force (extrusion pressure times the area). However, larger diameter substrates can be assembled from smaller substrates outside the die in order to make large segmented substrates [57].

In summary, honeycombs made from points, lines, and planes are assembled outside of their die; they are alternatives to direct honeycomb extrusion, a process in which the flow streams assemble and knit inside the die. Ultimately, profitability, speed, flexibility, and process-dependent properties (quality of knit and total knit area per unit volume of honeycomb) are the criteria for comparing these different honeycomb fabrication processes.

2.2.3 Composition

For paste extrusion of a honeycomb, the particles, binder, and processing equipment are designed to work together as a system of compromises and synergies to obtain a successful honeycomb product.

2.2.3.1 Paste

The direct extrusion of a honeycomb requires that the material be viscoplastically formed into a honeycomb, and requires that this shape and composition neither slump nor fracture during handling, drying, debinding, firing, finishing, assembling, and operating. There are two classes of materials for honeycomb extrusion: melts and pastes. Melts are densified and then shaped (like a glass melt). Pastes are shaped and then densified (like a ceramic paste). A melt is homogeneous. A paste is a composite in that it is a mixture of binder "liquid" and particles, a powder which is usually either too refractory or reactive to be economically melt-processed. Any powder (polymer, metal, glass, ceramic, etc.) can be extruded into a honeycomb provided it is compatible with its binder, a polymer solution.

Examples of paste extrusion are spaghetti [58] and pencil lead, the first high-tech ceramic [59]. For a paste, the shaping shear occurs in the binder phase, such that the particle assembly transforms from one shape to the next without fracture. The role of the binder is to saturate (no compressible gas phase) and to not seep/leak through the particulate pack. "Incompressibilility" and "no seepage" are the necessary conditions to create a hydrostatic condition for shear strain for extrusion for plastic shaping of the particle network (soil mechanics calls this the *critical state*) [60–64]. The other role of the binder is to build viscosity during mixing to create sufficient shear stress for deagglomeration of the particles (*viscous* processing [65, 66]) and then to create sufficient yield stress to maintain the shape of the wet-green extrudate upon exiting the die (Terminology: "wet-green" is as-extruded: undried and unfired, "dry-green" is dried and unfired, "brown" is debinded, and "fired" is the final sintered ceramic). The goal is to disperse the particles and then to agglomerate them into a single uniform agglomerate (the paste) and then to reshape/extrude it into a honeycomb.

2.2.3.2
Mixing

The formation of the paste is a transient event. Particles and liquid are brought together. The components are mixed and "sintered": liquid–solid interfaces displace/minimize the air–solid and air–liquid interfaces, expelling the air phase, coalescing/consolidating the free-flowing, air-dispersed ingredients into a saturated (no gas phase) plastic paste, a dough that sticks to itself. The transition from free-flowing powder to paste results in an increase in shear stress, and this works the paste to shear apart hard agglomerates and elongate liquid droplets. The paste is homogenized from being a poorly mixed composite of low and high viscosities and solids loadings to being a paste which is homogeneous in viscosity and composition. The mixing process breaks/shears the paste at its inhomogeneities and then reknits, so as to progressively remove the defect population, until the paste uniformly plastically deforms rather than fractures when its spaghetti extrudate is bent around a radius in an unconfined state (volume is not kept constant). The tighter the bend radius without fracture, the "longer" the paste is said to be. A paste that readily fractures is called "short". Some "long" pastes dilate (increase in volume), imbibe air so as to uniformly open the particle network in a show of fracture toughness, rather than undergoing brittle opening of one large crack in the unconfined state. Extrusion is a confined deformation process (constant volume), in which cracks cannot open up within the die, and thereby the fracture response to stress is suppressed to favor the plastic response.

In manufacturing, high-shear mixing and pressurization is accomplished with a twin-screw auger [67–69]. In the laboratory, the paste can be evaluated with a Brabender® mixer/rheometer and/or calendered (pasta maker: fold and roll again and again) to observe how the properties evolve with shear stress and strain as the paste is homogenized. The paste can be calendered into thin sheets, as thin as the honeycomb web, to observe flexibility, drying, and the population of mixing defects [70, 71]. Calendering is a good surrogate test for evaluating the honeycomb extrudability of a paste. A paste will calender well if it extrudes well as a honeycomb. But, a paste that calenders well does not necessarily knit well during honeycomb extrusion.

The thinness of a sheet which can be made by calendering is determined by the tensile cohesiveness of the paste and its peal adhesiveness to the roller, and for extrusion these properties relate to die-entrance yield stress and die-land yield stress [72], respectively. Honeycomb extrusion has the additional constraint that self-adhesion (knit) is also required. To simulate the honeycomb knit, sheets can be laminated in the calender. The goal is for the strength of lamination to be the same as the cohesiveness of the original tape. The quality of the knit is expected to be a function of thickness reduction (elongation) in the nip of the rollers, a function of particle rearrangement and accommodation at the lamination interface.

2.2.3.3
The Binder

C.F. Binns described the paste process as "making ropes out of sand" [73]. To do this, a binder is required. The binder is usually a solution of a high molecular weight polymer. The required molecular weight and concentration increase with particle size. A good starting point for 5 μm particles is a 15 % solution of a polymer with a molecular weight greater than 200 000. The polymer concentration can be lower if the particles have a natural plasticity, like clay as opposed to glass powder. A good starting point for solids loading (volume fraction of the paste occupied by non-deformable particles times 100) is 50 vol% for a typical powder [72, 74, 75]. The binder can also contain immiscible liquids [76], surface active agents like lubricants, and others. Composition adjustments can quickly be evaluated with the calender method described above.

The rheology of the paste can be described as Hershel–Bulkley (a shear-thinning paste with a yield stress), and its Benbow parameters [72] can be measured by capillary rheometry. The Benbow parameters can be used to describe the honeycomb extrusion flow, which can be used to quantify the rheological and tribological effect of each component of the paste [63, 64, 77].

The target yield stress of a paste depends on many factors, for example, workable extrusion pressure and necessary wet strength (the necessary yield stress of the wet-green honeycomb extrudate to prevent deformation during handling). There are trade-offs: pastes with higher yield stresses result in higher extrusion pressures, and this limits the possible combinations of maximum cell density, open frontal area, diameter, and extrusion rate for the paste–extruder–die system. The combination of high cell density and high open frontal area is a challenging target. For the current

Corning product line, the maximum combination is the 900/2 Celcor® substrate [11]: 900 channels per square inch (csi) with a 2 mil web (nominally, a 50 μm thick web with a 800 μm pitch) with 88 % OFA .

The binder must also enable sufficient lateral paste flow and knit between the pins in the honeycomb die. Webs or corner intersections (depending on the location of the feed hole) contain the knit lines that run the axial length of the honeycomb. The binder contributes to the quality of this knit. An overlubricated stiff paste may calender well, but does not knit, and thus would result in a loose bundle of ribbed spaghetti exiting the extrusion die instead of a honeycomb.

The patent literature covers a wide variety of binder systems for paste extrusion: aqueous, nonaqueous, and combinations [72, 74, 76, 78–81]. Of these, that based on Methocel® polymer (methylcellulose and hypromellose, manufactured by Dow Chemical Co.) are the most fascinating in that their aqueous polymer solutions gel reversibly on heating [82, 84]. Methocel is insoluble in hot water and soluble in cold water; it is a unique commercial polymer. The consequence is that the wet-green extrudate stiffens upon heating (the batch can be cooled during mixing to prevent gelation from the heat from the work of mixing), even before it dries. And, it dries very quickly, because the water is no longer osmotically held by this polymer once the temperature is above its gel point. This is very attractive for manufacturing.

2.2.4
Thermal Processing

Once the particles are assembled into the shape of a honeycomb, a series of extraction processing steps are executed to transform the particle assembly into a sintered ceramic. Water is removed, then polymer, then contaminants and inorganic decomposition products, then surface area. The loss of surface area, the loss of porosity, is due to transition from porous to nonporous ceramic. Surface area is traded for strength.

2.2.4.1
Diffusion: Drying and Debinding

After the honeycomb shape is extruded, a series of subtraction processes are preformed that requires chemical and thermal diffusion and exchange. The honeycomb geometry is perfect to aid these processes. The great advantage of the honeycomb geometry is that its diffusion distances can be effectively short, because honeycomb channels are like diffusion superhighways/arteries connecting to the capillaries of web porosity. Thus, a honeycomb with a large volume dries and debinds like a thin self-supporting tape. Diffusion times that would normally scale to the square of half the thickness of an object [85] can be made to scale to the square of the radius of the channel or of half the web thickness because of high open frontal area and channels that are easily flowed through. The optimized "fast-light-off" design of the automotive catalytic substrate [86] is beneficial for the flow-through reactions required for its own fabrication: drying, debinding, and firing of large honeycombs.

About 50 vol% of the wet-green honeycomb web is binder (water, polymer, etc.), but on the basis of the entire volume of the substrate, it is only 10 vol% binder (for a honeycomb with 80% open frontal area), and less than 2 vol% polymer for the entire object. Such a content of polymer is very low for such a large ceramic object; it has almost the same polymer volume fraction and diffusion distance as is found in the famous "Norton Sand Filled Balloon" demonstration [87].

2.2.4.2
Melt Manipulation

Cordierite ($2\,Al_2O_3 \cdot 2\,MgO \cdot 5\,SiO_2$) honeycombs are used as substrates for ceramic automotive catalyst supports and diesel particulate filters (DPF). Most cordierite honeycombs are extruded from precursor powders that react on firing to become cordierite. There are many precursors and resulting microstructures; however, the "talc path" to cordierite is a good demonstration of the use of melt manipulation to make a unique porous ceramic [88].

For the talc path to cordierite, the cordierite precursors are talc, quartz, kaolin, calcined kaolin and alumina. During the reaction path to cordierite (1400–1430 °C), at about 1355–1370 °C the so-called talc melt event occurs [89]. It is a misnomer for a series of low-melt eutectics, but it picturesquely describes how the whole talc particle becomes part of a liquid phase that undergoes capillarily drainage into the surrounding, rigid, porous microstructure. The talc particle melts and drains away into its porous surroundings where it noninstantaneously reacts to form cordierite, and a relic pore of the same size and shape as the talc particle is left behind.

To demonstrate the power of the talc melt event, the precursor paste was divided into two pastes: a talc paste and a non-talc paste (quartz, kaolin, calcined kaolin, and alumina). Core-clad spaghetti was coextruded with a talc paste core and non-talc paste cladding; their proportions were such that together they would react to become cordierite. The wet-green core-clad spaghetti was bundled and isopressed to make a honeycomb precursor: an array of talc filaments in a matrix of non-talc. This composite billet was reduction extruded. The resulting extrudates were rebundled, isopressed, and reduction-extruded again. On firing, the array of talc filaments melted and were capillarily drained into the non-talc porous matrix where they reacted to form cordierite, and a cordierite honeycomb resulted (Fig. 5). Had the distribution been reversed (an array of non-talc filaments in a talc matrix), the honeycomb structure would have collapsed during the talc melt event. This demonstrates the need to engineer the non-talc distribution to percolate and sinter throughout the honeycomb in such a way as to form a strong enough skeleton to absorb and to hold/support the talc melt (which is a liquid and not load-bearing) without losing the honeycomb shape while reacting to form load-bearing cordierite. This same technique can be used to make silicon carbide honeycombs [43].

In the standard random mix of cordierite precursors, during the talc melt event, the honeycomb web is a porous non-talc ceramic, and its pore size distribution

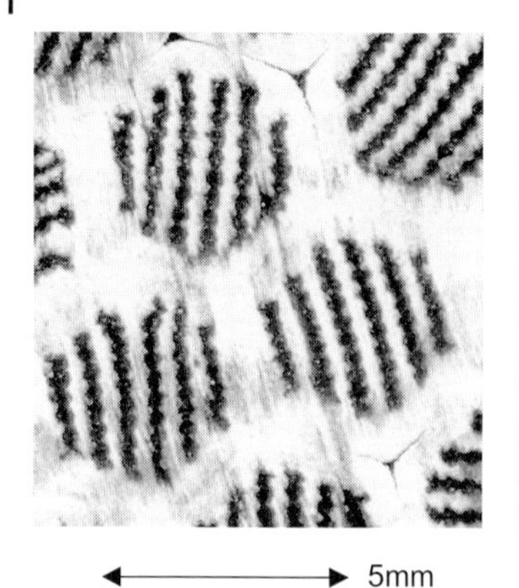
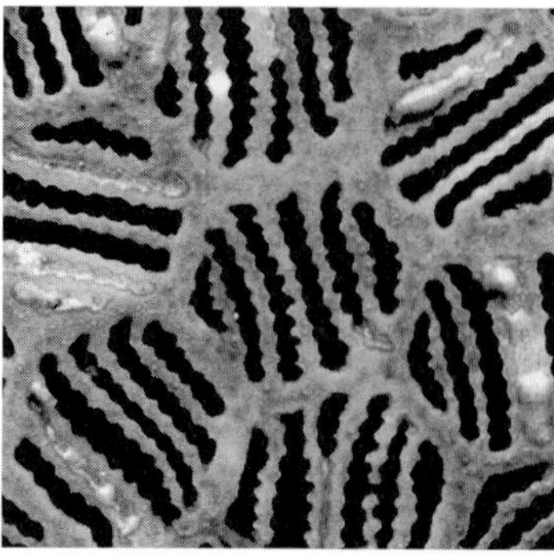

Fig. 5 Talc/non-talc structures: Wet-green talc and non-talc spaghetti were bundled together in proportions that react to form cordierite on firing. The bundle was reduction-extruded, bundled again, isopressed, and reduction extruded. The left photo is a cross section of the dry-green billet (the black area is talc paste, and the white area non-talc paste). The right photo was taken after firing. Note how relic channels were created in the original positions of the talc. The talc melt event caused the talc to form a liquid that was capillarily drained into the porous non-talc region, where it reacted to form cordierite.

redistributes the talc melt with capillary diameter gradients and a surface tension strong enough to resist gravity, causing redistribution of mass and its complement, porosity. On completion of the firing reaction, the webs are cordierite with about 26% porosity, down from about 50% porosity just after binder burnout. Some shrinkage occurs during firing, and it is not necessarily isotropic [90].

2.2.4.3
Sinter Shrinkage Manipulation

The cordierite honeycomb has porous ceramic webs. These webs can be made even more porous and permeable by arranging the precursor particles to pack more openly, and this can be accomplished by adding pore formers to the batch. In diesel particulate filter (DPF) production (Chapter 5.2) [18], graphite particles are added to the batch to create a fired-web porosity of about 49% (13 μm mean diameter) for high web permeability (Section 4.2) and "flow through the wall" design (alternate channels are plugged to force the exhaust flow through the web so as to remove entrained carbon soot particles; Chapter 5.2). The cordierite precursor particles pack around the graphite particle, after which the graphite particle is oxidized/burnt away leaving a relic pore defined by its adjacent precursor particles. Platy graphite is used as a pore former, because its shape has a high aspect ratio, which has a low percolation point for creating a microstructure with high permeability for the DPF application.

Talc and graphite are both platy and form platy relic porosity, and hence the way in which these two particle size distributions percolate together through the paste results in a pore network of high permeability and strength.

During honeycomb extrusion, under shear, the platy particles align their faces with the side faces of the pins of the honeycomb die, which results in a degree of coparallelness between the webs and the platy particles [88]. This can result in beneficial anisotropic cordierite crystal growth for microcracking to achieve a low coefficient of thermal expansion (CTE; e.g., $0.3 \times 10^{-6}\,K^{-1}$), and this alignment can result in beneficial anisotropic drying and firing shrinkage for increasing OFA [90].

Ceramic processing is about the reduction of surface area without loss of shape. In special cases of "enhanced surface diffusion", sintering can occur without loss of porosity [91–93], but in most cases some porosity must be traded in sintering for strength.

High porosity can be obtained by addition of pore formers, open packing particles, and shrinkage inhibitors [49], but some of the porosity is lost to shrinkage during firing. This shrinkage can decrease web thickness, width, and length, and the shrinkage distribution depends on particle arrangements, orientations, and constraints [93–95]. In a special case, a decrease in porosity can result in an increase in open frontal area and an increase in cell density during firing [90].

There is a practical limit to the maximum cell density that can be directly extruded. Everything in honeycomb extrusion process works against the direct extrusion of a honeycomb with higher cell density:

- The higher the cell density, the larger the drag area between the extrudate and the die.
- The higher the cell density, the thinner the webs, and the higher the flow intolerance to inhomogeneities in the batch and the die.
- The higher the cell density, the thinner the webs, and the faster the die wears out of tolerance.

Thus, when the cell density of the honeycomb can be increased without increasing the cell density of the honeycomb die, it is commercially significant. However, converting porosity into higher cell density is problematic because pore formers can be expensive and this transient high porosity is fragile and prone to damage and distortion during pronounced shrinkage during drying, debinding, and sintering.

2.2.5
Post-Extrusion Forming

Post-extrusion forming denotes shape-changing steps that can be performed on the honeycomb after it has been extruded. The most familiar one is the checkerboard plug pattern on the faces of a honeycomb that makes it into a "flow through the wall" filter, such as a diesel particulate filter (Section 5.2) [96, 97]. Less familiar post-extrusion forming operations are reduction extrusion and hot draw for making honeycomb funnels (convergent channels) and ultrahigh cell density honeycombs.

These processes allow one to get around some of the cell density limits imposed by the paste and die technology. The following are some examples of how honeycombs can be viscoplastically altered by reduction extrusion [5] and by hot draw down [50, 93] to increase cell density.

2.2.5.1
Reduction Extrusion

Reduction extrusion entails ram extruding (Fig. 4) a billet of paste through a reduction die (frustum-shaped die) to plastically to reduce the diameter and to increase the length of the billet while maintaining its volume. When the billet is a composite of more than one paste of equivalent plasticity, the cross-sectional pattern of this billet can be reduced in a self-similar manner by reduction extrusion, and this process is popularly known as MFCX (microfabrication by coextrusion) [27, 48].

If a wet-green honeycomb extrudate is reloaded into an extruder barrel and ram-extruded through a reduction die, it plastically deforms in an undesirable manner: the webs buckle and the channels collapse and it ultimately passes through the die as a solid plug. However, if the channels are backfilled with an incompressible, non-seeping material of about the same rheology as the webs, then the honeycomb plastically reduces through the reduction die without its channels collapsing. There is creep flow through the reduction die, so the flow lines (channels) do not cross. The reduction occurs in almost a self-similar manner. The channel filaments in the die show the flow path in the same detail as a FLUENT® flow model trace of a homogeneous paste (Fig. 6). The volume and number of channels are conserved from the entrance to the exit of the reduction die. Channel continuity is maintained through the billet, die, and extrudate. The resulting novel products are honeycomb funnels and honeycombs with higher cell density than can be made by direct honeycomb extrusion.

Reduction extrusion of the wet-green honeycomb can be described from another perspective: the wet-green honeycomb is used as an extrudable mold for plast-casting. Plast-casting is like gel-casting, but in plast-casting the viscous slurry sets to become an extrudable plastic paste [34, 99, 100] rather than an elastic body like in gel-casting [101, 102]. Here, the wet-green honeycomb is used as an extrudable mold to cast, in its matrix of plastic webs, an array of extrudable filaments, the goal being a homogeneous plasticity of a heterogeneous (composite) structure.

The following is an example of the reduction extrusion of a paste-extruded honeycomb [6, 103]: The extruded honeycomb is dried (it now has an open porosity from the loss of more water volume than shrinkage volume during drying) and then trimmed to fit a mold in which it is reconstituted by submersion in hot water (90 °C). During reconstitution the web porosity is saturated with hot water, which is a nonsolvent for the methylcellulose binder (Section 2.2.3.3). While hot, saturated, and still in the mold, the hot channel water, but not the pore water, is drained and replaced with a hot liquid microcrystalline wax (90 °C), which is very fluid. The wax easily fills the channels under only gravitational pressure and is immiscible with the pore water. The backfilled honeycomb is cooled, and the water in the pores redissolves the methylcellulose binder as the liquid wax in the channels crystallizes/soli-

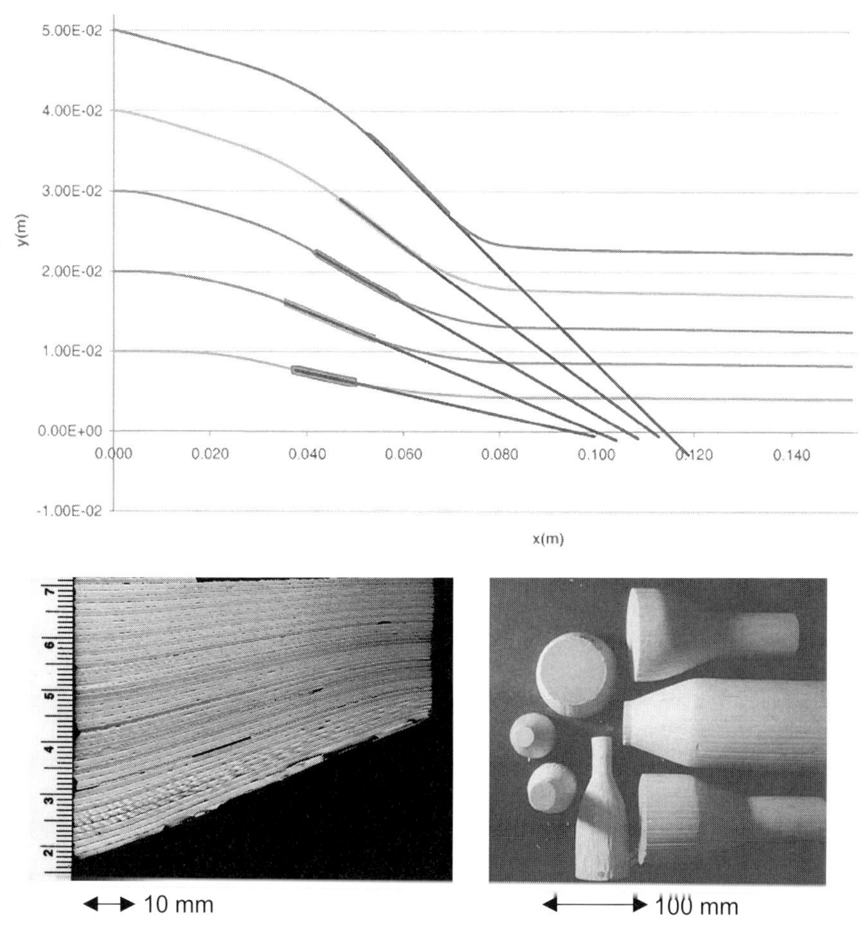

Fig. 6 FLUENT® program was used to model paste flow through a frustum, and its flow trace (top) is understandably similar to the cross section (bottom left) of a honeycomb funnel (bottom right) made by reduction extrusion. Note that both the flow trace and the channels do not converge to a common point (lines have been superimposed on the flow trace to show this lack of convergence to a point).

difies. The webs go from being elastic to being plastic while the channel filaments go from being viscous to being plastic (microcrystalline wax is pliably plastic at room temperature, whereas paraffin wax is brittle). Once at room temperature, the paste matrix and wax filament array are uniformly plastic. The process depends on immiscibility or osmotic balance between the channel/array volume and the web/-matrix volume, and on the dramatic change in rheology of these two volumes coinciding and resulting in a sufficiently similar plasticity at room temperature. This wax-backfilled honeycomb billet is ram-extruded through a reduction die, after which the resulting reduction-extruded honeycombs are dewaxed, dried, vacuum pyrolyzed [104], and fired.

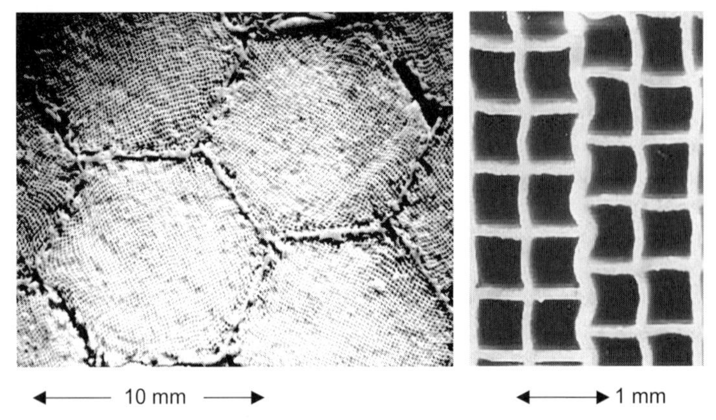

Fig. 7 Segmented reduction extrusion. The left photo is an example of a polycrystalline honeycomb made by bundling and reduction extrusion of previous reduced substrates which had a thick skin on the exterior of the honeycomb. The photo on the right is an example of skinless segmented honeycombs also made by bundled reduction extrusion. Note the misregistration between the honeycomb grids and the knit of the webs at the "grain" boundary.

The process is surprisingly robust, and can easily increase the cell density of a honeycomb from 60 cells per square centimeter to 250 and higher. Reduction extrusion trades honeycomb diameter for cell density. Backfilled honeycombs can be bundled together to increase diameter and channel count; however, the channel lattice will now be "polycrystalline". This bundle can be reduction extruded, too (Fig. 7). The honeycomb can even be re-extruded through the honeycomb die to create two tiers of honeycomb channels (Fig. 8). Care must be taken to not reduce too far, because the potential bulk density and the web thickness of the resulting honeycomb can quickly become incredibly small. "To go too far" is to lose the finest honeycomb structure to the point at which it becomes just a random mix of wax and ceramic paste.

Fig. 8 Celcor® squared: 1000 μm macrochannels with macrowebs composed of 100 μm microchannels with 23 μm microwebs. This object was made by extruding a backfilled honeycomb through a honeycomb die.

For direct paste extrusion of a honeycomb, a rule of thumb is that an extruded web should have at least ten particles spanning its width, otherwise flow disruptions (log jams) will occur in the die. Reduction extrusion does not have this problem, since the smallest features are defined by soft boundary conditions (wax filament array) and not by hard boundary conditions (metal honeycomb die). But a soft boundary condition has its own problems: reduction extrusion decreases the channel diameter and web thickness, but does not correspondingly decrease the size of the microstructure: the particles in the paste remain the same size before and after the reduction die.

Reduction extrusion of the honeycomb increases the cell density but loses some of the open frontal area, either due to shrinkage of the wax during solidification (ca. 15 vol%), and/or due to slip planes [105, 106], and/or due to intrusion of the wax during dilation or during roughening of the web by reorientation of platy particles. For reduction extrusion, the web roughness seems to scale to the largest particle in the paste and the wax backfill, the talc particle (ca. 8 μm) and microcrystalline wax crystal (ca. 18 μm), respectively. So, as the web thickness reduces to become comparable to the size of the largest precursor particle (and the fired mean pore size of ca. 2 μm), considerable roughening of the web occurs. This is expected since the paste can no longer be approximated as a continuum at this level of scrutiny. Its graininess/microstructure and degree of mixedness do not reduce as its honeycomb geometry does during reduction extrusion (Fig. 9) [107].

For greater reduction to higher cell densities, finer web-paste and channel-fill microstructures are required, and this is an incentive to use nanoparticles [108]. The ultimate goal is to have no microstructure.

2.2.5.2 Hot Draw Reduction

A paste of glass particles can be honeycomb extruded, and its particle size distribution of the glass can be eliminated by viscous sintering so that, ideally, there is no microstructure left to scale to the original glass particle size distribution. This glass honeycomb can be hot drawn down (Figs. 9 and 10) to make a honeycomb fiber. Ideally, this fiber can be drawn down to a submicrometer web thickness (see Fig. 2) smaller than the original particles from which it was made [98]. However, glass has its own problems. Glass powder, made by comminution (ball mill, attritor, jet mill, etc.) has contaminants that scale to the glass particle size distribution. This wear debris is often a nonglass which does not viscously hot draw; hence, the debris particle stays the same size as the web shrinks around it. Ultimately, these stones limit the diameter down to which the fiber can be drawn. Superfine glass powders which are built up (e.g., pyrolysis sootlike Cab-o-sil® fumed silica, and sol colloids like Ludox® colloidal silica) rather than broken down do not have grinding debris.

Other sources of microstructure are leaching (alkali depletion of the glass powder or web surface), devitrification, and coarse porosity resulting from unintended pore formers. The pore size distribution in the debinded particle network can be bimodal as the result of a pore-former size distribution in addition to the normal interstices

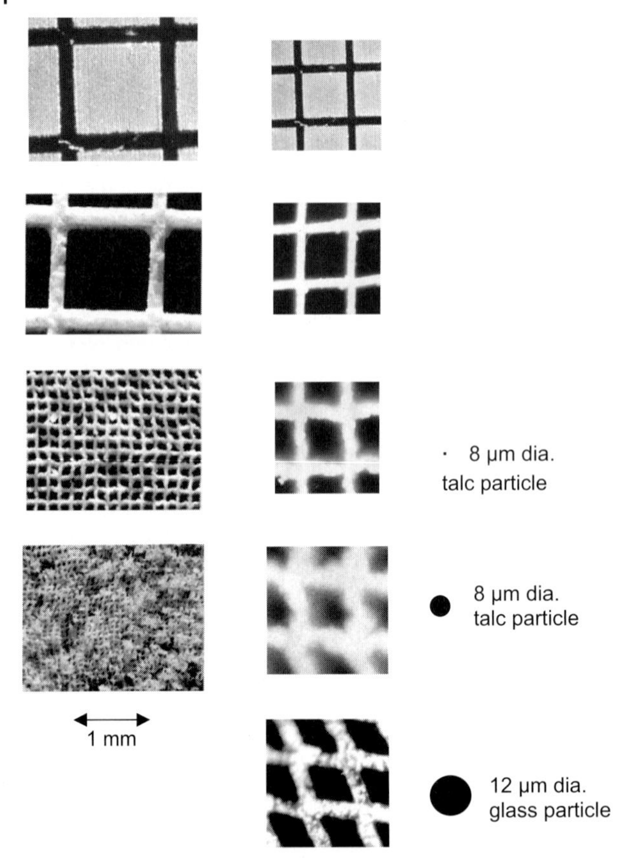

Fig. 9 Microstructural limits of reduction: The left column of photos was taken at the same magnification, and the center column of photos are the same as the left, but were magnified differently so that they have the same size of channel image. The right column is at the same magnification as the center and has a circle that represents the biggest particle in the batch (8 μm diameter talc or 12 μm diameter glass). The top row of photos are of the exit face of the die from which the lower photos were made. The middle group of photos is of cordierite. Note for these cordierite honeycombs how the apparent open frontal area decreases and how the webs roughen as the web thickness approaches the particle size of the paste. The exception is the bottom photo; this honeycomb was made from a glass paste which was viscously sintered to remove all trace of its original particles and hot drawn down to have smooth webs thinner than the diameter of the original particles (12 μm) from which it was made.

of the glass particle size distribution. A coarse pore size distribution can result from packing of the glass particles around unintended pore formers like gas bubbles, metal shavings, lint, insoluble or partially dissolved polymer particles, and agglomerates. The relic pores left by these pore formers can result in pores much larger than the normal interstices of the packed primary particles, and these large relic pores become bubbles as the much finer pore size distribution in the surrounding

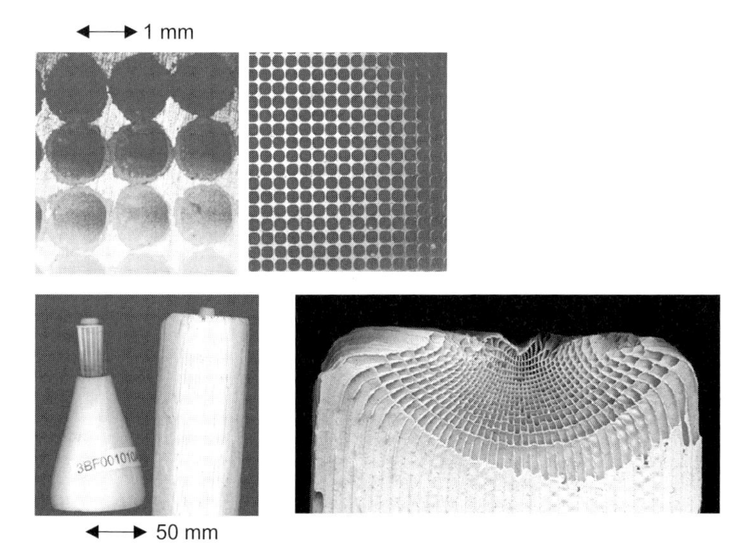

Fig. 10 The funnel on the left (the one with the metal collar) is an ink reservoir; it is a Pyrex® honeycomb funnel that was formed by hot drawing, and its geometry was dictated by soft boundary conditions; that is, the temperature profile. The top photos are of the macro- (left) and microfaces (right) of this funnel, and they were taken at the same magnification. The funnel on the right is extremely abrupt, and it was made by reduction extrusion through a square-entry die, a hard boundary condition. The photo on the bottom right is an off-axis cross section of this funnel. The hard boundary conditions of reduction extrusion die can create a more extreme axial gradient than can hot draw.

matrix is sintered away first. These remaining large bubbles require much more time and temperature to collapse than did the much smaller native interstices (submicrometer). The bubbles are visible, since the glass is clear, and this can make the glass webs opaque.

When designing a paste, it is necessary that the debinding and sintering temperatures are sufficiently separated to avoid carbon and gas entombment with the onset of closed porosity. The debinding temperature can be decreased by using [74]:

- A polymer with a lower burnout temperature [109].
- A catalyst [110].
- A more reactive atmosphere [111].
- A particle network with higher permeability (pore size distribution created by particle size distribution and packing efficiency).
- A multicomponent/multistage binder-removal process [112].

The time and temperature before the onset of closed porosity can be increased by selecting a more viscous glass, larger particle size, and higher initial porosity and permeability. The goal here is to not make a foamed glass.

The surface area of the glass particles must be cleaned and the interstitial gases evacuated or exchanged with helium before closing the open porosity so as to not have a nondiffusing atmosphere in the bubble, because a nondiffusing atmosphere would resist the viscous collapse of the bubble [113–115]. Some large bubbles can persist, because the larger their diameter, the longer the time and/or the higher the temperature needed to collapse even a vacuum bubble. It is better to prevent or eliminate unintended pore formers during batching and mixing rather than to eliminate their conesquences during sintering. However, if a "fix" is still needed, hot isopressing of the closed porosity with argon gas (nondiffusing) can collapse the bubbles at a lower temperature and in a shorter time.

The sintering time and temperature schedule must be such as to avoid phase separation and devitrification in the glass, which affect the viscosity. Devitrification/precipitation can grow crystallites sufficiently large to prevent useful draw-down. Potassium borosilicate glass is particularly resistant to devitrification, and fused silica is not. The fused-silica strategy is to sinter for a shorter time at a lower temperature by starting with cleaner and smaller particles and pores.

Viscous sintering of a glass powder is a careful balancing act for processing of a honeycomb of glass particles. Sintering consumes fastest the surfaces with the smallest radii of curvature. The art and science of ceramic processing is to reduce surface area without losing the shape of the object, without turning the mass into a sphere or a puddle. The goal is for the pore size distribution of the packed glass particles to be much smaller than the web thickness to minimize sintering away of the channel surface area (which would result in a loss of open frontal area) and of the honeycomb shape (rounding of the channels and gravitational slump) during viscous sintering to zero web porosity.

In the forming of a glass honeycomb by paste extrusion, it is interesting to track the surface area throughout the process: surface area is created by pulverization/-comminution or precipitation to create a glass powder, some surface area is lost in sintering to create a nonporous honeycomb preform, and then surface area is regained in hot draw to create a honeycomb fiber. Here, the random surface area of the powder is eventually recreated as the nonrandom surface area of the honeycomb fiber which defines the microfilament array of air.

An example is the fabrication of a *glass honeycomb funnel* from potassium borosilicate glass (7761 Pyrex® glass, see Fig. 11 for viscosity) [52, 107].

Glass tubes were hot melt extruded from a glass melt tank. The tubes were crushed, dry ball-milled, and dry sieved to make a particle size distribution with an average particle size of about 12 μm (Fig. 12). The glass powder was dry blended with Methocel® powder, and liquids were added during the mulling process, whereby the powder was consolidated into paste chunks. The stiff paste was then evacuated and ram-extruded three times through a spaghetti die (an array of 3 mm diameter holes) for high-shear mixing, and then ram-extruded through a honeycomb die. The honeycomb extrudate was dielectrically dried enough to fix its shape and then convection-dried (70 °C). It was then trimmed (30-inch length) and vertically suspended in an electric furnace (700 °C max.) in a controlled atmosphere to debind and then to sinter. The length of the honeycomb shrank due to sintering while it lengthened owing to gravitational slump. This vertically suspended fired honeycomb ended with a slight axial taper due to the gravitational stress gradient along its length. This

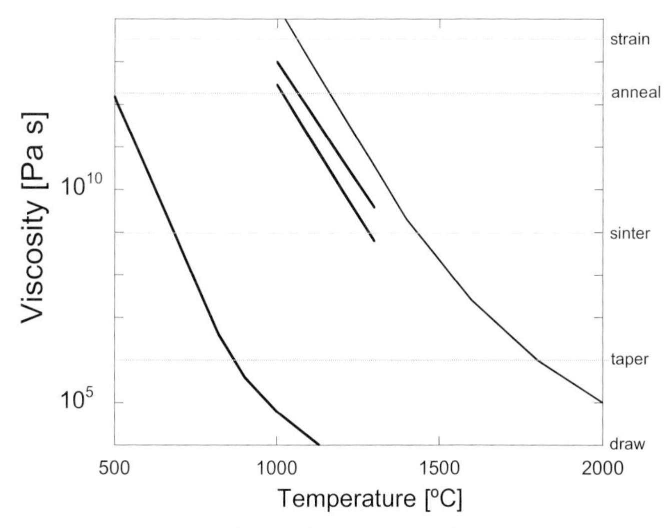

Fig. 11 Viscosity curves from the literature: Top to bottom: GE fused silica (low-hydroxyl) [116], Corning 7940 (HPFS® glass, ca. 0.16 wt% OH) [117], 7971 (ULE® glass, ca. 0.16 wt% OH) [117], Pyrex® 7761 (potassium borosilicate) glass [118].

glass honeycomb was then heated at the middle of its length and then drawn down to make a honeycomb funnel. A balance of temperature gradients and channel pressure [21, 22, 52] and axial tension controlled the draw geometry. The funnel was cut from the draw; it is the root and not the gob. The resulting honeycomb funnel was ground and polished to have a macro base with about 90 channels per square centimeter, a micro top with about 2000, and web thicknesses of about 200 and 20 μm, respectively (see Fig. 10).

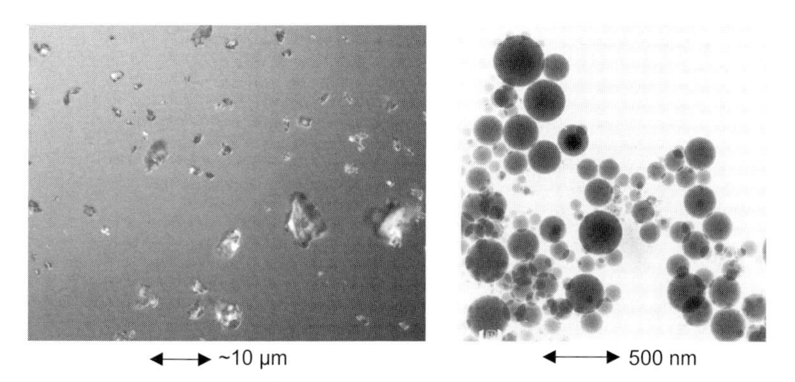

Fig. 12 Two powders: Pyrex® 7761 glass particles with a 12 μm mean diameter and spherical ULE® soot with a 0.2 μm mean diameter.

This honeycomb funnel is used as an ink reservoir for applications in combinatorial chemistry and biology [119]. The inks are loaded into the macro-array of openings from where they are conveyed through the channels by capillary action resulting from the axial diameter gradient of the funnel. The net force from surface tension is greater than that of gravity on the slugs of ink, so the inks move from the macro to the micro end of the funnel to form a micro array of ink menisci. A matching pin array dips into this menisci array, picks up and transfers a drop array, and prints it on a slide.

This same Pyrex® preform can be drawn down further to make a honeycomb fiber (see Fig. 9). However, the extent to which this Pyrex preform can be hot-drawn is limited by the stone, grinding, and metal wear debris from all the previous Pyrex processing steps.

An example at the other end of the glass spectrum is the production of *high-purity fused silica honeycomb* by paste-extrusion of high-purity fused silica (HPFS®) soot powder [120–125]. This work was an interdepartmental effort at Corning Inc. to merge existing technologies of soot generation, honeycomb extrusion, beneficiation, and sintering for a cost-effective manufacture of clear honeycomb structures [126, 127].

The strategy was to use existing materials (standard HPFS soot and binder, about 10% Methocel solution [120–125]) and processes (mixing and extrusion equipment) to create clear honeycombs without making any significant changes to the current processes, and so this approach differed substantially from the literature on silica paste extrusion. Current approaches in the literature use the least amount of polymer [127–130] to make a simple final shape: a rod or a tube. These objects have much thicker diffusion distances (ca. 1 mm) than the honeycomb web (ca. 0.1 mm) described here, and so the literature objects are more difficult to dry, debind, and clean. This honeycomb experiment was to determine whether the easy-cleaning thinness of the honeycomb webs was sufficient to compensate for the rest of the process, so as to allow the production of clear honeycombs from the existing infrastructure.

Silica soot is made by flame combustion of silica precursors. It is an aerosol process that grows a particle up to the desired diameter. Unlike the above Pyrex glass powder, it is a clean process. The precursors are made by digesting silica particles to synthesize silicon tetrafluoride [131], silicon tetrachloride [132], and now environmentally friendly organosilicon precursor molecules. These are purified and flame hydrolyzed to make pyrogenic silica soot particles [133]. Other compositions can be made, too (e.g., ULE® 7–8% titania glass [134]). The flame combustion process controls the nucleation and growth (coalescence of molten particles), and can generate small soot particles with a desired shape (fractal to spheroid) and size (0.1–1.0 μm diameter) without grinding. The soot powder production process differs from the commercial process for making HPFS and ULE glass, which is a glassmaking process in which the molten/viscous soot particles are directly deposited and sintered, layer by layer, to make boules for making fibers [135] and blanks for making lenses and mirrors [133, 134]. In soot powder production, the soot particles are quenched to prevent consolidation and collected in a bag house. The bagged soot particles are free of hard agglomerates. Silica soots are also commercially available from Cabot (Cabosil® powder) and Degussa (Aerosil® powder); alumina and carbon soots are also available.

The soot particles used in this honeycomb experiment were beautiful spheres (Fig. 12). These soot particles were easily made into an extrudable paste and were plastically extruded into the shape of a honeycomb [123–126]. The honeycomb was dried, debinded, and purified by a beneficiation with hot chlorine gas, so that only silica remained [91, 92, 115].

The goal was to purify the honeycomb with the aid of all of its accessible surface area. The thin porous webs of the honeycomb geometry were perfect for chlorine beneficiation for the removal of iron and sodium resulting from the extrusion process: wear debris from the mixing equipment and ash from the binder, respectively. The honeycomb geometry is optimal for this beneficiation process because the diffusion length is half a web thickness, which is very small compared to the radius of the honeycomb substrate or boule. The honeycomb channels are like "superhighways" to the torturosity of the porosity of the webs. The hot chlorine gas reacts/scavenges for rogue cations to form volatile substances such as iron chloride, sodium chloride, and so on. It is important to minimize the trace sodium content to prevent devitrification of the silica during sintering [115, 136].

The debinding process window was a bit larger for HPFS soot powder than it was for the above Pyrex example. Here, binder removal ($< 600\,°C$) occurred at a much lower temperature than surface area removal (sintering, $> 1200\,°C$), and so there was sufficient temperature difference to carry out the purification step between these two processes: between the opening/growing of porosity by drying and debinding and the closing/shrinking of porosity by viscous sintering in fast diffusing helium [113, 114]. These processes were very dependent on permeablility.

Once the honeycomb was sufficiently purified, the temperature was ramped up to sinter the silica particles to full density and make a clear honeycomb. The sinter process was fast to avoid devitrification [137, 138] and gravitational slump. Fast is not a problem for silica glass, because of its extremely low CTE ($0.5\ ppm\ K^{-1}$), and it is extremely resistant to thermal shock.

Large HPFS® and ULE® honeycombs ($120 \times 120mm \times 25mm$) have been sintered. They were made by this paste extrusion process (see Figs. 4 and 13) as an alternative to the present production method of water-jet cutting of a honeycomb from a solid blank of glass in order to make lightweight cores for telescope mirrors [14, 121]. These honeycombs were made from a soot with a particle size distribution which is favorable to both crack-free drying (a problem if the particles are too small) and slump-free firing (a problem if the particles are too large). The same particle surface curvature that drives sintering also drives cracking during drying.

Smaller particles sinter faster to full density at a lower temperature than do larger particles, because the energy released from the reduction of surface area and curvature drives the sintering process [115]. In the final stage of sintering to full density, the closed pore size distribution should scale to the original particle size distribution, to the original interstitial size distribution. Sintering time t_{sint} is a function of particle size d, particle packing, surface energy γ, and viscosity η (see Fig. 11). The sintering model is rate-controlled by the energy released by the reduction of surface area (shrinkage) and consumed by viscous dissipation [93, 115, 139, 140]:

$$t_{sint} = \tfrac{3}{4}\, d\eta/\gamma. \quad (1)$$

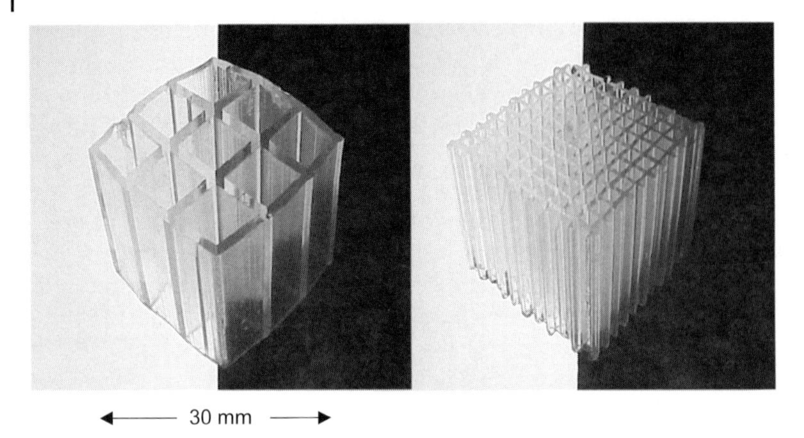

Fig. 13 High purity fused silica honeycombs.

The other viscous processes which occur simultaneously are loss of channel area (open frontal area, OFA) and gravitational slump. Roughly, the time dependence for slump is a function of gravity g, density ρ, viscosity η, and height h of the honeycomb (Fig. 14). The slump time t_{slump} is the time required to thicken the web at the base of the honeycomb by 5 % ($e = 0.05$) [141]:

$$t_{\text{slump}} = (4\,e\eta)/(\rho g h). \tag{2}$$

The other time-dependent process is devitrification: nucleation and growth [117, 137, 138]. The goal is to sinter the honeycomb before the nucleation and growth of crystobalite can occur. This requires low-temperature sintering with small particles to avoid both slump and/or devitrification [115].

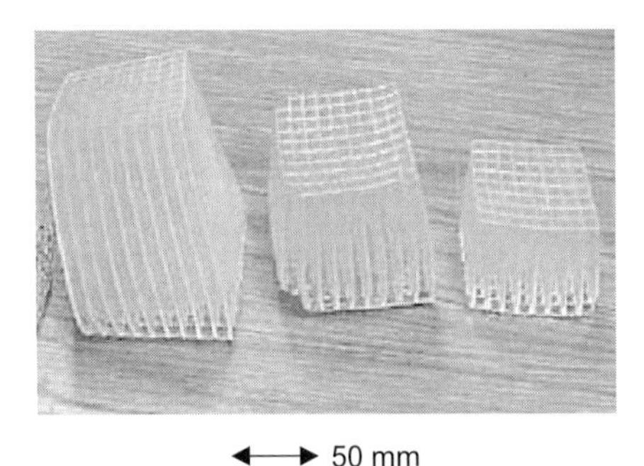

Fig. 14 Slump as a function of height for HPFS® glass consolidated in helium. As expected, shorter honeycombs have less gravitational pressure and less slump flow.

Figure 15 shows sinter time (Eq. 1), induction time [117] and slump time (Eq. 2) as a function of temperature. It defines a region of process space in which it is pos-

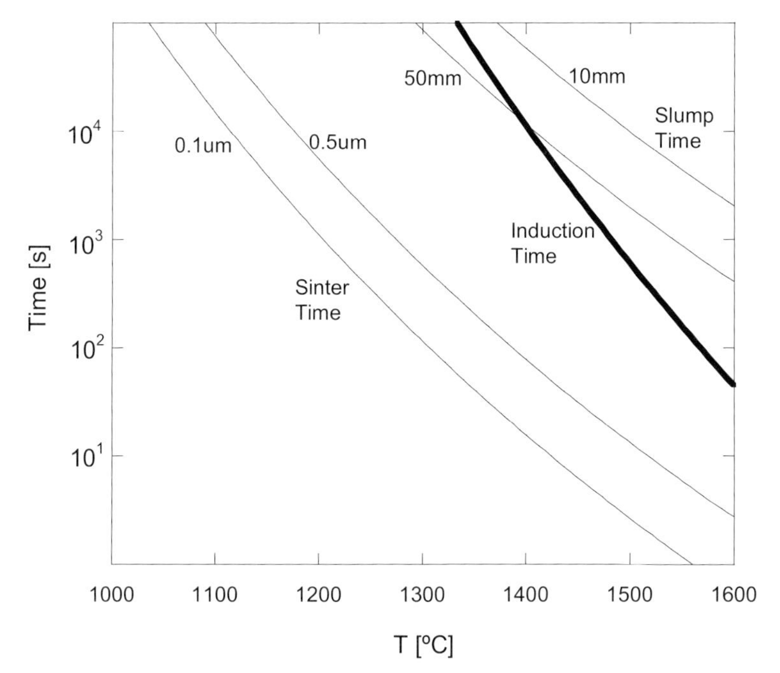

Fig. 15 Sintering, induction, and slump times as a function of temperature for HPFS® glass as extrapolated from Mazurin's data on Corning 7940 glass (ca. 0.16 wt% OH) [117]. Two sintering curves are shown: 0.1 and 0.5 μm particle size *d* (Eq. 1). Two slump curves are shown: 10 and 50 mm honeycomb height *h* (Eq. 2). The slump time was calculated for a 5% thickening at the base ($e = 0.05$). The sintering, slump, and devitrification times would be longer for drier glass and for more viscous glass, such as the GE fused silica in Fig. 11.

sible to sinter silica soot honeycombs clear without bubbles, crystals, and slump. Decreasing the particle size, temperature, and height/ thickness of the honeycomb puts the process in a more optimal space. But this is only true if there are no unintended pore formers that result in pores and consequential bubbles which are much larger than the soot particles. To remove these bubbles requires a higher temperature and longer time than is predicted from the size of the soot particles, and this higher temperature and longer time result in some slumping of the honeycomb during sintering. Get rid of the unintended pore formers during batching and mixing [65, 66, 69], and bubble-free sintering will occur at a lower temperature and in a shorter time with less slump and less devitrification (crystallites).

Clear honeycombs have been made by this paste extrusion process (see Figs. 13 and 14), and these have demonstrated the robustness of the honeycomb geometry (thin porous webs and fast channels) for the removal of water, polymer, sodium, iron, and finally the surface area of porosity.

Currently, efforts in soot extrusion are directed at making a bubble-free honeycomb which is free of microstructure but not free of channels. This honeycomb is the preform from which a honeycomb fiber can be hot drawn. In the previous example we were trying to minimize slump from bulk viscous flow due to gravity. In this example we want slump – in particular, hot viscous draw – in order to make a honeycomb fiber. The fiber drawing process (> 1900 °C) requires more thermal processing time than does simple sintering, so it has a greater probability of devitrification, which causes the honeycomb to fail to draw. Devitrification can be prevented by melting all of the crystobalite nuclei in the preform, and this is done by ramping its temperature to above the melting point of crystobalite (ca. 1725 °C).

Currently, silica *photonic crystal fibers* are routinely hot drawn from preforms made by tube bundling and fusing. The resulting honeycomb fibers have channel diameters on the order of a micrometer with web thicknesses of less than 200 nm (see Fig. 2). These are some of the finest and highest aspect ratio objects ever made. However, the same will be possible for preforms made by paste extrusion, too [98]. This is our current focus. After all, no one wants to bundle tubes if they do not have to, and it would be satisfying to demonstrate:

- The surface area lost in sintering is regained in hot draw.
- A submicrometer web thickness that can be made from submicron soot particles in the roundabout way of paste extrusion.
- A preform made by paste extrusion with no irreductible microstructure, a preform which has only the reducible synthetic meso → micro → nanostructure of channels.

2.2.6
Summary

Our quest for honeycombs with higher cell densities began with the porous cordierite honeycomb with a few hundred channels per square centimeter and ended with a fused silica photonic crystal fiber with almost a billion channels per square centimeter. Along the way, the following observations were made:

- Ceramic and glass honeycombs can be made by paste extrusion, which is an effective way to mechanically create porosity: geometric porosity and geometric surface area. The porous honeycomb geometry is beneficial to its own fabrication (drying, debinding, beneficiation, and firing).
- The paste-extruded honeycomb is a convenient way to arrange particles for easy access for fast chemical and thermal diffusion. Large honeycomb objects can be made with very short effective diffusion distances.
- Nonporous honeycombs start off as being porous when made by paste extrusion. The process of sintering to full density follows the path of open porosity to closed porosity to no porosity. The initial porosity gives access to the surface area of the particles for beneficiation and gas exchange before the porosity closes, after which corrections cannot be easily made.
- Creating a porous honeycomb preform with a microstructure much finer than its web thickness is a prerequisite for reduction extrusion and hot draw.

Sintering a glass honeycomb to have no microstructure is prerequisite for the hot drawing of a honeycomb fiber.
- A glass honeycomb can be hot drawn down to make a honeycomb funnel and a honeycomb fiber, the limiting case being a photonic crystal fiber.

This chapter described the manufacturing of honeycombs by paste extrusion, but in so doing, it necessarily described the manufacturing of porous ceramic and glass microstructures, because to make a good ceramic or glass honeycomb by paste extrusion, one must have a composition and process to make a good porous microstructure, which is a prerequisite to obtaining the final web microstructure of the honeycomb.

An excellent set of tables describing ceramic honeycombs can be found on the web at www.dieselnet.com/tech/dpf_wall-flow.html#intro [96] and http://customer2.corning.com/environmental/ [11].

Acknowledgements

The work presented in this chapter was carried out at Corning Incorporated with a wide cross section of expertise: P. Bardhan, D. Beall, N. Borrelli, C. Booker, A. Buchtel, P. Cimo, T. Dannoux, G. Dillon, J. Fajardo, C. Fekety, M. Fischer, E. Funk, D. Gokey, E. Hale, L. Holleran, J. Humphrey, C. He, K. Hrdina, T. Johnson, K. Koch, R. Layton, R. McCarthy, G. Merkel, A. Olszewski, P. Oram, S. Rajamma, A. Roselstad, D. St. Julien, E. Sanford, N. Venkataraman, J. Wang, M. Wasilewski, J. M. Whalen, D. Witte, J. Wight, S. Wu, R. Wusirika, K. Zaun.

References

1 S. Gulati, *Ceramic Catalysts Supports for Gasoline Fuel* in *Structured Catalysts and Reactors*, A. Cybulski, J.A. Moulijn (Eds.), Marcel Dekker, New York (1996), Chap. 2.
2 R.J Farrauto, R.M. Heck, *Catalytic Air Pollution Control, Commercial Technology*, John Wiley & Sons, Inc., New York, 1995.
3 S.T. Gulati, US Patent 4,323,614, 1982.
4 G.E. Cunningham, G.D. Lipp, L.S. Rajnik, US Patent 5,238,386, 1993.
5 Y. Ichikawa, T. Kondo, M. Miyazaki, M. Shirai, US Patent 6,656,564, 2003.
6 D. St.Julien, J. Wight, S-H Wu, K.E. Zaun, US Patent 6,299,958, 2001.
7 www.whatman.com
8 P. Russell, *Science* **2003**, *299*, 358–362.
9 B. Laprade, R. Starcher, "The 2 Micron Pore Microchannel Plate, the Development of the World's Fastest Detector," Burle Electro Optics, Sturbridge MA (4/3/01). This article is located at the website: http://www.burle.com/cgi-bin/byteserver.pl/pdf/2micron2.pdf
10 Philips, "Micron-Size Honeycomb Manufacturing Process Enabling a Breakthrough in Multichannel Layered Structures". This article is at the website: http://www.yet2.com/app/list/techpak?id=30774&sid=90&abc=0
11 The Corning website for Celcor® information: www.customer2.corning.com/environmental/.
12 M. Edwards, T. W. Hobbs, Current Fabrication Techniques for ULE® and Fused Silica Lightweight Mirrors in Space Telescopes and Instruments V, SPIE Proceedings Vol. 3356, P.Y. Bely, J. B. Breckinridge (Eds.), SPIE, 1998, 3356–3341.
13 M. Lampton, "Optical Space Telescope Assembly," Space Sciences Laboratory, University of California Berkeley. This article is

located at the website: www.snap.lbl.gov/pubdocs/Lampton.pdf.

14 C.M. Smith, N. Venkataram, M.T. Gallagher, D. Muller, *Nature* **2003**, *424, 8/7/03*, 657–659.

15 Precision Micro Extrusion, Coors Tech Brochure G0201 8510-1072 rev.A, 2002.

16 R. Bagley, US Patent 3,905,743, 1975.

17 T. Palucka, *Invention & Technology*, **2003**, *19[3]*, 22–31

18 W.H. Pitcher, US Patent 4,329,162, 1982.

19 H.P. Hood, M.E. Norderg, US Patent 2,221,709, 1940.

20 M. Raney, US Patent 1,563,587, 1925.

21 S.B. Dawes. M.T. Gallagher, D.W. Hawtof, N. Venkataraman, US Patent Application 2003/0230118, 2003.

22 J.C. Fajardo, M.T. Gallagher, J.A. West, N. Venkataraman, US Patent Application 2003/0231846, 2003.

23 D.D. Walz, US Patent 3,616,841, 1971.

24 D.D. Walz, US Patent 3,946,039, 1976.

25 General Electric, GB Patent 916,784, 1963.

26 K. Schwartzwalder, H. Somers, A.V. Somers, US Patent 3,090,094, 1963.

27 A.T. Crumm, J.W. Halloran, *J. Am. Ceram. Soc.* **1998** *81* [4] 1053–1057.

28 M.E. Badding, J.F. Wight, US Patent 6,551,735, 2003.

29 J.K. Cochran, K.M. Hurysz, K.J. Lee, M. Liu, W.L. Rauch, T.H. Sanders, "Extruded SOFC Stacks, Processing and Performance," (CB-S2-11-2004), 28th International Conference & Expo on Advanced Ceramics & Composites, Cocoa Beach, FA (1/25/04).

30 M.J.Cima, J.S.Haggerty, E.M.Sachs, P.A. Williams, US Patent 5204,055, 1993.

31 S.S. Crump, US Patent 5,121,329, 1992.

32 P.D. Calvert, J. Cesarano, US Patent 6,027,326, 2000.

33 X. Zhao, J.R.G. Evans, M.J. Edirisinghe, J-H. Song, *J. Am. Ceram. Soc.*, **2002**, *85* [8], 2113–2114.

34 J. Cesarano, J.A. Lewis, J.E. Smay, *Langmuir* **2002**, *18*, 5429–5437.

35 J. Cesarano, J.A. Lewis, J.E. Smay, B.A. Tuttle, *J. Appl. Phys.* **2002**, *92[10]*, 6119–6127.

36 J. Cesarano, L.R. Evans, R.M. Ferrizz, J.E. Miller, J.N. Stuecker, "Novel Monolithic Supports for Catalytic Combustion,"18th North American Catalysis Society Meeting, Cancun, Mexico, June 1–6, 2003.

37 J. Cesarano, R.M. Ferrizz, E. Lindsey, J.E. Miller, J.N. Stuecker, "Robocast Monoliths for Catalyst Support and Diesel Particulate Traps," 28th International Cocoa Beach Conference and Expo on Advanced Ceramics and Composites, CB-S3-49-2004.

38 J.F. Bianchi, F. Gonzales, G. Muench, W.C. Pfefferle, S. Roychoudhury, "Development and Performance of Microlith™ Light-off Preconverters for LEV/ULEV," SAE Technical Paper Series 971023, SAE International, Warrendale PA, 1997.

39 J-P. Stringaro, US Patent 5,240,663, 1993.

40 D. St.Julien, US Patent 5,525,291, 1996.

41 J-A. Bruhn, *Designs in Miniature: The Story of Mosaic Glass*, The Corning Glass Museum, Corning, NY, 1995.

42 I.M. Lachman, US Patent 5,053,092, 1991.

43 L.J. Levy, US Patent 5,774,779, 1998.

44 A. Barda, M. Griffith, J.W. Halloran, C.V. Hoy, *J. Am. Ceram. Soc.* **1998**, *81* [1], 152–158.

45 Information at the website: http://www.schott.com/fiberoptics/english/products/imagin/flexiblecomponents/leached.html.

46 G.J. Fine, US Patent 4,913,518, 1990.

47 L. Curtis, US Patent 3,589,793, 1971.

48 A. Barda, G.A. Brady, J.W. Halloran, G.E. Hilmas, D. Popovic, S. Somers, G. Zywicki, US Patent 5,645,781, 1997.

49 Kriven, W., Lee, S-J., *J. Am. Ceram. Soc.* **2001**, *84[4]*, 767–774.

50 Blodgett, K.B., *J. Am. Cer. Soc.* **1951**, *34[1]*, 14.

51 H. Rauscher, US Patent 4,112,170, 1978.

52 N.F. Borrelli, A.R.E. Carre, T.L.A. Dannoux, B. Eid, D. Root, R.R. Wusirika, US Patent 6,350,618, 2002.

53 R.Z. Hollenbach, US Patent 3,112,184, 1963.

54 F.J. Sergeys, US Patent 3,755,204, 1973.

55 C.L. Kehr, US Patent 3,660,217, 1972.

56 R. Brück, M. Reizig, European Patent 1285153A1, 2003.

57 R.L. Frost, R.D. McBrayer, V.K. Purjari, US Patent 4,381,815, 1983.

58 Croce, J.D., *Ultimate Pasta*, D.K. Publishing, Inc., New York, 1997.

59 H. Petroski, *The Pencil: A History of Design and Circumstance*, Knopf, New York, 1992.

60 M. Janney, "Plasticity of Ceramic Particulate Systems," Doctorate Dissertation, University of Florida, FL (1982).

61 G. Onoda, M. Janney, Application of Soil Mechanics Concepts to Ceramic Particulate Processing in Advances in Powder Technolo-

gy, G.Y.Chin (Ed.), American Society of Metals, OH, 1981, pp. 53–74.
62 J. Reed, J. Wight, *Am. Ceram. Soc. Bull.* **2001**, *80* [4], 31–35; *80* [6], 73–76.
63 J. Reed, J. Wight, *J. Am. Ceram. Soc.*, **2002**, *85* [7], 1681–1688.
64 J. Reed, J. Wight, *J. Am. Ceram. Soc.*, **2002**, *85* [7], 1689–1694.
65 K. Kendall, N.M. Alford, S.R. Tan, J.D. Birchall, *J. Mater. Res.* **1986**, *1* [1], 120–123.
66 J.W. Cotton, *Ceram. Ind.*, **1993**, June, 60–61.
67 R.W. Gardner, D.L. Guile, M. Lynn, US Patent 4,551,295, 1985.
68 E. Ford, D.L. Guile, L.R. Quatrini, US Patent 5,213,737, 1993.
69 N.A. Golomb, C.J. Malarkey, US Patent 6,375,450, 2002.
70 M. Inoue, Z. Kato, J.-Y. Kim, K. Saito, N. Uchida, K. Uematsu, "Development of Direct Observation Method for Internal Structures in Silicon Nitride Granule and Green Body," *Trans. MRS Jpn.* **1990**, 192–199.
71 Z. Kato, J.-Y. Kim, M. Miyashita, N. Uchida, K. Uematsu, *J. Am. Ceram. Soc.* **1991**, *74* [9], 2170–2174.
72 J. Benbow, J. Bridgwater, *Paste Flow and Extrusion* in *Oxford Series on Advanced Manufacturing*, J.R. Crookall, M.C. Shaw, N.P. Suh (Eds.), Clarendon Press, Oxford, 1993.
73 C.F. Binns, *The Potter's Craft, a Practical Guide for the Studio and the Workshop*, 2nd Ed., D.Van Nostrand, New York, 1922
74 R.M German, *Powder Injection Molding*, Metal Powder Industries Federation, Princeton NJ, 1990.
75 J. Reed, *Introduction to the Principles of Ceramic Processing*, Wiley Interscience, New York, 1995.
76 D. Chalasani, M. Fischer, C. Malarkey, K. McCarthy, B. Stutts, M. Zak, US Patent 6,080,345, 2000.
77 G.Dillon, "Analysis of the Effect of Sodium Stearate on the Extrusion Behavior of a Cordierite Body," Master Thesis, New York State College of Ceramics, Alfred, NY, 1992.
78 M.K. Faber, T.D. Ketcham, D. St. Julien, US Patent 5,458,834, 1995.
79 J.M. Barnard, R.E. Johnson, K.A. Wexell, US Patent 5,574,957, 1996.
80 I.M. Lachman, L.A. Nordlie, US Patent 4,631,267, 1986.
81 K. Gadkaree, D. Hickman; Y.L. Peng, T. Tao, US Patent 6,228,803, 2001.
82 N. Sarkar, *J. Appl. Polym. Sci.* **1979**, *24*, 1073–1087.
83 F.A. Cantaloupe, R.I. Frost, L.M. Holleran, US Patent 3,919,384, 1975.
84 R.W. Gardner, D.L. Guile, US Patent 4,551,295, 1985.
85 C.J. Geankoplis, *Transport Processes and Unit Operations*, 2nd ed., Allyn and Bacon, Inc., Newton, MA, 1983.
86 Information at website: www.corning.com/environmentaltechnologies/ auto_emissions_magazine/archived_issues/Fall1998/article3.asp.
87 F.H. Norton, *J. Am. Ceram. Soc.* **1948**, *31* [8], 236–241.
88 I.M. Lackman, R.M. Lewis, US Patent 3,885,977, 1975.
89 I.M.Lachman, *Porosity in Extruded Cellular Ceramics* in *Advances in Ceramics, Vol 9, Forming of Ceramics*, The Amer. Ceram. Soc., Columbus OH, 1984, pp. 201–211.
90 D.M. Beall, G. Merkel, US Patent 6,506,336, 2003.
91 J. Lee, D.W. Readey, *Microstructure Development in Fe_2O_3 in HCl Vapor* in *Materials Science Research Vol.16, Sintering and Heterogeneous Catalysis*, G.C. Kuczynski, A.E. Miller, G.A. Sargent (Eds.), Plenum, New York, 1984, pp. 145–147.
92 D.W. Readey, T. Quadir, *Microstructure Evolution in SnO_2 and CdO in Reducing Atmosphere* in *Materials Science Research, Vol.16, Sintering and Heterogeneous Catalysis*, G.C. Kuczynski, A.E. Miller, G.A. Sargent (Eds.), Plenum, New York, 1984, pp. 159–171.
93 W.D. Kingery, H.K. Bowen, D.R. Uhlmann, *Introduction to Ceramics*, 2nd ed. John Wiley & Sons, New York, 1976, p. 494.
94 K.R. Mikeska, D.T. Schaefer, US Patent 5,254,191, 1993.
95 A.H. Kumar, B.J. Thaler, A.N. Prabhu, US Patent 5,876,536, 1999.
96 www.dieselnet.com/tech/dpf_wall-flow.html#intro.
97 P. Stobbe, H.G. Petersen, S.C. Sorenson, J.W. Høj, "A New Closing Method for Diesel Particulate Filters," SAE Technical Paper Series 960129, SAE International, Warrendale, PA, 1996.
98 N.F. Borrelli, J.F. Wight, R.R. Wusirika, US Patent 6,260,388, 2001.

99 R. Clasen, US Patent 4,682,995, 1987.
100 D.A. Fleming, P. Hubbauer, D. W. Johnson, J. B. MacChesney, T. E. Stockert, F. W. Walz, US Patent 6,080,339, 2000.
101 N.S. Bell, L. Bergstrom, W.S. Sigmund, *J. Am. Ceram. Soc.* **2000**, *83* [7], 1557–1574.
102 J. B. Macchesney, European Patent 1,172,339, 2002.
103 A.B., Buchtel, D.A. Earl, L., Holleran, *J. Mater. Sci. Letters*, **2001**, *20*, 1759–1761.
104 S.K. Robinson, "MIM/Ceramic Part Debinding Methods," Industrial Heating website posting 2/14/02 http://www.industrialheating.com/CDA/ArticleInformation/features/BNP_Features_Item/0,2832,72164,00.html.
105 H. Hodgkinson, *Claycraft* **1962**, *362]*, 42–48.
106 F. Moore, *Rheology of Ceramic Systems*, Maclaren and Sons, London,1965, pp. 70–71.
107 J. Wight, "Celcor Funnels: Channeling from the Macro to the Micro," ACerS Annual Meeting, St.Louis (5/02).
108 M. Allahverdi, E. Niver, R. Riman, A. Safari, "Photonic Band Gap (PBG) Structures via Micro-Fabrication by Co-Extrusion (MFCX)," Nanotechnology for Photonic Materials and Devices, NJIT Kick Off Meeting, 11/19/2002.
109 M. Hurley, *Ceram. Ind.*, **1995**, Nov., 51–44.
110 D.C. Krueger, US Patent 5,531,958, 1996.
111 J. ter Maat, H. Wohlfromm, *Int. Ceram.* **1998**, *[2]*, 35–39.
112 L. Bowie, C.L. Kehr, D. Wayne, US Patent 3,351,495, 1967.
113 J.E. Shelby, *J. Am. Ceram. Soc.* **1972**, *55* [4], 195–197.
114 P.K Onorato, D.R. Uhlmann, M.C. Weinberg, *J. Am. Ceram. Soc.* **1980**, *63* [3], 175–180.
115 E.M., Rabinovich, *J. Mater. Sci.* **1985**, *20*, 4259–4297.
116 GE Website: http://www.gequartz.com/en/thermal.htm.
117 O.V. Mazurin, L.K. Leko, L.A. Komarova, *J. Non-Cryst. Solids* **1975**, *18*, 1–9.
118 Corning 7761 data sheet, Corning Inc., Corning NY.
119 D. St. Julien, "Inorganic Tools for Bio-Discovery," Bioceramics Symposium, Am. Ceram. Soc. 103th Annual Meeting and Expo, April 22–25, 2001.
120 Engelhard Industries, GB Patent 1,010,702, 1965.
121 R. Clasen, B. Schmidl, US Patent 4,816,051, 1989.
122 T. Yagi, T. Satoh, Y. Koinuma, K. Yoshida, US Patent 5,314,520, 1994.
123 G. Kar, K.E. Hrdina, J. Wight, C. Yu, US Patent 6,468,374, 2002.
124 G. Kar, K.E. Hrdina, J. Wight, C. Yu, US Patent 6,548,142, 2003.
125 G. Kar, K.E. Hrdina, J. Wight, C. Yu, US Patent 6,479,129, 2002.
126 D.Gokey, K.Hrdina, J.Wight, "Paste Extruded Glass Honeycombs," Glass & Optical Materials Division of The American Ceramic Society Fall Meeting (10/2003).
127 J.E. Pierson, US Patent 3,782,982, 1974.
128 R.D. Shoup, W.J. Wein, US Patent 4,059,658, 1977.
129 R. Clasen, US Patent 4,682,995, 1987.
130 D.A. Fleming, P. Hubbauer, D.W. Johnson, J.B. MacChesney, T.E. Stockert, F.W. Walz, US Patent 6,080,339, 2000.
131 R.N. Secord, US Patent 2,886,414, 1959.
132 L.S. Belknap, US Patent 3,145,083, 1964.
133 J.F. Hyde, US Patent 2,272,342, 1942.
134 M.E. Nordberg, US Patent 2,326,059, 1943.
135 D.B. Keck, R.D. Maurer, US Patent 3,775,075, 1973.
136 P.P. Bihuniak, *J. Am. Ceram. Soc.* **1983**, *66*, C-188.
137 D. Uhlmann, *J. Non-Cryst. Solids* **1972**, *7*, 337.
138 A.K. Varshneya, *Fundamentals of Inorganic Glasses*, Academic Press, Boston, 1993, p. 55.
139 J.K. Mackenzie, R. Shuttleworth, *Proc. Phys. Soc.* **1949**, *LXII*, 12-B, 833–852.
140 Scherer, G.W., *J. Am. Ceram. Soc.* **1977**, *60* [5], 236–246.
141 Y. M. Stokes, "Very Viscous Flows Driven by Gravity with Particular Application to Slumping of Molten Glass", Doctorate Thesis, University of Adelaide, Dept. of Applied Mathematics, July 1998.
142 N. Borrelli, A. Douglas, J. Fajardo, M. Gallagher, K. Koch, J. Wight, "Fabrication Approaches for Glass-Guiding and Air-Guiding PCFs," American Ceramic Society Annual Meeting, St.Louis, MO, (4/30–5/3/00).

2.3
Three-Dimensional Periodic Structures

Jennifer A. Lewis and James E. Smay

2.3.1
Introduction

Three-dimensional (3D) periodic structures comprised of interconnected cylindrical rods may find widespread technological application as advanced ceramics [1, 2], sensors [3, 4], composites [5, 6], tissue engineering scaffolds [7], and photonic materials [8, 9]. Robotic deposition processes are direct-write techniques utilizing the extrusion of particle-filled inks and are capable of assembling such structures with submillimeter precision. Examples include fused deposition [4, 10] and Robocasting [11–13]. In both techniques, the extruded ink forms a continuous filament that is patterned in a layer-by-layer sequence to assemble complex 3D architectures. The primary distinction between these two approaches lies in ink design and post-deposition processing. Robocasting utilizes concentrated colloidal gels as inks [12, 13], whereas fused deposition uses particle-filled, polymeric inks [10]. Process control of and geometries attainable by either technique depend intimately on the dynamic viscoelastic properties of the ink during extrusion and patterning. The principal advantage of using colloidal gel-based inks is that the as-formed structures do not have high organic content and therefore are not subject to the lengthy binder-removal process typically associated with fused deposition. While fused deposition is a viable technique and other methods [14] (e.g., 3D printing [15], stereolithography [16]) may produce similar structures, this chapter focuses on the manufacturing of 3D periodic structures by extrusion-based robotic deposition using colloidal gel-based inks. First, a brief overview of this technique is provided. This is followed by detailed description of the design of colloidal gel-based inks, ink flow during deposition, and shape retention of the as-deposited features. Finally, examples of the types of 3D structures that can be produced by this assembly route are provided.

2.3.2
Direct-Write Assembly

Robotic deposition, illustrated schematically in Fig. 1, employs an ink delivery system mounted on a z-axis motion control stage for agile printing on a moving x–y platform. Coordinated three-axis motion is controlled by a computer program,

Cellular Ceramics: Structure, Manufacturing, Properties and Applications.
Michael Scheffler, Paolo Colombo (Eds.)

ISBN: 3-527-31320-6

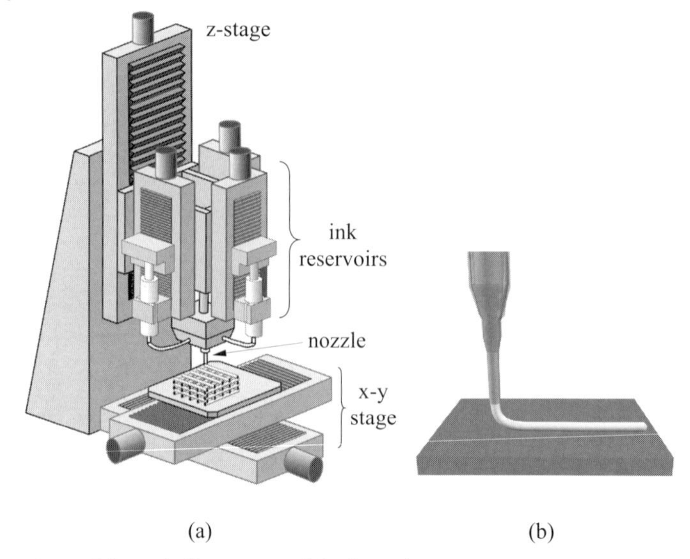

Fig. 1 Schematic illustrations of a) robotic deposition apparatus and b) ink extrusion through a cylindrical nozzle onto a substrate.

which allows for the design and assembly of complex 3D structures in a layer-by-layer deposition scheme. The ink(s) are housed in multiple reservoirs affixed to the z-axis stage and extruded through a cylindrical nozzle oriented vertically above the x–y stage. The composition of the extruded filament can be varied by mixing various inks (shown schematically in Fig. 1a), or different ink compositions can be deposited discretely from different nozzles. Upon exiting the nozzle, the cylindrical rod (or filament) of ink bends by 90° to lie parallel to the x–y plane, as shown in Fig. 1b. The volumetric flow rate of the ink is proportional to the x–y table speed, which is typically 1–10 mm s^{-1}. A multireservoir ink delivery system facilitates compositional grading along a given rod, between rods in a given layer, or between layers. Fused deposition uses a similar motion-control platform; however, the ink delivery system requires a heated nozzle to liquefy the particle-filled polymer ink during extrusion.

Direct-write assembly techniques allow any conceivable 2D pattern to be printed within a layer. Upon solidification, the layer then serves as a platform for the deposition of subsequent layers to give the desired 3D architecture. In the original conception of Robocasting, concentrated colloidal suspensions were utilized for the construction of space-filling solid components. These inks solidified after deposition by a drying-induced pseudoplastic-to-dilatent transition [11]. Space-filling components benefited from this initial ink fluidity, which allowed for the creation of a continuous structure. To fabricate self-supporting spanning structures, the ink design [12, 13] was changed to utilize the inherent viscoelastic properties of colloidal gels to enable the assembly of 3D periodic structures.

2.3.3 Colloidal Inks

Colloidal inks developed for direct-write assembly of 3D periodic structures with spanning features must satisfy two important criteria. First, they must exhibit a well-controlled viscoelastic response, that is, they must be able to flow through a deposition nozzle and then "set" immediately to facilitate shape retention of the deposited features even as they span gaps in the underlying layer(s). Second, they must contain high volume fractions of colloid to minimize drying-induced shrinkage after assembly, that is, the particle network must be able to resist compressive stresses arising from capillary tension [17]. These criteria require careful control of colloidal forces to first generate a highly concentrated, stable dispersion followed by inducing a system change (e.g., ΔpH, ionic strength, or solvent quality) that promotes a fluid-to-gel transition [13].

Colloidal inks are produced by first preparing an aqueous dispersion of colloidal particles (e.g., ceramic, polymer, semiconducting, or metal colloid). Polyelectrolyte species such as poly(acrylic acid), PAA, are typically added as dispersants to provide suspension stability. PAA is a linear polymer that has one ionizable carboxylic acid group per monomer unit. Under appropriate conditions of pH and ionic strength, it is fully ionized (or electrostatically charged), and this allows it to provide the electrosteric stabilization necessary to create highly concentrated colloidal suspensions with a solids loading of about 50 vol%. By altering solution conditions such as decreasing pH or increasing ionic strength, the colloidal stability can be reduced to cause the desired fluid-to-gel transition and yield a concentrated, colloidal gel-based ink, as illustrated schematically in Fig. 2.

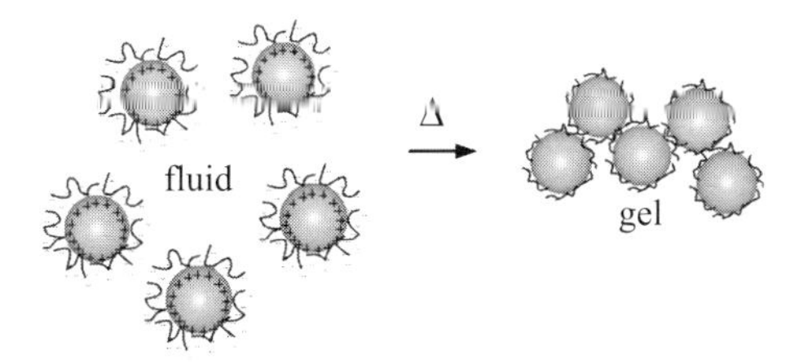

Fig. 2 Schematic illustration of the fluid-to-gel transition that occurs for a polyelectrolyte-stabilized colloidal dispersion (fluid phase) upon changing pH or increasing ionic strength of the solution to induce attractive interactions between colloidal particles (gel phase).

Colloidal gels consist of a percolating network of attractive particles capable of transmitting stress above a critical volume fraction ϕ_{gel}, known as the gel point. If the colloid volume fraction ϕ within the gelled ink is held constant, the elastic properties of the ink can be controlled by altering the strength of the interparticle attractions according to the scaling relationship [18] given by:

$$y = k\left(\frac{\phi}{\phi_{gel}} - 1\right)^{x} \tag{1}$$

where y is the elastic property of interest (shear yield stress τ_y or elastic modulus G') k is a constant, and x is a scaling exponent (ca. 2.5). The equilibrium mechanical properties of colloidal gels are governed by two parameters: ϕ, which is representative of the interparticle bond density, and ϕ_{gel}, which scales inversely with bond strength. As the interparticle forces are made more attractive, colloidal gels (of constant ϕ) experience a significant and controllable rise in their elastic properties [18–26] (see Fig. 3). In addition to the ability to control equilibrium elastic properties, colloidal gels have a dynamic ability to break down (when the applied shear stress τ exceeds τ_y) and rebuild network structure (under quiescent conditions, $\tau \rightarrow 0$). This makes colloidal gel-based inks well suited to fulfill the requirements of flow through the deposition nozzle while maintaining the elasticity necessary to promote shape retention.

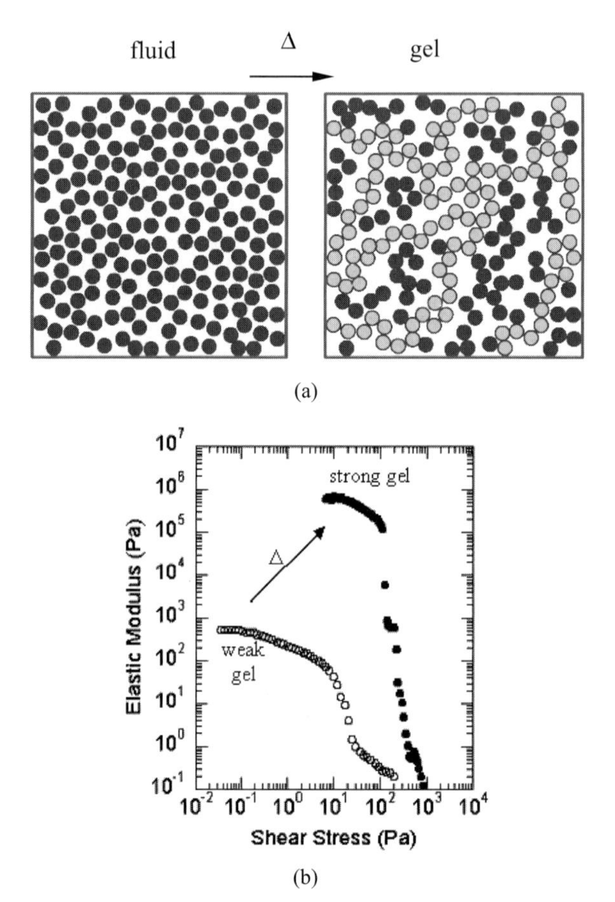

Fig. 3 a) Schematic illustration of attractive particle network that forms upon gelation and b) log–log plot of elastic modulus as a function of shear stress for both a weak and strong colloidal gel. (Adopted from Ref. [13].)

2.3.4
Ink Flow during Deposition

During direct-write assembly, the concentrated colloidal-gel-based ink must flow through a fine deposition nozzle (ca. 100 μm to 1 mm in diameter) at the volumetric flow rate required to maintain a constant deposition speed ν of 1–10 mm s^{-1}. The shear rate profile that the ink experiences during the deposition process depends on the nozzle diameter, the deposition speed, and the rheological properties of the ink [7, 12]. Recently, the flow behavior of viscous Newtonian fluids within a cylindrical deposition nozzle has been modeled by Baer et al. [27]. The computed shear rate profile across the nozzle cross section is shown in Fig. 4 for a 100 Pa·s fluid flowing

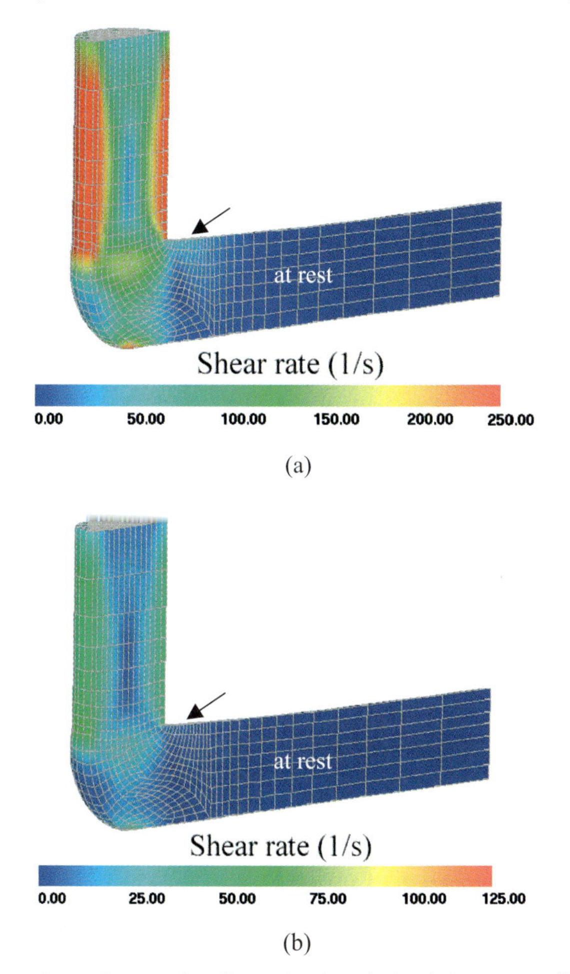

Fig. 4 Contour plots illustrating the calculated shear rate profile within the deposition nozzle (arrow denotes end of nozzle) for a viscous Newtonian fluid (100 Pa·s viscosity) deposited at a table speed of 5 mm s^{-1} through cylindrical nozzles of diameter of a) 250 μm and b) 1.37 mm. (Adopted from Ref. [27].)

through nozzles of diameter 0.25 and 1.37 mm at a constant deposition speed of 5 mm s^{-1}. As expected, the fluid experiences a maximum shear rate (or stress) along the nozzle walls. The maximum values observed were approximately 250 s^{-1} in the 250 μm nozzle and 50 s^{-1} in the 1.37 mm nozzle under these conditions. In both profiles, it is evident that the observed shear rate (or stress) experienced by the fluid decreases towards the center of the nozzle to a nearly stress-free state. Immediately upon exiting the nozzle, the ink experiences a 90° bend and all regions within the filament undergo some deformation. However, once the ink filament is deposited onto the underlying layer (or substrate), it returns to a quiescent state.

The rheological behavior of colloidal gels is complex, and can be described by the Hershel–Bulkley model [28]:

$$\tau = \tau_y + K\dot{\gamma}^n \tag{2}$$

where n is the shear thinning exponent and K is the viscosity parameter. Ink flows through the cylindrical nozzle when a pressure gradient (ΔP) is applied in the axial direction producing a radially varying shear stress τ_r parallel to the nozzle axis given by:

$$\tau_r = \frac{r\Delta P}{2\ell} \tag{3}$$

where r is the radial position within the nozzle (i.e., $r = 0$ at the center axis and $r = R$ at the nozzle wall). The flow rate can be calculated from the constitutive equation (Eq. 2) and the applied stress field given by Eq. (3).

A Hershel–Bulkley material flowing through a fine nozzle may develop a three-zone velocity profile consisting of 1) an unyielded core of radius r_c moving at constant velocity surrounded by 2) a yielded shell experiencing laminar flow, and, possibly, 3) a particle-depleted slip layer at the nozzle wall [29, 30]. The slip layer is a thin fluid layer with thickness $\delta \ll R$, between the nozzle wall and the bulk ink. The volumetric flow rate Q as a function of applied pressure is found by summing the contributions from slip at the nozzle wall and the integrated velocity profile in the core-shell region:

$$Q = \pi R^2 (v_s + f(\tau_s)) \tag{4}$$

where v_s is the slip velocity, and $f(\tau_s)$ the integrated velocity profile of the core and shell region. The slip velocity is determined by:

$$v_s = \frac{\left(\tau_R^4 - \tau_s^4\right)}{4\tau_R^3 \eta_c} \tag{5}$$

where τ_R is the shear stress at the nozzle wall, τ_s the shear stress at the slip layer/gel interface, and $\eta_c = 30$ mPa·s. The function $f(\tau_s)$ is given by

$$f(\tau_s) = \frac{R\alpha^{m+1}}{m+1}\left(\frac{\tau_s}{K}\right)^m \left(\frac{\tau_s}{\tau_R}\right)\left[1 - \frac{2\alpha}{m+2} + \frac{\alpha^2}{(m+2)(m+3)}\right] \tag{6}$$

where $m = 1/n$, n is the Hershel–Bulkley shear thinning exponent, K the viscosity parameter $\alpha = (1-\tau_y/\tau_s)$, and τ_y is the gel yield stress. These material parameters can be used to calculate a pressure gradient required for a desired flow rate (or deposition speed) using Eqs. (3)–(6), while the shear rate profile within the nozzle is derived from Eqs. (2) and (3). Under the condition of no slip boundary ($\nu_s = 0$), Eq. (6) reduces to the Buckingham–Reiner relationship for an ideal Bingham fluid or the Hagan–Poiseuille relationship for a Newtonian fluid [31].

The shear rate profiles for a representative colloidal ink were calculated at varying deposition speeds under slip and no-slip boundary conditions by using Eq. (2), where the slip layer thickness was taken as the mean particle diameter. These profiles are plotted in Fig. 5, where the abscissa is the reduced radial position within the nozzle r/R. The no-slip boundary condition (solid lines) led to a wall shear rate of 100–400 s^{-1} for deposition speeds of 2–8 mm s^{-1}. No core region is apparent on this scale except for the 2 mm s^{-1} profile, where $r_c/R \approx 0.01$. Introduction of the slip layer (dashed curves) reduced the wall shear rate to between 0.8 and 5 s^{-1}, with the core region expanding to $r_c/R \approx 0.1$–0.3 with decreasing deposition speed. These calculations indicate that the no-slip condition leads to high shear rates throughout the extruded filament and thereby alters the ink structure. In contrast, the ink structure experiences significantly less change during extrusion under the slip boundary condition.

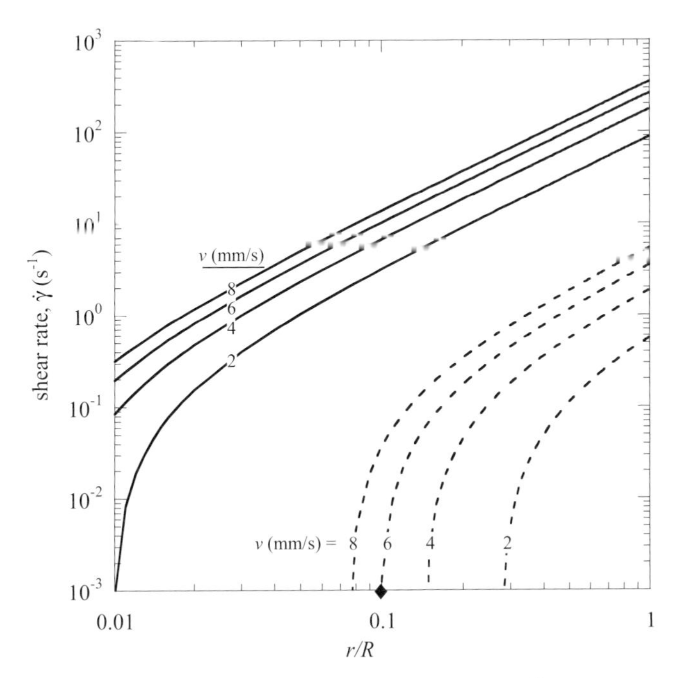

Fig. 5 Calculated shear rate profiles for a representative colloidal ink ($\tau_y = 100$ Pa) in a 200 μm diameter capillary geometry as a function of varying deposition speed. Solid lines assume no-slip condition, and dashed lines assume slip boundary of 0.65 μm. (Adopted from Ref. [12].)

2.3.5 Shape Evolution of Spanning Filaments

To create 3D periodic structures by direct-write assembly, filaments are deposited in the form of a linear array of rods aligned along the x- or y-axis such that their orientation is orthogonal to the underlying layer. During deposition, the extruded filament must span gaps in the underlying layer, as it first anchors to one of the supporting filaments, then traverses the unsupported gap region, and finally anchors to an adjacent supporting filament. Despite the dynamic nature of this process, the spanning filament can be viewed to be reasonably stress-free in the initial state and to deform to the equilibrium shape allowed by the elastic properties of the filamentary elements (or rods) when anchored to the second support [12].

A static, simply supported beam is an idealized model of the spanning phenomena that captures the essential body forces and ink properties, which influence the structural evolution of spanning filaments after deposition. In this model, the spanning filament is represented as a beam of circular cross section that is simply supported at its ends (see Fig. 6b). Such a beam deflects in proportion to the distributed load of its own weight as a function of distance from the supports given by [32]

$$\delta z = \frac{Wy}{24EI}\left(2Ly^2 - y^3 - L^3\right) \tag{7}$$

where W is the distributed load ($= 0.25[\rho_{\text{eff}}]g_o\pi D^2$), ρ_{eff} the effective ink density ($\rho_{\text{ink}}-\rho_{\text{fluid}}$), and ρ_{fluid} the density of the deposition medium (e.g., air or liquid), g_o the gravitational constant, y the position along the rod, E the Young's modulus of the gel ($E=(1+2\nu)G'$), $\nu = 0.5$ is Poisson's ratio for the gel [18], and I the area moment of inertia of the circular cross section ($=\pi D^4/64$). Thus, for a specific span distance, the modulus and/or effective moment of inertia may be chosen as degrees of freedom to capture the elastic behavior of the beam and deflections predicted by Eq. (7).

To demonstrate the influence of ink rheology on shape evolution, V-shaped test structures were deposited by direct-write assembly, as shown in Fig. 6, and their deflected shape measured by noncontact laser profilometry. The height profiles observed for test structures produced from representative colloidal inks of varying elastic properties are shown in Fig. 7. The color scale representing height data was limited to a range from 0 to 200 μm corresponding to the diameter D of a single filament. The maximum rod deflection occurred midway between the inner supports and increased with span length ($L = 0.53$ mm near the apex of the structure to $L = 2.60$ mm at the base of the triangular support). Test structures assembled from the weakest colloidal gel experienced severe deformation even at modest span lengths (see Fig. 7a), whereas those assembled from the strongest ink exhibited maximum deflections of less than $0.25D$ for spanning distances up to 2 mm ($10D$; see Fig. 7b). Such observations provide important ink-design guidelines. Spanning-filament deflection is directly related to the elasticity of the colloidal gel in a form similar to Eq. (7). By tailoring ink rheology, 3D periodic structures comprised of spanning filaments can be created by direct-write assembly.

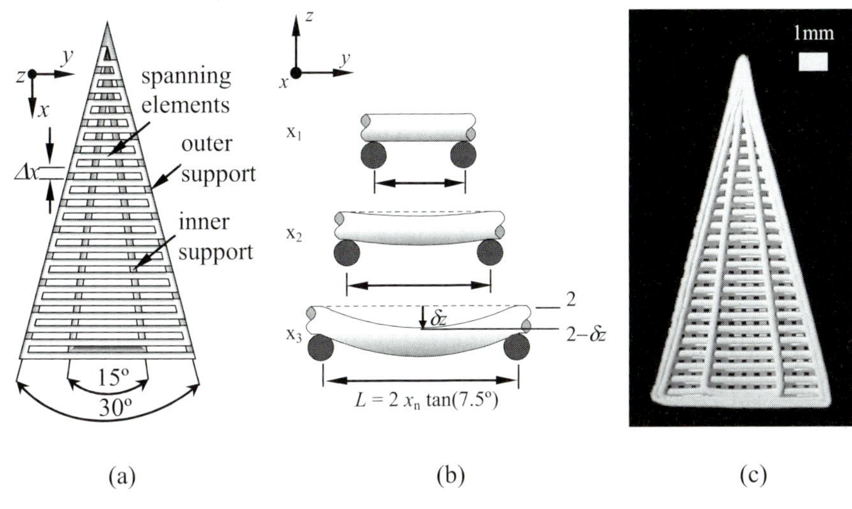

Fig. 6 a) Schematic top view of V-shaped test structure highlighting the inner and outer support structures and spanning elements (marker layer is not shown). b) Illustration of selected spans demonstrating the reference height of 2 mm and the variation of span length L between the inner supports as a function of x position. c) Top view of a dried V-shaped test structure assembled from a concentrated colloidal ink at 6 mm s^{-1} deposition speed. (Adopted from Ref. [12].)

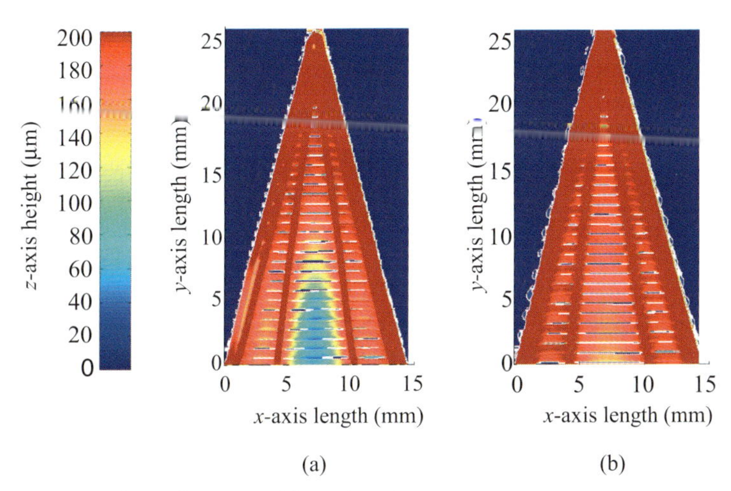

Fig. 7 Height profiles acquired for V-shaped test structures assembled from a) weak ($G'(\infty) = 1$ kPa, $\tau_{yield} = 30$ Pa) and strong ($G'(\infty) = 10$ kPa, $\tau_{yield} = 100$ Pa) colloidal gel. (Adopted from Ref. [12].)

2.3.6
Direct-Write Assembly of 3D Periodic Structures

Robotic deposition is a facile approach for fabricating 3D periodic structures from a wide variety of materials, including structural, functional, and bioactive ceramics. 3D periodic lattices of varying geometry, filament (rod) diameter, lattice spacing, and composition have been produced. Recent work in the area of piezoelectric ceramic/polymer composites will be highlighted to illustrate the structural features accessible by this technique.

3D periodic lattices consisting of alternating layers of parallel filaments with a 90° rotation between layers were assembled from a concentrated colloidal ink at a deposition speed of 6 mm s^{-1}. The filament diameter D was held constant within any single structure while the spacing between filaments within a layer was varied. As an example, the rod spacing L varied from 300 μm to 1.2 mm when using a 200 μm nozzle and was fixed at 1.21 μm when using a 400 μm nozzle. To illustrate that the extruded filament need not be deposited in a straight line, radial arrays were assembled ($D = 200$ μm) by sequential deposition of layers with alternating patterns of concentric rings and a circular array of radially oriented rods. The first layer was an array of five equally spaced concentric rings between an inner and outer diameters of 3.8 and 9.7 mm. In the second layer, an array of radially oriented rods was deposited between the inner and outer radii. The angular spacing between the rods was 4.68° such that the arc length between adjacent rods varied from 0.38 to 0.19 mm on going from the outer to inner radius. The pore architecture can be easily designed by varying the deposition pattern and the spacing between filaments.

Representative 3D periodic lattices and a radial array assembled from a colloidal gel-based ink are shown in Fig. 8. The 3D lattice structures in Fig. 8a and b result if filaments aligned parallel to the x-axis are stacked in simple columns, and the filaments aligned parallel to the y-axis are stacked in a similar pattern. The 3D lattice in Fig. 8c results from an alternative pattern in which x-axis-aligned filaments are stacked with a 1/2 rod spacing offset in alternating layers. High-quality interlayer bonding is observed at the junction points between filaments in adjacent layers. These junctions strengthen the cellular ceramics, facilitating both green-body handling and post-densification processing. In fact, the cross-sectional images in Fig. 8 were obtained by carefully slicing the structure with a diamond saw, but no embedding in a support material was required. The representative radial array shown in Fig. 8c has an overall cylindrical symmetry. The visible top layer is an array of radial lines, and the underlying layer is a series of concentric rings. The rings maintained their circular shape during deposition despite the changing arc length between supports provided by the radial lines in the previous layer. In addition, the deposited rods maintained a circular cross section and spanned the gaps in underlying layers with minimal deflection.

The 3D periodic structures can be represented by unit cells, as highlighted in Fig. 8a and c. Schematic illustrations of unit cells for both simple and "face-centered" stacking patterns are shown in Fig. 9, along with accompanying illustrations of their pore space. For the simple stacking pattern, the unit cell has a height c defined as twice

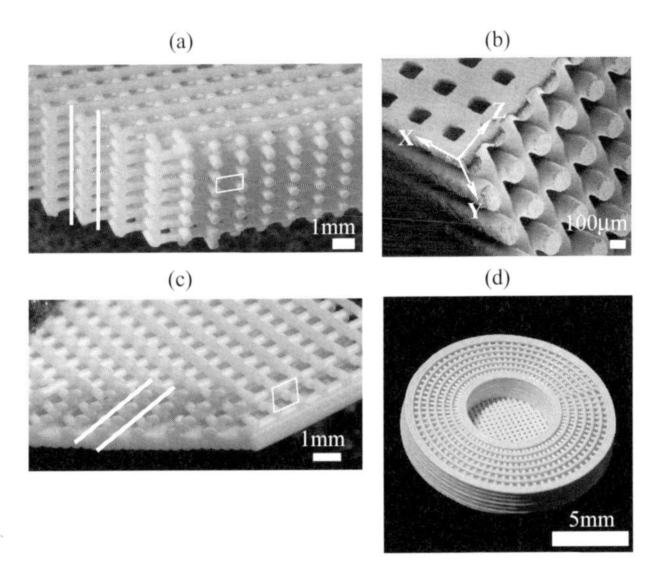

Fig. 8 Optical and SEM images of 3D periodic lattices in a)–c) and a radial array in d) comprised of cylindrical rods (ca. 250 μm in diameter) interconnected in all three dimensions [12]. (Reprinted with permission from Smay, J.E., Cesarano III, J., and Lewis, J.A., *Langmuir* **2002**, *18*, 5429–5437. Copyright 2002 American Chemical Society.)

the filament diameter, $c = 2D$. For the face-centered stacking pattern, the unit cell repeat height is $c = 4D$. In the lateral directions, the unit cell dimensions a, b are determined by the spacing of filaments within a layer. For the 3D lattices shown in Fig. 8, the unit cell lengths are identical ($a = b$); however, this is not required. Therefore, both simple and face-centered stacking patterns can be produced with tetragonal ($a = b \neq c$) or orthorhombic ($a \neq b \neq c$) unit cells if alternating filament layers are perpendicular. Also note that the filament orientation between layers need not be restricted to 90° rotations.

Within any single layer the solid volume fraction Γ_p is defined by:

$$\Gamma_p = \frac{\pi}{4}\frac{D}{L} \tag{8}$$

with $1-\Gamma_p$ equal to the volume fraction of porosity. The number of pores per inch (ppi) or pores per meter (ppm) is simply the reciprocal of the filament spacing. Given the ease with which the filament spacing can be varied by robotic deposition, the ppi can be different in all three dimensions. The surface area (SA) to volume (V) ratio is another important feature of cellular ceramics, and is given by:

$$SA/V = \frac{\pi}{L} = \frac{4\Gamma_p}{D} \tag{9}$$

where SA/V is given in units of inverse length (m^{-1}). Note that Eq. (9) neglects the effect of the junction regions created due to rod–rod interfacial bonding. This ratio

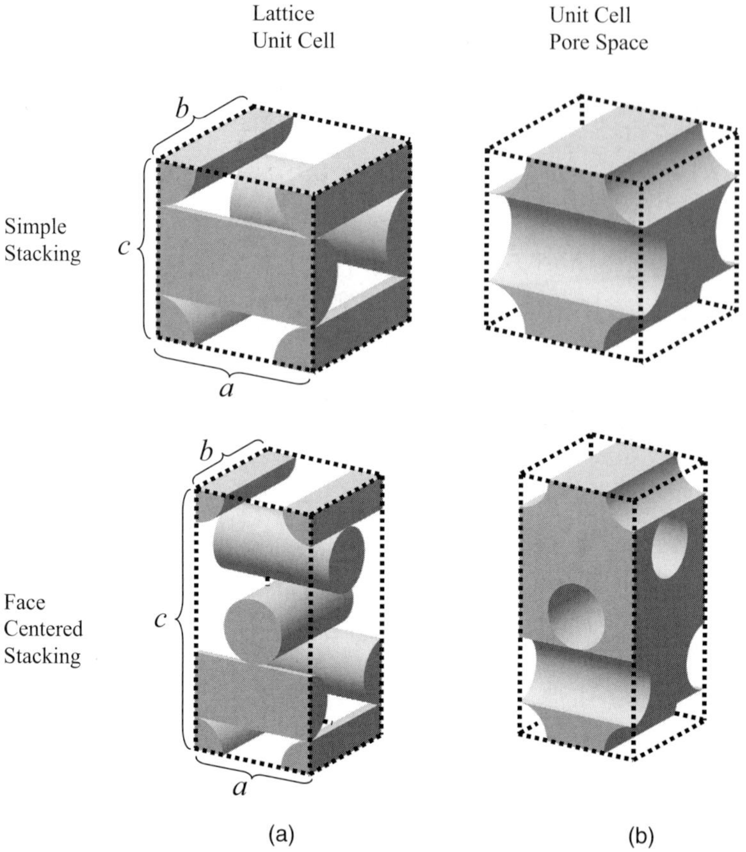

Fig. 9 Schematic illustrations of the a) unit cells and b) corresponding pore space for simple (top row) and face-centered (bottom row) stacking patterns.

SA/V increases as the lateral dimension of the unit cell decreases; however, in the limit that $L \leq D$, the pores become isolated and SA/V decreases dramatically.

The 3D periodic structures shown in Fig. 10 illustrate systematic variation of unit cell size. Each structure is comprised of a simple stacking pattern in which the c dimension of the unit cell is held constant. The lateral unit cell dimensions varied from 230 μm (sample a) to 937 μm (sample h). For these samples, the filament diameter is $D \approx 160$ μm, such that in the lateral directions, Γ_p varies from 0.55 to 0.13, ppi varies from 110 in^{-1} to 27 in^{-1}, and SA/V varies from 136 cm^{-1} to 33 cm^{-1}. For the radial array shown in Fig. 8d, the calculation of Γ_p depends on radial position. In this case, Γ_p in the radially oriented layer varied linearly from 0.85 at the inner radius to 0.4 at the outer radius.

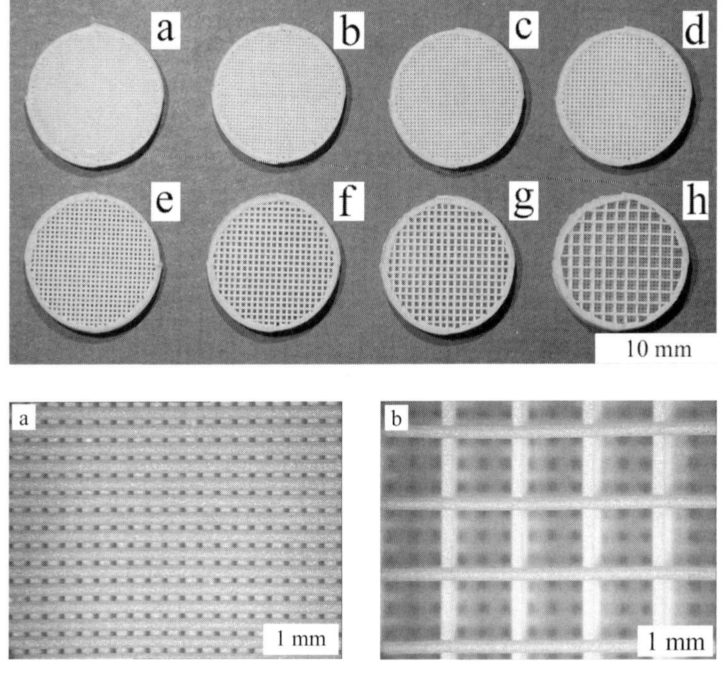

Fig. 10 Optical images of 3D periodic structures with varying lattice constant, and corresponding higher magnification views of structures a) and h) with respective solids volume fractions of 0.17 and 0.70. (Adopted from Ref. [33].)

2.3.7 Summary

Direct-write assembly processes such as Robocasting and fused deposition provide remarkable control and flexibility in designing cellular ceramics with 3D periodic structures. As demonstrated above, use of colloidal gel-based inks allows one to create regular 3D patterns of varying lattice geometry, lattice constants, and rod diameter. Such 3D structures can be utilized directly or can serve as a constituent in 3D interpenetrating composites (e.g., ceramic–polymeric [33], ceramic–metal [34], or ceramic–ceramic composites) formed by infilling their intervening pore space with a second phase.

References

1 Lewis, J.A., *J. Am. Ceram. Soc.*, **2000**. *83*, 2341–2359.
2 Tohver, V., J.E. Smay, A. Braem, P.V. Braun, and J.A. Lewis, *Proc. Natl. Acad. Sci. USA*, **2001**. 98, 8950–8954.
3 Tressler, J.F., S. Alkpu, A. Dogan, and R.E. Newnham, *Compos. Part A (Appl. Sci. Manuf.)*, **1999**. *30*, 477–482.
4 Allahverdi, M., S.C. Danforth, M. Jafari, and A. Safari, *J. Eur. Ceram. Soc.*, **2001**. *21*, 1485–1490.
5 Rao, M.P., A.J. Sanchez-Herencia, G.E. Beltz, R.M. McMeeking, and F.F. Lange, *Science*, **1999**. *286*, 102–105.
6 Soundararajan, R., G. Kuhn, R. Atisivan, S. Bose, and A. Bandyopadhyay, *J. Am. Ceram. Soc.*, **2001**. *84*, 509–513.
7 Chu, T.M., J.W. Halloran, S.J. Hollister, and S.E. Feinberg, *J. Mater. Sci. Mater. Med.*, **2001**. 12, 471–478.
8 Joannopoulos, J.D., Villeneuve, P.R., and S. Fan, *Nature*, **1997**. *386*, 143–149.
9 Vlasov, Y.A., X.A. Bo, J.C. Sturm, and D.J. Norris, *Nature*, **2001**. *414*, 289–293.
10 Agarwala, M., A. Bandyopadhyay, R. van Weewn, A. Safari, S.C. Danforth, N.A. Langrana, V.R. Jamalabad, and P.J. Whalen, *Am. Ceram. Soc. Bull.*, **1996**. *75*, 60–66.
11 Cesarano III, J., Segalman, R., and Calvert, P., *Ceram. Ind.*, **1998**. *148*, 94–102.
12 Smay, J.E., Cesarano III, J., and Lewis, J.A., *Langmuir*, **2002**. *18*, 5429–5437.
13 Smay, J.E., G. Gratson, R.F. Shepard, J. Cesareno II, and J.A. Lewis, *Adv. Mater.*, **2002**. *14*, 1279–1283.
14 Rice, R., *Ceramic Fabrication Technology*, Marcel Dekker, New York, 2003.
15 Sachs, E., M. Cima, P. Williams, D. Brancazio, and J. Cornie, *J. Eng. Ind. Trans. ASME*, **1992**. *114*, 481–488.
16 Griffith, M.L. and Halloran, J.W., *J. Am. Ceram. Soc.*, **1996**. *79*, 2601–2608.
17 Guo, J.J. and Lewis, J.A., *J. Am. Ceram. Soc.*, **1999**. *82*, 2345–2358.
18 Channell, G.M. and Zukoski, C.F., *AICHE J.*, **1997**. *43*, 1700–1708.
19 Rueb, C.J. and C.F. Zukoski, *J. Rheol.*, **1997**. 197–218.
20 Buscall, R., P.D.A. Mills, J.W. Goodwin, and D.W. Lawson, *J. Chem. Soc. Faraday Trans. I*, **1988**. *84*, 4249–4260.
21 Grant, M.C. and Russel, W.B., *Phys. Rev. E (Stat. Phys. Plasmas Fluids, Relat. Interdisc. Top.)*, **1993**. 47, 2606–2614.
22 Russel, W.B., *J. Rheol.*, **1980**. *24*, 287–317.
23 Shih, W.Y., Shih, W.-H., and Aksay, I.A., *J. Am. Ceram. Soc.*, **1999**. *82*, 616–624.
24 Shih, W.-H., W.Y. Shih, S.-I. Kim, J. Liu, and I.A. Aksay, *Phys. Rev. A*, **1990**. *42*, 4772–4779.
25 Shih, W.-H., J. Liu, W.Y. Shih, S.I. Kim, M. Sarikaya, and I.A. Aksay, *Mater. Res. Soc. Symp. Proc.*, **1989**. *155*, 83–92.
26 Sonntag, R.C. and Russel, W.B., *J. Colloid Interface Sci.*, **1987**. *116*, 485–489.
27 Morissette, S.L., J.A. Lewis, J. Cesarano III, D.B. Dimos, and T. Baer, *J. Am. Ceram. Soc.*, **2000**. *83*, 2409–2416.
28 Herschel, W.H. and Bulkley, R., *Kolloid Z.*, **1926**. *39*, 291.
29 Buscall, R., McGowan, J.I., and Morton-Jones, A.J., *J. Rheol.*, **1993**. *37*, 621–641.
30 Yilmazer, U. and Kalyon, D.M., *J. Rheol.*, **1989**. *33*, 1197–1212.
31 Hunter, R.J., *Foundations of Colloid Science*, Oxford University Press, New York, 1992.
32 Shigley, J.E. and Mischke, C.R., *Mechanical Engineering Design*, McGraw Hill, New York, 1989.
33 Smay, J.E., J. Cesarano III, B.A. Tuttle, and J.A. Lewis, *J. Appl. Phys.*, **2002**. *92*, 6119–6127.
34 Marchi, C.S., M. Kouzeli, R. Rao, J.A. Lewis, and D.C. Dunand, *Scripta Mater.*, **2003**. *49*, 861–866.

2.4
Connected Fibers: Fiber Felts and Mats

Janet B. Davis and David B. Marshall

2.4.1
Introduction

Porous fibrous ceramics are used for numerous and diverse applications including catalyst supports, hot-gas filters, composite reinforcement, biomaterials, and acoustic and thermal insulation. The optimal composition and microstructure is different for each application, and a complete description of the processing methods for all fibrous ceramics is clearly beyond the scope of this chapter.

Instead, a brief description will be given of several forms of fiber products that are commercially available and how these are used to produce a specific class of materials, namely, thermal insulation used to protect reusable launch vehicles. In this way, a number of fiber-forming methods and typical post-forming operations for fiber products pertain. The examples of thermal protection systems were selected because their performance requirements are exacting and their successful development has necessitated innovative processing approaches.

Both rigid and flexible passive insulation systems consist of refractory oxide fibers. To date, however, only a limited number of fiber types have been used as primary constituents of heat shields. Oxide fiber production methods will first be described in general terms and then specific examples relevant for thermal protection systems will be presented in detail. The same format will be followed in the discussion of fiber consolidation methods. Generally available fiber forms will be discussed prior to detailed descriptions of processing methods for thermal protection tiles and blankets. The latter will include a description of the system fabrication methods rather than those used only to form the insulation component, as many of the key properties contributing to their implementation are associated with surface modifications and construction methods. The evolution of these systems will be presented, as will ongoing work to improve their durability and thermal performance and simplify their production.

Cellular Ceramics: Structure, Manufacturing, Properties and Applications.
Michael Scheffler, Paolo Colombo (Eds.)

ISBN: 3-527-31320-6

2.4.2
Oxide Fibers

Several types of fibers have been used to produce insulation for applications ranging from furnace linings to thermal protection systems for space vehicles. Glass fibers were among the earliest available synthetic fibers and are generally processed from the melt [1–4]. Melts can be formed directly from mineral ores prior to spinning or may be homogenized through a prior melting cycle to form marbles. Molten glass is passed through a series of heated orifices to form individual filaments that may be subjected to one or more drawing methods to modify the fiber diameter and quench the amorphous filament. Melt spinning or blowing processes can be used to form fibers of various diameters ($< 1\,\mu m$ to $10\,\mu m$) and compositions. These compositions are, however, limited to glass formers that have the appropriate rheological properties to form continuous strands when melted [5]. The primary constituent is typically silica, but other chemical components are necessary to decrease the melting temperature and melt viscosity to allow homogenization and fining. Processes that can be used to draw the molten glass into filaments include rotary air attenuation (similar to the method used to make cotton candy), steam or air attenuation, and flame attenuation. Bundles of strands or filaments can be collected and wound together to produce strands suitable for textile methods such as weaving [6].

Of available glass fibers, high-purity, small-diameter silica fibers have been the primary constituent of all types of rigid tile insulation and have been used in a flexible mat form in thermal protection blankets. These require specialized processing to form small-diameter filaments and to purify the fibers. Their processing is described in more detail in Section 2.4.2.1.

The limitations imposed by melt spinning arise from the restricted set of chemical compositions that melt at temperatures low enough for practical processing and have the proper viscosity for fiber production. Other refractory oxide fibers of Al_2O_3, ZrO_2, mullite, and yttrium aluminum garnet (YAG) were developed by using alternative approaches [5]. These are typically produced from a liquid precursor solution that can be made viscous through the use of additives and/or by increasing their concentration. The fibers are converted to their final crystalline form by pyrolysis and sintering. These fibers, too, can be produced with a wide range of diameters depending on the drawing method. Both small- ($\sim 3\,\mu m$) and large-diameter ($\sim 11\,\mu m$) fibers produced from precursor solutions have been used to produce heat-shield materials. The two most prevalent compositions used are based on alumina–silica and alumina–borosilicate compositions. The drawing methods used for small- and large-diameter fibers are quite different, and both will be described.

2.4.2.1
Melt-Blown Silica Fibers

Of the various glass fibers available, only those with the highest thermal stability are used for aerospace insulation applications. These are typically high-purity silica fibers (e.g., Q-fiber, Johns Manville Corporation) formed by using a flame attenuation process suited to producing small-diameter filaments. They require composition refinement to prevent crystallization during use, which can compromise both the thermal and mechanical performance of the heat-shield insulation.

The fibers are blown from high-silica sand which is melted once to form marbles that are subsequently remelted and drawn through furnace bushings consisting of a sievelike arrangement of orifices to produce coarse fibers. These are again melted in a second attenuation step to form several smaller diameter fibers. This second step uses a high-temperature gas flame, typically impinging at right angles to the primary fibers (Fig. 1a) [7]. The resulting amorphous fibers exhibit a range of diameters (Fig. 1b) with an average diameter of 0.75–1.6 µm [8]. They are propelled by high-velocity gas through a forming tube and collected as an entangled mass on a conveyor belt. In addition to the fibers, inclusions or "shot" which have a different morphology and can be of different chemical composition are also collected in the mat [9].

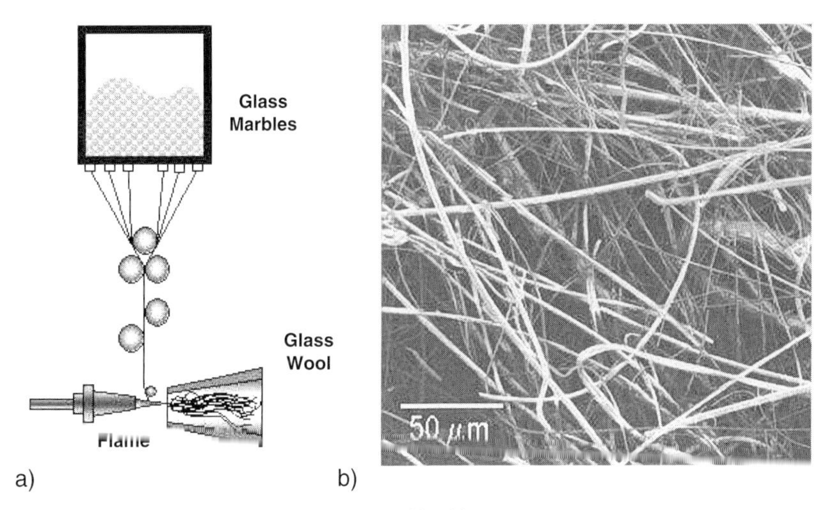

Fig. 1 a) Schematic of the flame-attenuation fiber-blowing process. b) Scanning electron micrograph of high-purity silica fibers (Q-fiber, Johns Manville).

Alternatively, blown fibers can be collected on a perforated rotating drum and aligned as they are removed as an agglomerated mass to pass through an orifice and be spooled onto a second drum rotating faster than the first. This process is called drafting, and the tension produced by the differential drum speeds aligns the fibers in a well-defined bundle [10]. Drafting cycles can be repeated until the desired degree of fiber alignment is achieved. The drafted fiber bundles can then undergo additional processes such as twisting to form continuous strands suitable for textile processing such as weaving.

As-drawn fibers contain impurities such as sodium that are beneficial in the melting and spinning processes but detrimental in service since they promote fiber crystallization. To eliminate such impurities, the fibers are subjected to an acid leaching process. The process starts with glass fiber and then any extraneous mono- or divalent ions are leached from the glass to leave an open pore structure. Depending on the purity of the starting raw materials the resulting silica content after leaching can be quite high (99.5+%) [8]. All subsequent processing steps are performed with the goal of minimizing contamination.

2.4.2.2 Blown Alumina–Silica Fibers

Fibers with a high volume fraction of alumina (e.g., Saffil, J&J Dyson) are also used as components of both flexible and rigid thermal protection systems [8]. Saffil fibers are made by using a solution-precursor spinning process especially developed to manufacture small-diameter fibers. The fibers are made from an aqueous solution comprising aluminum oxide chloride, a silica sol (added to control the crystallization behavior), and a nonionic, soluble, high molecular weight organic polymer (added to control viscosity). The viscosity of the composition is tailored to the fibrizing method (blowing) and is typically less than 10 poises [11].

During blowing, the solution is extruded through an aperture into a high-velocity gas stream. The extruded liquid stream is drawn down by the action of the gas stream to reduce its diameter. A diameter reduction factor of about 20 is usual for this type of blowing process. For very fine fibers the viscosity of the extruded composition must be carefully maintained during this step. This means that loss of solvent from and/or gelling of the composition must be controlled by manipulation of velocity, temperature, and particularly the drawing atmosphere. To minimize the loss of solvent from the composition, air with a high relative humidity is typically used. This fiber-forming method typically results in a product with a lower shot content than the melt-spinning process described above.

The fiber can be dried further after attenuation in the gas stream and is then exposed to ammonia vapor or a basic amine atmosphere to cause gelling of the precursor and to preserve the fiber shape. Fibers are then subjected to a hydrothermal treatment after being collected in the form of a loose mat [11]. It is during low-temperature hydrothermal heat treatment that the aluminum salt is decomposed to an aluminum hydroxide. The precise mechanism whereby ammonia or a basic amine assists the process is not fully understood, but it is believed that the release of acid anions is assisted by the formation of a soluble substance that is more easily removed in hydrothermal treatment. In general, the use of ammonia or a basic amine in conjunction with hydrothermal treatment gives fibers with a higher BET surface area and smaller grain size. The fibers are subjected to a final high-temperature heat treatment step to fully crystallize the fiber and remove residual porosity. This heat treatment is accompanied by a number of phase changes involving transition aluminas with a progression from η-Al_2O_3 to γ-Al_2O_3 to δ-Al_2O_3 to θ-Al_2O_3 and finally the stable form α-Al_2O_3 [5]. Careful control of the heat-treatment parameters is required to control the fiber microstructure. A fine, uniform grain size and high density are desired for good mechanical performance but this is difficult to achieve in practice in pure alumina fibers. Therefore, silica is added as a second phase to stabilize the transition alumina forms to allow pore removal and inhibit

Fig. 2 Scanning electron micrograph of alumina–silica fibers (Saffil, J&J Dyson).

crystal growth. The resulting fibers (Fig. 2) comprise approximately 95 wt% Al_2O_3 and 5 wt% SiO_2 and are relatively fine (~ 3 µm) and have short staple lengths (2–4 cm) [5].

2.4.2.3
Drawn Alumina–Borosilicate Fibers

Large diameter continuous oxide fibers are typically produced by extruding precursor solutions of appropriate viscosity through an orifice and drawing them to the desired diameter [12]. Commercially available fibers are produced for the most part from aqueous solution precursors based on alumina. Viscosity control is the key to the process. The solution must not shear-thin or change viscosity during extrusion. Furthermore, to produce continuous fibers the drawing and gravitational forces must balance the surface tension, viscosity, and inertial drag of the fiber as it is accelerated during attenuation [13]. This requirement establishes an appropriate range of viscosity so that fibers are not subject to necking (viscosity too low) or capillary fracture (viscosity too high) during drawing. Once formed, the fibers are subjected to heat treatments to pyrolyze the precursor to remove organics and water and heat-treated to higher temperatures to obtain the desired microstructure and crystalline phase. The removal of fugitive components without forming defects becomes increasingly difficult as the fiber diameter increases and ultimately limits the maximum diameter of fibers that can be processed from solution precursors or sols (typically ~ 20 µm).

Of the family of larger diameter continuous oxide fibers (Nextel, 3M Co.), the most widely used in thermal protection systems is Nextel 312 which contains 14 wt% boria. These fibers are produced from a mixture of aluminum carboxylates and colloidal silica mixed with organic and boria-containing additives. The precursor is

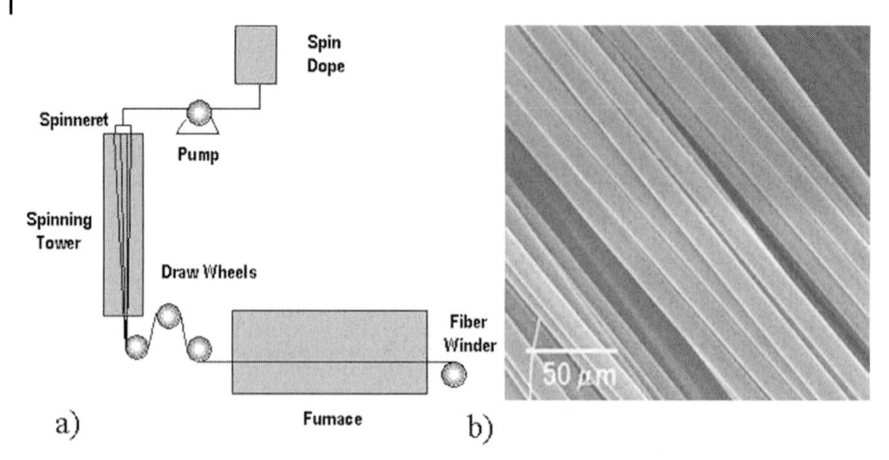

Fig. 3 a) Schematic illustration of the dry spinning process. b) Scanning electron micrograph of continuous aliminoborosilicate fibers (Nextel 312, 3M Co.).

formed as a dilute solution that is filtered prior to being concentrated by solvent evaporation to form a viscous spin dope suitable for fiber drawing. A dry-spinning process is used in which the spin dope is pumped via a metering pump through a spinneret. The fiber diameter is controlled by varying the pumping rate and drawing speed during spinning. The fibers are then heat-treated to pyrolyze, crystallize, and sinter the fibers. The dry-spinning process is shown schematically in Fig. 3a [12].

The composition of the formed fibers is 62% Al_2O_3, 24% SiO_2, and 14% B_2O_3, and the predominant crystalline species after heat treatment are aluminum borate and mullite. The fibers are approximately 11 µm in diameter (Fig. 3b) and each roving or tow comprises 390 individual filaments. The continuous fibers can be woven to produce fabrics or chopped to shorter lengths and combined with low-conductivity oxide fibers to form insulation tiles.

2.4.3
Fiber Product Forms

Fibers produced by the methods described above can be used to produce a variety of three-dimensional porous bodies. Several examples are given although a number of additional processing approaches exist. These are summarized in Fig. 4 [5].

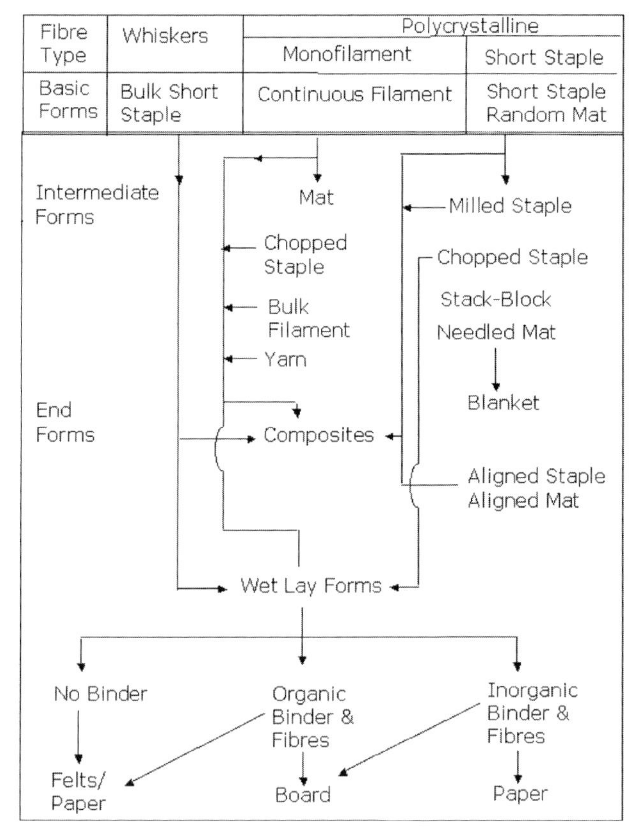

Fig. 4 Fiber forms.

2.4.3.1 Continuous Monofilaments

Monofilaments of continuous blown or drawn fibers can be used to produce composite reinforcements or fiber preforms by weaving, knitting, or braiding. The fibers are typically collected as aligned fiber bundles or roving. Continuous fiber bundles can also be chopped to short lengths of several millimeters and mixed with discontinuous fibers to form mats or molded felts or blocks.

2.4.3.2 Fiber Mat

Mats are formed by collecting staple fibers that are randomly oriented in the length and breadth directions on a moving belt. The structure is built up in the thickness direction as layers of such deposits are formed atop one another. This results in the physical properties differing in the in-plane and through-thickness directions. It

Fig. 5 Commercially available oxide fiber forms include a) mat, b) board, c) bulk fibers, d) paper, and e) felt.

also produces a loose flexible mat with a striated appearance and two-dimensional layers that can be easily separated from one another (Fig. 5a). This allows the weight of the mat per unit area to be easily adjusted. Mats can be used in their as-deposited form or processed into rigid board by using organic or inorganic binders (Fig. 5b). They can also be further processed into bulk or milled fiber.

Mat properties can be modified by several techniques. Needled mats are produced by incorporating organic nonwoven fiber scrim cloths between layers of the mat and pushing an array of needles through the assembly in the through-thickness direction. The needles force the organic fibers and some of the inorganic fibers to follow their trajectory. This results in a fraction of the fibers interpenetrating several of the two-dimensional mat layers. The process can be repeated to change the area fraction of interpenetrating fibers. The organic fibers can then be removed by heat treatment if desired.

Mats can also be reconfigured as "stacked blocks" for use as insulation in applications where delamination of the two-dimensional fiber layers is especially problematic. For this configuration, strips of insulation are cut, turned 90° to their original orientation, and reassembled. While useful for furnace insulation, this product has not been incorporated in thermal protection systems.

2.4.3.3
Bulk Fiber

Mats are shaped into bulk form in a shredding or chopping machine that produces three-dimensional fiber agglomerates (Fig. 5c). Chopping can produce a range of fiber lengths. Further reduction in the fiber length, if required, can be achieved by high-energy milling processes such as hammer milling. Bulk fibers are typically processed into insulation components by dispersing the fiber agglomerates in a liquid and then removing the liquid (usually water) by filtration to leave a shaped body. Depending on the method used for consolidation, products can range from paper to felt to board.

Ceramic paper is formed by blending short bulk inorganic fibers with binders and a dispersing medium (Fig. 5d). Several types of fiber can be included depending on the final product requirements and the necessary fiber bonding characteristics. Binders can be organic or inorganic and are added to achieve a specific concentration relative to the fiber content. The binders must be uniformly distributed to produce a suitable strength, modulus, and other mechanical properties. The organic binders impart strength in the green state to allow ease of handling. After forming, the paper can be densified by compressing or calendering.

Felts are produced in much the same way as paper excepting the final densification steps (Fig. 5e). They, too, may contain organic fibers or binders to improve their handling strength but are often available in flexible binderless forms. Their density and homogeneity depend on the length of the fibers, their degree of agglomeration, and the process used to mix the fibers in the dispersing medium. Rigid boards can also be produced with bulk fiber bonded together with second-phase fibers or binding agents. These can be made rigid at low temperatures by using organic binders and can be compressed to form insulation with a range of densities. Inorganic binding agents can be incorporated to promote fiber bonding during thermal excursion such that the boards retain their shape and rigidity in service at elevated temperatures.

2.4.4
High-Performance Insulation for Space Vehicles

Space Shuttle thermal protection systems are a special class of high-performance insulation. They must withstand high temperatures, severe thermal shock and gradients, high acoustic loads during launch, and structural deflection due to aerodynamic loads, and they must remain dimensionally stable and be reusable. Furthermore, they must satisfy these requirements while being as light as possible and inexpensive to produce, install, and maintain. When the Shuttle program began, the primary heat-shield candidates included replaceable ablator panels, re-radiative panels produced from carbon–carbon composites, metals, and oxide insulation [8]. Oxide insulation was selected for all areas except control surfaces that experience the highest temperatures.

Originally, most surfaces of the Space Shuttle orbiters were protected by insulation tiles [14]. These rigid tiles were primarily comprised of silica fibers and a silica bonding agent, and their success was dependent on developing purification methods for both constituents and modifying their surface characteristics to control the optical properties and impart additional durability. Tile compositions and processing approaches have changed somewhat since the first systems were developed. These changes were made to improve the temperature resistance and mechanical performance. Even with such improvements, however, they remain inherently expensive to produce and maintain, and this has led to the development of alternative systems.

In an effort to decrease the overall cost of space-vehicle heat shields, additional concepts including flexible insulation systems and/or thermal protection blankets have been implemented [15]. Initially, these were also comprised primarily of silica and were used on vehicle surfaces that experience lower peak temperatures during vehicle reentry. Their temperature capability has, however, been improved through the use of new constituents and processing methods, and they are now being considered for higher temperature areas [16].

2.4.4.1
Rigid Space Shuttle Tiles

The two major components of preliminary Space Shuttle heat shields were carbon–-carbon hot structure used to protect vehicle leading edges and oxide fibrous insulation tiles used to protect the remaining vehicle surfaces [14]. The tile system was selected over competing approaches such as ablative or metallic reradiative panels because it was reusable and of a relatively simpler design. Furthermore, the insulation thickness could be tailored in a straightforward manner to meet local thermal performance requirements [8].

All of the rigid insulation heat shield materials are formed by a slurry casting method and incorporate bulk fiber. The earliest tiles comprised silica fibers and a silica binder [17]. The first system was of lower density and thermal capability than the all-silica systems developed subsequently. By incorporating additional fiber types and bonding agents, improvements in the tile strength and thermal capacity were achieved. The evolution of these systems and their processing methods will be described.

Although the primary focus is the fibrous insulation block used to produce thermal protection tiles, two additional key components will also be discussed. These are an outer (exposed) surface coating that provides the necessary optical properties and enhances the surface durability, and an inner coating that provides adequate mechanical properties for bonding the tiles to the underlying vehicle structure. Without these components and the processing approaches developed to co-process them with the insulation system, the tiles would not have been adequate for service.

Lockheed developed the first Space Shuttle heat shield tiles in the early 1960s. These *all-silica tiles* are designated LI-900, they have a low density of 0.15 g cm^{-3} (compared to 2.2 g cm^{-3} for fused quartz) and are comprised wholly of silica. The primary constituent is the high-purity amorphous fiber described above. Before the fiber is used, however, it undergoes additional processing steps to stabilize the microstructure against shrinkage and to remove nonfibrous inclusions, which would cause density variations and devitrification of the fibers.

These additional processing steps are essential since crystalline silica fibers would exhibit polymorphic phase transformations during thermal cycling that would compromise the performance of the heat shield. Due to the slight variability in glass melt composition and purity and the resulting effect on reproducibility of insulation properties, a number of fiber lots are blended together prior to use. The fibers are also heat-treated to eliminate any porosity that may have formed during the purification process (acid leaching). The blended, heat-treated bulk fiber is then cleaned by using a hydrocyclone to remove glass shot and dried [18].

The colloidal silica binder must also be of high purity so as to adequately bond the silica fibers together to form rigid insulation without contaminating them. Of particular concern are devitrifying agents such as sodium. Contamination sources include processing silica sols from sodium silicate or using NaOH to prevent gelation. The Na^+ ions are typically removed by a series of column deionization cycles involving prolonged storage of the sols and/or heat treatments between cycles to facilitate the migration of Na^+ to sol particle surfaces for removal. These cycles are repeated until the Na_2O content reaches an acceptably low level, as determined by heat treatment and X-ray diffraction to determine the amount of crystalline phase formed (typically less than 300 ppm) [19].

After the constituents are chemically refined, the tile insulation blocks are formed by using a slurry casting method. Casting generally results in preferential orientation of fibers perpendicular to the pressing direction. This, in turn, produces anisotropy in the thermal and mechanical properties of the insulation billet. Hence, tile properties are often reported for both the through-thickness (parallel to the pressing direction) and in-plane (perpendicular to the pressing direction) orientations. When bonded to the vehicle, the in-plane orientation is parallel to the surface.

To form the slurry, predetermined weights of fibers and deionized water are combined in a low-shear mixer. After the fibers are dispersed, the mixture is transferred to a casting tower and de-aired. The water is removed by a combination of pressing from the top surface and applying a vacuum to the bottom surface of the mold. Water is removed until the billet reaches a predetermined thickness and the water makes up approximately 80 % of the billet weight. The colloidal silica binder is then pumped into the top of the mold while vacuum is again applied to the bottom so that the binder displaces the water. The binder is then gelled to ensure a uniform distribution after drying [18].

The billets are dried in a convection oven or microwave drier or both and have a density of approximately 0.11 g cm^{-3}. Once dry, the billets are sintered at 1290 °C until they reach the desired density of 0.15 g cm^{-3}. They are then machined to shape, heat-treated to remove organic contaminants, and coated. Although most tiles are nominally 6 × 6-inch squares (except for close-out tiles), no two tiles on an orbiter are exactly alike. The service heat load determines the thickness of each tile and the curvature of each tile's underside is matched to the contour of the Shuttle's skin at the exact point the tile is to be bonded. This custom machining adds considerable expense to the system.

The inner coating is generally the same for each of the tile systems discussed and is required since tiles cannot withstand airframe load deformation. Therefore, strain isolation is necessary between the tiles and the orbiter structure. This isolation is provided by a compliant pad. These are thermal insulators made of Nomex felt material and are bonded to the tiles with a high-temperature silicon adhesive. The same adhesive is then used to bond the assembly to the vehicle surface [14].

The bond assembly introduces stress concentrations in the tile. This results in localized failure of the tile just above the bond line [20]. To solve this problem, the inner surface of the tile is densified to distribute the load more uniformly. The densification material comprises a Ludox ammonia-stabilized silica binder mixed with silica slip particles and a silicon tetraboride colorant added to facilitate visual inspection of the penetration depth in the tile. Several coats of the pigmented slurry are brushed onto the tile surface that is to be bonded to the strain isolator pad. The tile is then air-dried for 24 h before the coating is sintered. The densification coating penetrates the tile to a depth of approximately 3 mm, and the strength and stiffness of the tile and bonding system are increased sufficiently for reliable attachment to the vehicle [21].

Two top surface coatings are used for the LI-900 tiles to give the tiles different heat-rejection capabilities. Tiles on the lower or windward vehicle surface have a black coating for high emittance at high temperature and the tiles on the upper surface (now mostly replaced by blankets; see Section 2.4.4.2) have a white coating to limit in-orbit vehicle temperature. The black tiles are coated on the top and sides with a mixture of powdered silicon tetraboride and borosilicate glass frit in an alcohol carrier with a methylcellulose prebinder [22]. The viscosity and component particle sizes are measured prior to application. This coating slurry is sprayed onto the top and side surfaces of tiles. The slurry is applied to achieve a targeted weight based on coated area that will result in a dense coating approximately 300 μm thick after heating to a temperature of 1230 °C [18]. The white coating contains no silicon tetraboride but is applied in the same manner and sintered to a slightly lower temperature. The black tiles of this type are used to protect vehicle surfaces that reach temperatures of less than 1200 °C and the white tiles for areas that reach temperatures of less than 650 °C. After the ceramic coatings are sintered, the tiles are waterproofed with a silane vapor. Dimethylethoxysilane is injected into each tile through an existing hole in the surface coating with a needleless gun.

For some areas of the vehicle, the strength of the low-density LI-900 is inadequate. A higher density tile system, LI-2200, with a slightly different processing scheme was developed by NASA, Ames, for those areas. Although these insulation billets are also produced by slurry casting, they do not contain colloidal silica as a binder. They do, however, contain silicon carbide powder as an emittance aid [18].

Several other details of the insulation billet-processing scheme were changed to reach higher densities. The silica fibers were not heat-treated prior to casting, and a more aggressive slurry mixing process was employed. The fibers were mixed with water, silicon carbide powder, and ammonium hydroxide (all in preweighed amounts) in a V-blender with an intensifier bar. The size of the fiber agglomerate was modified during this step to allow a higher packing density during casting. The casting process itself was unchanged, although the degree of fiber compaction was increased. After drying, the green density of the billet was approximately 0.21 g cm^{-3}, and after firing to 1300 °C the final density was approximately 0.35 g cm^{-3}. The tiles were machined and the surface coatings applied as previous described.

Following the development of all-silica tiles, other compositions were investigated. These *blended-fiber tiles* generally consisted of insulation blocks formed by blending two or more types of fibers to impart additional specific strength. This benefit, however, was often achieved at the expense of other desirable properties such as low thermal conductivity or dimensional stability at elevated temperatures. In most cases parametric studies were performed to characterize the effects of varying the composition and processing parameters on the microstructure and physical properties of the insulation. The results of a few of the published studies of this type are summarized herein. Several groups, however, continue to conduct proprietary research on these systems using fibers and binders other than those discussed in this summary, and future improvements may be expected.

The first of the blended-fiber systems, developed at NASA, Ames, was termed fibrous refractory composite insulation (FRCI) [18, 23, 24]. In these insulation materials, high-purity silica fibers were combined with aluminoborosilicate fibers. Both fiber types are heat-treated prior to blending. The silica fibers were preconditioned by dispersing them in a mixture of deionized water and hydrochloric acid ($pH \approx 3$) while at the same time bubbling nitrogen gas through the slurry to sediment out shot and other contaminants. The fibers were treated in this manner for a couple of hours and then rinsed with deionized water. Afterward, the fibers were thermally treated as described above for LI-900 tiles.

The relatively larger diameter (~ 11 µm) aluminoborosilicate fibers are available as rovings that are cut to lengths of approximately 3 mm prior to use. They are then heated to at least 1090 °C to promote crystallization. The stiffer fully crystalline fibers have been observed to disperse more uniformly during fiber blending than amorphous forms [25]. The resulting mullite content is determined prior to use. A number of tile compositions have been investigated with the aluminoborosilicate fibers making up 20–80 wt% of the billet. Of these, the composition with 20% aluminoborosilicate fibers and a density ranging from 0.12 to 0.22 g cm^{-3} has been produced in a production capacity (FRCI-20-12).

The fiber slurry is prepared in several steps. The aluminoborosilicate fibers are first mixed with a portion of the silica fibers in a high-shear mixer. This mixture is then combined with the remaining silica fibers and silicon carbide powder in a V-blender with an intensifier bar. The slurry pH is adjusted with ammonium hydroxide to aid dispersion. After mixing, the slurries are transferred to a casting tower and de-aired.

Rapid removal of the water during casting is required to prevent sedimentation, which results in composition and density gradients in the final billet. During pressing, the fibers become aligned perpendicular to the pressing direction.

After drying, the tile billets are fired to 1315 °C to promote fiber-to-fiber bonding. The bonding in this system is effected by the formation of borosilicate glass at fiber junctions with the boron fluxing agent provided by the aluminoborosilicate fibers. The difference in fiber bonding between silica tiles and blended fiber tiles is shown in Fig. 6. During sintering, the heating rate is chosen to permit relatively uniform temperatures to be achieved throughout the tile. A faster heating rate causes nonuniform sintering and may result in cracking. Longer or higher temperature firing results in more initial shrinkage during sintering but less subsequent shrinkage in service. Increased firing temperatures may lead to crystallization of silica fibers.

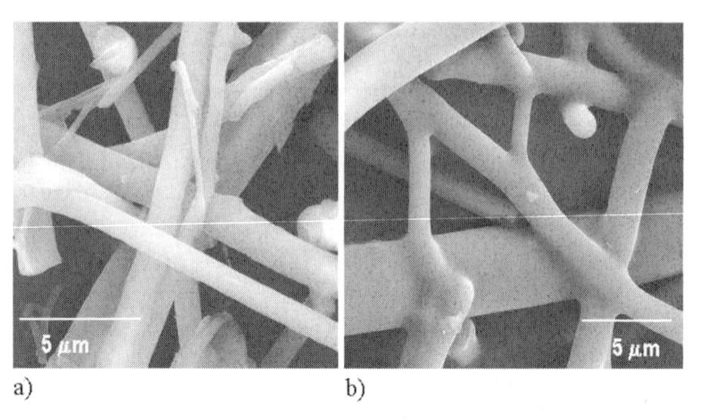

Fig. 6 After sintering, the fibers bonded together in a) LI-900 silica tiles and b) alumina-enhanced thermal barrier blended-fiber tiles appear different.

In general, blended-fiber tiles can be produced with higher through-thickness strength, lower density, and lower thermal conductivity than LI-2200. FRCI-20-12 (20 % nominal aluminoborosilicate fibers with a nominal density of 12 lb ft^{-3}) tiles have, in fact, been successfully used in place of LI-2200 on Shuttle orbiters.

Further modifications to the blended-fiber tile systems include the incorporation of small-diameter alumina fibers to increase the temperature to which the tiles can be exposed without dimensional instability or slumping. These alumina-enhanced thermal barrier (AETB) systems are of nominal composition 20 % small-diameter alumina fibers (Saffil), 12 % large diameter aluminoborosilicate fiber (Nextel 312), and 68 % silica fiber (Q-fiber) [25]. They are processed similarly to FRCI insulation and in two nominal densities, 0.13 and 0.19 g cm^{-3}. The lower density version has successfully been flown on the Shuttle base heat shield, replacing LI-900 tiles, and the higher density version has been flown as an experimental material.

The improved dimensional stability of AETB over FRCI tiles has been attributed to the presence of the more refractory alumina fiber as well as a reduction in the concentration of aluminoborosilicate fiber resulting in a lower boria content and less softening of the silica fibers. The alumina fibers, however, also increase the thermal expansion coefficient of the tile, which, in turn, necessitates modification of the surface coating composition. The effect of various additives was investigated, as was the effect of changing the coating morphology from a thin glass shell to thicker tile-infiltrating glaze. These combined modifications have been quite successful at preventing coating spallation and undesirable tile shrinkage during sintering. These coatings (toughened uniform fibrous insulation or TUFI) penetrate several millimeters into the tile surface and also improve impact resistance [26].

To produce the tile-infiltrating coating slurry, the glass frit, fluxing agent, and emittance agent are separately milled to reduce their particle size before they are mixed together. The glass frit comprises a small amount of boron oxide added to a commercially available, relatively pure, acid-leached borosilicate glass available under the name Vycor. The coating contains a silicon tetraboride fluxing agent that is oxidized exothermically to produce a high boron oxide borosilicate glass flux. The resultant multicomponent heterogeneous glass encapsulates the molybdenum disilicide emittance agent. To adjust the sintering compatibility to that of the AETB insulation, approximately 20 wt% molybdenum disilicide is incorporated. All of the components are milled in ethanol and then mixed together in a Kendall mixer or equivalent to blend the individual slurries together. No methycellulose is added to increase the viscosity of this coating composition, and this in conjunction with milling (which increases the surface area of the particles and decreases the viscosity of the fluxed glass composition during sintering) facilitates penetration of the coating into the substrate.

The slurry is sprayed onto the insulation block with an airbrush or spray gun in several applications. A preweighed amount of slurry is applied and this effectively impregnates the outer surface of the tile and creates a porous graded composite. After spraying, the substrate is dried overnight at room temperature or for several hours at temperatures up to about 70 °C. After drying, the coating is sintered in a preheated furnace for about 90 min at about 1200–1260 °C. The fired coating is porous and black in color.

Tile systems that have been used to date on Space Shuttle orbiters have been described in the preceding paragraphs. Advanced concepts for next-generation vehicles are discussed at the end of this chapter. The tile compositions and processes used to make them have evolved since the earliest systems were developed. Improved strength and dimensional stability have been realized through the incorporation of additional types of fibers. As these additions were made, however, more aggressive mixing approaches were needed to ensure adequate homogeneity. In the filter pressing process, sedimentation due to mass differences between different fiber types and fiber agglomerate sizes is detrimental and leads to nonuniform through-thickness density and tile warping. Mixing the slurry until a rapid liquid-removal process is employed minimizes this problem. As a final check to ensure that each tile is uniform and homogeneous, witness specimens are also usually taken from various locations on the billets and tested for strength, density, and other properties.

General properties of rigid tile insulation systems are summarized in Table 1.

Table 1. General properties of rigid tile insulation systems.

Density	120–350 kg m^{-3}
In-plane thermal conductivity	0.067–0.1 W m^{-1} K^{-1}
Through-thickness thermal conductivity	0.047–0.075 W m^{-1} K^{-1}
In-plane tensile strength	0.47–3.2 MPa
Through-thickness tensile strength	0.16–0.69 MPa
In-plane tensile modulus	0.17–0.22 GPa
Through-thickness tensile modulus	0.048–0.11 GPa

2.4.4.2
Flexible Insulation Blankets

Flexible blankets are an alternative and less expensive fiber-based insulation system. Rigid tiles are inherently expensive because 1) tile size is limited by strain mismatch with the vehicle; 2) tiles cannot be directly bonded to the vehicle and a compliant material must be placed beneath each tile to isolate it from the structure; and 3) each tile must be precisely machined and installed to a unique shape and position to maintain the smooth surface profile required for aerodynamic performance. Flexible insulation blankets have replaced most of the white LI-900 tiles on the Space Shuttle orbiters. These are compliant structures that can be used in larger sizes than tiles and do not require strain-isolation pads. They are bonded with silicone adhesive directly to the vehicle structure at a reduced installation cost. Currently, they are used on the upper sidewalls of the orbiter's fuselage, sections of the payload bay doors, most of the vertical stabilizer and rudder speed brake areas, the outboard and aft sections of the upper wing, parts of the elevons, and around the observation windows. Improvements to the temperature capability and durability of these systems may allow their use on additional vehicle surfaces, as discussed below.

Flexible blankets (advanced flexible reusable surface insulation or AFRSI) comprise an insulation layer sandwiched between two woven ceramic fabric sheets and sewn together with silica thread (Astroquartz). The low-density insulation, or batting, is processed from the same high-purity amorphous silica fiber used in making tiles (Section 2.4.4.1). Bulk fiber is formed into a binderless felt (Q-felt) that is manufactured by a water deposition process. This form is produced to achieve a specified weight per area, and the density and thickness may vary slightly from one sheet to the next. It is sandwiched between an outer woven high-temperature fabric produced from a waterproofed high-purity silica textile fibers (Astroquartz) and an inner woven fabric produced with lower temperature capability glass fibers (S-glass). After the composite is sewn with silica thread, it has a quilt-like appearance [8, 15].

The blankets require a coating to rigidize the surface to minimize aerodynamically induced abrasion. The coating composition, referred to as C-9 in the literature, is a mixture of ammonia-stabilized collodial silica (Ludox) and high-purity ground silica filler particles [27]. The coating is applied in two steps with the first application comprising colloidal silica and 2-propanol slurry that penetrates the waterproofed fabric. The second coat contains the silica filler powder but no 2-propanol and sits atop the surface of the fabric. The density of the assembled AFRSI blanket is approximately 0.1 g cm^{-3} and it varies in thickness from 1.1 to 2.4 cm. The appropriate thickness is determined by the heat load the blanket encounters during vehicle reentry, and thicker blankets are used to protect hotter areas. The blankets are cut to the required shape and bonded directly to the orbiter by RTV silicone adhesive 0.5 cm thick. The thin glue line reduces weight and minimizes thermal expansion during temperature changes.

A blanket waterproofing agent is typically applied before each flight. Compared to tiles, the AFRSI blankets have a much rougher surface and therefore an increased propensity to force a laminar boundary layer into turbulence. This increased roughness can also produce markedly amplified local heating where boundary layers are relatively thin, such as the windward side of most reentry vehicles. Improving the surface smoothness of these materials is key to their use on additional vehicle surfaces.

The original AFRSI constructions have been modified to increase their thermal capabilities. These improvements involved substituting a woven Nextel 312 fabric and sewing thread for the Astroquartz materials used in the first construct, and incorporating alumina mat insulation in place of quartz felt. These higher temperature blankets are used for the dome heat shields that surround the base of the main engine nozzles [16, 28]. Additional improvements, not yet incorporated on vehicles, have also been explored and are discussed in the Section 2.4.4.3.

2.4.4.3 Innovations in Thermal Protection Systems

Several potential improvements to the performance of thermal protection tiles and blankets may result from recent or on-going research aimed at reducing the thermal conductivity of the insulation or improving the durability of the overall system. The first category of improvements can be achieved by incorporating additives that modify heat conduction processes or by combining more than one type of insulation in a single component. The second category of improvements typically involves integrating new materials and surface treatments.

Heat transfer through fibrous insulation occurs by a combination of conduction, convection, and radiation. Reducing contributions from any of these mechanisms may decrease the thermal conductivity of tile or blanket insulation and enhance its performance. At high temperatures, the radiative component becomes increasingly important. A number of attempts have been made to modify radiative heat transfer in fibrous insulation by incorporation of reflective particles [29] or sheets embedded in the insulation, or by coating individual fibers with a reflective material by using sol–gel and other processes [30]. These have met with varying degrees of success. The most widely used example is commercially available multilayer insulation (MLI) blankets, in which highly reflective metal foils are sandwiched in a predetermined stacking sequence throughout a flexible fibrous ceramic mat. Although this construction is effective in decreasing the thermal conductivity of the insulation, it also reduces the maximum use temperature.

Another approach for decreasing the thermal conductivity of tile and blanket insulation materials is to reduce gas convection through the insulation by incorporation of aerogels [31, 32]. The fine pore structure of these materials, typically on the order of 50 nm or so, is very effective at reducing convective heat flow. In addition, the weight penalty is very small. Processing techniques based on supercritical drying have been successfully used to incorporate aerogels of various compositions in both rigid and flexible fibrous insulations. The primary limitation of the approach lies in the poor high temperature stability of the aerogels themselves. They sinter and densify at temperatures well below the peak surface temperatures of thermal protection components. For this reason, they are likely to be of use only when embedded in insulation near the cooler face.

The embedded-aerogel approach to optimizing the performance of heat-shield components is similar to other methods explored to combine more than one type of insulation material in a single component. Multilayer tile material has been produced from layers of AETB tiles of two different densities [33]. The insulation layers were bonded together by a high-strength, high-temperature alumina or silica binder having a coefficient of thermal expansion similar to that of the insulation layers. In this way, the surface exposed to the highest peak temperature can comprise higher density tile that is dimensionally stable, while the embedded portion of the insulation, which reaches lower temperatures, can comprise lower density tile that is lighter and of lower thermal conductivity. This approach provides a means of optimizing several performance parameters if successful.

In addition to improving heat shields by decreasing the conductivity of the insulation, they can be improved by using new construction methods or more effective protective surface treatments. For example, changing the form of the insulation used in thermal protection blankets from a flexible mat to a rigid board significantly improves the resulting surface smoothness, as needed for better aerodynamic performance [34]. The rigid insulation board allows more uniform tensioning of threads during the sewing process and better control of the fabric tension along blanket edges. The boards can be processed with organic binders that are removed by heat treatment to restore the requisite flexibility of the component subsequent to fabrication.

Flexible reusable insulation requires a coating on the outer woven sheet that infiltrates and stiffens the fabric to provide an aerodynamic surface. The coating must act as a "high-temperature starch" without embrittling the fabric. Since the coated fabric layer is essentially a thin ceramic matrix composite in which the infiltrated coating is the matrix, the requirements for blanket durability are the same as those for damage tolerance in structural ceramic composites: a weak bond is needed between the matrix and the fibers to prevent embrittlement. The upper surface of the orbiter is protected by blankets that consist of silica-based fabric, insulation, and coating. At temperatures above about 700–800 °C, the silica-based coating bonds strongly to the fibers, embrittling the outer fabric and limiting its lifetime. Development of more refractory blanket fabrics and compatible coatings with temperature capability up to the range of 1000–1200 °C would allow use of blankets on additional vehicle surfaces. This concept has been explored, and several improved blanket systems have been developed [35–38]. In general, varying the fabric/coating combina-

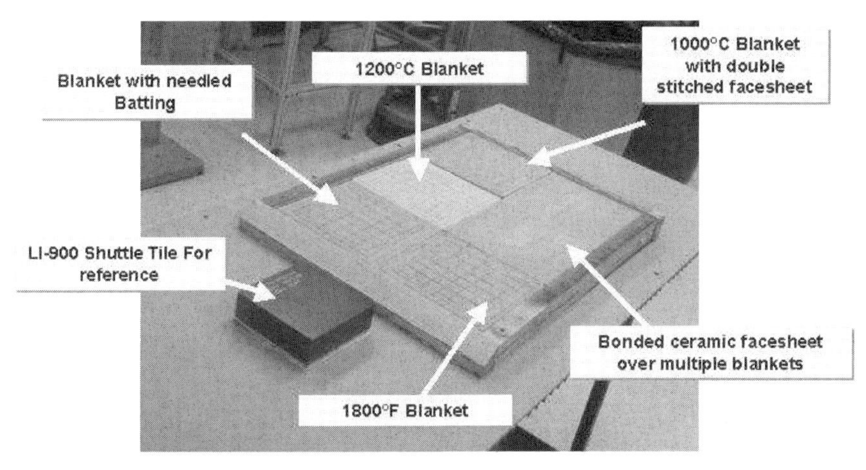

Fig. 7 A family of improved thermal protection blankets has recently been demonstrated.

tion affects the thermal capability and durability of the system. Suites of new flexible insulation systems with tailored properties now exist, and examples are shown in Fig. 7. In some cases, these materials can be used up to temperatures of 1200 °C with no loss of durability.

An alternative approach that has been taken to produce durable blanket surfaces is to incorporate a woven hybrid aluminoborosilicate/Inconel wire/braze alloy wire outer fabric [39]. A thin Inconel foil is brazed onto the surface of the blanket after the sewing operation. No waterproofing agent is applied to the durable advanced flexible reusable surface insulation (DurAFRSI), since the superalloy outer surface is waterproof. Compared to tile-type TPS, the installed cost of DurAFRSI is relatively low, but it has a more highly catalytic surface and a lower emissivity, both of which can lead to amplified surface heating. Additionally, its maximum use temperature is limited to below 1000 °C.

Improvements in surface treatment for tiles include combining the two types of glass coatings described in Section 2.4.4.1 such that the first coating penetrates into the tile and the second thin dense glass shell covers and seals the surface. These double-coated tiles have been shown to be more impact resistant than when either coating is used separately [38]. Tiles have also been protected with ceramic composite surfaces [38]. These are typically produced by using matrices derived from preceramic polymer precursors. The composite is laminated directly atop the tile by conventional polymer composite processing methods. The precursor is then converted to an oxide ceramic matrix in a heat treatment step that also bonds it to the insulation block. The surface durability and maximum use capability depend on the composite system used and its construction details (fabric weave, number of fabric layers, etc.). In general, these can be produced with a higher specific impact resistance than any of the glass-coated systems. An example of such a tile undergoing impact testing is shown in Fig. 8.

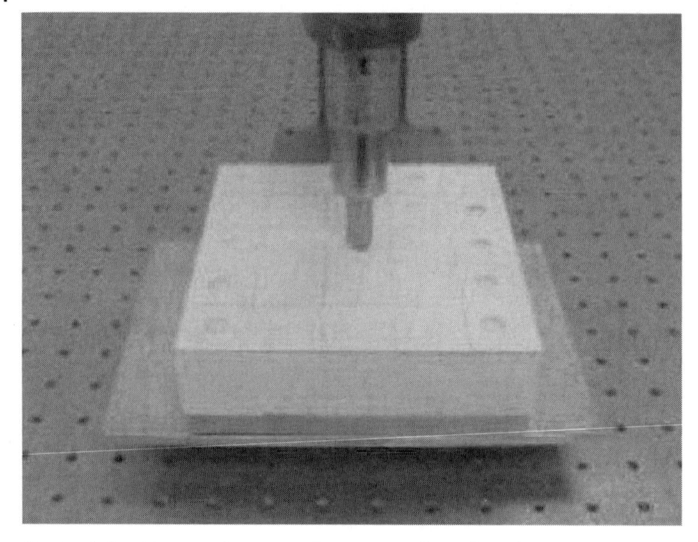

Fig. 8 A durable ceramic composite covered tile undergoing impact testing.

2.4.5 Summary

Ceramic fiber felts and mats have utility for a broad range of applications. The optimal composition and microstructure is different for each case. Common to many, however, are the forms of fiber products that are used. This chapter summarized many of the processes used to produce typical fiber forms and provide some detail regarding a specific class of materials, namely, thermal insulation used to protect reusable launch vehicles.

References

1 "The Origin of Rock Wool," *Stone*, **1936**, *57*.
2 J. Player, U.S. Patent 103,650, 1870.
3 H. Lang, *Chem. Metall. Eng.*, **1933**, *27*, 365.
4 *The Handbook of Glass Manufacture*, Vol. II, F. V. Tooley (Ed.), Ashlee Publishing, New York, 1984.
5 *Handbook of Composites: Strong Fibers, Vol. 1*, W. Watt and B.V. Perov (Eds.), Elsevier Science Publishers, Amsterdam, 1985, p. 115.
6 *Engineered Materials Handbook: Ceramics and Glasses*, Vol. 4, S.J. Schneider, Jr. (Ed.), ASM International, Metals Park, 1991, p. 402.
7 *Overview of Fiberglass*, Johns Manville Corporation Product Brochure.
8 J. Cleland and F. Iannetti, *Thermal Protection System of the Space Shuttle*, NASA Contractor Report 4227, 1989.
9 W.C. Millar and T.A. Scripps, *Am. Ceram. Soc. Bull.*, **1982**, *61*, 711.
10 *Glass Engineering Handbook*, E.B. Shand (Ed.), McGraw-Hill Book Co., New York, Toronto, London, 1959.
11 J.S. Kenworthy, M.J. Morton, and M.D. Taylor, U.S. Patent 3,950,479, 1976.
12 K.A. Karst, and H.G. Sowman, U.S. Patent 4,047,965, 1977.

13 *Ceramic Fibers and Coatings: Advanced Materials for the Twenty-First Century*, National Materials Advisory Board Publication NMAB-494, National Academy Press, 1998.
14 L.J. Korb, C.A. Morant, R.M. Calland, and C.S. Thatcher, *Am. Ceram. Soc. Bull.*, **1981**, *60*, 1188–1193.
15 H. Goldstein, D. Leiser, P. Sawko, H. Larson, C. Estrella, M. Smith, and F. Pitoniak, *NASA Tech. Briefs*, **1985**, *9*, 107–108.
16 S.A., Chiu, W.C., Pitts, AIAA Paper 91-0695, **1991**.
17 J.D. Buckley, G. Strouhal, and J.J. Gangler, *Am. Ceram. Soc. Bull.*, **1981**, *60*, 1196–1200.
18 R.P. Banas, E.R. Gzowski, and W.T. Larsen, *Ceram. Eng. Sci. Proc.*, **1983**, *4*, 591–610.
19 E. Bahnsen, S. Garofalini, and A. Pechmen, "Accelerated Purification of Colloidal Silica Sols", *NASA Tech Briefs MSC-16793*, **1978**.
20 D.J. Green, J.E. Ritter, Jr., and F.F., Lange, *J. Am. Ceram. Soc.*, **1982**, *65*, 141–46.
21 A.M. Lovelace, R.L. Dotts, and J.W. Holt, U.S. Patent 4,338,368, 1980.
22 H. Goldstein, V.E. Katvala, and D.B. Leiser, "High-Temperature Glass and Glass Coatings", *NASA Tech Briefs ARC-11051*, **1977**.
23 R.A. Frosch, D.B. Leiser, H. Goldstein, M. Smith, U.S. Patent 4,148,962, 1970.
24 D.B. Leiser, M. Smith, and H. Goldstien, *Am. Ceram. Soc. Bull.*, **1981**, *60*, 1201–1204.
25 D.B. Leiser, M. Smith, and D.A. Stewart, *Ceram. Eng. Sci. Proc.*, **1985**, *6*, 757–768.
26 D.B. Leiser, M. Smith, R.A. Churchward, and V.W. Katvala; U.S. Patent 5,079,082, 1992.
27 D. Mui and H.M. Clancy, *Ceram. Eng. Sci. Proc.*, **1985**, *6*, 793–805.
28 D.J. Rasky, 25th International Conference On Environmental Systems, No. 951618, 1995.
29 Grunet, W.E., Notaro, F. and Reid, R.L., AIAA 4th Thermophysics Conference, **1969**, AIAA-69-605.
30 D.D. Hass, B.D. Prasad, D.E. Glass, K.E. Wiedemann, NASA Contractor Report 201733, 1997.
31 J. Ryu, U.S. Patent 6,068,882, 2000.
32 S.M. White, U.S. Patent Application 20020061396, filed 1997.
33 R. A. DiChiara, Jr. and F. K. Myers, U.S. Patent 6,607,851, 2003.
34 A. Barney, C. Whittington, B. Eilertson, B., and Z. Seminski, US Patent 6,652,950B2, 2003.
35 P.A. Hogenson, US Patent 5,626,951 A1, 1997.
36 J.B. Davis, D.B. Marshall, K.S. Oka, R.M. Housley, and P.E.D. Morgan, *Compos. Part A*, **1999**, *30*, 483–488.
37 J.B. Davis, D.B. Marshall, P.E.D. Morgan, K.S. Oka, A.O. Barney, and P.A. Hogenson, AIAA Manuscript No. AIAA-2000-0172, 2000.
38 S.J. Scotti, C. Clay and M. Rezin, AIAA Manuscript No. AIAA-2003-2697, 2003.
39 D. Rasky, and D. Kourtides, U.S. Patent Application ARC 12081-1, filed 1998.

2.5 Microcellular Ceramics from Wood

Heino Sieber and Mrityunjay Singh

2.5.1 Introduction

Wood is a natural composite material developed through long-term genetic evolution. It exhibits an anisotropic, porous morphology with excellent strength at low density, high stiffness, elasticity, and tolerance to damage on the micro- and macroscales [1, 2]. In contrast to most technically advanced man-made materials, the morphology of wood is characterized by a complex, hierarchical anatomy. The microstructural features of wood range from the milli- (growth-ring patterns) via micro- (tracheidal-cell patterns) down to the nanometer scale (molecular cellulose fiber and membrane structures of cell walls). The most distinct feature of the wood structure, however, is the open porous system of the tracheidal cells, which provide the transportation path for water and nutrients in the living wood and form a uniaxial pore structure. The mechanical properties of plant materials like wood and palms are determined by this open cellular structure, which results in pronounced anisotropy [3].

The heterogeneous tissues of wood are made by different types of cells. The wood cell walls are made of biopolymers such as cellulose, hemicellulose, pectin, and proteins. Several cellulose macromolecules are bundled to fibrils, and lignin serves as a binder between the cellulose microfibrils and increases the mechanical strength. The average elemental composition of wood is about 50 wt% C, 43.4 wt% O, 6.1 wt% H, 0.2 wt% N, and 0.3 wt% ash consisting mainly of oxides of K, Ca, Mg, Na, Si, P, and Al [4].

The morphology and the arrangement of the cells vary between coniferous wood, deciduous wood, and wood-forming plants like palms (Fig. 1). Coniferous woods have a relatively uniform structure interrupted by seasonal rings. It consists of 90–95 % tracheids, which are long and slender cells (diameter up to 50 μm, length up to a few millimeters) that are tapered at the ends and form a nearly monomodal pore morphology. Deciduous woods are less homogeneous. They contain additionally tracheary cells (diameter up to 500 μm, length less than 1 mm) which form long tubes of a few centimeters. Tracheids and tracheary cells are oriented in the direction of the trunk axis. Cells arranged in radial direction (rays) and pores in the cell walls create a three-dimensional open-porous network for transportation. The axial microstructure of deciduous woods and palms is characterized by a multimodal porous morphology with pore diameters between a few and a few hundred micrometers.

Cellular Ceramics: Structure, Manufacturing, Properties and Applications.
Michael Scheffler, Paolo Colombo (Eds.)

ISBN: 3-527-31320-6

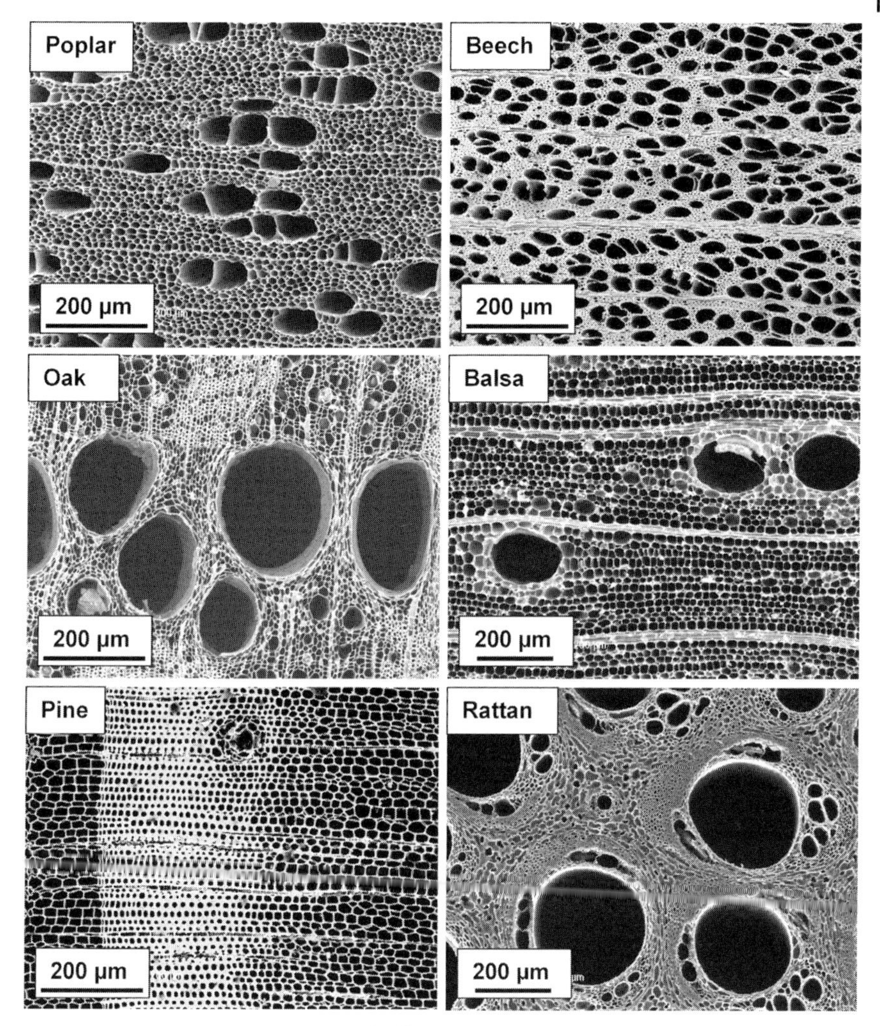

Figure 1 SEM images: microstructure of different wood plants.

The cellular anatomy of naturally grown wood provides an attractive template for the design of microcellular ceramic materials with hierarchically ordered pore structures on different length scales. In recent years, different technologies for transforming native wood materials into cellular ceramics and ceramic composites has attained particular interest due to unidirectional pore structures on the microscopic level that can not be produced by conventional ceramic processing technologies [5, 6]. The inherent cellular and open-porous morphology of the bioorganic materials is easily accessible to liquid or gaseous infiltrants of different compositions. By using high-temperature pyrolysis (carbonization) and subsequent infiltration-reaction processes, the bioorganic structures can be converted to biomorphous ceramics within reasonable time and maintain the morphological features of the native template.

Wood preforms allow the manufacture of cellular ceramics with cell diameters in the micrometer range. Porous ceramics with cellular structures on a larger scale can be manufactured by conversion of wood-derived materials such as cellulose fiber felts, preprocessed papers, or corrugated cardboard structures. Ohzawa et al. [7] and Almeida, Streitwieser et al. [8] infiltrated cellulose-fiber paper preforms with SiC by pressure-pulsed chemical vapor infiltration for high-temperature filter applications. Sieber et al. [9, 10] used macrocellular templates of corrugated cardboard with cell diameters of a few millimeters for reactive conversion to SiC/mullite-based ceramic composites. The processing of cellulose-fiber papers filled with Si powder yields highly porous SiC ceramics with fibrous morphology [11, 12].

Typically, the fabrication of cellular ceramics and ceramic composites from wood templates involves two processing steps: preparation of porous biocarbon (C_B) templates and subsequent conversion to carbide (reactive techniques) or oxide (molding techniques) ceramic structures. While the reactive techniques involve transformation of the C_B template into carbide phases by solid/liquid or gas-phase reactions, the molding technique reproduces the microstructural morphologies of the wood template by coating of internal surfaces (Fig. 2). The latter techniques have mainly been applied for manufacturing microcellular, oxide-based ceramics. Both processing routes are discussed in detail in following sections.

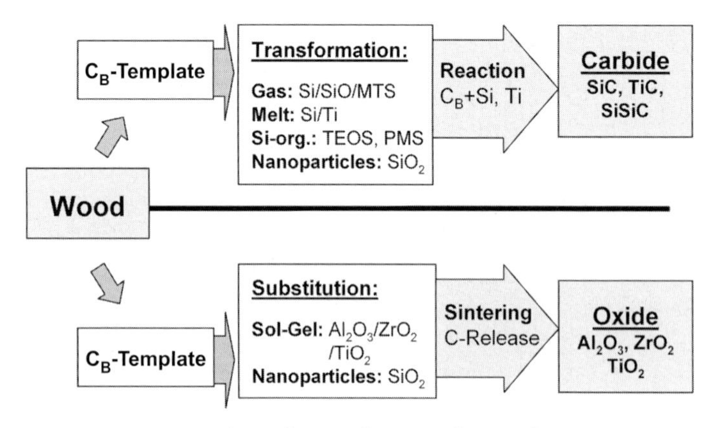

Figure 2 Processing scheme for manufacturing of biomorphous carbide and oxide ceramic from wood (MTS: methyltrichlorosilane, TEOS: tetraethylorthosilicate, PMS: polymethylsiloxane).

2.5.2
Fabrication of Porous Biocarbon Templates

The preparation of biocarbon (C_B) template structures by pyrolysis of native wood materials is carried out above 700 °C. During pyrolysis, a slow heating rate of about 1 °C min^{-1} must be applied up to about 500 °C to completely decompose the polyaromatic wood polymers cellulose, hemicellulose, and lignin to carbon. Afterwards, a higher heating rate of about 5 °C min^{-1} up to peak temperature can be used [13, 14].

The formation of carbon template structures from wood is a thermal degradation process [15]. It can be divided in three steps:

- Release of water, which results in "dehydrocellulose", followed by elimination of CO and CO_2 above 260 °C.
- Pyrolytic decomposition of hemicellulose, cellulose, and lignin between 260 and 500 °C, associated with volatilization of aliphatic acids, carbonyl compounds, and alcohols.
- Breakdown of the –C–C– chains in the biopolymer structures and formation of graphitic carbon structures above 600 °C.

The weight loss during pyrolysis of the wood preforms is almost complete at about 600 °C and results in a final weight loss in the range of 70–80 wt % [13, 16]. The weight losses of various kinds of wood can differ and are related to the molecular composition of hemicellulose, cellulose, and lignin in the wood. Due to its aromatic structure, lignin yields a much smaller weight loss (ca. 55 wt %) than cellulose (ca. 80 wt %) [16].

The weight loss leads to anisotropic shrinkage of the wood materials during pyrolysis by about 20 % in the axial, 30 % in the radial, and 40 % in the tangential direction on average [13, 16, 17]. It can be explained by decomposition and rearrangement of the oxygen-bonded glucose units of the cellulose fiber during thermal treatment and thus depends on the relative orientation of the cellulose fiber to the wood axis [18]. After pyrolysis, the total porosity of the C_B template is 20–25 % higher than that of the native (dried) wood [13, 16]. The weight loss and shrinkage during pyrolysis of pinewood are exemplarily shown in Fig. 3.

Despite the large weight loss and anisotropic shrinkage during pyrolysis, the fine microstructural features of the cellular wood anatomy on the submicrometer scale are retained with high precision in the C_B template. By using slow heating rates during pyrolysis, even large wood species can be carbonized into C_B template structures maintaining the morphology of the natural wood anatomy and without cracking of the material [13].

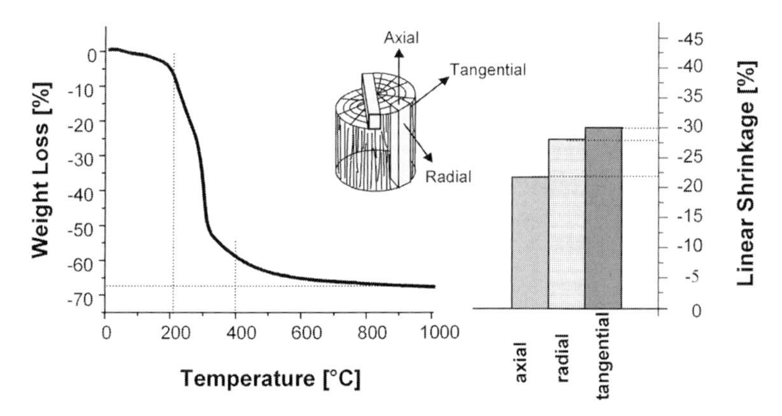

Figure 3 Weight loss and linear shrinkage during pyrolysis of pinewood (from [17]).

2.5.3
Preparation of Carbide-Based Biomorphous Ceramics

Carbide-based biomorphous ceramics can be fabricated by reactive conversion of the porous C_B template. To manufacture SiC-based ceramics, different reactive processing routes were applied [19], such as infiltration with Si melts, $Si/SiO/CH_3SiCl_3$ vapors, or sols containing SiO_2 precursors (Fig. 4). Infiltration with SiO_2 precursor sols at ambient temperature is an easier processing route with only one high-temperature reaction step, but it often results in low mechanical stability of the obtained SiC material. Examples are the infiltration of organosilcon mono- or polymers into charcoal and their conversion to highly porous SiC and SiOC ceramics [20, 21]. Herzog et al. developed a processing route with infiltration of nanosized SiO_2 sols into wood chars [22]. Repeated infiltration and carbothermal reduction of the C_B template yields beech- and pinewood derived SiC ceramics with porosities of 65–75 %.

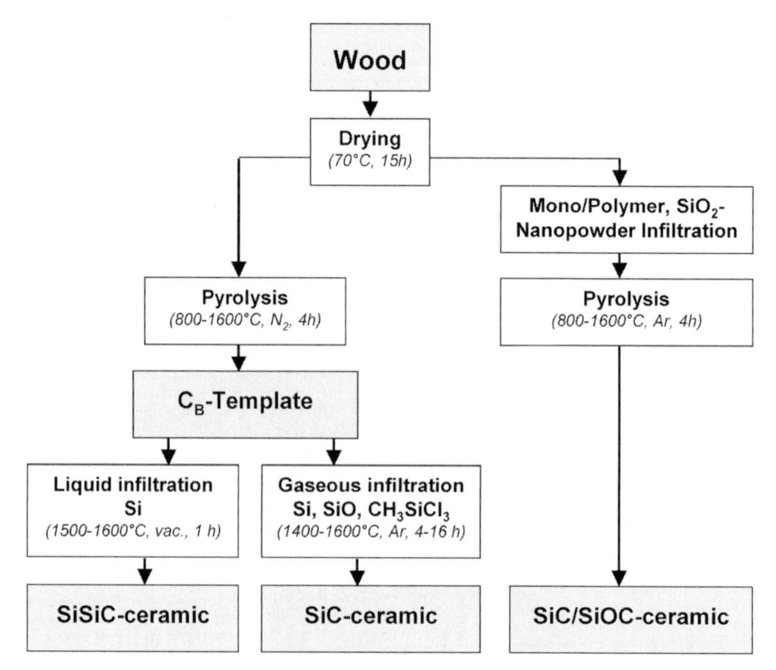

Figure 4 Processing scheme for manufacturing microcellular SiC ceramics from wood (after [19]).

Besides the synthesis of biomorphous SiC-based ceramics, a few reports also deal with the conversion of bioorganic materials to TiC-based ceramics. The reactive infiltration of carbonized wood structures with pure Ti vapor in vacuum at 1600 °C yields a thin surface layer of TiC of few 100 μm in depth [23]. Sun et al. prepared highly porous, biomorphous TiC ceramics by infiltration of wood char with tetrabutyl titanate and high-temperature reaction in inert atmosphere [24]. A more promising approach for synthesis of mechanically stable, porous, biomorphous TiC ceramics is chemical vapor infiltration-reaction (CVI-R) of carbonized wood specimens with $TiCl_4$ at about 1200 °C [25].

2.5.3.1
Processing by Silicon-Melt Infiltration

SiSiC ceramic composites with microcellular morphology and excellent mechanical stability can be prepared by infiltration of liquid Si into the carbonized wood templates, similar to conventional liquid-silicon infiltration (LSI) processing [26]. Byrne and Nagle [27], Greil et al. [16], Martinez-Fernandez et al. [28, 29], and Singh et al. [5, 30, 31] converted different kinds of wood structures by spontaneous infiltration with an Si melt to microcellular SiSiC composites. Similarly, Shin and Park [32] used charcoal for fabrication of biomorphous SiSiC composites. After reactive infiltration with liquid Si at 1450–1600 °C, the carbon of the wood char is transformed into β-SiC. The small pores in the C_B template, up to a pore diameter of approximately 50 μm, are filled with residual Si, and an SiSiC ceramic–ceramic composite is formed. The Si content of the final cellular SiSiC composite, as well as the total porosity and pore size distribution, depends on infiltration parameters and the anatomy of the wooden preforms. To fabricate more porous biomorphous SiSiC ceramics, the specimens can be subjected to stoichiometric or near-stoichiometric Si infiltration in which only the Si required for conversion of the biocarbon cell walls is supplied [33], or the unconsumed Si can be removed afterwards by heat treatment [34] or chemical etching [16].

Figure 5a shows redwood-derived microcellular SiSiC ceramics after spontaneous Si-melt infiltration. The pores walls were converted to β-SiC, while the pore volumes were filled with residual Si after processing. To remove the excess Si remaining in the pores, the samples were heat-treated at 1550 °C for 4 h (Fig. 5b). After initial infiltration at 1450 °C for 30 min, the geometrical density was around 2.4 g cm^{-3}, and it decreased to 2.0 g cm^{-3} after heat treatment at 1550 °C [34].

Biomorphous SiSiC ceramic composites contain about 1–5 vol % of unconverted carbon after processing, due mainly to local density variations in the C_B template. Especially in denser areas of the C_B template (e.g., latewood areas in pine char), the carbon density can exceed the critical limit of about 0.97 g cm^{-3}, which results in clogging of the pores during Si-melt infiltration and inhibits further SiC formation [35]. The microstructure of the biomorphous SiSiC ceramic composites is characterized mainly by β-SiC grains with a grain size in the micrometer range (up to ca. 10 μm) formed by intimate contact of the SiC phase with the Si melt. Additionally, layers of a nanograined β-SiC phase with average grain diameters of less than 100 nm, formed only between the coarse-grained SiC phase and the unconverted carbon regions, are observed [36, 37]. While the coarse-grained SiC phase is due to dissolution and recrystallization processes of the carbon and SiC in contact with the Si melt, the nanograined SiC phase originates from gas-phase and solid-state reactions of C and Si. The evolution of the different SiC phases during infiltration of liquid Si into carbonized wood templates is a combined infiltration/reaction process influ-

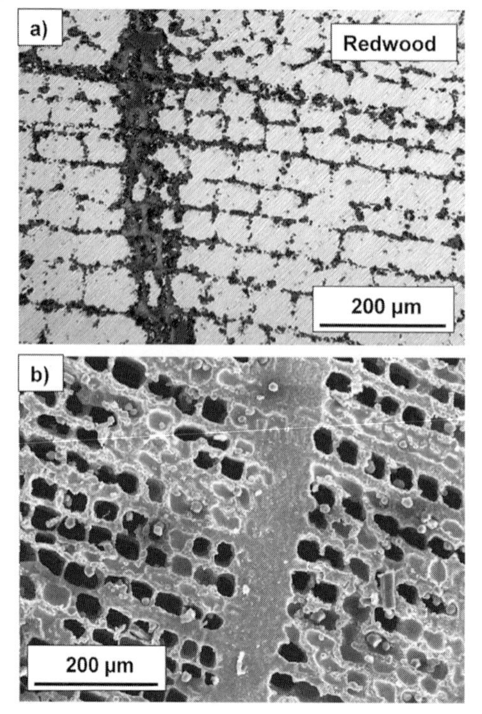

Figure 5 SEM images: microcellular SiSiC ceramic derived from redwood a) after Si-melt infiltration, and b) heat-treated at 1550 °C to remove excess silicon (reprinted from [34], with permission by CPRC, Hanyang University, Korea).

enced by the local carbon morphology, as well as by time-dependent dissolution, recrystallization, and SiC grain-growth processes [38].

The microcellular, biomorphous SiSiC ceramics exhibit excellent but anisotropic mechanical properties, due to the unidirectional cellular pore morphology of the wood template [16, 29, 33, 39–41]. The microstructure–strength correlation in porous materials is described in the literature by two main approaches: cellular materials with porosities greater than 70 %, in which the strength is mostly related to the bending strength of the cell walls [42], and systems with lower porosities, in which the strength is assumed to be only dependent on the minimum solid area perpendicular to the applied stress [43]. While the cellular model of Gibson and Ashby exhibits good agreement for the compressive strength of low-density biomorphous SiSiC [33], the model cannot differentiate between different topologies of the materials and thus, for a wide variety of biomorphous SiSiC microstructures, the anisotropic behavior and porosity dependence fit better with the assumption of the minimum-solid-area approach [29].

The mechanical properties of porous, biomorphous SiSiC ceramic composites can be retained up to high temperatures until melting of Si starts [33]. Potential applications of microcellular, biomorphous SiSiC ceramics are filters, microreactor devices, and absorbers for high temperatures [37], as well as medical implants [44].

2.5.3.2 Gas-Phase Processing

In contrast to Si-melt infiltration, highly-porous, single-phase biomorphous SiC ceramics with a cellular morphology on the micrometer scale can be manufactured by using Si-containing vapors as reactants. During processing, the Si-containing vapor penetrates the pores of the C_B template and reacts with the carbon to form β-SiC. Different reactive Si-containing vapor species such as Si [17, 45, 46], SiO [47] and CH_3SiCl_3 (methyltrichlorosilane, MTS) [48] can be used, and the porosity of the final SiC ceramics depends on the reaction products formed:

$$C_{B(s)} + Si_{(g)} \rightarrow SiC_{B(s)} \tag{1}$$

$$2\,C_{B(s)} + SiO_{(g)} \rightarrow SiC_{B(s)} + CO_{(g)} \tag{2a}$$

$$3\,C_{B(s)} + 2\,SiO_{(g)} \rightarrow 2\,SiC_{B(s)} + CO_{2(g)} \tag{2b}$$

$$C_{B(s)} + (1+x)\,CH_3SiCl_{3(g)} + 2\,H_{2(g)} \rightarrow SiC_{B(s)} + x\,SiC_{(s)} + 3(1+x)\,HCl_{(g)} + CH_{4(g)} \tag{3}$$

where x denotes the amount of additionally deposited SiC from CVI processing.

The specific volume change associated with the reaction of the wood-derived carbon with the Si-containing vapors to give β-SiC is given by:

$$\Delta V/V_C = (V_{SiC} - V_C)/V_C = \frac{n_{SiC}\,M_{SiC}\,\rho_C}{n_C\,M_C\,\rho_{SiC}} - 1 \tag{4}$$

where n, M, and ρ denote the number of moles, the molecular weight, and the densities of the indicated phases. The density of SiC is $\rho_{SiC} = 3.21$ g cm^{-3}. Depending on the density of the carbon after pyrolysis (graphite: $\rho_C = 2.26$ g cm^{-3}, vitreous carbon: $\rho_C = 1.44$ g cm^{-3}) and the kind of reaction (Eqs. 1–3), specific volume changes occur (Fig. 6).

The measured skeletal density of the wood char after pyrolysis at 1600 °C is 1.4–1.6 g cm^{-3}, and the specific surface area of a few tens of m^2/g indicates a pronounced porous strut morphology [14, 15, 47]. The reaction of the C_B template struts with Si vapor (Eq. 1) results in 1:1 conversion of porous carbon to β-SiC and thus in an increase in volume during SiC formation together with densification of the porous carbon strut. CVI processing with gaseous organosilicon precursors (Eq. 3) yields a much denser SiC ceramic due to the deposition of additional SiC. In contrast, processing with SiO vapor results in more porous cellular materials due to the release of carbon by CO and CO_2 evaporation during the reaction (Eqs. 2a and 2b).

Figure 7 shows biomorphous SiC ceramics from pinewood after Si-vapor infiltration. In contrast to Si-melt infiltration, no Si was deposited during processing and a highly porous, single-phase SiC ceramic was formed. The final microcellular SiC

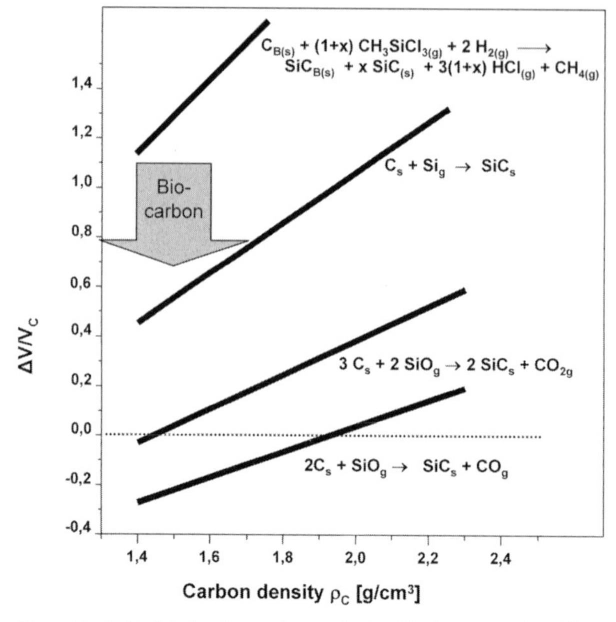

Figure 6 Calculated volume change during SiC formation by different reactions as a function of initial carbon template density (modified from [45]).

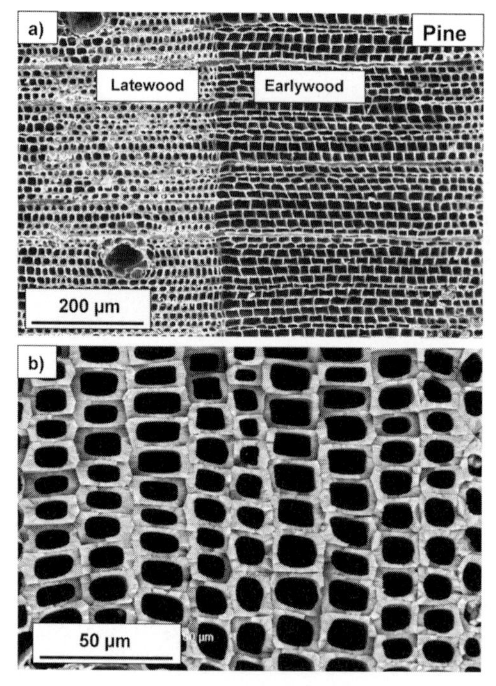

Figure 7 SEM images: microcellular SiC ceramic derived from pinewood after Si-gas infiltration at 1600 °C for 4 h (see also [17]).

ceramics completely reproduce the porous wood morphology down to the micrometer level. The β-SiC grains formed in the reaction are in the submicrometer range and form a nearly dense strut morphology. The microstructure, porosity, and mechanical properties of the biomorphous SiC ceramics after gas-phase infiltration depend on the kind of wood as well as on the infiltrating and reacting vapor species. For instance, SiC ceramics derived from pinewood and different Si-containing vapors exhibit bending strengths in the axial direction between 4 and 21 MPa [49] (compression strength 70–151 MPa [50]) for a porosity in the range between 60 and 80 % (Tab. 1).

Table 1 Properties of biomorphous SiC-based ceramics from pinewood prepared by different technologies [16, 33, 49, 50]. Compression strength was measured in the axial direction at room temperature; radial and tangential strengths are lower.

	Material	Density [g cm^{-3}]	Bending strength [MPa]	Compression strength [MPa]	Porosity [vol %]	Specific surface area [m^2 g^{-1}]	Refs.
C_B template	C_B	0.31	15–18	–	78	50	16
Si-melt infiltration* (1450 °C/4 h, vacuum)	SiSiC	1.2–1.7	–	50–120	35–45	–	33
SiO-vapor infiltration** (1600 °C/8 h, Ar)	SiC_{SiO}	0.6	4	70	80	16.1	49, 50
Si-vapor infiltration** (1600 °C/4 h, Ar)	SiC_{Si}	1.0	13	120	70	3.3	49, 50
MTS-vapor infiltration** (1250 °C, 1600 °C/1 h, Ar)	SiC_{MTS}	1.2	21	151	60	0.5	49, 50

* Compression strength was measured at 1150 °C [33].
** Bending strength was measured by biaxial ball-on-ring testing [49].

2.5.4 Preparation of Oxide-Based Biomorphous Ceramics

Only a few investigations have focused on the synthesis of microcellular oxide ceramics with biomorphous microstructures. Metal alkoxides and metal chlorides can be used as precursors which can form a macromolecular oxide network through hydrolysis and condensation on the inner cell walls of the porous structure of native plants and C_B templates. Subsequently, the biocarbon is burned out by heating in air, and a highly porous oxide ceramic or ceramic composite is formed. Further annealing leads to consolidation and sintering of the highly porous structure and results in microcellular, biomorphous oxide ceramics of different compositions.

The technique was used by Yermolenko et al. for the formation of ZrO_2 fibers by oxidizing hydrated cellulose fibers impregnated with a zirconium salt [51]. Padel and Padhi [52, 53] manufactured Al_2O_3 and TiO_2 fibers by infiltration of natural sisal, jute, and hemp fibers with $AlCl_3$ and $TiCl_4$, respectively. Ota et al. [54] produced biomorphous oxide ceramics by infiltration of Japanese wood species with titanium isopropoxide. After high-temperature treatment in air the wood structures were converted into porous TiO_2 ceramics. Shin et al. [55] synthesized biomorphous SiO_2 ceramics from wood by a surfactant-templated sol–gel process. Rattan palm preforms were converted to biomorphous Al_2O_3, ZrO_2, and mullite ($Al_6Si_2O_{13}$) ceramics by a sol–gel process with metal alkoxides, metal oxide chlorides, and SiO_2 nanopowders [56]. Singh and Yee [57] infiltrated jelutong wood with ZrO_2 sol for manufacturing monoclinic ZrO_2 ceramics with biomorphous structure. Highly porous, biomorphous Al_2O_3, TiO_2, and ZrO_2 ceramics were prepared from pinewood and from cellulose fiber preforms by a sol–gel process with metal alkoxides [58–60].

As an example, the manufacture of microcellular oxide ceramics from pinewood will be described [58]. The process starts with infiltration of the dried native pine specimen with liquid organometallic precursors. Low-viscosity, stable oxide sols were prepared for the infiltration process. The properties of the different sols (e.g., concentration, viscosity, stability) had to be adjusted for optimal infiltration behavior into the pinewood preforms. After sol infiltration the samples were dried in air to form in situ gels of the respective oxide. This procedure was repeated several times to increase the precursor content in the raw samples. Then the samples were pyrolyzed at 800 °C to carbonize the wood template and decompose the organometallic precursor into the respective oxide. Subsequent infiltrations were performed into the pyrolyzed pinewood templates. For consolidation and sintering into microcellular oxide ceramics, the specimens were annealed in air at temperatures up to 1550 °C. The processing scheme is summarized in Fig. 8. The final oxide ceramics are pseudomorphous to the initial pinewood. Additional infiltration processes, followed by repeated annealing, were performed to increase the density of the final materials and to fill the pores left by burning out the C_B template.

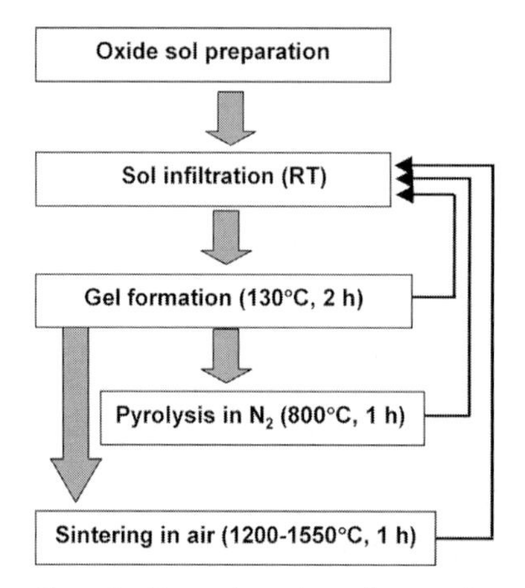

Figure 8 Processing scheme for manufacturing microcellular oxide ceramic from wood (after [58]).

The technique was used for conversion of pinewood samples to biomorphous Al_2O_3, TiO_2, and ZrO_2 ceramics. After processing, the biocarbon of the pine chars was completely burned out to give brightly colored biomorphous oxide ceramics maintaining the macro- and microstructural features of the biological template. However, due to Ca impurities in the native pinewood, small amounts of mixed Ca oxides were also observed in the microcellular oxide ceramics [58].

Figure 9 illustrates the cellular microstructure in the axial direction of biomorphous oxide ceramics derived from pinewood. The initial cellular anatomy was reproduced in the ceramic products. In the biomorphous Al_2O_3 ceramics, burning the carbon out of the pinewood cell walls left small holes in the earlywood regions, where the cell walls of the specimens were not completely consolidated. Due to the smaller cell diameter and thicker cell walls in the earlywood region, the morphology is characterized by a hollow-fiber structure, separated by voids of up to 5 μm in size. In contrast to Al_2O_3, the cell walls of pinewood were replaced by dense layers of TiO_2 in the biomorphous TiO_2 ceramics. Most of the vessels are kept open after conversion to ceramic. The microstructure of biomorphous ZrO_2 ceramics looks similar to that of biomorphous Al_2O_3 and TiO_2 ceramics. The size of the Al_2O_3, TiO_2, and ZrO_2 grains after sintering is about 3–5 μm. Similar results were obtained by using rattan palms as bioorganic template structures [59, 60] or cellulose fiber felts [61].

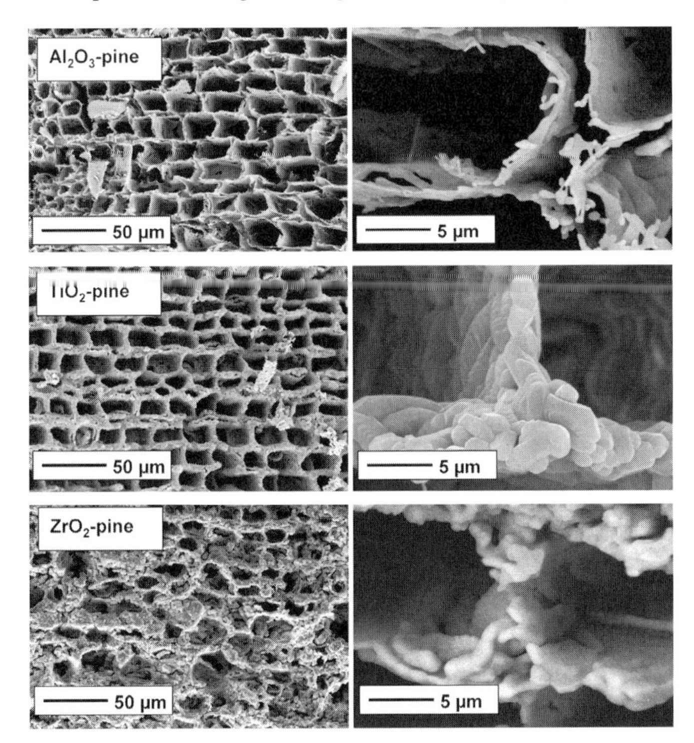

Figure 9 SEM images: microcellular Al_2O_3, TiO_2, and ZrO_2 ceramics derived from pinewood after annealing at 1550, 1200, and 1500 °C, respectively, for 1 h in air (see also [58]).

Table 2 summarizes the material properties of biomorphous oxide ceramics made from pinewood. The porosity was calculated from the difference between the geometrical and skeletal densities measured by He pycnometry. The skeletal densities of biomorphous Al_2O_3 and TiO_2 ceramics are very close to the theoretical density of Al_2O_3 (3.97 g cm^{-3}) and TiO_2 (4.26 g cm^{-3}), and they indicate only small amount of closed porosity. Due to the content of monoclinic ZrO_2 (theoretical density 5.68 g cm^{-3}) and impurities, the skeletal density of the biomorphous ZrO_2 is only 5.81 g cm^{-3} (theoretical density 6.7 g cm^{-3}).

Table 2 Density and porosity values of biomorphous oxide ceramics derived from pinewood by infiltration with different organometallic precursors (after [59]).

Oxide	Density [g cm^{-3}]		Porosity [%]
	geometrical	skeletal	
α-Al_2O_3	0.22	3.93	94
Rutile TiO_2	0.86	4.26	80
Cubic ZrO_2	1.11	5.81	81

2.5.5 Summary

Microcellular carbide and oxide ceramics with cell diameters from a few up to several hundred micrometers can be manufactured from bioorganic structures by utilizing the open-porous morphology of wood materials and applying different bioorganic-to-ceramic conversion technologies. Due to the unidirectional pore morphology in the native wood templates, biomorphous ceramics exhibit pronounced anisotropy in physical properties such as penetrability, heat transfer, electrical conductivity, and mechanical stiffness, and are thus suitable for different applications as high-temperature or corrosion-resistant materials.

Biological templates from wood and wood-derived materials are available on a large scale, and they are cheap and regenerable. Wood and products derived therefrom, such as paper and cardboard, can be easily machined and preformed into complex three-dimensional shapes of different porosities and cellular morphologies. The large variety of natural plant tissues, the established forming and shaping technologies for wooden materials, and the utilization of different conversion technologies facilitate the manufacture of microcellular ceramics and ceramic composites with tailored microstructures and compositions.

Acknowledgements

Financial support from the Volkswagen Foundation under contract I/73 043 is gratefully acknowledged. H.S. wants to thank C. Zollfrank, C.R. Rambo, E. Vogli, J. Cao, and P. Greil for many helpful discussions during preparation of biomorphic ceramics.

References

1 Gibson, L.J., *Met. Mater.* **1992**, *8*, 333–348.
2 Greil, P., *J. Eur. Ceram. Soc.* **2001**, *21*, 105–118.
3 Lucas, P.W., Darvell , B.W., Lee, P.K., Yuen, T.D.B., Choong, M.F., *Phil. Trans. R. Soc. London* **1995**, *B348*, 363–385.
4 Fengel, D., Wegener, G., *Wood: Chemistry, Ultrastructure, Reactions*, de Gruyter, Berlin, New York, 1983.
5 Singh, M., *Ceram. Sci. Eng. Proc.* **2000**, *21*, 39–44.
6 Sieber, H., Hoffmann, C., Kaindl, A., Greil, P., *Adv. Eng. Mater.* **2000**, *2*, 105–109.
7 Ohzawa, Y., Hshino, H., Fujikawa, M., Nakane K., Sugiyama, K., *J. Mater. Sci.* **1998**, *33*, 5259–5264.
8 Almeida Streitwieser, D., Popovska, N., Gerhard, H., and Emig, G., *J. Eur. Ceram. Soc.*, in press.
9 Sieber, H., Kaindl, A., Schwarze, D., Werner J.P., Greil, P., *cfi/Ber. DKG* **2000**, *77*, 21–24.
10 Sieber, H., Schwarze, D., Kaindl, A., Friedrich H., and Greil, P., *Ceram. Trans. 108*, Eds.: J.P. Singh, N.P. Bansal, and K. Niihara, The American Ceramic Society, **2000**, pp. 571–580.
11 Fey, T., Sieber H., and Greil, P., *High Temperature Ceramic Matrix Composites – HTCMC-5*, Eds.: M. Singh, R.J. Kerans, E. Lara-Curcio, and R. Naslain, The American Ceramic Society, **2004**, 419–423.
12 Rusina, O., Kirmeier, R., Molinero, A., Rambo, C.R., Sieber, H., *Ceram. Trans. 166*, Ed.: N. P. Bansal, The American Ceramic Society, **2004**, 171–178.
13 Byrne, C.E., Nagle, D.C., *Carbon* **1997**, *35*, 259–266.
14 Byrne, C.E., Nagle, D.C., *Carbon* **1997**, *35*, 267–173.
15 Klemm, D., Philipp, B., Heinze, T., Heinze, U., Wagenknecht, W., *Comprehensive Cellulose Chemistry*, Vol. I, Wiley-VCH, Weinheim, 1998, pp. 107–125.
16 Greil, P., Lifka, T., Kaindl, A., *J. Eur. Ceram. Soc.* **1998**, *18*, 1961–1983.
17 Vogli, E., Sieber H., Greil, P., *J. Eur. Ceram. Soc.* **2002**, *22*, 2663–2668.
18 Tang, M.M., Bacon, R., *Carbon* **1964**, *2*, 211–220.
19 H. Sieber, *High Temperature Ceramic Matrix Composites – HTCMC-5*, Eds.: M. Singh, R.J. Kerans, E. Lara-Curcio, and R. Naslain, The American Ceramic Society, **2004**, 407–412.
20 Ota, T., Takahashi, M., Hibi, T., Ozawa, M., Suzuki, S., Hikichi Y., Suzuki, H., *J. Am. Ceram. Soc.* **1995**, *78*, 3409–3411.
21 Zollfrank, C., Kladny, R., Sieber H., Greil, P., *J. Eur. Ceram. Soc.* **2004**, *24*, 479–487.
22 Herzog, A., Klingner, R., Vogt, U., Graule, T., *J. Am. Ceram. Soc.* **2001**, *87*, 781–793.
23 Sieber, H., Rambo C.R., Benes, J., *Ceram. Eng. Sci. Proc.* **2003**, *24*, 135–140.
24 Sun, B., Fan, T., Zhang, D., Okabe, T., *Carbon* **2004**, *42*, 177–182.
25 Sieber, H., Zollfrank, C., Almeida, D., Gerhard, H., Popovska, N., *Key Eng. Mater.* **2004**, *264–268*, 2227–2230.
26 Hillig, W.B., *J. Am. Ceram. Soc.* **1988**, *71*, C96–C99.
27 Byrne, C.E., Nagle, D.E., *Mater. Res. Innovat.* **1997**, *1*, 137–144.
28 Martinez-Fernandez, J., Valera-Feria F.M., Singh, M., *Scripta Mater.* **2000**, *43*, 813–818.
29 Martínez-Fernández, J., de Muñoz, A., Arellano-López, A.R., Varela-Feria, F.M., Domínguez-Rodríguez, A., Singh, M., *Acta Mater.* **2003**, *51*, 3259-3275.
30 Singh, M., *Adv. Mater. Proc.* **2002**, *160*, 39–45.

31 J.A., Salem, Singh, M., *J. Eur. Ceram. Soc.* **2002**, *22*, 2709–2717.

32 Shin, D.-W., Park, S.S., *J. Am. Ceram. Soc.* **1999**, *82*, 3251–3253.

33 Varela-Feria, F.M., Martinez-Fernandez, J., Arellano-Lopez, A.R., Singh, M., *J. Eur. Ceram. Soc.*, **2002**, *22*, 2719–2725.

34 Singh, M., Yee, B.M., *J. Ceram. Proc. Res.* **2004**, *5*, 121–126.

35 Behrendt, D.R., Singh, M., *J. Mater. Synth. Proc.* **1994**, *2*, 117–124.

36 Zollfrank, C., Sieber, H., *J. Eur. Ceram. Soc.* **2004**, *24*, 495–506.

37 de Arellano-López, A.R., Martinez-Fernández, J., González, P., Domínguez, C., Fernández-Quero, V., Singh, M., *Int. J. Appl. Ceram. Technol.* **2004**, *1*, 56–67.

38 Zollfrank, C., Sieber, H., *J. Am. Ceram. Soc.* **2005** *88(1)*, 51–58.

39 Varela-Feria, F.M., López Pombero, S., Martínez-Fernández, J., de Arellano López, A.R., Singh, M., *Ceram. Eng. Sci. Proc.* **2001**, *22*, 135–145.

40 Qiao, G., Ma, R., Cai, N., Zhang, C., Jin, Z., *Mater. Sci. Eng. A* **2002**, *323*, 301–305.

41 Munoz, A., Martinez-Fernandez, J., Singh, M., *J. Eur. Ceram. Soc.* **2002**, *22*, 2727–2733.

42 Gibson, L.J., Ashby, M.F., *Cellular Solids, Structure and Properties*, Pergamon Press, New York, **2000**.

43 Rice, R.W., *J. Mater. Sci.* **1996**, *31*, 102–118.

44 González, P., Serra, J., Liste, S., Chiussi, S., León, B., Pérez-Amor, M., Martínez-Fernández, J., de Arellano-López, A. R., Varela-Feria, F. M., *Biomaterials* **2003**, *24*, 4827–4832.

45 Sieber, H., Vogli, E., Greil, P., *Ceram. Eng. Sci. Proc.* **2001**, *22*, 225–230.

46 Qian, J., Wang J., Jin, Z., *Mater. Chem. Phys.* **2003**, *82*, 648–653.

47 Vogli, E., Mukerji, J., Hoffmann, C., Kladny, R., Sieber H., Greil, P., *J. Am. Ceram. Soc.* **2001**, *84*, 1236–1240.

48 Sieber, H., Vogli, E., Müller, F., Greil, P., Popovska, N., Gerhard H., Emig, G., *Key Eng. Mater.* **2002**, *206–213*, 2013–2016.

49 Greil, P., Vogli, E., Fey, T., Bezold, A., Popovska, N., Gerhard H., Sieber, H., *J. Eur. Ceram. Soc.* **2002**, *22*, 2697–2707.

50 Vogli, E., Ph. D. Thesis, University of Erlangen-Nuernberg, Germany, **2003**.

51 Yermolenko, I.N., Vityaz, P.A., Ulyanova T.M., Fyodorova, I.L, *Sprechsaal* **1985**, *118*, 323–325.

52 Patel, M., Padhi, B.K., *J. Mater. Sci.* **1990**, *25*, 1335–1343.

53 Patel, M., Padhi, B.K., *J. Mater. Sci. Lett.* **1993**, *12*, 1234–1235.

54 Ota, T., Imaeda, M., Takase, H., Kobayashi, M., Kinoshita, N., Hirashita, T., Miyazaki, H., Hikichi, Y., *J. Am. Ceram. Soc.* **2000**, *83*, 1521–1523.

55 Shin, Y.S., Liu, J., Chang, J.H., Nie, Z.M., Exarhos, G., *Adv. Mater.* **2001**, *13*, 728–732.

56 Sieber, H., Rambo, C., Cao, J., Vogli, E., Greil, P., *Key Eng. Mater.* **2002**, *206–213*, 2009–1012.

57 Singh M., Yee, B.M., *J. Eur. Ceram. Soc.* **2004**, *24*, 209–217.

58 Cao, J., Rambo, C., Sieber, H., *Ceram. Int.* **2004**, *30*, 1967–1970.

59 Cao, J., Rambo, C.R., Sieber, H., *J. Porous Mater.* **2004**, *11*, 163–172.

60 Rambo, C., Cao J., Sieber, H., *Mater. Chem. Phys.* **2004**, *87*, 345–352.

61 Cao, J., Rusina, O., Sieber, H., *Ceram. Int.* **2004**, *30*, 1971–1974.

2.6 Carbon Foams

James Klett

2.6.1 Introduction

Carbon is a unique element in that in one of its allotropes is extremely thermally conductive and another is insulating. It has the ability to span orders of magnitudes in properties simply by how it is processed into a structure and the degree of heat treatment. One such structure is a foam made from carbon or graphite. Typical process sequences include blowing, carbonizing, and then thermally treating the foam. Autoclave processes are also feasible. Foams can be blown into a mold for net-shape composites or machined into intricate parts. Analytical models of graphitic foams have predicted them to have a compression modulus of approximately 2 GPa with a density of about 0.1 g cm^{-3} [1, 2]. Few, if any, other foams or core materials have a density and compression modulus near these theoretical values. Graphitic foams have been produced with various processes and precursors. The resultant foam properties are dependent on the process path and precursor. Some foams have a very high thermal conductivity, while others exhibit greater structural integrity. The creation of these foams has begun to foster novel thinking on how to design structures and thermal-management systems with these "alternative" carbon materials.

2.6.2 History

The first carbon foam, developed in the late 1960s, was reticulated vitreous carbon foam. Ford [3] reported on carbon foams produced by carbonizing thermosetting organic polymer foams through a simple heat treatment. Then, Googin et al. [4] at the Oak Ridge Atomic Energy Commission Laboratory reported the first process dedicated to controlling the structure and material properties of carbon and graphitic foams by varying the precursor material (partially cured urethane polymer). In the several decades following these initial discoveries, many researchers explored a variety of applications for these materials [5–14] ranging from electrodes to insulating liners for temperatures up to 2500 °C. Reticulated carbon foams have been used as the template for many of the metal and ceramic foams currently used in industry.

Cellular Ceramics: Structure, Manufacturing, Properties and Applications.
Michael Scheffler, Paolo Colombo (Eds.)

ISBN: 3-527-31320-6

In the 1970s, research focused primarily on producing carbon foams from alternative precursors. For example, Klett [8] at Sandia National Laboratories produced the first carbon foams from cork, a natural cellular precursor. Others worked on various changes to processing and precursors in an attempt to modify properties and reduce cost. One technique produced a reticulated vitreous carbon foam, commonly called glassy carbon foam [15, 16]. Other researchers also produced carbon foams by carbonization of thermoplastic or thermoset foams, such as phenolic resins and polyurethanes. For example, syntactic foams were obtained by molding a mixture of hollow plastic microspheres with a binder followed by carbonization; such products have a closed porosity of greater than 90 %. A process for the preparation of carbon foam from polyurethanes has been described, as has a process starting from phenolic foams. The majority of these "glassy" carbon foams were used for thermal insulation, although some structural applications were found.

In the early 1990s, researchers at the Wright Patterson Air Force Base (WPAFB) Materials Laboratory pioneered mesophase pitch-derived graphitic foams, specifically for replacing expensive 3D woven fiber performs in polymer composites and as replacements for honeycomb materials [1, 2, 17–22]. Their work was centered on developing a highly structural material that was lightweight. Concurrently, Ultramet Corp. performed research on RVC foam and used chemical vapor deposition (CVD) as a templating technique to place pyrolytic graphite on the glassy carbon ligaments, producing three-dimensional carbon structures with high-modulus ligaments.

With the goal of producing very inexpensive carbon foams, researchers at West Virginia University developed a method that used coal as a precursor for high-strength foams with excellent thermal insulation properties [23–26]. Then, in 1997, Klett [27–37] at the Oak Ridge National Laboratory (ORNL) reported the first graphitic foams with bulk thermal conductivities greater than $40\,W\,m^{-1}\,K^{-1}$ (recently, conductivities over $180\,W\,m^{-1}\,K^{-1}$ have been measured [38]). By combining an open cellular structure with a thermal conductivity to density ratio (κ/ρ) of greater than $345\,W\,m^{-1}K^{-1}\,g^{-1}\,cm^{3}$ (compared to 45 for copper), this material presents a unique opportunity to radically change the approach to solving many heat-transfer problems. This graphite material has been examined for the core of heat-transfer devices such as radiators and heat sinks, evaporative coolers, and phase-change devices.

This chapter discusses the various processing methods for several major types of carbon foams. This is not all-inclusive but is a cross-section of the many fundamental techniques for producing carbon and graphite foams.

2.6.3
Terminology

Before embarking on a review of carbon and graphitic foams, it is necessary to review the terminology used in the carbon community. The first terms that are critical to understanding these materials are carbon, graphite, and graphitization. These have been defined in many circles [39–45], but the most relevant definitions were developed by the carbon and graphite committee of ASTM subcommittee D02.F0 and are listed below.

2.6.3.1 Carbon

Element number 6 of the periodic table, with the electron ground state $1s^2 2s^2 2p^2$. In carbon and graphite technology, an artifact consisting predominantly of the element carbon and having limited long-range order.

2.6.3.2 Graphite

An allotropic crystalline form of the element carbon, occurring as a mineral, commonly consisting of a hexagonal array of carbon atoms but also known in a rhombohedral form. In carbon and graphite technology, a material consisting predominantly of the element carbon that possesses extensive long-range three-dimensional crystallographic order, as determined by X-ray diffraction studies.

The presence of long-range order is usually accompanied by high electrical and thermal conductivity within the hexagonal plane. This also results in a material having relatively easy machinability when compared to non-graphitic materials. The use of the term graphite without reporting confirmation of long-range crystallographic order should be avoided, as it can be misleading.

2.6.3.3 Graphitization

Graphitization is a solid state transformation of thermodynamically unstable non-graphitic carbon into graphite by thermal treatment. The degree of graphitization is a measure of the extent of long-range 3D crystallographic order, as determined by diffraction studies alone. The degree of graphitization significantly affects many properties, such as thermal conductivity, electrical conductivity, strength, and stiffness.

A common, but incorrect, use of the term graphitization is to indicate a process of thermal treatment of carbon materials above 2200 °C regardless of any resultant crystallinity. The use of the term graphitization without reporting confirmation of long-range 3D crystallographic order determined by diffraction studies should be avoided, as it can be misleading.

The term graphitization is quite often used to indicate a heat treatment to very high temperatures. Many assume that the common heat treatment referred to as graphitization of a carbon part necessarily yields a graphitic structure. To avoid confusion, the term should not be used without confirmation of extensive crystallinity in diffraction studies. Other studies such as TEM are suggestive of, but do not confirm, long-range 3D crystallinity.

2.6.3.4
Foam

According to Gibson and Ashby, a cellular material is a material made up of interconnected struts or plates, and a foam is a subset of cellular materials made up of polyhedral cells, giving a 3D structure [46]. At some point, with increasing density, the material resembles a solid with pores rather than a foam. Gibson and Ashby suggest that one way to delineate between a foam and a porous solid is to look at the stress–strain response under compression loading [46]. A relative density (r.d.) of approximately 0.3 seemed to be the transition point between a foam and a porous solid. Recently, graphitic foams have been formed with relative densities higher than 0.3 that exhibit the same fundamental compressive failure modes as lower density foams (see Fig. 1). In addition, MER Corp. [47] reported carbon and graphite

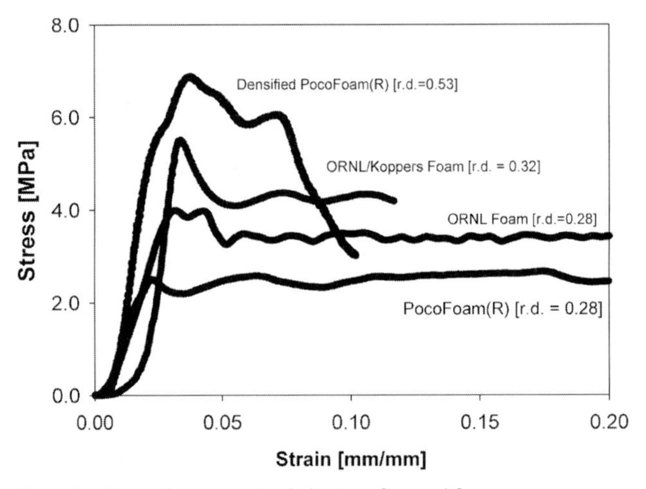

Figure 1 Plots of compression behavior of several foam structures showing a change in compression failure modes as the relative density increases above 0.5.

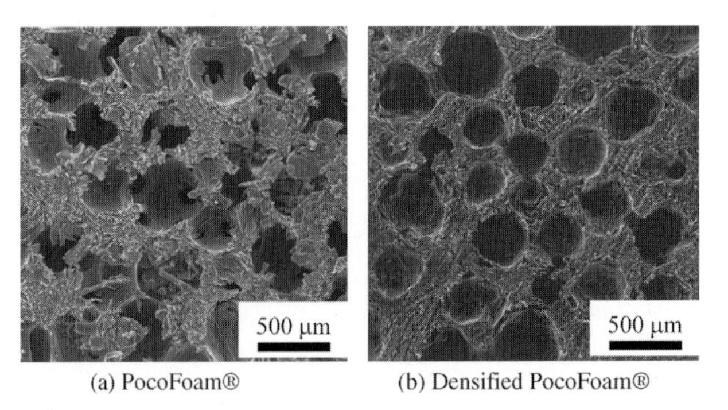

Figure 2 SEM images of standard PocoFoam and PocoFoam densified to a relative density of 0.53.

foams with densities up to 0.8 g cm^{-3} (r.d. = 0.35), and Stiller et al. [25] reported foams with densities up to 1.0 g cm^{-3} (r.d. = 0.44). Knippenburg et. al. reported a foam with r.d. greater than 0.3 [6]. As can be seen in Fig. 1, when the relative density of a foam is sufficiently high, the failure mode changes significantly, and the plateau is lacking (here, the graphite foam was densified with mesophase pitch, yielding a structure with interconnecting pores, but very large ligaments; see Fig. 2).

Clearly, most researchers classify their materials as foam when it exhibits a structure in accord with the first definition of Gibson and Ashby, but are not in agreement when it comes to a specific cutoff of relative density. In addition, the polymer industry does not use relative density in its definition of foam. Because of these inconsistencies, ASTM defined a foam without using relative density as a delineation mark. Instead ASTM agreed on the definitions given below based on the 3D cellular structure commonly assumed to be the essence of a foam. This chapter attempts to adhere to the ASTM definitions. However, some of the patents referenced in this section were granted prior to the ASTM and Gibson and Ashby definitions.

Carbon foam is a porous carbon product containing regularly shaped, predominantly concave, homogeneously dispersed cells which interact to form a three-dimensional array throughout a continuum material of carbon, predominantly in the non-graphitic state. The final result is either an open- or a closed-cell product.

Graphite foam is a porous graphite product containing regularly shaped, predominantly concave, homogeneously dispersed cells which interact to form a three-dimensional array throughout a continuum material of carbon, predominantly in the graphitic state. The final result is either an open- or a closed-cell product.

2.6.4
Foaming Processes

Foams can be fabricated by many different techniques outlined in the next section. There are three general types of precursors for foams: 1) thermosetting, 2) thermoplastic, and 3) hydrocarbon vapors. However, since the process of chemical vapor deposition (CVD) of carbon requires a foam substrate to begin with, this is included in the discussions of thermosetting and thermoplastic precursors.

2.6.4.1
Thermosetting Precursors

The first carbon foams were prepared by heat-treating existing polymeric foams, such as those from polyurethane, polyester, or phenolics [3]. Then, Googin et al. at the U.S. Atomic Energy Commision's Oak Ridge plant adjusted the precursor to tailor the resulting foam properties [4]. By mixing furfuryl alcohol with urethane-forming compounds prior to foaming, the resulting *glassy carbon foam* had drastically different properties than the converted polyurethane foam. In addition, graphite powder was added to the mix to result in the first "graphitic" foam. While the foam contained graphitic structures, the resulting product was an excellent thermal insulator for high-temperature furnaces. In fact, this foam was specifically designed for high-temperature stability to facilitate longer furnace runs before replacing the insulation pack. Hence, this is an example of graphitic foams with very low thermal conductivities.

Following years of research on variations of the above type of process, Knippenburg et al. developed a significant improvement. Reticulated polyurethane foams [6] were stabilized with gaseous oxygen to prevent the structure from swelling, softening, or collapsing during processing. Alternatively, the foams could be immersed in a poly(vinyl alcohol) solution at 80 °C. Subsequent evaporation of the water from the solution coats the ligaments with a thin film of poly(vinyl alcohol) which does not react or dissolve in the impregnating agent in the next step. Next, the ligaments are coated with a dilute solution of phenolic resins in ethyl alcohol. After drying the solvent from the foam, the phenolic resin is cured. Many coating steps can be performed to regulate the final foam properties (density, strength, porosity, etc.).

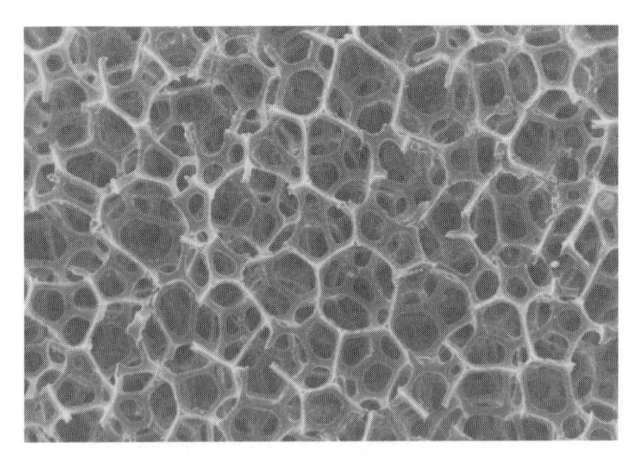

Figure 3 Reticulated foam structure (reticulated carbon foam made by Ultramet Corp.). Photo courtesy of Ed Stankewitz, Ultramet Corp.

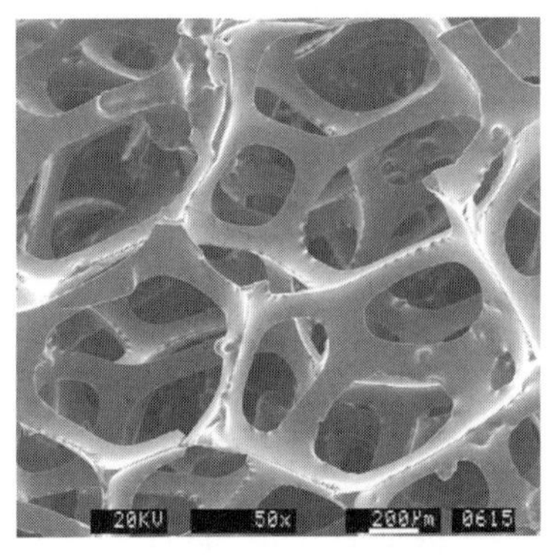

Figure 4 Reticulated carbon foam (RVC) made by ERG Corp [41].

Finally, the coated polyurethane foam is pyrolyzed to convert the structure to a glassy carbon foam typically referred to as *reticulated vitreous carbon foam*. This type of process (with many variations) is practiced by Ultramet Corp. which markets the foam as Ultrafoam. For specific applications, Ultramet produces foams with a controlled aspect ratio of cells near unity and with billet thicknesses greater than 20 cm (Fig. 3). A similar process is patented by Energy Research and Generation, Inc. (ERG) [48], which markets a reticulated vitreous carbon foam under the tradename RVC (Fig. 4), which is limited to 10 cm in thickness.

By varying the starting foam structure (linear cell count, ligament thickness, etc.) and varying the coating process, foams with the same overall density but drastically different structures (i.e. different cell count, ppi, strengths, thermal conductivities, and permeability factors) can be made.

Another process for producing a glassy carbon foam is described in ref. [49]. This process carbonizes a polystyrene-based low-density foam to yield a carbon foam. First, an inverse emulsion of a styrene monomer in water is prepared and polymerized to produce a styrene foam. The cells of the styrene foam are filled with an aqueous resorcinol–formaldehyde solution, which is cured to form a gel which coats the pores of the styrene foam. Subsequent heating to elevated temperatures converts the gel-filled polystyrene foam to a carbon foam.

Hayward et al. [50] combined expanded graphite with phenolic resins prior to foaming. First, a flexible graphite (e.g., Grafoil® gasket material) is chopped into small particles and thermally shocked to expand the graphitic planes into a very low density structure (i.e., exfoliation). A phenolic resin is mixed with the expanded graphite, and application of heat and pressure produces a foamed structure with both graphitic and glassy regions. This product was shown to be an excellent thermal insulator, despite being partially graphitic.

Another process for the production of carbon foam was developed by Simandl et al. at the US DOE Y-12 plant in Oak Ridge [51]. This process variation is unique because a polyacrylonitrile (PAN) precursor was used instead of a glassy carbon precursor (like phenols and urethanes). This process utilizes a phase inversion of polyacrylnitrile in a solution of an alkali metal halide in a solvent such as propylene carbonate, tetramethylene sulfone, or mixtures of both. Typically, gels of polyacrylonitrile produce a structure incapable of self-support when converted to a carbon material. However, the alkali metal, in low concentrations, promotes solubilization and unraveling of the highly crystalline, helical PAN molecules in solution and then maintains these extended molecules for the desired strut formation of the gel during phase inversion. Thus, a stable carbon foam can be produced. This carbon foam, after heat treatment at elevated temperatures (> 2400 °C), was found to be an excellent thermal insulator at densities up to 0.3 g cm^{-3}.

Another process for producing foam is the *lost-foam technique*. While this technique is more widely used for metal foams, it has been adapted for production of carbon foams. In general, the process begins by producing a solid structure with interconnected porosity from a salt. Then the precursor for the foam is infiltrated into the interconnecting porous structure and cured. At the appropriate stage in curing, the salt is dissolved out of the polymer foam. The polymer foam is then carbonized to form carbon foam. One such technique [52] describes such a product as a machinable and structurally stable, low-density, microcellular carbon foam. In this process, pulverized sodium chloride is first classified to improve particle size uniformity. The particles are cold pressed into a compact having internal pores and then sintered. The sintered compact is submerged in a phenolic polymer solution to uniformly fill the pores of the compact with phenolic polymer. The compact is then heated to pyrolyze the phenolic polymer to carbon in the form of a foam. Finally, the sodium chloride is leached away with water to give a carbon foam suitable for catalyst supports or high-temperature insulation.

2.6.4.2
Thermoplastic Precursors

Carbon foams from pitches have been under development for several decades [9, 53]. Recently, the processes have been investigated for large-scale manufacture. As material becomes available, data are being generated to support realization of potential applications for theses carbon foams. Traditionally, pitches are cheaper than synthetic polymers but can be difficult to process.

The first process for the manufacture of pitch foams was developed by Bonzom et al. [9, 53]. In this process *petroleum-derived pitches* are expanded at elevated temperature by a pore-forming agent in a mold. The pitches used by Bonzom et al. were residues from petroleum cracking, subjected to thermal aging to increase the ratio of aromatic to aliphatic componets. Typically, the pitch had a Kraemer–Sarnow (KS) softening point between 70 and 210 °C. The initial pressure was selected such that, at the decomposition temperature of the porogenic agent, gases would not evolve (i.e., the initial pressure is higher than the pressure of normal pyrolysis under atmospheric conditions). After the decomposition of the porogenic agent, the pressure was lowered to expand the pitch (thermodynamic flash). The temperature could also be adjusted between the decomposition and decompression stages to control growth rate of the cells by adjusting the viscosity of the pitch. After expansion, the pitch requires oxidative stabilization to render it infusible and prevent melting. Bonzom originally showed that these pitch foams could be used as floor coverings, as insulators, or as surfaces for the collection of solar energy. Further heat treatment to 2400 °C was performed to convert the pitch foam to carbon foam. Since the precursor pitch is isotropic in nature, oxidative stabilization rendered the structure nongraphitizable. Therefore, the foam was not converted to a graphitic structure and remained primarily a noncrystalline carbon material. The carbon foams could be used as thermal insulation, catalyst supports, or as filters for corrosive products. In addition, because of its noncrystalline nature, the carbon foam can be activated to produce a highly adsorbent material for separation of liquids and gases.

Another method, developed at West Virginia University in the late 1990s by Stiller et. al., uses *coal-derived pitch*, an even cheaper precursor than the petroleum-derived pitches used by Bonzom et al. [9]. The process begins by hydrogenating and deashing bituminous coal in a specialized reactor at temperatures between 325 and 450 °C at a hydrogen overpressure of about 3.4–17.2 MPa for between 15 and 90 min. Tetralin can be employed as a proton-donating agent. After the reactor has cooled, the contents are removed and, if necessary, the tetralin is separated by distillation. Then the product is deashed according to ref. [54]. The resulting deashed hydrogenated coal can be extracted with tetrahydrofuran (THF), and the solution filtered to remove inorganic matter. Under these hydrogenating conditions, more than one-half of the coal mass is rendered soluble in THF. The THF portion contains all of the asphaltenes, or coal-derived pitch precursors, and oils. When extraction is complete, the THF can be evaporated for recycling, and the recovered coal-derived pitch precursor separated by employing a suitable solvent, such as toluene. The toluene-soluble fraction is generally referred to as "oils", and the remainder as the asphaltene fraction or coal-derived pitch-precursor fraction. After separating the asphaltenes from the oils by decanting, the asphaltenes are foamed by heating at about 325–500 °C and pressures up to 100 MPa under inert conditions (argon or nitrogen). The pressure is maintained at a generally constant level by means of a ballast tank and throttle valve. Pyrolysis of the asphaltenes produces volatile components which generate a foam at these high pressures and increased pitch viscosity. The foam is then soaked for up to 8 h to sufficiently coke it and render it infusible. After coking, the foamed material is calcined at a temperature substantially higher than the coking temperature, between 975 and 1025 °C, to remove any residual volatile material. The carbonized foam is then subjected to graphitization temperatures of up to 2600 °C to further heat-treat the material and increase its strength (see Fig. 5). Despite these temperatures, the material is not generally graphitized into a highly crystalline structure, but is only partially graphitic with small crystalline regions dispersed among more noncrystalline regions. Variations of this process have been commercialized by Touchstone Research Lab.

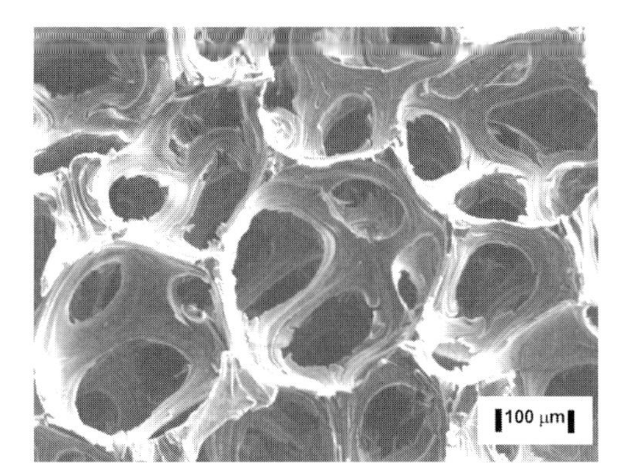

Figure 5 Foams produced from coal-derived pitches by West Virginia University (sample courtesy of Al Stiller).

In the late 1980s, research demonstrated that production of fibers from *mesophase pitch* could result in graphitic fibers with very high moduli. The unique discotic nematic liquid-crystal precursor, mesophase pitch, resulted in highly oriented graphitic structures aligned along the fiber axis. During spinning, the mesophase crystals align along the fiber axis and, after heat treatment at very high temperatures (2400–3000 °C), convert to graphite. The high modulus of the graphene planes give the fibers high modulus, which results in high-stiffness composites (a requirement for many aerospace applications). This graphitic alignment also results in efficient thermal transport. If such a highly structured material could be formed into a 3D network of interconnected ligaments, the resulting material would be an alternative preform for composite materials, use of which would eliminate the tedious lay-up of fiber tows in the desired directions. Hagar et al. [17] began work on this theory in the early 1990s and developed a process to utilize mesophase pitch as the precursor to produce a graphite foam.

Ideally, the foam can be modeled as graphitic carbon fibers interconnected to a macroscopically isotropic reinforcement. Unfortunately, the foams produced with this technique were not reticulated as desired, but had a more spherical structure (Fig. 6). The "fiber" or strut structure was really a somewhat triangular cell wall and therefore did not exactly microscopically resemble a fiber (Fig. 7) The graphitic planes tend to align along the length of the strut and follow the triangular surface of the cell wall. This structure is similar to that found in noncircular fibers, as seen by Edie et. al. [55], where the orientation of the graphene planes is more planar. At the node (the point where the cell walls meet) the graphitic planes come together into a somewhat randomly oriented connection.

During foaming, biaxial stresses from the growing bubbles align the pre-graphitic mesophase planes along the cell walls to provide the ideal structure for good

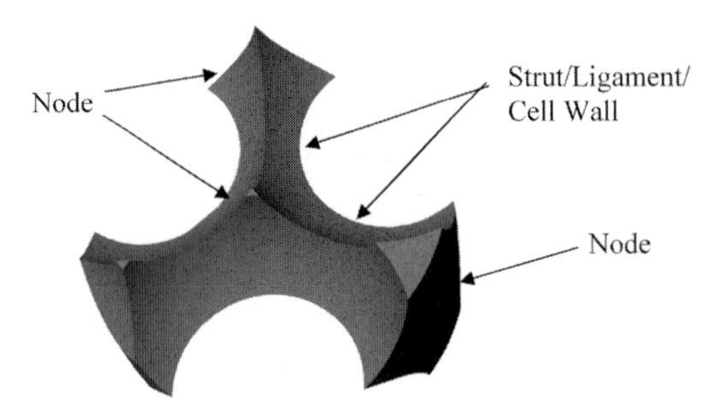

Figure 6 3D representation of foam structure.

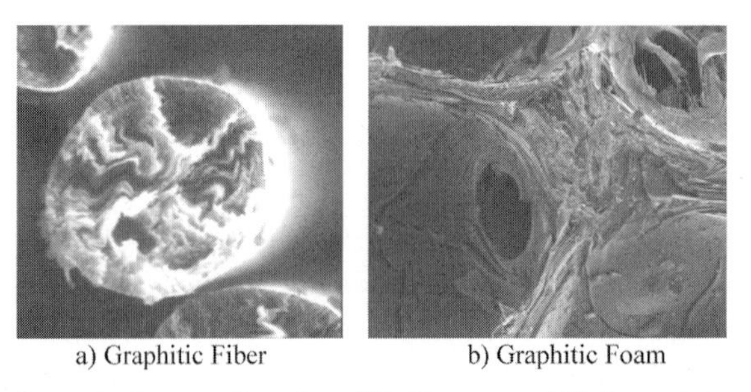

Figure 7 Cross section of graphitic fiber vs graphitic foam strut.

mechanical and thermal properties. As in film blowing (e.g., plastic bags), the biaxial stresses appear to impart better orientation in the liquid crystal than normal shear in fiber spinning. This may result in better orientation of the graphitic planes along the struts in mesophase pitch-derived foam than that seen in round fibers. This alignment is observable in Figure 7, as evidenced by the striations in the strut and node surfaces.

There are several techniques for producing mesophase pich-derived graphitic foams. The first, by Hagar et al. [1, 2, 17–20] and later by Kearns [21, 22], was a traditional blowing technique in which the mesophase is softened, saturated with a blowing agent (gas), and then flashed (blown) into a foam. The process requires oxidative stabilization of the foam to prevent remelting during carbonization and graphitization.

This process produces a foam with a microcellular structure and uniform pore size (Fig. 8). First, a quantity of a pitch is pressed in a mold to provide a pressed article. Then the article is placed in a pressure vessel and covered with an inert gas and pressurized to about 200–500 psi. The furnace is heated about 10–40 °C above the melting temperature of the pitch. Following a short hold time to allow equilibrium to be reached, additional inert gas is added to obtain a final pressure in the pressure vessel of about 1000–1500 psi. The furnace is held constant for about 10–40 min and then quickly vented to atmospheric pressure, thereby resulting in a thermodynamic flash of the dissolved gases in the precursor. The resulting expansion provides a porous foam due to the high viscosity of the mesophase precursor at these temperatures. After removal from the furnace, the foams are stabilized with oxygen at an elevated temperature near the softening point of the precursor. The maximum temperature at which the stabilization is carried out should not markedly exceed the temperature at which the pitch foam softens. In general, stabilization can be carried out by subjecting the porous foam to an oxygen or air atmosphere for about 8–24 h at about 150–260 °C, preferably about 150–220 °C, or until a weight gain of about 5–10 % is achieved. Slow cooling (ca. 0.1–5 °C min^{-1}) of the stabilized porous pitch foam to ambient temperature is necessary. This is a critical step because a fast cooling rate may result in thermal stresses and cracking of the sample (dependent on billet size). The porous pitch foam can then be converted to a porous carbon foam by heating to 900–1100 °C. The porous carbon foam can then be converted to a porous graphitic foam by heating to graphitization temperatures (> 2400 °C). This process has been licensed by MER Corporation, Tucson, Arizona (www.mercorp.com).

An advantage of this process is that resulting graphitic foam has a very low density of about 0.1 g cm^{-3} and very high specific strength. However, thermal and electrical transport are limited by the processing conditions. First, the foaming step is performed at a temperature not much higher than the melting point of the pitch. During biaxial extension of the mesophase domains, the very high viscosity limits the mobility of the mesophase domains and prevents formation of a highly oriented structure parallel to the cell walls. In the fabrication of mesophase fibers, the precursor first travels through a narrow capillary designed to induce very high shear rates which can reorient the domains in a very viscous material. However, in the normal

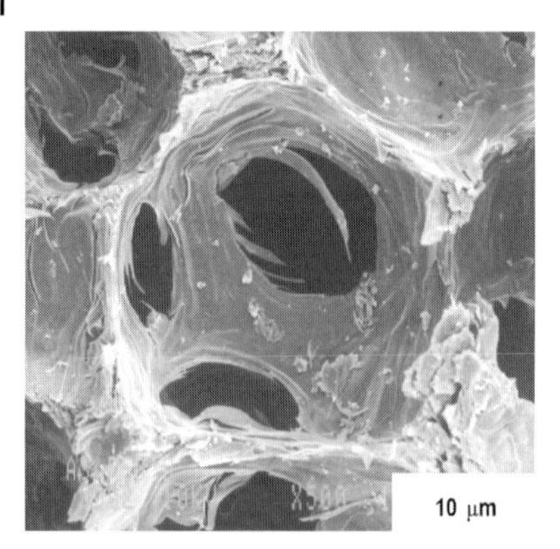

Figure 8 Microcellular carbon foams produced by Kearns (photo courtesy of Dave Anderson, University of Dayton Research Institute).

blowing process, it is expected that the stresses imparted to the mesophase domains are significantly lower than those occurring in carbon fiber production. Hence, the mesophase domains do not reorient as well as desired. Next, stabilization of the porous foam is intended to render the surface layer infusible so that the porous structure of the foam is maintained during subsequent treatments. Additionally, the stabilization locks in the domain size of the mesophase regions, thus limiting the crystal size of the resultant graphitic structure [56]. Thus, during post-stabilization heat treatments, any reorganization of the mesogens into a pre-graphitic and eventually a graphitic structure is limited to the original domain size. Hence, domain size of the original mesophase becomes the maximum crystallite size after graphitization [56]. This helps the mechanical strength, as small crystallite sizes (or grain boundaries) in flaw-limited ceramics, such as foams, tend to increase strength. However, it severely limits the thermal and electrical transport properties of foams produced with this method.

Another process for producing graphitic foams was developed at Oak Ridge National Laboratory (ORNL) by Klett et al. [21, 22, 29–31, 57–61]. This process requires fewer steps and overcomes some of the deficiencies of the Hagar–Kearns processes. It results in a foam that is denser, significantly more thermally and electrically conductive, and structurally weaker. In fact, they were the first to produce carbon foams sufficiently conductive to be considered for heat sinks and other thermal-management applications, rather than thermally insulating or structural applications. The ORNL process is relatively simple and is free of the oxidative stabilization traditionally required for processing of pitches and mesophases.

First, a mesophase pitch precursor is heated in an oxygen-free environment to about 50 °C above its softening point. Once the pitch has melted, the pressure is elevated and the temperature is raised at a controlled rate. While the pitch is molten, low molecular weight species begin to evolve. These volatile gases form bubbles at nucleation sites on the bottom and sides of the crucible and rise to the top, beginning to orient the mesophase crystals in the vertical direction. With time, a significant amount of the mesophase crystals are oriented vertically.

At high temperatures the mesophase begins to pyrolyze (polymerize) and create additional volatile species. This pyrolysis, which can be very rapid and is dependent on the precursor, is accompanied by an increase in the molecular weight of the precursor which, in turn, increases the melt viscosity of the liquid mesophase. As the rapid evolution of gases progresses, the increase in viscosity tends to capture the bubbles in place, forcing the material to foam in the unrestrained z direction. As the temperature is further increased, the foamed mesophase continues to pyrolyze, further increasing the viscosity of the material until it has sufficiently cross-linked and is rendered infusible.

While the foam synthesis process is rather simple, the morphological changes occurring during processing are complex. There is a delicate relationship between the viscosity–temperature behavior, melting temperature, and pyrolysis temperature of the mesophase pitch. Initially, the pyrolysis gases develop at a temperature such that the viscosity is sufficient to result in a stable foam. Premature evolution of pyrolysis gas causes the pitch to froth and results in foam with a significant density gradient (which may be desirable in some applications). If the gases are evolved too late, when the pitch viscosity is high, the bubbles may not be uniform, and cracking can occur due to thermal stresses. If the pyrolysis gases are evolved very slowly, as for certain high-melting Conoco pitches, the pores tend to be smaller [60].

Bubble formation is closely related to the autoclave operating pressure and temperature. Typically, the higher the autoclave gas pressure, the higher the temperature at which gas evolution occurs and the smaller the resulting pores. However, depending on the unique rheological properties of the starting pitch, the cell walls have different thicknesses, the bubble sizes can be dramatically different, and the mechanical and thermal properties can be affected. Unlike some other foaming techniques, such as slurry-derived metallic foams, the resultant properties of the graphitic foam (e.g., bubble size, ligament size, relative density, thermal and mechanical properties) are not independent properties. They are all dependent on the precursor's melt viscosity, pyrolysis temperature, and other pitch rheological properties.

The foamed mesophase is carbonized by heating to between 600 and 1000 °C to yield a relatively pure carbon foam. In this state, the foam is an excellent thermal insulator with a bulk thermal conductivity of about 1.2 W m^{-1} K^{-1} at a density of 0.5 g cm^{-3}. Because the carbonized foam was formed with a mesophase that was not oxidatively stabilized during the pyrolysis/carbonization stages, the mesophase crystals are not inhibited and can grow to very large sizes. Consequently, when the carbon foam is converted to a highly graphitic foam by heat treatment above 2800 °C under an argon purge, the resultant graphite crystals are highly aligned and significantly larger than those found in mesophase pitch-derived carbon fibers (similar to needle coke). Hence, the ligaments of the graphite foam produced with this method are more thermally conductive than even the best mesophase pitch-based graphite fibers.

A typical resultant mesophase foam is illustrated in Fig. 9. The foam typically exhibits uniformly shaped bubbles with a normal distribution. The average pore size, orientation, and distribution are determined primarily by the pitch viscosity and processing pressure during foaming. Additionally, the mesophase-derived foam has a preferred orientation of crystals in the z direction with accompanying anisotropy of properties in the xy plane even though bubble shape may not appear anisotropic.

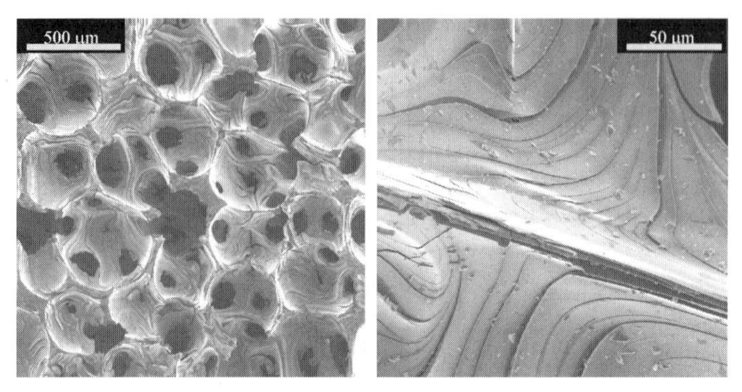

Figure 9 SEM images of mesophase pitch-derived foams.

This material differs from the carbon and graphite foams produced previously. It has a predominately spherical porosity with smaller openings between the cells (Fig. 10). Highly oriented sheets of graphite parallel to the cell walls yield thermal conductivities in the cell walls as high as 1640 W m^{-1} K^{-1} [59]. At a relative density of approximately 0.3, this results in a bulk apparent thermal conductivity of more than 180 W m^{-1} K^{-1}. Hence, a graphite foam can be produced with a bulk thermal conductivity almost equivalent to those of dense aluminum alloys, but at one-fifth the weight. This process has been licensed by Poco Graphite, Inc. of Decatur, TX (www.poco.com).

To examine the unique graphitic structure, the ORNL foams were examined by TEM. Figure 11 shows a typical TEM image of a cell wall. Unlike many synthetic graphitic materials, the graphene planes here are highly aligned and relatively defect

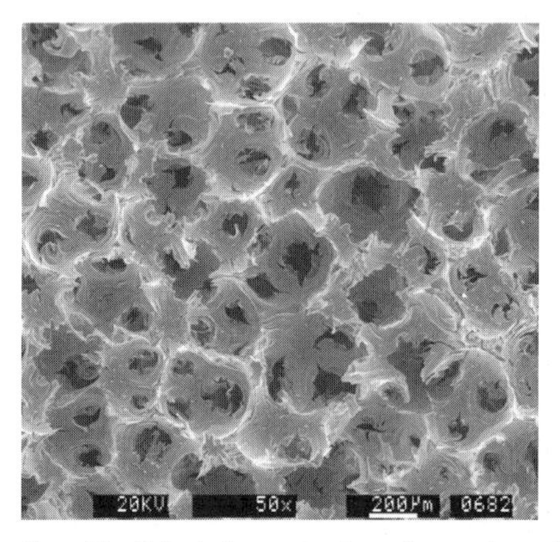

Figure 10 Spherical pore structure of mesophase pitch-based graphite foams.

free over extremely long ranges. More importantly, there do not appear to be any bifurcations of the basal planes, which would result in large mean free paths. In addition, there are several areas (one shown enlarged) where the structure is aligned in true 3D order such that both the 002 and the 110 planes are visible. The 002 planes appear as black and white bands and are easily resolved. The 110 planes are harder to identify and run perpendicular to the 002 planes (point A in Fig. 11). The TEM observations support the supposition that the graphite crystals in the foam ligaments are highly aligned and will exhibit extremely high thermal conductivity. This is confirmed by the X-ray diffraction pattern of the foam, which shows significant 002, 100, 101, and 110 peaks (Fig. 12). Interestingly, this highly crystalline structure does not develop until the foams are heat-treated to very high temperatures. Figure 13 illustrates the change in the X-ray diffraction pattern as the foams are heat-treated from 1000 to 2800 °C. Peak development begins around 2200 °C, but significant crystalline structure only develops above 2500 °C. The development of long-range crystal structure is significant because the graphite foams developed with this technique did not go through an oxidative stabilization stage. Oxidative

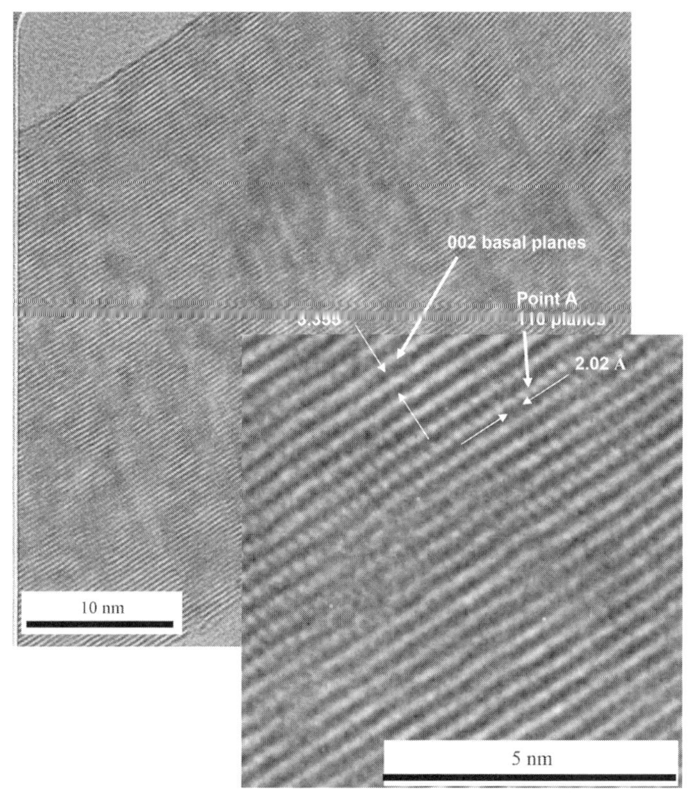

Figure 11 TEM images (JEOL 4000EX, accelerating voltage 400 kV) of graphite foam ligament illustrating the highly ordered nature of the structure [41].

stabilization would have minimized the domain size of the mesophase during formation and thus limited the size of the crystals in the final heat treatment. Hence, while the crystals are not developed in the foaming stage, the ability to form large crystals in later heat treatments is regulated in this stage.

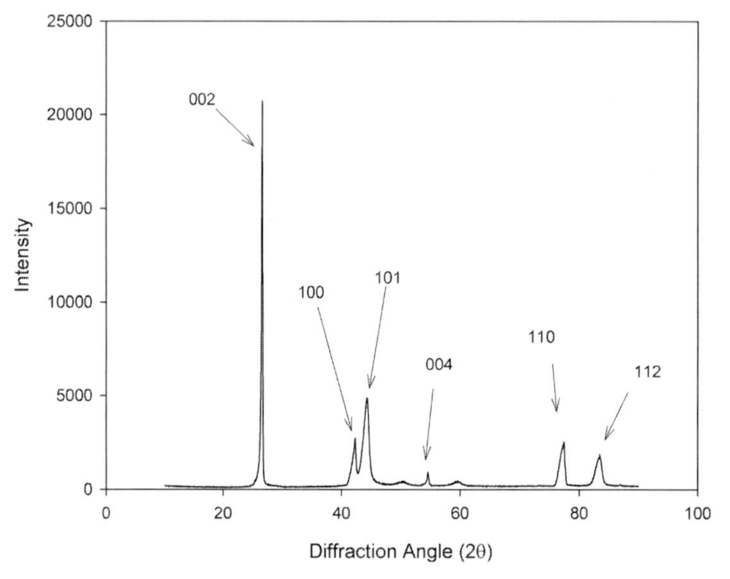

Figure 12 X-Ray diffraction pattern of highly graphitic foams.

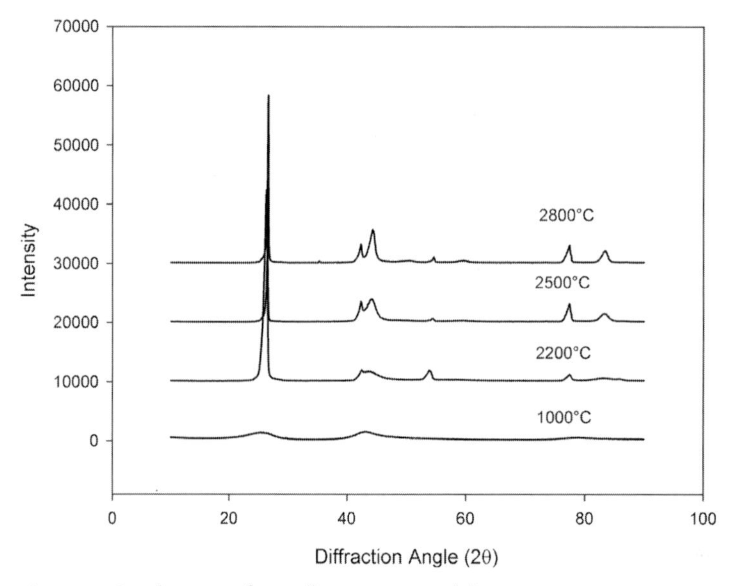

Figure 13 Development of crystalline structure with heat treatment temperature.

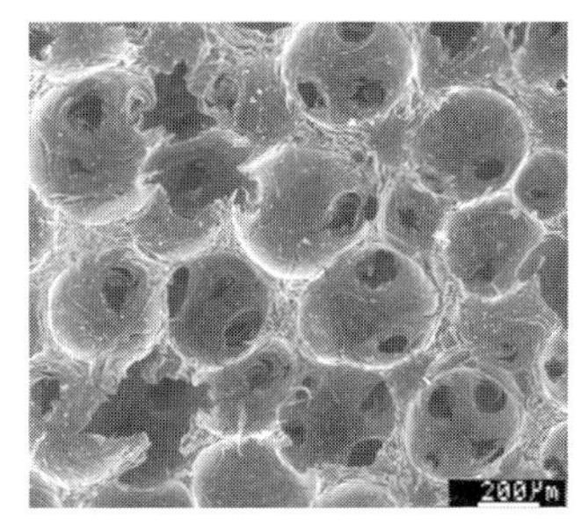

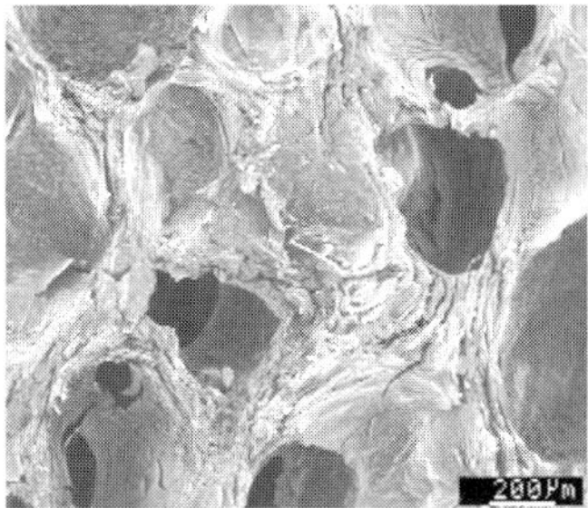

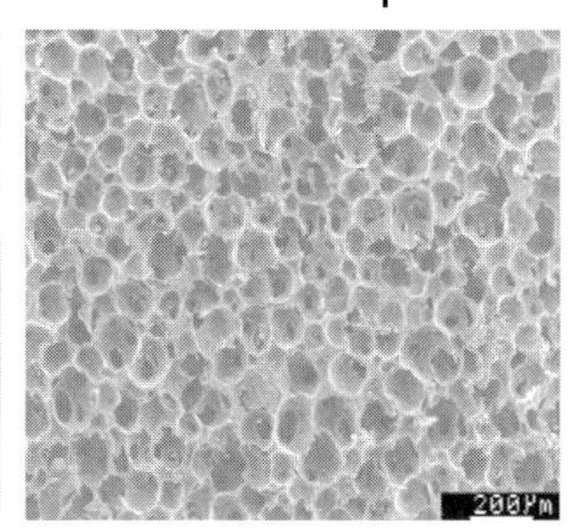

Figure 14 Images of thermally conductive graphitic foams made from synthetic and petroleum- and coal-derived mesophases.

The ORNL process can be used with different mesophase precursors to form different foams without changing process parameters. Figure 14 shows scanning electron micrographs of foams made with the same operating conditions and a synthetic mesophase (AR mesophase), a petroleum mesophase (Conoco Dry Mesophase), and a coal-derived mesophase (Koppers mesophase). The strikingly different morphologies of the resultant foams are due to the viscosities of the precursor at the temperature when the pyrolysis gases evolve. It is feasible to blend the three pitches, even with isotropic pitches, to produce a hybrid foam with a tailored pore structure.

2.6.5 Properties of Carbon and Graphite Foam

Because of the wide range of structures developed and specific precursors used to fabricate carbon and graphite foams, there is an equally wide range in mechanical and thermal properties. The advantage is that virtually any desired material property within the range can be fabricated. The disadvantage is that customers usually want the product customized to their individual needs, which increases costs. Unfortunately, overall engineering is as important to heat transfer with graphite foams as is material tailorability. First, there are foams with very low pressure drops (reticulated carbon foams) and ones with very high pressure drops (ORNL foams, see Fig. 15). However, selecting a specific foam for heat- (or mass-) transfer applications where fluids flow through the foams is more complicated than selecting a pore size to minimize pumping power. In the case of heat transfer, the pumping power must be balanced with heat transfer. In addition, there are engineering solutions to the design of the system that can be utilized to reduce pressure drop without seriously affecting heat transfer. Such a design is used frequently with low permeability filters for HEPA vacuums by corrugating the structure. Corrugation, or pleating, can reduce the pumping power by a factor of more than 10, but not significantly decrease heat transfer.

Table 1 lists several properties of various types of carbon and graphite foams. This is not all-inclusive, just representative of what is available commercially. Many specific properties can be obtained by tailoring various processing conditions.

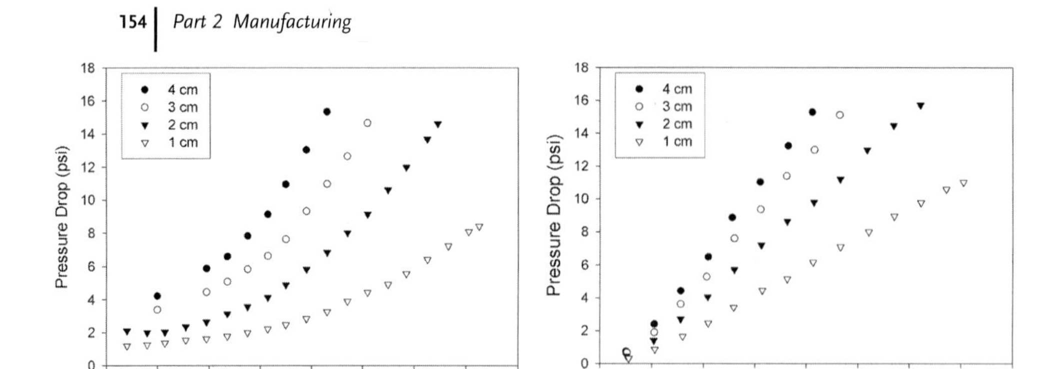

Figure 15 Pressure drops as a function of air and water velocity through Pocofoam®, as measured at ORNL.

An interesting aspect of carbon is its electrical conductivity. The thermal insulating foams exhibit high electrical resistivity, and some are even used with ohmic heating for furnace elements. However, as the structure becomes more graphitic, the foams become electrically conductive. The mechanisms that scatter phonons and interfere with thermal transport also scatter electrons and interefere with electrical transport [62]. Hence, there is a relationship between electrical resistivity and thermal conductivity in the thermally conductive foams, similar to that found in mesophase derived carbon fibers. Figure 16 illustrates that foams with high thermal conductivity ($\lambda > 100\,\text{W m}^{-1}\,\text{K}^{-1}$), exhibit an electrical resistivity of less than $10\,\mu\Omega\cdot\text{m}$, which is nearing that of semiconductors.

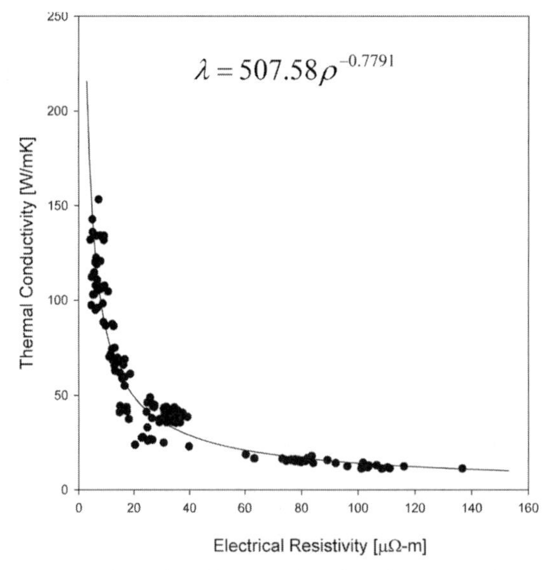

Figure 16 Thermal conductivity of graphitic foams as a function of electrical resistivity.

Table 1 General properties of various experimental and commercial graphite foams (data from manufacturer's website unless otherwise specified).

	Average bulk density [g cm^{-3}]	Compressive strength [MPa]	Compressive modulus [MPa]	Tensile strength [MPa]	Shear strength [MPa]	Room-temperature average z-plane thermal conductivity [W m^{-1} K^{-1}]
Carbon foams						
Touchstone C-Foam® 17 [52]	0.272	3.4	138	1380	–	0.25
Touchstone C-Foam® 25 [52]	0.40	13.8	551	3450	–	0.75
Ultramet Ultrafoam [16]	0.042	0.625	–	0.810	0.29	0.085
WPAFB [pers.comm.]	0.16**	–	–	125**	–	0.25*
ERG Foam [15]	0.049	0.482	62	0.345		< 0.15
ORNL Carbon Foam A	0.54*	5.3	348	–	–	2.0
Graphite foams						
ORNL Graphite Foam A	0.57*	2.1*	144*	0.7*	1.87**	180*
ORNL Graphite Foam B	0.59*	5*	180*	–	–	134*
ORNL Graphite Foam C	0.70*	5.1*	413*	–	–	187*
PocoFoam®	0.61*	2.66*	135*	–	–	182*
PocoHTC® [53]	0.9	5.89	–	–	–	245
WPAFB foams [52]	0.16**	–	–	2.0**	1.43**	17*

* Measured at ORNL. ** Measured at WPAFB.

2.6.6 Summary

Carbon foams are a very broad class of materials. In addition, they can exhibit some of the most widely varied properties and can be made with a multitude of processes. They can be everything from thermally insulating to thermally conductive, structural or nonstructural, or reticulated or spherical. Their uses include catalyst supports, filters, gas diffusers, and templates for other metallic and ceramic foams. Recently, the low-density, high-strength foams developed by the US Air Force offer the opportunity for replacing very expensive fiber weaves in 3D structural applications. Furthermore, the development of foams with high thermal conductivity opens even more applications for these unique materials. Now, a material which can be thermally insulating and used as a heat shield can simply be heat-treated to very high temperatures and converted to a thermally conductive material which can readily transfer heat. These applications range from computer cooling to engine cooling to residential coolers. The ability to transfer heat at a rate equivalent to aluminium but at one-fifth its weight has significant applications if the systems utilizing this technology can be engineered properly to capture the full potential of the foams.

References

1 Hager J.W. and Anderson D.P. Idealized Ligament Formation and Geometry in Open-Celled Foams. Proceedings of 21st Biennial Conference on Carbon, 1993, 102.
2 Hager J.W., *Mater. Res. Soc. Symp. Pro.* **1992**; *270*: 41–46.
3 Ford W., US Patent 3,121,050. 1964.
4 Googin J., Napier J., and Scrivner M., US Patent 3,345,440. 1967.
5 Cowlard F.C. and Lewis J.C., *J. Mater. Sci.* **1967**; *2*: 507–512.
6 Knippenberg W.F. and Lersmacher B., *Phillips Tech. Rev.* **1976**; *36* (4): 93–103.
7 Noda T., Inagaki M., and Yamada S., *J. Non-Cryst. Solids* **1969**; *1*: 285–302.
8 Klett R.D., US Patent 3,914,392. 1975.
9 Bonzom A., Crepaux A.P., and Montard A.-M.E.J., US Patent 4,276,246. 1981.
10 Ettinger B. and Wolosin S., US Patent 3,666,526. 1972.
11 Luhleich H., Nickel H., and Dias F., US Patent 3,927,187. 1975.
12 Marek R. and Udichak W., US Patent 3,922,334. 1975.
13 Franck H.-G., et al., US Patent 3,784,487. 1974.
14 Vinton C. and Franklin C., US Patent 3,927,186. 1975.
15 ERG Product Literature, 2000.
16 Ultramet Product Literature: Ultrafoam, 1998.
17 Hager J.W. and Anderson D.P., Progress in Open-Celled Foams. Proceedings of 40th International Sampe Symposium, Anaheim, California, 1995.
18 Hager J.W. and Lake M.L., *Mater. Res. Soc. Symp. Proc.* **1992**; *270*: 29–34.
19 Hager J.W., Newman J.W., Johannes N., and Turrill F.H., US Patent 6,013,371. 2000.
20 Sandhu S.S. and Hager J.W., *Mater. Res. Soc. Symp. Proc.* **1992**; *270*: 35–40.
21 Kearns K. Graphitic Carbon Foam Processing. Proceedings of 21st Annual Conference on Composites, Materials, and Structures, Cocoa Beach, Florida, 1997, 835–847.
22 Kearns K. M., US Patent 5,868,974. 1999.
23 Stiller A.H., Stansberry P.G., and Zondlo J.W., US Patent 5,888,469. 1999.
24 Stiller A.H., Yocum A., and Plucinski J., US Patent 6,183,854. 1999.
25 Stiller A.H., Stansbery P., and Zondlo J., US Patent 6,346,226. 2001.
26 Stiller A.H., Stansbery P., and Zondlo J., US Patent 6,241,957. 1999.
27 Klett J. High Thermal Conductivity, Mesophase Pitch-Derived Carbon Foam. Proceedings of The 1998 43rd International SAMPE Symposium and Exhibition, Part 1 (of 2), Anaheim, California, USA, 1998, 745–755.
28 Klett J., *J. Compos. Manuf.* **1999**; *15* (4): 1-7.
29 Klett J., US Patent 6,033,506. 2000.
30 Klett J., US Patent 6,261,485. 2001.
31 Klett J., US Patent 6,287,375. 2001.
32 Klett J., US Patent 6,344,159. 2002.
33 Klett J., US Patent 6,387,343. 2002.
34 Klett J., US Patent 6,398,994. 2002.
35 Klett J. and Burchell T., US Patent 6,399,149. 2002.
36 Klett J. and Burchell T.D., High Thermal Conductivity, Mesophase Pitch Derived Carbon Foam. Proceedings of Science and Technology of Carbon, Strasbourg, France, 1998.
37 Klett J., Walls C., and Burchell T. D., High Thermal Conductivity Mesophase Pitch-Derived Carbon Foams: Effect of Precursor on Structure and Properties. Proceedings of Carbon '99, Charleston, SC, 1999.
38 Klett J.W., McMillan A., and Gallego N., Carbon Foam for Electronics Cooling, National Laboratory Fuel Cell Annual Report, 2002.
39 Fitzer E., Kochling K.H., Boehm H.P., and Marsh H., *Pure Appl. Chem.* **1995**; *67* (3): 473–506.
40 ICCTC, *Carbon* **1982**; *20*: 445.
41 ICCTC, *Carbon* **1983**; *21*: 517.
42 ICCTC, *Carbon* **1985**; *23*: 601.
43 ICCTC, *Carbon* **1986**; *24*: 246.
44 ICCTC, *Carbon* **1987**; *25*: 317.
45 ICCTC, *Carbon* **1987**; *25*: 449.
46 Gibson L.J. and Ashby M.F., *Cellular Solids: Structure and Properties*. Pergamon Press, New York, 1988.
47 www.mercorp.com, 2004, MER Corp. Website.
48 Vinton C.S. and Franklin C.H., US Patent 4,022,875. 1977.
49 Fung-Ming K., US Patent 4,992,254. 1991.
50 Hayward T.P., US Patent 5,582,781. 1996.
51 Simandl R.F. and Brown J.D., US Patent 5,300,272. 1994.

52 Hopper R.W. and Pekala R.W., US Patent 4,806,290. 1989.

53 Maricle D. L., US Patent 4,125,676. 1978.

54 Stiller A.H., Sears J.T., and Hammack R.W., US Patent 4,272,356. 1981.

55 Edie D.D., Fox N.K., Barnett B.C., and Fain C.C., *Carbon* **1986**; *24* (4): 477–482.

56 Rouzard J. N. and Oberlin A., *Carbon* **1989**; *27* (4): 517–529.

57 Edie D. D. and Dunham M. G., *Carbon* **1989**; *27* (5): 647–655.

58 White J. L. and Sheaffer P. M., *Carbon* **1989**; *27*: 697–707.

59 Klett J., McMillan A.D., Gallego N.C., and Walls C.A., *J. Mater. Sci.* **2004**; *39*: 3659–3676.

60 Klett J., Hardy R., Romine E., Walls C., and Burchell T., *Carbon* **2000**; *38* (7): 953–973.

61 Klett J., Klett L., Burchell T., and Walls C., *Society of Automotive Engineers Technical Paper Series* 2000; (00FCC-117).

62 Lavin J. G., Boyington D. R., Lahijani J., Nysten B., and Issi J.-P., *Carbon* **1993**; *31* (6): 1001–1002.

2.7
Glass Foams

Giovanni Scarinci, Giovanna Brusatin, Enrico Bernardo

2.7.1
Introduction

Glass foam has a unique combination of properties: it is lightweight, rigid, compression-resistant, thermally insulating, freeze-tolerant, nonflammable, chemically inert and nontoxic, rodent- and insect-resistant, bacteria-resistant, and water- and steam-resistant. Moreover, glass foam facilitates quick construction and has low transport costs, it is easy to handle, cut, and drill, and is readily combined with concrete. This combination of properties makes glass foam practically irreplaceable both in construction (e.g., for the insulation of roofs, walls, floors, and ceilings under hot or cold conditions) and in many other fields [1].

Glass foam is generally obtained by the action of a gas-generating agent (termed gasifier or foaming agent, mostly carbon or carbonaceous substances), which is ground together with the starting glass to a finely divided powder. The mixture of glass powder, foaming agent, and occasionally other mineral agents is heated to a temperature at which the evolution of gas from the foaming agent occurs within a pyroplastic mass of the softened glass particles undergoing viscous flow sintering. The evolved gas leads to a multitude of initially spherical small bubbles which, under the increasing gas pressure, expand to a foam structure of polyhedral cells that, after cooling of the glass, constitute the pores in the glass foam.

The properties of finished foamed glass products depend strongly on the type and quantity of the added foaming agents, on the initial size of the glass particles, and on the firing schedule. The result is a glass foam with high compressive strength and dimensional stability, characterized by a density of only 0.13 to 0.3 g cm^{-3} [2].

2.7.2
Historical Background

The production of glass foam dates back to the 1930s, when major research activity was conducted throughout the industrialized countries. Owing to the many patents granted in the same period, it is uncertain who the first inventors of glass foam were. The patented processes can be divided into two fundamental types: manufacturing of glass foam by the above-described sintering of finely ground glass powders

Cellular Ceramics: Structure, Manufacturing, Properties and Applications.
Michael Scheffler, Paolo Colombo (Eds.)

ISBN: 3-527-31320-6

with a suitable foaming agent, and the direct introduction of fluids (air, CO_2, water vapor) into the molten glass.

The reports of Kern [3] and Kitaigorodski [4] can be considered to be the first examples of glass foam from a sintering process. Kern mixed finely ground amorphous silica with up to 20 wt% of combustible material, such as coal, lignite, or wood, and "plasticizing agents" (hydrochloric acid, chloride solution, or a NaOH solution). Heating the mixtures to 1500 °C resulted in highly porous silica glass articles. Kitaigorodski began industrial production of glass foam in 1932, on the basis of technologies developed at the Mendeleev Institute of Moscow. The starting mixture of finely powdered glass and $CaCO_3$ as foaming agent was heated in steel molds up to about 850 °C; after cooling to 600–700 °C, the foamed glass was demolded in the form of blocks and annealed in tunnel kilns. Later, anthracite and carbon black were also employed as foaming agents, as alternatives to $CaCO_3$. The final density was about 0.3 g cm^{-3}, and the pore size around 5 mm; the thermal conductivity ranged from 0.06 to 0.08 W m^{-1} K^{-1}.

The direct introduction of gases into molten glass was the subject of intense research activity, mainly in France (Saint Gobain) and the USA (Pittsburgh Plate Glass and Corning Glass Works) soon after the first experience with the sintering approach. In 1934 Long [5] demonstrated the feasibility of a "spongelike glass" based on the production of a glass with a notable amount of dissolved gases (mainly water) from a mixture of silica, borax, and zinc oxide. After fusion, the glass material was cooled slowly to the softening temperature, which ranged from 500 to 700 °C, and maintained at that temperature until the entrapped gases evolved, leaving a multitude of bubbles and resulting in a cellular body with a density of about 1.25 g cm^{-3}. To achieve a density comparable with that reported for the sintering process, the volume of bubbles in the glass was dramatically increased by injection of a gas or a vapor. In 1940 Lytle [6] described the production of glass foam by forcing carbon dioxide, steam, or air, continuously or intermittently, into molten glass: the glass was formed in a column furnace, and the gases were introduced at the bottom of the column. The foamed glass, still fluid, was poured into molds moving on a belt conveyor or laminated through two belts. A number of patents were granted concerning the same technology in the following years [7–10], with slight improvements, such as subjecting the glass with entrapped gas bubbles to a negative pressure (depressurization) to make the bubbles expand. Peyches [11] reported the foaming of glass by local overheating of glass on the electrodes of Joule-heated glass melters, which resulted in the evolution of dissolved gases; such action could be enhanced by the injection of gas into the glass directly through the surface responsible for overheating. Other modifications concerned the mechanism of gas injection into the molten glass: Powell [12] reported the formation of bubbles from the thermal decomposition of solid foaming agents such as sodium nitrate and sodium sulfate, introduced into the molten glass while casting into molds.

World War II boosted the research, development, and production of glass foam in the USA, since it was largely used as nonflammable thermal insulation in the internal walls, floors, and ceilings of ships and submarines [4]. The most important results were patented by Pittsburgh Plate Glass and Corning Glass Works, later merged as Pittsburgh Corning Corporation; the first large production plant for glass foam in the USA started up in Port Allegany, PA, in 1943. The patents referred mainly to the sintering of finely powdered glass with $CaCO_3$ and carbon as alternative foaming agents [13–16]. The use of sheet glass, that is, a glass with a conventional chemical composition instead of a specific composition, was mentioned for the first time in Ref. [15]. Reference [16] reported the use of negative pressure as a foaming aid. The currently employed methods for glass foam manufacturing are given in Ref. [17], in which carbon black is reported as foaming agent for conventional window glass or borosilicate glass; in the same patent particular attention was dedicated to the presence of oxygen-releasing agents, such as SO_3 in the glass composition or Fe_2O_3 and Sb_2O_3 as additives to the mixture of glass powders and foaming agents. Willis [18] developed a sort of compromise between the sintering and foaming approaches by using entrapped gases: glass powders were sintered under pressure to cause gases to permeate into the interstices, and subsequently heated to expand the bubbles.

2.7.3 Starting Glasses

The first glass foams, dating back to the 1930s, were developed from specially formulated pristine glasses [19], especially in the case of direct introduction of gases into the molten glass. This process is at present hardly used, since it requires more sophisticated industrial plants, and it is much more energy-intensive than the currently used sintering process, which is carried out at much lower temperatures and can employ large quantities of powdered recycled glass. In some cases the substitution of pristine glass with waste glass is almost complete, so that glass foam production can be considered an effective way of recycling a great number of glass products (container and flat glass, borosilicate glass for the chemical industry, and, more recently, even fluorescent lamp glass [20]). Tens of millions tons of waste glass are generated every year worldwide, a large percentage of which is in the form of crushed solids with a wide range of particle sizes in a mixture of various colors. Moreover, the glass is contaminated with metallic and nonmetallic fragments (plastics, ceramics, aluminum, iron, paper, organic substances, and so on). The costs of color sorting and impurity removal are high, especially if a significant portion of the glass is finely divided. The use of recycled glass in the production of new glass articles is consequently barely profitable; in addition, new articles produced with introduction of impure recycled glass may contain defects and thus may exhibit poor mechanical properties.

Besides waste glass, other vitrified solid wastes are being generated in increasing quantities, for instance, fly ashes from burning of coal or municipal solid waste (MSW), slags of various kinds, sewage sludge ashes, and so on. The use of fly ashes, calcined together with sand and other minor constituents (borax and sodium salts), as the basis material for glass foam manufacturing dates back to the late 1960s [21]; much literature was dedicated more recently to this raw material [22, 23]. The vitrification of a number of inorganic wastes is particularly advantageous [24], since the obtained waste glass generally has high chemical durability and can be safely disposed of in landfills, with no risk of pollution. However, the high costs of landfilling, together with those of vitrification, lead to the need for applications for the obtained glass, for example, manufacturing profitable products like glass foam [25]. Such a prospect could also be profitable for glasses from end-of-life cathode ray tubes (CRTs) [26, 27], direct recycling of which is particularly difficult, so that they are mainly disposed of in landfills.

Some research was performed on manufacturing glass foams from natural materials like clays [28–31]. Cowan et al. [28] described the foaming of montmorillonite clay with carbon or organic compounds as foaming agents and Na_2O as sintering "flux". Seki et al. [31] reported the foaming of mixtures of silicate glass, volcanic ashes (volcanic glass materials), and waterglass. Finally, some research focused on the use of slags directly as raw materials (without previous vitrification) [32].

2.7.4 Modern Foaming Process

The main industrial process currently employed for glass foam manufacture is sintering of powdered glass admixed with suitable agents. The main parameters of this process are described in what follows.

2.7.4.1 Initial Particle Size of the Glass and the Foaming Agent

The starting glass must be ground and sieved to a grain size size of less than 0.4 mm, otherwise the foaming process is almost completely halted [33]. There is a definite relation between the fineness of the initial particles and the pore diameter of the obtained glass foam. For example, Fig. 1a [34] shows the microstructures of foams obtained from soda-lime glass cullet with 5 wt% SiC as foaming agent with variation of the glass powder size as indicated. Figure 1b shows the pore diameter of these foams as a function of the glass powder size [34].

Likewise, the particle size of the foaming agent affects the cell size [35] and the foaming behavior. A sample containing coarse-grained SiC (74–78 μm) hardly foams at 950 °C, while a sample containing fine-grained SiC (4–7 μm) exhibits a very large increase in volume at lower temperatures [36]. Moreover, precompaction of the initial mixture is also very important for the resulting foam structure: precompacted samples always have a much more uniform structure than loose powder mixtures, probably due to earlier onset of closed porosity.

The microstructural homogeneity of the foams is limited if the glass and foaming agent powders have very different dimensions, as can be seen by comparing the morphology of glass foams obtained with the carbon content of fly ashes (not previously calcined; Fig. 2a) and carbon black (Fig. 2b) as foaming agent [37]. In the first case, the starting dimensions of the foaming agent and the glass powder are

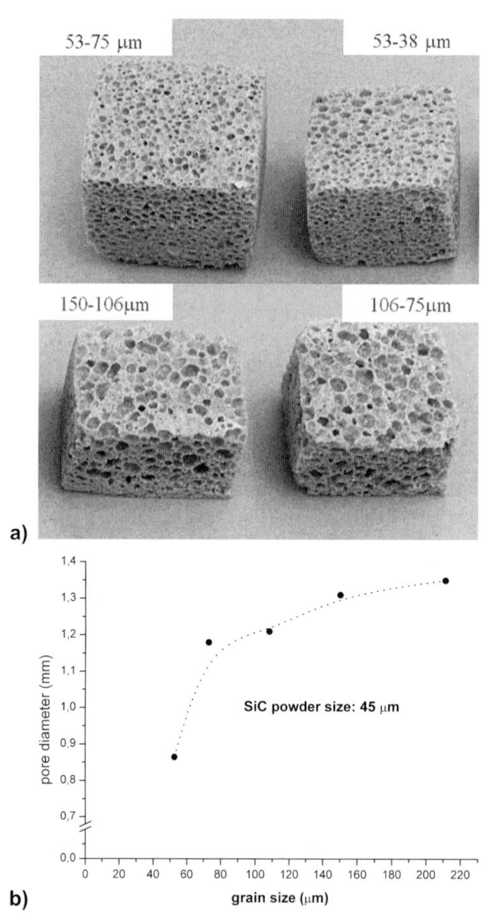

Fig. 1 a) Influence of glass powder grain size on the microstructure of the glass foam obtained with 5 wt % SiC (45 μm) as foaming agent. The range of glass powder sizes is indicated. b) Pore diameter of foam produced with SiC as a function of glass powder size.

notably different (1–20 μm and 75–150 μm, respectively), and the inhomogeneity of cell size distribution is probably due to agglomerating effects of fly ash, while in the second case they have about the same size. Another innovative strategy [37] to obtain homogeneous size distribution of pores is the use of a foaming agent homogeneously dispersed in the form of a solution, as in the treatment of finely ground soda-lime glass cullet by introduction of polymethylmethacrylate (PMMA) dissolved in CH_2Cl_2, which, after drying, produces a polymeric layer on the glass granules. The actual foaming agent is formed by pyrolysis of the polymer at 850 °C in an N_2 atmosphere for 5 min, followed by oxidation of the pyrolysis residues at the same temperature in air for 5 min. The layer of PMMA on the glass granules is transformed into a layer of carbonaceous products on pyrolysis of the polymer and thus leads to a very homogeneous distribution of pores, as illustrated in Fig. 2c.

A given foaming process leads to a precise density, corresponding to a certain global volume of pores. This volume, however, may be distributed in a limited number of large pores or in a great number of small pores. Such different distributions can

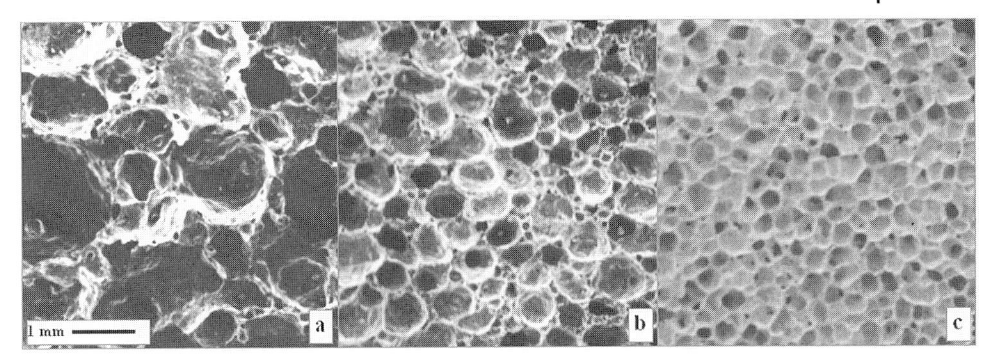

Fig. 2 SEM images of the microstructures of glass foams obtained by using the carbon content of a) fly ash (not previously calcined), b) carbon black, and c) PMMA pyrolysis residues as foaming agents and with comparable apparent densities of 0.28, 0.284, and 0.29 g cm^{-3}, respectively.

have a major influence on the performance and the properties of the glass foam. Generally, the lower the foam density, the lower the thermal conductivity (and the better the thermal insulation properties), without any evidence in the literature of dependence on cell size. On the other hand, the smaller the cell size, the higher the compressive strength of the product. The crushing strength is a function of the inverse square root of the cell size [38]. Ketov [39] observed a strong influence of the size of the initial glass particles on the density of the product. Decreasing the size of the particles to less than 25–30 μm leads to a final glass foam with a density of 160 kg m^{-3}, a thermal conductivity of 0.035 W m^{-1}/K^{-1}, and a compression strength of 1.3 MPa.

2.7.4.2 Heating Rate

The heating rate must be accurately controlled, because it is a very important factor for optimizing the glass foam product [23]. Owing to the fineness of the starting powders and the consequent high degree of dispersion, the foaming mixture contains a large amount of air entrained in the interstices between the particles of the glass and the gasifier. As a result, the foaming mixture has low thermal conductivity, so the rate of heating to the foaming temperature must not be too high. The larger the sample, the slower the heating rate should be to maintain a uniform temperature distribution in the poorly conductive material. When the temperature rise is rapid (i.e., 40 °C min^{-1}), large cracks develop throughout the glass mass. Heating rates of 5–10 °C min^{-1} are usually unproblematic. On the other hand, too a slow heating rate is also undesirable, since prolonged isothermal heating at high temperature could lead to premature generation of gases (i.e., before glass powder sintering). Also the temperature gradient in the furnace should be kept small (±5 °C); otherwise, inhomogeneous foams are obtained.

2.7.4.3
Foaming Temperature

Proper selection of the maximum temperature for the foaming process is of basic importance. Glass viscosity (strongly dependent on temperature) and the foaming temperature are strictly related. The optimum foaming temperature must be selected by considering, on the one hand, the maximum foam stability, which is controlled by viscosity, and, on the other hand, the internal cell structure, characterized by homogeneous and regular shape and size of the pores and by the minimum thickness of their separating walls [23]. If the selected foaming temperature is too high, the melt viscosity is too low (i.e., $< 10^3$ Pa·s), and controlling the structure becomes difficult, because bubbles rise to the top of the mold (as in fining of glass) and consequently the bubble distribution is far from uniform. On the contrary, if the temperature is too low, the glass viscosity is too high, and gas expansion is difficult and little increase in volume oocurs. In this case, the formation of the separating walls does not go to completion, and residual open porosity results, so that water absorption of the glass foam in service increases.

At an optimum selected temperature, the increase in volume can be remarkable. Koese [40] reported the volume changes in a pressed soda-lime glass/8 wt% SiC pellet during heating and cooling in air. In the sintering range, from about 600 to 750 °C, he found a slight contraction, due to sinter shrinkage of the compacted glass powder. Above 750–800 °C the foaming agent became active and caused a strong increase in volume (about 700%) by a gas-generating reaction. Porosity increased at higher firing temperatures and the compressive strength was lowered.

2.7.4.4
Heat-Treatment Time

There is a definite dependence of glass foam density on the duration of heat treatment. During the period of gas release, the density continuously decreases down to a minimum value. When this stage finishes, a gradual process of destruction and collapse of the foam by coalescence of the pores begins. Thus the surface energy of the system is decreased by reduction of the specific surface area of the walls of the cells. This process leads to a new increase in density [37] (Fig. 3). Hence, it is necessary to calculate precisely the time of heat treatment and to remove the glass foam from the hot zone prior to the beginning of the coalescence process.

2.7.4.5
Chemical Dissolved Oxygen

A very important role on the foaming capability of glasses is played by chemical dissolved oxygen, which is related to the redox potential of the glass and can be released by reduction reactions. It is well known that a characteristic equilibrium always exists in an alkali silicate glass melt between tri- and divalent iron with oxygen evolution. This phenomenon is of great importance in the foaming process, because the

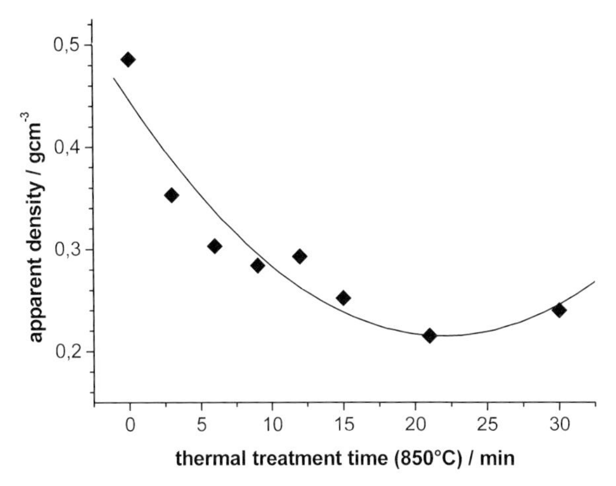

Fig. 3 Apparent density of foams obtained with carbon black as foaming agent as a function of thermal treatment at 850 °C.

reduction of Fe^{III} to Fe^{II} is presumably the gas-forming reaction, since oxygen is used to oxidize carbon (the most common foaming agent) in the pyroplastic glass. The equilibrium position depends on various parameters, such as temperature, heating rate, glass composition, and furnace atmosphere. The onset of dissociation is usually at a temperature above 1100 °C, but the presence of dispersed activated carbon catalyzes the oxidation/reduction process at much lower temperatures. The reduction of sodium sulfate, introduced during fabrication of most common glasses as fining agent, oxidizes carbon to CO:

$$Na_2SO_4 + 4C \rightarrow Na_2S + 4CO.$$

The absence of sulfur-containing species in the off-gas indicates that sodium sulfate has been reduced to sulfide, and Na_2S dissolved in the glass [41].

2.7.4.6
Cooling Rate

Rapid cooling from the foaming temperature to a temperature slightly higher than the annealing range can be used to lock in – through the corresponding increase in viscosity – the obtained structure and the evolved gas. In any case, it is necessary to adopt a very low rate of cooling through the glass transformation range to eliminate any residual stress. During cooling the pressure in the cells drops, and this leads to some shrinkage and development of tensile stresses between the layers [23].

The annealing rate depends on several factors, among which are the chemical composition of the glass (and consequently its expansion coefficient) and the glass foam structure (in particular, number and size of the pores, thickness of the cell walls, type of foaming agent, shape and dimensions of the foamed products). Unlike the foaming process, which is governed mainly by the composition of the foaming mixture, annealing can be controlled only by the duration of the process. If the walls of the cells intercommunicate (i.e., open porosity) the thermal conductivity is higher, and thus the cooling rate can be up to 2 °C min^{-1} instead of 0.5–0.7 °C min^{-1}.

2.7.5 Foaming Agents

In general, any glass (preferably in powdered form) can be transformed into a foam by addition of suitable substances (foaming agents) which generate gaseous products by decomposition or reaction at temperatures above its softening (Littleton) temperature (corresponding to a viscosity of $10^{6.6}$ Pa·s). If gas generation occurs at temperatures below the softening point, the glass powder has not yet sintered to closed porosity, and the gaseous products cannot be retained by the mass, and if the gas generation takes place when the glass viscosity is too low, the gaseous products are released from the melt, as in the fining of glass melts.

It follows that the most convenient viscosity range for optimizing the foaming process for development of maximum of porosity and minimum apparent density is 10^5 to 10^3 Pa·s. This viscosity is similar to that used in hot drawing operations. For a standard sheet soda-lime glass composition, it corresponds to the temperature range of 800–1000 °C.

The interdependence of viscosity and foaming temperature is very sensitive to any change in chemical composition of the raw materials, especially the type and amount of foaming agent. The same type of foaming agent can differently influence the structure and properties of different glasses, and the same glass can be differently influenced by the addition of different foaming agents. Moreover, decomposition of many foaming agents leaves a finely distributed ash which can influence some properties of the glass, such as viscosity and crystallization tendency. Typically, fineness and quantity of foaming agent strongly influence the size of the final pores; therefore, the type, size, and optimum concentration of the foaming agent must be accurately chosen.

2.7.5.1 Foaming by Thermal Decomposition

Foaming agents that undergo thermal decomposition include carbonates such as $CaCO_3$ and Na_2CO_3. The foaming of glass with added $CaCO_3$ is possible, as stated by Koese [40], only with wet glass particles, because partial leaching of alkali metal oxides from the glass surface causes adhesion of the particles at lower temperatures at which $CaCO_3$ can act as a foaming agent. During the gradual softening of the sintered glass under continuous heating, the calcium or sodium carbonate particles decompose to the oxide with simultaneous release of gaseous carbon dioxide. The oxide is incorporated into the molten glass mass and acts as a glass modifier, thus altering the viscosity of the molten glass. The CO_2 gas which is released is trapped in the viscous glass mass and its pressure is gradually increased, forcing the molten glass mass to expand. On subsequent cooling the molten glass mass is solidified and a cellular structure is formed.

As foaming agents, $CaCO_3$ and Na_2CO_3 cause a different amounts of volume expansion [42]. A maximum expansion of over 450% was obtained when 2 wt% of $CaCO_3$ was added to glass powder obtained from recycled colorless soda-lime glass, whereas a maximum volume expansion of only 90% was achieved with 5 wt% of Na_2CO_3 as foaming agent. Such a large difference is probably attributable, to different characteristics of the glass melt due to incorporation of the residual oxide component from decomposition of the carbonate compounds. Both the viscosity and the surface tension of a silicate glass melt can be altered by the addition of CaO or Na_2O. In particular, the modification of the viscosity of the glass melt is dependent on the miscibility of the oxide system. Based on these considerations, it is conceivable that the addition of small amounts (1–3 wt%) of CaO to the silicate glass melt will give rise to a small degree of phase separation which consequently results in the formation of a glass melt with high surface tension, but low viscosity, thus permitting a greater volume expansion of the molten mass. On the other hand, the addition of 2–7 wt% of Na_2O to the glass melt will cause a larger degree of phase separation which leads to the formation of a glass melt with low surface tension, but higher viscosity, thus limiting the volume expansion of the molten mass.

As an alternative to carbonates, calcium sulfate (e.g., gypsum) can be employed, the foaming gas being SO_2, which has an advantageous thermal conductivity, lower than that of CO_2, but requires more control, since it is known to be noxious [43, 44].

2.7.5.2 Foaming by Reaction

Softened glass can be foamed by a chemical reaction between the interparticle atmosphere and a suitable agent. Carbon dioxide is commonly produced in the softened glass mass by oxidation of carbon-containing foaming agents, such as pure carbon, SiC, sugar, starch, or organic wastes (excreta [45]). The foam gas present in the closed pores is mostly a mixture of CO_2 and CO. Since the main glass foam manufacturer, Pittsburgh Corning, is known to employ carbon [17], foaming by oxidation can be considered the most important foaming process.

The foaming activity is strongly influenced by the type of carbon (coke, anthracite, graphite, carbon black) and the glass composition (sulfate content). Foaming occurs at temperatures in the range 800–900 °C, and the evolution of gaseous species is due, in addition to the oxidation of carbon by the atmosphere, to secondary reactions of carbon with some constituents of the glass, such as H_2O, alkali, and sulfates, so that foaming can even be performed in nonoxidizing atmospheres. The formation of CO/CO_2 by the reduction of oxides in the softened glass alters the chemical composition and viscosity of the system; like in the case of carbonates, which leave a certain amount of residual oxides to be dissolved in the glass, foaming cannot be attributed only to the formation of gas bubbles in an unmodified glass, but depends on the superposition of the physicochemical interactions in the glass on gas evolution. The foam products obtained by oxidation of carbon are usually fine-pored. The addition of carbon to the glass mass is in the range of 0.2–2 wt%.

The chemical reaction which leads to the development of gaseous products may be expressed as [46]:

$$\text{glass-}SO_4^{2-} + 2\,C \rightarrow \text{glass-}S^{2-} + CO + CO_2.$$

The obtained glass foam contains a fair quantity of hydrogen sulfide trapped in its cells and, moreover, it can evolve this gaseous product if the sulfides contained in the glass come into contact with atmospheric moisture when the foam is ruptured. This drawback of several of the existing glass foam technologies limits the possible fields of its applications, because of the toxicity of hydrogen sulfide.

However, as emphasized by Demidovich [4], the evolving gas may be generated by oxidation of carbon by steam:

$$C + H_2O \rightarrow CO + H_2,$$
$$C + 2\,H_2O \rightarrow CO_2 + 2\,H_2.$$

Only chemically fixed water (as hydroxyl groups on the glass surface) can be a source of steam at the foaming temperature. Recently, Ketov [47] has had success with the introduction of hydroxyl groups at the stage of powder preparation. The main role in the hydration of the dispersed glass is played by the exchange of the surface sodium ions of the material for the H^+ ions of the water environment. A layer of polysilicic acids appears on the glass surface as a result of ion exchange and hydrolysis. Formation of polysilicic acids in the system results in evolution of steam at temperatures above 500–600 °C, at which glass is close to the pyroplastic state.

The employment of organic compounds (hydrocarbons or organic wastes) instead of carbon leads to the need to control the "carbon yield" of the compounds: Cowan et al. [28] found that sodium, calcium, and iron acetates, or other sodium salts like sodium lauryl sulfate, sodium oleate, sodium stearate, and so on, are much more effective as foaming agents than sugar, starch, and hydrocarbons that easily vaporize or sublime, since a certain amount of char, available for foaming, is formed on heating. For example, in the case of sodium acetate, the formation of free carbon is described by the following reaction:

$$2\,NaC_2H_3O_2 \rightarrow Na_2CO_3 + CO + 2C + 3\,H_2.$$

Silicon carbide (SiC) is also considered a very effective foaming agent, capable of giving uniform, controlled and precise cell sizes in the glass foam [48]. Although carbon is preferred for manufacturing the most important type of glass foams (i.e., blocks and shapes), SiC is known as a foaming agent for other commercial applications. The foaming activity of SiC is generally reported for higher temperatures than for carbon (950–1150 °C), and the reactions of SiC with the atmosphere and with the constituents of the glass are much more complex than those of carbon. Bayer and

Koese [36] showed a number of thermodynamically possible reactions between SiC and the gas (Tab. 1).

Table 1 Possible reactions of SiC in various atmospheres

Reaction	ΔH/kJ at 1000 K	ΔH/kJ at 1500 K
$SiC + \frac{1}{2} O_2 \rightarrow SiO + C$	–134.7	–179.08
$SiC + \frac{1}{2} O_2 \rightarrow CO + Si$	–157.3	–204.60
$SiC + O_2 \rightarrow SiO + CO$	–333.5	–412.3
$SiC + O_2 \rightarrow SiO_2 + C$	–656.1	–575.7
$SiC + O_2 \rightarrow CO_2 + Si$	–350.6	–357.3
$SiC + 2\,O_2 \rightarrow SiO_2 + CO_2$	–1048.1	–970.7
$SiC + 2\,CO \rightarrow SiO_2 + 3\,C$	–258.6	–91.2
$SiC + 3\,H_2O \rightarrow SiO_2 + CO + 3\,H_2$	–279.9	–327.2
$SiC + 4\,H_2O \rightarrow SiO_2 + CO_2 + 4\,H_2$	–281.6	–316.3

Silicon carbide produces a certain amountvof carbonaceous products, which in turn are capable of foaming, and residual silicon oxide, to be incorporated in the glass. The release of silicon oxides from the foaming agent may result in partial crystallization of the glass, with precipitation of cristobalite, particularly pronounced in borosilicate glasses. The large volume changes due to the polymorphic transformation of such phases on cooling of foams may lead to microcracking and mechanically weak samples. Addition of kaolin to the glass powder (sometimes > 10 wt %) is effective in reducing cristobalite precipitation, since it increases the content of Al_2O_3, useful as devitrification inhibitor [36].

Secondary foaming effects due to reduction of oxides present in the foaming of glass with C and SiC can be regarded as a starting point for innovative foams, which although commercially unavailable are widely described in the literature. The key feature of such foaming is the lack of a reaction with the furnace, thus avoiding the risk of a differential access of comburent oxygen to C and SiC, and resulting in a rather homogeneous and fine-porous microstructure. Bayer [49] pointed out the efficiency of SiC or Si_3N_4 together with easily reducible compounds. In particular, Si_3N_4 reacts with transition metal oxides (MnO_2, Fe_2O_3, Co_2O_3, NiO, CuO) with the evolution of N_2 and N_2O, leading to a very large expansion (volume increase by a factor of about 10) and low bulk density (100 kg m^{-3}).

The foaming activity of nitrides was recently investigated for manufacturing of glass foams from CRT glasses. Mear et al. [27] used AlN together with Fe_2O_3, and thus obtained a pronounced expansion and a very fine microstructure.

The foaming agents normally employed are usable only below 850–900 °C and are hence not suitable for use with high-viscosity glasses. The patent by Camerlinck [50] pertains to the use of a foaming agent whose utilization temperature is on the order of 1050 °C and which can be consequently used when high-temperature foaming is desirable (e.g., if the starting glass is highly viscous or is to be converted to a glass ceramic, so it is important to operate above the devitrification temperature to avoid premature crystallization). This foaming agent is a mixture of finely powdered SnO_2

and SiC, with a slight excess of SiC. At about 1050 °C SnO_2 is reduced by SiC with evolution of CO_2, which acts as foaming agent.

2.7.6 Glass Foam Products

Although it has been known for a long time and has many potential applications and excellent properties, only relatively small quantities of glass foam are produced industrially, and the number of glass foam manufacturers is very limited. The main reasons for the lack of growth of the foamed glass industry in the past were probably cost and demand. However, while in the past glass foam was manufactured from a purposely produced glass (with corresponding raw materials, plant, and energy costs), now the growing availability of waste glass and a better knowledge of the process parameters make it possible to produce glass foam at reduced cost. Furthermore, the flammability of organic insulation and the toxicity of its combustion products have created the need for new inorganic materials. Foamed glass from waste is therefore expected to become more and more important in the near future.

Currently, there are three main types of glass foam (and, consequently, different types of glass foam processes) [44]:

Loose glass foam aggregate (for totally replacing natural aggregate in concrete and for applications as ground insulation, foundation piles, floor and roof insulation, backfill insulation), obtained by continuous production of sheets of glass foam. Due to commercial confidentiality there is little detailed information about the process [35, 51, 52].

A mixture of powdered waste glass (from container and/or flat glass) and foaming agent (reduced to a particle size ranging from 75 to 150 μm, first by a hammer mill and subsequently by a ball mill) is continuously fed onto a moving belt that passes through a furnace heated to the foaming temperature (between 700 and 900 °C). After cooling, the slabs of glass foam, of width up to 2 m and thickness up to 100 mm, can be broken up to form loose aggregate, graded into different ranges of particle size depending on the application. The feedstock can be dry or wet, but it is believed that the wet route will not produce completely closed cells and hence result in lower thermal insulation.

Glass foam blocks and shapes, up to sizes on the order of 1200 × 600 × 160 mm, are generally manufactured by a continuous process, though a batch process can also be employed. Possible applications of glass foam blocks are in precast concrete panels, concrete bricks, piping insulation, storage vessel wall insulation, block paving, but mainly floor and roof insulation. Pittsburgh Corning Corporation of Pittsburgh, PA, has developed and marketed a product known as Foamglas Insulation Systems. Although in this case, too, due to commercial confidentiality, the process used is not completely clear, it is possible, on the basis of some US patents [53–57], to suppose that the foam is manufactured in individual blocks contained in molds and passed through a furnace. After foaming, it is necessary to subject the blocks to prolonged annealing to reduce internal strains and obtain a stable product. With larger blocks, increasing difficulties arise with heat transfer, due to the high thermal insulation capability of both the feedstock and the product. Then it is likely that the core will not achieve sintering temperature before the outside cells start to collapse. The molded glass foam can be cut and, if necessary, machined to the required shape.

In addition, Pittsburgh Corning has patented [58] an innovative method of glass foam manufacturing by means of a fluidized bed. Gas is passed through the glass powder as it is sintered to produce the glass foam. However, it is not known if this process is currently being used.

The foams for block and shape manufacturing exhibit a very low density (100–170 kg m^{-3}), so that the porosity is well above 90 %; the thermal conductivity is consequently very low (0.04–0.05 W m^{-1} K^{-1}), and glass foams can thus be regarded as a valid alternative to polymer foams. Consequently, blocks and shapes can be considered the most important type of glass foam.

Spherical pellets are produced in a pelletization process. The finely ground glass and the foaming agent are formed into spheres and then fed into a rotary furnace where the granules soften and the foaming agent exerts its action. The spheres of glass foam are then annealed and cooled, and can be used in the manufacture of blocks, panels, or slabs, as they can be subsequently sintered.

Ducman et al. [33] found that with the addition of waterglass (up to 50 wt %, in the form of droplets) to a powdered waste glass (NaCa silicate waste glass, obtained by crushing bottles), after firing of the raw granules at 805 °C for 1 min, foamed granules of high porosity can be obtained.

Table 2 lists selected properties of commercial glass foam products [44].

Table 2 Typical properties of commercial glass foam products

Density	0.1–0.3 g cm^{-3}
Porosity	85–95 %
Crushing strength	0.4–6 MPa
Flexural strength	0.3–1 MPa
Flexural modulus of elasticity	0.6–1.5 GPa
Coefficient of thermal expansion	8.9×10^{-6} K^{-1}
Thermal conductivity	0.04–0.08 W m^{-1} K^{-1}
Specific heat	0.84 kJ kg^{-1} K^{-1}
Thermal diffusivity at 0 °C	$(3.5–4.9) \times 10^{-7}$ m^2 s^{-1}
Sound transmission loss at normal frequency	28 dB/100 mm

2.7.7 Alternative Processes and Products

A relatively rich literature is dedicated to special glass foam products from processes alternative to those industrially employed or from innovative raw materials. A summary of these "alternative" processes and products is listed below.

2.7.7.1 Foams from Evaporation of Metals

Munters [59] reported that zinc or cadmium powder could be introduced into a molten mass of glass with subsequent evaporation. After condensation of the vapors, fast cooling of the obtained glass foam created a vacuum inside the cells. The pores, being evacuated and covered with a heat-reflecting metallic layer, constituted a highly insulating object owing to the Dewar principle.

2.7.7.2 High-Silica Foams from Phase-Separating Glasses

Elmer et al. [60] patented a method for manufacturing high-silica refractory glass foam. The key point of the invention was the application of a phase-separating borosilicate glass. After the leaching of the silica-poor phase, the resulting microporous high-silica glass was finely ground and impregnated with an aqueous solution of boric acid; the powders were sintered with the formation of a number of bubbles due to the large amount of water entrapped during the impregnation stage. Johnson [61] later patented a much simpler process based on the direct foaming of a phase-separating glass by using alkali metal carbonates and sulfates; the leaching stage was performed on the foamed product. In both applications the composition of the borosilicate glass had to be designed to result in spinodal decomposition to two interconnected phases, which is essential for properly leaching out the silica-poor phase.

2.7.7.3 Microwave Heating

Microwave heating has been developed for industrial applications such as drying and sintering of ceramics. Although glass compositions are generally transparent to microwave radiation at room temperature, at higher temperatures, in excess of 500 °C, the glass structure relaxes [62] and absorption of microwave radiation increases rapidly with consequent rapid volumetric heating of the glass. Hence, it should be possible to realize a continuous furnace for production of glass foam by using dual heating: conventional electric heating in a preheat zone to take the feedstock up to 500 °C, and microwave radiation to reach the required foaming temperature of 800–900 °C. In this way the processing time would be shorter due to rapid uniform heating. The homogeneous absorption of microwave radiation throughout the glass mass is believed to cause homogeneous foaming.

In recent laboratory-scale experiments, glass foam samples containing short metallic fibers as reinforcing elements were produced by microwave heating [63]. The glass chosen for this investigation was a soda-borosilicate glass, which coupled with microwaves above 600 °C. Stainless steel fibers (10 vol%) were added to enhance the structural integrity of the foams and increase fracture toughness.

2.7.7.4
Glass Foam from Silica Gel

A new method of producing glass foam for refractory thermal insulation material has been patented by Lee [64]. A glass foam having a closed-pore structure is produced by using porous silica gel, so that heat resistance and corrosion resistance can be greatly enhanced. The method comprises the following steps of 1) preparing a silica gel according to a sol–gel method from an alkali metal silicate, which is obtained by fusing silica and alkali; 2) exposing the silica gel to wet air to absorb a certain amount of moisture (the amount of absorbed moisture can be adjusted to regulate the foaming rate); 3) adding a certain amount of silica gel to a heat-resistant mold to be shaped; 4) calcining the mold at a constant rate up to 980–1300 °C (at least to the softening temperature of glass) in a tunnel or shuttle kiln, whereby the silica gel is fused into a glass foam; 5) cooling the fused glass foam.

2.7.7.5
High-Density Glass Foam

A recent invention by Hojaji et al. [65] refers to large high-density glass foam tiles which can be used as a facade on both exterior and interior building walls. These heavy (0.45–1.15 g cm^{-3}) glass foam tiles are capable of absorbing a substantial portion of the shock wave caused by an explosion, and thus may be used on the critical surfaces of buildings at risk of terrorist attacks, in combination with concrete, steel, or other high-strength building materials. Moreover, these tiles also have the advantage of being more resistant to earthquakes. (Prior work by the inventors and others [22] developed methods for making glass foam tiles of a wide a variety of densities.)

2.7.7.6
Partially Crystallized Glass Foam

Ketov [39] achieved formation of crystalline structure in foamed silicate, with the consequence that the system forms a rigid skeleton at the softening temperature of the glass, thus preventing destruction of the foam structure. In this case, it was possible to observe the absence of any increase of density with time after reaching its minimum value at the foaming temperature.

This important result was obtained by the addition of various substances promoting glass crystallization. Partial crystallization of glass does not result, however, in changes of the specific volume of the substance and in material cracking. The crystalline phases should not exhibit any pronounced change in volume on cooling; crystal phases subject to such changes, due to polymorphic transformations, are undesirable.

2.7.7.7
Foaming of CRT Glasses

The foaming of CRT glasses with $CaCO_3$ is also interesting. Glass foams were prepared [26] from panel glass and mixtures of panel glass and funnel and neck glasses, by dry mixing of fine glass powders with $CaCO_3$. While the panel glass is a barium-–strontium glass, the funnel and neck glasses, both employed in the rear parts of the CRTs (the neck being the envelop of the electron gun), are lead silicate glasses. It is known that lead-containing wastes are particularly difficult to recycle since lead is a noxious heavy metal.

The foaming of CRT glasses by $CaCO_3$ at 725 °C for 15–30 min led to lightweight articles with densities in the range 0.18–0.4 g cm^{-3} and fine-pored (pore diameter ca. 100 μm) and homogeneous microstructure (Fig. 4). Several openings between adjacent cells are clearly visible, that is, a fraction of the foam is open-celled. The XRD spectra revealed slight formation of wollastonite $CaO \cdot SiO_2$, a desirable crystal phase since it does not exhibit volume changes due to polymorphic transformations on cooling of glass foams.

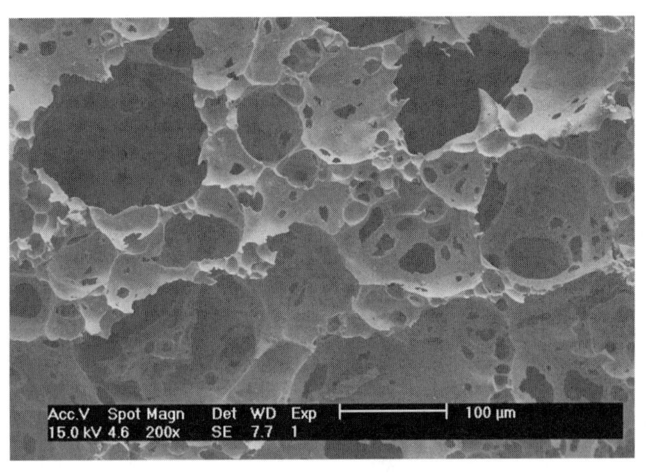

Fig. 4 SEM image of the microstructure of glass foam obtained from CRT glasses by using $CaCO_3$ as foaming agent.

The crushing strength of the foams (up to 5 MPa at a density of 0.4 g cm^{-3} and a porosity of 85 %) is remarkable when compared to the strength predicted [66] for such low density and partially open celled microstructure. Some preliminary chemical tests (acid attack) demonstrated that the heavy metal release of the foams was negligible and, above all, independent of the chemical formulation (ratio of panel/-lead-rich glass) of the glass cullet employed. Therefore, the foaming of CRT glasses by $CaCO_3$ thermal decomposition appears to be a promising way of treating heavy metal containing glasses, since foaming does not depend on oxidation/reduction processes, which could cause the precipitation of metallic colloids by reduction of easily reducible oxides like PbO. Moreover, the relatively low temperature required (CRT glasses have a low softening point) prevents the volatilization of the PbO.

2.7.8 Summary

For more than seventy years glass foam production technology has been established as a valid method both for obtaining products having a unique combination of properties (with particular reference to high insulation capability, very low density, excellent resistance to fire, water impermeability, high compression strength, and dimensional stability) and for recycling growing quantities of glass waste. The continuously decreasing price owing to improvement of process technologies and the growing introduction of wastes as raw materials, as well as the advantages coming from energy saving through a better insulation capability, will certainly increase the application possibilities for this material in the near future.

References

1 Mc Lellan, G.W., Shand, E.B., *Glass Engineering Handbook*, McGraw-Hill Book Co., New York, 1984, Chap. 19.

2 Pfaender, H.G., *Schott Guide to Glass*, Chapman & Hall, London, 1992, pp. 186–187.

3 Kern, L., US Patent 1,898,839, 1933.

4 Demidovich, B.K., *Production and Application of Glass foam* (in russian), Ed. Nauka i Tekhnika, Minsk, 1972, p. 301.

5 Long, B. (Saint Gobain, France), US Patent 1,945,052, 1934.

6 Lytle, W.O. (Pittsburgh Plate Glass, USA), US Patent 2,215,223, 1940.

7 Miller, R.A. (Pittsburgh Plate Glass, USA), US Patent 2,233,631, 1941.

8 Fox, J.H., Lytle, W.O. (Pittsburgh Plate Glass, USA), US Patent 2,261,022, 1941.

9 Black, H.R. (Corning, USA), US Patent 2,272,930, 1942.

10 Fox, J.H., Lytle, W.O. (Pittsburgh Plate Glass, USA), US Patent 2,354,807, 1944.

11 Peyches, I. (Saint Gobain, France), US Patent 2,658,096, 1953.

12 Powell, E.R., US Patent 3,133,820, 1964.

13 Haux, E.H. (Pittsburgh Plate Glass, USA), US Patent 2,191,658, 1940.

14 Lytle, W.O. (Pittsburgh Plate Glass, USA), US Patent 2,322,581, 1943.

15 Owen, W. (Pittsburgh Plate Glass, USA), US Patent 2,310,457, 1943.

16 Owen, W. (Pittsburgh Corning, USA), US Patent 2,401,582, 1946.

17 Ford, W.D. (Pittsburgh Corning, USA), US Patent 2,691,248, 1954.

18 Willis, S. (Corning, USA), US Patent 2,255,236, 1941.

19 Akulich, S.S., Demidovich, B.K., Sadchenko, N.P., Voznesensky, V.A., Proceedings XIth International Congress on Glass, Prague, 1977, vol. 5, pp. 483–491.

20 Byung Il, K., Korean Patent NKR 2003026078 (2003).

21 D'Eustachio, D., Johnson, H.E. (Pittsburgh Corning, USA), US Patent 3,441,396, 1969.

22 Hojaji, H., US Patent 4,430,108, 1984.

23 Hojaji, H., *Mater. Res. Soc. Symp. Proc.*, **1988**, *136*, 185–206.

24 Gutmann, R., *Glastech. Ber. Glass Sci. Tech.* **1996**, *69, 285–299.*

25 Colombo, P., Brusatin, G., Bernardo, E., Scarinci, G., *Curr. Opin. Solid State Mater. Sci.* **2003**, *7, 225–239.*

26 Bernardo, E., Scarinci, G., Hreglich, S., Proc IVth Int. Congress "Valorisation and Recycling of Industrial Wastes (VARIREI)", L'Aquila, Italy, 2003.

27 Mear, F., Yot, P., Cambon, N., Liautard, B., *Verre* **2003**, *9, 72–77.*

28 Cowan, J.H., Rostoker, D. (Corning, USA), US Patent 3,666,506, 1972.

29 Rostoker, D. (Corning, USA), US Patent 3,793,039, 1974.

30 Kurz, F.W.A., US Patent 3,874,861, 1975.

31 Seki, Y., Nakamura, M., US Patent 3,951,632, 1976.

32 Shabanov, V.F., Pavlov, V.F., Pavlov, J.V., Pavlova, N.A., Russian Patent RU 2,192,397 C2 (2002).

33 Ducman, W., Kovacevic', M., *Key Eng. Mater.* **1997**, *132–136, 2264–2267.*

34 Brusatin, G., Scarinci. G., Bernardo, E., unpublished results.

35 Solomon, D., Rossetti, M., US Patent 5,516,351, 1996.

36 Bayer, G., Koese, S., *Riv. Staz. Sperim. Vetro* **1979**, *9, 310–320.*

37 Brusatin, G., Scarinci, G., Proceed. IVth Int. Congress "Valorisation and Recycling of Industrial Wastes (VARIREI)", L'Aquila, Italy, 2003.

38 Morgan, J., Wood, J., Bradt, R., *Mater. Sci. Eng.* **1981**, *47, 37–42.*

39 Ketov, A.A., Proc. Int. Symp. Recycling and Reuse of Glass Cullet, Dundee, Scotland, 2001, p. 1–7.

40 Koese,V.S., Bayer, G., *Glastechn. Ber.* **1982**, *55, 151–160.*

41 Steiner, A., Beerkens, R.G.C., Proc. I.C.G. Ann. Meeting, July 1–6, 2001, Edinburgh, Scotland.

42 Low, N.M.P., *J. Mater. Sci.* **1981**, *16, 800–808.*

43 Lynsavage, W., *Ceram. Bull.* **1951**, *30, 230–231.*

44 Hurley, J., Glass Research and Development Final Report: a U.K. Market Survey for Glass Foam, published by: WRAP (The Waste and Resources Action Programme), The Old Academy, 21 Horsefair, Banbury, Oxon OX16 0AH, www.wrap.org.uk.

45 Mackenzie, J.D., US Patent 3,811,851, 1974.

46 Shill, F., *Glass Foam (Production and Applications)* (in Russian), Ed. Literatura Po Stroitelstvu, Moscow, 1965, p. 307.

47 Ketov, A.A., Proc. Int. Symp. Advances in Waste Management and Recycling, Dundee, Scotland, 2003, pp. 695–704.

48 Brusatin, G., Scarinci, G., Zampieri, L., Colombo, P., Proc. XIXth Int. Congress on Glass, Edinburgh, Scotland, 2001, vol. 2, pp. 17–18

49 Bayer, G., *J. Non-Cryst.Solids* **1980**, *38–39, 855–860.*

50 Camerlinck, P., (Saint Gobain, France), US Patent 3,975,174 (1976).

51 Kraemer, S., Seidle, A., Mayer, R., Streibl, L., US Patent 3,473,904, 1969.

52 Malesak, J., US Patent 3,607,170, 1971.

53 King, W.C., Medvid, R.J. (Pittsburgh Corning, USA), US Patent 3,959,541, 1976.

54 Rostoker, D. (Pittsburgh Corning, USA), US Patent 4,119,422, 1978.

55 Kirkpatrick, J.D. (Pittsburgh Corning, USA), US Patent 4,198,224, 1978.

56 Kijowski, J., Miller, G.D. (Pittsburgh Corning, USA), US Patent 4,571,321, 1986.

57 Linton, R.W., Orlowski, A.W. (Pittsburgh Corning, USA), US Patent 4,623,585, 1986.

58 Smolenski, C., European Patent EP 0,294,046, 1988.

59 Munters, C.G., US Patent 2,012,617, 1931.

60 Elmer, T.H., Middaugh, H.D. (Corning, USA), US Patent 3,592,619, 1971.

61 Johnson, J.D., US Patent 3,945,816, 1976.

62 Knox, M., Copley, G., *Glass Tech.*, **1996**, *38*, 91–96.

63 Minay, E.J., Veronesi, P., Cannillo, V., Leonelli, C., Boccaccini, A.R., *J. Eur. Ceram. Soc.* **2004**, *24, 3203–3208.*

64 Lee, Y.W., Korea, P.N., WO 0187783, Int. Appl. N. PCT/KR01/00602 (2001).

65 Hojaji, H., Buarque de Macedo, P.M., US Patent Application 20030145534A1, 2003.

66 Gibson, L.J., Ashby, M.F., *Cellular Solids: Structure and Properties*, 2nd ed., Cambridge University Press, 1999, p. 213.

2.8 Hollow Spheres

Srinivasa Rao Boddapati and Rajendra K. Bordia

2.8.1 Introduction

Hollow spheres have found applications in such diverse areas as refractory thermal insulation, lightweight composites, fiber-optic sensors, laser-fusion targets, encapsulation, and gas and chemical storage. Here, attention is restricted to ceramic hollow spheres. Advancements in processing, properties, and applications of solid and hollow spheres up to 1994 were discussed as a focused symposium at the 1994 fall meeting of the Materials Research Society held in Boston [1]. The focus of this chapter is on processing of ceramic hollow spheres, although a general overview of properties and applications of hollow spheres is also given. A variety of processing methods have been developed to produce hollow spheres over the years: spray techniques such as spray drying [2–4], spray pyrolysis [5–11], and the coaxial-nozzle method [12–15], the sacrificial-core method [16–26], the sol–gel/emulsion method [27–30], and layer-by-layer deposition on colloidal templates [31–36]. Each method has its own advantages and disadvantages. Spray techniques are most suitable for bulk production and larger diameter spheres. On the other hand, solution- or colloid-based techniques yield hollow spheres of controlled morphology and size, but they are not suitable for bulk production. All techniques without exception use some solvent, either for holding the ceramic particles (slurry or colloid) or for dissolving the salts or chemical compounds to be coated onto the core particles. After drying the droplets, the dried hollow spheres are sintered at high temperatures (> 400 °C, depending on the particle size) to improve their mechanical properties. Before the 1990s, most studies on inorganic hollow spheres were confined to hollow glass spheres, which were made by blowing molten glass, driven by research into targets for inertial-confinement fusion [13]. The two most important parameters of hollow spheres are sphere size and aspect ratio (the ratio of sphere diameter to wall thickness), which govern their properties. Various processing methods employed to fabricate ceramic hollow spheres are described first, followed by their properties and applications.

Cellular Ceramics: Structure, Manufacturing, Properties and Applications.
Michael Scheffler, Paolo Colombo (Eds.)

ISBN: 3-527-31320-6

2.8.2
Processing Methods

The processing of ceramic hollow spheres (CHS) can be broadly classified into five different categories: 1) sacrificial-core method, 2) layer-by-layer deposition, 3) emulsion/sol–gel method, 4) spray and coaxial-nozzle methods, and 5) reaction methods. Other methods such as spray drying and aerosol-based techniques can also be grouped with the nozzle techniques, as these methods employ a droplet-generation technique as a primary means for generating hollow spheres. In the coaxial-nozzle technique, a slurry containing ceramic particles or a precursor solution flows through a narrow orifice while an inert gas flows through an inner nozzle concentric with the outer nozzle. The sacrificial-core method uses a sacrificial core such as polystyrene spheres onto which the shell material of the required hollow sphere is coated and later heated to high temperatures to remove the core and sinter the shell material. A variation of the sacrificial-core method is the layer-by-layer (LbL) deposition technique. Although it can be viewed as a variation of the sacrificial-core method, it gives greater flexibility in controlling the shell wall thickness of hollow spheres by deposition of individual layers of polyelectrolyte and hollow-sphere material. The emulsion/sol–gel method utilizes the principle of limited or no miscibility between two liquids in forming an emulsion. The emulsified globules of one of the liquids acts as a template for coating with the hollow-sphere material, and the liquid in the core is either extracted or evaporated during subsequent heating to higher temperatures. Occasionally, polymer particles have also been used in the emulsion/sol–gel method as a template for making hollow ceramic spheres. Each of the techniques mentioned above is described and discussed in detail in the following subsections.

2.8.2.1
Sacrificial-Core Method

The sacrificial-core method is one of the most widely used for making ceramic hollow spheres (CHS). In this method a spherical polymer particle (core) is coated with a slurry or solution of ceramic material, the core–shell composite is dried, and the core is removed by chemical dissolution or by heating to a temperature at which the polymer decomposes. Sometimes the polymer sphere is negatively or positively charged, and this enhances adsorption between the sphere and the material to be coated (ceramic particle in colloidal form or ceramic-producing salt in solution). The thus-formed hollow spheres are heated to higher temperatures to sinter the particles in the wall and impart adequate mechanical strength and rigidity to the hollow sphere. A number of ceramic materials have been processed into hollow-sphere form by using this technique: Y_2O_3 [16, 37], $Pb(Zr_{0.52}Ti_{0.48})O_3$ [28, 38], SiO_2 and Fe_3O_4 [17], hematite [20, 24], SiO_2 [18], TiO_2 [23, 25, 39, 41, 42], TiO_2 and SnO_2 [40], silver/TiO_2 [43], and BN [44]. CHS have also been made by using combustible cores [26], mechanofusion [45], the reactant itself as a core [19, 44], vesicles as cores [21], and micellized diblock copolymer as template [46]. Yang et al. [23] modified the sacrificial-core technique to produce tunable-cavity hollow spheres from core–shell gel template particles which were prepared by inward sulfonation of polystyrene particles with concentrated sulfuric acid. Yin et al. [22] and Aoi et al. [47], functionalized interior surfaces of TiO_2 hollow spheres with silver and platinum nanoparticles, respectively. This kind of functionalization of the interior surface of the CHS eliminates the step of diffusing the substance to be immobilized or functionalized into the interior of the hollow sphere. Along similar lines, inorganic cores such as SiO_2 have been used to fabricate polymer hollow spheres by coating polystyrene nanospheres onto the SiO_2 cores and later removing the SiO_2 by etching [48]. Among the different template methods used to make CHS, the catanionic (a mixture of cationic and anionic surfactants) vesicle template method gives very small diameters (60–120 nm) and wall thicknesses (1–2 nm). Diameter and wall thickness of hollow spheres are the two most important parameters which control their properties. In some applications (inertial-confinement fusion targets), the surface roughness also plays an important role. The wall or shell of a CHS is formed by coating the ceramic material from a solution onto a core by different means: sol–gel, precipitation, hydrolysis of ceramic-producing salts, mechanofusion, and so on. The timescale of this process is on the order of hours or, in some cases, days [19]. The method is amenable to producing CHS having a diameter ranging from a few tens of nanometers to millimeters. The diameter of the hollow spheres is controlled by the diameter of the core or template particle, and the wall thickness, in most cases, is controlled by the concentration of the solution used for coating the cores and the exposure time.

2.8.2.2 Layer-by-Layer Deposition

Layer-by-layer deposition has been used to make hollow spheres of different ceramic materials [31–36, 49, 50]. In this technique, colloidal core particles are coated with alternating layers of polyelectrolytes and charged ceramic particles. The charge on the polyelectrolytes and the ceramic particles are opposite in nature. As a result, the ceramic particles in a solution are adsorbed onto the polyelectrolyte layer by electrostatic interaction. A schematic depicting various stages of the process is shown in Fig. 1. A polymer sphere such as negatively charged polystyrene is used as core. Adsorption of positively charged polyelectrolytes such as poly(diallyldimethylammonium chloride), PDADMAC, onto the polystyrene in solution lead to a polyelectrolyte layer with excess positive charge, onto which negatively charged ceramic particles (mostly nanosize) are adsorbed by electrostatic interaction. By repeating the process of alternating adsorption of polyelectrolyte and ceramic particle layers, the required wall thickness of the hollow sphere is obtained. Sometimes, three or five alternating layers of PDADMAC and polystyrenesulfonate (PSS) are coated onto the polystyrene or polymer core particles to provide a uniformly charged and smooth polyelectrolyte surface [32, 34, 35, 50]. The cavity or void size of the hollow sphere is governed by the diameter of the polystyrene or polymer core particle and shrinkage during sintering. The core is removed by thermal or chemical means, and the

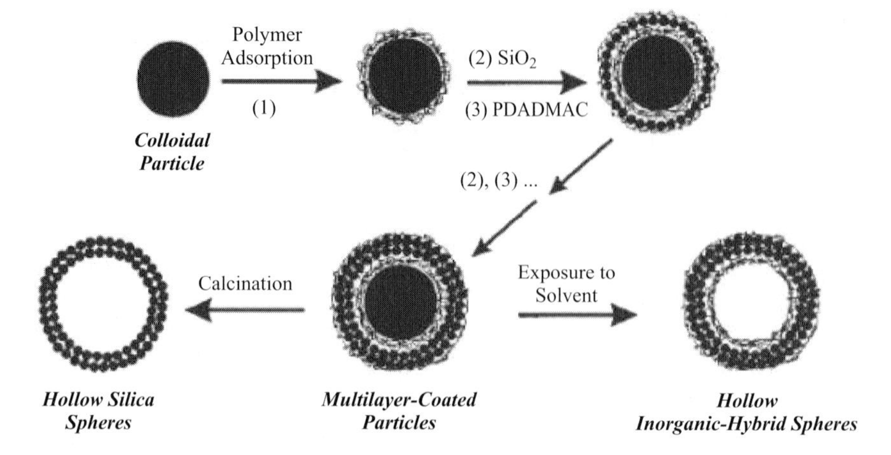

Fig. 1 Schematic of layer-by-layer deposition method [Reprinted with permission from F. Caruso, R.A. Caruso, and H. Möhwald, *Science*, **1998**, *282*, 1111–1114. Copyright (1998) AAAS].

resulting green hollow spheres are generally sintered at temperatures above 400 °C. If thermal heating is employed to remove the core material, the sintering step is combined with the core-removal step. In the case of SiO_2/PDADMAC layer pairs, after sintering, the diameter of hollow spheres shrank by about 5–10% [33]. The number of polyelectrolyte and ceramic particle layers deposited on the core controls the wall thickness of the hollow spheres. This method has been used to obtained hollow spheres of silica [31, 35, 36], titania [34], silica/Fe_3O_4 composite [32], laponite [36], and zeolite [49, 50] (see Tab. 1). Figures 2–4 show the magnetic (Fe_3O_4) and Fe_3O_4/SiO_2 composite hollow spheres prepared by depositing ceramic nanoparticle layers in alternation with polyelectrolyte layers [each layer contains PDADMAC/PSS/PDADMAC (PE_3)] on initially PE_3-coated polystyrene particles. This technique has been used extensively to coat planar substrates, colloidal particles, and self-assembled colloidal particles. More detailes can be found in Ref. [51]. The parameters governing successful generation of ceramic hollow spheres depend on the selection of core materials, polyelectrolytes, particle size, shape and concentration of ceramic nanoparticles in the solution or inorganic molecular precursors, core removal, and sintering of the hollow spheres. Although it is possible to control the wall thickness of the hollow sphere by the number of cycles used to deposit polyelectrolyte and ceramic layers, the excess polyelectrolyte must be removed each time the ceramic layer is adsorbed or coated onto the polyelectrolyte, which is time-consuming. In addition, repeated use of centrifugal forces for this purpose can disrupt the integrity of the ceramic layer. When ceramic precursor is used instead of a ceramic colloidal particle for coating or infiltrating the polyelectrolyte-coated polymer core, at least certain minimum number of layers of ceramic precursor is required to bring about a mechanically stable shell [34]. Separate preparation of ceramic colloidal particles is required for coating a ceramic layer onto the polyelectrolyte, unless inorganic molecular precursors are used, as the process is mainly driven by electrostatic

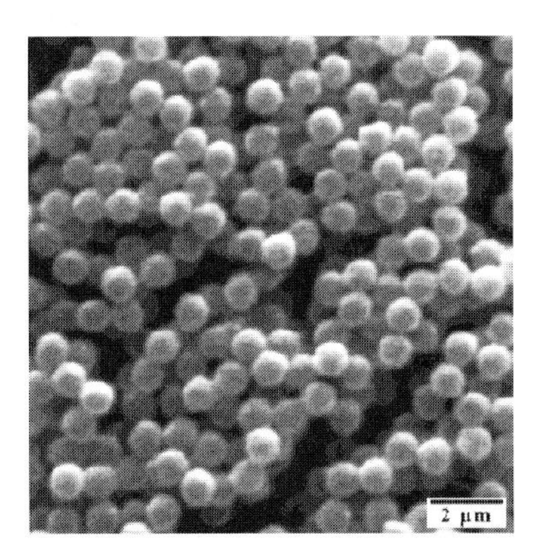

Fig. 2 SEM micrograph of magnetic hollow spheres prepared by depositing five Fe_3O_4 nanoparticle layers in alternation with PDADMAC/PSS/PDADMAC on PE_3-coated polystyrene particles, followed by calcination at 500 °C [Reprinted with permission from F. Caruso et al., *Chem. Mater.*, **2001**, *13*, 109. Copyright (2001) American Chemical Society].

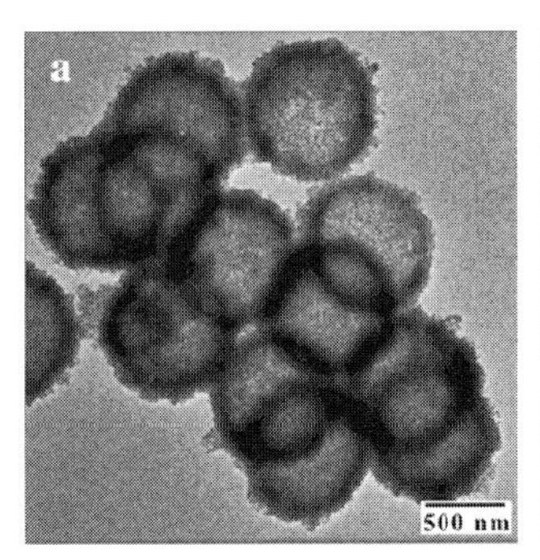

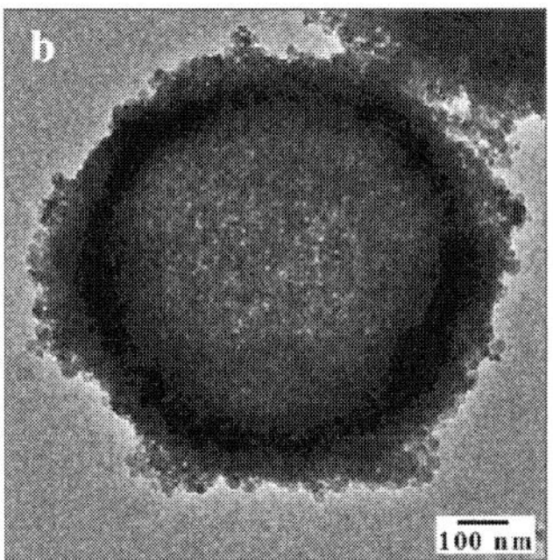

Fig. 3 TEM images of magnetic hollow spheres prepared by calcination (500 °C) of PE_3-modified PS particles coated with inner and outermost SiO_2 nanoparticle layers and three intermediate Fe_3O_4 nanoparticle layers (each inorganic nanoparticle layer was separated by PE_3). a) Low- and b) high magnification image [Reprinted with permission from F. Caruso et al., *Chem. Mater.*, **2001**, *13*, 109. Copyright (2001) American Chemical Society].

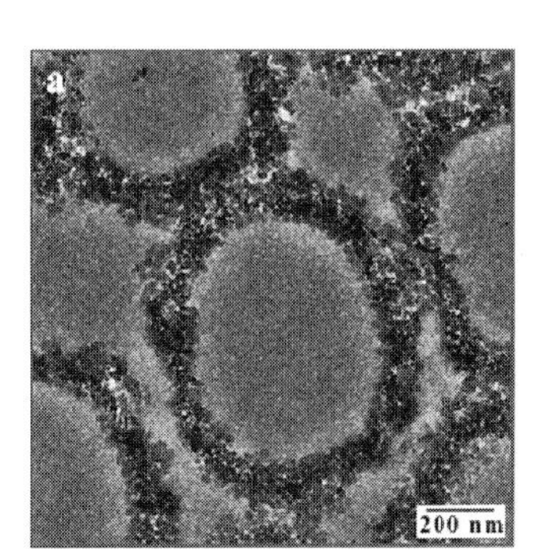

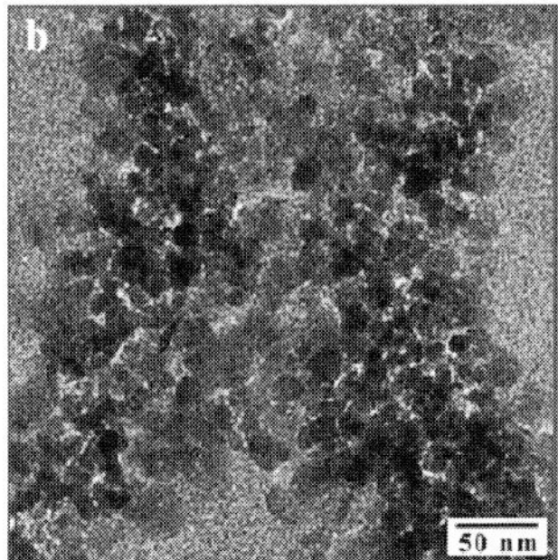

Fig. 4 TEM images of cross sections of the magnetic hollow spheres displayed in Figure 3. a) Low- and b) high-magnification image [Reprinted with permission from F. Caruso et al., *Chem. Mater.*, **2001**, *13*, 109. Copyright (2001) American Chemical Society].

interaction. Another major shortcoming is maintaining individual coated particles as well as separation of coated particles without disturbing their geometry. Centrifuging is used to separate the coated particles and is ineffective if the particle size is smaller than 100 nm. Filtration is used to separate the coated particles when the particles are smaller than 100 nm, although they lead to frequent clogging of pores of the filter and hence frequent replacement of filters and attendant cost.

Table 1 Ceramic hollow spheres produced by layer-by-layer deposition.

Ceramic hollow sphere	Wall material (ceramic nanoparticles or precursors)	Polyelectrolyte	Ref.
SiO_2	25 nm SiO_2 particles	PDADMAC	31, 33, 35, 36
TiO_2	titanium(IV) bis(ammonium lactate) dihydroxide	PDADMAC	36
Fe_3O_4	Fe_3O_4 sol (particle size 12 nm)	PDADMAC and polystyrenesulfonate	32, 36
Laponite $[Na_{0.7}(Si_8Mg_{5.5}Li_{0.3})O_{20}(OH)_4]$	laponite nanoplatelets	PDADMAC	36
Zeolite	nanosilicalite	PDADMAC, PDADMAC/PSS/ PDADMAC/PSS/ PDADMAC	49, 50

2.8.2.3
Emulsion/Sol–Gel Method

A water-extraction sol–gel process, patented by Sowman [52], has been used to prepare hollow ceramic spheres. Because water extraction and the dispersion were carried out simultaneously, control of the particle size was a challenge, however, and the product had a broad particle size distribution. The emulsion/sol–gel method developed by Liu and Wilcox, Sr., [27, 53], is an improvement on the water-extraction sol–gel technique. In this method (Fig. 5), hollow ceramic spheres are prepared by forming an emulsion of aqueous oxide/hydroxide sols by dispersing them in an organic liquid using a suitable surfactant followed by water extraction from the emulsified sol globules. The process parameters controlling both aspect ratio and size of hollow microspheres are the colloid concentration, water-extraction rate, and droplet size. Colloid concentration affects the initial droplet size as it influences the viscosity of the colloid and thereby partly controls the droplet size during the emulsion step. The effect of colloid concentration on the aspect ratio of hollow spheres depends on the critical concentration for gelation of a particular sol. At a fixed droplet size and water-extraction rate, dilute colloids lead to the formation of hollow spheres with lower aspect ratios than those from concentrated colloids, as the shrinkage of the droplet is inversely proportional to colloid concentration. Wall thickness is controlled by the water-extraction rate, and at a fixed colloid concentration and droplet size, higher water-extraction rate leads to quick formation of shell membranes. As a result, higher aspect ratio hollow spheres were obtained due to reduced shrinkage and thinner shell wall. The droplet size affects whether a droplet will form a hollow or a solid sphere. Liu and Wilcox [27] reported that the microspheres smaller than 1 μm were found to be solid in most cases. They attributed this to the small diffusion distance, which led to a homogenous colloid concentration, rather than a gradient in colloid concentration from the surface to the interior of the droplet, and hence homogenous gelation rather than surface gelation. The advantage of this method, compared to that of Sowman [52], is that the emulsion and water-extraction steps are separate, and hollow-sphere size can be controlled during the water-extraction step independent of emulsion step. By using this technique hollow spheres of SiO_2 for removal of heavy metal ions from wastewater [54], Si/Al composite oxide [55], SiO_2 for preparation of hollow gold spheres [56], and SiO_2 from tetraethoxysilane (TEOS) [57] have been maufactured. Yang and Chaki [28] prepared lead zirconate titanate (PZT) hollow spheres by coating polyacrylamide latex microspheres (20–100 μm) in an emulsion with PZT sol and subsequently burning out the polymer core. In this case, polymer spheres are used as templates, unlike in usual emulsion-based techniques in which oil droplets are used as templates. The PZT sol was prepared from lead nitrate $Pb(NO_3)_2$, zirconium *n*-butoxide $Zr(C_4H_9O)_4$, and titanium isopropoxide $Ti(OC_3H_7)_4$ and the molar ratio of the chemical compounds used for processing was such that the cation ratio in the final compound was maintained as $Pb(Zr_{0.52}Ti_{0.48})O_3$. The zirconium and titanium alkoxides were poured into $Pb(NO_3)_2$ solution and stirred to obtain PZT sol. The polyacrylamide spheres were dispersed in water-free toluene by stirring followed by addition of NH_3. The resultant mixture was stirred so that an emulsion of NH_3 droplets was formed. The droplets of NH_3 which came in contact with polymer spheres were adsorbed, and this was responsible for gelation of PZT sol when it came in contact with NH_3. The diameter of the PZT hollow spheres ranged from 20 to 95 μm.

Collins et al. [58] reported the formation of TiO_2 hollow spheres in nonaqueous emulsions such as oil in formamide from titanium alkoxide precursors. Among the tested precursors, titanium(IV) ethoxide alone gave intact hollow spheres, whereas titanium(IV) propoxide, although successful to some extent, often gave very thin shells which were susceptible to damage. In contrast, titanium(IV) butoxide failed to produce intact hollow spheres due to the reduction in the rates of hydrolysis and condensation because of the large alkyl chains. Wu et al. prepared mesostructured lead titanate by an oil-in-water emulsion-mediated route using titanium butoxide, lead acetate, and dodecylamine in 1-butanol as the source of TiO_2, PbO, and template for mesostructure in the shell wall, respectively [30]. It was shown that the presence of KOH as a mineralizer is essential in the appropriate concentration (i.e., at molar ratios of K_2O/H_2O of 4.0/1780 to 10.0/1780) for producing mesostructured hollow spheres. Sun et al. prepared hollow silica spheres with multilamellar structure in the shell using poly(ethylene oxide)–poly(propylene oxide)–poly(cthylene oxide) block copolymer based emulsion in combination with sodium silicate [29]. Recently, Liu et al. produced Cu_2O hollow spheres using a multiple-emulsion sys-

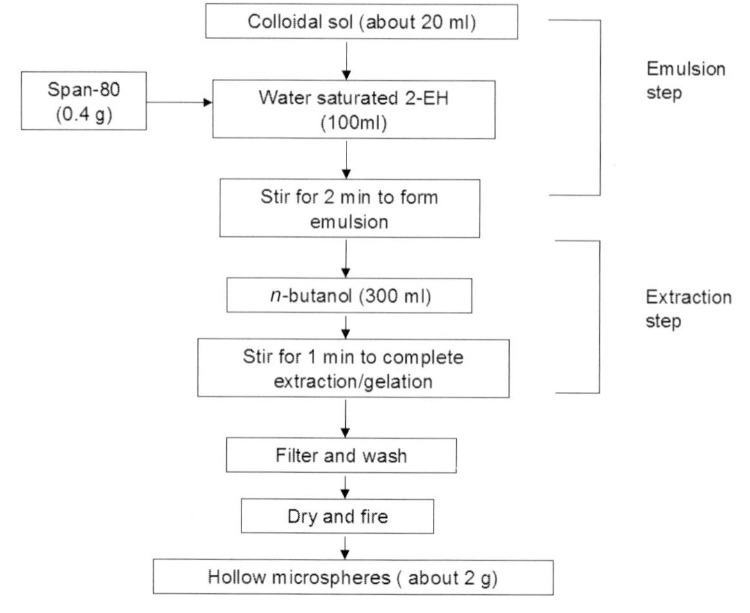

Fig. 5 Flowchart of different steps involved in emulsion preparation and water extraction (after J. G. Liu and D. L. Wilcox, Sr. [27]).

tem (oil/water/oil) as a template [59] (Fig. 6). In this technique, the outer oil acts as a dispersing medium, the inner oil droplets as a template, and the reactants are dispersed in the interim water layer. The individual droplets in the emulsion act as a template and play a space-limiting role and at the same time assist prevent agglomeration of product particles. However, the droplets did coalesce due to Brownian motion when the emulsion was heated in a microwave oven.

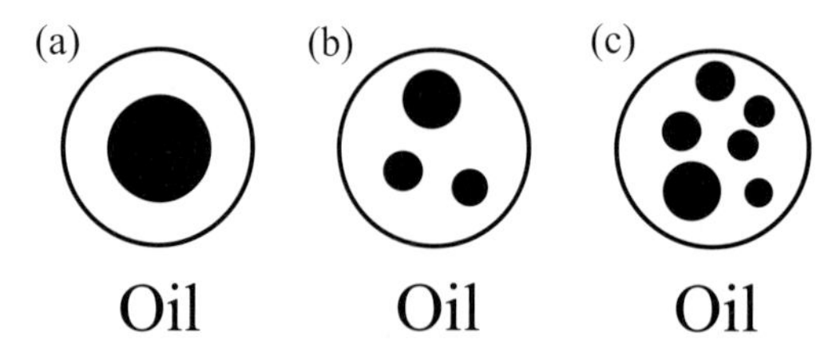

Fig. 6 Three different modes of existence of water droplets in oil/water/oil emulsions (black stands for oil inside the water droplet, and white for the water phase; water droplets are surrounded by oil). a) An oil droplet in a water droplet, b) several oil droplets far from each other in a water droplet, and c) many oil droplets close to each other in a water droplet (after H. Liu et al. [59]).

2.8.2.4 Spray and Coaxial-Nozzle Techniques

Hollow spheres have been prepared by a host of droplet generation techniques using solutions of the salts of the target material or slurries of the target material as the starting materials. The solution or slurry is either sprayed through a hot chamber [13, 60] or through a coaxial nozzle [12, 13, 61–63] set up to generate spherical droplets. Torobin used the coaxial nozzle technique to make hollow porous microspheres as substrates and containers for catalysts [61, 64, 65].

In a spray pyrolysis process, hollow spheres were generated by atomizing the inorganic precursor in solution form, evaporating the solvent during flight, and precipitating the inorganic salt on the surface. The remaining solvent was removed by drying, and the precipitated salt was pyrolyzed at higher temperatures (> 250 °C) to form the desired ceramic phase in the shell of a hollow sphere. These pyrolyzed spheres or particles were sintered at higher temperatures to achieve sufficient mechanical strength. Messing et al. described in detail various steps involved in spray pyrolysis and also the conditions required for producing solid and hollow particles, fibers, thin films, and composite particles [66, 67]. Pratsinis and Vemury reviewed the formation of particles in gases and described the advantage of processing particles in gases [68]. Most of the spray pyrolysis methods differ in the way the droplets are produced. Hollow spheres of zircon ($ZrSiO_4$) were prepared by mechanically mixing a zirconium salt and a silica sol in proportions corresponding to the zircon composition and spraying the aqueous solution as a mist into a preheated (150 °C) chamber [5]. The effect of calcination temperature on the phase evolution and the size and morphology of the hollow spheres was investigated. Tartaj et al. synthesized silica-coated γ-Fe_2O_3 hollow spheres by aerosol pyrolysis of a methanol solution containing iron ammonium citrate and silicon ethoxide [7]. Aerosol techniques have been used to produce fine metal and metal oxide particles [8], as well as for encapsulation of oil droplets with metal oxide [9] and mesostructured SiO_2 [69]. Pyrolysis of aerosol droplets was employed to generate hollow spheres of CuO from copper sulfate pentahydrate [10], and of $NiFe_2O_4$ from nickel and iron nitrates [6]. Jokanovic et al. used ultrasonic spray pyrolysis for producing TiO_2 hollow spheres [11]. Tani et al used emulsion combustion to prepare hollow spherical particles of Al_2O_3, TiO_2, ZrO_2, and Y_2O_3 from metal precursors [70]. Spray drying is one of the most versatile methods of powder processing [71]. In spray drying (see Fig. 7), water- or organic-based slurries of ceramic particles are sprayed in the form of droplets into a chamber containing hot air or other, inert gases. Spray drying (atomizing and drying) leads to a large variety of particle shapes depending on the processing parameters and slurry characteristics: solid, elongated, pancake, donut-shaped, needlelike, or hollow particles. The effect of spray-drying conditions on the morphology of a mesoporous hollow silica particle is shown in Fig. 8. It has been shown in the case of Al_2O_3 slurry that a flocculated slurry leads to solid particles, whereas a dispersed suspension leads to hollow particles [72]. Spray drying has been applied successfully to produce hollow spheres of TiO_2 [4], hydroxyapatite [3], and SiO_2 [2]. The success of most of the spray pyrolysis and spray drying techniques in generating hollow spheres depends on a number of factors: evaporation rate, concentration of the solution, and droplet size. Formation of hollow particles in spray drying can take place in four different ways [71]:

1) Formation of surface layer and blowing of the droplet as the evaporating liquid inside the droplet expands.
2) Diffusion of solids back into the droplet is slower than evaporation of the liquid.
3) Liquid along with the solids flows to the surface due to capillary action.
4) Air or any other gas contained in the droplet leaves the space it occupied when the liquid in the surrounding film evaporates and forms a porous layer.

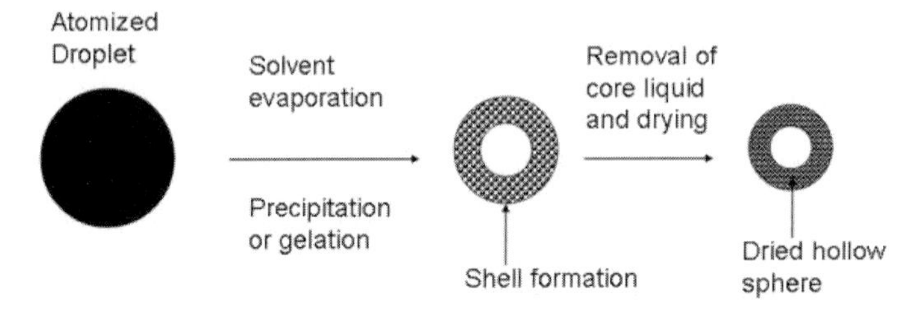

Fig. 7 Schematic of spray drying.

When a solution is used for generating hollow spheres, the water or other organic solvents inside the liquid droplets act as a blowing agent, whereas in a coaxial-nozzle method a gas is blown through a nozzle concentric with an outer nozzle to generate spherical droplets containing gas. Although the methodology of converting these spherical droplets into hollow spheres differs among the individual techniques, the principle is same for generating spherical droplets containing a liquid or a gas. Both liquid-droplet generation [60, 73] and sol–gel techniques [74, 75] were used to produce hollow glass and silica spheres. The liquid-droplet process used for making

Fig. 8 Scanning electron micrograph of calcined mesoporous silica hollow spheres produced by spray drying. This particular morphology arises from the formation of a rigid crust during drying, before all the solvent has diffused out of the interior of the droplet [Reprinted with permission from P. J. Bruinsma et al., *Chem. Mater.*, **1997**, *9* [11], 2510. Copyright (1997) American Chemical Society.]

hollow glass spheres for fusion targets needs special mention as it is capable of making spheres with smooth surfaces and uniform wall thickness. A great deal of effort has gone into making these hollow spheres meet stringent product specifications set by the fusion targets. In a coaxial-nozzle technique, similar to the liquid-droplet generation technique, ceramic hollow spheres are generated by passing a slurry of ceramic particles through an outer nozzle while a gas flows through the inner nozzle, and as a result spherical hollow (occupied by a gas; see Fig. 9) drops are formed due to surface tension and hydrodynamic forces. The size of the hollow sphere and wall thickness are determined by the diameter of the outer nozzle, the spacing between the inner and outer nozzles, and the slurry concentration. This process is capable of producing monosized hollow spheres with uniform wall thickness. Spheres with tight tolerances have been produced due to the uniformity of processing (variations in diameter of less than ±5%, spherodicity of ±4%, and bulk density variations of ±6%) [15]. The capabilities of different processing techniques and aspect ratios of hollow spheres produced by various techniques are given in Tab. 2.

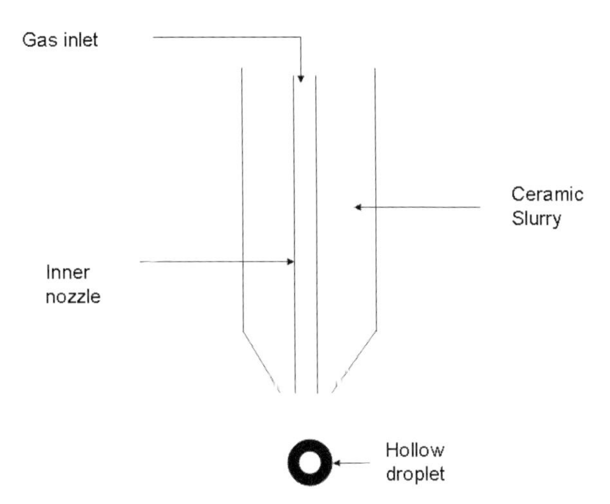

Fig. 9 Schematic of generation of hollow droplets by the coaxial-nozzle technique.

Table 2 Size and aspect ratio of ceramic hollow spheres produced by different techniques.

Fabrication technique	Range of hollow sphere diameter, nm	Range of aspect ratios
Sacrificial core	60–2 000 000	5–60
Layer-by-layer deposition	200–1 500	10–30
Emulsion/sol–gel	300–110 000	2–20
Spray or coaxial nozzle	100–6 000 000	2–100
Reaction-based	50–50 000	4–20

2.8.2.5 Reaction-Based and Other Methods

The interest in ceramic hollow spheres has increased tremendously in the last few years, as is evident from the host of new techniques developed for producing them [76–83]: radio-frequency (RF) plasma [76], assembly during reaction [77], gelation and hydrothermal treatment [78], chemical conversion of hollow templates (carbon nanotubes) [79], self-assembly [80], solid-state reaction [81], oxidation [82], and solution chemistry [83]. The mechanism of formation of hollow spheres from hollow templates is straightforward, whereas that of hollow-sphere formation in reaction-based methods remains to be investigated.

2.8.3 Cellular Ceramics from Hollow Spheres (Syntactic Foams)

Cellular ceramics have been made by sintering ceramic hollow spheres [84–86]. Sintered hollow glass spheres have been fabricated by sintering hollow glass spheres between 500 and 550 °C for various times [84]. The relative density (density of foam/density of the hollow-sphere wall material) of the resultant foams was in the range of 9–24 %. It was observed that the sintering of hollow spheres is a complex process, as there was no systematic correlation between the wall thickness of the spheres before and after sintering (thickness decreased in some cases and increased in others). Al_2O_3 foams with 10 % relative density have been produced by a two-step process [86]: sacrificial cores were coated with Al_2O_3 powder slurry followed by a second coating which connected the coated particles. The compact was dried, calcined at 600 °C, and sintered in air at 1600 °C for 1 h. Cylindrical foams with diameters in the range 2–4 mm have been made by this technique. The compressive strength of the individual hollow spheres, made from different cores, was in the range of 25–31 MPa. The compressive strength of foams made of hollow spheres was in the range of 1–4 MPa.

2.8.4 Properties

Very limited information is available on the properties of ceramic hollow spheres. Green [84] and Green and Hoagland [85] studied the mechanical properties of lightweight ceramics made by sintering hollow glass spheres. In addition, they proposed a micromechanical model, based on shell theory, to predict the elastic modulus and fracture toughness. The data predicted by the model were in reasonable agreement with the experimental data, especially in the nonlinear variation of elastic modulus and fracture toughness as a function of density. Chung et al. studied the compressive mechanical behavior of hollow spheres made of alumina by both experiments and finite-element modeling [87]. The sphere strength (the load per maximum cross-sectional area of the sphere at fracture) was shown to be proportional to the square of the relative density (density of sphere/density of sphere wall material), as given in Eq. (1)

$$\sigma_s = C\sigma_o \left(\frac{\rho_s}{\rho_o}\right)^2, \tag{1}$$

where σ_s is sphere strength, C a constant depending on the type of loading, σ_o the tensile strength of the wall material, ρ_s the density of the hollow sphere, and ρ_o the density of the sphere wall material.

The value of C was dictated by the stress analysis and fracture criterion used for different types of loading, and it is an order of magnitude larger for contact loading (0.774) than for concentrated loading (0.0275). The value of the constant determined from the experimental results was lower by a factor of three, and the discrepancy has been attributed to the imperfections in the sphere such as nonuniformity of wall thickness, imperfect sphere geometry, flaws, tails, and so on. The sphere strength increased from about 3 MPa to about 10 MPa when the relative density was increased from 0.05 to 0.14.

The optical properties of titania (TiO_2) hollow spheres used to make photonic crystals have been studied [88]. In a photonic band gap crystal, an electromagnetic wave cannot pass though it in any direction in a particular frequency range. The band gap of the photonic crystals can be controlled by controlling the wall thickness as well as the connectivity between the hollow spheres. Liu and Wilcox, Sr., modeled the dielectric behavior of cordierite ($2\,MgO \cdot 2Al_2O_3 \cdot 5\,SiO_2$) reinforced with silica–alumina hollow spheres over a frequency range of 1 kHz to 1 MHz [89]. The dielectric constant of cordierite decreased from 5.7 to 3.6 when the hollow sphere loading was increased from 0 to 48 vol% (equal to 37 vol% porosity). The Bruggeman effective medium theory was used to predict the dielectric constant of the composite, and the model predictions were in good agreement with the experimental results.

2.8.5 Applications

Cochran described potential applications of ceramic hollow spheres in different areas [90]. These range from individual spheres used as under water hydrophones, ultrasonic imaging, drug-delivery capsules, artificial cell carriers, and targets for inertial-confinement fusion to collections of hollow spheres in the form of syntactic foams, fillers in lightweight composites, photonic crystals, refractories, kiln furniture, energy-absorbing structures, and catalyst supports. Newnham et al. have extensively investigated the possibility of using ceramic hollow spheres, mainly lead zirconate titanate (PZT) and lead titanate, as single-element transducers and an array of transducers for exposimetry and tissue ablation, medical imaging, hydrophones, and underwater flat-panel arrays [15, 91–96]. The smaller the transducer the better would be the resolution. Ceramic hollow spheres have been investigated for reducing the thermal conductivity of materials used in radiant burners [14]. Cenospheres, hollow spheres largely composed of silica and alumina, are waste from coal-fired power plants and are being used for variety of applications such as automobile bodies, tires, insulating materials, road construction materials, landfill stabilizers, and so on [97]. Hollow porous microspheres have been developed for hosting catalysts [98]. Hollow spheres of SiC and SiO_2 have been investigated for possible use in preventing hypervelocity impact perforation, based on the superior energy-absorbing capability of hollow spheres [99]. In addition, hollow spheres have been investigated for various potential applications: cell carriers [100], for making colloidal crystals [101], for making ceramic foams [86], IR-transparent components [102], and hollow glass sphere/alumina composites [103].

2.8.6 Summary

The techniques developed for fabrication of ceramic hollow spheres have been divided into five main categories: sacrificial core, layer-by-layer deposition on a colloid template, emulsion/sol–gel, spray or hollow-droplet generation, and reaction-based methods. The essential features, advantages, and disadvantages of each method have been discussed. The properties of hollow spheres, especially compressive mechanical behavior of individual spheres, have been described. Ceramic hollow spheres have potential applications in medical imaging, underwater hydrophones, energy-absorbing structures, cell carriers, photonic band-gap crystals, thermal insulation, and syntactic foams.

References

1 D. L. Wilcox, Sr., M. Berg, T. Bernat, D. Kellerman and J. K. Cochran, Jr., Hollow and Solid Spheres and Microspheres: Science and Technology Associated with their Fabrication and Application in D. L. Wilcox Sr., M. Berg, T. Bernat, D. Kellerman and J. K. Cochran, Jr. (eds.) *Mater. Res. Soc. Proc.*, **1994**, *372*, 3–13.

2 P. J. Bruinsma, A. Y. Kim, J. Liu and S. Baskaran, *Chem. Mater.*, **1997**, *9*, 2507–2512.

3 P. Luo and T. G. Nieh, *Biomaterials*, **1996**, *17*, 1959–1964.

4 M. Iida, T. Sasaki and M. Watanabe, *Chem. Mater.*, **1998**, *10*, 3780–3782.

5 S. S. Jada, *J. Mater. Sci. Lett.*, **1990**, *9*, 565–568.

6 A. M. Gadalla and H.-F. Yu, *J. Mater. Res.*, **1990**, *5*[12], 2923–2927.

7 P. Tartaj, T. Gonzalez-Carreno and C. J. Serna, *Adv. Mater.*, **2001**, *13*[21], 1620–1624.

8 M. Ramamurthi and K. H. Leong, *J. Aerosol Sci.*, **1987**, *18*[2], 175–191.

9 L. Durand-Keklikian and R. E. Patch, *J. Aerosol Sci.*, **1988**, *19*[4], 511–521.

10 C. Roth and R. Köbrich, *J. Aerosol Sci.*, **1988**, *19*[7], 939–942.

11 V. Jokanovic, A. M. Spasic and D. Uskokovic, *J. Colloid Interface Sci.*, **2004**, *278*, 342–352.

12 L. B. Torobin, US patent no. 4,777,154, **1988**.

13 C. D. Hendricks, A. Rosencwaig, R. L. Woerner, J. C. Koo, J. L. Dressler, J. W. Sherohman, S. L. Weinland and M. Jeffries, *J. Nucl. Mater.*, **1979**, 85–86, 107–111.

14 A. T. Chapman, J. K. Cochran, T. R. Ford, S. D. Furlong and D. L. McElroy, Insulation Materials: Testing and Applications, 2nd Volume, ASTM STP 1116, R. S. Graves and D. C. Wysocki (eds.), American Society for Testing and Materials, Philadelphia, **1991**, pp. 464–475.

15 R. Meyer, Jr., H. Weitzing, Q. Xu, Q. Zhang, R. E. Newnham and J. K. Cochran, *J. Am. Ceram. Soc.*, **1994**, *77*[6], 1669–1672.
16 N. Kawahashi and E. Matijevic, *J. Colloid Interf. Sci.*, **1991**, *143*[1], 103–110.
17 H. Bamnolker, B. Nitzan, S. Gura and S. Margel, *J. Mater. Sci.*, **1997**, *16*, 1412–1415.
18 G. Zhu, S. Qiu, O. Terasaki and Y. Wei, *J. Am. Chem. Soc.*, **2001**, *123*, 7723–7724.
19 M. Shao, D. Wang, B. Hu, G. Yu and Y. Qian, *J. Cryst. Growth*, **2003**, *249*, 549–552.
20 Z. Huang and F. Tang, *Colloid Polym. Sci.*, **2004**, *182*, 1198–1205.
21 H.-P. Hentze, S. R. Raghavan, C. A. McKelvey and E. W. Kaler, *Langmuir*, **2003**, *19*, 1069–1074.
22 Y. Yin, Y. Lu, B. Gates and Y. Xia, *Chem. Mater.*, **2001**, *13*, 1146–1148.
23 Z. Yang, Z. Niu, Y. Lu, Z. Hu and C. C. Han, *Angew. Chem. Int. Ed.*, **2003**, *42*, 1943–1945.
24 H. Shiho and N. Kawahashi, *J. Colloid Interf. Sci.*, **2000**, *226*, 91–97.
25 G. C. Li and Z. K. Zhang, *Mater. Lett.*, **2004**, *58*, 2768–2771.
26 G. K. Sargeant, *Brit. Ceram. Trans.*, **1991**, *90*[4], 132–135.
27 J. G. Liu and D. L. Wilcox, Sr., *J. Mater. Res.*, **1995**, *10*[1], 84–94.
28 X. Yang and T. K. Chaki, *Mater. Sci. Eng.*, **1996**, *B39*, 123–128.
29 Q. Sun, P. J. Kooyman, J. G. Grossmann, P. H. H. Bomans, P. M. Frederik, P. C. M. M. Magusin, T. P. M. Beelen, R. A. van Santen and N. A. J. M. Sommerdijk, *Adv. Mater.*, **2003**, *15*[3], 1097–1100.
30 M. Wu, G. Wang, H. Xu, J. Long, F. L. Y. Shek, S. M.-F. Lo, I. D. Williams, S. Feng and R. Xu, *Langmuir*, **2003**, *19*, 1362–1367.
31 F. Caruso, R. A. Caruso and H. Möhwald, *Science*, **1998**, *282*, 1111–1114.
32 F. Caruso, M. Spasova, A. Susha, M. Giersig and R. A. Caruso, *Chem. Mater.*, **2001**, *13*, 109–116.
33 F. Caruso, *Chem. Eur. J.*, **2000**, *6*[3], 413–419.
34 F. Caruso, X. Shi, R. A. Caruso and A. Susha, *Adv. Mater.*, **2001**, *13*[10], 740–744.
35 F. Caruso, R. A. Caruso and H. Möhwald, *Chem. Mater.*, **1999**, *11*, 3309–3314.
36 R. A. Caruso, A. Susha and F. Caruso, *Chem. Mater.*, **2001**, *13*, 400–409.
37 M. Jaeckel and H. Smigilski, US patent no. 4,917,857, **1990**.
38 X. Yang and T. K. Chaki, *J. Mater. Sci.*, **1996**, *31*, 2536–2567.
39 A. Imhof, *Langmuir*, **2001**, *17*, 3579–3585.
40 Z. Zhang, Y. Yin, B. Gates and Y. Xia, *Adv. Mater.*, **2000**, *12*[3], 206–209.
41 S. Eiden and G. Maret, *J. Colloids Interf. Sci.*, **2002**, *250*, 281–284.
42 Y. Kobayashi, S. Gu, T. Kondo, E. Mine, D. Nagao and M. Konno, *J. Chem. Eng. Jpn.*, **2004**, *37*[7], 912–914.
43 C. Song, D. Wang, G. Gu, Y. Lin, J. Yang, L. Chen, X. Fu and Z. Hu, *J. Colloids Interf Sci.*, **2004**, *272*, 340–344.
44 L. Chen, Y. Gu, L. Shi, Z. Yang, J. Ma and Y. Qian, *Solid State Commun.*, **2004**, *130*, 537–540.
45 T. Kato, H. Ushijima, M. Katsumata, T. Hyodo, Y. Shimizu and M. Egashira, *J. Am. Ceram. Soc.*, **2004**, *87*[1], 60–67.
46 H. Huang, E. E. Remsen, T. Kowalewski and K. L. Wooley, *J. Am. Chem. Soc.*, **1999**, *121*, 3805–3806.
47 Y. Aoi, H. Kambayashi, E. Kamijo and S. Deki, *J. Mater. Res.*, **2003**, *18*[12], 2832–2836.
48 M. S. Fleming, T. K. Mandal and D. R. Walt, *Chem. Mater.*, **2001**, *13*, 2210–2216.
49 K. H. Rhodes, S. A. Davis, F. Caruso, B. Zhang and S. Mann, *Chem. Mater.*, **2000**, *12*, 2832–2834.
50 X. D. Wang, W. L. Yang, Y. Tang, Y. J. Wang, S. K. Fu and Z. Gao, *Chem. Commun.*, **2000**, 2161–2162.
51 G. Decher and J. B. Schlenoff, *Multilayer Thin Films: Sequential Assembly of Nanocomposite Materials*, Wiley-VCH, Weinheim, 2003.
52 H. G. Sowman, US patent No. 4,349,456, 1982.
53 D. L. Wilcox, J. G. Liu and J.-L. Loon, US Patent No. 5,492,870, 1996.
54 E. Bae, S. Chah and J. Yi, *J. Colloid Interf. Sci.*, **2000**, *230*, 367–376.
55 Y. Jiang, J. Zhao, H. Bala, H. Xu, N. Tao, X. Ding and Z. Wang, *Mater. Lett.*, **2004**, *58*, 2401–2405.
56 S. Chah, J. H. Fendler and J. Yi, *J. Colloid Interf. Sci.*, **2002**, *250*, 142–148.
57 S. Schacht, Q. Huo, I. G. Voigt-Martin, G. D. Stucky and F. Schuth, *Science*, **1996**, *273*[5276], 768–771.
58 A. M. Collins, C. Spickermann and S. Mann, *J. Mater. Chem.*, **2003**, *13*, 1112–1114.
59 H. Liu, Y. Ni, F. Wang, G. Yin, J. Hong, Q. Ma and Z. Xu, *Colloids Surf. A Physicochem. Eng. Asp.*, **2004**, *235*, 79–82.
60 C. Hendricks, *Glass Sci. Technol.*, **1984**, *2*, 149–168.

61 L. B. Torobin, US patent no. 5,212,143, 1993.
62 J. K. Cochran, Jr., US patent no. 4,867,931, 1989.
63 L. B. Torobin, US patent no. 5,397,759, 1995.
64 L. B. Torobin, US patent no. 4,743,545, 1988.
65 L. B. Torobin, US patent no. 4,637,990, 1987.
66 G. L. Messing, S.-C. Zhang and G. V. Jayanthi, *J. Am. Ceram. Soc.*, **1993**, *76*[11], 2707–2726.
67 G. L. Messing, T. J. Gardner and R. R., Ciminelli, *Sci. Ceram.*, **1983**, *12*, 117–124.
68 S. E. Pratsinis and S. Vemury, *Powder Technol.*, **1996**, *88*, 267–273.
69 Y. Lu, H. Fan, A. Stump, T. L. Ward, T. Rieker and C. J. Brinker, *Nature*, **1999**, *398*, 223–226.
70 T. Tani, N. Watanabe and K. Takatori, *J. Am. Ceram. Soc.*, **2003**, 86[6], 898–904.
71 K. Masters, Spray Drying Handbook, 5th ed., Longman Scientific & Technical, Essex, UK, **1988**, pp. 329–342.
72 G. Bertrand, C. Filiatre, H. Mahdjoub, A. Foissy and C. Coddet, *J. Eur. Ceram. Soc.*, **2003**, *23*, 263–271.
73 K. Kim and K. Y. Jang, *J. Am. Ceram. Soc.*, **1991**, *74*[8], 1987–1992.
74 M. Nogami, J. Hayakawa and Y. Moriya, *J. Mater. Sci.*, **1982**, *17*, 2845–2849.
75 R. L. Downs, M. A. Ebner and W. J. Miller in *Sol-Gel Technology for Thin Films, Fibers*, L. Klein, ed., Noyes Publications, New Jersey, **1988**, pp. 330–381.
76 Z. Károly and J. Szépvölgyi, *Powder Technol.*, **2003**, *132*, 211–215.
77 M. Yang and J.-J. Zhu, *J. Cryst. Growth*, **2003**, *256*, 134–138.
78 H.-P. Lin, Y.-R. Cheng and C.-Y. Mou, *Chem. Mater.*, **1998**, *10*, 3772–3776.
79 Y. Gu, L. Chen, Z. Li, Y. Qian and W. Zhang, *Carbon*, **2004**, *42*, 235–238.
80 T. He, D. Chen, X. Jiao, Y. Xu and Y. Gu, *Langmuir*, **2004**, *20*, 8404–8408.
81 J. Wang, J. Yang, Y. Bao and J. Sun, *Powder Technol.*, **2004**, *145*, 172–175.
82 Y.-L. Li and T. Ishigaki, *Chem. Mater.*, **2001**, *13*, 1577–1584.
83 K. Kosuge and P. S. Singh, *Microporous Mesoporous Mater.*, **2001**, *44–45*, 139–145.
84 D. J. Green, *J. Am. Ceram. Soc.*, **1985**, *68*[7], 403–409.
85 D. J. Green and R. G. Hoagland, *J. Am. Ceram. Soc.*, **1985**, *68*[7], 395–398.
86 I. Thijs, J. Luyten and S. Mullens, *J. Am. Ceram. Soc.*, **2003**, *87*[1], 170–172.
87 J. H. Chung, J. K. Cochran and K. J. Lee, Hollow and Solid Spheres and Microspheres: Science and Technology Associated with their Fabrication and Application in D. L. Wilcox Sr., M. Berg, T. Bernat, D. Kellerman and J. K. Cochran, Jr., eds., *Mater. Res. Soc. Proc.*, **1994**, *372*, 179–186.
88 J. G. Liu, D. L. Wilcox, Sr., *J. Appl. Phys.*, **1995**, *77*[12], 6456–6460.
89 R. Rengarajan, P. Jiang, V. Colvin and D. Mittleman, *Appl. Phys. Lett.*, **2000**, *77*[22], 3517–3519.
90 J. K. Cochran, *Curr. Opin. Solid State Mater. Sci.*, **1998**, *3*, 474–479.
91 O. M. Al-Bataineh, R. J. Meyer Jr., R. E. Newnham and N. B. Smith, Proc. IEEE 2002 Ultrasonics Symp., Munich, Germany, 8–11 Oct., 2002, *2*, 1473–1476.
92 S. Alkoy, J. K. Cochran, Jr. and R. E. Newnham, IEEE-ISAF Proc., Montreux, Switzerland, 1998, 345–348.
93 R. E. Newnham, S. Alkoy, A. C. Hladky, W. J. Hughes, D. C. Markley, R. J. Meyer and J. Zhang, Proceedings of MTS/IEEE Oceans 2001 Conference and Exhibition, Honolulu, HI, **2001**, *3*, 1529–1535.
94 R. Meyer Jr., S. Alkoy, R. Newnham and J. Cochran, Proceedings of the 1999 IEEE International Ultrasonics Symposium, Lake Tahoe, CA, 18–21 October **1999**, *2*, 1299–1302.
95 S. Alkoy, A. Dogan, A.-C. Hladky, P. Langlet, J. K. Cochran and R. E. Newnham, IEEE International Frequency Control Symposium, Honolulu, HI, 1996, 586–594.
96 G. Geiger, *Am. Ceram. Soc. Bull.*, **1994**, *73*[8], 57–61.
97 N. Wandell, *Am. Ceram. Soc. Bull.*, **1996**, *75*[6], 79–81.
98 L. B. Torobin, US patent no. 4,793,980, **1988**.
99 Y. Li, J. B. Li and R. Zhang, *Compos. Struct.*, **2004**, *64*, 71–78.
100 K.-L. Eckert, M. Mathey, J. Mayer, F. R. Homberger, P. E. Thomann, P. Groscurth and E. Wintermantel, *Biomaterials*, **2000**, *21*, 63–69.
101 P. Jiang, J. F. Bertone and V. L. Colvin, *Science*, **2001**, *291*, 453–457.
102 A. Krell, G. Baur and C. Daehne, *Proc. SPIE*, **2003**, *5078*, 199–207.
103 S. J. Wu and L. C. De Jonghe, *J. Mater. Sci.*, **1997**, *32*, 6075–6084.

2.9
Cellular Concrete

Michael W. Grutzeck

2.9.1
Introduction

Unlike conventional high-fired cellular ceramics discussed in earlier chapters, cellular concrete is one of the many ceramiclike materials that can be produced via a comparatively low temperature, chemically driven hydration reaction. Air-dried cellular concrete consists of two phases: a gaseous phase (gas bubbles, ca. 60 vol%) and void spaces/micropores (ca. 20 vol%) that were formally occupied by evaporable water, and a surrounding matrix phase consisting predominantly of calcium silicate hydrate formed by a dissolution/precipitation reaction that occurs during the reaction of Portland cement with water. The water has three functions. First it makes it physically possible to mix, aerate, and place the cellular material. Second, it acts as a solvent for the micrometer-sized anhydrous cement particles present in Portland cement. Third, it acts as a reagent that combines with the dissolved species in solution to form the insoluble hydrates that comprise the matrix material in cellular concrete. The hydrates that form, in descending order of importance are: calcium silicate hydrate, smaller amounts of calcium aluminate hydrates, and calcium hydroxide [1, 2]. Much like sintering causes a green ceramic body to develop its final properties, the root cause for hardening of cellular concretes is a series of complex hydration reactions.

The water consumed during hydration that becomes part of or closely associated with a newly formed hydrated phase is considered to be nonevaporable, because it occupies a lattice position or is chemically bound to a surface. Unlike excess water that remains in interstitial spaces and voids, water of hydration is normally not lost during drying at 105 °C [3]. Thus, total porosity of cellular concrete is defined as the volume of space occupied by the gas phase and evaporable water divided by the total volume of the sample [4].

In the strictest sense, cellular concrete is not a concrete. By definition, concrete contains (by volume) approximately one part Portland cement, two parts fine aggregate (sand-sized limestone or quartz), three parts course aggregate (centimeter-sized limestone or quartz), and enough water to make the combination workable during mixing and placing [5]. For this reason "cellular concrete" is somewhat of a misnomer. Cellular concrete is little more than a hardened Portland cement slurry that has been aerated prior to setting to give it a homogeneous void or cell structure con-

Cellular Ceramics: Structure, Manufacturing, Properties and Applications.
Michael Scheffler, Paolo Colombo (Eds.)

ISBN: 3-527-31320-6

taining 50–80 vol% or more of air bubbles, void spaces, and capillary porosity [6, 7]. (A slurry is a mixture of Portland cement and water. For sake of completeness, a mortar is a slurry that has been mixed with sand, and a concrete is a slurry that contains both sand and larger sized aggregate.)

Cellular concrete will undergo a process of setting and hardening much like its namesake. If the reaction is carried out below the boiling point of water, the calcium silicate hydrates that form are tens of nanometers in size; they have short-range order as observed by magic angle spinning nuclear magnetic resonance (MAS NMR), but only poorly organized long-range order [8]. It is suggested that the predominant calcium silicate hydrate that forms has a layerlike structure similar to certain clays (phyllosilicates) and is thus able to host interlayer water molecules [1–3]. These characteristics give the calcium silicate hydrates the distinction of being X-ray amorphous and are purportedly the cause of their gel-like properties. If cured above 100 °C in a pressurized vessel under saturated steam pressure, the calcium silicate hydrates that form during mixing and molding will begin to develop crystallinity, which makes them easier to analyze and understand in the thermodynamic sense, that is, their solubilities and phase equilibria are relatively straightforward [9].

The phases that form are determined by the bulk composition and final curing conditions of the mixtures, and each of the phases that form has a given set of characteristics. Some hydrates have desirable properties while others do not. For example, it has been repeatedly observed that a conventional cement paste cured significantly above 100 °C will develop calcium silicate hydrate phases that are unsound [1,2]. It is for this reason that autoclave-cured products are commonly augmented with 50 wt% or more of a silica source. This causes enough of a change in bulk composition that subsequent autoclave curing will lead to the production of a stronger phase [1, 2, 9]. Although phase relations at low temperatures are still uncertain, at this point in time enough data exist to allow one to piece things together and offer hypothetical versions of phase diagrams for the low-temperature phases occurring in the system $CaO–SiO_2–H_2O$ [1,8]. Phase compatibility and phase properties are important design tools that can be used to formulate cellular cements for different applications.

2.9.2 Types of Cellular Concrete

Curing temperature has the single greatest effect on the outcome of a given hydration reaction. If we adopt an arbitrary dividing line (e.g., ca. 100 °C) we can divide cellular concretes into two groups: one that consists of cast-in-place fill materials and precast masonry products that are cured at ambient or slightly elevated temperatures, and another that consists of precast masonry products and reinforced slabs and panels that are molded, precured to develop green strength, and then autoclaved at temperatures significantly higher than 100 °C in a pressurized, steam-heated environment. These two types of cellular concretes share a common name and have a similar appearance (see Fig. 1), but their chemistries are radically different.

Fig. 1 Cellular characters of the two cellular concretes are nearly the same. The one on the left is a cellular concrete made at ambient or slightly higher temperature (taken from A.J. Voton's web site, http://www.ajvoton.com/foam-concrete.html). The one on the right is a piece of aerated autoclaved concrete (AAC) made at 180 °C by Hebel GmbH.

2.9.2.1
Low Temperature Cured Cellular Concrete

Low temperature cured cellular concrete consists of gas bubbles and voids enclosed by a matrix that is, for the most part, identical to that associated with conventional hydrated cement products. In fact one can visualize a cellular concrete as being a mixture of a conventional Portland cement slurry or mortar mixed in a stationary mixer or ready-mix truck that has been blended with up to 80 vol% of air bubbles, voids, and micropores. Both foamed and unfoamed materials have the same phase composition; e.g. the calcium silicate hydrate in the cured products has a Ca/Si molar ratio close to 1.7. Because Portland cement is the main ingredient and other additives such as sand or gravel are essentially inert when cured at low temperatures, the calcium silicate hydrates that form in cellular concrete provide the same strength and impermeability to the matrix as would be expected to occur in a solid hydrated paste or mortar. The inclusion of a gas phase reduces the density of the material, making it lightweight, self-insulating, and soundproof, but at the same time the gas phase makes the cellular concrete less strong, more permeable to water vapor, and ultimately less durable than its more massive counterparts. Although not widely accepted at present in the USA it is anticipated that cellular concrete of all types will increase in popularity as its positive attributes become more widely known.

There are two categories of lightweight low-strength cellular concrete: those that are cast in place and those that are precast and cured in a mold or formwork with/without reinforcing (here collectively referred to as masonry). Wet densities of foamed slurries and mortar usually range from 100 to 900 kg m^{-3} for cement slurries and up to 1600 kg m^{-3} for mortars containing sand or fly ash. ASTM has two standards that are instructive in that they provide some background on cellular concrete as well as a description of appropriate test methods for making and testing hardened cellular concrete. ASTM C 796-97 (Standard Test Method for Foaming Agents for Use in Producing Cellular Concrete Using Preformed Foam) provides the reader with a composition and a means of producing foam and test samples; it is to be used in conjunction with ASTM C 869-91 (Standard Specification for Foaming Agents Used in Making Preformed Foam for Cellular Concrete). The former standard provides recipes for a Type I and a Type III Portland cement slurry that are designed to have wet densities within the 600–700 kg m^{-3} range when used with different preformed foaming agents, while the latter standard contains a table that lists the physical requirements of the test specimens themselves. Specifications for the mandated 0.58 water/cement ratio Type I and 0.64 water/cement ratio Type III cement slurries are as follows: compressive strength (1.4 MPa), tensile splitting strength (0.17 MPa), maximum water absorption (25 vol%), and maximum loss of air during pumping (4.5 vol %). Other variables that are determined are oven dry density and air content.

Cast-in-place cellular concrete is commonly used for geotechnical applications as a replacement for unstable soils, for example, nonslumping backfill for highway construction, especially around bridge approaches and bridge piers; backfilling of tunnels; and insulation of buried steam pipes and utility-containing culverts that may have to be reexcavated in the future [10]. When used as backfill, it is considerably more stable than clay-rich soil, which tends to creep with time, causing slabs cast on top of such soils to buckle and crack. When used as fill around steep bridge embankments it will not flow as would a clay rich soil. For example, densities of one set of foamed slurries studied by Cellular Concrete LLC ranged from 320 to 960 kg m^{-3}, and their associated strength ranged from 3.5 to 65 kg cm^{-2}. Production of a lightweight fill material is relatively simple. A surfactant/air entrainer or a preformed foam is added to a concrete mix in a ready-mix truck, allowed to mix to generate foam and/or distribute preformed foam evenly, and then poured into place where it sets and cures [10].

Lightweight masonry includes materials that are cast in place in formwork or precast in a factory and cured with low-pressure steam. Materials in this category include UL fire and FM rated roof decks, UL fire-rated soundproofing floor fill, leveling fill, cellular acoustic paneling, and impact/energy adsorbing block or panel [10]. Masonry formed by casting a slurry or mortar in formwork is commonly used in floor and roof systems worldwide. These serve as a foundation for the application of the final wearing course (e.g., compound roof, wood, or tile floor). Masonry products formed in a factory and then used elsewhere are less common in the USA. Although cellular concrete has great potential it tends to be less popular than equivalent Plaster of Paris based paneling or lightweight concrete made with lightweight aggregate (e.g., expanded perlite), but this trend may be changing. Cellular concrete is much more durable in wet places and thus better suited for bathroom and basement applications. In addition, cellular concrete masonry is able to adsorb energy caused by impacts of various types and adsorb traffic noise along highways when used as a sound wall. The cellular character and low strength of the material provides controlled energy adsorption as the cellular structure collapses. These masonry materials can be used for crash abatement at the bottom of hills or at the ends of airport runways as a means of stopping out of control vehicles and aircraft. Sound walls are another area where cellular concrete may find increased use once sensitivity to wet/dry and freeze/thaw damage of uncoated material are minimized.

2.9.2.2
Autoclave-Cured Cellular Concrete

The second type of cellular concrete is called autoclaved aerated concrete (AAC) because it is cured in a pressurized vessel of some type in the presence of free water at temperatures well above the boiling point of water. The most common curing chamber available to accomplish this is a steam-heated autoclave that is operated along the liquid/vapor curve of water at 180–190 °C. The development of final properties of reinforced precast AAC units such as a slab or a beam used for floors, walls, or roofs, or of a masonry product such as a block or a panel that is later assembled with thin-set mortar and used for exterior and interior walls or panels normally occurs in two steps: during precure and during final cure in an autoclave.

There are two main types of AAC used worldwide: those based on quartz flour and those based on Class F fly ash. Both are durable once coated with a waterproofing agent and both have the same physical and mechanical properties, but one is gray and the other white. In both cases, approximately 60 wt% or more of the siliceous material is blended with smaller amounts of Portland cement and lime and enough hot water to make a thin slurry. During the last minute of mixing, micrometer-sized aluminum metal flake is added to the mixture. The aluminum reacts with the caustic in the cement to form millimeter-sized hydrogen bubbles, which in turn cause the slurry to double or even treble in size. Slaking of the lime in the mixture and hydration of the cement causes the "cake" to thicken and harden in 45–60 min, that is, to develop green strength. The cake is strong enough at this point to be demolded and cut to size with wire saws. The blocks are then autoclaved at 180–190 °C and saturated steam pressure [11]. Although not very common in the USA, AAC is a very popular construction material used worldwide [12, 13].

RILEM has published a set of recommended test methods much like ASTM has done for cellular concrete [11]. There are a total of 27 such recommendations which cover the entire gamut of characteristics discussed below. ASTM is currently developing similar test procedures for AAC as part of its Committee C 27.60. To date, two have been accepted and published: ASTM C 1386-98 (Standard Specification for Precast Autoclaved Aerated Concrete (PAAC) Wall Construction Units) and ASTM C1452-00 (Standard Specification for Reinforced Autoclaved Aerated Concrete Elements). Work in progress includes a standard for measuring the modulus of elasticity of AAC.

Even though cellular concrete does not normally contain fine/coarse aggregate, the choice to call these aerated materials "concretes" was intentional. By associating them by name with conventional concrete it was implied that the aerated cellular cement pastes that comprise cellular concrete would have the same performance, albeit at a lower level, and the same durability as conventional concrete. Second, the use of concrete as a descriptor made these materials more acceptable to engineers, who are keenly aware of the fact that cement pastes shrink and crack and do not perform well in traditional engineering applications [5].

2.9.3 Per-Capita Consumption

If the per-capita consumption of cellular concrete masonry in all of its forms were ranked in descending order by geographical location, consumption in the USA would appear near the bottom of the list [12,13]. The regions appearing near the top would be Europe and Asia. One can only speculate as to the cause of this; more than likely it is a combination of economic and cultural factors. Cellular concrete products are currently more expensive in the USA than conventional concrete block; U.S. families tend to prefer wood framed houses over stucco-finished concrete block houses; and U.S. home owners tend to move every seven years or so and thus they consider price more so than long-term durability when choosing a residence.

Nevertheless, AAC is gaining increasing acceptance as a construction material for public and commercial buildings in the USA, because AAC is nearly maintenance free and thus more cost effective in the long term. Currently there are four U.S. producers and one Mexican distributor (Contec) based in Texas supplying the industry with AAC block and panel that is being used to construct factories, elementary schools, college dormitories, libraries, office buildings, hotels, and other commercial buildings. The history surrounding its introduction in the U.S. market is both interesting and telling; growing pains are clearly evident. Two successful German companies (Hebel and Ytong) built plants in the Southern part of the United States and for a period of ten years or so tried to establish an AAC presence in the U.S. They were not successful. There was little consumer acceptance. They subsequently sold their plants to U.S. companies who have continued to operate them. Ytong has become Aercon Florida LLC and Hebel has become Babb International in Georgia. ACCO Aerated Concrete Systems in Florida and E-Crete in Arizona were built by the current owners. At the same time EPRI constructed a mobile AAC plant that was used to manufacture on-site AAC block from the fly ash of ten coal burning power plants throughout the USA. They published three volumes containing analyses of the various fly ashes used, typical recipes, and a tabulation of the physical and mechanical properties of the AAC produced at each plant [14]. Open houses were held and samples were distributed, but very little interest was actually generated. In contrast, Mexico has the distinction of having the oldest continuously operating AAC plant in the Americas. Buildings made with their AAC have withstood countless earthquakes. Acceptance in Mexico is much higher than in the USA. Perhaps cellular concrete has been adopted by the Mexicans, Europeans, and Asians because it is both lightweight and self-insulating and is well suited as a replacement for solid stone and or cement block used in the stucco-finished housing common in these areas.

Whatever the cause may be, in general terms, the situation on a worldwide basis is very different from that in the USA. As acceptance by the commercial sector increases and the U.S. population is exposed to cellular concrete products and begins to adopt them, prices of cellular concrete masonry materials will come down and cellular concretes will be able to compete with wood in price. Given the current state of affairs for masonry, it is safe to say that the most commonly used cellular concretes in the USA at present are the cast-in-place lightweight geotechnical fill materials and controlled-strength fill materials used for construction purposes [6, 7].

2.9.4
Overview of Cellular Concrete

Cellular concrete is a composite material. The hydrous solid dictates its performance and properties; the gas phase dictates its specific gravity and ease of handling. The properties of the matrix phase can be manipulated almost infinitely by varying bulk composition, type of starting materials, water/cement ratio, and precuring and final curing conditions. The gas phase can be introduced in a number of ways during the mixing and precuring stage. Although the gas phase affects final properties of the composite, it is essentially considered to be inert filler.

2.9.4.1
The Gas Phase

The gas phase present in cellular concrete is introduced by using chemical air-entraining agents, pre-formed foams, and/or micrometer-sized aluminum flake. These substances are added to a Portland cement slurry during mixing to produce a stable array of gas bubbles in the slurry. The mechanisms of chemical air-entraining agents [15] are well known and are similar to what is discussed in earlier chapters of this book. Preformed foams are prepared by mixing a surfactant with water in a pressurized container and then adding it to the concrete. Such foams have densities in the 40–70 kg m^{-3} range [10]. Both foaming agents are normally used to foam cellular concretes that are cured at ambient temperatures or at temperatures well below 100 °C. The resulting cellular pastes and slurries can be cast in place, or can be precast and cured in the factory.

Aluminum flake reacts with the caustic solution that evolves during the hydration reaction to form hydrogen gas bubbles. The formation of portlandite $Ca(OH)_2$, which drives the reaction, is due to the fact that the calcium silicate and calcium aluminate hydrates that form during the hydration of Portland cement, that is, calcium silicate hydrate and the various calcium aluminate hydrates, contain less calcium that the starting materials. Because the reaction is a through-solution reaction, the excess calcium and hydroxyl ions remain in solution until concentrations significantly exceed the 20 mmol L^{-1} considered to be the solubility of $Ca(OH)_2$ in water, that is, the solution becomes supersaturated with respect to portlandite, which begins to precipitate as a separate phase. At this point the pH is approximately 12, which is caustic enough to make the reaction of aluminum proceed. Aluminum flake is not normally used to foam cast-in-place cellular concrete cured below 100 °C. Its use is restricted to applications that require production of foam after mixing and placing. For example, if one wishes to produce a reinforced masonry product containing reinforcing rebar or wire cages one has no choice but to introduce a thin slurry into the mold and allow it to rise and engulf the reinforcement. The process makes it possible to reinforce the product without introducing large air pockets along metal/cement contacts.

As one begins to experiment with the formulation of the extremely fluid AAC mixtures used by the industry, one soon arrives at the conclusion that the ability to produce a stable and homogeneous cellular structure with aluminum flake is actually more of an art than science. Viscosity of the slurry will dictate how the bubbles arrange themselves in space over time, but unlike many traditional cellular ceramics the viscosity of the slurry cannot be varied independently since water is a reactant and a certain minimal amount is necessary for mixing and placing and complete hydration of the starting materials. This makes the use of micrometer aluminum flake to foam these mixtures especially tricky, because the composition and viscosity of the solution phase varies while the aluminum flake is reacting. Chemical air-entraining mixtures are easier to use because the quantity of chemical added is usually correlated to the amount of foam that is produced. In both cases however ingredients do vary with time and thus a rigorous QA program must be in place to insure product uniformity.

2.9.4.2
The Matrix Phase

The matrix material that surrounds the air bubbles and voids is composed predominantly of calcium silicate hydrate. At low temperatures the hydrate is X-ray amorphous, difficult to study by methods designed for crystalline materials, and is easily altered or even decomposed under high vacuum. Thus, its structure is still unknown and the mechanisms of reaction responsible for its formation remain controversial. For this reason this phase is commonly referred to as "calcium silicate hydrate" and its chemical formula abbreviated as C–S–H [1,2], where C, S, and H are shorthand notations used by cement chemists to represent CaO, SiO_2, and H_2O, respectively. Written without hyphens this notation would imply that the hydrate had the composition $CaSiO_3 \cdot H_2O$ which is not correct; the hyphens in the formula signify that the composition of the hydrate is extremely variable, depending on the nature of the starting materials and composition of the coexisting solution phase.

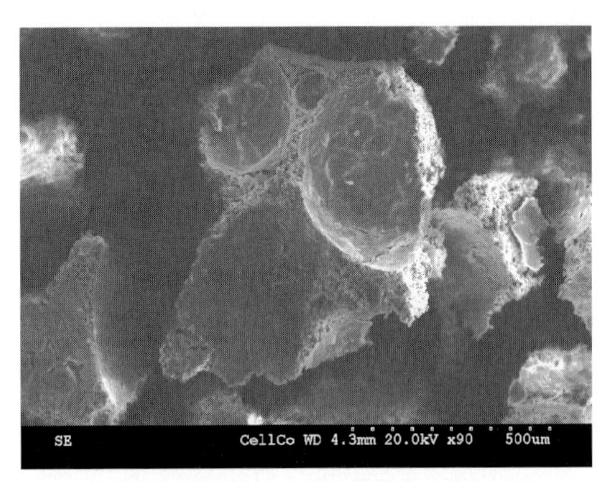

Fig. 2 Overview of a cellular concrete (ca. 425 kg m^{-3}) containing Portland cement, water, and a chemical foam that was cured at room temperature for weeks/months. The SEM micrograph shows that it consists of air bubbles dispersed in a C–S–H matrix (sample prepared by Cellular Concrete LLC).

C–S–H is a generic name for a substance commonly found in cement and concrete that behaves much like a solid solution in the classic phase-equilibrium sense, that is, it has a variable composition and were it represented on a phase diagram for the system $CaO–SiO_2–H_2O$ at 25 °C, for example, it would occupy a relatively large area rather than a specific point. The nominal composition of a water-saturated C–S–H is $Ca_{1.7}SiO_{3.7} \cdot 4\,H_2O$ [1]. The water of hydration is considered nonevaporable because it is held in place by hydrogen bonds and Coulombic forces in interlayers and on surfaces, and it requires temperatures well above 105 °C to be driven off. Conversely, water of mixing that is not consumed by the hydration reaction and now exists in interstitial cavities and in voids in between the hydrated and now hardened C–S–H particles can be removed by evaporative drying at ambient and slightly elevated temperatures. Richardson [16] has published a comprehensive article on the nature and morphology of C–S–H that forms in various C_3S and C_3S plus slag systems. The morphology of C–S–H is complex and varied depending upon its physical location with respect to the hydrating grains and the water-filled spaces around them. From electron micrographs (Figs. 2–5) it is clear that the C–S–H that grows into previously water filled spaces (outer product, Op) has a very high surface area consisting of fibrils and foils, as opposed to the inner product (Ip) hydrate that surrounds the hydrating grains themselves. The micrographs are presented in order of increasing magnification. Figure 2 shows the matrix/bubble arrangement in a low-temperature cellular concrete provided by Cellular Concrete LLC [10], Fig. 3 a magnified view of the interior surface of a bubble, Fig. 4 the microstructure of C–S–H that makes up the matrix, and Fig. 5 three TEM images of a single hydrating grain of C_3S, showing detail of inner and outer product C–S–H, taken from Richardson [16]. These provide a visualization of the geometry of the sample, the relation of its matrix microstructure to the bubbles in the sample, and a representation of the development of porosity in the matrix at various magnifications. Many authors have sug-

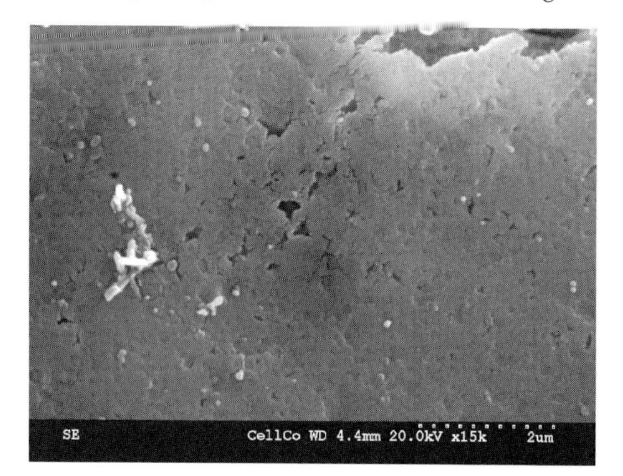

Fig. 3 Microstructure in bubble. The morphology seems to suggest that the water film in contact with the air has precipitated some hexagonal hydrate, probably $Ca(OH)_2$, which I assume has been carbonated, so that much of the individual detail usually associated with portlandite crystals is erased. The scale (denoted by 11 small dots in the labeling area) is 2 μm long.

Fig. 4 Microstructure in matrix between bubbles at two magnifications (scale: 5 μm and 20 μm). The rounded shapes represent C–S–H overgrowths on potentially unhydrated cores of clinker particles. As such the hydration reaction is only partially complete.

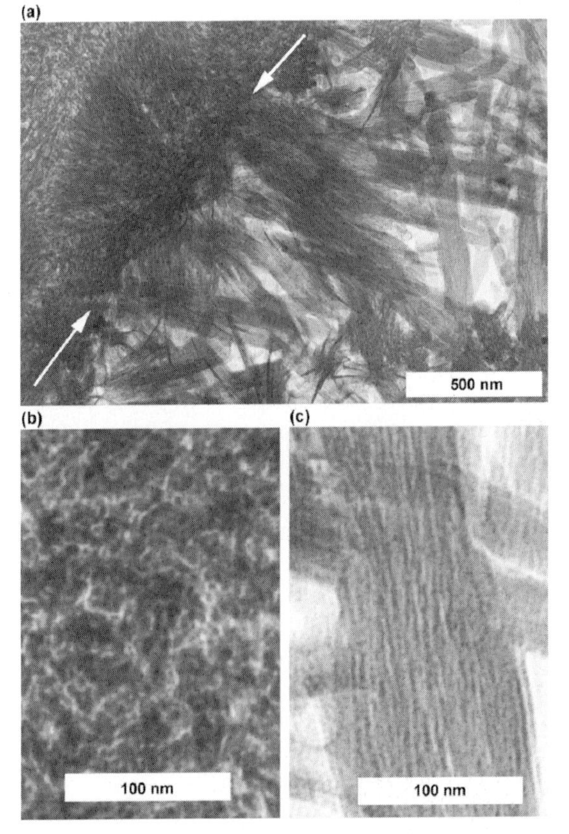

Fig. 5 a) TEM image showing Ip and Op C–S–H present in a hardened C_3S paste with $w/c = 0.4$ hydrated at 20 °C for eight years. White arrows indicate the Ip–Op boundary; the Ip is in the upper left of the micrograph. b) Enlargement of a region of Ip C–S–H. c) An enlargement of a fibril of Op C–S–H.

gested that the C–S–H that forms as outer product is a poorly crystalline analogue of a mineral known as tobermorite ($Ca_5Si_6O_{17} \cdot 5\,H_2O$) [1,2] and that the C–S–H has fractal geometry.

Although difficult to study, it is generally accepted that the outer-product C–S–H has a very high surface area (80–100 m^2/g) [17], a layerlike structure purportedly similar to those associated with tobermorite, but having more water ($Ca_5Si_6O_{17} \cdot 10\,H_2O$), and jennite ($Ca_9Si_6O_{21} \cdot 11\,H_2O$), exchangeable calcium ions, protons, and water molecules, and a variable Ca/Si molar ratio in the vicinity of 1.7 [1, 2, 16]. In addition, the C–S–H tends to lose its water of hydration gradually over a range of temperatures (100–200 °C), much like a gel would, rather than in discrete steps as a mineral would. As a result of these characteristics C–S–H has often been described as having "gel-like" behavior [3].

Very lightweight materials used as cast-in-place insulating material for underground steam and hot-water pipes or as geotechnical fill (controlled low-strength material) made with surfactants or preformed foams generally have densities in the 100–300 kg m^{-3} range [10]. Lightweight masonry that is prefabricated and cured below 100 °C must be strong enough to withstand normal cutting, transport, and handling. Materials such as lightweight masonry block, cellular paneling, and energy-absorbing materials have densities in the 200–600 kg m^{-3} range.

If these same materials were cured above 100 °C in a pressure vessel the previously X-ray amorphous C–S–H phases tend to crystallize as α-dicalcium silicate hydrate $Ca_2(HSiO_3)(OH)$. This phase has relatively little strength, so autoclaved concretes are usually formulated with large additions of Class F fly ash or quartz flour to achieve a Ca/Si molar ratio closer to 0.8, as found in tobermorite [11]. The tobermorite that forms has a distinctive morphology; it consists of intergrown bladelike crystals similar to those pictured in Figs. 6–9, again given in order of increasing magnification. Figure 6 shows matrix and bubbles in a sample of AAC made by

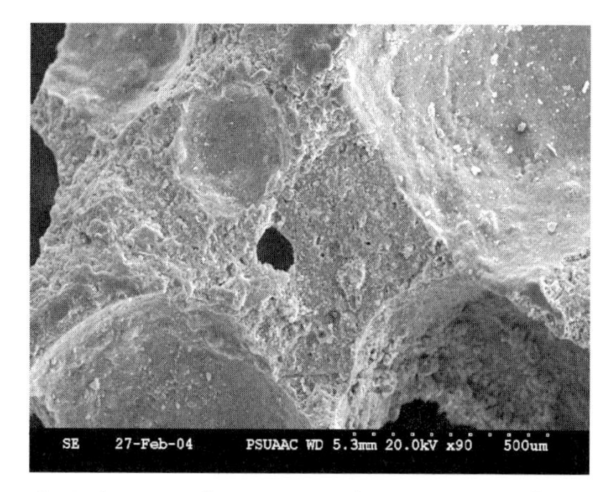

Fig. 6 Low-magnification image of quartz AAC manufactured by Asahi Chemical Co. Japan under licence to Hebel (scale: 500 μm). Voids formed by the bubbles and the tobermorite matrix between the bubbles can be seen.

Hebel, Fig. 7 the crystalline tobermorite that grows into open spaces inside bubbles, Fig. 8 the more compact tobermorite that grows in the matrix, and Fig. 9 TEM images taken from Ref. [16]. Further details on the nature of the crystallization process can be found in one of many articles by Mitsuda et al. [18], who have spent ten or more years determining the characteristics of tobermorite chemistry. It is easy to distinguish the two types of cellular concrete from one another in the SEM images, but not as easy in the TEM images. Tobermorite has a distinctive platy crystalline appearance. The suggestion that C–S–H could be tobermorite-like is most evident when comparing the TEMs. Here both crystals look similar, save for a distinctive striation on the C–S–H that is absent on the tobermorite.

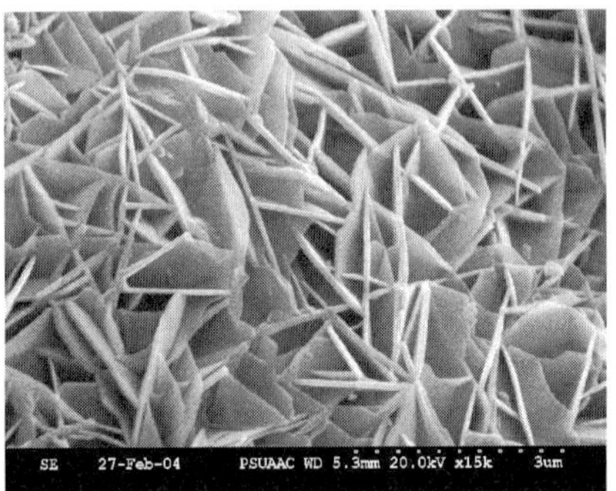

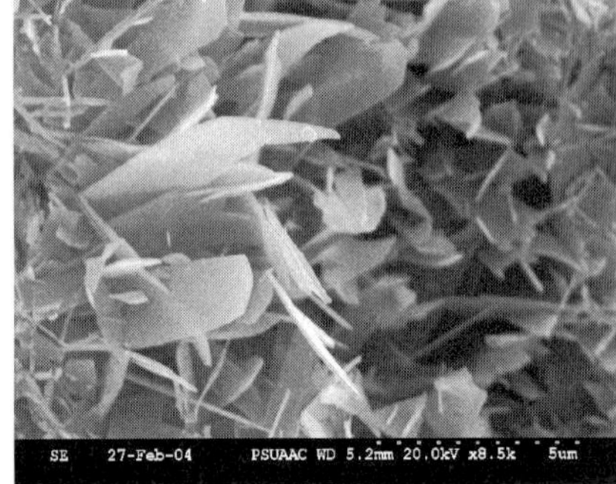

Fig. 7 Tobermorite crystals that grow into bubble voids. Plates are much better developed because they have free space to grow into (scale: 3 μm and 5 μm).

Fig. 8 Tobermorite that forms in the matrix has a less well defined microstructure, lower porosity, and hence appears more massive. (scale: 5 μm)

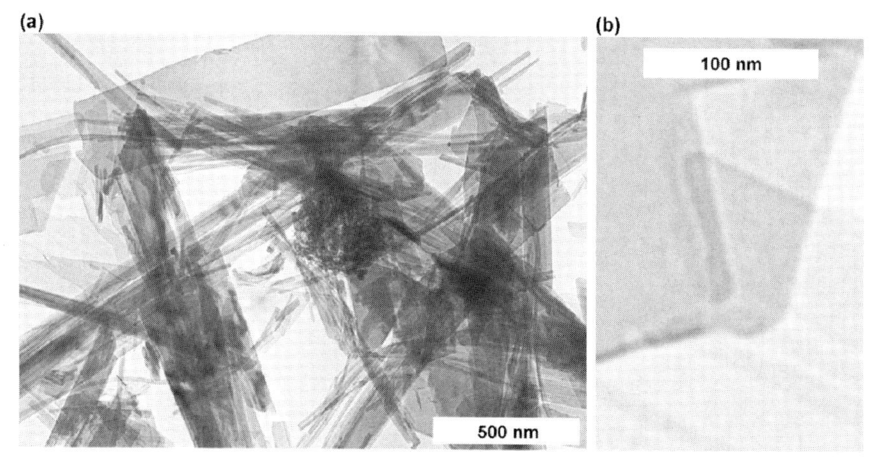

Figure 9 a) TEM image showing crystals of tobermorite.
b) Enlargement of overlapping crystals of tobermorite.

AAC formulations have bulk compositions that are significantly more silica rich than their low-temperature counterparts. Typical formulations based on quartz flour and fly ash are given in Tab. 1. Quartz flour AAC slurries normally contain a greater proportion of lime relative to Portland cement vis à vis an equivalent AAC made from Class F fly ash, which usually contains a greater proportion of cement and less lime.

Table 1 Formulation of a quartz AAC and a fly ash AAC (in wt %)

Ingredient	Quartz AAC		Fly ash AAC	
	w returns	w/o returns	w returns	w/o returns
Quartz flour	25.4	34.7	–	–
Dry class F fly ash	–	–	41.1	41.7
Dry returns	10.0	–	9.0	–
Portland cement	14.0	14.7	10.2	19.7
Quicklime	9.0	9.4	3.5	4.8
Al flake	0.05	0.04	0.04	0.07
Anhydrite (A)/gypsum (G)	1.8 (G)	1.8 (G)	1.7 (A)	1.7 (A)
Water	39.8	39.4	34.5	32.1

In both cases, the starting materials contain enough lime and Portland cement that the slurries will hydrate at room temperature and develop green strength, but at these temperatures the additives (quartz sand flour or Class F fly ash) will mostly act as inert fillers. However at elevated steam temperatures of 160–180 °C, these additives will combine with the lime-rich C–S–H phase in the green body to form tobermorite ($Ca_5Si_6O_{17} \cdot 10\,H_2O$). Densities of autoclave-cured masonry are in the

200–1000 kg m^{-3} range which is a direct consequence of varying the air content of the mixture [11].

Autoclave curing at 180–190 °C and saturated steam pressure causes the C–S–H formed at ambient temperatures in the green block to undergo varying degrees of crystallization. Sato and Grutzeck [19] proposed that crystallization occurs most easily from a lime-rich C–S–H precursor that forms during the initial setting of the cellular concrete. They suggest that poorly crystallized lime-rich C–S–H contains shorter silicate units [$(Si_2O_7)^{6-}$] that are more easily rearranged during higher-temperature curing than a comparable C–S–H already containing considerably longer silicate chains [$(Si_3O_9)_n^{6-}$]. Very fine silica such as a colloidal silica is so reactive (pozzolanic) that it causes tobermorite-like C–S–H with a Ca/Si ratio closer to unity to form in the green cake. This tends to inhibit further crystallization at autoclave temperatures. Conversely, the industry standard, a fine-grained silica sand (silica flour) is relatively unreactive. The C–S–H that forms tends to form around the cement grains and coat the quartz grains. As a consequence the C–S–H has a Ca/Si ratio near 1.7. Once this is autoclaved the C–S–H and remaining anhydrous ingredients combine to form a very crystalline tobermorite, similar to that in Figures 6–9. Clearly, small amounts of impurities as well as changes in silica source and granularity tend to alter the shape and size of the tobermorite crystals. The effect of such changes on transport properties is unknown.

The calcium aluminate hydrates that form and coexist with C–S–H are crystalline at all temperatures: $Ca_2Al_2O_5 \cdot 8\,H_2O$ which in shorthand becomes C_2AH_8, $Ca_4Al_6O_7 \cdot 13\,H_2O$ (C_4AH_{13}), and hydrogarnet $Ca_3Al_2O_6 \cdot 6\,H_2O$ (C_3AH_6) and are thus much easier to study than C–S–H by X-ray diffraction and electron microscopy. The final phase that forms in a hydrated cement is portlandite $Ca(OH)_2$ (CH). Portlandite forms as a byproduct of the decomposition and formation of C–S–H from tricalcium silicate Ca_3SiO_5 (C_3S) and dicalcium silicate Ca_2SiO_4 (C_2S). For more information on the nature of these and other intermediate phases found in cementitious materials the reader is referred to Taylor [1] and Lea [2].

2.9.5 Portland Cement

Because Portland cement and its hydration products are central to the production of cellular concrete, it is necessary to spend some time discussing its history, chemistry and properties. However, this is not an easy matter, because the chemistry of Portland cement hydration is daunting. Even after 100 years or so of commercial use in the USA there are still questions as to what actually happens in terms of dissolution, nucleation, and growth when Portland cement is hydrated [1, 2].

2.9.5.1
History

Portland cement was formulated, evaluated, and first commercialized in the UK in the mid 1800s. It was found that a mixture of certain argillaceous limestone could be fired at 1400–1500 °C to form a mixture of calcium compounds (phases) that would react with water, and given time would harden to form a rock-hard product much like the famed Portland stone quarried in the southern parts of England. Its name was meant to imply that Portland cement would be as good as or perhaps even better than Portland stone. As it turned out, it was as good as it was supposed to be and became immediately popular. Engineers began to use it to repair landmark structures and the rest is history. A revolutionary new material was born. The technology was brought to the USA in the early 1900 by entrepreneurs. It was referred to as a synthetic cement in contrast to what was then available in the USA, that is Rosendale and other natural cements made from argillaceous limestone deposits in Rosendale, NY and Louisville, KY. Once it was realized that Portland cement could be made from limestone and kaolinite clay, cement plants sprouted up all across the world [1,2].

2.9.5.2
Fabrication of Portland Cement

Like many commercially available cements the needed raw materials are often fired in a furnace of some type to dehydrate, vitrify, or otherwise make a substance that contains a large amount of latent free energy. In other words, once treated it will react with a solution of some sort and form hydrated solids that have the unique ability to intergrow and produce monolithic solids with useful physical and mechanical properties. Portland cement is currently made in rotary kilns equipped with a host of preheaters, coolers, recyclers, grinders, and filters needed to make them energy efficient and environmentally compliant. The kilns are fired at one end and fed with raw materials at the other. The limestone and clay move through the kiln and gradually become dewatered, decarbonated, sintered, and then react at 1450–1500 °C in the presence of a small amount of a liquid phase [2]. The now-agglomerated reaction product (5–20 mm) is called clinker. The clinker falls through a grate where it is cooled by incoming air. The clinker is then ground in ball mills containing steel balls and 3–5 wt% gypsum added to control premature setting. Grain size is such (ca. 20–50 μm) that surface areas typically are in the 2000–3000 $cm^2\ g^{-1}$ range. The Portland cement is bagged as it is or blended with a variety of active and, more recently, inert fillers. The active additives are known as pozzolanic materials. They are typically "glassy" in nature and are reactive with the lime in the pore solution that forms during hydration. Typically Class F fly ash and granulated blast furnace slag are added to the ground clinker. Natural pozzolans such as diamateous earth, Trass from Germany, pozzolana from Italy, and manufactured products such as metakaolinite and condensed silica fume (an industrial byproduct essentially equivalent to silica fume produced by various manufacturers such as Cabot Corporation in the USA and Degussa in Europe) can also be added [1,2]. Recently, there has been a move afoot to base standards governing the manufacture of Portland cement on performance rather than on chemistry alone. For example, clinker can be ground extremely fine, and this makes it more reactive in the first few days and weeks. Standards require a minimum compressive strength at 28 d. These cements react to such an extent that they are often stronger than they need to be. Therefore, the manufacturer can add 5–10 wt% limestone as inert filler. Unfortunately for the material scientist these "alterations" go unnoticed because the manufacturer is not required to label such ingredients on their bags.

Producing Portland cement clinker is a true "ceramic" process. In fact Portland cement clinker is the largest volume ceramic produced in the world. In the simplest case ground Portland cement clinker contains four major phases. Listed in order of decreasing abundance they are: tricalcium silicate (Ca_3SiO_5, C_3S, ca. 60 wt%), dicalcium silicate (Ca_2SiO_4, C_2S, ca. 20 wt%), tricalcium aluminate ($Ca_3Al_2O_6$, C_3A, ca. 10 wt%), and tetracalcium aluminoferrite ($Ca_4Al_2Fe_2O_{10}$, C_4AF, 10 wt%). Although C_3A is the most reactive phase (and the reason gypsum is added to the clinker during grinding; see Eq. 3 below) it plays a rather minor role in the overall property development of Portland cement based materials. By far the most important and consequently the most studied of the four phases is C_3S. When C_3S is mixed with water it undergoes dissolution, the solution reaches saturation/supersaturation, and nucleation begins, followed by rapid growth of highly insoluble hydrates.

2.9.5.3
Hydration

Once Portland cement is mixed with water the resulting viscoelastic mixture undergoes hydration and will harden with time. The anhydrous grains undergo surface dissolution, the solution that forms becomes supersaturated with respect to calcium ions, which in turn causes hydrated phases to nucleate and grow. The calcium silicate hydrates that form are collectively known as calcium silicate hydrate (C–S–H). Once the paste hardens it is called hydrated cement paste. The strength of the hardened cement paste will continue to increase until all of the anhydrous starting materials are consumed. This process can take as long as a year or considerably longer if particles are large and water content is small. Portland cement, like many cements can be mixed and then placed, since the reaction is not immediate (induction period) as in the case of mixing two homogeneous solutions such as Na_2SO_4 and $BaCl_2$ to instantaneously precipitate $BaSO_4$. Epoxies and other cements exhibit this same characteristic. In the case of Portland cement, the first C–S–H that precipitates from solution does so within minutes of mixing. Although it does not have the necessary morphology or layer structure needed to join the grains together [20], it does coat the grains and cause the reaction to become diffusion-controlled; it is suggested that this change is the cause of the induction period. The induction period of cement slurry is normally 3–5 h. The end of the induction period is signaled by a vigorous increase in the rate of heat evolution of the sample. The reaction that occurs is highly exothermic and has a very distinct footprint on the output of a conduction calorimeter. Figure 10 shows an example of a conduction calorimetric output for hydration of C_3S with water at 21 °C for 24 h [21].

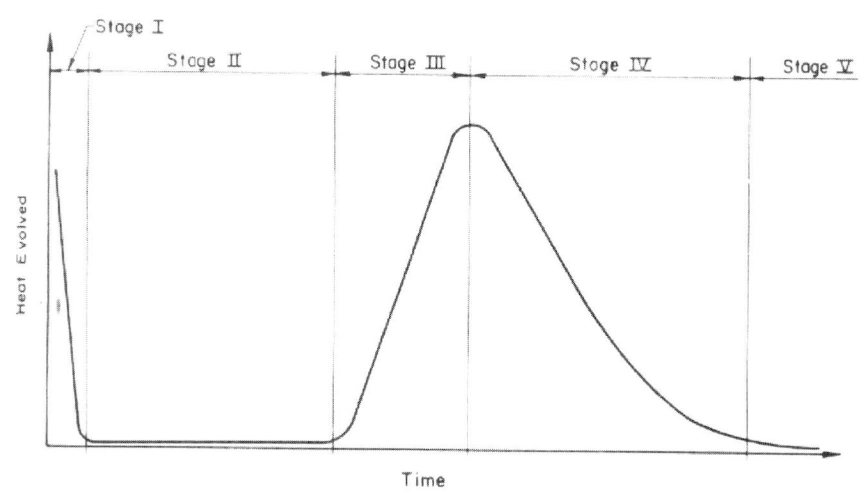

Fig. 10 Heat flux of hydrating tricalcium silicate as a function of time, determined by using an isothermal conduction calorimeter. Stage I = surface wetting, State II = induction period, State III = acceleratory period, State IV = deaccelatory period, State V = normal diffusion-controlled reaction. After Kondo and Daimon (published by Young [21]).

Portland cement continues to hydrate for days, months, and years. Water is important to the setting and curing process, and more harm than good comes from allowing a concrete to dry out. When Portland cement is hydrated the reaction is relatively slow and the development of the hydrated phases required to convert a slurry into a solid is much more gradual than the hydration of Plaster of Paris ($CaSO_4 \cdot 0.5\,H_2O$), for example, which goes to completion in approximately 15 min. Portland cement will harden in approximately 8 h. The first 3–4 h are considered the induction period. After entering what is called the acceleratory period, it normally takes 4-5 h for cement paste to harden to the touch. During the acceleratory period reaction rates accelerate as a second C–S–H having a layer structure begins to form. Compared to the initial sorosilicate-like C–S–H that forms initially, the surface area of the second hydrate is large, often greater than 100 m^2/g [17]. In addition, the C–S–H has a layer structure which contains Ca^{2+} ions, protons, and water molecules. Its morphology is often described as fibrillar to foil-like [16]. The developing hydrates intergrow and interlock, and when connectivities are great enough the hydrating grains become "glued" together, and thus cause the cement paste to harden. As larger quantities of C–S–H are formed water pockets become isolated from one another and the rate of reaction becomes diffusion-controlled. As the degree of joining increases so does the rigidity of the paste.

Hydration reactions for the four clinker phases in Portland cement are given below [5]:

$$2\,Ca_3SiO_5 + H_2O \rightarrow \text{C–S–H (nominally 3:2:8)} + 3\,Ca(OH)_2 \qquad (1)$$

$$2\,Ca_2SiO_4 + 5\,H_2O \rightarrow \text{C–S–H (nominally 3:2:8)} + Ca(OH)_2 \quad (2)$$

$$Ca_3Al_2O_6 + 3\,CaSO_4 \cdot 2\,H_2O + 26\,H_2O \rightarrow Ca_3Al_2O_6(CaSO_4)_3 \cdot 32\,H_2O \text{ (ettringite)} \quad (3)$$

$$Ca_3Al_2O_6(CaSO_4)_3 \cdot 32\,H_2O + 2\,Ca_3Al_2O_6 + 4\,H_2O \rightarrow 3\,Ca_3Al_2O_6(CaSO_4) \cdot 12\,H_2O \text{ (monosulfate)} \quad (4)$$

$$Ca_3Al_2O_6 + H_2O \rightarrow Ca_2Al(OH)_6 \cdot [OH,\ 3H_2O] + Ca_2Al(OH)_6 \cdot [Al(OH)_4 + 3\,H_2O] \rightarrow Ca_3Al_2(OH)_{12} \text{ (sulfate-free cement)} \quad (5)$$

$$Ca_4Al_2Fe_2O_{10} + Ca(OH)_2 + 14\,H_2O \rightarrow Ca_4(Al, Fe)OH_{12} \cdot [2\,OH \cdot 6\,H_2O] \quad (6)$$

In contrast Plaster of Paris ($CaSO_4 \cdot 0.5\,H_2O$, hemihydrate) combines with water to form a solid in approximately 15 min in a highly exothermic reaction (Eq. 7).

$$CaSO_4 \cdot 0.5\,H_2O + 1.5\,H_2O \rightarrow CaSO_4 \cdot 2\,H_2O \quad (7)$$

In this case the excess water will come off during drying because the gypsum crystals are much larger and do not have interlayers that are capable of containing chemically bound interstitial water. The only water molecules present are those within the structure itself, hydrogen-bonded to the Ca^{2+} ions in the gypsum lattice. Even the largest grains have been converted to gypsum and just a few remnant hemihydrate cores normally remain [22].

The currently accepted model for the structure of C–S–H was proposed by Taylor [1]. It proposes that the C–S–H that forms has a rudimentary layer structure, similar to that found in a phyllosilicate (i.e., clay), but without the hexagonal silicate structure of a phyllosilicate (i.e., C–S–H does not have Q_3 connectivity, as determined by ^{29}Si MAS NMR). Rather it is proposed that the C–S–H consists of a central layer of octahedrally coordinated Ca^{2+} ions (CaO_2^{2-} sheet) upon whose surface are bonded varying amounts of dimeric and longer polymeric silicate chains (dreierketten). Two such layers stacked next to each other would define an interlayer. The dreierketten have a repeat distance of three. The central silicate tetrahedron in a chain of three is the so-called bridging tetrahedron. It has two unpaired oxygen atoms that extend into the interlayer space. It is in this layer and in the vicinity of the oxygen ions that Ca^{2+} ions and protons can substitute for one another, leading to what can be considered to be a solid-solution series. The interlayer also contains varying amounts of molecular water. The Ca/Si ratio of the C–S–H will vary in response to the composition of the coexisting solution. Taylor proposed that the initial silicate units that attach themselves to the CaO_2 layer are predominantly dimeric ($Si_2O_7^{6-}$), but with time the silicate dimers tend to polymerize to form increasingly longer chains. Grutzeck has published a model for the reaction of C_3S and water which describes the earliest reactions in more detail [20]. It purports that there are two major C–S–H phases that form during the first 8–10 h of curing at ambient temperature. The first phase that forms has a sorosilicate-like structure with a Ca/Si ratio close to two that closely resembles a calcium-deficient jaffeite (tricalcium silicate hydrate $Ca_3Si_2O_7 \cdot 3\,Ca(OH)_2$ with loss of some $Ca(OH)_2$) similar in structure to a dimeric jennite C–S–H described by Richardson [16]. It is proposed that the sorosilicate-like C–S–H develops as a result of an equilibration process that occurs during the induction period, and that the initial hydrate formation during the first few minutes after mixing is a highly localized event that occurs in the vicinity of the anhydrous grains as a consequence of rapid dissolution and buildup of critically supersaturated concentration gradients around each grain. The C–S–H that precipitates is a highly disorganized, thermodynamically unstable monomeric hydrate that coats the grains and causes the reaction to become diffusion-controlled (i.e., causes the induction period). The hydrate now controls the solubility of the C_3S. It dissolves incongruently and thus begins to reorganize itself to form more stable dimers surrounded by double chains of octahedrally coordinated Ca atoms. The jaffeite structure contains molecular $Ca(OH)_2$, which is mobile and could conceivably be replaced by water molecules. It is proposed that the C–S–H will continue to exist until the coexisting solution phase becomes deficient in Ca ions, causing it to decompose. It forms via a phase transformation of the sorosilicate as a consequence of radical changes in coexisting pore solution chemistry as a consequence of $Ca(OH)_2$ precipitation. It is proposed that the jaffeite-like structure converts to a dimeric jennite-like structure, which has a layer structure that is similar in principle to that exhibited by a Ca-deficient tobermorite. Additional lime is lost by sorosilicate forming the jennite structure. With time jennite is joined by tobermorite-like C–S–H with gradually lengthening dreierkette as the system drives towards equilibrium. Although Grutzeck's model is empirical [20], it can be used to visualize a mechanism responsible for much of the kinetic data reported in the literature.

The detailed chemistry of Portland cement is far too complex and as such well beyond the scope of this chapter, but at the same time the general reaction is easy enough to visualize. The matrix phase in a cellular concrete forms as a result of hydrate formation around the entrained gas bubbles. The matrix is not glassy nor is it theoretically dense as in traditional ceramics. It is porous because the developing foils and fibrils that form on the surfaces of the anhydrous grains intergrow and interlock, and with time the hydrates replace both the anhydrous phases and the water that held the anhydrous starting materials in suspension during mixing and placing [20]. For further information on the chemistry of cement, see Refs. [1, 2, 5].

2.9.6
Properties of Calcium Silicate Hydrate in Cellular Concretes

Unlike high-fired ceramics, cellular concretes have an extremely large degree of latitude in terms of the ingredients that can be used to make them. They can be produced from a variety of materials with a variety of foaming agents. Cellular concretes can be manufactured from Portland cement, but they can also be manufactured from Portland cement mixed with waste materials such as Class F fly ash and ground blast-furnace slag. These additives are waste products that often have to be land-filled. Because of their pozzolanic character they actually aid the long-term performance of the cellular concrete. However, one should be aware of possible deleterious chemical interactions that might occur before proceeding with full-scale use of a given cellular concrete. One should at least evaluate the proposed combination experimentally. An example of a worst case scenario follows. A contractor was using a cellular concrete fill material as a base material for an asphalt-covered parking lot in Texas. The fill specifications included Portland cement and Class F fly ash. The contractor noted that a nearby coal burning power plant was literally giving its fly ash away so he used it in formulating his mixture. The mixture was allowed to cure for a day or so and then was coated with asphalt. Everything went smoothly.

However, the contractor neither understood nor realized that the fly ash was the product of flue gas desulfurization, and in addition to fly ash it also contained a great deal of gypsum and hannebachite. After sitting in the Texas sun for a month or so the sulfate and sulfide reacted with the calcium aluminate and the calcium silicate hydrates in the fill material and additional water adsorbed from the ground to form the expansive minerals known as ettringite and/or thaumasite. The asphalt covering heaved and buckled and had to be replaced.

2.9.6.1
Cast-in-Place or Precast Cellular Concrete

The strength of a low-temperature cellular concrete will decrease as the amount of porosity (void space occupied by air bubbles and evaporable water) increases, as proposed in 1972 by Hoff [4], who also provided an equation for calculating strength as a function of theoretical porosity of preformed foamed cellular concretes made with Type III and Class G oil-well cements. More recent data taken from a Cellular Concrete LLC brochure on cellular concrete [10] was used to produce the graphs in Figs. 11 and 12. The trends as a function of mixture density are similar to those published by Hoff. Curve-fitting programs were used to fit the data points to a curve. The equations are given, as is the goodness of fit.

The trend of the strength versus dry density curve of the cured cellular concrete tends to follow the same trend as that of AAC (see Fig. 13). However, the fact that the strength curve tracks the lower limit of the AAC curve suggests that a properly formulated low-temperature cellular concrete is not as strong as an a similarly formulated autoclaved equivalent. Hoff suggests that 280 kg m^{-3} may be the practical lower limit for such products cured at ambient temperature. Lower densities lead to such low-strength products that they tend to break when handled. Thermal conductivity increases with increasing density (Fig. 12), as does the permeability to and ability to absorb water. Water adsorption is on the order of 14–20 wt% after long-term immersion. The layer of matrix between each bubble is quite impermeable because it consists of a relatively dense C–S–H microstructure. Matrix thickness between walls in the cellular concretes pictured in Fig. 1 are on the order of 0.1 mm. Adding coarse-grained quartz or limestone sand to the mixture increases its density without effecting its resistance to water-penetration and freeze/thaw resistance. Adding a similar amount of Class F fly ash will compromise the isolated cellular structure

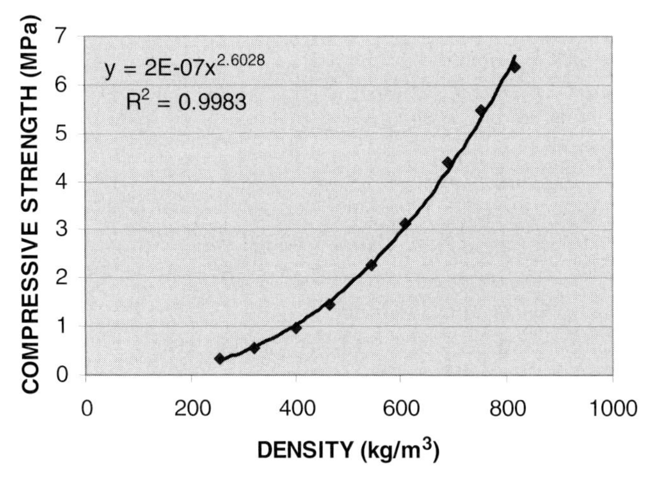

Fig. 11 Variation in compressive strength with density of low-temperature cellular concrete.

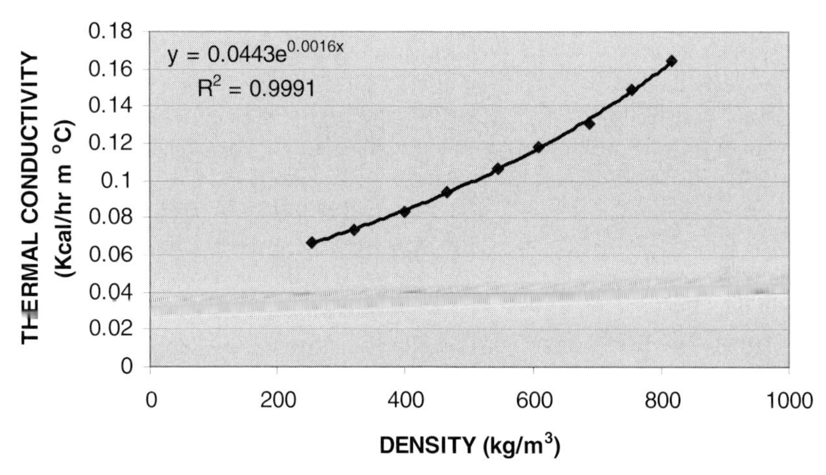

Fig. 12 Variation in thermal conductivity with density of low-temperature cellular concrete.

somewhat in the short term. Both water adsorption and freeze/thaw behavior are significantly degraded [6]. It is suggested that the relative percentage of interfaces between the sand and the cement paste are relatively small when compared to a finer grained material such as fly ash, which has a much larger surface area. It is proposed that the fly ash is a reactive pozzolan, and as each fly ash sphere dissolves it develops a relatively porous interface with the matrix C–S–H, similar to those observed in a normal cement/fly ash paste [23]. The random association and potential overlap of these interfaces throughout the matrix reduce the size of the wall separating individual bubbles. In the worst case bubble connectivity could increase as more fly ash is added. Cellular concretes containing fly ash are generally more prone to freeze/thaw due to their higher permeability to water [6, 7].

The recipes used by the various manufactures of cellular masonry are proprietary, and references to cellular concrete in the open literature are limited. However, patent literature on this subject is abundant, mostly concentrated in the Russian patent office. The microstructure of cellular concrete is unimpressive. The C–S–H that forms is noncrystalline with a microstructure containing porosity and an occasional large crystal of $Ca(OH)_2$. SEM images of C–S–H in a cellular concrete cured at low temperature are given in Figs. 3 and 4. For comparison, the microstructure of a Class F fly ash AAC sample is shown in Figs. 7 and 8.

2.9.6.2
Autoclaved Aerated Concrete (AAC)

When mixed, autoclave-cured cements are normally much more fluid than cellular concretes. This is somewhat misleading, however, because they also contain CaO, which is slaked soon after mixing and thickens the mixture after 5–6 min. Metal powders used to foam AAC formulations are normally aluminum-based. The aluminum powder is ball-milled, and particle size is critical in achieving a specified bubble size. Envision the following: A micrometer-sized powder of thin aluminum flakes is introduced into an AAC slurry at the very last minute of mixing to provide the necessary hydrogen bubbles to make the slurry take on a highly cellular character. The flakes must be thin rather than round to provide a fine edge where the gas bubbles originate. The making of a particular size of bubble is necessary to minimize the merging of bubbles into very large ones that will move upward and be lost to the atmosphere. The viscosity of the mixture must be such that it begins to thicken during gas evolution and thus captures the bubbles before they migrate upward in the mix.

AAC has environmental advantages. Because it is tobermorite-based it does not need as much lime-containing ingredients to produce as does cellular cement paste. In addition it consists of 80 % void space and only 20 % solids, so 4 m^3 of AAC can be produced from 1 m^3 of starting material [11]. Final curing occurs in a steam-heated autoclave, so emissions of CO_2, CO, and NO_x during curing are minimal when compared to high firing of ceramics in a kiln. Furthermore, waste materials such as Class F fly ash and ground granulated blast-furnace slag, as well as trimmings from the green block during sample sizing and condensate from the boiler, can be recycled into new AAC mixtures [11, 23–26].

Typically the ingredients are mixed and the slurry is poured and allowed to harden for 45–60 min, at which point the material has enough green strength to be demolded and cut to size with wire saws. The slurry normally contains ball-milled quartz flour (10–20 μm), Class F fly ash, or micrometer-sized copper mine tailings (the three silica sources used in the USA at present), Portland cement, vertical-kiln-fired lime that is slow to slake, recycled trimmings, and additional water if necessary. The silica sources are mixed with water maintained at about 60 °C in constantly stirred thousand-gallon open tanks equipped with overhead stirrers. Waste material is similarly slurried in another tank. It consists of the waste material trimmed from the green cakes prior to autoclaving. The mixer is charged with known quantities of raw and recycled materials, to which are added the Portland cement and then the lime in a series of timed steps. Typical formulations are given in Tab. 1. Mixing is continuous and quite strong, forming a vortex in the middle of the mixer. Some manufactures add gypsum to their mix at this point, followed closely by a preslurried dose of aluminum powder. After pouring the mixture begins to develop the alkalinity needed to for the aluminum powder react. Ideally this occurs after the lime has slaked and the mixture has thickened slightly, making bubble retention easier. The mix can expand two- or threefold. Gas bubbles and capillary pores occupy as much as 80 % of the void space in the mixture. Densities can range from about 200 to 1000 kg m^{-3}, and associated strengths increase with density. As density increases so does the amount of matrix material present; in this case the amount and kinds of aluminum added control the outcome. One hour after being poured into a massive preheated (38–40 °C) steel mold coated with an oily release agent, the AAC has evolved into a cake. In some cases the mold is not attached to the base and is slightly tapered to facilitate its upward removal by using a lift. When casting reinforced panels in molds that are up to 6 m or more in length, the mold has a removable side and the beam is tipped out of the mold and manipulated so that the direction of rise is horizontal prior to cutting with piano-wire saws. Autoclaves are heated with superheated steam. Commercial autoclaves are manufactured by a variety of companies. Essentially they are large steel cylinders 40–50 m or more in length and nearly 3 m in diameter. Ideally they have doors at each end that allows the manufacturer to roll the green cake into one end of the autoclave while removing the cured material from the other. The AAC is cast on small steel carts mounted on railway tracks. An autoclave can take many cycles of green cakes. Because a heating cycle takes about 24 h for heat/cure/cool, cakes tend to accumulate. Some will sit for many hours while the last made enter the autoclave. Like a ceramic (and depending on whether or not the piece is reinforced with wire cage, the cake is preheated for up to 6–8 h as the temperature and pressure are raised to 180 °C. Curing lasts for 8 (without metal reinforcement) to 12 h (with metal reinforcement), and cooling takes 6–8 h, too. The manufacturer can adjust the density of the AAC by adding or subtracting Al flake, which in turn affects the amount of air included in the sample.

The final AAC material consists of approximately 80 vol % void space and 20 vol % crystalline tobermorite [27]. Unlike Portland cement cured at room temperature, AAC consists of the crystalline calcium silicate hydrate mineral known as 1.1 nm tobermorite ($Ca_5Si_6O_{17} \cdot 5\,H_2O$) [18, 28]. Although it is suggested that this mineral forms a poorly crystalline analogue at room temperature called C–S–H, this has never been proven [1].

Three classes of AAC based on compressive strength are commonly manufactured: low-, medium-, and high-strength. Table 2 summarizes the general characteristics of these classes of AAC [11]. As is the case for low-temperature cellular concrete, the density of AAC is also directly related to strength. Rilem [11] has published a plot of AAC strength versus dry density, reproduced here as Fig. 13. Other density-related properties that similarly increase with declining strength are thermal and acoustic insulating value [11].

Table 2 Classification of AAC according to characteristic compressive strength [11]

Property	Low	Medium	High
Compressive strength/MPa	< 1.8	1.8–4.0	> 4.0
Young's modulus/MPa	< 900	900–2500	> 2500
Density/kg m^{-3}	200–400	300–600	500–1000
Thermal conductivity, dry/W m^{-1} K^{-1}	< 0.10	0.6–0.14	> 0.12

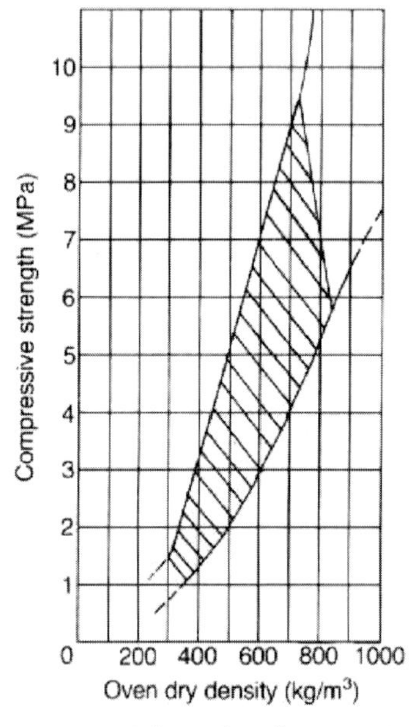

Figure 13 Relationship of compressive strength to dry density. For any two equivalent density (dry) AAC and low temperature cellular concrete samples the expected strength of the autoclaved sample should always be higher [11].

AAC has mechanical properties that depend on its water content. After autoclave curing, AAC contains about 25–35 % (as much as 45 % in very low density materials) water by weight of the dry material [11]. Strength at this point is at its lowest. Some manufacturers test their samples at this point and base their formulations on meeting wet strength and the requirements of their class. Others test their samples when they are dried to about 10 wt % moisture. Strength increases as water is lost by the sample. The average water content of AAC after 1–3 years of service reaches an "equilibrium" value of about 3.5 wt %. Ninety percent of AAC samples in use have less than 5 wt % moisture content. Figure 14 depicts the relative compressive strength of an AAC sample as it dries [11]. The insulating value increases with the degree of drying [29–32] of AAC, and thus it is proposed that the degree of soundproofing will also increase as a sample dries.

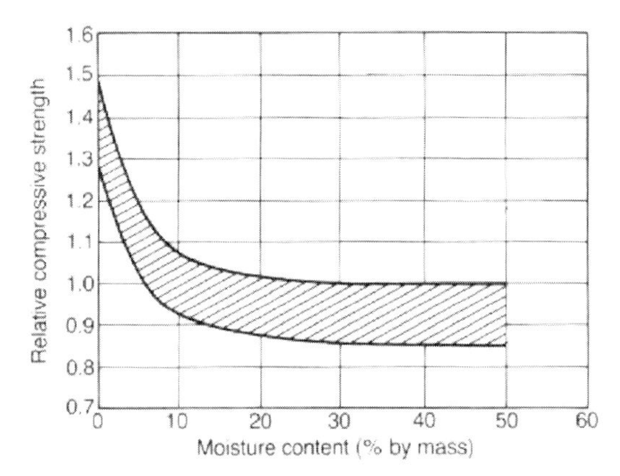

Fig. 14 Moisture content versus relative strength [11].

Strength and gas permeability are extremely sensitive to direction of rise [29]. Bubbles tend to rise somewhat, and the upper portion of the AAC is commonly slightly less dense than the rest of the sample. For this reason testing is always done perpendicular to direction of rise. Direct tensile strength is 15–35 % of compressive strength, and modulus of rupture (MOR) of AAC is approximated by Eq. (8), where f_{ct} is compressive strength (in MPa) [11]:

$$\mathrm{MOR} = 0.27 + 0.21 f_{ct}. \tag{8}$$

The shear strength is approximately 20–30 % of the compressive strength [11].

The use of fracture mechanics as a predictive tool for cellular concrete, AAC, and concrete products in general is still in its infancy, although attempts have been made with a variety of models [11] with a fair degree of success and controversy. In terms of linear elastic fracture mechanics, the toughness of AAC can be compared and contrasted by measuring the critical strain energy release rate G_C (in N m^{-1}) and critical stress intensity factor or fracture toughness K_{IC} (in MN m$^{-3/2}$) [30–32]. These parameters can be equated to each other for "plane-stress conditions" by Eq. (9)

$$K_{IC} = \sqrt{E G_C} \tag{9}$$

where E is the elastic modulus [11]. Representative fracture toughness data versus density are given in Fig. 15.

Nonlinear fracture mechanics have also been used to calculate toughness (fictitious crack model), but in this instance curve fitting (strain softening) and area measurement under the entire force–crack mouth opening displacement diagram are used to derive values [11, 33–35] which do in fact provide additional insight for well-defined ceramic systems, but for cement-based cellular concrete systems, the results should be viewed with some caution when comparing samples from different sources prepared in different ways, because of the uncertainty introduced by

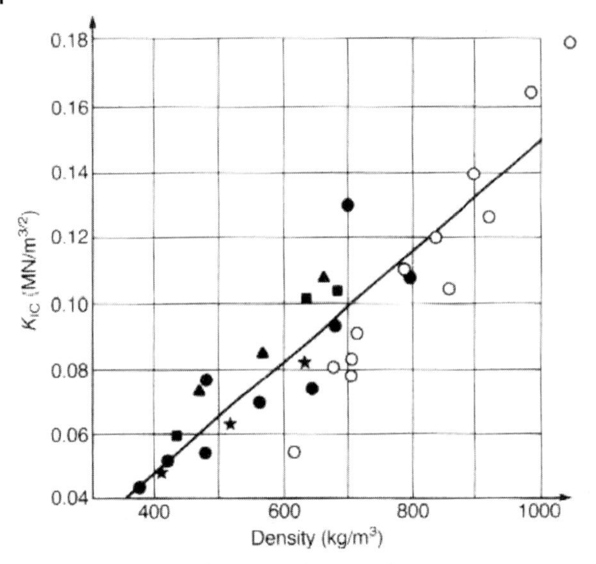

Fig. 15 Fracture toughness as a function of density. The normal block produced in the USA has a density of 400–600 kg m^{-3}. Its fracture toughness is approximately 0.05–0.08 MN m$^{-3/2}$. Adding zeolites increases this value slightly [11].

multiple cracking [36]. On the other hand, if all measurements are carried out on the same samples as a function of a single variable such as AAC rising direction, then the results of nonlinear fracture mechanics are informative, and in this case illustrated that the brittle character of the samples does vary as a function of the rising direction of the AAC cake [35].

Toughness can be increased by judicious use of fiber reinforcement. Fibers can be introduced during mixing [36] or can be grown in situ during hydrothermal curing [37]. Asbestos-reinforced cement is a perfect example of the benefit derived from such a process. In a post-asbestos world, however, fiber reinforcement has never been the same. Glass, steel, cellulose, and plastic fibers have been introduced into all mannerer of cementitious material with varying degrees of success [36]. There is an ongoing effort to find a mineral replacement that can be used in lieu of asbestos. For example, Low and Beaudoin [38] used fibrous wollastonite to reinforce C–S–H pastes, and Grutzeck has grown zeolites in Class F fly ash AAC by augmenting the mix with significant amounts of sodium hydroxide [37]. To date just a slight increase has been documented (unpublished data).

AAC has an interlocking microstructure. Although it is able to float, it is not impermeable. Water will enter the structure, but relatively slowly [39–41]. The microstructure shown in Figs. 7 and 8 contains openings that are in the range of 0.03 μm in size, which compares to the pore size in the "green" block of close to 6 μm [28]. The coefficient of water permeability K ranges from 10^{-12} to 10^{-13} m^2, while gas permeability for a dry AAC sample is approximately 10^{-14} m^2 [11]. A typical absorption/desorption isotherm for water in AAC is given in Fig. 16. Note that the plot is extremely nonlinear. Its shape is reminiscent of those of clays [42] and could be related to the inclusion of liquid water in interlayers at very high humidities.

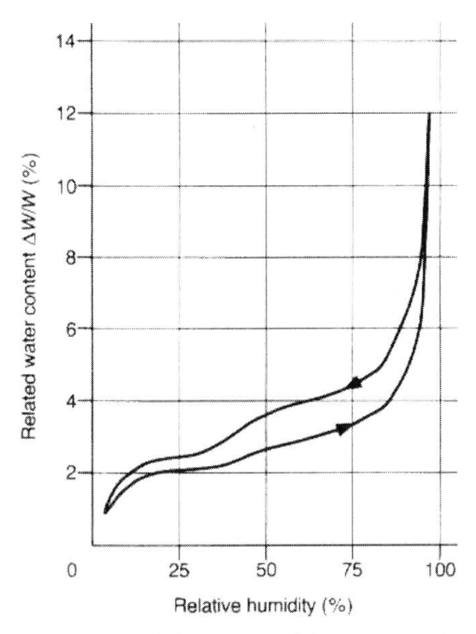

Fig. 16 Typical absorption and desorption isotherm of AAC [11].

Due to the sensitivity of AAC to water, the material is always coated with stucco or some type of waterproofing before it is exposed to water or high humidity [6, 7, 11]. Durability is not a problem if the AAC is continually frozen (arctic climate) or air-dried (arid climate), but it is an issue in temperate climates where conditions cycle many times during a season. Unexpectedly, the resistance of uncoated AAC to freeze/thaw and wet/dry cycling is not impressive compared to air-entrained ordinary Portland cement (OPC) concrete, even though it contains 50 vol % air bubbles.

2.9.7 Durability of Cellular Concrete

Although bubbles and voids in AAC are not interconnected, bubbles and pockets of evaporable water are normally isolated from one another by a rim of hydration products. The thickness of these rims and their relation to each other and how they grow together to form the matrix ultimately determines the permeability of the cellular concrete. As the density of AAC is decreased (air content is increased) the thickness of the matrix wall decreases and thus resistance to water ingress also decreases. In all cases water will slowly permeate the material with time, but it will do so more quickly in low-density material [43, 44]. Since water can freeze in bubbles filled with water and expand and cause cracks or shrink during drying, uncoated AAC is not considered durable when exposed to freeze/thaw and wet/dry cycling as occur in temperate climates. Cellular concretes exposed to environmental wet/dry and freeze/thaw cycles must be coated to prevent deterioration [11, 43, 44]. Unfortunately this sensitivity will increase the cost of AAC materials used in large-volume applications such as sound walls along highways and thus deter its use.

Moisture transport in AAC is not yet fully understood. Modeling studies have been carried out and models proposed [11, 39–41]. Interestingly, freeze/thaw resistance is related to bubble size and the water content of the air bubbles. A conventional AAC sample undergoing 100 freeze/thaw cycles lost 12 wt% of surface material, while a companion sample with very small bubbles lost 3 wt% [6]. Hasegawa [41] and Senbu and Kamada [43] both reported that freezing of water in matrix pores caused little damage. What did cause damage, however, was the presence of water in the large pores (ca. 40% filled). When these froze the AAC sample began to spall. Compared to a conventional concrete that does not contain bubbles, freezing that occurs in the paste pores normally causes damage by expansion when the water freezes. However, if it is air-entrained (surfactant is included that provides 3–5% air bubbles in the mix), the freezing water will expand into the air bubbles and damage will be averted. In AAC the damage is similarly averted if water content is low. But once the air bubbles become partially filled, then freeze/thaw damage will occur [43].

Higher porosity leads to an overall increase in permeability and thus an increase in susceptibility to the effects of freezing and thawing. Permeability and freeze/thaw damage increases as the density of the composite decreases. It is known that the addition of a pozzolanic material such as Class F fly ash and /or a reactive silica source to a Portland cement paste will increase long-term performance. The smaller particles usually fit between the hydrating cement grains and help the hydration products to "bridge the water-filled gap" between cement particles in the paste. These additives also provide a more tortuous path, since they react with the excess lime to produce more C–S–H. Reactivity of these additives (pozzolana) at ambient temperatures is limited in the short term; therefore, they often reduce early strength, but then enhance it over the long term. Finally, the degree of reactivity is generally correlated with temperature. Cellular concrete cast and cured at ambient temperatures matures over a period of weeks to months. Pozzolanic additives are usually considered relatively inert. Autoclaved concrete cured above 100 °C matures in a day. In this instance the reactivity of the additives is comparable to that of the cement itself. Coating these materials with waterproofing materials and/or paints or stucco prevents ingress of water and thus makes them more durable [41, 44]. Many companies worldwide deal in such coating materials.

2.9.8 Summary

Cellular concrete is an umbrella term for a variety of cast-in-place structural fill and precured masonry materials. These materials are generally light in weight, soundproof, and resistant to failure due to bacterial, insect, and fire damage. Cellular concrete is used throughout the world to varying degrees. In Europe and Asia this technology has been generally adopted for construction of housing. Its use in the USA for this purpose is limited but now on the rise. Cellular concrete used as low-strength fill and soil-replacement materials continues to be the most commonly used product in the USA. Alternative uses such as aircraft arresting materials and sound barriers along highways are applications that could become more widespread as costs are reduced. Cellular concrete can be made by chemical reactions that can compete with more conventional high-fired materials. The area is one with great promise and should become more popular in the future as wood and energy costs escalate and limiting CO_2 production becomes more important to the USA.

Cellular concrete occupies a rather small segment of the much larger literature on concrete. The literature normally pertains to the use of concrete in civil and infrastructure applications. Rather than delve into the mechanical performance data in greater detail than I have and instead focus more fully on the underlying science, I now refer the reader to those journals and texts that deal with the engineering aspects of concrete for further information. I also suggest that the reader refer to the extensive bibliographies produced by Houst and Wittmann [45] and Tada [46]. These summarize the literature on AAC for the period 1949–1992. Having the knowledge set forth in this chapter as a guide, one should now be better able to decide what is important and what is not.

Acknowledgement

The support of National Science Foundation (NSF GRANT CMS-9988543) is gratefully acknowledged.

References

1 Taylor, H.F.W., *Cement Chemistry*, 2nd ed., Thomas Telford Publishing, London 1997, 459 pp.
2 Lea, F.M., *The Chemistry of Cement and Concrete*, 3rd ed., Edward Arnold Publishers, London 1970, 727 pp.
3 Powers, T.C., Brownyard, T.L., *J. Am. Concr. Inst.*, **1947**, *43*, 101–992.
4 Hoff, G.C., *Cem. Concr. Res.*, **1972**, *2*, 91–100.
5 Young, J.F., Mindess, S., Gray, R.J., Bentur, A., *The Science and Technology of Civil Engineering Materials*, Prentice Hall, Upper Saddle River, NJ 1998, 384 pp.
6 Pospisil, J.D., *Testing and Evaluation of Cellular Concrete for Freeze-Thaw Durability*, MS Thesis in Civil Engineering, The Pennsylvania State University, University Park, PA 1998, 130 pp.
7 Tikalsky, P.J., Pospisil, J., MacDonald, W., *Cem. Concr. Res.* **2004**, *34*, 889–893.
8 Grutzeck, M.W., Benesi, A., Fanning, B., *J. Am. Ceram. Soc.* **1989**, *72*, 665–668.

9 Roy, D.M., Harker, R.I., 4th ISCC Washington 1960, Vol 1, 1962, p. 196; Buckner, D.A., Roy, D.M., Roy, R., *Am. J. Sci.* **1960**, *257*, 132–147; Harker, R.I., Roy, D.M., Tuttle, O.F., *J. Am. Ceram. Soc.* **1962**, *45*, 471–473, Pistorius, C.W.F.T., *Am. J. Sci.* **1963**, *261*, 79–87; Harker R.I., *J. Am. Ceram. Soc.* **1964**, *47*, 521–529; Roy, D.M., Johnson, A.M., Proceedings Symposium Autoclaved Calcium Silicate Building Products, London 1965, Soc. Chem. Indust., London 1965, p. 114; Kalousek, G.L., *J. Am. Concr. Inst.* **1954**, *25*, 365–378; Luke, K., Taylor, H.F.W., *Cem. Concr. Res.* **1984**, *14*, 657–662.

10 Cellular Concrete LLC, Roselle Park, N.J. (http://www.cellular-concrete.com) supplied a sample of ambient temperature cured cellular concrete which I used to obtain the SEMs presented in Figures 2–5, as well as a CD containing Technical Bulletins of various kinds that I used to augment the text and prepare the plots in Figures 11 and 12.

11 Aroni, S., de Groot, G.J., Robinson, M.J., Svanholm, G., Wittman, F.H. (Eds.), *Autoclaved Aerated Concrete, Properties, Testing and Design, RILEM Recommended Practice, RILEM Technical Committees 78-MCA and 51-ALC*, E & FN Spon, London 1993, 404 pp.

12 W. Dubral, "On Production and Application of AAC Worldwide" in *Advances in Autoclaved Aerated Concrete*, F.H. Wittmann (Ed.), A.A. Balkema, Rotterdam, 1992, pp. 3–9.

13 G. Båve, "Regional Climatic Conditions, Building Physics and Economics" in *Autoclaved Aerated Concrete, Moisture and Properties*, F.H. Wittmann (Ed.), Elsevier Scientific Publishing Co., Amsterdam, 1983, pp. 1–12.

14 Electric Power Research Institute (EPRI), *Environmental and Physical Properties of Autoclaved Cellular Concrete, Vols. 1–3*, Prepared by the University of Pittsburgh, R.D. Neufeld and L.E. Vallejo (Principal Investigators), EPRI, Pleasant Hill, CA, 1996.

15 Ramachandran, V.S., *Concrete Admixtures Handbook*, Noyes Publications, Park Ridge, NJ, 1984, 626 pp.

16 Richardson, I.G., *Cem. Concr. Res.* **2004**, *34*, 1733–1777.

17 Grutzeck, M., LaRosa-Thompson, J., Kwan, S., "Characteristics of C–S–H Gels" in Proceedings of the 10th International Congress on the Chemistry of Cement, Vol. 2, Gothenburg, Sweden, June 2–6, 1997, H. Justnes (Ed.), Amarkai AB and Congrex Göteborg AB, Göteborg 1997, paper 2ii067, 10 pages.

18 T. Mitsuda, T. Kiribayashi, K. Sasaki, H. Ishida, "Influence of Hydrothermal Processing on the Properties of Autoclaved Aerated Concrete" in *Autoclaved Aerated Concrete*, Wittmann, F.H. (Ed.), A.A. Balkema, Rotterdam, 1992, pp. 11–18.

19 H. Sato, M.W. Grutzeck, "Effect of Starting Materials on the Synthesis of Tobermorite" in Materials Research Society Symposium Proceedings, Advanced Cementitious Systems: Mechanisms and Properties, December 2–4, 1991, Glasser, F.P., McCarthy, G.J., Young, J.F., Mason, T.O., Pratt, P.L. (Eds.), Materials Research Society, Pittsburgh, 1992, pp. 235–240.

20 Grutzeck, M. W., *Mater. Res. Innov.* **1999**, *3*, 160–170.

21 Young, J. F., *Cem. Concr. Res.* **1972**, *2*, 415–433.

22 NIST Web Site http://visiblecement.nist.gov/plaster.html

23 Grutzeck, M.W., Roy, D.M., Scheetz, B.E., "Hydration Mechanisms of High-Lime Fly Ash" in Portland Cement Composites in Effects of Fly Ash Incorporation in Cement and Concrete, Proceeding Symposium N, Materials Research Society Annual Meeting November 1981, Diamond, S. (Ed.), Materials Research Society, University Park, PA, pp. 92–101

24 N. Kohler, "Global Energetic Budget of Aerated Concrete" in *Autoclaved Aerated Concrete, Moisture and Properties*, Wittmann, F.H. (Ed.), Elsevier Scientific Publishing Co., Amsterdam, 1983, pp. 13–26.

25 D. Hums, "Ecological Aspects for the Production and Use of Autoclaved Aerated Concrete" in Autoclaved Aerated Concrete, Wittmann, F.H. (Ed.), A.A. Balkema, Rotterdam, 1992, pp. 271–275.

26 J. Lutter, "New Research on the Primary Energy Content of Building Materials" in *Autoclaved Aerated Concrete*, Wittmann, F.H. (Ed.), A.A. Balkema, Rotterdam, 1992, pp. 277–281.

27 Isu, N., Teramura, S., Ido, K., Mitsuda, T., "Influence of Quartz Particle Size on the Chemical and Mechanical Properties of Autoclaved Lightweight Concrete" in *Autoclaved Aerated Concrete*, Wittmann, F.H. (Ed.), A.A. Balkema, Rotterdam, 1992, pp. 27–34;

Isu, N., Teramura, S., Ishida, H., Mitsuda, T., *Cem. Concr. Res.* **1995**, *25*, 249–254.

28 Mitsuda, T., Sasaki, K., Ishida, H., "Influence of Particle Size of Quartz on the Tobermorite Formation" in *Autoclaved Aerated Concrete*, Wittmann, F.H. (Ed.), A.A. Balkema, Rotterdam, 1992, pp. 19–26; Isu, N., Teramura, S., Ido, K., Mitsuda, T., "Influence of Quartz Particle Size on the Chemical and Mechanical Properties of Autoclaved Lightweight Concrete" in *Autoclaved Aerated Concrete*, Wittmann, F.H. (Ed.), A.A. Balkema, Rotterdam, 1992, pp. 27–34.

29 Wägner, F., Schober, G., Mörtel, H., *Cem. Concr. Res.* **1995**, *25*, 1621–1626.

30 Millard, W.R., "The Thermal Performance of European Autoclaved Aerated Concrete" in *Autoclaved Aerated Concrete*, Wittmann, F.H. (Ed.), A.A. Balkema, Rotterdam, 1992, pp. 83–88.

31 Frey, E., "Recent Results on Thermal Conductivity and Hydroscopic Moisture Content of AAC" in *Autoclaved Aerated Concrete*, Wittmann, F.H. (Ed.), A.A. Balkema, Rotterdam, 1992, pp. 89–92.

32 Lippe, K.F., "The Effect of Moisture on the Thermal Conductivity of AAC" in *Autoclaved Aerated Concrete*, Wittmann, F.H. (Ed.), A.A. Balkema, Rotterdam, 1992, pp. 99–102.

33 Zhou, W., Feng, N., Yan, G., "Fracture Energy Experiments of AAC and its Fractal Analysis" in *Autoclaved Aerated Concrete*, Wittmann, F.H. (Ed.), A.A. Balkema, Rotterdam, 1992, pp. 135–140.

34 Slowik, V., Wittmann, F.H., "Fracture Energy and Strain Softening of AAC" in *Autoclaved Aerated Concrete*, Wittmann, F.H. (Ed.), A.A. Balkema, Rotterdam, 1992, pp. 141–145; Alvaredo, A.M., Wittmann, F.H., "Influence of Fracture Energy on Failure of AAC-Elements" in *Autoclaved Aerated Concrete*, Wittmann, F.H. (Ed.), A.A. Balkema, Rotterdam, 1992, pp. 147–151.

35 Trunk, B., Schober, G., Helbling, A.K., Wittmann, F.H., *Cem. Concr. Res.* **1999**, *29*, 855–859.

36 Bentur, A., Mindess, S., *Fibre Reinforced Cementitious Composites*, Elsevier Applied Science, London, 1990, 449 pp.

37 Grutzeck, M.W., Kwan, S., DiCola, M., *Cem. Concr. Res.***2004**, *34*, 949–955.

38 Low, N.M.P. and Beaudoin, J.J., *Cem. Concr. Res.* **1994**, *24*, 250–258.

39 Daïan, J.-F., Bellini da Cunha, J.A., "Experimental Determination of AAC Moisture Transport Coefficients under Temperature Gradients" in *Autoclaved Aerated Concrete*, Wittmann, F.H. (Ed.), A.A. Balkema, Rotterdam, 1992, pp. 105–111.

40 Wittman, X., Sadouki, H., Wittmann, F.H., "Determination of Hydral Diffusion Coefficients of AAC – A Combined Experimental and Numerical Method" in *Autoclaved Aerated Concrete*, Wittmann, F.H. (Ed.), A.A. Balkema, Rotterdam, 1992, pp. 113–118.

41 Hasegawa, T., "Investigation of Moisture Contents of Autoclaved Lightweight Concrete Walls in Cold Districts" in *Autoclaved Aerated Concrete*, Wittmann, F.H. (Ed.), A.A. Balkema, Rotterdam, 1992, pp. 125–132.

42 Grim, R.E., *Clay Mineralogy*, McGraw-Hill Book Company, New York, 1968, 596 pp.

43 Senbu, O., Kamada, E., "Mechanism of Frost Deterioration of AAC" in *Autoclaved Aerated Concrete*, Wittmann, F.H. (Ed.), A.A. Balkema, Rotterdam, 1992, pp. 153–156.

44 Hama, Y., Kamada, E., Tabata, M., Watanabe, T., "Frost Resistance of Increased Density Autoclaved Aerated Concrete" in *Autoclaved Aerated Concrete*, Wittmann, F.H. (Ed.), A.A. Balkema, Rotterdam, 1992, pp. 157–163.

45 Houst, Y., Wittmann, "Bibliography of Autoclaved Aerated Concrete" in *Autoclaved Aerated Concrete, Moisture and Properties*, Wittmann, F.H. (Ed.), Elsevier Scientific Publishing Co., Amsterdam, 1983, pp. 325–369.

46 Tada, S., "Bibliography on Autoclaved Aerated Concrete 1982–1992" in *Autoclaved Aerated Concrete*, Wittmann, F.H. (Ed.), A.A. Balkema, Rotterdam, 1992, pp. 325–360.

Part 3
Structure

Cellular Ceramics: Structure, Manufacturing, Properties and Applications.
Michael Scheffler, Paolo Colombo (Eds.)

ISBN: 3-527-31320-6

3.1
Characterization of Structure and Morphology

Steven Mullens, Jan Luyten, and Juergen Zeschky

3.1.1
Introduction and Theoretical Background

3.1.1.1
The Importance of Foam Structure Characterization

Ceramic foams, also named cellular ceramics or porous ceramics, offer a series of unique properties because of their combination of a highly porous cellular structure and a tailored composition of the strut material. Porous ceramics are promising candidates for a wide range of engineering applications. As the internal structure of these materials can vary widely, there is a diversity of fields in which these products are already being used or in which they can be of substantial benefit [1].

When considering the internal, three-dimensional architecture of foam materials, one can use a number of characterization parameters such as cell size distribution and cell morphology (for example anisotropy), window opening, strut thickness, shape and length, interconnectivity, porosity, and type of porosity (open versus closed). Figure 1 shows some of these structural parameters for the open and closed cell units.

In the last few years, several innovative manufacturing routes have led to a wide diversity of ceramic foam materials, with respect to the materials of which they are composed, specific structure, and mechanical strength. The window of properties and characteristics for each technology can be tailored to some extent by means of

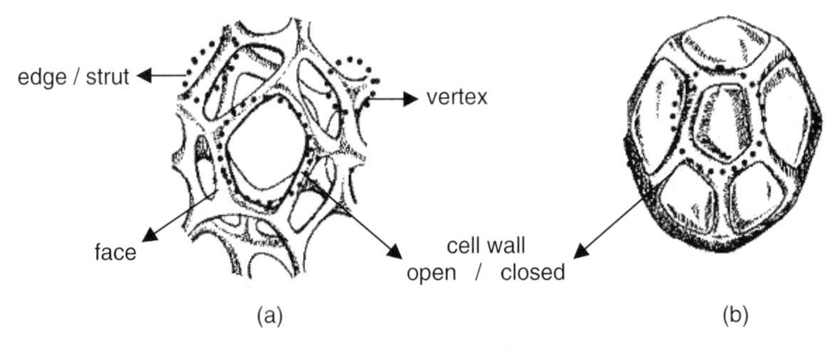

Figure 1 Some components of the structure of a cell unit in a) open and b) closed foams.

Cellular Ceramics: Structure, Manufacturing, Properties and Applications.
Michael Scheffler, Paolo Colombo (Eds.)

ISBN: 3-527-31320-6

the experimental conditions during processing, depending on the desired properties for a specific application.

The ability to tailor the pore structure is exemplified by comparing the strut morphology of ceramic foams produced by replication of reticulated polyurethane and by direct foaming of a ceramic suspension with a gelling agent [3]. Figure 2a shows the triangular-shaped strut with a hole in the center that results from pyrolysis of the polymer skeleton. Clearly, this feature will affect properties such as the mechanical behavior of the foam. In contrast, direct foaming of a ceramic suspension produces a material with dense struts (Fig. 2b).

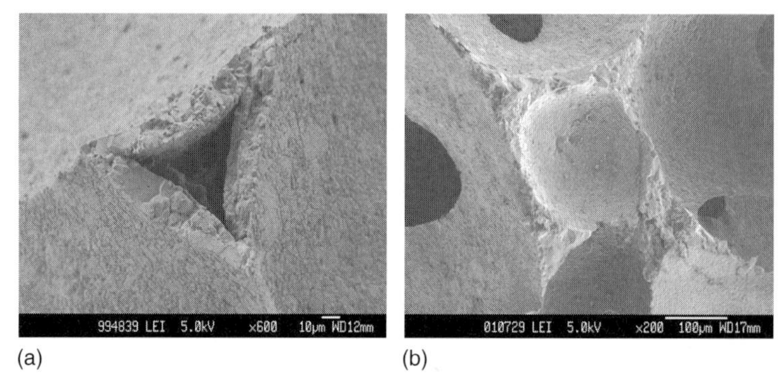

Figure 2 Strut morphology for a foam made by a) replication of reticulated polyurethane foam and b) direct foaming of a ceramic suspension.

As the manufacturing and processing parameters determine the ultimate properties of the foam structure and thus its domain of application, accurate and quantitative characterization of the morphology and the internal architecture of the foam is essential both for product development, manufacturing, and end-use.

3.1.1.2
Structure-Dependent Properties

Ceramic foams significantly extend the range of properties of materials that are available to engineers, because they have an unique combination of characteristics. Table 1 lists some of the properties of ceramic foams directly related to their typical micro- and macrostructure [1].

The relative density, defined as the foam density divided by the density of the solid material, is one of the most important structural properties that governs the general behavior of a ceramic foam. The relative density of common foams usually ranges between 0.5 and 0.30.

The thermal conductivity varies from 0.1 to 1 W m^{-1} K^{-1}. Heat transfer can be further reduced by decreasing the cell size while maintaining a low density [4]. The electrical resistivity increases when an electrically conductive material is foamed [5], and the dielectric constant of dielectric foams increases with increasing density [6].

Table 1 Some typical properties of ceramic foams and their relative values.

Low	High
Relative density	Specific Strength
Thermal conductivity	Permeability
Dielectric constant	Thermal-shock resistance
Thermal mass	Porosity
	Specific surface area
	Hardness/wear resistance
	Resistance to chemical corrosion
	Tortuosity of flow path

As the mechanical properties (strength, Young's modulus, etc.) of a foam material often limit its applicability – especially for low-density foams – it is a parameter of fundamental importance. The mechanical properties are strongly dependent on the structure of the foam [7, 8]. Figure 3 illustrates the evolution of the three-point bending strength of reaction-bonded aluminum oxide foams (RBAO), made by replication of reticulated polyurethane foam, and gel-cast alumina foams, as a function of relative density [9].

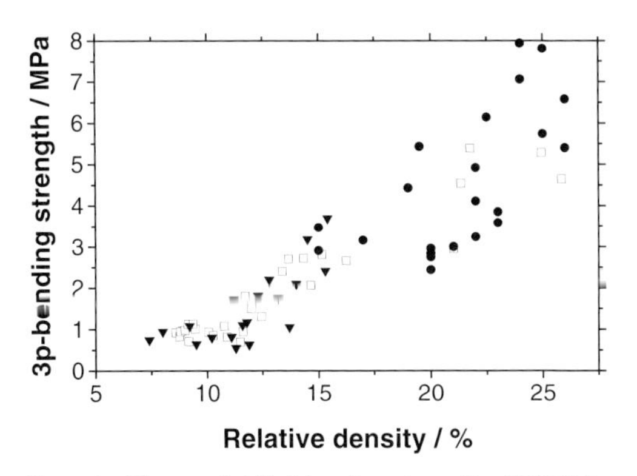

Figure 3 Three-point (3p) bending strength of RBAO foams manufactured by replication of reticulated polyurethane foam (□: pre-oxidized RBAO, ●: long-chain-treated RBAO) and gel-cast alumina foams (▼) as a function of relative density.

Important characteristics for filter applications include thermal-shock resistance, permeability, and specific surface area (catalytic filters). Thermal-shock resistance of ceramic foams is good to excellent, due to the low coefficient of thermal expansion [$(1–9) \times 10^{-6}$ K^{-1}]. The Darcian permeability of reticulated (open-cell) foams lies in the range of 10^{-11} to 10^{-7} m^2, primarily dependent on the pore size distribution and the total porosity. Strong nonlinear increases in Darcian permeability occurs with increasing porosity (80–90 %) and increasing cell size [10–12].

3.1.1.3
Parameters Describing the Structure of the Foams

The complete description of the complex, three-dimensional internal architecture of ceramic foams requires several structural parameters. Depending on the specific application, other characteristics may be relevant. Some important structural parameters are:

1) Cell size and its distribution
2) Strut thickness and its distribution
3) Strut shape and morphology (e.g., dense or hollow struts)
4) Cell window opening
5) Fractional density
6) Degree of anisotropy (of porosity, of pore size, graded materials, etc.)
7) Surface to volume ratio.

Although several models describing the internal relationship between these parameters and the properties have been proposed, only the most important will be considered here.

One approach commonly used to relate the structural parameters of the foam and the mechanical behavior involves the development of a micromechanical model. This implies the assumption of a unit-cell geometry and a deformation mode within the struts. The failure of a single strut generally constitutes the failure of the unit cell and thereby of the bulk foam. It is assumed that the mechanical behavior of the unit cell is representative of the bulk structure.

The mechanical behavior of cellular ceramic materials has been described by the model of Gibson and Ashby [13]. The complications encountered in trying to identify a suitable unit cell that is representative of the complex macrostructure of real foams led them to consider a simple geometry for this unit cell (Fig. 4).

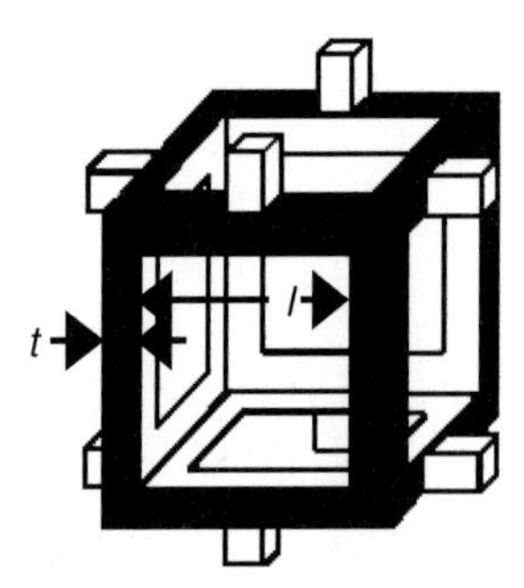

Figure 4 The unit cell as defined in the model of Gibson and Ashby.

The open-cell foam is modelled as a cubic array of individual members of length l and thickness t. The cells meet at the midpoint of the struts of the adjoining cell. Applying a load on the structure will cause bending moments on the cell walls. The mechanical analysis is substantially simplified by this cell geometry, and this allows the derivation of a general set of expressions for the mechanical behavior. However, assigning a single unit cell to these complex macrostructures may be a major oversimplification [14, 15]. Gibson and Ashby have shown that by using this unit cell the relative density of most three-dimensional cellular solids can be related to the macrostructure by the following expressions for open- and closed-cell foams:

$$\rho_{rel} = C_1 \cdot \left(\frac{t}{l}\right)^2 \quad \text{(open cells)}, \tag{1}$$

$$\rho_{rel} = C_2 \cdot \left(\frac{t}{l}\right) \quad \text{(closed cells)}. \tag{2}$$

These expressions should be accurate for relative densities less than 0.3. The parameters C_1 and C_2 are constants characterizing the cell geometry. Ashby and Gibson derived 0.333 and 0.766 for C_1 and C_2, respectively.

During the foaming process, cell elongation in the direction of foam expansion can occur. Therefore, in many cases, characterization in three dimensions is necessary for an accurate description of cell size. Zhang and Ashby [16] argued that the cell geometry best representing isotropic foams, and at the same time capable of filling space, is a tetrakaidecahedron, also known as the Kelvin cell. This truncated octahedron was considered the best unit cell for partitioning space into cavities of equal volume while minimizing the interfacial area (Fig. 5a) until in 1993 Weaire and Phelan discovered a foam structure containing two different types of cavities of equal volume and with a smaller surface area than the Kelvin foam [17]. The unit cell consists of eight polyhedra (two dodecahedra and six 14-hedra) and is replicated in a cubic lattice. Figure 5b shows a dodecahedron at the front upper right; the other visible cells are 14-hedra.

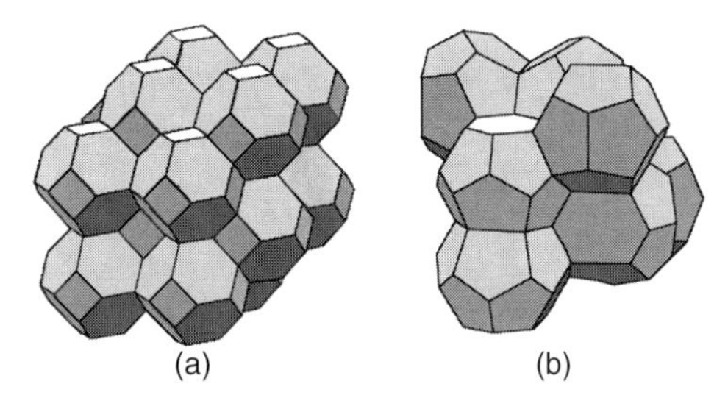

Figure 5 Foam structure composed of a) tetrakaidecahedra (Kelvin cells) and b) the Weaire–Phelan unit cell consisting of eight polyhedra.

At high relative densities (> 0.3), the cellular structure is lost and identification of a repeating unit cell becomes more problematic.

A theoretical model by Peng et al. [18] correlates the overall porosity with the structural parameters of a single cell. Under the condition of equally sized spherical foam cells having a dense packing and thus a mean pore coordination number of

12, the pore volume fraction V_p can be expressed in terms of the ratio of window size to cell size k

$$V_p = \frac{\pi}{\sqrt{2}} \cdot \left(\frac{3}{1-k^2} - \frac{3}{5} \cdot \left(\frac{1}{\sqrt{1-k^2}} \right)^3 - 1 \right). \tag{3}$$

This equation holds for fractional porosities above a critical value of $V_p = 0.74$. However, the model assumptions are rarely found in practice. For example, the window diameter strongly depends on the fabrication method and varies over a wide range for foams with the same fractional density.

3.1.2
Characterization of Foam Pore Structure

Although resolving the structure of cellular foams has been the subject of scientific research for some time, no simple standard experimental technique or procedure to determine some of the most common structural parameters has been identified so far. Although most of the characterization techniques have proven their value in different fields of scientific research for several years, their potential in investigating the parameters that describe the shape, isotropy, and interconnectivity of single cells, as well as the isotropy and fractional density of the whole component, in highly porous materials has yet to be explored.

The following section of this chapter describes a series of analytical characterization techniques commonly used in the determination of some of the most relevant structural parameters of porous materials. For each technique, both the underlying principle and a general interpretation of the measured data are briefly reviewed. Also, the experimental specifications are listed in greater detail.

For comparing the various techniques, a set of porous alumina samples was manufactured. More details on the manufacturing route are given in Section 3.1.2.1. Because of the nondestructive nature of some characterization techniques, the same set of samples could be applied for different techniques. The bar-shaped ceramic foam samples were cut in half: one part was sliced and analyzed by image analysis, and the other was successively used for the Visiocell measurements, permeability measurements (capillary flow porometry), and micro computer tomography. As all the reported measurements originate from the same batch of samples, the influence of process variations is eliminated. This approach enables a more reliable comparison of the techniques discussed below. Section 3.1.2.2 ends with a short overview of other techniques that are either well known for characterizing porous samples or are relatively new in this research field. Section 3.1.2.3 summarizes the various techniques and the corresponding pore sizes for this set of samples.

3.1.2.1
Sample Preparation

Most of the analytical techniques were applied to a series of porous alumina samples, which were all produced by replication of reticulated polyurethane foam. The advantages of this method include the wide diversity in mean cell size of the polyurethane sponges, the narrow pore size distribution, and the relatively large window of processing parameters, which enable a broad range of structures for the ceramic foams. Processing is described in Chapter 2.1 and detailed in Ref. [2].

Four polyurethane foams were selected (kindly supplied by Recticel Co., Belgium). The reaction-bonded aluminum oxide suspension used to coat the polymeric sponges contained a high concentration (30–35 vol%) or a low concentration (20 vol%) of ceramic/metal powder. After drying, the polyurethane was pyrolyzed at 600 °C in air, and the resulting samples were sintered for 1 h at 1750 °C. This procedure resulted in eight ceramic foam samples, with expected mean pore sizes ranging from 400 to 1800 μm.

Table 2 lists information about the samples that were investigated. The sample identification contains the approximate mean cell size of the polyurethane foam and the concentration of reaction-bonded suspension used to coat the polyurethane foam (e.g., PU-620-20: sample made starting from a polyurethane foam with an approximate pore diameter of 620 μm and coated with a suspension containing 20 vol% of aluminum/alumina powder).

Table 2 Overview of the investigated samples.

Sample number	Sample identification	Cell size PU foam* [μm]	Shrinkage after sintering** [%]	Relative density [%]
1	PU-620-20	580	29	11.7
2	PU-620-30	580	21	15.9
3	PU-890-20	860	28	11.8
4	PU-890-35	860	23	20.9
5	PU-1330-20	1230	28	11.5
6	PU-1330-35	1230	23	22.9
7	PU-1900-20	1850	27	10.6
8	PU-1900-35	1850	23	18.1

* Pore size determined by Visiocell (see Section 3.1.2.2).
** Volume shrinkage measured by geometry difference after sintering.

3.1.2.2
Characterization Methods

The different methods can be classified according to some inherent aspects of the technique itself:

1) Nondestructive and destructive methods are distinguished according to whether the foam is irreversibly modified by the test itself (e.g., mercury intrusion porosimetry) or by sample preparation (e.g., thin slices used in image analysis). Capillary flow analysis, liquid extrusion porosimetry, and micro computer tomography are examples of nondestructive characterization techniques.
2) Microscopic or macroscopic techniques according to the size of the area of interest.
3) The dimensionality of the analysis: transferring two-dimensional data to the representation of the three-dimensional internal structure is problematic for geometrically complex pore shapes and requires in many cases foreknowledge and assumptions on shape, distribution, and isotropy. The step from spatial morphology to planar sections, as in two-dimensional characterization techniques like image analysis, involves a great loss of information.

Ideally, because many properties of the cellular material are directly related to the architecture of the sample, a direct, noninvasive investigation of the three-dimensional internal structure of the material is preferred. However, because no single characterization technique can provide a complete overview of all structural parameters, a combination of methods is in many cases necessary. Prior knowledge of the sample structure can help in selecting the appropriate technique or combination of complementary techniques to give a realistic description of the sample, depending on the desired information for a specific application.

3.1.2.2.1 **The PPI Method**

Principle

The size of a cell is one of the key parameters in the design of a foam. For many applications, foam performance is directly influenced by cell size. For more than 25 years, the cellular structure and the cell size have been defined by the unit ppi (pores per inch) [19]. The number of pores is counted over a standard length of one inch. The linear intercept method was primarily used when manual measurements were required to obtain data, because they could be performed by drawing random lines on images of sections. Figure 6 examplifies this method on a micrograph of a polyurethane foam. With modern computer-based instruments, it is usually easier to measure the intersection areas which are made by a sampling plane with the object.

Interpretation of measured characteristics

This unit can be a little confusing and subjective because of:

1) The unclear definition of a “pore”: it can be a window or the full section of the cell.
2) The size of the pore: in case of a cell window, the size depends on the viewing angle. For the cell section, the size depends on the location of the section (top, middle, or bottom).

3) The data scatter broadly within a sample: counting from A to B or from C to D in Fig. 6 leads to significant differences.
4) Analogous to other characterization techniques, dimensional reduction is an important issue: the cell is a three-dimensional feature of the foam, while ppi is a reduction of the volume to a linear, one-dimensional count of a undefined unit (the pore). Under the assumption of spherical cells that are relatively uniform in size, a correction factor can be derived [20]:

$$D_{\text{sphere}} = \frac{t}{0.616} \quad (4)$$

where D_{sphere} is the average sphere diameter and t the average cell chord length.

Because of the problems described above, the ppi measure has not been universally accepted as an international unit. Methods to evaluate it, such as direct counting, pressure drop, and three-point methods have not been recognized. Visual comparison of the cell structure with test samples (or images of cell structures) with known ppi is an alternative method. The relation between pressure-drop data and ppi value is commonly applied in the characterization of polymer foams. For ceramic foams, this relation is not straightforward, as the presence of cell windows, which are almost completely absent in reticulated polymer foams, greatly affects the measured pressure drop [21].

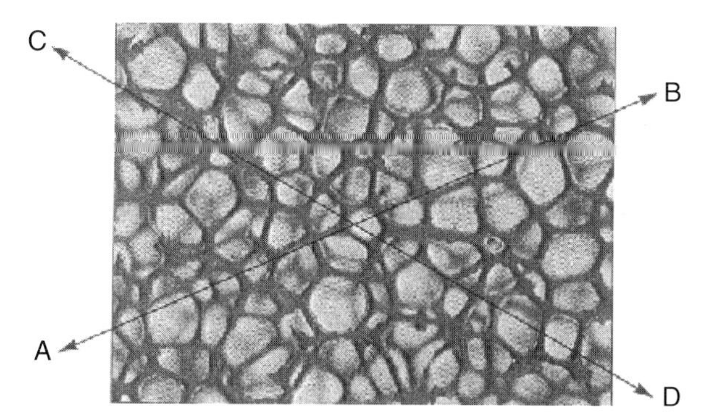

Figure 6 Illustration of the principle and the uncertainty of the ppi method.

As a result of the different measurements and definitions, each foam manufacturer has its own reference scale, and a foam defined as 80 ppi by one producer could be defined as 110 ppi by another, and this leads to confusing and contradictory specifications. Owing to lack of standardization, this method cannot fulfil the specification requirements of the new high-tech developments and applications. Therefore, it is becoming increasingly less relevant in present-day characterization and will not be discussed further.

3.1.2.2.2 **Visiocell**

Principle

Visiocell is a three-dimensional method based on light micrographs, originately developed by the company Recticel to characterise polyurethane foams [19]. The basis of this technique is an image of a horizontal cut of a foam sample (for polyurethane samples, the cut is made perpendicular to the foam-rise direction) taken with a magnifying camera. A representative cell is selected by identifying its approximately circular shape comprising ten struts and one or two small pentagon(s) in its center (Fig. 7a). These pentagons are the underside and/or upper side window(s). The actual measurement is performed by superimposing a calibrated ring, printed on transparent paper. The ring that most closely fits the cell in the image indicates the cell size in the polyurethane foam. The cell diameter is in this case defined as the average between the internal and external circle of the ring.

Specifications

The accuracy of this technique is about 2 %. An accurate correlation between the cell diameter by the Visiocell method and the ppi scale is impossible due to the inaccuracy of the ppi measurement (Fig. 7b). The nature of the Visiocell measurement only permits the determination of regular cell structures with cell sizes larger than 450 µm. The use of Visiocell in the characterization of ceramic foams is limited as its starts from the assumption of a uniform distribution of spherical pores. Therefore, it can only be used for a fast estimate of cell size.

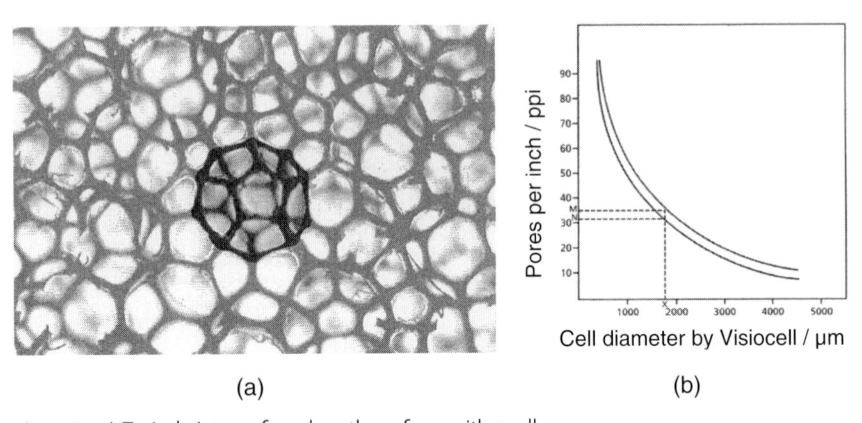

Figure 7 a) Typical picture of a polyurethane foam with a cell diameter of approximately 1000 µm and b) the correlation and margin of error between ppi and Visiocell results.

Results

The four polyurethane foams and the resulting eight ceramic foams were investigated with this method. Starting from the pore size of the polyurethane foam and taking into account the shrinkage of the different samples, one can calculate an "expected" value for the pore size. There is a relatively good agreement between the calculated pore size and the pore size measured by Visiocell, as shown in Fig. 8. The deviation from the calculated values is 2–7 %.

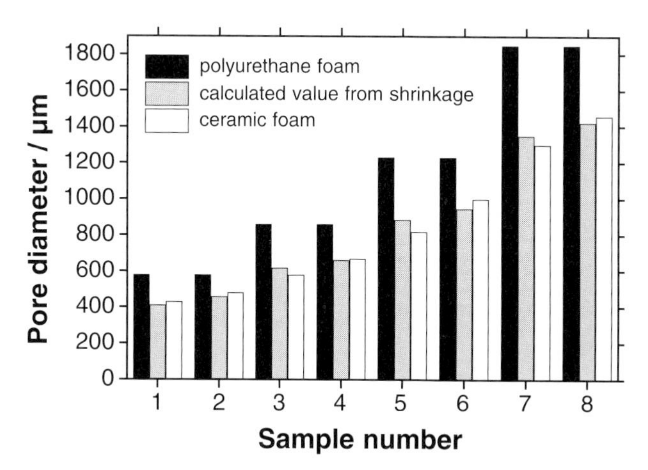

Figure 8 Visiocell measurement of the pore diameter for the polyurethane foams, the expected pore diameters calculated from the shrinkage, and the Visiocell measurements of the resulting alumina foams.

3.1.2.2.3 **Image Analysis**

Image analysis has proven its value in different fields of scientific research [22–25]. Common application fields include geology (simulation of permeability of porous stone, soil studies, composition of coal or minerals), medicine (characterization of bone structure, analysis of medical images), biology (cell biology by selective staining, root quantification, particle counting), and materials science (grain size distribution, microstructure of composite materials). Image analysis uses a series of image-processing routines to extract meaningful numerical information starting from an image. The results from an image analysis for a porous material can include several important pore characteristics like the number of intersections, the perimeter, the pore size distribution, the porosity, and the shape factor of the pores [26].

Principle

The image analysis routine typically requires several sequential processing steps [27, 28]. Although these steps can differ depending on the kind of image and on the information one wants to obtain, a general outline of the procedure is described in the following.

1) *Image acquisition and input:*

The conventional approach for image acquisition is microscopy. As the magnification of the image is governed by the pore size of the sample, this can be done by

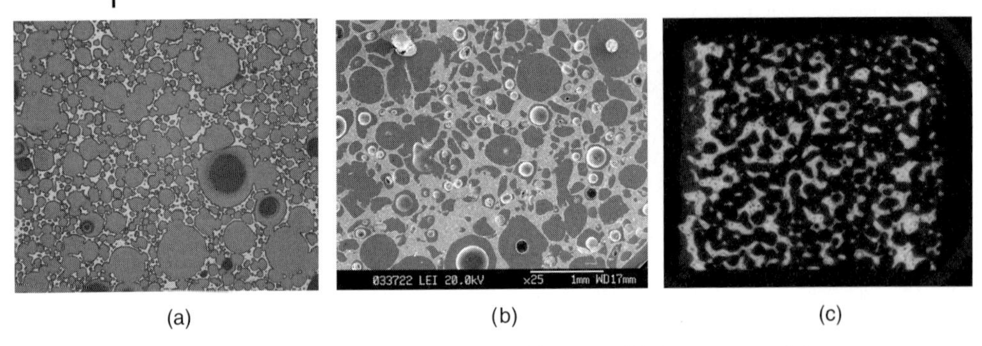

Figure 9 a) Optical micrograph of an SiC foam manufactured by direct foaming, imbedded in a resin, b) SEM image of a porous alumina material made by incorporating organic fillers, imbedded in a resin, and c) section by computerized X-ray microtomography of a foam made by replication of reticulated polyurethane foam.

light or scanning electron microscopy (SEM). Careful preparation of the sample is required to obtain sufficient difference in brightness or contrast between ceramic material and the substance in the pore (air or imbedded material). One solution to the contrast problem consists of cutting thin slices, infiltrating the cellular material with a black resin, and finally polishing the plane of interest. Image analysis can also be performed on images from other sources, such as the two-dimensional slices from computerized X-ray microtomography. Figure 9 shows some images of porous ceramics obtained by optical microscopy (a), scanning electron microscopy (b), and micro computer tomography (c). The porous ceramics were made by gel casting (a), by incorporation of organic fillers (b), and by replication of reticulated polyurethane (c). In these images, the ceramic material is represented by the light areas, and the resin (or air) by darker areas.

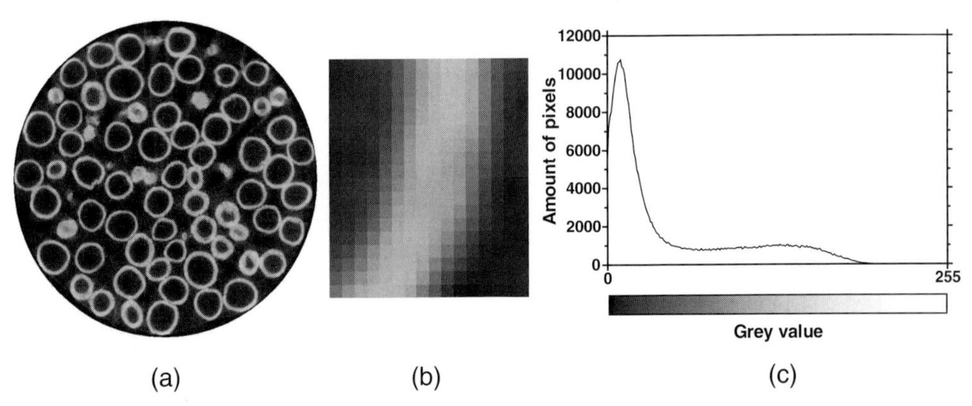

Figure 10 a) Circular region of interest of a computerized X-ray microtomography slice of a hollow-sphere material (705 × 705 pixels) [26], b) magnification of a boundary region, and c) distribution of the gray-scale values in the image.

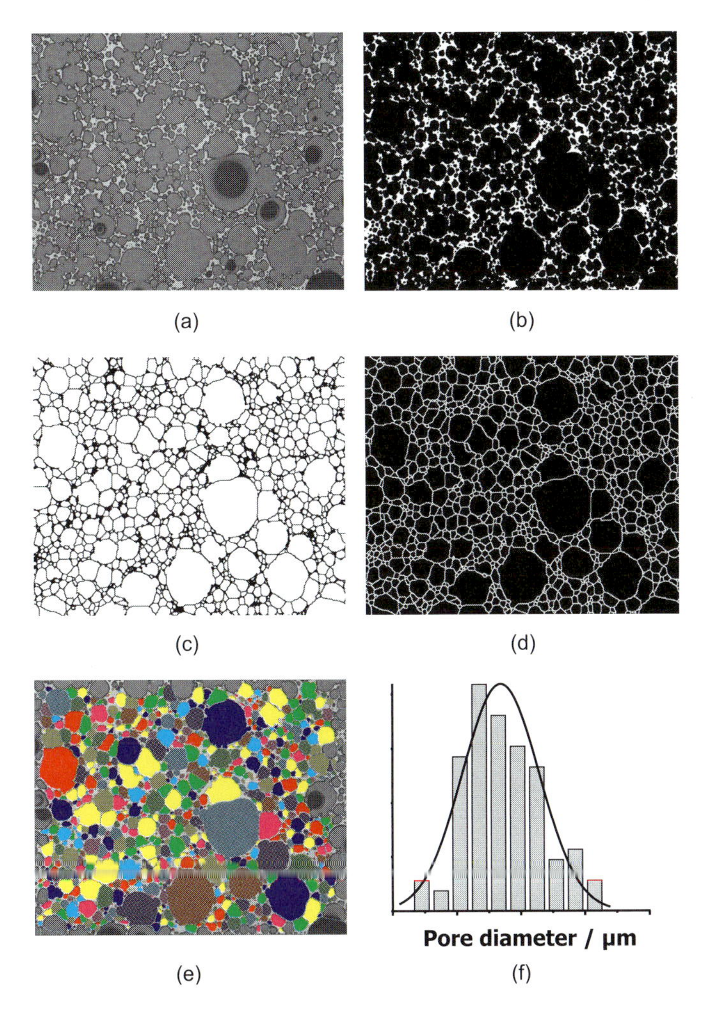

Figure 11 Overview of the successive steps in the image analysis of a SiC foam: a) image acquisition, b) image thresholding (binary image), c, d) skeletonization, watershed, and distance algorithms, e) pore identification and measurement of pore diameter, and f) histogram representation of pore diameter distribution.

2) *Image enhancement:*
Before image analysis is performed, some preliminary image enhancement is usually required. The selection of the region of interest (ROI), which is possible for a variety of geometrical or random shapes, excludes edge effects and incomplete representation of certain features by frame limitations. Image defects are suppressed by other functions, such as smoothing or median filtering for noise reduction, improvement of contrast, correction for nonuniform image illumination, low-pass filtering for correction of shading, and stray light effects. If necessary, image details can be improved to further enhance the visibility of features of interest.

3) *Image thresholding:*

The next step, probably one of the most important in image analysis, is threshold segmentation, in which pixels that represent the actual feature of interest are identified in the image by analyzing relative pixel intensity. This can be an interactive manual procedure in which the operator decides the best boundary threshold, or an automatical procedure based on the gray-scale histogram. The histogram is a graph of the distribution of red, green, blue, gray scale, hue, saturation, and/or lightness values in an image as a function of the number of pixels at each value. The lightness values of the image can range from black to white (gray values 0 to 255). In the ideal case, a clear separation of the pixel intensity of the feature of interest exists. Several threshold algorithms are available, depending on the type of image, the nature of the boundary, and so on.

In the case of porous materials, only two material phases, air and solid, are of interest. However, most gray-scale images do not have a sharp boundary at the interface between the two different phases. Instead, there is a boundary region over a distance of several pixels in which the gray-scale changes gradually from one gray level to the other. This feature is illustrated in Fig. 10, which shows the unsharp, diffuse boundary in an image of a collection of sintered alumina hollow spheres [29], obtained by micro computer tomography, and the corresponding gray-scale histogram. The maxima at gray value of 10 (dark) and 130 (light) correspond to air and ceramic material respectively. The minimum in the gray-scale histogram can be set as the threshold for this the image. Small changes in the threshold parameters alter the size of the regions of interest (i.e., the ceramic material) by changing the specific location in the boundary region where the pore is defined in the binary image. Inconsistent human judgements can and often do lead to operator-dependent measurements.

4) *Binary image processing:*

When the thresholding procedure leaves artefacts behind, the use of combinations of erosion and dilation, which are binary image processing functions that compare pixels to their immediate neighbors, permits selective correction of the binary image details. The Euclidian distance function generates a distance-transformed image by assigning values to pixels within features that measure the distance to the nearest background point. It can be regarded geographically as a "landscape" with gray-scale hills and valleys. This distance map is used as input for the watershed function. This algorithm is a strong tool for separating touching convex shapes by flooding the valleys and finding the dividing lines between the different valleys in the "landscape". However, if the overlap is too great, this method will fail, and adjacent features are considered as being one. Thus, for samples with very low density, manual correction to identify the individual pores is laborious but indispensable. Skeletonization, also known as thinning, determines the skeleton, one pixel wide, of the white regions of the formerly defined binary image.

5) *Measurements:*

Once the features have been unambiguously identified by the routine, different measurements can be performed: counting features (with the possibility of eliminating features based on their size or shape), size measurements (length, perimeter, area), shape measurements (circularity), and intensity (topology, information on gray scale). The generated data are classified and can be analyzed statistically.

Figure 11 gives an overview of these successive steps in the image analysis routine for measuring the pore size distribution of a SiC foam manufactured by gel casting.

Alternative image analysis routines are based on granulometry: the pores are assumed to be particles whose sizes can be determined by passing them successively through meshes with increasing size and collecting what remains after each pass. Certainly for pores with more irregular shapes, this approach is advantageous, as no assumptions about the shape factor are made.

Interpretation of measured characteristics

Image analysis can quantitatively extract important structural parameters for porous materials. However, apart from problems related to sample preparation (embedding, slicing, contrast, etc.), other issues concerning the interpretation of the data must be addressed.

The pore size distribution and other structural properties of the porous material are part of a three-dimensional pore space. The image analysis data are based on only two-dimensional random sections. As the intersections through the individual pores are randomly oriented in space, the pore size distribution measured in this way will not completely represent the actual distribution. Inevitably, the reduction from spatial morphologies to their planar sections results in most cases in a great loss of information. Only in the case of materials with well-defined pore shape and uniform pore size distribution does an analytical solution for this reduction exist.

Several methods and techniques have been proposed for obtaining three-dimensional information from two-dimensional images of the micro- and macrostructure. The scientific and mathematical field dealing with the relationships between the data from two-dimensional images (either sections or projections) and the three-dimensional reality which they represent is called *stereology* [30–32]. Although a complete overview of the stereological approach is outside the scope of this chapter, some general principles are described below.

Two aspects must be taken into account when converting the two-dimensional data to the three-dimensional representation of the porous sample [33]. The cut-section effect deals with the fact that the intersection plane rarely cuts through the center of each pore. As a result, the pore size distribution will be broader, even if the real distribution is monodisperse. For polydisperse distributions, the problem becomes even more complex, as smaller features are less likely to be cut by a plane than larger features. This is known as the intersection-probability effect. Additional errors may originate from the presence of preferred spatial orientations and shape variations (e.g., elongated pores).

Mathematical theories to correct the two-dimensional data are mainly based on randomly distributed spheres. Assuming a monodisperse distribution of spheres, the intersection probability effect can be resolved by stating:

$$n_v = \frac{n_a}{D} \tag{5}$$

where n_v is the total number of spheres per unit volume, n_a the total number of spheres per unit area, and D the diameter of the spheres. This equation can be modified to apply to a distribution of other shapes.

The cut-section effect can be resolved analytically only for spheres. The function describing the probability of a random intersection of a sphere rises to a maximum near the diameter of the sphere. Hence, the mean intersection length is close to the true three-dimensional size of the object. However, depending on the true three-dimensional shape and the preferred orientation of the pores, other sectional shapes will be produced [33]. This poses a serious limitation in calculating real-life samples, as deviations from perfect spheres and irregular pore forms will have substantial influence on the results.

Saltikov proposed a method of unfolding a population of intersection lengths into the true length using a function of the intersection lengths. This method works well for spheres and spherelike shapes. Large errors are introduced when more complex shapes are present [34, 35].

Shape-related parameters can be included in some more advanced methods like an extended Schwarz–Saltikov approach or iterative solutions (the program StripStar, written by R. Heilbronner, University Basel or the program CSDCorrections, written by M. Higgins, Université du Québec a Chicoutimi). More information concerning the underlying principles of stereology and conversion programs can be found elsewhere [33, 36].

Another option for extracting three-dimensional data is serial sectioning. Apart from being time consuming and laborious, this method is rarely applicable to large enough volumes of material to give statistically meaningful data. Moreover, the section spacing in the z direction limits the lower pore size detection [33, 37]. The degree of anisotropy can be estimated by comparing the parameters of the porous material in x, y, and z directions.

Specifications

The eight foam samples were embedded in a resin and cut into thin sections. Images were digitally recorded under an Axioplan II microscope (Carl Zeiss Vision GmbH), equipped with a digital camera Axiocam (magnification 25×, 1300 × 1030 pixels, 4.068 μm/pixel). For each samples three to five images were recorded, and the data averaged. Image analysis was performed with the KS400 routine version 3.0 (Carl Zeiss Vision GmbH), which allows the application of user-defined macros. Powerful freeware programs such as ImageJ (National Institutes of Health; http://rsb.info.nih.gov/ij/) and Imagetool (UTHSCSA; http://ddsdx.uthscsa.edu/dig/itdesc.html) are available on the internet.

Results

1) *Porosity:*

The determination of the two-dimensional porosity of the RBAO ceramic foam sample is a relatively straightforward procedure provided the image has been thresholded correctly. The porosity is calculated as the count of pixels that represent pores on a given surface [28, 37].

Figure 12 compares the density measured by image analysis and the geometrical density (percentage of the theoretical density). This can be regarded as an evaluation of the threshold process, as large and consistent differences would point to incorrect identification of pore boundaries.

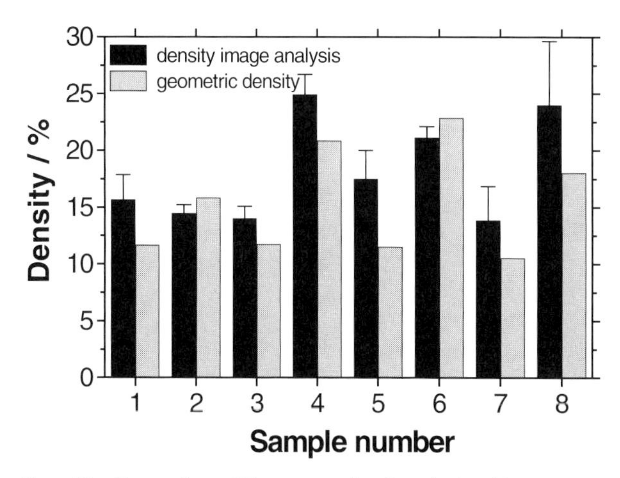

Figure 12 Comparison of the average density calculated by image analysis and the geometrical density for the eight ceramic foam samples.

2) *Pore size distribution:*
The results of the image analysis routine are parameters such as the area, the equivalent diameter, and the circularity of each pore. Pore size distributions can be derived from the calculation of the area of the individual pores in the thresholded image [22]. In this study, each pore area in the intersection plane was determined by counting the number of pixels in each pore, multiplied by the area of one pixel. Assuming spherical pores, the diameter of a circle with equivalent area D_{eq} can be calculated:

$$D_{eq} = 2 \cdot \sqrt{\frac{\text{area}}{\pi}}. \tag{6}$$

The equivalent diameter can be classified in a histogram function by using a suitable bin size which can be fitted to an appropriate function. Another option is to present the pore size distribution as the cumulative pore fraction as a function of equivalent pore diameter. Depending on the number of pores in one image, this corresponds to 100 (for the largest pore size) to over 1100 pores that were analyzed. The calculation of the cumulative curve is based on the summation of all measurements. Small artefacts remaining in the image were removed by imposing a minimum equivalent pore diameter.

This procedure is exemplified in Fig. 13 for sample 2. Starting from a microscopic image, a histogram classification of the equivalent pore diameter (as calculated from Eq. 6) is obtained. The cumulative counts of four separate measurements (Fig. 13b) shows good agreement. The histogram function and the cumulative graph with black dots represent the average of the four measurements, corresponding to the analysis of 631 pores.

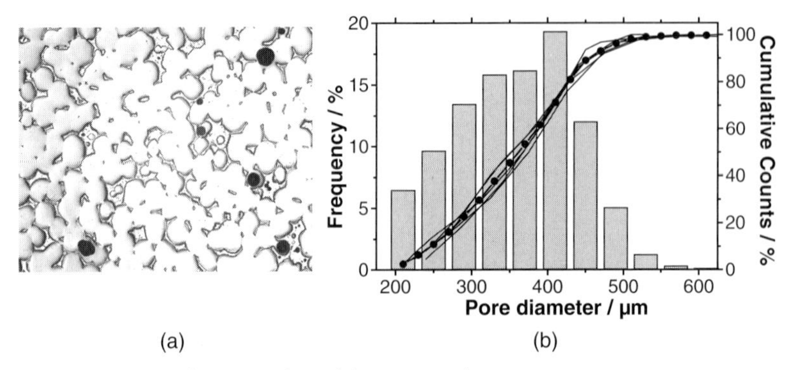

Figure 13 a) Typical micrograph used for image analysis and b) the cumulative curves and histogram classification of the equivalent pore diameter for sample 2.

Figure 14 presents the overview of the cumulative counts for the eight investigated samples. The pore size distributions cover the range from about 150 to 1700 µm. The wide range of pore sizes results from the combination of the pore size of the polyurethane foam as the starting material, the composition of the ceramic suspension used to coat the foam, and from other processing parameters, such as the distance between the rolls used to remove the excess ceramic slurry after wetting. The jagged curve for sample 8 is directly related to the small number of analyzed pores (100).

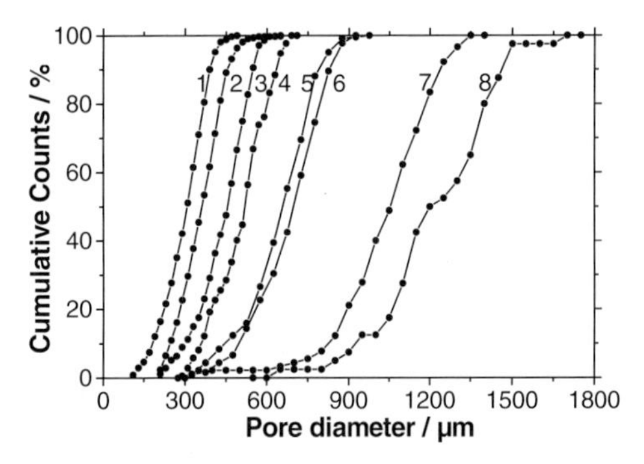

Figure 14 Overview of the cumulative curves of equivalent pore diameter for the eight investigated samples.

Differentiation of the cumulative curves over discrete intervals yields the area number density for each size interval (i.e., the number of pores in each bin divided by the total area measured). Figure 15 presents the area number density as a function of the equivalent pore diameter for samples 1, 3, 5, and 7; the mean pore diameter increases with increasing pore size of the original polyurethane foam.

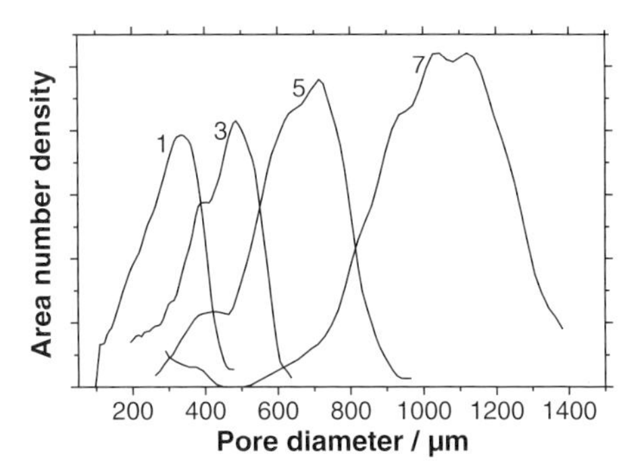

Figure 15 Area number density by differentiation of the cumulative curves as a function of the equivalent pore diameter for samples 1, 3, 5 and 7.

Assuming that the investigated samples are characterized by a random orientation of spherical pores with a relatively narrow cell size distribution, a simple stereological correction factor can be applied to determine the true three-dimensional mean pore diameter. It can be shown that the average sphere diameter D_{sphere} is larger than the average circular segment diameter D_{circ} due to random truncation of the cells with respect to depth at the plane of the specimen surface. The relation can be expressed by the following equation [20, 22]:

$$D_{sphere} = \frac{D_{circ}}{0.785} \, . \tag{7}$$

Figure 16 compares the mean pore diameters measured by the Visiocell method and the image analysis results. When a uniform distribution of spheres is assumed, the largest diameter in the pore size distribution corresponds to a section in the center of the sphere. Because of the limited number of measurements in this bin size, the d_{90} value of the cumulative curve is regarded as the largest diameter.

Applying the simple stereological correction factor from Eq. (7) results in a mean pore diameter that is in good agreement with the other methods. Differences between the d_{90} value and the corrected d_{50} value range between 1 and 7 %. This indicates that the basic assumptions for Eq. (7) are valid for these kinds of samples. For most samples, the difference between the d_{90} value from image analysis and the Visiocell method range between 3 and 5 % (except for samples 1 and 6, with differences of 9 and 17 %, respectively).

Alternatively, the two-dimensional slices from micro computer tomography (μ-CT) can be used as input for the image analysis routine. As the difference in linear attenuation coefficient between air and ceramic material is fairly large, the image contrast is high. Figure 17a shows a slice of a μ-CT scan of sample 8. The corresponding identification of the individual pores by the image analysis routine is presented in Fig. 17b.

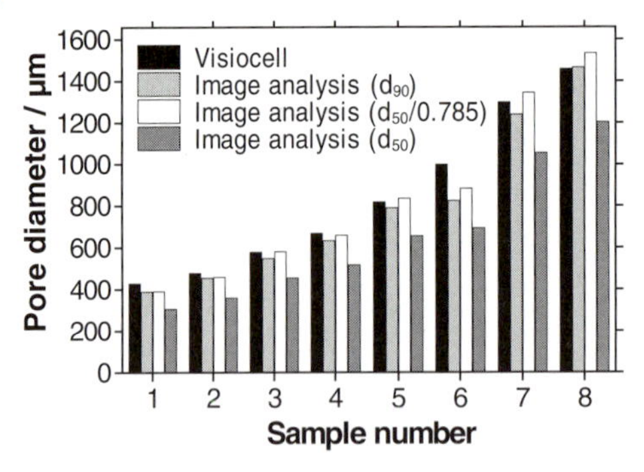

Figure 16 Comparison of the equivalent pore diameter for the eight investigated samples, measured by the Visiocell method and image analysis: the d_{90} value, three-dimensional calculation according to Eq (7) ($d_{50}/0.785$), and d_{50} value.

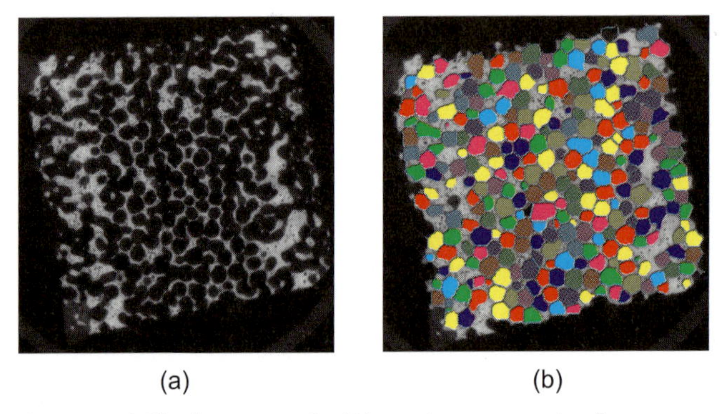

Figure 17 a) Slice by computerized X-ray microtomography of sample 8 and b) identification of the pore diameter in the image-analysis routine.

Image analysis was performed on μ-CT slices of samples 5, 6, 7, and 8. The measured data are the average of analyses on two randomly chosen μ-CT slices. This corresponds to about 500 analyzed pores. Figure 18 compares the d_{90} and d_{50} values in the cumulative curves for analysis of images obtained by light microscopy and slices obtained by tomography. The μ-CT-derived data are consistently lower for all investigated samples, both for the d_{90} and the d_{50} value. The differences in d_{90} value are about 4–8 % (17 % for sample 8). This might be due to difference in setting the threshold of the optical image and the thresholding in the microtomography procedure.

3) *Shape factor:*

The shape of the pores is of interest in many applications. Moreover, if stereological corrections are applied, information on the shape of the features is in most cases necessary. Depending on the expected form of the intersection (circular, elliptical, triangular, or rectangular), the aspect ratio and the degree of roundness must be estimated. When performing image analysis on thin slices that are cut in x, y, and z directions, the differences in intersection forms provide information on the anisotropy and the general shape of the pore channel.

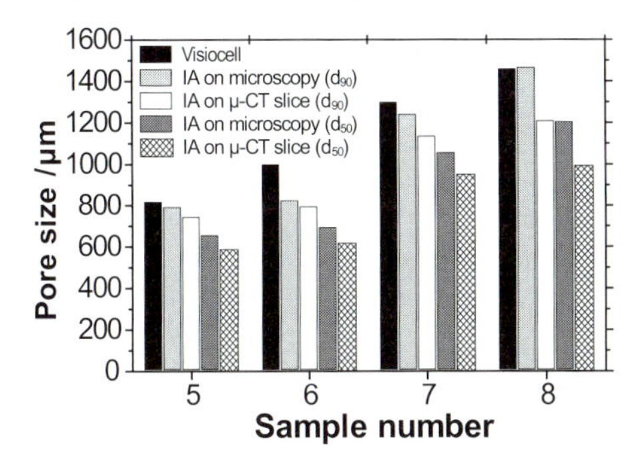

Figure 18 Comparison of the pore size determined by the Visiocell method, d_{90} values from image analysis on optical microscopy images and on microtomography (μ-CT) slices and d_{50} values from image analysis on microscopy optical images and on μ-CT slices.

As these samples can be described by a sphere-like pore model, a circularity factor of each pore in the sample can be calculated. The circularity factor F_{circ} represents the roundness of the region and can be defined as

$$F_{\text{circ}} = 4\pi \cdot \left(\frac{\text{area}}{\text{perimeter}^2} \right). \tag{8}$$

For a perfect circle, the shape factor is 1. More elongated polygonal intersections lower the circularity factor. Figure 19 presents the shape factor for all pores in a micrograph of sample 1 (total number of pores 167). The mean circularity for this image is 0.74, with a standard deviation of 0.1.

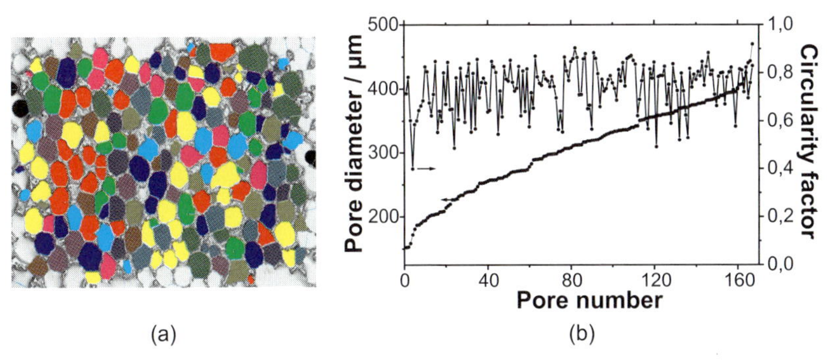

Figure 19 a) Identification of pores by image analysis on microscopic image of ceramic foam and b) ascending list of the equivalent diameter of each pore in the image of sample 1 and the corresponding circularity factor.

3.1.2.2.4 Capillary Flow Porometry

The structure of porous materials is often very complex because of the nature of the porosity (closed porosity, blind pores, and through-pores), the pore shape, the surface area, and the isotropy. For applications like filter media, the fluid or gas flow characteristics, the barrier properties, and the efficiency of the process are governed by the combination of all aspects of the pore structure.

Capillary flow porometry, also known as extrusion flow porometry, can be applied to determine some important characteristics of porous materials [38, 39]:

- Distribution of the constricted part of a pore channel.
- Flow distribution curve (bubble point, mean diameter).
- Liquid and gas permeability.
- Surface area.

Principle

The sample is placed in a wetting liquid to fill the pores of the sample. As the liquid/solid surface free energy γ_{ls} is lower than the solid/gas surface free energy γ_{sg}, pores are spontaneously filled by the liquid. When a nonreacting gas increasingly pressurises the sample, the liquid is removed from the pores and a gas flow through the pores results. To displace the liquid from the pores, the work done by the gas must be equal to the increase in surface free energy. This can be expressed by the following equation:

$$p\,\mathrm{d}V = (\gamma_{sg} - \gamma_{ls})\,\mathrm{d}S \tag{9}$$

where p is differential pressure, $\mathrm{d}V$ the increase in volume of gas in the pore, and $\mathrm{d}S$ the increase in solid/gas surface area (or the corresponding decrease in solid/liquid interfacial area).

The pore cross section can be quite complex. The diameter of a pore at any location along the pore path can be defined as the diameter of a cylindrical opening that has the same $\mathrm{d}S/\mathrm{d}V$ ratio as the actual pore. This is illustrated in Fig. 20.

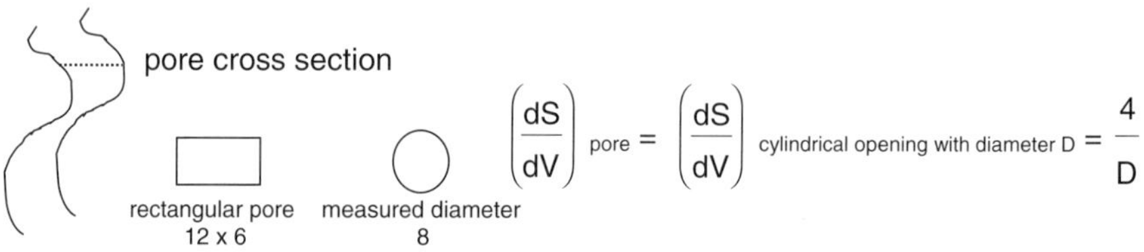

pore cross section

$\left(\frac{dS}{dV}\right)$ pore $= \left(\frac{dS}{dV}\right)$ cylindrical opening with diameter D $= \frac{4}{D}$

rectangular pore 12 x 6 | measured diameter 8

Figure 20 Schematic representation of pore cross section and the definition of pore diameter in capillary flow analysis.

Taking into account the relation between the contact angle θ and surface tension γ, the relation between the pore diameter D and the differential pressure needed to displace the liquid from the pore can be expressed as:

$$p = 4\gamma \cdot \left(\frac{\cos\theta}{D}\right). \tag{10}$$

For exact measurements, information on the contact angle is necessary. For low surface tension wetting fluids the contact angle approaches zero. Figure 21 gives a schematic representation of the principle of this method. When the pressure is gradually increased, the liquid is removed first from the pores with the largest diameter, and a gas flow through the pore results. Higher pressures are needed to empty the smaller pores.

The presence of pores is detected by measuring the flow rate at a given applied differential pressure with flow meters. The flow curve as a function of the differential pressure is measured for both the wet and the dry sample.

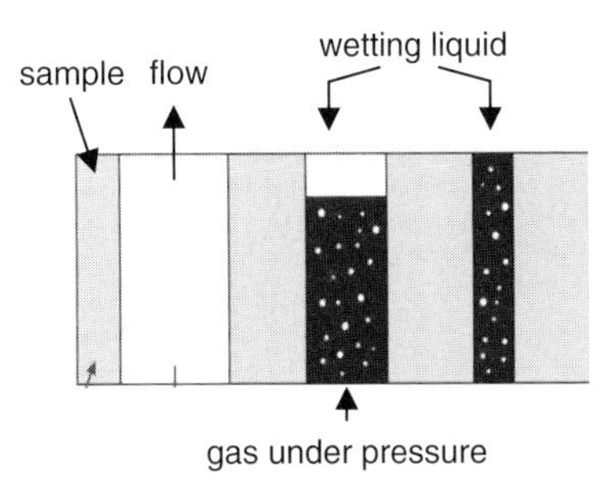

Figure 21 Schematic view of the principle of capillary flow analysis.

Interpretation of measured characteristics
The measured differential pressures and flow rates are used to calculate a number of *pore characteristics* of the sample.

1) The measurement of the *constricted pore diameter distribution* is based on the complete removal of a wetting liquid from a pore with a certain diameter at a distinct differential pressure; it can be calculated from Eq. (10). Figure 22 shows the schematic view of a through-pore channel. To measure a gas flow through the pore, the differential pressure must be large enough to remove the liquid from the most constricted part of the pore. Until this pressure is reached, the gas displaces liquid in the pore up to the smallest diameter in the pore channel. Thus, the differential pressure that permits flow through a pore corresponds to the displacement of the liquid at the most constricted part of the pore. The flow meter detects the pore by sensing the increase in flow rate at a certain differential pressure. The calculated pore diameter is then the diameter of the through-pore at its most constricted part.

The bubble point, being the largest constricted through-pore diameter, corresponds to the lowest pressure at which gas flow through the wet sample is detected. The mean constricted through pore diameter is calculated by the intersection of the wet flow curve and the half-dry flow curve. This mean pore diameter corresponds to half of the flow through pores larger than the mean value. A flow distribution curve as a function of the pore diameter can be calculated as well in terms of the function f:

$$f = -\left(\frac{df_w}{df_d}\right) 100/dD \tag{11}$$

where f_w and f_d are the flow rates through wet and dry samples, respectively, at the same differential pressure. The area under the curve in a certain pore range is the percentage flow through that pore size range. A typical result of a capillary flow analysis of a ceramic foam is presented in Fig. 23.

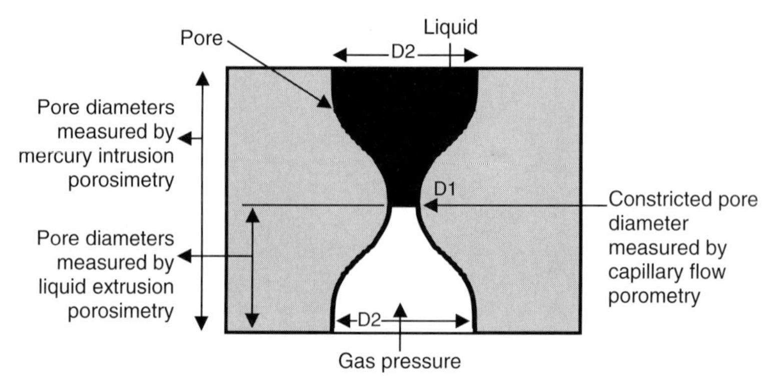

Figure 22 Schematic view of a through-pore channel and the pore diameter measured by capillary flow analysis (in comparison with liquid extrusion and mercury intrusion porosimetry).

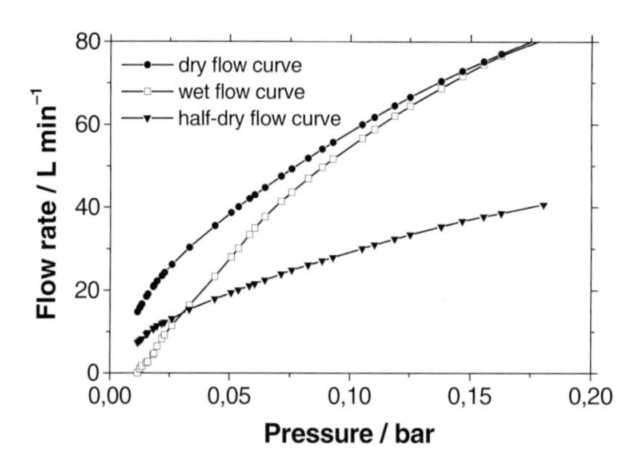

Figure 23 Capillary flow analysis of sample 1, showing the dry flow curve, the wet flow curve, and the half-dry flow curve.

2) The gas flow curve through the dry sample permits the calculation of the *gas permeability* using Darcy's law [40]:

$$F = k \cdot \left(\frac{A}{\mu \cdot l}\right) \cdot (p_i - p_o) \tag{12}$$

where F is the volume flow rate at the average pressure per unit time, k the permeability, μ the viscosity of the fluid, A the cross-sectional area of the porous sample, l the thickness of the porous sample, p_i the inlet pressure, and p_o the outlet pressure.

Analogously, *liquid permeability* for a variety of liquids and solutions can be measured by using the corresponding wet flow curve.

3) The flow rate data enable calculation of the *surface area* by using the Kozeny–Carman relation [41, 42]. Accurate measurements are possible when the specific surface area is less than 10 $m^2\ g^{-1}$ and the distribution of pore diameters is relatively narrow. The agreement with gas adsorption measurements is reasonably good if the sample is free of blind pores.

Specifications

In this study, a PMI Capillary flow porometer, type CFP-1200-A was used (Porous Materials Inc., USA). The specifications according to the manufacturer are summarized in Tab. 3. The measurable pore size stretches over a broad range from 13 nm to 500 μm. The low pressure needed to remove the liquid enables the characterization of mechanically weaker structures (e.g., polymer foams) and avoids distortion of the pore structures.

Table 3 Specification of the capillary flow analysis equipment.

Property	Specification
Pore size range	13–500 μm
Sample diameter	1–6 cm
Pressure range	0–35 bar
Pressurizing gas	dry, compressed air, nonflammable, noncorrosive
Pressure transducer range	0–35 bar
Resolution	1 in 20 000
Accuracy	0.15 % of reading
Mass flow transducer range	10 ml min^{-1} to 500 l min^{-1}

Results

Figure 24a presents pore diameter and filter flow (ratio of the wet and dry flow curves) as a function of the applied pressure for sample 4. From these raw data, the bubble point can easily be identified as the pore diameter that corresponds to the

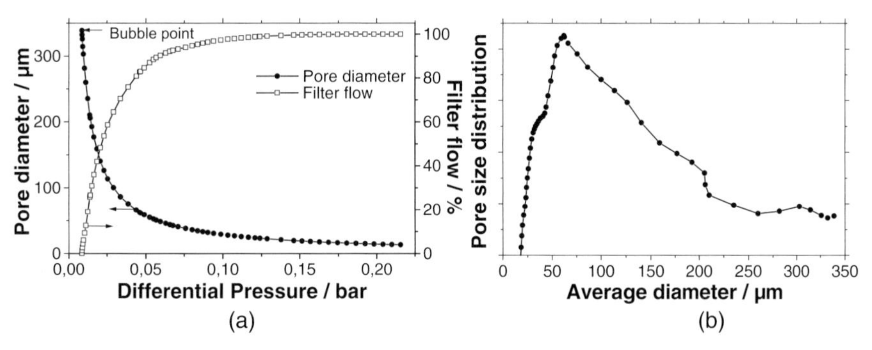

Figure 24 a) Relation between pore diameter and filter flow as a function of differential pressure and b) size distribution of the most restricted part in the pore channels for sample 4.

lowest pressure needed to initiate flow through the wet sample. For this sample, the bubble point is 338 μm, and 90 % of the filter flow occurs for pores larger than 49 μm. A pore size distribution is obtained by the differentiation of the filter flow as a function of the pore diameter (Fig. 24b).

Figure 25 presents an overview of the filter flow curves for the eight samples. Although the pore diameters of the different samples vary over a relatively broad range (from 200 to 1700 μm), the pore sizes as measured by capillary flow analysis are distributed over a much smaller range (ca. 20 to 700 μm). This can be explained by the presence of semiclosed windows and other restricted pore channels that cannot be detected by image analysis or other techniques based on microscopy.

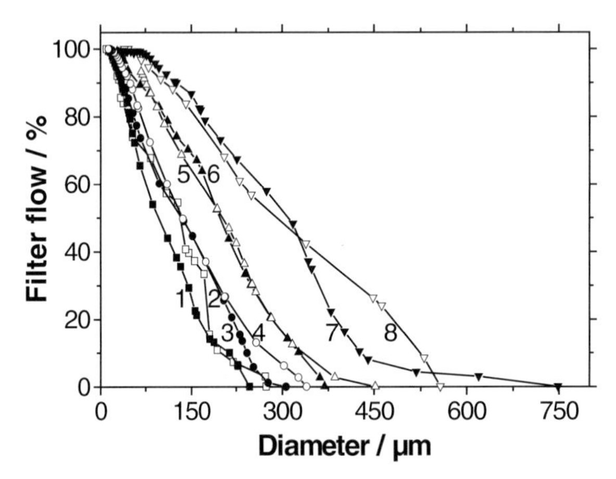

Figure 25 Filter flow curves measured by capillary flow analysis for the eight samples.

Comparing the information of the capillary flow analysis with the pore diameter determined by the Visiocell measurements clearly shows the difference between the two techniques regarding the measured property (Fig. 26). As capillary flow analysis only measures the most restricted part of the pore channel, mean pore flow diameter and the bubble point are lower than the average diameter of the pore size determined by the Visiocell technique.

Although a general increase in the restricted pore diameter is seen with increasing cell size of the polyurethane foam, this trend is not followed to the same degree. When coating with more viscous suspensions, the formation of semiclosed cell windows cannot be avoided, as exemplified by the lowering of the restricted pore channel diameter for samples 6 and 8 compared to samples 5 and 7, respectively.

The mean flow pore diameter and the bubble point are critical parameters of porous materials that are used in applications such as filtration or porous bone replacements. For the latter, bone ingrowth requires a minimum pore channel diameter that enables migrating biological material like cells or veins to penetrate the porous implant. The combination of the mean pore diameter and the most restricted part of the pore channel play a critical role in the ability for complete ingrowth.

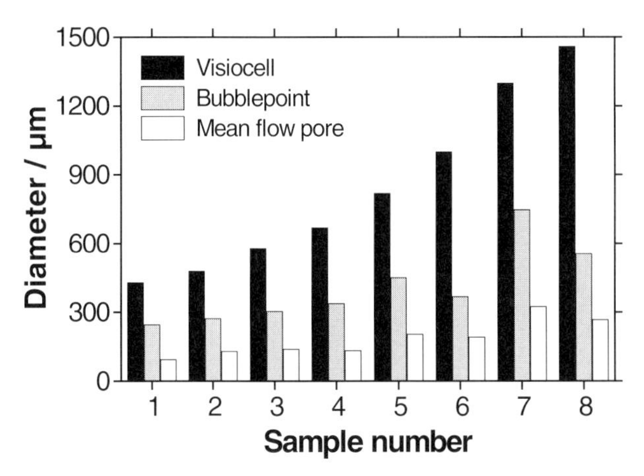

Figure 26 Comparison of the pore diameter by the Visiocell method, bubble point, and mean flow pore diameter measured by capillary flow analysis.

3.1.2.2.5 Micro Computer Tomography

Knowledge of the structure and morphology of highly porous ceramic materials is essential for various applications. Structural properties such as fractional density, which corresponds to the overall porosity, the mean size and size distribution of the foam cells and the surrounding struts, the shape of the cells, and the isotropy of the individual cells influence many properties. For example, gas permeability is increased by large windows interconnecting cells and decreased by small cell diameters.

Several approaches have been used to evaluate the structure, such as magnetic resonance imaging and optical analysis. Tomography facilitates three-dimensional imaging of complex structures. Optical tomography was used to evaluate transparent aqueous foams: light was transmitted through a sample and detected by a camera with a plane detector [43, 44]. The drawback of light optical analysis is that it cannot be used for nondestructive evaluation of the three-dimensional structure of nontransparent ceramics. However, X rays transmit through most ceramic materials, and thus X-ray tomography can be used to characterize porous ceramic materials [45].

Principle

X-ray micro tomography is based on the evaluation of stacked two-dimensional layers of the sample that are scanned sequentially [46] (Fig. 27a). An X-ray source, usually a microfocus X-ray tube, emits a fan-shaped beam of intensity I_0 that passes through the specimen. The intensity I is detected by a line array of length x. The relative intensity I/I_0 of the transmitted beam depends on the amount and nature of the material being analyzed. Between two recordings the sample is slightly rotated. After a full rotation up to several hundred recordings have been taken. The intensity data can be displayed as a sinugram (Fig. 27b).

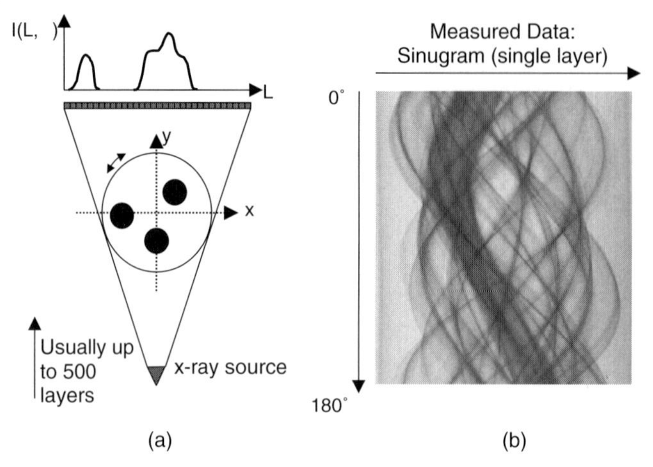

Figure 27 a) Measurement schematic of X-ray tomography and b) raw intensity data displayed as sinugram.

The intensities I and I_0 with and without material and can be described by a line integral

$$I = I_0 \cdot \exp\left(- \int_L \mu(x, y) \; dx \; dy \right) \tag{13}$$

where μ is the material-dependent X-ray absorption coefficient and x is the position on the line L between the X-ray source and the detector; $\mu(x,y)$ is called the object function. During the measurement the sample rotates around the center axis with an angle θ

$$\begin{pmatrix} s \\ t \end{pmatrix} = \begin{pmatrix} \cos\theta - \sin\theta \\ \sin\theta - \cos\theta \end{pmatrix} \cdot \begin{pmatrix} x \\ y \end{pmatrix} \tag{14}$$

where s and t are the coordinates of the rotating system, and x,y the coordinates of the fixed system. The X-ray beam is parallel to the s-axis in the fixed system. If t_0 is the distance between the beam and the s-axis, the points that are on the beam can be described according to

$$x \; \sin\theta + y \; \cos\theta = t_0 \; . \tag{15}$$

Equation (13) can be transformed using the Dirac $\delta(x)$, which has the value of infinity for $x = 0$, zero elsewhere, and a total integral of one. Equation (15) can be described as a line integral after normalization by I_0 and inverting:

$$l\,(\theta, t_0) = \int_{-\infty}^{+\infty} \int_{-\infty}^{+\infty} \mu(x, y) \cdot \delta(x \; \sin\theta + y \; \cos\theta - t_0) \; dx \; dy = \ln \frac{I_0}{I(\theta, t_0)}. \tag{16}$$

This projection of $\mu(x,y)$ on the line integrals $l(\theta,t_0)$ is equivalent to the Radon transformation. A projection P_θ is defined as the line integrals $(-\infty < t < +\infty)$ for a fixed value of the angle θ. A usual measurement can make use of up to several hundred projections.

For a fixed projection angle θ the object function can be calculated from all beams that pass through a point (x_0,y_0):

$$\mu(x_0, y_0) = \int_0^\pi \mu_\theta(x_0, y_0)\, d\theta = \int_0^\pi P_\theta\, (x_0\, \sin\theta\, + y_0\, \cos\theta)\, d\theta\,. \tag{17}$$

As only a limited number M of projection angles are measured, the integral must be replaced by a sum:

$$\mu(x_0, y_0) = \sum_\theta P_\theta(x_0\, \sin\theta + y_0\, \cos\theta) \cdot \frac{\pi}{M}\,. \tag{18}$$

Points that are not passed by any beam are calculated by interpolation of measured adjacent points. These mathematically extensive Radon backprojection calculations can be carried out in several different ways that are mathematically equivalent but differ in the image quality. The Radon backtransformation, as well as the necessary execution of low-pass, ramp, and window filters to achieve artefact-free two-dimensional images of the sample, is performed by high-performance computers. An example of a two-dimensional reconstruction of a polymer-derived ceramic foam is given in Fig. 28a. Stacking successive two-dimensional reconstructed layers facilitates three-dimensional reconstruction of the volume of interest of the measured specimen (Fig. 28b).

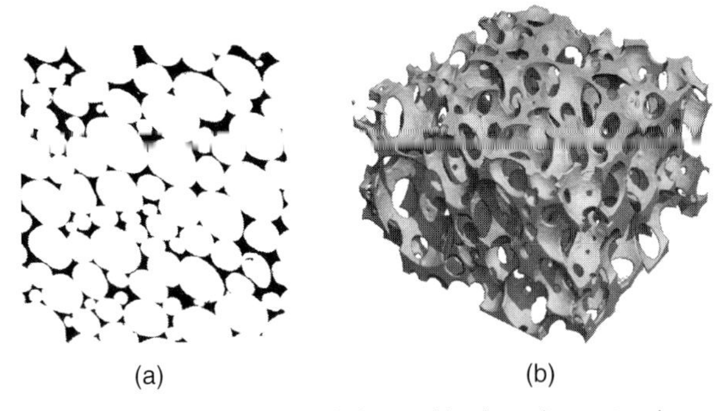

Figure 28 a) Two-dimensional slice and b) three-dimensional reconstruction of an open-cell ceramic foam derived from preceramic polymer.

Measured characteristics

The morphometric *structure model index* (SMI) is calculated from the three-dimensional tomographic data. The SMI was introduced for quantification of bone microarchitecture and can be used in foam structure analysis to characterize the shape and anisotropy of cellular materials. The SMI is calculated according to [47]

$$\mathrm{SMI} = 6 \cdot \left(\frac{V}{S^2}\right) \cdot \left(\frac{dS}{dr}\right) \tag{19}$$

where S is the strut surface in a volume V, and dS/dr the surface-area derivative. The foam surface area is rendered by triangulating the surface of a specific volume of interest (VOI) with the marching cubes method [48] and the volume is defined by setting up polyhedra inside the VOI that match the bounding surface triangles. An SMI of 4 describes spherical pores; any deviation from spherical cell geometry results in lower values. A rod cell structure is described by an SMI of 3, and lenticular pores have an SMI of 0. Negative SMIs result from cells with concave surfaces, which are common if coalescence of pores occurs.

The surface/volume ratio, mean cell size, and strut thickness were calculated by the distance transformation (DT) method [49]. The connectivity density (CD) was computed from CT data by using the Conn–Eulor principle [50]: reconstructed two-dimensional CT data images of two neighboring slices are compared by the Boolean EXCLUSIVE–OR operator. The result is superimposed onto the original pictures and analyzed. All new bridges (B: new connections), holes (H), and islands (I) are counted and CD is calculated:

$$\mathrm{CD} = -\frac{\Sigma H + \Sigma I - \Sigma B}{2h \cdot A} \tag{20}$$

where h is the distance between the two slices ($h = 38\,\mu\text{m}$), and A the image area. In the case of determining CD of the pores, B and I denote cell windows and cells, respectively, and H, which corresponds to isolated strut material completely surrounded by pores, is unlikely to occur. The value is identical if CD is calculated for the strut material, where B denotes new struts, H cells, and I isolated struts.

A brief overview of CT data-evaluation parameters is given in Fig. 29.

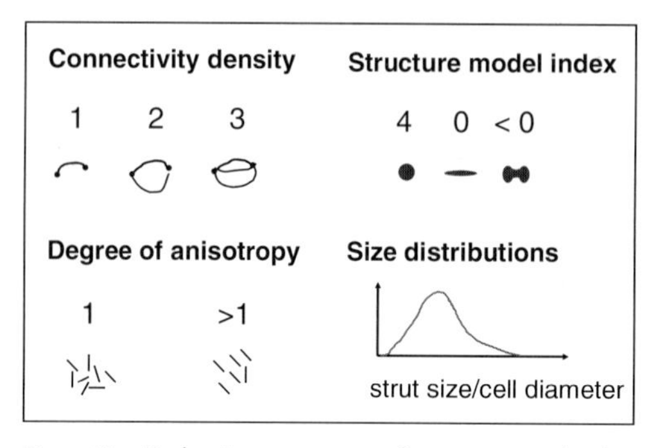

Figure 29 Evaluation parameters from tomography data.

Comparison of ceramic foams from different fabrication methods

Different fabrication methods for ceramic foams result in ceramics with unique structures. Due to the fabrication process, reticulated foams show hollow struts that can be clearly seen both by image analysis and two-dimensional tomography. These foams are characterized by pronounced anisotropy (degree of anisotropy 1.18 compared to 1.08 for polymer-derived ceramic foams) and irregularly shaped foam cells (SMI $\ll 0$). Conversely, sintered hollow spheres have a low degree of anisotropy and a moderate SMI. The low SMI is due to the fact that not only is the interior of the sphere taken into account (which would result in an SMI of ca. 4), but also the irregularly shaped gaps between the cells. The polymer-derived ceramic foams that were foamed by in situ formation of gas bubbles had the highest SMI corresponding to nearly spherical foam cells with high interconnectivity. A summary is given in Fig. 30.

Preparation method	Polymer derived ceramic foam	Reticulated ceramic foam	Sintered hollow spheres
2D image	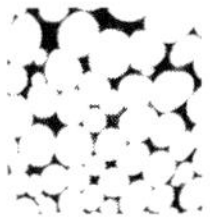	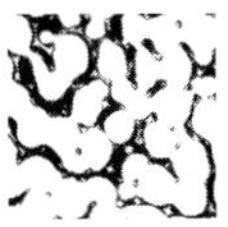	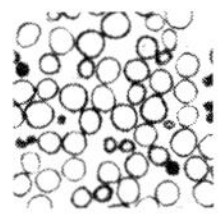
3D image	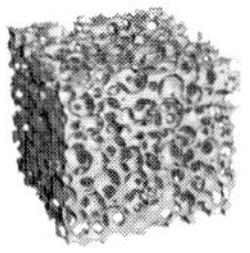	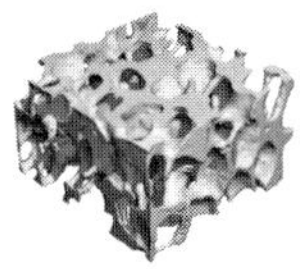	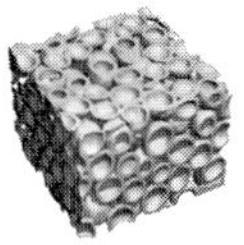
Structure model index	0.95	-2.30	-0.47
Degree of anisotropy	1.08	1.18	1.03
Connectivity density	5.2	0.2	0.3

Figure 30 Different structural parameters of ceramic foams from different preparation methods.

Figure 31 shows the dependence of strut size and cell window distribution on the preparation method. Foams made from sintered hollow spheres and the polymer precursor have narrow strut thickness distributions. The bimodal distribution for the foam from sintered hollow spheres likely results from two adjacent spheres that are attached to each other (0.45 mm) and thus result in a strut thickness of twice the thickness of the sphere shell (0.25 mm). The reticulated ceramic foam, in contrast, shows a wide distribution of strut thickness. The cell diameter of the ceramic foams differ both in mean value and distribution (Fig. 31b). The polymer-derived ceramic foams have a narrow distribution with a peak at 0.9 mm. The foam made from sintered hollow spheres shows a wider bimodal distribution with peaks at 0.7 and 1.4 mm. The distribution of the reticulated foam has a bimodal curve with a maximum of 1.3 mm.

Specifications

The resolution of tomography is physically limited by the wavelength and energy of the radiation. High-energy synchrotron radiation provides a higher resolution than conventional X-ray sources, which is desirable for the evaluation of microcellular foams with cell diameters in the micrometer and submicrometer range. However, challenging technological difficulties limit the resolution, and the physical limits of

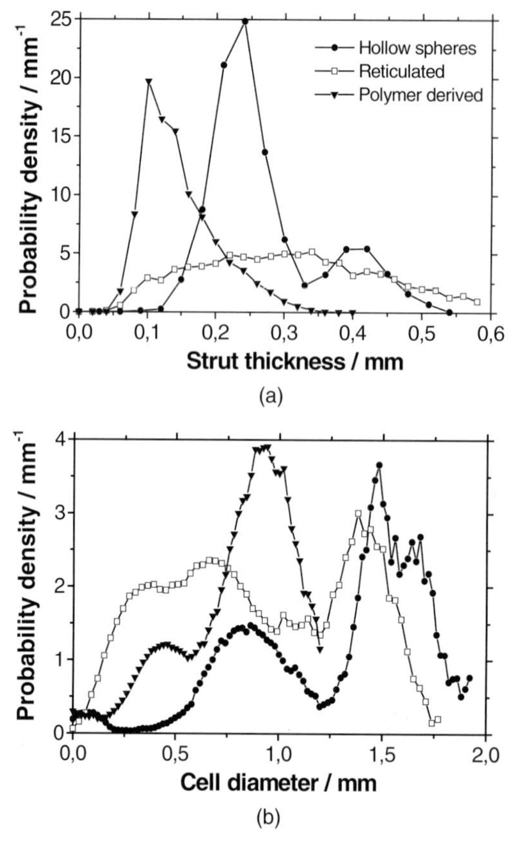

Figure 31 a) Strut thickness and b) cell diameter distribution of different ceramic foams: polymer-derived ceramic foam [49], reticulated ceramic foam [50], and cellular ceramic from sintered hollow spheres [51].

X-ray tomography have not yet been reached. The resolution is set by the size of the detector array and the number of projections taken. Common line arrays have around 1000 detector elements. Additional projections that are carried out with translation of the X-ray source and the detector will increase the resolution. Any additional projection will increase the calculation effort. A VOI containing $1000 \times 1000 \times 1000$ volume elements results in a raw data set of several gigabytes. Handling this much data is time-consuming and requires high-performance computers. The resolution of the X-ray tomograph in this study (μCT 40, Scanco Medical AG, Bassersdorf, CH) scales linearly with sample size, resulting in a resolution of 20 μm for a sample diameter of 20 mm and a best resolution of 5 μm for samples of 5 mm in diameter. High resolution X-ray tomographs that scan large volumes of interest with a best resolution below 2 μm are available at prices in accordance to this performance. In the near future, resolutions better than 400 nm will be achieved by nano CT tomographs. However, the VOI size will be limited to about 1 mm^3.

Results

1) The two-dimensional *sectional images* made by micro computer tomography offer the opportunity for a visual inspection of the structure in a nondestructive manner. Parameters like homogeneity of the sample, especially the thickness of deposition in the interior of the foam and edge effects can easily be observed from these kinds of images. Figure 32 shows slices through the middle of the sample (scan 150 of about 400 slices) for samples 1, 2, 7, and 8.

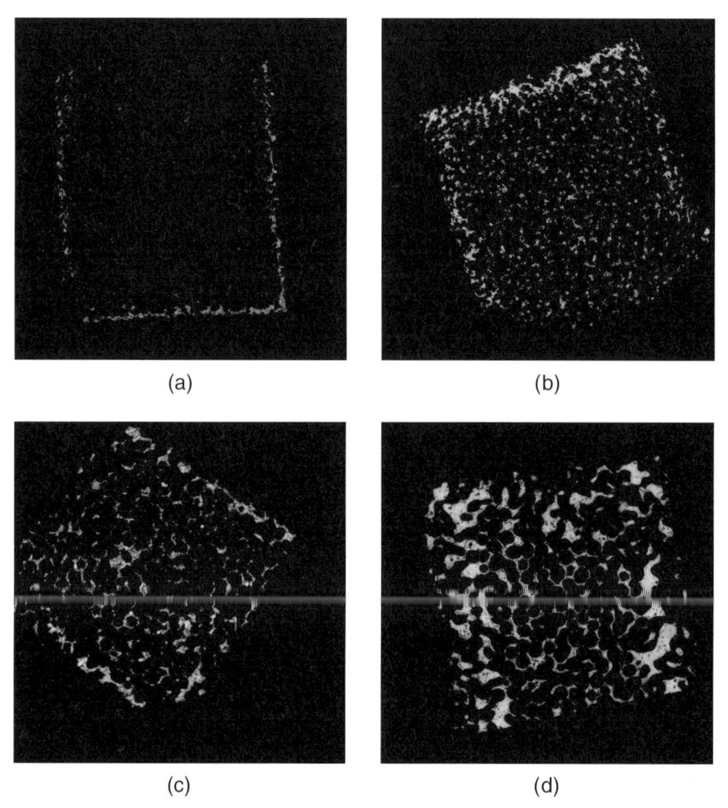

Figure 32 Two-dimensional slices made by micro computer tomography for a) sample 1, b) sample 2, c) sample 7, and d) sample 8.

2) The *pore size distribution* of the ceramic foams fabricated by replication of reticulated polyurethane is presented as a series of cumulative curves in Fig. 33. The trend of increasing pore diameter for the eight samples, as measured by image analysis and the Visiocell method, is not followed by the micro computer tomography analysis. A complete comparison between the different characterization techniques for these samples and possible fundamental reasons for disagreement are discussed in Section 3.1.2.3.

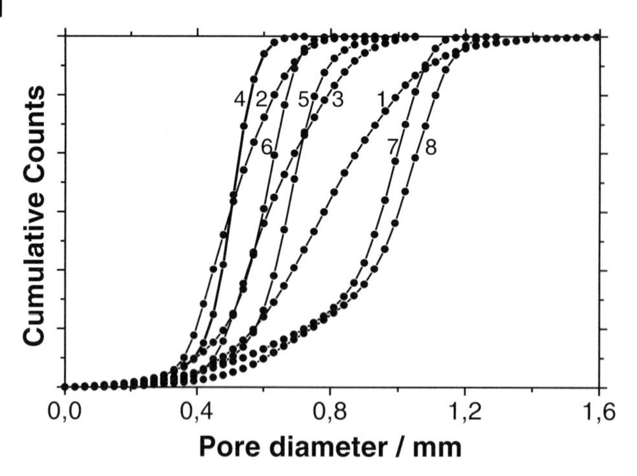

Figure 33 Cumulative pore diameter curves for the eight samples measured by micro computer tomography.

3) The *strut thickness distributions* for samples 1, 2, 7, and 8 are compared in Fig. 34a. Both the polyurethane skeleton and the composition of the suspension that was used to coat the polymer skeleton have a substantial influence on the strut thickness distribution. With increasing cell size of the polyurethane foam, thicker struts are obtained in the ceramic replica when coated with equal concentrations of ceramic suspension. Coating the polyurethane foam with a more viscous suspension by increasing its solids loading leads to higher mean strut thickness and a broader distribution. Figure 34b presents the mean strut thickness for all samples and the width of the distribution. Visual inspection of the two-dimensional slices confirms this trend (Fig. 31). Especially in filter applications, strut morphology and dimensions are important parameters.

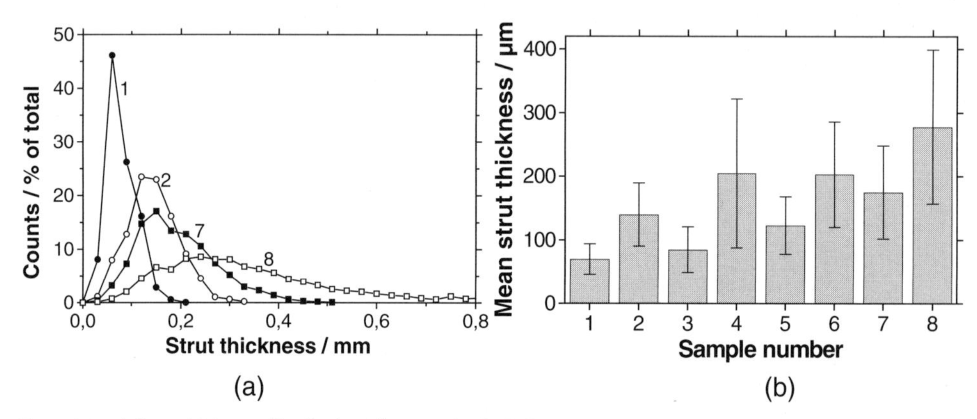

Figure 34 a) Strut thickness distributions for samples 1, 2, 7, and 8 and b) mean strut thickness and width (shown by error bars) of the distribution for all samples.

3.1.2.2.6 **Other Techniques**

Mercury porosimetry is based on the intrusion of mercury into the pore volume under pressure. The pressure required to force mercury into pores gives the pore diameter and the liquid-intruded volume gives pore volume and pore volume distribution. However, pores larger than a few hundred micrometers are hard to detect [52, 53].

Analogous to capillary flow analysis (see Section 3.1.2.2.4), *liquid extrusion porosimetry* is based on the extrusion of a wetting liquid from the porous sample that is placed on a membrane. The characteristics of the membrane are such that its largest pore is smaller than the smallest pore in the sample. By gradually increasing the pressure of a nonreactive gas above the sample, the liquid is removed from its pores. However, because of the smaller pore size in the membrane, the pressure is not sufficient to remove the liquid from these pores. As a result, no gas flows through the membrane. Instead, the displaced liquid from the pores of the sample merely passes through the liquid-filled pores of the membrane. By measuring the liquid volume extruded from the sample, this technique has the ability to determine the pore volume distribution, the pore diameter, and the liquid permeability [54, 56].

The *BET gas adsorption technique* (Brunauer–Emmett–Teller) is based on the principle of measuring the amount of adsorbed gas on the pore surface as a function of the gas pressure below the equilibrium vapor pressure. Measured characteristics are the pore diameter and the combined volume of blind pores and through-pores. The measurable pore size ranges from 0.5 nm to 1 μm. This technique is considered to be one of the most accurate in determining the surface area of samples with high specific surface area [54].

NMR micro-imaging is a three-dimensional, nondestructive technique that can be used to characterize foam materials. Measurement with sufficient resolution is still very time consuming for materials with small pores. Nevertheless, with technical improvements, it could become a promising technique [57–59].

In *ultrasound imaging*, ultrasonic excitation causes a reflected and backscattered signal in the sample that can be detected for depth-resolved cross sections. This method of nondestructive imaging and noninvasive detection of porosity has been applied to bone samples and aluminum castings. Detailed surface and subsurface images can be obtained, with a resolution from 1 to 100 μm, depending on the frequency of the acoustic microscope. More experimental work has to be done to verify the value of this technique in the characterization of cellular ceramics [60, 61].

Although the theoretical concepts of confocal microscopy date from several decades ago, the actual applicability of this imaging technique is a relatively new development due to the recent advances in optical and electronic technology. Especially in the biology-related sciences, *laser scanning confocal microscopy* has proven its value in the visualization of micro-organisms after staining with fluorescing dyes (osteoblast cells adhering to porous substrates, protein immobilization, etc.) [62, 63]. Also in geology, the technique is widely applied in the determination of the pore geometry in porous sandstones or for understanding the failure mechanisms in geomaterials [64]. Confocal imaging rejects the out-of-focus information by placing a pinhole in front of the detector. This confocal pinhole is what gives the system its confocal property, by rejecting light that did not originate from the focal plane of the microscope objective. Confocal imaging can only be performed with pointwise illumination and detection, which is an important advantage of using laser scanning microscopy. This ability to reject light from above or below the focal plane enables the confocal microscope to perform depth discrimination and optical tomography. A true three-dimensional image can be processed by taking a series of confocal images at successive depths in the specimen. Moreover, the submicrometer resolution in both lateral and axial directions makes it a powerful tool in three-dimensional mapping of pore architecture.

3.1.2.3 Comparison of Methods

Figure 35 gives an overview of the mean pore diameter of the eight samples, investigated with several techniques. Because capillary flow analysis measures the restricted pore diameter and not the pore diameter as such, these results are not included in the comparison.

The values obtained by the Visiocell method are meaningful, as the main conditions for a meaningful interpretation of the data are fulfilled for these samples. Image analysis was performed both on microscopic images and on two-dimensional

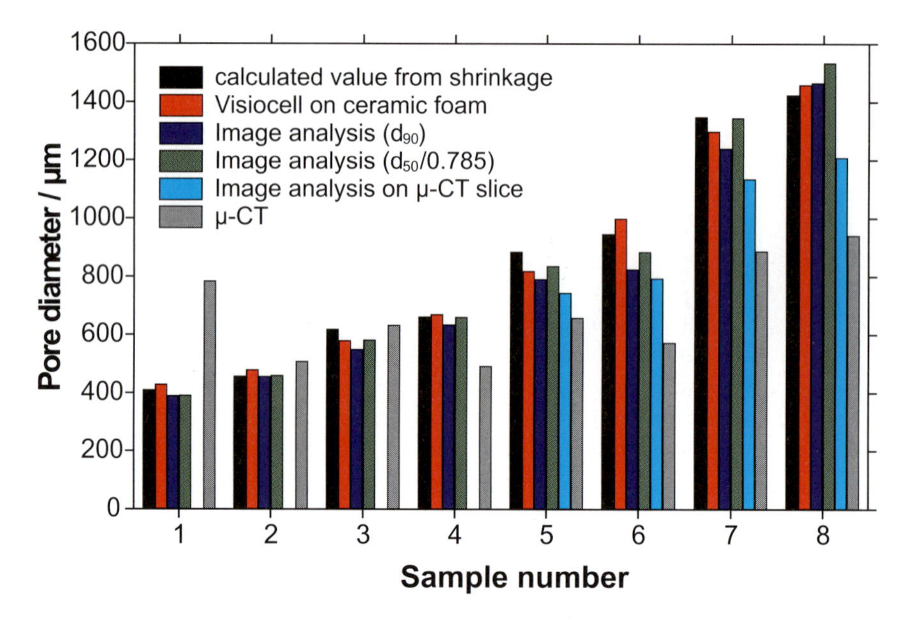

Figure 35 Comparison of the measured pore diameter of the polyurethane foam from the expected value of the ceramic foam taking into account the shrinkage, the pore diameter by Visiocell, the d_{90} value by image analysis on microscopy images, the three-dimensionally corrected d_{50} value by image analysis on microscopy images, the d_{90} value by image analysis on micro computer tomography images, and the three-dimensional mean pore diameter from micro computer tomography.

slices obtained by micro computer tomography. By assuming a relatively narrow distribution of spherical pores, a stereological correction factor to convert the two-dimensional data to the three-dimensions of the actual pore space can be applied to the mean pore diameter. A three-dimensional characterization of the pore architecture was performed by micro computer tomography, from which the mean pore diameter was obtained.

The data from Visiocell and image analysis are in reasonable agreement when corrected for the dimensional aspect of the latter technique. This can be done either by selecting the largest two-dimensional diameter (the d_{90} value) or by applying a stereological factor to the mean pore diameter. For these samples, the differences between the two methods are 4–7 %.

The mean pore diameters measured by micro computer tomography are for almost all samples in discordance with the results from Visiocell or image analysis. Part of the explanation probably involves differences in threshold procedure, as image analysis on the micro computer tomography slices does not completely match the results on the microscopic value. The image analysis results for the micro computer tomography images are consistently lower than the results from microscopy images. In both methods, the threshold procedure and possible deviations from the actual pore boundary structure will have a significance influence on the measurement.

Thresholding differences cannot account fully for the complete mismatch of the results. Other features come into play. The assumptions made on the shape factor of the individual pores may affect the calculations, both for image analysis and micro computer tomography. With regard to the SMI values of these samples, micro computer tomography detects the pore structures more as rodlike pores, as opposed to the sphere-like porosity in image analysis. Another factor that may contribute to this issue is the fact that every single pore is taken into account equally in the SMI calculation and the pore diameter distribution in the mico computer tomography procedure. This contrary to image analysis, in which artefacts are filtered by imposing a minimum pore area. If small, very irregular pores are present, this might influence the SMI value. Unfortunately, the micro computer tomography software does not enable a visual verification of the identified pores.

A general overview of some of the most common techniques for analysing pore architecture and the specifications and capabilities of each technique are summarised in Tab. 4.

3.1.3 Summary

Due to the complex nature of the three-dimensional pore space in most porous ceramics and the wide variety of internal structures, no single analytical technique so far provides a complete and accurate description of the morphology.

Table 4 Comparison of the characterization techniques.

	Mercury intrusion	Liquid extrusion	Capillary flow	Image analysis	Visiocell	Computer tomography	Gas adsorption
Dimension of measurement							
	3D	3D	3D	2D	3D	3D	3D
Pore diameter range (in μm)							
	300–0.03	2000–0.05	500–0.013	depends on image input	> 450	> 5	1–0.0005
Capabilities and interpretation of measurement for different pore parameters							
Pore diameter	based on volume; many diameters	based on volume; many diameters	most constricted diameter	yes	only for spherical pores	yes	many diameters
Pore volume	yes, through- and blind pores	yes, through- and blind pores	no	assumptions necessary	only for spherical pores	yes	yes, through- and blind pores
Pore nature	combination of through- and blind pores	through- and converging/diverging pores	only through-pores	no	no		combination of through- and blind pores
Permeability	no	liquid	gas and liquid	no	no	no	no
Surface area	of blind and through-pores	of through-pores only	of through-pores only	yes	no	yes	of blind and through-pores
Strut thickness	no	no	no	yes/no	yes	yes	no

In the last few years, new developments in the production of foam materials and analysis techniques have occurred. Several innovative processing routes have led to a large diversity of ceramic foam materials. Substantial progress in characterization methods has led to faster and more accurate measurements. These developments necessitate identifying and implementing a standardized procedure for analyzing the pore structure of this class of materials, because of the significant effect of structure on their properties and uses.

Several experimental techniques have been reviewed with regard to their application as characterization techniques for cellular ceramics. A meaningful comparison of the techniques is difficult and not straightforward, as each technique probes the pore structure differently. Moreover, the dimensional aspect of the measurement is of key significance in the interpretation of the results. Presently, combining information from multiple techniques is the only way one can obtain an overall picture of the real three-dimensional architecture.

References

1 Colombo, P., *Key Eng. Mater.* **2002**, *206–213*, 1913–1918.
2 Klempner, D., Sendijarevic, V.: *Polymeric Foams and Foam Technology*, Hanser Gardner Publications, Cincinnati, USA, 2004.
3 Luyten, J., Cooymans, J. , De Wilde, A., Thijs, I., *Key Eng. Mater.* **2002**, *206–213*, 1937–1940.
4 Sepulveda, P., dos Santos, W., Pandolfelli, V., Bressiani, J., Taylor, R., *Am. Ceram. Soc. Bull.* **1999**, *82(2)*, 61–66.
5 Sepulveda, P., *Am. Ceram. Soc. Bull.* **1997**, *76(10)*, 61–65.
6 Strom, L., Sweeting, T., Norris, D., Norris, J., *Mater. Res. Soc. Symp. Proc.* **1995**, *371*, 321–326.
7 Ashby, M., *Metall. Trans.* **1983**, *14A*, 1755–1769.
8 Brezny, R., Green, D., Chap. 9 in *Material Science and Technology, Vol.11*, Cahn, R., Haasen, P., Kramer, E.J., Swain, M. (Eds.), Wiley-VCH, Weinheim, 1991, pp. 453–516.
9 Luyten, J., Mullens, S., Cooymans, J., De Wilde, A., Thijs, I., *Adv. Eng. Mater.* **2003**, *5(10)*, 715–718.
10 Innocentini, M., Salvini, V., Pandofelli, V., Coury, J., *Am. Ceram. Soc. Bull.* **1999**, *82(2)*, 78–84.
11 Philipse, A., Schram, H., *J. Am. Ceram. Soc.* **1991**, *74(4)*, 724–732.
12 Innocentini, M., Sepulveda, P., Salvini, V., Pandolfelli, V., *J. Am. Ceram. Soc.* **1998**, *81(12)*, 3349–3352.
13 Gibson, L., Ashby, M., *Cellular Solids: Structure and Properties 3*, Pergamon Press, New York, 1988.
14 Will, J., *Porous Support Structures and Sintered Thin Films Electrolytes for Solid Oxide Fuel Cells*, PhD thesis, SFIT Zurich, 1998.
15 Brezny, R., *Mechanical Behavior of Open Cell Ceramics*, PhD thesis, Penn State Univ., 1990.
16 Zhang, J., Ashby, M., *Theoretical Studies On Isotropic Foams*, Report # CUED/C-Mats/ TR 158, Cambridge University Engineering Department, April 1989.
17 Weaire, D., Phelan, R., *Philos. Mag. Lett.* **1994**, *69(2)*, 107–110.
18 Peng, H., Fan, Z., Evans, J., *Mater. Sci. Eng. A* **2001**, *303*, 37–45.
19 Wallaeys, B., Mortelmans, R., *Recticel Brochure Technical Foams – The Guide 2000*, Recticel Co., Wetteren, 2000.
20 ASTM Standards, *Standard Test Method for Cell Size of Rigid Cellular Plastics*, ASTM D 3576, 1977, 919–922.
21 Richardson, J., Peng, Y., Remue, D., *Appl. Catal. A Gen.* **2000**, *204*, 19–32.
22 Lock, P., Jing, X., Zimmerman R., Schlueter, E., *J. Appl. Phys.* **2002**, *92(10)*, 6311–6319.
23 Pierret, A., Capowiez, Y., Belzunces, L., Moran, C., *Geoderma* **2002**, *106*, 247–271.

24 Liu, Z-Q., Austin, T., Thomas, C., Clement, J., *Comput. Biol. Med.* **1996**, *26(1)*, 65–76.
25 Guild, F., Summerscales, J., *Composites* **1993**, *24(5)*, 383–393.
26 Cárcel, A., Ferrer, C., Determination of Pore Size "Distribution and Uniformity in Closed Cell Metal Foams from 2D-Image Analysis" in Proceedings of the International Conference on Cellular Metals and Metal Foaming Technology, Banhart, J. (Ed.), MIT Publications, Berlin, 2003.
27 J. Russ in Roadmap Guide to Image Analysis, Workshop Notes (http://www.drjohnruss.com/).
28 Landis, E., Petrell, A., Nagy, E., *Concr. Sci. Eng.* **2000**, *2(8)*, 162–169.
29 Thijs, I., Luyten, J., Mullens, S., *J. Am. Ceram. Soc.* **2003**, *87(1)*, 170–172.
30 Underwood, E., *Quantative Stereology*, Addison-Wesley Publishers, Reading, 1970.
31 Royet, J. *Prog. Neurobiol.* **1991**, *37*, 433–474.
32 Russ, J., Dehoff, R., *Practical Stereology*, Plenum Press, New York, 1999 (http://www.Practical-Stereology.org).
33 Higgins, M., *Am. Mineral.* **2000**, *85*, 1105–1116.
34 Cruz-Orive, L., *J. Microsc.* **1983**, *131(3)*, 265–290.
35 Berger, A., Roselle, G., *Am. Mineral.* **2001**, *86*, 215–224.
36 Heilbronner, R., Bruhn, D., *J. Struct. Geol.* **1998**, *20(6)*, 695–705.
37 Moreau, E., Velde, B., Terribile, F., *Geoderma* **1999**, *92*, 55–72.
38 Jena, A., Gupta, K., Sarkar, P., *Am. Ceram. Soc. Bull* **2003**, *82(12)*, 9401–9406.
39 Jena, A., Gupta, K., *J. Power Sources* **2001**, *96(1)*, 214–219.
40 Fernando, J., Chung, D., *J. Porous Mater.* **2002**, *9(3)*, 211–219.
41 Kozeny, J., Carman, P., *Ber. Wien. Akd.* **1927**, *136*, 271–278.
42 Carman, P., *Trans. Inst. Chem. Eng.* **1937**, *15*, 150–166.
43 Monnereau, C., Adler, M., *J. Colloid Interface Sci.* **1998**, *202*, 45–53.
44 Thomas, P., Darton, R., Whalley, P., *Ind. Eng. Chem. Res.* **1998**, *37*, 710–717.
45 Sasov, A., *J. Microsc.* **1987**, 147, 169–187.
46 Elmoutaouakkil, A., Salvo, L., Maire, E., Peix, G., *Adv. Eng. Mater.* **2002**, *4*, 803–807.
47 Hildebrand, T., Rüegsegger, R., *Comput. Methods Biomech. Eng.* **1997**, 15–23.
48 Lorensen, W., Cline, H., *Comput. Graphics* **1987**, *21(4)*, 163–169.
49 Hildebrand, T., Rüegsegger, R. *J. Microsc.* **1997**, *185*, 67–75.
50 Gundersen, H., Boyce, R., Nyengaard, J., Odgaard, A., *Bone* **1993**, *14*, 217–222.
51 Zeschky, J., Goetz-Neunhoeffer, F., Neubauer, J., Lo, S., Kummer, B., Scheffler, M., Greil, P., *Composites Sci. Technol.* **2003**, *63*, 2361–2370.
52 Schwartzwalder, K., Somers, A., U.S. Patent 3,090,094, 1963.
53 Luyten, J., Mullens, S., Cooymans, J., DeWilde, A., Thijs, I., "Ceramic Foams: Synthesis and Characterization" in Proceedings of Shaping II, 2nd International Conference on Shaping of Advanced Ceramics (Ghent), Luyten, J., Erauw, J-P. (Eds.), VITO, Mol, 2002.
54 Jena, A., Gupta, K., *Materialprüfung* **2002**, *44(6)*, 243–245.
55 Banhart, J., *Progr. Mater. Sci.* **2001**, *46*, 559–632.
56 Jena, A., Gupta, K., *Ceram. Eng. Sci. Proc.* **2002**, *23(4)*, 277–284.
57 Szayna, M., Voelkel, R., *Solid State Nucl. Magn. Reson.* **1999**, *2*, 99–102.
58 Caprihan, A., Clewett, C., Kuethe, D., Fukushima, E., Glass, S., *Magn. Reson. Imag.* **2001**, *19*, 311–317.
59 Beyea, S., Caprihan, A., Glass, S., Giovanni, A., *J. Appl. Phys.* **2003**, *94(2)*, 935–941.
60 Callé, S., Remenieras, J-P., Bou Matar, O., Defontaine, M., Patat, F., *Ultrasound Med. Biol.* **2003**, *29(3)*, 465–472.
61 Maev, R., Sokolowski, J., Lee, H., Maeva, E., Denissov, A., *Mater. Charact.* **2001**, *46(4)*, 263–269.
62 Damaskinos, S., Dixon, A., Ellis, K., Diehl-Jones, W., *Micron* **1995**, *26(6)*, 493–502.
63 Ljunglöf, A., Thömmes, J., *J. Chromatogr. A* **1998**, 813(2), 387–395.
64 Fredrich, J., *Phys. Chem. Earth (A)* **1999**, *24(7)*, 551–561.

3.2
Modeling Structure–Property Relationships in Random Cellular Materials

Anthony P. Roberts

3.2.1
Introduction

Today new and advanced materials are constantly being developed. The central goal is to develop materials with superior properties to those currently available, or to achieve similar properties at a reduced cost or weight. Since materials synthesis and extensive empirical testing are both time-consuming and expensive, it is important to guide the process as far as possible by theoretical modeling. Composite and porous materials have found broad application in virtually every field of science and engineering because they offer combinations of properties not available in their constituent materials. A key problem is to understand and quantify the relationship between the internal structure of these materials and their properties. The resulting structure–property relationships are used for designing and improving materials, or conversely, for interpreting experimental relationships in terms of microstructural features.

Ideally, the aim is to construct a theory that employs general microstructural information to make accurate property predictions. Since the properties of advanced composites and porous materials are dependent on their complex internal structure, it has proved extremely difficult to develop accurate structure–property relationships in all but the simplest of cases. Two main problems must be solved in order to generate accurate structure–property theories. First, an appropriate model of microstructure is needed; second, the properties of the microstructure must be accurately evaluated. Often these are contradictory goals; a realistic structural model generally prohibits the use of analytical techniques to predict properties, and conversely, the requirement of making an accurate property prediction generally dictates an over-simplified model of structure.

This chapter surveys past results and recent progress in the related fields of measuring and modeling microstructure, and predicting the properties of random materials. Attention is restricted to the influence of the shape of the solid–pore morphology on material properties, rather than the role of the internal microstructure of the solid struts (which is taken as given). We also deal only with the rigidity (stiffness) of porous materials, rather than more complex mechanical properties such as failure stress.

Cellular Ceramics: Structure, Manufacturing, Properties and Applications.
Michael Scheffler, Paolo Colombo (Eds.)

ISBN: 3-527-31320-6

No single model or simulation can attempt to reproduce the rich variety of structures that can occur in porous materials in general, or even in the subset defined by cellular ceramics. A brief introduction is provided to the main classes of statistical models that have been used to represent porous materials. These include Boolean, level-set Gaussian, coalescing bubble, and Voronoi tessellation models. The statistical characterization of porous materials and how this information can be used to "tune" models so that they better mimic real materials are also discussed. This offers the possibility of statistically reconstructing three-dimensional models from experimentally measured two-dimensional images. Technical aspects of rigorously predicting composite properties have been reviewed in two recent monographs [1, 2]. A simplified summary of these theoretical results is provided for general porous materials, as well as results more directly relevant to cellular materials. Recent practical advances in computational theories which are pertinent to cellular ceramics are reviewed.

3.2.2
Theoretical Structure–Property Relations

The deformation of composite materials under stress is governed by the equations of elasticity. In principle it is therefore necessary to solve these equations within a representative element of a heterogeneous material to determine its macroscopic behavior [1–3]. All of the methods described below follow this approach, although other theoretical approaches bypass the elasticity equations altogether. The minimum solid area models, reviewed in Part 4, are an example. An example of cellular ceramic and a simple model of bubble growth and coalescence are shown in Fig. 1. To illustrate theories discussed below computational estimates of Young's modulus as a function of density for this model of "expanding bubbles" are plotted in Fig. 2.

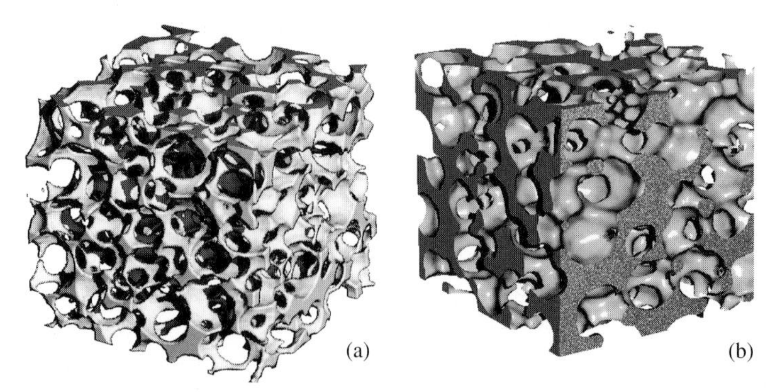

Fig. 1 a) An example of a CT scan of a cellular ceramic. Image courtesy of J. Zeschky, see also Ref. [50]. b) A simple "expanding-bubbles" model generated by dilating spherical pores seeded at the centers of a dense hard sphere pack.

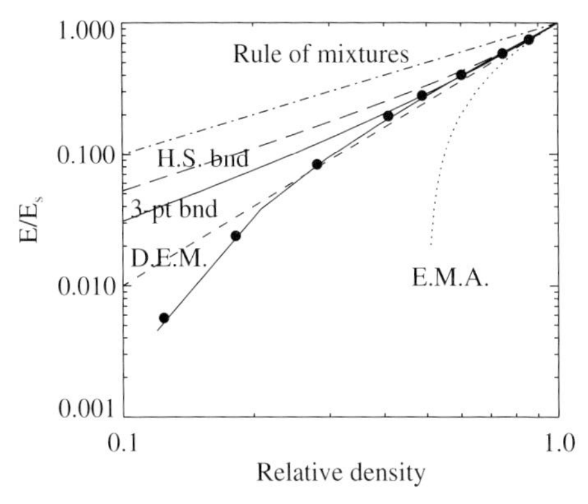

Fig. 2 A comparison of five theories with finite-element data for the expanding-bubbles model shown in Fig. 1. The predictions and bounds are described in the text. The solid line is a fit to the data based on Eq. (13).

The simplest approach to estimating the properties of a composite material is to use the rule of mixtures, which states that the Young's modulus is

$$E = pE_1 + (1 - p)E_2 \quad (1)$$

where p is the volume fraction of phase one and E_1 (E_2) is the modulus of phase 1 (2). This formula is exact for a material composed of slabs of each material, but only for stress parallel to the direction of the slabs. For a porous material $p = \rho/\rho_s$ corresponds to the relative density and we take $E_1 = E_s$ (and $E_2 = 0$). The rule predicts a relative modulus of $E/E_s = p$, which significantly overestimates the effective modulus (see Fig. 2).

The best known theories for more realistic models apply to low concentrations of spherical inclusions in an otherwise homogeneous matrix [3]. For randomly distributed dilute spherical pores of concentration 1–p, the relative Young's modulus is given by the simple formula

$$\frac{E}{E_s} = 2p - 1. \quad (2)$$

Here attention is restricted to the case where the Poisson ratio of the solid structure is $\nu_s = 0.2$. Although algebraically quite complex, the result only differs by a few percent [4] if $0 < \nu_s < 0.4$, which is generally the case. Note that the result applies to a solid with a few remote bubbles, that is, the incipient stages of foam formation.

There are two main approximate methods of extending Eq. (2) to finite concentrations of inclusions. The first method has been called the self-consistent or effective-medium approximation (E.M.A.) [3, 5, 6]. For general ν_s and phase properties the formula cannot be expressed in closed form, but for hollow inclusions and a solid Poisson's ratio of $\nu_s = 0.2$ the result is given again by Eq. (2), which is therefore expected to be useful outside the strict dilute limit. However, as can be seen in

Fig. 2, the result is inaccurate for the expanding-bubbles model with relative densities below $p = 0.8$. The second common generalization of the dilute result to finite porosity is provided by the differential effective medium (D.E.M.) theory (see Ref. [7] for a review). Again the result is only expressible in closed form if $\nu_s = 0.2$ and the inclusion phase is hollow. This gives

$$\frac{E}{E_s} = p^2 . \tag{3}$$

The differential result provides a very good approximation for the expanding-bubbles model for relative densities $p > 0.3$ (see Fig. 2). For intermediate densities, this is probably more fortuitous than an indication of a good model: At low relative densities the actual microstructure of the bubble-growth model is quite different from that "built-in" [1] to the differential result [8].

To go beyond the spherical-inclusion approximation two relevant strands of literature have emerged for porous materials. The fully general approach incorporates arbitrary microstructural information, typified by variational bounding methods. An alternative approach focuses on periodic structures.

The rigorous bounding methods are dealt with in several comprehensive reviews [1–3, 9]. For high-contrast (e.g., porous) composites the bounds tend to be so far apart that they lack predictive power. Indeed, for porous materials the lower bound is zero, but there is evidence that the upper bound E_u can sometimes provide a reasonable approximation [1], as well as a useful theoretical upper limit. The most commonly used result is the Hashin–Shtrikman bound (H.S. bnd) for isotropic materials. For porous solids the bound simplifies to

$$\frac{E_u}{E_s} = \frac{p}{1+C(1-p)} . \tag{4}$$

The result is only weakly dependent on ν_s via the constant C, and $C = 1$ for $\nu_s = 0.2$. Note that the rule of mixtures exceeds the bound (which does not apply to anisotropic materials). A key theoretical advantage of the bounding approach is that it can be extended to incorporate arbitrary microstructural information. For example the three-point (3-p) bounds of Milton and Phan-Thien [10] for $\nu_s = 0.2$ are of the form of Eq. (4) with [4]

$$C = \frac{33\eta+7\zeta}{5\zeta(9\eta-\zeta)} . \tag{5}$$

The parameters η and ζ involve multidimensional integrals of the three-point correlation functions of the microstructure. These have been tabulated for a wide range of materials [4, 9, 11, 12]. Although progress has been made for many realistic classes of models, it is quite a challenging task to actually derive the correlation functions for a given statistical model (e.g., no results exist for Voronoi tessellations). The improvement of the three-point bounds over the Hashin–Shtrikman bounds is shown in Fig. 2. The three-point bound provides a reasonable approximation for $p \geq 0.4$, but significantly overestimates the modulus of the bubble model for lower densities. Note that the bound has been evaluated for uncorrelated pore centers, which does not exactly match the expanding-bubbles model. Nevertheless, the bound for the bubble model is expected to be slightly greater than the bound for uncorrelated pores, because the struts are more uniform in the former model (and hence provide greater stiffness). In general it has been shown that the three-point bound does not provide accurate predictions for low-density ($p < 0.4$) porous materials [4].

The most effective approach for low-density cellular materials has been to model the properties of periodic arrays by approximating the behavior of single cells with thin-beam theories. The results provide qualitative predictions of foam properties and illustrate the basic mechanisms of deformation in cellular materials. First consider a simple cubic array of uniformly spaced intersecting aligned struts. From elementary considerations, the Young's modulus is $E/E_s \approx p/3$ for uniaxial compression along a strut axis. The linear dependence of modulus on density is typical of model foams that contain straight-through struts that traverse the extent of the sample; longitudinal compression or tension is the only mode of deformation [13–15]. Since most foams do not contain straight-through struts, beam bending comes into play [14–17]. In this case a quadratic dependence of the modulus on density is observed. The most commonly used result for open-cell foams is [18]

$$\frac{E}{E_s} \approx Cp^2; \qquad \nu \approx \frac{1}{3} \tag{6}$$

where the prefactor $C \approx 1$ and Poisson's ratio have been empirically determined. This semi-empirical formula broadly describes data obtained for many different types of foams.

Zhu et al. [15] and Warren and Kraynik [19] derived analytic results for an open-cell tetrakaidecahedral model packed in a body-centered cubic array. The model is shown in Fig. 3a. The results of Zhu et al. for Young's modulus and Poisson's ratio for strain parallel to the $\langle 100 \rangle$ axis are

$$\frac{E_{100}}{E_s} = \frac{2C_z p^2}{3(1+C_z p)}; \qquad \nu_{12} = \frac{1}{2}\left(\frac{1-C_z p}{1+C_z p}\right) \tag{7}$$

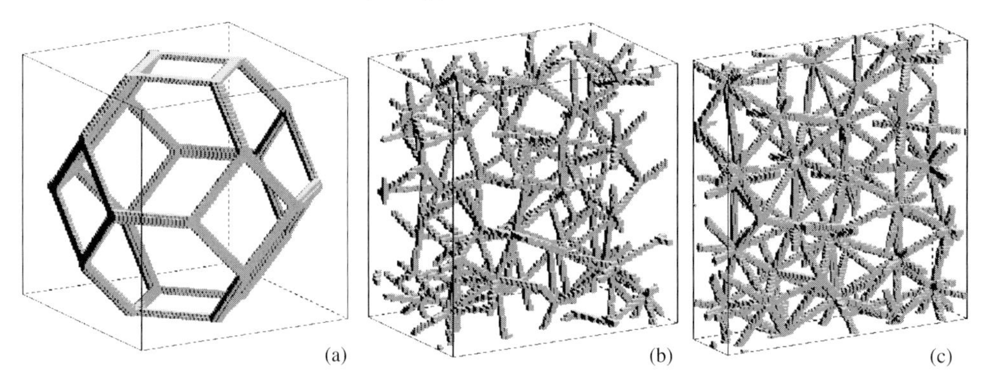

Fig. 3 Models of cellular solids. a) The periodic tetrakaidecahedral model [28]. b) and c) are node-bond models with coordination numbers (average number of connections per node) of 5.5 and 12, respectively [42].

where $C_Z = 8\sqrt{2}IA^2$ depends on the cross-sectional area A and the second moment of the area I. For equilateral triangles $C_Z = 1.09$ [15], and for cylindrical beams $C_Z = 0.900$. Note that Poisson's ratio depends on orientation. The notation ν_{12} corresponds to expansion measured in the $\langle 010 \rangle$ or $\langle 001 \rangle$ directions. Note that the foam is relatively stiff under uniform compression, with the bulk modulus given by $K/K_s = {}^1/_9 \cdot p$, and $\nu_{12} \to 0.5$ as $p \to 0$.

Several methods have been proposed to derive analytic predictions for random isotropic foams. A typical result, which performs an isotropic average of randomly placed long thin (i.e., straight-through) struts, was derived by Christensen [13]

$$\frac{E}{E_s} = \frac{1}{6}p; \quad \nu = \frac{1}{4}. \tag{8}$$

who noted that the results are equivalent to those of Gent and Thomas [20]. In the low-density limit, the same results were derived for a rotationally averaged simple cubic structure [14]. The absence of bending in these models is indicated by the linear dependence of Young's modulus on density.

Warren and Kraynik [14] derived analytic results for the properties of a foam comprised of isotropically oriented tetrahedrally arranged struts. The geometry can be visualized as a node located at the center of a tetrahedron with equilateral faces, in which the four struts (separated by an angle of 109.5°) connect the central node to the vertices. The results are

$$\frac{E}{E_s} = \frac{C_W p^2 (11+4C_W p)}{10+31C_W p+4C_W^2 p^2}; \quad \nu = \frac{1}{2}\frac{(1-C_W p)(10+8C_W p)}{10+31C_W p+4C_W^2 p^2} \tag{9}$$

where $C_W = 18I/\sqrt{3}\,A^2$. For struts of equilateral triangular cross section $C_W = 1$, while for a circular cross section $C_W \approx 0.827$. As expected from the definition of the model, beam bending is the primary mode of deformation for uniaxial compression. However, Eq. (9) imply $K/K_s = {}^1/_9 \cdot p$, that is, bending is not activated under pure compression. Like the tetrakaidecahedral model, Poisson's ratio of the model therefore tends to 0.5 at low densities.

Christensen [13] derived a result for a closed-cell material comprised of randomly located and isotropically oriented large intersecting thin plates. The results are

$$\frac{E}{E_s} = \frac{2(7-5\nu_s)}{3(1-\nu_s)(9+5\nu_s)}p; \quad \nu = \frac{1+5\nu_s}{9+5\nu_s}. \tag{10}$$

where the subscript "s" indicates the solid phase. The linear dependence ($n = 1$) of modulus on density is typical for cellular materials with straight-through elements. In this case, cell-wall stretching is the only mechanism of deformation.

Analysis of more complex closed-cell foams is very difficult, but computational results [21–25] have been obtained for the closed-cell tetrakaidecahedral foam shown in Figure 3a. Simone and Gibson [25] recently found that Young's modulus is nearly equal (within 10%) for loading in the $\langle 100 \rangle$, $\langle 111 \rangle$, and $\langle 110 \rangle$ directions. For the density range $0.05 < p < 0.20$, their results for the $\langle 100 \rangle$ direction were fitted with the formula

$$\frac{E_{100}}{E_s} \approx 0.315p + 0.209p^2 \tag{11}$$

which is consistent with the result $E_{100}/E_s \approx 0.33\,p$ obtained by Renz and Ehrenstein [24]. For the case where the face thickness is 5 % of the edge thickness, Mills and Zhu [23] found $E_{100}/E_s \approx 0.06\,p^{1.06}$ in the density range $0.015 < p < 0.1$.

For closed-cell materials Gibson and Ashby proposed the semi-empirical formula

$$\frac{E}{E_s} \approx \phi^2 p^2 + (1-\phi)p; \qquad \nu \approx \frac{1}{3} \tag{12}$$

where ϕ is the fraction of solid mass in the cell-edges (the remaining fraction $1-\phi$ is in the cell faces). The first term of Eq. (12) accounts for deformation in the cell edges. Note that the case $\phi = 1$ corresponds to the semi-empirical formula for open-cell solids (i.e., Eq. 6). The second term corresponds to stretching deformation in the cell faces. The result provides good agreement with data for closed-cell foams when $0.6 \leq \phi \leq 0.8$ [18]. However, some foams have $\phi < 0.5$ [23], in which case Eq. (12) violates the Hashin–Shtrikman upper bound [26]. The idea of partitioning stiffness into contributions from the "open-cell" skeleton and "closed-cell" faces was shown not to work for materials with an increasing number of missing faces [26]. This is because the interaction between microstructure is inherently nonlinear.

The above models generally assume a relatively simple microstructure in order to "solve" the equations of elasticity; that is, they prioritize solution of the second problem of property prediction over the first. The bounding methods incorporate arbitrary microstructure, but are limited to particular classes of models for which high-order correlation functions are available and are not guaranteed to provide an accurate predictive method. An alternative which has been made viable by the increasing speed of computers is to use the finite element method to estimate properties of realistic models [27]. This leads to new questions about designing accurate models, but also allows other pertinent questions to be tackled. For example, how do the various theories for porous/composite materials perform and how does the effect of disorder alter the properties of solids. The latter means the effect of randomness, irregular or missing struts, and missing faces. Moreover, the properties of noncellular porous materials such as aerogels and bone can also be studied by this method.

3.2.3 Modeling and Measuring Structure

There are three broad approaches to modeling random materials. The first is to simulate material formation from first principles. Generally, detailed geometrical, physical, and chemical elements of the relevant processes must be accurately modeled, making simulation quite complex. Whole fields of research are devoted to some of the many classes of porous and composite materials. For example, determining minimal surface energy configurations of "single cells" of periodic foams [28, 29] is a significant undertaking. Other well-known examples for which detailed simulations are undertaken include spinodal decomposition, bone growth, aerogel formation, and sphere packing. Since these processes are all relevant to modeling porous and cellular ceramics it is impossible to develop a unified simulation scheme for all materials of interest. A second approach is to employ and adapt a variety of realistic statistical models that appear to mimic the structure observed in actual materials. A quantitative correspondence between the model and real structures can be made by choosing the model to match statistical information about the material. For example, the model may be chosen to match the volume fraction and correlation function (equivalently, small-angle scattering profile) of a sample. A third method that has recently become viable is to directly compute the properties of a digital model obtained from a micro computer tomography scan [30]. This section focuses on the second approach, as random models are particularly convenient for generating computational structure–property relationships.

The most common models of random cellular solids are generated by Voronoi tessellation of distributions of "seed points" in space. Around each seed is a region of space that is closer to that seed than any other. This region defines the cell of a Voronoi (or Dirichlet) tessellation [31]. Placing a solid wall at each face of these cells gives a closed-cell tessellation. An open-cell tessellation results if only the edges where two cell-walls intersect are defined as solid. For several different random (e.g., Poisson) distributions of seed points, the average number of faces per cell falls in the range 13.7–15.5 [32].

The Voronoi tessellation can also be obtained [31] by allowing spherical bubbles to grow with uniform velocity from each of the seed points. Where two bubbles touch, growth is halted at the contact point, but allowed to continue elsewhere. In this respect the tessellation is similar to the actual process of liquid-foam formation [33]. Of course, physical constraints such as minimization of surface energy will also play an important role in shaping the foam. Depending on the properties of the liquid and the processing conditions, the resultant solid foam will be comprised of open and/or closed cells. Figure 4 shows a summary of this process.

The amount of order in the Voronoi tessellation depends on the order in the seed points. If regular arrays are used, ordered anisotropic foams will result. Indeed the open-cell models used by Warren and Kraynik [19], Zhu et al. [15] and Ko [16] turn out to be equivalent to Voronoi tessellations of the body-centred cubic (bcc, Fig. 3a), face-centred cubic, and hexagonal close-packed lattices. If a purely random (Poisson) distribution of points is used, highly irregular isotropic foams containing a wide size distribution of large and small cells will result.

It is worth noting that the tessellation of the bcc array (the tetrakaidecahedral cell model discussed above) is a reasonable approximation to the foam introduced by Lord Kelvin [19, 22, 28]. The cells of the Kelvin foam are uniformly shaped, fill space, and satisfy Plateau's law of foam equilibrium (three faces meet at angles of 120°, and four struts join at 109.5°). For this to be true, the faces and edges are slightly curved [28], unlike those of the tetrakaidecahedral cell model.

While the importance of the open- and closed-cell tessellations is clear, a related model appears relevant to certain types of cellular ceramics. Figure 1 shows a porous ceramic in which the bubbles have coalesced to an open-cellular structure with cir-

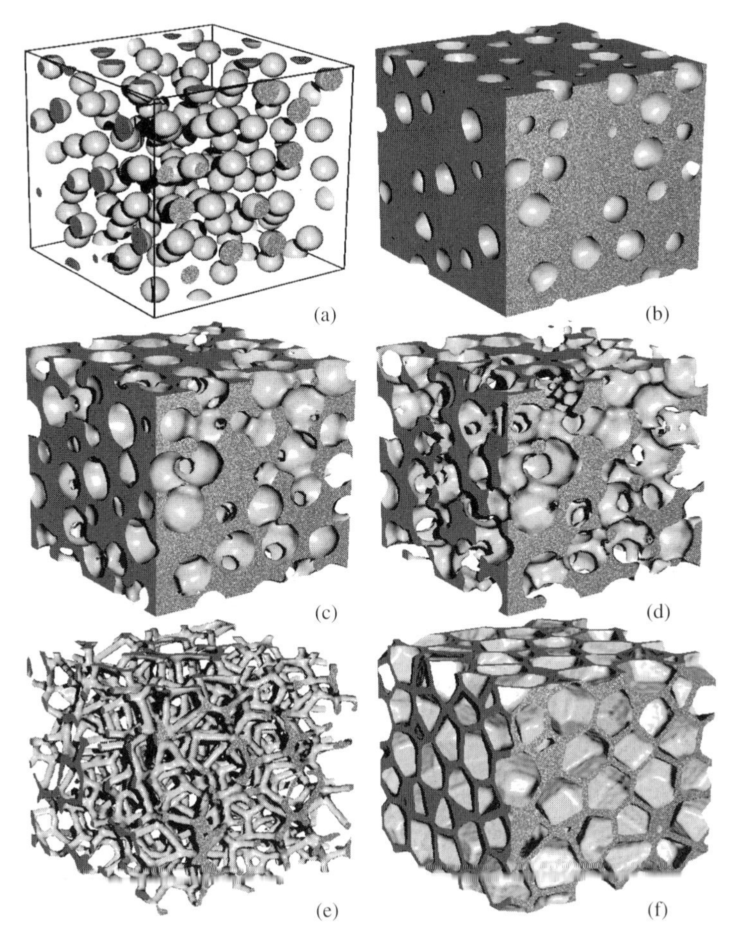

Fig. 4 The expanding-bubbles model (a–d) and the open- (e) and closed-cell (f) Voronoi tessellations. Image a) shows the seed points of the models. In the expanding-bubbles model the pores grow without changing shape, which mimics the structure seen in Fig. 1a. The tessellations are generated from the same seed points by using the standard Voronoi rules.

cular throats and less regular struts than the uniform connectors of conventional open-cell Voronoi tessellations. A similar microstructure was observed by Colombo and Hellman [34]. This type of structure was mimicked by inverting the solid and pore phases of a distribution of hard spheres of unit radius. The relative density is decreased (increased) by simply allowing the spheres to grow (shrink). For clarity, this is referred to as the expanding-bubbles model, but it is just the inversion of a cherry-pit model [1], and is similar to the microstructure underlying minimum solid area models (Part 4). The model is shown in Fig. 4c and d.

Of course not all cellular solids are formed by bubble-expansion processes. The intricate structure of trabecular bone is one example, while sol–gel-derived materials are another important exception. Figures 5 and 6 show a graphical summary of a range of other statistical models that have been studied. The Boolean models involve placing solid objects or pores at uncorrelated points in space. Boolean models [1, 35] have been studied in great detail because it is relatively easy to determine statistical measures of the models. It is possible to generate other realistic microstructure models by using the level-cut Gaussian random field (GRF) scheme [36–38]. One starts with a Gaussian random field $y(\mathbf{r})$ which assigns a (spatially correlated) random number to each point in space. The random field can be generated by a number of methods, including summing sinusoids with random coefficients. The distribution of coefficients entirely determines the statistical properties of the resultant field (see for example, Refs. [37–39]). A two-phase solid–pore model can be defined by letting the region in space where $-\infty < y(\mathbf{r}) < \beta$ be solid, while the remainder corresponds to the pore space (Fig. 6a). This type of structure is reminiscent of that generated in spinodal decomposition. An interesting two-cut GRF model [38] can be generated by defining the solid phase to lie in the region $-\beta < y(\mathbf{r}) < \beta$ (Fig. 6b). This type of structure, observed in microemulsions, has formed the basis for biomimetic silica materials [40]. Open- and closed-cell models can be obtained from the two-cut version by forming the intersection (Fig. 6c) and union (Fig. 6d) sets of two statistically independent two-cut GRF models [41]. The open-cell model has been used to model aerogels [41], while the closed cell model is similar to the "wavy" pores seen in some foamed polymers [26].

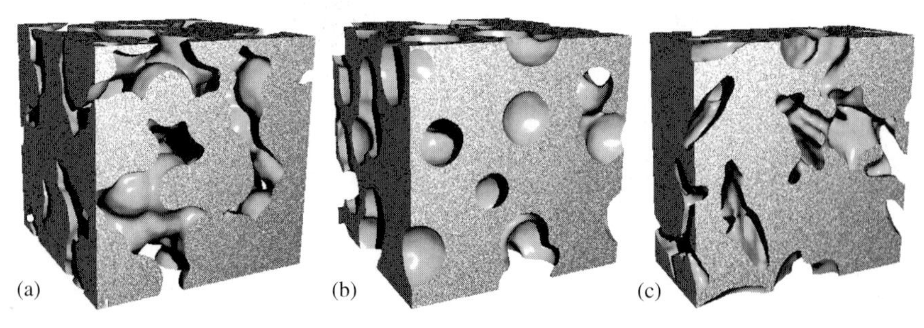

Fig. 5 Boolean models of porous media. a) Overlapping solid spheres, b) spherical pores, and c) oblate spheroidal pores (aspect ratio four). After Ref. [4].

Figure 3 shows two "node-bond" network models which were studied to determine the influence of coordination number on properties [42]. The nodes are defined as the centers of hard spheres. Two nodes are connected by a strut if they are closer than a specified distance. If a value of 1.1 times the original sphere diameter is used a loosely connected structure results. The average number of connections that a node has is 5.5. If the interconnection distance is increased to 1.5 the coordination number increases to 12.

While statistical models capture particular qualitative features of many different materials it is natural to ask how well a model is able to mimic a given microstructure. Since there are many degrees of freedom in these models, it is possible to tune them to some extent to match measured information; for example, the specification

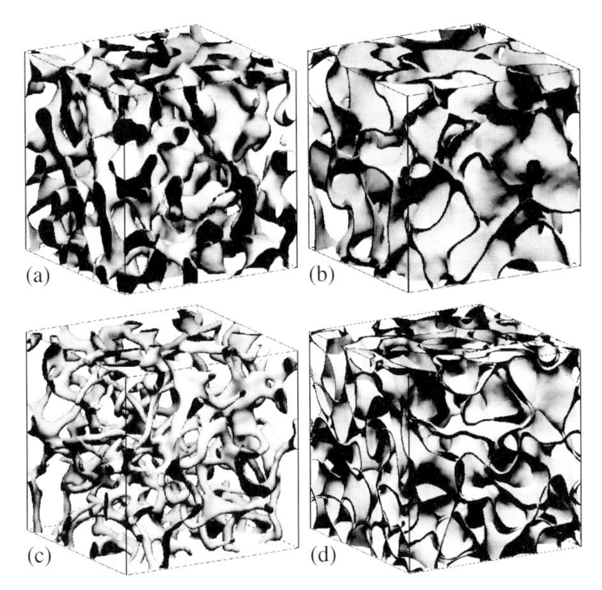

Fig. 6 Three-dimensional Gaussian random field (GRF) models. a) Single-cut, b) two-cut, c) open-cell intersection, and d) closed-cell union. After Ref. [4].

of seed-point distributions in tessellations and network models, the size and shape distribution in Boolean or sphere-pack models, and the choice of correlation functions in Gaussian level-set models. What types of measurable information might be used to choose a model for a particular material? If the goal is to develop structure–property relations, are there particular characteristics of structure that play a key role in determining properties? Even for "simple" properties like conductivity and elasticity there seems to be no simple answer to this question.

As for structure–property relationships there are two approaches to the problem of characterizing structure depending on whether the methods apply to arbitrary structures or particular classes of models. We will focus largely on the former methods, but mention the very detailed literature on the morphology of foams (see, for example, Refs. [29, 32]). This research focuses on well-defined cells with polyhedral facets. Key geometrical questions that have been answered include estimates of the number of cell faces, the surface area of cells, and their edge length. The results depend on the spatial distribution of cells, which may be randomly chosen or generated by using algorithms that mimic froth formation. The shape of cell struts and thickness of cell walls can also be predicted by using physical principles for idealized foam models [29].

Although the correlation between cell features, such as the average number of cell faces, and foam properties is not known, detailed morphological studies of equilibrated foams could be used to tune Voronoi tessellation models. For example, the random sequential adsorption (RSA) method of generating a sphere pack whose centers are subsequently used as the seed-point distribution of a tessellation results in a model in which the number of edges per polyhedral face is about 14.9, whereas equilibrated foams have a value of around 13.7 [29] and are more monodisperse. The random sequential adsorption algorithm does not provide very dense packings [32], which in turn leads to elongated and unusual cell shapes in the tessellation process. It is possible to model more monodisperse foams with a denser sphere pack. The tessellation shown in Fig. 4f is quite monodisperse and was obtained by removing 2% of the spheres from a bcc array and "thermally" shaking the array until an isotropic system was obtained [26]. I conjecture that the average number of edges per cell of this model will be closer to 13.7. Note that there are many types of cellular solids and foams that do not satisfy the strict assumptions of polyhedral facets. In these cases more general characterization methods are necessary.

The most obvious way of characterizing a general porous or composite material is by using the hierarchy of correlation functions of one of the phases. The first-order function is the probability that a point lies within the phase, which is just the volume fraction or relative density. The second-order function is the probability that two randomly chosen points a distance r apart both lie in one phase, and so forth. The rate of decay and presence of peaks in the correlation function indicate the degree and nature of "correlation" in the composite. Examples of normalized autocorrelation functions are shown in Fig. 7. The data are for the open- and closed-cell tessellations shown in Fig. 4e and f, based on a dense hard-sphere pack, while the third model is a closed-cell tessellation based on a Poisson distribution. The Fourier transform $I(q)$ of the correlation function, which corresponds exactly with small-angle scattering intensity, is more revealing (see Fig. 8). The peak around $q = 0.25$ corresponds to a correlation at wavelength $2\pi/q = 25$, which is the average cell diameter of the models. If an uncorrelated (Poisson) seed distribution is used the subsidiary peaks are far weaker (Fig. 8). The correlation function clearly contains important

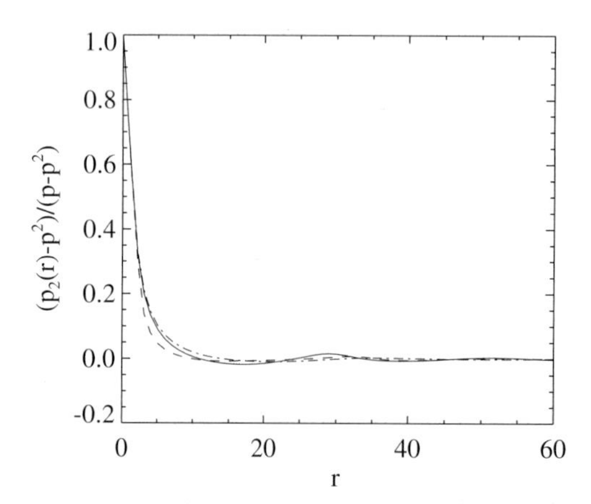

Fig. 7 The normalized two-point correlation does not appear to provide a strong signature of structure. The dashed and solid lines are for the open- and closed-cell tessellations shown in Figs. 4e and f, respectively. The dot-dashed line is a closed-cell tessellation of a Poisson distribution.

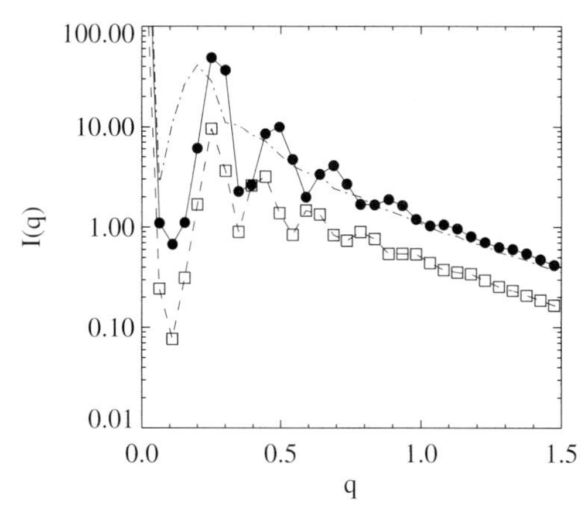

Fig. 8 The Fourier transform (scattering intensity) of the correlation functions for the open- (dashed) and closed-cell (solid) tessellations. The tessellation based on a completely random (Poisson) seed distribution (dot-dashed) only shows a scattering peak at the average cell size.

morphological information, but like most measures it does not appear to have a direct link with properties: the open- and closed-cell models have similar scattering profiles but very different properties, while two closed-cell tessellations with quite different scattering profiles will have similar physical properties.

The implied weak dependence of properties on the correlation function is borne out by the rigorous property-prediction methods, which prove that the properties of isotropic materials do not explicitly depend on the two-point function [1, 3]. Indeed, rigorous prediction methods indicate that third-order functions are needed to distinguish between composites, but the differentiation between materials with quite different properties is not strong. For example the bounds are continuous over a percolation transition. This implies that “higher order” information is critical in determining properties, but even accurate measurement of the third-order function is difficult.

A range of other methods of characterizing microstructure have been identified. These include surface–surface and surface–volume correlation functions, which are important in flow problems [1], and the chord-length probability function which measures the probability that a line segment in a specified phase has given length. Given a three-dimensional structure it is also possible to define a pore size distribution. The actual definition of pore size is complicated by the generally irregular shape of pores and their interconnections. Other measures which have received attention are the lineal-path function and coarseness functions, which measure the variance of (say) porosity as a function of window size. These functions and interrelations between them have been reviewed by Torquato [1].

An important goal of research in characterizing microstructure is to improve the match between model and real materials. This matching has been termed “statistical reconstruction”, where one tunes a model to match measured information. Two avenues have been pursued in this field: model-based and model-independent approaches. Both methods can be based on two-dimensional information, which allows a three-dimensional model to be built up from a two-dimensional image (provided the model is isotropic). Model-based approaches use quantitative statistical measures to select a statistical model with appropriate morphological parameters to match structure. Examples include matching the correlation functions of level sets of Gaussian random fields [37, 43] to an experimentally measured correlation function. A problem with this approach [44] is that the two-point function is practically nonunique, that is, many models were able to reproduce a given correlation function. To solve this problem the chord distribution was successfully employed [44] to distinguish between several candidate models. More recent work uses integral geometry (volume fraction, surface area, integral mean curvature, and genus) as a basis for choosing the parameters of polydisperse Boolean models [30]. The latter models also provided good predictions of material properties.

As noted above, model-independent methods can be used to "reconstruct" materials [45]. To match microstructure these models flip voxels from one phase to another until the statistical properties of the digital model match those of the composite. Two advantages of this approach are that first, the microstructure does not have to be "close" to a model (as in the model-based approach), and second the microstructure measures employed (e.g., the correlation functions) need not be available in a simple form. Apart from being computationally intensive, a disadvantage of not imposing some structure via a model is that fine details, such as thin struts, which tend not to be strongly emphasized by arbitrary microstructure measures (such as chord distributions and correlation functions) may not appear in the digital reconstruction.

The possibility of modeling a given composite, given statistical information about its microstructure, is a relatively new field. While there has been excellent progress, the generality and robustness of the method are not yet known. Near the percolation threshold, where just a few small connections dictate the properties, it is unlikely that statistical methods will be able to reconstruct accurate models for structure–property predictions. On the other hand, materials are not generally produced at or near this threshold. In part the requirements for modeling are reduced by the availability of micro-CT scans, but for low-contrast and/or nonporous materials, as well as materials with submicrometer structure, modeling will remain a key component of understanding properties. Moreover, once a model is chosen, parameters of the model, such as porosity and surface area, can be varied to study the influence of microstructure on properties and how they may be optimized for a given application.

3.2.4 Computational Structure–Property Relations

As discussed in Section 3.2.3 considerable effort is required to generate models which quantitatively match the statistical properties of a given material. An expedient alternative is to simply choose representative parameters for each model, and calculate theoretical structure–property relationships for each "class" of models. This approach has been undertaken for a range of materials [4, 26, 42, 46]. The elastic properties of the digitized models have been calculated by the finite-element method [27]. Instead of tabulating the results at discrete intervals simple two-parameter curves were fitted to the data. For medium- to high-density solids the equation

$$\frac{E}{E_s} = \left(\frac{p-p_0}{1-p_0}\right)^m \tag{13}$$

was used to describe the data. Note that p_0 is a fitting parameter and does not necessarily correspond to a geometrical percolation threshold. For example, for the expanding-bubbles model shown in Fig. 1b, $m = 1.61$ and $p_0 = 0.088$ for volume fractions $p \geq 0.12$. The formula would not be accurate if extrapolated to lower fractions.

At lower densities data for solids which remain interconnected are well described by a the power law

$$E/E_s \approx Cp^n \tag{14}$$

which can be used to extrapolate the data below the computed range. The fitting parameters for the models shown in Figs. 3–6 are reported in Tab. 1. In all cases the solid Poisson's ratio of $\nu_s = 0.2$ was used.

Table 1 Simple structure–property relations for different model porous materials. For $p > p_{min}$, the data can be described by $E/E_s = [(p-p_0)/(1-p_0)]^m$ to within a few percent. For the last seven models (which do not have a finite percolation threshold), the data can be extrapolated by using the power law $E/E_s = Cp^n$ for $p < p_{max}$. The Poisson's ratio can be approximately described by Eq. (15) with parameters ν_1 and p_1. The values of ν_1 and p_1 for the last four models are estimated by extrapolating data obtained at one Poisson's ratio ($\nu_s = 0.2$).

Model	$p < p_{max}$			$p > p_{min}$			ν	
	n	C	p_{max}	m	p_0	p_{min}	ν_1	p_1
Solid spheres				2.23	0.348	0.50	0.140	0.528
Spherical pores				1.65	0.182	0.50	0.221	0.160
Oblate pores				2.25	0.202	0.50	0.166	0.396
Expanding bubbles				1.61	0.088	0.12		
"Cherry-pit" spheres				1.38	0.568	0.60		
Single-cut GRF				1.64	0.214	0.30	0.184	0.258
Two-cut GRF	1.58	0.717	0.50	2.09	–0.064	0.10	0.220	–0.045
Open-cell GRF	3.15	4.200	0.20	2.15	0.029	0.20	0.233	0.114
Closed-cell GRF	1.54	0.694	0.40	2.30	–0.121	0.15	0.227	–0.029
Closed-cell tessellation	1.19	0.563	0.30	2.09	–0.140	0.15	0.28	0
Open-cell tessellation	2.04	0.930	0.50	2.12	–0.006	0.04	0.50	0
Bond model, $\bar{z} = 5.5$	1.81	0.535	0.35	4.27	–0.445	0.25	0.25	0
Bond model, $\bar{z} = 12$	1.29	0.376	0.25	2.80	–0.198	0.10	0.25	0

In general it is expected that the Young's modulus of a porous material will depend on volume fraction, microstructure, and Poisson's ratio of the solid phase ν_s. However, it has been recently proven [47–49] that E is independent of ν_s in two dimensions, and that E is practically independent (to within a few percent) of ν_s in three dimensions [4]. In contrast, the bulk and shear moduli do depend quite significantly on ν_s. To calculate the bulk and shear moduli it is therefore necessary to estimate the Poisson ratio of the porous material. The following simple formula has been found to provide a rough fit of the data [42]

$$\nu = \nu_s + \frac{1-p}{1-p_1}(\nu_1 - \nu_s)\ . \tag{15}$$

The fitting parameters are listed in Tab. 1. Figure 9 shows the typical "flow-diagram" behavior of the Poisson's ratio which tends to become independent of Poisson's ratio of the solid phase as the density decreases [4]. This is because the microstructure dictates the amount of macroscopic lateral expansion of a body subjected to uniaxial stress at low densities. To see this consider the extreme case of diamond-shaped cell made of four struts which is uniaxially loaded from corner to corner. The unloaded corners will bow out and thus cause a much more significant lateral expansion than the contribution of the "internal" expansion of the struts.

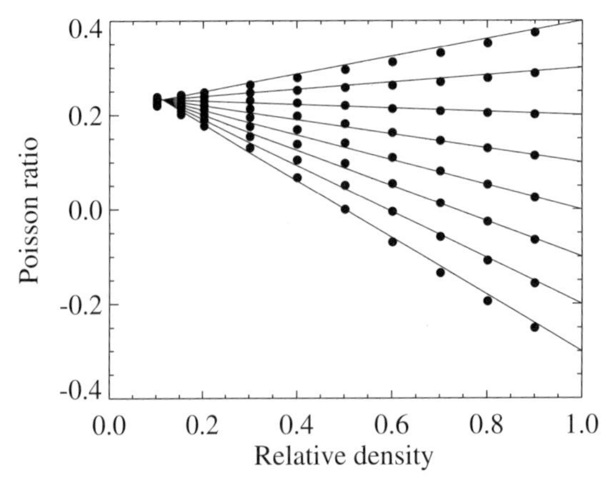

Fig. 9 Typical behavior of Poisson's ratio of a porous material as a function of density and Poisson's ratio of the solid skeleton. The lines represent a best fit to Equation (15). Finite-element data is for the open-cell GRF model shown in Fig. 6c.

In general most models [4] have a Poisson's ratio of around $\nu = 0.2$ at low densities, with the interesting exception of the random open-cell Voronoi tessellation, which has a ratio of $\nu \approx 0.5$ [42]. Physically, this interesting behavior corresponds to a material which preserves volume under uniaxial compression; the compression is matched by lateral expansion. Equivalently, the material has much higher bulk modulus than shear modulus. The computational results are well predicted by the Warren–Kraynik model (Eq. 9). Essentially, the bending,mode of deformation is not activated when a near-tetrahedral joint is uniformly loaded, but under shear or uniaxial loading, bending is activated. This implies $K \propto p$, but $E \propto p^2$, so that $\nu = 1/2 - E/6K \rightarrow 0.5$ as $p \rightarrow 0$. To my knowledge this behavior has not been observed in real foams, although the prediction is robust to variations in seed distribution [42] and even to missing struts. Around 15 % of the struts must be deleted to reduce the Poisson's ratio to around 0.35 [42].

Examples of comparisons between computational theories and experimental data are shown in Figs. 10–12. Figure 10 plots data for SiOC ceramic foam [34] against results for the open-cell tessellation and the expanding-bubbles model. The latter is used because of the close similarity between SiOC and the model, at least at intermediate volume fractions. For relative density $p > 0.2$ both models are consistent with the data, while for $0.1 < p < 0.2$ the open-cell tessellation provides a much better match. This indicates that SiOC has a more structurally efficient morphology than that provided by the expanding-bubbles model.

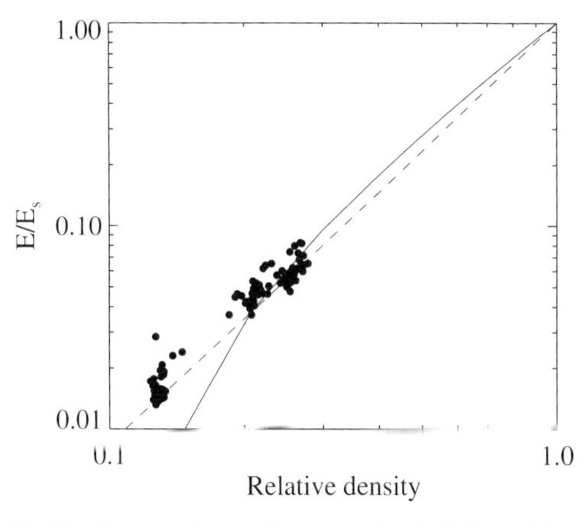

Fig. 10 A comparison of the expanding bubble model (——) and open-cell (– – –) Voronoi tessellation model with data from SiOC foams (•) [34].

Figure 11 compares data from open-cell foams with various computational theories. The models are able to span the data observed in real systems, although it is also necessary to check that the geometry of the models is correct by using micrographs or micro-CT images. Data for a closed-cell glass foam are shown in Fig. 12. The closed-cell Voronoi tessellation provides an excellent match with the data. For this review I also computed the theoretical elastic stiffness of a ceramic foam/Mg alloy composite studied by Zeschky et al. [50]. The foam ceramic support (fraction $p = 0.18$) has a morphology similar to that shown in Fig. 1 and is infiltrated with magnesium. Under the authors' assumption that the moduli of the ceramic and metal are $E_s = 190$ GPa and $E_m = 37$ GPa, the expanding-bubbles model this gave a prediction of 51 GPa, while for the open-cell tessellation the stiffness was slightly lower at 49 GPa. The estimates are about 10% lower than the measured value of 56 GPa. This either indicates that the foam structure is stiffer than predicted by either of the models, or that the estimate of the ceramic or metal stiffness is too low.

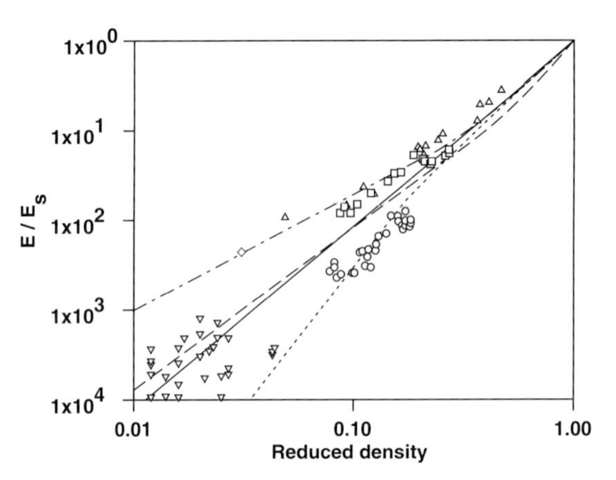

Fig. 11 Young's modulus of open-cell foams. The data is for alumina [54] (○, $E_s = 380$ GPa, $\rho_s = 3970$ kg/cm^3), rubber latex obtained by Lederman [55] (□) and Gent and Thomas [20] (△), open-cell foams [17] (▽), and reticulated vitreous carbon [13] (◇, $E_s = 6.9$ GPa). The lines correspond to the four open-cell FEM theories: high (– - –) and low (– – –) coordination number foams, open-cell Voronoi tessellation (——), and the open-cell Gaussian random field model (- - -). After Ref. [42].

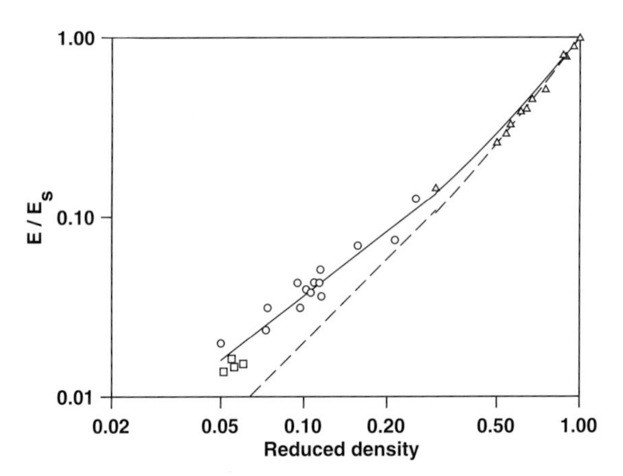

Fig. 12 Young's modulus of foamed glasses with closed cells. The data is from Morgan et al. [56] (□), Zwissler and Adams [57] (○, $E_s =$ 69 GPa [58]) and Walsh et al. [59] (△) ($E_s = 75$ GPa). The solid line corresponds to the closed-cell Voronoi tessellation. Results for the closed-cell GRF model (dashed line) are shown for comparison. After Ref. [26].

3.2.5 Summary

A summary of structure–property relationships for porous materials has been provided. The two equally important aspects of the problem are the development of realistic microstructural models and accurate estimation of their properties. Analytical theories have been derived for particular microstructures in high- and low-density limits. These theories are useful for demonstrating the effect of particular variations of structure on properties. For example, the dilute-pore (i.e., high-density) theories can be extended to ellipsoidal inclusions, while the thin-beam theories for low-density periodic cells can elucidate the role of strut shape in properties. A disadvantage of simple structural models is that they do not accurately describe many materials; most cellular solids are not composed of periodic arrays of well-defined beams connected at nodes. Moreover, it is not clear how accurate these theories are for intermediate densities. Thus, if disagreement between theory and experiment is found, it is not clear if the microstructural model is at fault, or the approximation used to estimate its properties.

Computational theories for statistical models avoid this problem; deviations between experimental data and prediction definitely correspond to a difference in structure, which can be informative. In this approach it is implicitly assumed that material microstructure can be divided into a number of recognizable "classes", and that the structure–property relationships among members of the same class will be similar. For example, there are many different types of sintered materials, but it is hypothesized that their properties can be estimated by the structure–property relationship of overlapping solid spheres (see Fig. 5a). Similarly I conjecture that foams made up of networks of tetrahedral elements can be effectively modeled by the Voronoi tessellation with a seed-point distribution given by the centers of random hard spheres. Of course there will be differences depending on which seed distribution is employed to generate the tessellation, or whether the model has been allowed to equilibrate in some way. However, it is assumed that these differences are minor compared to those found between classes of models. The models can be refined and adapted to study variants of these problems. For example, the effect of missing cell faces has been studied [4], and it is possible to model polydisperse foams by using generalized tessellation schemes [51].

Methods of making the models more accurate have been discussed. These "statistical reconstruction" procedures rely on matching statistical properties of the model to those of the real material. Since many statistical properties of isotropic materials can be measured in two-dimensions it is possible to reconstruct a three-dimensional medium from an image. If three-dimensional information is available, from micro-CT scans, for example, then other measures such as integral geometric techniques can be employed [52]. If a micro-CT image is available and only a single property estimate is required, then the finite-element techniques can be used directly on the image [53].

This chapter dealt with the linear-elastic properties of heterogeneous materials. Other macroscopic properties such as conductivity, diffusivity, and fluid permeability can be treated in an analogous way. A major goal of future studies in this field will be the application of finite-element techniques to the study of yield and fracture in porous and composite materials.

References

1 Torquato, S., *Random Heterogeneous Materials. Microstructure and Properties*, Springer-Verlag, New York, 2002.
2 Milton, G.W., *The Theory of Composites*, Cambridge University Press, Cambridge, 2002.
3 Hashin, Z., Analysis of Composite-Materials – A Survey, *J. Appl. Mech.* **1983**, *50*, 481–505.
4 Roberts, A.P. and Garboczi, E.J., Computation of the Linear Elastic Properties of Random Porous Materials with a Wide Variety of Microstructure, *Proc. Roy. Soc. Lond. A* **2002**, *458*, 1033–1054.
5 Hill, R., A Self-Consistent Mechanics of Composite Materials, *J. Mech. Phys. Solids* **1965**, *13*, 213–222.
6 Budiansky, B., On the Elastic Moduli of Some Heterogeneous Materials, *J. Mech. Phys. Solids* **1965**, *13*, 223–227.
7 McLaughlin, R., A Study of the Differential Scheme for Composite Materials, *Int. J. Eng. Sci.* **1977**, *15*, 237–244.
8 Milton, G.W., Correlation of the Electromagnetic and Elastic Properties of Composites and Microgeometries Corresponding with Effective Medium Theory, in Johnson, D.L. and Sen, P.N. (eds.), *Physics and Chemistry of Porous Media*, American Institute of Physics, Woodbury, NY, USA, , pp. 66–77.
9 Torquato, S., Random Heterogeneous Media: Microstructure and Improved Bounds on Effective Properties, *Appl. Mech. Rev.* **1991**, *44*, 37–76.
10 Milton, G.W. and Phan-Thien, N., New Bounds on Effective Elastic Moduli of Two-Component Materials, *Proc. Roy. Soc. London A* **1982**, *380*, 305–331.
11 Roberts, A.P. and Knackstedt, M.A., Structure-Property Correlations in Model Composite Materials, *Phys. Rev. E* **1996**, *54*, 2313–2328.
12 Jeulin, D. and Savary, L., Effective Complex Permittivity of Random Composites, *J. Physique I* **1997**, *7*, 1123–1142.
13 Christensen, R.M., Mechanics of Low Density Materials, *J. Mech. Phys. Solids* **1986**, *34*(6), 563–578.
14 Warren, W.E. and Kraynik, A.M., The Linear Elastic Properties of Open-Cell Foams, *J. Appl. Mech.* **1988**, *55*, 341–346.
15 Zhu, H.X., Knott, J.F., and Mills, N.J., Analysis of the Elastic Properties of Open-Cell Foams with Tetrakaidecahedral Cells, *J. Mech. Phys. Solids* **1997**, *45*, 319–343.
16 Ko, W.L., Deformations of Foamed Elastomers, *J. Cell. Plast.* **1965**, *1*, 45–50.
17 Gibson, L.J. and Ashby, M.F., The Mechanics of Three-Dimensional Cellular Materials, *Proc. Roy. Soc. Lond. A* **1982**, *382*, 43–59.
18 Gibson, L.J. and Ashby, M.F., *Cellular Solids: Structure and Properties*, Pergamon Press, Oxford, **1988**.
19 Warren, W.E. and Kraynik, A.M., Linear Elastic Behavior of a Low Density Kelvin Foam with Open Cells, *J. Appl. Mech.* **1997**, *64*, 787–794.
20 Gent, A.N. and Thomas, A.G., The Deformation of Foamed Elastic Materials, *J. Appl. Polym. Sci.* **1959**, *1*, 107.
21 Grenestedt, J.L. and Tanaka, K., Influence of Cell Shape Variations on Elastic Stiffness of Closed Cell Cellular Solids, *Scripta Mater.* **1999**, *40*(1), 71–77.
22 Grenestedt, J.L., Effective Elastic Behavior of Some Models for "Perfect" Cellular Solids, *Int. J. Solids Struct.* **1999**, *36*, 1471–1501.
23 Mills, N.J. and Zhu, H.X., The High Strain Compression of Closed-Cell Polymer Foams, *J. Mech. Phys. Solids* **1999**, *47*, 669–695.
24 Renz, R. and Ehrenstein, G.W., Calculation of Deformation of Cellular Plastics by Applying the Finite Element Method, *Cell. Polym.* **1982**, *1*, 5–13.
25 Simone, A.E. and Gibson, L.J., Effects of Solid Distribution on the Stiffness and Strength of Metallic Foams, *Acta Mater.* **1998**, *46*(6), 2139–2150.
26 Roberts, A.P. and Garboczi, E.J., Elastic Properties of Model Random Three-Dimensional

Closed-Cell Cellular Solids, *Acta Mater.* **2001**, *49*, 189–197.

27 Garboczi, E.J. and Day, A.R., An Algorithm for Computing the Effective Linear Elastic Properties of Heterogeneous Materials: Three-Dimensional Results for Composites with Equal Phase Poisson Ratios, *J. Mech. Phys. Solids* **1995**, *43*, 1349–1362.

28 Weaire, D. and Fortes, M.A., Stress and Strain in Liquid and Solid Foams, *Adv. Phys.* **1994**, *43*(6), 685–738.

29 Kraynik, A.M., Foam Structure: From Soap Froth to Solid Foams, *Mater. Res. Bull.* **2003**, *4*, 275–278.

30 Arns, C.H., Knackstedt, M.A., and Mecke, K.R., Reconstructing Complex Materials via Effective Grain Shapes, *Phys. Rev. Lett* **2003**, *91*, 215506-1.

31 Stoyan, D., Kendall, W.S., and Mecke, J., *Stochastic Geometry and its Applications*, Wiley, Chichester, 2nd ed., **1995**.

32 Oger, L., Gervois, A., Troadec, J.P., and Rivier, N., Voronoi Tesselation of Packings of Spheres: Topological Correlation and Statistics, *Phil. Mag. B* **1996**, *74*, 177–197.

33 Van der Burg, M.W.D., Shulmeister, V., Van der Geissen, E., and Marissen, R., On the Linear Elastic Properties of Regular and Random Open-Cell Foams Models, *J. Cell. Plast.* **1997**, *33*, 31–54.

34 Colombo, P. and Hellmann, J.R., Ceramic Foams from Preceramic Polymers, *Mater. Res. Innovat.* **2002**, *6*, 260–272.

35 Serra, J., *Image Analysis and Mathematical Morphology*, Academic Press, London, **1988**.

36 Cahn, J.W., Phase Seperation by Spinodal Decomposition in Isotropic Systems, *J. Chem. Phys.* **1965**, *42*, 93–99.

37 Adler, P.M., *Porous Media*, Butterworth-Heinemann, Boston, **1992**.

38 Berk, N.F., Scattering Properties of a Model Bicontinuous Structure with a Well Defined Length Scale, *Phys. Rev. Lett.* **1987**, *58*, 2718–2721.

39 Roberts, A.P. and Teubner, M., Transport Properties of Heterogeneous Materials Derived from Gaussian Random Fields: Bounds and Simulation., *Phys. Rev. E* **1995**, *51*, 4141–4154.

40 McGrath, K.M., Dabbs, D.M., Yao, N., Aksay, I.A., and Gruner, S.M., Formation of a Silicate L-3 Phase with Continuously Adjustable Pore Sizes, *Science* **1997**, *277*, 552–556.

41 Roberts, A.P., Morphology and Thermal Conductivity of Model Organic Aerogels, *Phys. Rev. E* **1997**, *55*, 1286–1289.

42 Roberts, A.P. and Garboczi, E.J., Elastic Properties of Model Random Three-Dimensional Open-Cell Solids, *J. Mech. Phys. Solids* **2002**, *50*, 33–55.

43 Quiblier, J.A., A New Three-Dimensional Modeling Technique for Studying Porous Media, *J. Colloid Interface Sci.* **1984**, *98*, 84–102.

44 Roberts, A.P., Statistical Reconstruction of Three-Dimensional Porous Media from Two-Dimensional Images, *Phys. Rev. E* **1997**, *56*, 3203–3212.

45 Yeong, C.L.Y. and Torquato, S., Reconstructing Random Media. II. Three Dimensional Media from Two Dimensional Cuts, *Phys. Rev. E* **1998**, *58*, 224–233.

46 Roberts, A.P. and Garboczi, E.J., Elastic Properties of Model Porous Ceramics, *J. Am. Ceram. Soc.* **2000**, *83*(12), 3041–3048.

47 Day, A.R., Snyder, K.A., Garboczi, E.J., and Thorpe, M.F., The Elastic Moduli of Sheet Containing Spherical Holes, *J. Mech. Phys. Solids* **1992**, *40*, 1031–1051.

48 Cherkaev, A.V., Lurie, K.A., and Milton, G.W., Invariant Properties of the Stress in Plane Elasticity and Equivalence Classes of Composites, *Proc. R. Soc. Lond. A* **1992**, *438*, 519–529.

49 Thorpe, M.F. and Jasiuk, I., New Results in the Theory of Elasticity for Two-Dimensional Composites, *Proc. Roy. Soc. Lond. A* **1992**, *438*, 531–544.

50 Zeschky, J., Lo, J.H., Scheffler, M., Hoeppel, H.-W., Arnold, M., and Greil, P., Polysiloxane-Derived Ceramic Foam for the Reinforcement of Mg Alloy, *Z. Metallkd.* **2002**, *93*, 812–818.

51 Richard, P., Oger, L., Troadec, J.P., and Gervois, A., Tessellation of Binary Assemblies of Spheres, *Physica A* **1998**, *259*, 205–221.

52 Arns, C.H., *The Influence of Morphology on Physical Properties of Reservoir Rocks*, Ph.D. thesis, Petroleum Engineering, UNSW, **2002**.

53 Saadatfar, M., Knackstedt, M.A., Arns, C.H., Sakellariou, A., Senden, T.J.S., Sheppard, A.P., Sok, R.M., Steininger, H., and Schrof, W., Polymeric Foam Properties Derived From 3D Images, *Physica A* **2004**, *1*, 131–136.

54 Hagiwara, H. and Green, D.J., Elastic Behavior of Open-Cell Alumina, *J. Am. Ceram. Soc.* **1987**, *70*(11), 811–15.

55 Lederman, J.M., The Prediction of the Tensile Properties of Flexible Foams, *J. Appl. Polym. Sci.* **1971**, *15*, 693–703.

56 Morgan, J.S., Wood, J.L., and Bradt, R.C., Cell Size Effects on the Strength of Foamed Glass, *Mater. Sci. Eng.* **1981**, *47*(1), 37–42.

57 Zwissler, J.G. and Adams, M.A., Fracture Mechanics of Cellular Glass, in R.C. Bradt, A.G. Evans, D.P.H. Hasselman, and F.F. Lange (eds.), *Fracture Mechanics of Ceramics*, Plenum Press, New York, vol. 6, **1983**, pp. 211–241.

58 Green, D.J., Fabrication and Properties of Lightweight Ceramics Produced by Sintering of Hollow Spheres, *J. Am. Ceram. Soc.* **1985**, *68*(7), 403.

59 Walsh, J.B., Brace, W.F., and England, A.W., Effect of Porosity on Compressibility of Glass, *J. Am. Ceram. Soc.* **1965**, *48*(12), 605–608.

Part 4
Properties

Cellular Ceramics: Structure, Manufacturing, Properties and Applications.
Michael Scheffler, Paolo Colombo (Eds.)

ISBN: 3-527-31320-6

4.1
Mechanical Properties

Roy Rice

4.1.1
Introduction

Mechanical properties of cellular ceramics play an important role in their various uses. This is clearly so for the more limited cases in which the cellular component serves primarily or exclusively a mechanical function. However, mechanical properties are also typically quite important for most, if not all, cases where the primary or exclusive function(s) of a cellular component is nonmechanical such as thermal insulation, filtration, catalysis, or controlled combustion (e.g., burner). This arises since such applications commonly entail various mechanical stresses which must be survived for the cellular component to continue to satisfactorily serve its desired nonmechanical functions.

The mechanical properties addressed are, in order of treatment, primarily elastic moduli (mainly Young's modulus) and Poisson's ratio, fracture toughness, tensile (flexure) strength, compressive (crushing) strength, and thermal stress/shock resistance. Hardness, erosion, and wear, for which there is limited or no literature, are addressed little or not at all. (Sonic velocities are not directly covered, but their dependence on porosity is directly derivable from that of the elastic properties [1].) The focus of this survey is on nominally room temperature behavior, but thermal stress/shock resistance of cellular materials is addressed. Note that while average property trends are addressed as a function of average porosity and its average character, local porosity extremes often play a role. The extent of this role of such porosity variations varies with the property, generally being lesser for elastic properties and greater for fracture properties, especially tensile fracture [1].

Besides the basic properties of the material of which the cellular body is made, the primary determinant of the mechanical, and many of the pertinent nonmechanical, properties is the amount and character of the porosity. Thus, the focus is on the porosity dependence of mechanical properties, primarily the reduction of most mechanical properties as the amount (i.e., volume fraction) of porosity P increases and as a function of pore character. Differences due to basic material properties at theoretical density (i.e., at $P = 0$) are addressed by normalizing (i.e., dividing) property values for a given material at any value of P by those at $P = 0$. However, effects of pore/cell size are noted later, as are property differences due to the character of

Cellular Ceramics: Structure, Manufacturing, Properties and Applications.
Michael Scheffler, Paolo Colombo (Eds.)

ISBN: 3-527-31320-6

the porosity and some of the important trade-offs between mechanical and nonmechanical properties needed for important types of applications. The important and developing topic of porous sandwich components consisting of a porous core between two denser layers of the same or other material is noted here (see refs. [2, 3] for more information), but not addressed. However, much of what is presented here is valuable input for development of such sandwich bodies.

Turning to the porosity dependence of mechanical properties, this is addressed by models for such dependence based on the amount and character of the porosity, which also reflects aspects of component fabrication. These models are summarized in the next section, where the common distinctions between the two main types of cellular solids are followed. The first are honeycomb structures, which consist of tubular pores, often aligned with one another, as are commonly derived by green-body extrusion of ceramics. There are other methods of fabricating tubular porous structures of the same or similar tubular pore structures as obtained from extrusion, such as tape processing and, more recently and more diverse, rapid prototyping/solid free-form fabrication (SFF) [4, 5]. However, these structures would be included under honeycomb structures. The other basic type of cellular ceramics are foam structures. Again, while these are commonly made by foaming, hence their name, related structures are made by other fabrication routes that produce similar pore structures and are thus included under the heading of foamed materials.

Before proceeding to modeling of the porosity dependence of mechanical properties two general related issues should be noted. First, such modeling involves simplifying the pore structure of given bodies to a more uniform, idealized one. Second, there are challenges in characterizing the porosity of real bodies due to variations in the type, character, and amounts of porosity in a given body. Thus, the focus, as noted earlier, is on basic property trends as a function of basic porosity trends. Comparison of such modeling with measured results is basic to improving understanding and modeling, as well as porosity characterization. Other important factors in such improvements are comparison of modeling and property results of differing fabrication/processing factors for the same as well as differing properties whose porosity dependences may or may not be similar. However, note that where electrical and thermal conductivity of bodies are determined mainly or totally by conduction through the solid as opposed to the pore phase, the dependences of the conductive property on porosity are the same as for the elastic moduli, as shown theoretically [1, 6] and experimentally [1].

4.1.2
Modeling the Porosity Dependence of Mechanical Properties of Cellular Ceramics

4.1.2.1
Earlier Models

While there are a number of models and approaches [1, 2, 7–15], four prominent models are considered here to provide perspective as well as both practical and functional utility. The first is an older, lesser used, model originally derived for elastic moduli, mainly Young's modulus E. The first model has both empirical and some analytical origins, with the latter being more rigorous in terms of the mechanics, and much less so in terms of the microstructure, which is generally taken as consisting of essentially homogeneously distributed, often spherical, pores to give homogenous properties [1]. This approach commonly results in the form:

$$E/E_0 = (\rho/\rho_0)^n = (1 - P)^n, \quad (1)$$

where E/E_0 is the value of E normalized by that at $P = 0$, ρ the density at any P and, ρ_0 the value at $P = 0$, and n an empirically determined exponent that is generally in the range of >0.5 to about 4, most commonly around 1–3. This expression has also been used for other mechanical properties, especially strengths, based on the rationale that their values are often determined by the porosity dependence of E [1].

Besides the serious neglect of specific microstructural effects in both its derivation and use (i.e., pore character and values of n are not correlated), it suffers two other deficiencies. First, it does not recognize the intrinsic limitation that different types of porosity have on the upper limit of the amount of that type of porosity that can exist in a body that is still a coherent solid (i.e., one that can sustain at least some stress) as opposed to simply being a pile or clump of individual powder grains (see Fig. 1 and discussion of Minimum Solid Area models below). Thus, though often not adequately recognized, each porous body that behaves as a solid body has a critical value P_C of the volume-fraction porosity P at which key physical properties, such as elastic moduli, of the body must go to zero, so that it no longer acts as a solid body [1, 7–10]. The values of P_C depend directly on the character of the porosity, ranging from slightly less than 1 for bodies with regular tubular pores (aligned with the stress direction) or spherical pores to typical values of 0.6–0.25 and potentially as low as about 0.1 in partially densified powder compacts. The P_C values can be readily calculated for Minimum Solid Area porosity models discussed below (and possibly from percolation models), or estimated by extrapolatiing key properties such as elastic moduli for porous bodies to zero. For porous bodies of partially densified powder, P_C values are the true green densities (i.e., without intentional or unintentional binder materials, e.g., water, after binder burnout).

A potential solution to inadequate recognition of P_C values is to normalize the values of P for a given body by dividing them by their P_C values (which then effects the value of n in fitting data to Eq. (1)), but this has thus far shown at best limited consolidation or improved coherence of individual real or model porous structures [10]. The other basic deficiency in use of Eq. (1) is that it by itself does not recognize that $n = 1$ is the upper limit of achievable properties [1], and thus that allowable values are $n < 1$.

Though there are serious questions about this model, it is included since it has been used a fair amount, and its use, limitations, and questions are important factors in the field of porosity effects on properties. Further, though it has been used primarily for partially sintered rather than cellular bodies, this model may have application to the latter (e.g., since it has a common form of some cellular models discussed next) and thus possibly aid in evaluating the transition in porosity dependence from lower degrees of porosity, where it has been mostly used, to the higher degrees of porosity of typical cellular solids.

4.1.2.2
Gibson–Ashby Models

In contrast to the above model, the models of Gibson and Ashby [2, 11–15] pay more attention to the porous microstructure, but end up with very similar mathematical relations. Specifically, these models, referred to here as G–A models, idealize the structure of various basic tubular cells for honeycombs and various box cells and their stacking and interconnection for foams (Fig. 1A). They then scale the mechanical behavior with the cell-structure parameters and the interconnection of the cells for various mechanical properties. This yields the desired properties in proportion

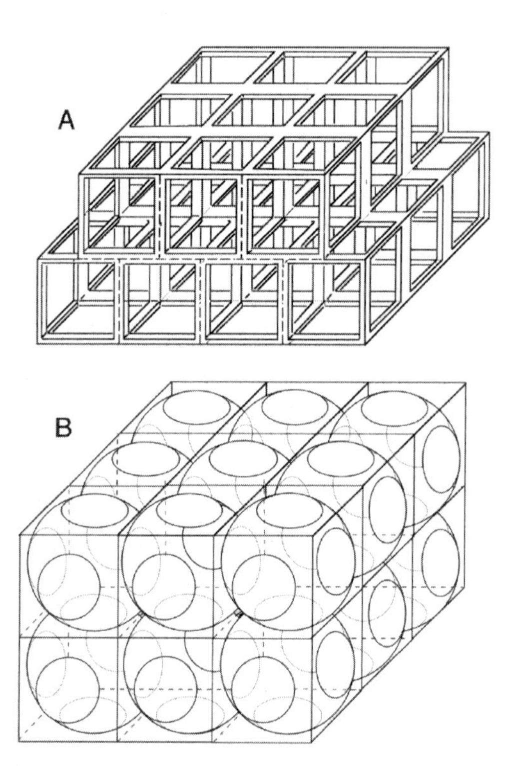

Fig. 1 Examples of idealized pore models used for G–A and MSA property modeling. A) A model of an open-cell foam structure in which unit pores often approach cubic character as shown (or parallelpiped character, not shown here but often used in G–A-type models to reflect the anisotropy that is common in many foam structures) [2, 11, 15]. This particular model shows a systematic shift from simple cubic packing of the cells, which is illustrative of changes in packing and related cell-to-cell joining that need to be considered in both types of models. B) Models of simple cubic packing of uniform spherical pores (or spherical particles) in a cubic cell structure commonly used as a reasonable model or random packing in MSA models [1, 7–10]. After Rice [8], published with permission of *J. Mater. Sci.*

to key microstructural parameters, with the constants of proportionality determined by fitting property data to their appropriate models. An important factor in this work of Gibson and Ashby [2] and others is that, while there can be important differences for different honeycomb or foamed materials, there are many close similarities within each of these types of cellular bodies for various polymer, metal, and ceramic materials, which reinforces trends for bodies for which there are fewer data (i.e., ceramics). A limitation of these cellular models is that there is less basis for relating the properties at high P to those at low P, and the transition in structure and behavior as P progresses from low to high values or the opposite. Thus, both the foregoing model and the following two modeling approaches may be of use in addressing such transitions and extrapolations of properties at high P to those at low P and vice versa. However, other than noted in this chapter, little consideration has been given to such cross-correlation of these models.

4.1.2.3 Minimum Solid Area (MSA) Models

Minimum Solid Area (MSA) models cover a broader range of pore structures ranging from low porosity to the medium and high porosity of cellular materials. However, much of MSA modeling is complementary to the above cellular modeling of Gibson and Ashby and others in that it also focuses on similar microstructures and their evolution as a given type of pores increases or decreases over the allowable range of P. A particular type of porosity reflecting idealizations of the pore structure resulting from each of several various basic processing parameters is assumed for a given model. Primary types of pores in MSA models are tubular pores from green body extrusion or tape casting and lamination, spherical pores from gas bubbles generated in sintering or melting, cubic or other polyhedral pores formed as intragranular pores in polycrystalline bodies (or single crystals), and pores between packed spherical particles that are being bonded to one another, especially by sintering. As with the preceding cellular models, MSA modeling assumes first that the body can be represented as a dense uniform packing of identical cells such that the effects of the pore in a cell reflect the effects of the porosity in the entire body (Fig. 1). Each cell consists of either a pore whose center is also the center of the cell surrounded by the solid phase or a spherical particle whose center is also the center of the cell surrounded by the pore phase (Fig. 1B). The cell geometry is determined by the dense (i.e., space-filling) packing or stacking of the individual cells (note that only two dimensional cells are needed for tubular pores). Most commonly, simple cubic packing is used for both its simplicity and as a good approximation for random packing of the solid and pore phases, but some denser packings have been used, and others offer possibly important extensions of this method. Once a representative cell structure has been selected based on the type and scale of the porosity being modeled, a complete evaluation of the range of feasible porosity can be made by incrementally changing the size of the pore or spherical particle to the cell size. This is basically a simulation of the porosity changes in a variety of foam and sintered bodies, but over their complete range of porosity, thus providing extrapolation from one degree or type of porosity to another.

The basic assumption of the relation of the cell structure to mechanical properties (and others, e.g., thermal and electrical conductivity) is that the property scales directly with the ratio of the MSA of a cell normal to the stress (or conductive flux) to the cross-sectional area of the cell in the same plane of the MSA in the cell. For example, the MSA for sintering particles is the area of a sintered neck with an adjacent particle, while that for a spherical pore is the area of surrounding solid material in the equatorial plane of the cell normal to the stress direction. Dividing such MSA values by the cross-sectional area of the cell in the same plane as the MSA gives the relative MSA, with which properties are assumed to directly correlate. While this assumption may be modified by further (e.g., finite-element) analysis, it has proven to be suitable for substantial useful evaluations [1, 7–9].

The output of MSA modeling over the entire allowable range of porosity is a curve of the relative property at any P up to the value of P_C for that porosity type, which is defined, at least approximately, if not exactly, by each specific model. Some MSA models, mainly those for tubular pores aligned parallel or perpendicular to the stress, result in simple equations of the form of Eq. (1), commonly with $n = 1$. However, most MSA models are generally not presented as an equation, since the individual calculations of each MSA value are complex due to physical requirements such as mass conservation in sintering and changes at key values of P (e.g., where the transition between open and closed porosity begins and the P_C value, both of which are valuable and often unique outputs of MSA modeling). Instead, the output of MSA modeling is generally a plot of the MSA values for the particular porosity type selected for analysis, normalized by the corresponding cell cross-sectional area, usually on a log scale, versus P (usually on a linear scale; Fig. 2).

Various earlier investigators used the MSA concept for a particular pore type, then tried to generalize results for the one pore structure to more general pore structures. It was later recognized that the MSA approach provided a variety of models for various common individual pore structures that collectively give a much more accurate picture of the porosity dependence of mechanical and related properties [1]. Thus, though only a limited number of MSA models have been developed, they cover a broad range of pore character and property behavior. Thus, they show that bodies with tubular or spherical pores have greater property retention and extend to higher P values, as opposed to the much more rapid decrease of properties and more limited range of allowable P values of bodies with porosity from partial sintering of particulate/powder compacts (Fig. 2). The intervening middle region of P values is where curves for bodies with combinations of these two types of porosity should occur. Linear combinations of the limited data for bodies with such mixed porosities support this, as does the recent demonstration that the curves of Fig. 2 can be normalized to one master curve by plotting them versus normalized P (i.e., versus P/P_C).

While, as noted above, the results of MSA modeling are generally not given as an equation or a family of equation, but as one or more plots, there are clearly three stages to the decrease in relative MSA, hence property, values on the typical semilog plot as P increases. These are first an approximately linear decrease for about one-

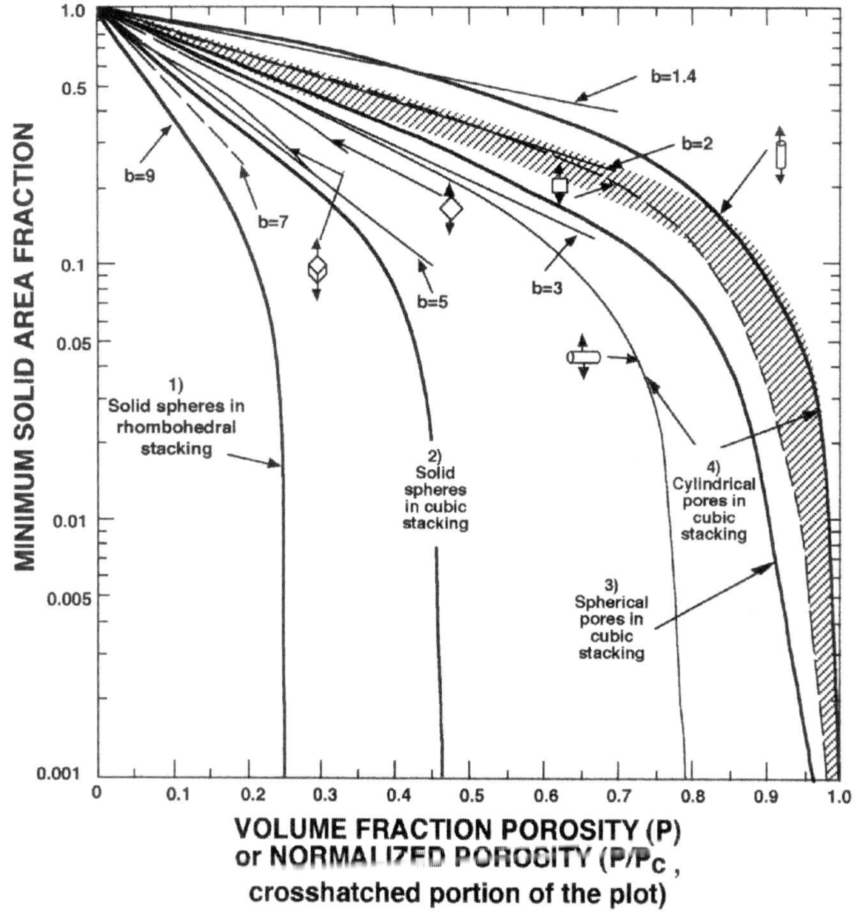

Fig. 2 Semilog plots of the MSA, which is also the relative Young's modulus or other pertinent mechanical properties normalized by their values at $P = 0$ versus the volume fraction porosity P for common basic MSA models. Note the essential consolidation of these individual models essentially to a single plot (crosshatched area) when the porosity is normalized by P_C, that is, the horizontal scale is P/P_C. After Rice [10], published with permission of *J. Mater. Sci.*

fourth to one-half of the allowable P range, followed by an accelerating rate of relative MSA decrease and then by a steeper rate of decrease to P_C. While each of these three regions can be approximated by equations, the initial approximately linear decrease in a semilog MSA plot is readily approximated by the familiar exponential relation of the form:

$$E/E_0 = e^{-bP} \tag{2}$$

where e is the natural logarithm base, and b the slope of the linear trend of data on a semilog plot. (Note that b varies inversely with P_C and is thus another indicator of

P_C values.) The values obtained for the various slopes of the different MSA models (Fig. 2) are consistent with slopes for the same plots of experimental data for real bodies in which the dominant pore character is similar to that of a particular MSA model. This agreement between extensive experimental data and MSA models supports the use of such models [1, 7–10].

4.1.2.4
Computer Models

The fourth, and potentially very diverse, method of modeling uses computers. A promising example is extension of MSA models, which is in the development stage, but the focus is on the more established use of the finite-element method (FEM) since there has been more work on this. While there are may possibilities, there are also challenges, an important one being whether the analysis is a simpler two-dimensional (2D) or a more rigorous three-dimensional (3D) analysis. Agarwal et al. [16] demonstrated that such 3D analysis of spherical pores in a glass matrix produced reasonable results by idealizing the structure as consisting of identical spherical pores uniformly spaced in the matrix and limited in total pore volume fraction to $P < 0.5$ (to limit interactions with adjacent pores), so the analysis could be done by solving the asymmetric problem of a representative cell consisting of a pore imbedded in a section of the matrix. Their results are in reasonable agreement with the MSA model for simple cubic packing of identical spherical pores [1] (see Figs. 2 and 8).

Recently, Roberts and Garboczi [17] reported FEM analyses of ceramics with pores between randomly packed identical spherical particles in various stages of sintering or of randomly packed overlapping spherical or ellipsoidal pores in a densifying ceramic matrix for values of $P \leq 0.5$. They also used a basic cubic cell structure of nominally identical particles, but consisting of substantially more than one randomly packed particle or pore per cell. Results were found to depend on the size of the cell and the pixels relative to one another. However, it was found that results from three suitable sizes could be extrapolated to an essentially "true" value for each degree of a given porosity. Results were shown to be reasonably consistent with some typical literature results and with the equation

$$E/E_0 = (1 - P/P_C)^n, \tag{3}$$

with $n = 2.23$ and an approximate P_C of 0.65. The value of P_C is only approximate since computational time increases rapidly as P_C is approached, so only an approximation can be made. However, the P_C values obtained are too high for the type of porosity involved to an extent which indicates modeling problems that need further attention [10].

More recently, the computer program OOF (Object-Oriented Finite Element Analysis of Real Material Microstructures) for simulating the effects of microstructure on physical properties has been used to evaluate effects of porosity on mechanical properties of ceramics [18]. In an attempt to introduce some real pore character into the simulation this evaluation used random cross sections of actual specimens and hence of their three-dimensional pore character (elongated spheroids in this case) from which to generate a pore structure for analysis of the mechanical effects of the pores. However, they used a 2D analysis which is much less demanding of computational capacity. Thus, the pore structure generated for their 2D analysis is tubular, which is clearly different from the starting 3D pore structure. The generated tubular pores also have varying diameters whose size distribution is not that of the actual pores themselves, but of the intersections of the original pores with the plane on which their evaluation took place. The results of this evaluation were consistent with that for tubular pores, which is of some use, but caution is indicated in use of this technique, since many of the 2D structures and issues are amenable to 2D analytical solutions, but are of limited applicability to 3D modeling [19].

4.1.3 Porosity Effects on Mechanical Properties of Cellular Ceramics

4.1.3.1 Honeycomb Structures

Honeycomb structures are almost exclusively anisotropic in their mechanical properties. The Young's modulus for honeycombs with straight tubular pores aligned parallel with an applied uniaxial stress is found readily by several methods (e.g. MSA and others) to simply be:

$$E/E_0 = 1 - P, \tag{4}$$

that is, Eq. (1) with $n = 1$, which holds for any straight tubular pores, regardless of the (constant) cross sectional geometry, and any packing of the pores. Note that this dependence is also the upper bound for porosity dependence of elastic properties (e.g., this represents the highest stiffness attainable for any given amount of porosity). This $1-P$ dependence is also the upper limit for all but some atypical toughness behavior and most, if not all, strength behavior [1]. Solutions for the same porosities stressed normal to the direction of aligned tubular pores are all substantially lower in property levels and have lower P_C values depending on the stacking of the aligned pores (Fig. 3). The properties and their porosity dependence also depend on pore cross-sectional shape. Those of circular cross section are apparently isotropic in any plane normal to the aligned pores, while pores of other basic geometrical cross-sectional shapes can range from isotropic, or nearly so, E values, to quite anisotropic modulus in a plane normal to that of alignment. Thus, for stressing normal to aligned close-packed pores of square cross section but parallel to the square pore walls gives half the values of E as for stressing parallel to the aligned pores, but the same dependence on P (i.e., varying as $1-P$), while stressing along diagonals of the square pore cross section results in greater porosity dependence. This and other variations noted by Gulati [22] are shown in Fig. 4 for aligned tubular pores of trian-

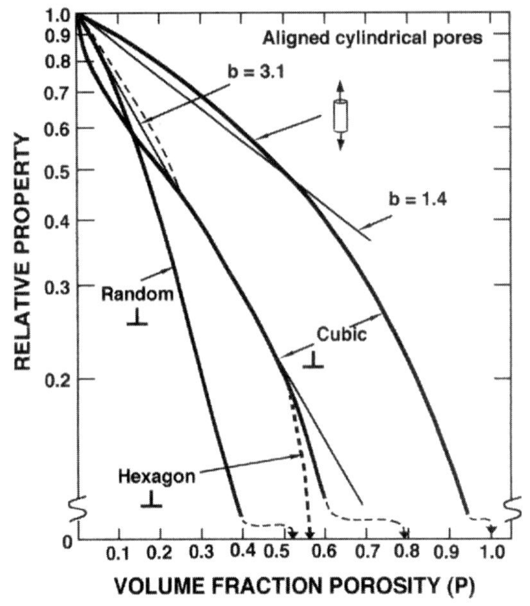

Fig. 3 Semilog plot of relative Young's modulus or other pertinent mechanical property for aligned cylindrical pores versus volume fraction porosity *P* for MSA and computer [20, 21] models of such cubically packed pores aligned with the stress axis (which is the upper limit of properties obtained with pores present). Results for simple cubic, random, and hexagonal stackings with the stress axis normal to that of the tubular pores are shown. (Only cubic packing was treated by MSA models, while all three stackings were calculated by 2D computer modeling.) Solid curves are for small specimen to pore size ratios (ca. 3/1); dashed lines and arrows are respectively for increasing and large specimen to pore size ratios. After Rice [1], published with permission of Marcel Dekker, Inc.

gular and square cross section. Note that the basic difference in behavior parallel and perpendicular to that of aligned tubular pores of circular cross section is corroborated by evaluation by Francl and Kingery [23] for thermal conductivity data for alumina (determined by conduction through the solid phase rather than radiation or convection through the pore phase).

Strength dependence on porosity is often similar to that of *E*, but often somewhat greater (i.e., lower strengths) since it is more affected by extremes of the porosity in the body. However, close-packed aligned pores of square cross section stressed in tension (flexure) show a linear dependence on cell wall strength σ_f and on $1-P$, while tensile strength for stressing normal to one set of cell walls and parallel to the others shows the same dependences on σ_f and *P*, but with only about one-half the tensile strength. Much more limited information is available (mainly or only for close-packed hexagonal cells) for bodies under biaxial flexure (tension) [24].

Crushing strengths are often important (e.g., for automotive exhaust catalyst supports) and have received some evaluation for densely packed aligned tubular pores stressed parallel to the axis of the aligned pores, or in two directions normal to the axis of alignment (one normal to half of the cell walls and parallel to the remaining

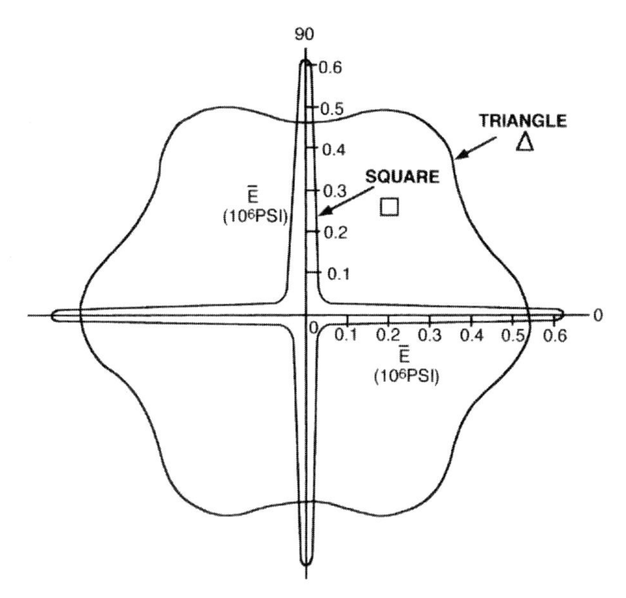

Fig. 4 Linear plot of the Young's modulus normal to densely packed aligned tubular pores with square or triangular cross sections showing, respectively, high anisotropy and near isotropy with high E. Note that densely packed hexagonal pores result in even more isotropic behavior, but at very low levels of E (slightly greater than the minimum for tubular pores with square cross sections) due to inherently low stiffness of such shaped pores. Data after Gulati [1, 22], published with permission of Marcel Dekker, Inc.

half of the cell walls, and the other parallel with the a diagonal direction of the square cross sections of the cells). While all three tests show crushing strengths linearly dependent on cell wall strength, they have greater dependence on P [e.g., as $(1-P)^2$] [24].

Thermal stress and shock failure has been investigated, especially for automotive exhaust catalyst supports, for which both closed-form and finite-element methods have been used [22].

4.1.3.2 Foams and Related Structures

More experimental and modeling evaluations are available for foams and related structures, especially along the lines of Gibson and Ashby and others. Modeling of open-cell foams is easier and results in simpler equations such as:

$$E/E_0 \sim (1-P)^2 \tag{5}$$

$$G/G_0 \sim 3/8\,(1-P)^2 \tag{6}$$

where G is the shear modulus at any P, G_0 is its value at $P = 0$, and the symbol ~ reflects the fact that constants of proportionality have been determined empirically.

(Note that the exponents are dependent on the packing and interconnection structure of the cells; the value of 2 is for one simple, commonly used structure [1, 2, 14].) Poisson's ratio ν is approximately constant regardless of P (i.e., the value at $P = 0$, commonly ca. 1/3). Equations for several other mechanical properties such as fracture toughness (normalized), tensile (flexure) strength, crushing (compressive) strength, and hardness all again involve proportionality constants (ca. 0.65 for each of these cases), depend linearly on strut strength, and all have a porosity dependence of $(1-P)^{1.5}$. (Note that while direct measurement of strut strengths is often not feasible, it can be obtained in some cases where fracture mirrors on fractured struts are clear enough to correlate with their fracture strengths, as discussed below. Also the indicated equivalence of tensile and compressive strengths is generally not born out by experiment, consistent with compressive strength being significant higher than tensile strength at low porosity [1].)

Modeling the properties of closed-cell foams and the transition from open- to closed-cell structure is done by incorporating a parameter φ, the fraction of solid material in the cell edges (struts), and the portion of solid material in the cell walls $1-f$, which are straightforward in definition but difficult to determine in real bodies. However, much closed-cell behavior is similar to that of open-cell foams [2, 11].

The above modeling results provide useful, at least general, guidance, as shown by the more limited data on ceramics and by the applicable data for plastic and metal foams, but there are still uncertainties. For example, data of Zwissler and Adams [25] for foamed glasses is cited as being consistent with the G–A open-cell model for Young's modulus, fracture toughness, and flexural strength [2]. This is true in a broad sense, but there is some uncertainty (Fig. 5). Thus, while most of the limited data fall within the scatter band of other foam data, some is near or outside this band, and there are some indications that the slopes (i.e., exponent values) may increase as P decreases (see also Fig. 8), and the average trends may differ measurably from those predicted by the applied model.

One source of the uncertainty are the measured values themselves and their extrapolation to $P = 0$. Linear extrapolation of the Young's modulus data to $P = 0$ gives $E_0 \approx 35$ GPa, about half the expected value. Doubling the measured values to be consistent with values at $P = 0$ still leaves the data in rough agreement with the model. However, even greater discrepancy is indicated in the extrapolation of tensile (flexure) strengths to $P = 0$, which gives a value of about 11 MPa, which may be as much as 5–10 times lower, but still be consistent with the model if so corrected. On the other hand, the fracture toughness values extrapolate to a typical value of about 0.8 $\text{MPa}\cdot\text{m}^{-1/2}$, consistent with values for bulk silicate glasses. However, this apparent consistency could be misleading, since there can be significant discrepancies in extrapolating fracture toughness values of some porous bodies owing to incompletely understood effects of porosity and some methods of toughness measurements [1, 26]. This issue of measurement accuracy will be addressed further below.

The above data on glass foam of Zwissler and Adams [25] has been shown to agree as well or better with the MSA model for cubically stacked spherical bubbles than with the Gibson and Ashby model, regardless of whether the data are corrected for differences between extrapolated values at $P = 0$ or not [1]. The more extensive

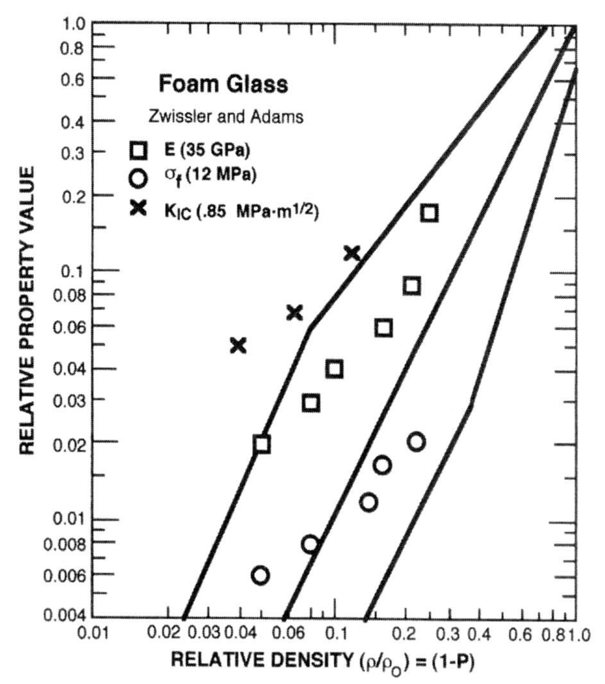

Fig. 5 Log-log plot of relative (i.e., normalized) Young's modulus E, fracture toughness K_{IC}, and flexure strength σ_f, that is, values for porous bodies divided by their values at $P = 0$ (data on foam silicate glass of Zwissler and Adams [25]) versus log $(1-P)$ (volume fraction porosity). Note again the equivalence of ρ/ρ_0, for example, as shown in Eq. (1). Approximate values of the data for each property extrapolated to $P = 0$, given in parentheses, shows some consistencies and inconsistencies of extrapolated and known values at $P = 0$. Note the mean and approximate bounds of values (solid lines) from surveys of Gibson and Ashby [2], and that the data are near to or within these limits (but less so if they are low due to possible test effects). While the approximate slopes of the $1-P$ dependence are not grossly inconsistent with model values, they are probably of lower slope (with changes to higher slopes at lower P values, similar to those shown in Fig. 8).

range of data for sintered SiO_2 foam of Harris and Welsh [27] also shows good agreement for the same MSA model (Fig. 6), and data for various porous SiO_2 bodies from sol–gel processing show agreement with and transitions between various MSA models pertinent to different stages of pore character and amount [1] (Fig. 7).

More recent and detailed studies on polycrystalline ceramic foams are overall consistent with the basic trends of foam models and the data trends noted above for the glass-foam results of Zwissler and Adams [25] (Fig. 5). Thus, Hagiwara and Green [28] found in their study on some alumina open-cell foams that the exponent for the porosity dependence of elastic moduli was about 2, as predicted in Eq. (5), but that the proportionality constants between the elastic properties and their porosity dependences were often lower than those given in the literature (e.g., a value of ca. 0.14 was found for G instead of the expected value of ca. 0.4). These differences were attributed to variations in cell-strut microstructure. Subsequently, Brezny and

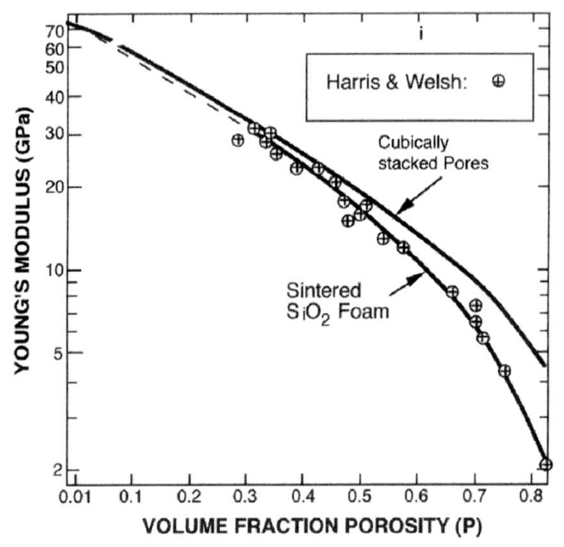

Fig. 6 Semilog plot of relative Young's modulus of sintered SiO_2 foam of Harris and Welsh versus the volume fraction porosity *P* and the MSA model for cubic stacking of ideal spherical pores (see also Fig. 8), which is a reasonable approximation for random packing of pores [1]. Note the general agreement of the model and the data. Some increasing deviation of the data below model values may reflect possible changes in cell structure and especially somewhat lower P_C values for real as opposed to ideal materials [1]. After Rice [8], published with permission of *J. Mater. Sci.*

Green [29], in a study on bending strength, fracture toughness, and compressive strength of some similar and other alumina-based open-cell foams, showed that the observed exponent for the porosity dependences was about 1.5, as expected, but the proportionality factor was found to be 0.13–0.23 instead of 0.65. More recently Vedula et al. [30] summarized the exponents for the porosity dependences to typically be about 2 and 1.5, respectively, for elastic moduli and fracture toughness, but that the exponent for compressive and tensile strengths could range from slightly less than the predicted value of 1.5 to greater than 2. Colombo and Modesti [31] reported that the porosity exponents for flexural strength and elastic moduli were each about 1 for quite open cell foams of SiOC (produced from foamed preceramic polymers), that is, lower than the respective expected values of 1.5 and 2.0. On the other hand, foams with less open cells gave respective values of 2 and 3.5 (i.e., somewhat to substantially higher than expected). This shows that general trends of Gibson–Ashby-type models may differ somewhat to substantially from data.

Studies (e.g., by Green et al.) show that much of the above variations in open-cell, mainly alumina-based, foams were due to effects of cell struts. Thus, the major source of variations of elastic moduli in their open-cell alumina foams was the microstructure of the cell struts, not the varying limited contents of closed cells [28]. (They also confirmed that Poisson's ratio was nominally constant at about 0.2, reasonably consistent with predictions of its being about 0.3 given the data scatter.) Strengths of the foams were found to correlate with those of the struts, which commonly

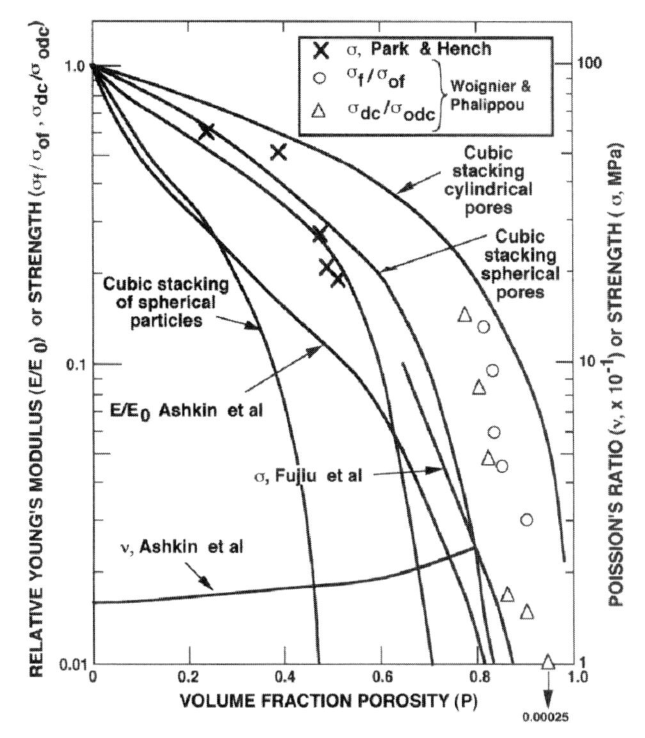

Fig. 7 Plot of absolute or relative mechanical properties versus volume fraction porosity P for bodies made from partially to fully sintered SiO_2 gels [38–42]. These bodies reflect transitions from stacked sintered particles to arrays of bubble-type pores, as shown by the various models. See also Fig. 8 for the same data in a different plot. Note the limited magnitude and modest change of Poisson's ratio ν, reasonably consistent with expectations from G–A models. After Rice [8], published with permission of *J. Mater. Sci.*

had low Weibull moduli (e.g., 1–3) [29, 32]. These trends were also shown to be true for glassy carbon foams, which gave particularly clear fractographic results [33].

Thermal shock resistance has also been addressed with more complex and less defined results. Tests with alumina-based open-cell foams showed that neither the macro thermal stresses across the complete foam specimen nor the local stresses across individual struts were the source of failure. This was shown to be due to heating up of quench fluid as it infiltrated the specimen, which strongly increased with increasing cell size [34]. Cell-size effects and limited dependence on density were confirmed in further experiments. Correlations between maximum thermal strains and degree of thermal stress damage (measured nondestructively) were shown [30].

Measurements of compressive (crushing) strength, performed on a few ceramic foams, gave some more complex and different results. Tests by Dam et al. [35] on open-cell alumina foams gave differing results according to whether the loading rams had stiff or compliant loading surfaces, with the former leading to significant discrepancies with G–A models, and the latter to better agreement. The use of compliant loading surfaces appeared to reduce general damage accumulation with increased loading. An increase in both Young's modulus *E* and crush strength with decreasing cell size was observed, contrary to G–A models. Later strength measurements on glassy carbon foam [36] showed both bending (tensile) and compressive (crushing) strengths that increased with decreasing cell size, but not *E* or fracture toughness. These results are more consistent with those of Morgan et al. [37] on foam glass, which showed a pronounced increase in compressive, tensile, and flexure strengths with increasing inverse square root of cell size (fracture toughness and *E* were essentially independent of cell size).

Structures the same as or similar to those generated by foaming can be produced by other fabrication methods and give added insight to the mechanical behavior as a function of porosity. SiO_2 bodies produced by sol–gel processing that range from zero to various high degrees of porosity are an example of this [38–42]. Figure 7 shows agreement with MSA models and indicates changes in pore structure with increasing porosity based on comparison of their property data (especially *E* data of Ashkin et al [38, 39]). Thus, the data are consistent with the MSA model for pores

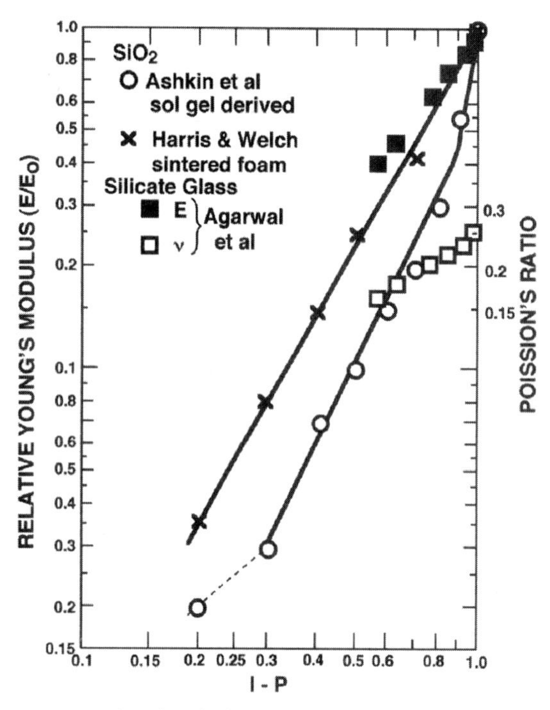

Fig. 8 Log-log plot of relative Young's modulus of SiO_2 made by sol–gel processing by Ashkin et al. [38, 39] (see also Fig. 7) and by sintering of fused quartz by Harris and Welsh [27] versus 1–*P* (see also Fig. 6). Note that both sets of data, while approximately consistent with G–A-type and related models using a constant value of the exponent *n* (e.g., ca. 2), shows some variation to higher *n* values at low *P*, consistent with the major change in pore structure indicated in Fig. 7. Note also for reference *E* and *ν* data for 3D finite-element analysis by Agarwal et al. [16] for spherical pores in glass.

between cubic packings of uniform spherical pores at low porosity, with a transition to one of cubic packing of uniform spherical pores at higher porosity. Such pore transition is consistent with the basic gel structure being one of strings of fine beads which pack closely at low porosity (high solids contents) but transition to outlining nominally spherical pores at lower densities [1]. For comparison of MSA and G–A models some of the previously shown Young's modulus data for SiO_2 are replotted as log E versus log $(1-P)$ in Fig. 8. While the sintered-foam data [27] show an essentially linear log-log plot, the data for the sintered sol–gel shows a bi- or trilinear plot consistent with the transitions in pore structure noted in conjunction with Fig. 7. Further, the two indicated changes in slope of the data for the sintered sol–gel are consistent with the physical requirements that the lower limit of $P = 0$ should give $E/E_0 = 1$ and that at the upper limit of P (i.e., $P = P_C$) the plot should give $E/E_0 = 0$. Note also that data of Agarwal et al. [16] for a silicate glass from 3D computer modeling show reasonable agreement with those of the sintered SiO_2 foam.

Another informative case of foamlike structures made by techniques other than foaming is that of bodies made by bonding ceramic beads, or especially balloons, together. Ceramic balloons can be made with relatively thin walls, but with more uniform wall thickness and resultant properties than are common in foaming, for which significant tapering to thinner walls with increasing distance from foam cell struts is observed [43]. This greater uniformity of wall thickness of ceramic balloons translates into substantially greater strengths of low-density foamlike bodies made by bonding such balloons together [1, 44, 45]. The strength to mass ratio of bodies formed by bonding balloons together can be improved by limiting bonding mainly to the contact areas between adjacent balloons.

Green [46] has made foamlike bodies by sintering compacts of glass balloons. Green and Hoagland [47] modeled the strength of these bodies.

4.1.4 Discussion

4.1.4.1 Measurement–Characterization Issues

Adequate measurement of pertinent properties and characterization of microstructure are necessary for a sound understanding of the mechanical behavior of cellular materials. This is particularly true of ceramics because of their brittle behavior and resultant sensitivity to stress concentrations. This need is shown by Dam et al. [35], whose tests on open-cell alumina foams showed more consistent results using compressive loading rams with compliant versus stiff loading surfaces. This is supported by Green's observation [46] that elastic moduli determined by compression loading of bodies of sintered glass balloons showed progressive deviations from values from ultrasonic or tensile measurements. Brezny and Green [48] concluded from their evaluation of scale effects on mechanical measurements on glassy carbon foams that specimen dimensions should be at least 15–20 times the cell size. Huang and

Gibson [12] concluded that valid fracture toughness measurements on brittle foams requires cracks whose size is greater than 10 times the cell size. Again note that the Young's modulus and tensile (flexure) strengths of Zwissler and Adams [25] for foamed glasses extrapolated to only about half the values expected at zero porosity, while the fracture toughness extrapolated to about the value expected for zero porosity. While extrapolation of properties to $P = 0$ is best done with models known to fit the type of porosity involved, linear extrapolations can raise questions for more detailed evaluation. Brezny and Green [49] have suggested procedures for uniaxial tensile and compressive strength testing.

Both the above examples and the basic nature of many porous bodies, including most, if not all, cellular ones, strongly suggest that much more work is needed to properly assess their mechanical, especially failure behavior. A key component of this which has been paid even less attention are effects of surface finishing; most surface finishing is presumably by machining, but is of limited or no specification. There has been limited or no comparison of, for example, as-fired versus as-machined specimens, which might be significant, especially for specimens of glass-based bodies.

Another basic limit of much data is the incomplete characterization of the microstructure, especially that of the porosity. Values of the volume fraction porosity P are usually given, but many reflect only the dominant source of porosity, and no indication of the statistical or spatial variations of P values are given. However, of greater concern is the almost total lack of characterization of the pore structure, such as the shape of pores and how they are spatially arrayed relative to each other and the solid structure.

4.1.4.2
Impact of Fabrication on Microstructure

The fabrication method used to make a cellular body obviously plays a major role in the amount, type, and character of the pore structures introduced, as shown for example by comparing foams with bodies made by bonding balloons together. The latter have more uniform pore wall thickness and a broader range of achievable wall thicknesses, but they are more limited in achieving thin and tapered wall thickness. However, the characterization of bonded balloon structures is probably simpler and more accurate than that of foams (and they can have significant advantages in the size, shape, and practicality of making porous bodies [1]). Similarly, there are trade-offs between extrusion and tape lamination, and to some extent with solid free-form fabrication, that impact costs, sizes, and shapes of bodies, as well as of resultant pore structures [43]. The limited development of foaming of glass and other ceramics, while producing reasonable foam cellular bodies, produces cell variations and often some overall anisotropy of the cellular body and its structure due to differing expansions in the rise direction versus the lateral direction during foaming. (Such anisotropies in foaming are common to forming many materials, including foamed foods such as cakes, pastries, and breads. Property effects of such anisotropies on properties have been modeled [1, 2].)

There can also be trade-offs in performance due to other microstructural factors in different fabrication methods. Thus, two effects have been identified in the commercial fabrication of open-cell ceramic foams by ceramic slip coating of open-cell polyurethane (PU) foams. First, such replication typically results in hollow ceramic struts due to the incomplete penetration of the PU foam structure by the ceramic slip and subsequent PU removal on firing the ceramic. The hollow struts that result from this fabrication generally do not reduce properties and can actually be of some benefit owing to limited reductions in density that give somewhat better property to mass ratios. However, another common strut effect in such ceramic foam fabrication, namely cracks in the ceramic struts due to differences in drying shrinkage and thermal expansion of the PU and the ceramic, lowers the resultant properties.

Other effects more specific to fabrication with a given material can occur, as illustrated by the unexpected benefits found by Lachman et al [1, 43, 50] in their engineering production of cordierite auto exhaust catalyst monoliths by extrusion. They selected clay as one of the main raw materials for both its lower cost and its easier extrudability. Extrusion resulted in considerable alignment of the clay particles, which in turn resulted in preferred orientation of the cordierite formed from them during firing. This preferred orientation of the resultant cordierite resulted in improved thermal stress/shock resistance of the cordierite honeycomb.

4.1.4.3
Porosity–Property Trade-Offs

Many important applications of cellular and related materials require various aspects of pore structures that place greater limits on the mechanical properties of these materials [1]. Thus most catalyst applications require high surface areas, but the highest surface areas are obtained with bodies made of lightly sintered fine powders, which results in most mechanical properties being lower and thus requires compromises in pore structure. Similarly, many applications require substantial flow or permeability through the porous structure. This is favored by higher P values and tubular versus equiaxial pores, which in turn may require some compromises in pore structure to limit reductions in both mechanical behavior and surface area. Thus, many applications are best met by controlled amounts of differing porosities in a given body

The above differences in different porosity effects are important reasons for better characterization of pore structures, as well for modeling effects of more than one type of porosity in a given body. Some characterization of dual porosities in a body is already done to some extent where effects of extruded tubular holes or foam cells and the pores in the extruded honeycomb walls or foam cell walls/struts are handled separately. Some more general combination of effects of different porosities in the same body has been demonstrated in MSA models [1], and it may be feasible in some other models, but further development is needed.

4.1.5 Summary

Substantial progress has been made in test methodology, data generation, and modeling and understanding of porosity effects on mechanical properties of cellular ceramics. Thus, some property measurements must be made with compliant test surfaces, and others require a sufficient number of pore cells across the test specimens. However, there is need for a broader range of property measurements, such as effects of various specimen surface finishes on strengths. Valuable methods of porosity characterization are available, in particular stereology, but are often used little or not at all, although they are sorely needed for more comprehensive characterization.

In modeling, which is a major factor in understanding mechanical behavior, substantial progress has also been made. Four modeling approaches – power law, Gibson–Ashby (G–A), Minimum Solid Area (MSA), and computer modeling – were considered in this chapter. All have some value, and use of two or more models for comparison and building a data base is recommended.

There has been considerable modeling of mechanical properties (and characterizing the porosity) of ceramic honeycomb structures due to development of tubular monoliths for catalytic, heat-exchanger, and filter applications. Though complicated to varying extents by anisotropic property behavior of honeycombs, their modeling is somewhat better developed, since much of it is amenable to 2D modeling.

Good progress has also been made in the study of 3D foam cellular structures, in part due to their development in other materials. However, noted cautions with the power-law modeling approach for 3D pore structures need to be addressed. G–A-type and MSA models are both valuable due to their explicit focus on microstructural factors. The power law and MSA modeling approaches, which cover a broader porosity range, should provide useful comparison with G–A models (which focus on cellular structures and have a substantial data base for such structures). More attention to broader ranges of types and amounts of porosity than commonly found in cellular bodies will also be useful, including extrapolations to P_C and $P = 0$, to better understand the behavior of cellular bodies. MSA models covering many of the structures of interest can aid in addressing transitions in pore character and properties between less and more porous bodies with generic types of porosity. Computer modeling holds substantial promise but needs to better address the issue of effectiveness of obtaining and validating 3D results. MSA modeling with computers may be particularly advantageous in modeling effects of combinations of porosity in bodies. Ultimately, computer modeling may be the only route to analyzing the behavior of more complex porous bodies with variations in amount, character, or both of the porosity, and is likely to be a valuable tool in evaluating trade-offs between important properties with significantly different porosity dependences.

References

1 Rice, R.W., *Porosity of Ceramics and its Effects on Properties*, Marcel Dekker, New York, 1998.
2 Gibson, L.J., Ashby, M.F., *Cellular Solids, Structure & Properties*, Cambridge University Press, Cambridge, England, 1999.
3 Van Voorhees, E.J., Green, D.J., *J. Am. Ceram. Soc.*, **1991**, *74* (11), 2747–2752.
4 Bose, S., Sugiura, S., Bandyopadhyay, A., *Scripta Mater.*, **1999**, *41* (9) 1009–1014.
5 Hattiangadi, A., Bandyopadhyay, A., *J. Am. Ceram. Soc.*, **2000**, *83* (11), 2730–2736.
6 Kachanov, M., Sevostianov, I., Shafiro, B., *J. Mech. Phys. Solids*, **2001**, *49*, 1–25. (See also Sevostianov, I., Kachanov, M., *J. Mech. Phys. Solids*, **2002**, *50*, 253–282.)
7 Rice, R.W., "The Porosity Dependance of Physical Properties of Materials – A Summary Review" in *Porous Ceramic Materials, Fabrication, Characterization, Applications*, Liu, D.-M., ed., Trans Tech Publ., Zurich, Switzerland, 1996, pp. 1–19.
8 Rice, R.W., *J. Mater. Sci.*,**1996**, *31*, 102–108.
9 Rice, R.W, *J. Mater. Sci.*,**1996**, *31*, 1509–1528.
10 Rice, R.W, *J. Mater. Sci.*, **2005**, *40*, 983–989.
11 Gibson, L.J., *Mater. Sci. Eng.*, **1989**, *A110*, 1–36.
12 Huang, J.S., Gibson, L.J., *Acta Metall. Mater.*, **1991**, *9* (7), 1627–1636.
13 Gibson, L.J., Ashby, M.F., *Proc. R. Soc. London*, **1982**, *A382*, 43–59
14 Gent, A.N., Thomas. A.G., *Rubber Chem. Tech.*, **1963**, *36*, 597–610.
15 Huber, A.T., Gibson, L.J., *J. Mater. Sci.*, **1988**, *23*, 3031–3040.
16 Agarwal, B.D., Panizza, G.A., Broutman, L.J., *J. Am. Ceram. Soc.*, **1971**, *54* (12), 620–624.
17 Roberts, A.P., Garboczi, E.J., *J. Am. Ceram. Soc.*, **2000**, *83* (12), 3041–3048.
18 Cannillo,V., Leonelli, C., Manfredini, T., Montorsi, M.A." Boccaccini, A., *J. Porous Mater.*, **2003**, *10*, 189–200.
19 Kachanov, M., Tsukrov, I., Shafiro, B., "Effective Moduli of Solids with Cavities of Various Shapes" in *Micromechanics of Random Media*, Ostoja-Starzewski, M., Jasiuk, I., eds., *Appl. Mech. Rev.*, 47, (1, part 2) ASME Book. No. AMR 139S170, 1994, pp S151–S174.
20 Day A.R., Snyder, K.A., Garboczi, E.J., Thorpe, M.F., *J. Mech. Phys. Solids*, **1992**, *40* (5), 1031–1051.
21 Snyder, K.A., Garboczi, E.J., Day, A.R., *J. Appl. Phys.*, **1992**, *72* (12), 5948–5955.
22 S. T. Gulati, "Effects of Cell Geometry on Thermal Shock Resistance of Catalytic Monoliths", Soc. Auto. Eng. Congress and Exposition Paper 750171, (Detroit, MI), 2/1975.
23 Francl, J., Kingery, W.D., *J. Am. Ceram. Soc.*, **1954**, *37* (2), 99–106.
24 S. T. Gulati, A. Schneider, "Mechanical Strength of Cellular Ceramic Substrates", in Proc. Ceramics for Environmental Protection ('Enviceram '88) Cologne, Germany, 12/7-9/1988, Deutsche Keramische Gesellschaft e.V., Köln, Germany.
25 J. G.Zwissler, J.G., Adams, M.A., "Fracture Mechanics of Cellular Glass", in *Fracture Mechanics of Ceramics*, Bradt, R.C., Evans, A.G., Hasselman, D.P.H., Lange, F.F., eds., Plenum Press, New York, 1983, pp. 211–242.
26 Rice, R.W., *J. Mater. Sci.*, **1996**, *31*, 1969–83.
27 Harris, J.N., Welsh, E.A., "Fused Silica Design Manual, I", Georgia Institute of Technology (Atlanta, Ga) for Naval Ordinance Systems Command Contract N 00017-72-C-4434, 5/1973.
28 Hagiwara, H., Green, D.J., *J. Am. Ceram. Soc.*, **1987**, *70* (11), pp 811–815.
29 Brezny, R., Green, D.J., *J. Am. Ceram. Soc.*, **1989**, *72* (7), 1145–1152.
30 Vedula, V.R., Green, D.J., Hellman, J.R., *J. Am. Ceram. Soc.*, **1999**, *82* (3), 649–656.
31 Colombo, P., Modesti, M., *J. Am. Ceram. Soc.*, **1999**, *82* (3), 575–578.
32 Brezny, R., Green D.J., Dam, C.Q., *J. Am. Ceram. Soc.*, **1989**, *72* (6), 885–889.
33 Brezny, R., Green, D.J., *J. Am. Ceram. Soc.*, **1991**, *74* (5), 1061–1065.
34 Orenstein, R.M., Green, D.J., *J. Am. Ceram. Soc.*, **1992**, *75* (7), 1899–1905.
35 Dam, C.Q., Brezny, R., Green, D.J., *J. Mater. Res.*, **1990**, *5* (1), 163–171.
36 Brezny, R., Green, D.J., *Acta Metall. Mater.*, **1990**, *38* (12), 2517–2526.
37 Morgan, J.S., Wood, J.J., Bradt, R.C., *Mater. Sci. Eng.*, **1981**, *47*, 37–42.
38 Ashkin, D., Haber, R.A., Wachtman, J.B., *J. Am. Ceram. Soc.*, **1990**, *73* (11), 3376–81.
39 D. Ashkin, "Properties of Bulk Microporous Ceramics," Ph.D. Thesis, Rutgers University, 1990.

40 Park, S.C., Hench, L.L., "Physical Properties of Partially Densified SiO_2 Gels", *Sci. Ceram. Chem. Proc.*, **1986**, 168–172.

41 Fujiu, T., Messing, G.L., Huebner, W., *J. Am. Ceram. Soc.*, **1990**, *73* (1), 85–90.

42 Woignier, T., Phalippou, J., *J. Non-Cryst. Solids*, **1988**, *100*, 404–408.

43 R. W. Rice, *Ceramic Fabrication Technology*, Marcel Dekker, New York, 2003.

44 Chung, J.H., Cochran, J.K., Lee, K.J., "Compressive Mechanical Behavior of Hollow Ceramic Spheres" in *Spheres and Microspheres: Synthesis and Applications*, Wilcox, D.L., Berg, M., Bernat, T., Cochran, J.K., and Kellerman, D., eds., *MRS Proc.* **1995**, *372*, 179–86.

45 P.R. Chu, "A Model for Coaxial Nozzle Formation of Hollow Ceramic Spheres" PhD Dissertation, Materials Science & Eng., Georgia Inst. Tech, Atlanta, GA, June 1991.

46 Green, D.J., *J. Am. Ceram. Soc.*, **1985**, *68* (7), 403–409.

47 Green, D.J., Hoagland, R.G., *J. Am. Ceram. Soc.*, **1985**, *68* (7), 395–398.

48 Brezny, R., Green, D., *J. Mater. Sci.*, **1990**, *25*, 4571–4578.

49 Brezny, R., Green, D.J., (Journal?(**1993**, *76* (9), 2185–2192.

50 Lachman, I.M., Bradley, R.D., Lewis, R.M., *Am. Ceram. Soc. Bull.*, **1981**, *60* (2), 202–205.

4.2
Permeability

Murilo Daniel de Mello Innocentini, Pilar Sepulveda, and Fernando dos Santos Ortega

4.2.1
Introduction

Fluid flow through cellular ceramics is a key topic in many engineering applications. Due to their unique pore geometry that provides very low pressure drop and the ability to operate in extreme conditions, cellular ceramics are the ideal substitutes for conventional porous media used in high-temperature processes involving filtration, fluid mixing, chemical reaction, or mass transfer. As the efficiency of these processes depends primarily on the permeability of the porous media employed, attention in this chapter is drawn to concepts of permeability evaluation and modeling of porous ceramics.

Quantification of permeability is initially discussed on the basis of the equations most applied in the literature. Concepts of flow through granular and fibrous media are then introduced, as they compose the basis of most correlations currently applied to cellular ceramics. Following this, experimental permeability data for honeycomb and foamlike ceramics gathered from the literature are discussed and related to structural features such as porosity, cell size, and, whenever possible, to processing variables. The most recent and innovative models for predicting permeability parameters of cellular ceramics are described and, in cases when no reliable modeling is possible, the reader is referred to other well-established theories available in the literature. The final section concerns criteria and practical examples to establish the correct flow regime through porous media and thus help to choose an optimum permeability model for a given application.

4.2.2
Description of Permeability

Permeability is regarded as a macroscopic measure of the ease with which a fluid driven by a pressure gradient flows through the voids of a porous medium. Thus, permeability is neither a property of the fluid nor a property of the porous medium, but reflects the effectiveness of the interaction between them [1].

Permeability of a porous medium is commonly expressed through parameters that are derived from two main equations: Darcy's law, Eq. (1), and Forchheimer's equation, Eq. (2),

Cellular Ceramics: Structure, Manufacturing, Properties and Applications.
Michael Scheffler, Paolo Colombo (Eds.)

ISBN: 3-527-31320-6

$$-\frac{\mathrm{d}P}{\mathrm{d}x} = \frac{\mu}{k_1} v_s, \qquad (1)$$

$$-\frac{\mathrm{d}P}{\mathrm{d}x} = \frac{\mu}{k_1} v_s + \frac{\rho}{k_2} v_s^2, \qquad (2)$$

where $-\mathrm{d}P/\mathrm{d}x$ is the pressure gradient along the flow direction, μ and ρ are respectively the absolute viscosity and the density of the fluid, and v_s is the superficial fluid velocity, defined by $v_s = Q/A$, where Q is the volumetric flow rate and A is the exposed surface area of the porous medium perpendicular to flow direction, and k_1 and k_2 are usually known as Darcian and non-Darcian permeabilities. These parameters incorporate only the structural features of the porous medium and therefore are considered constant even if the fluid or the flow conditions are changed. Note that although both parameters are referred to as "permeabilities", they are dimensionally distinct: k_1 is expressed in dimensions of square length [e.g., m^2, perm (10^{-4} m^2), and Darcy (10^{-12} m^2)], whereas k_2 is expressed in dimensions of length (typically m).

The basic difference between Eqs. (1) and (2) is the type of dependence that the fluid pressure assumes when related to the fluid velocity. Darcy's law, which is derived from experiments conducted at very low velocities, considers only the viscous effects on the fluid pressure drop and establishes a linear dependence between the pressure gradient and the fluid velocity through the porous medium. Forchheimer's equation, on the other hand, considers that the pressure gradient displays a parabolic trend with an increase in fluid velocity due to the contributions of inertia and turbulence. Although Darcy's law has been widely applied in the literature because of its simplicity, Forchheimer's equation has yielded more realistic and more reliable permeability parameters [2]. In fact, accurate use of Darcy's law and Forchheimer's equation depends very much on a correct determination of the flow regime within porous media. This controversial topic is discussed further in Section 4.2.5.

The integration of the pressure gradient $-\mathrm{d}P/\mathrm{d}x$ through the porous medium depends on the fluid compressibility. In contrast to liquids, gases and vapors expand along the flow path, and thus fluid velocity at the exit of the porous medium becomes higher than at the entrance, and this affects the pressure drop profile. This effect can be taken into account in Eqs. (1) and (2) by considering the fluid as an ideal gas. Table 1 presents integrated forms of Darcy's law and Forchheimer's equation according to the fluid compressibility.

The compressibility effect is more pronounced for low-porosity materials, such as rocks, bricks, tiles, concrete, and cast refractories, but it may also be considerable for highly porous media, depending on the thickness of the sample and on the pressure or velocity applied. Thus, permeability assessment by gas-flow experiments is more likely to generate errors if the compressibility effect is neglected [3].

Table 1 Permeability equations integrated according to fluid compressibility.*

Equation	Incompressible fluids		Compressible fluids	
Darcy	$\frac{\Delta P}{L} = \frac{\mu}{k_1} v_s$	(3)	$\frac{P_i^2 - P_o^2}{2\,P\,L} = \frac{\mu}{k_1} v_s$	(4)
Forchheimer	$\frac{\Delta P}{L} = \frac{\mu}{k_1} v_s + \frac{\rho}{k_2} v_s^2$	(5)	$\frac{P_i^2 - P_o^2}{2\,P\,L} = \frac{\mu}{k_1} v_s + \frac{\rho}{k_2} v_s^2$	(6)

* Where: $\Delta P = P_i - P_o$, L = medium thickness along the flow direction, P_i = absolute fluid pressure at the medium inlet, P_o = absolute fluid pressure at the medium outlet, P = absolute fluid pressure for which v_s, μ, and ρ are measured or calculated (usually P_i or P_o).

4.2.3
Experimental Evaluation of Permeability

Experimental evaluation of permeability parameters consists of a test in which a fluid is forced to flow through a porous sample in stationary regime. The sample is generally a disk or cylinder of thickness L and exposed area A, laterally sealed between two chambers. Absolute fluid pressure at the inlet (P_i) and outlet (P_o) of the sample are measured and recorded as a function of flow rate Q or fluid velocity ($v_s = Q/A$). If Forchheimer's equation is used, the collected data set is fitted according to the least-squares method to a parabolic model of the type: $y = ax + bx^2$, where y is either $P_i - P_o$ for liquids or $(P_i^2 - P_o^2)/2\,PL$ for gases and vapors, and x is the velocity v_s. For gas flow, velocity should be also corrected for the actual average temperature and pressure through the sample. The permeability parameters are then calculated from the fitted constants a and b, respectively, by $k_1 = \mu/a$ and $k_2 = \rho/b$. Table 2 provides equations that are useful for estimating the density ρ and the viscosity μ of water or common gases according to test temperature and pressure.

Several apparatuses for permeability evaluation are described in the literature [5]. The best configuration will depend mostly on the features of the sample. Figure 1 illustrates two typical setups used for ceramic foams. For highly porous materials, liquids are preferred as the testing fluid to provide a higher contribution of the inertial term ($\rho v_s^2/k_2$) and therefore reliable fitting of k_2 values. For consistent data analysis, samples from different batches should have their permeability parameters fitted within a similar pressure–velocity range [6]. If the sample is anisotropic, attention should be given to the flow direction. Generally, the experiment must be carried out with the sample mounted in the same direction as its regular usage.

Table 2 Useful equations for estimating the density and viscosity of water and common gases at different temperatures and pressures.*

Liquid water:

$$\rho = 1.4887\times 10^{-5}T^3 - 5.7544\times 10^{-3}T^2 + 1.0541\times 10^{-2}T + 1000.1 \quad (7)$$

$$\mu = \frac{0.1}{2.1482\left[(T-8.435)+\sqrt{8078.4+(T-8.435)^2}\right]-120} \quad (8)$$

Temperature range: 0–100 °C

Common gases:

$$\rho = \frac{PM}{\mathcal{R}(T+273)} \text{ (ideal gas law)} \quad (9)$$

$$\mu = \mu_o\left(\frac{T+273}{273}\right)^{1.5}\left(\frac{273+C_n}{T+273+C_n}\right) \text{ (Sutherland equation)} \quad (10)$$

Gas	M/kg mol^{-1}	$\mu_o/10^{-5}$Pa·s	C_n	Temperature range/°C
Ammonia (NH_3)	0.017	0.831	503	20–300
Carbon dioxide (CO_2)	0.044	1.38	254	20–280
Carbon monoxide (CO)	0.028	1.66	101	20–280
Chlorine (Cl_2)	0.071	1.23	350	20–500
Ethane (C_2H_6)	0.030	0.861	252	25–300
Ethylene (C_2H_4)	0.028	0.839	225	20–300
Hydrogen (H_2)	0.002	0.848	138	20–825
Methane (CH_4)	0.016	1.00	164	20–500
Nitrogen (N_2)	0.028	1.66	105	20–825
Nitrogen dioxide (NO_2)	0.046	1.78	128	20–250
Nitrogen monoxide (NO)	0.030	1.36	260	20–280
Oxygen (O_2)	0.032	1.92	125	15–830
Propane (C_3H_8)	0.044	0.75	290	20–300
Standard air	0.029	1.73	125	15–800

* Eq. (7) is a polynomial fitting of data given in Ref. [4]. In Eqs. (7)–(10), T is given in °C and P in Pascal for ρ and μ in SI units. R is the ideal gas constant (8.314 J mol^{-1} K^{-1}).

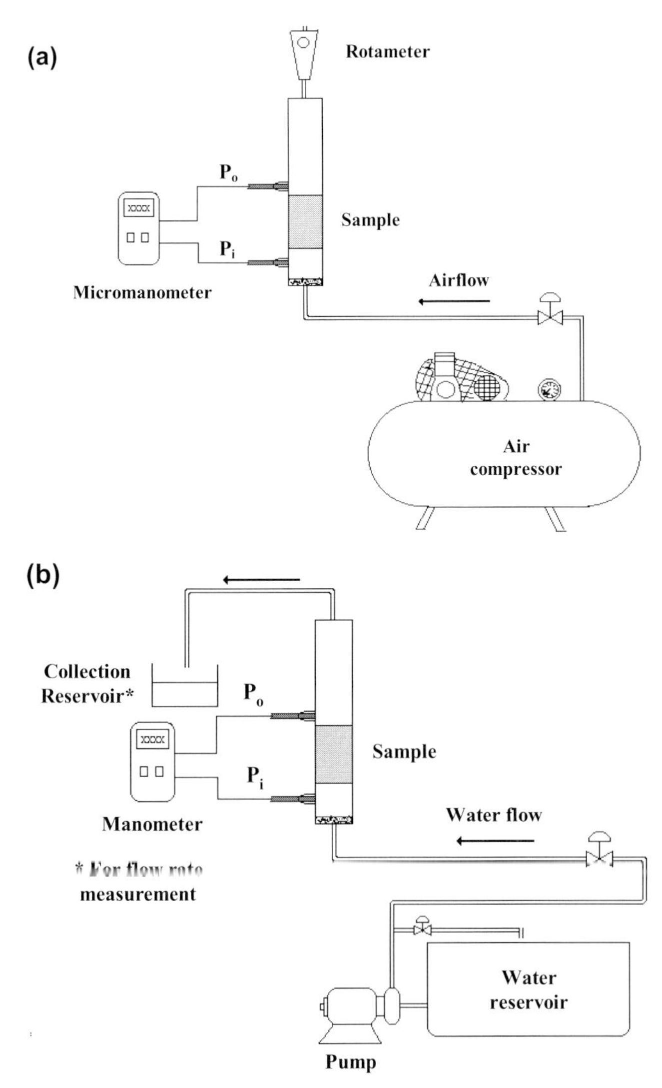

Fig. 1 Schematics of equipment commonly used for permeability evaluation of porous ceramics. a) Air-flow test. b) Water-flow test.

4.2.4
Models for Predicting Permeability

Permeability modeling based on structural parameters of porous media has gained special attention in the literature over the last few decades. The various models can be divided into the following categories [7]: 1) phenomenological models; 2) models based on conduit flow: i) geometrical models, ii) statistical models, and iii) models utilizing the complete Navier–Stokes equation; and 3) models based on flow around submerged objects.

The complexity of flow patterns and the difficulty to mathematically describe the pore structure have precluded the use of the aforementioned approaches to reliably predict permeability parameters for a general porous medium. For engineering purposes, valuable permeability modeling is available in the literature only for special classes of porous media, such as unconsolidated granular media (sand filters, fixed beds, etc.), consolidated granular media (ceramic filters, polymer membranes, bricks, concretes, mortars, etc.), and fibrous media (bag, candle, and cartridge filters). The permeability of cellular ceramics has been only recently addressed in the literature, and in many cases, models used in the determination of permeability constants are still derived from correlations originally developed for other classes of porous media.

The aim of this section is to give a brief overview of the main useful equations relating the permeability parameters to the different types of porous structures, as applied in the literature.

4.2.4.1 Granular Media

Most correlations currently applied to predict permeability of porous ceramics were developed in the 1900s on the basis of data for flow through unconsolidated granular media (i.e., a bed of loose particles) obtained for viscous flow [8–10] and inertial flow [11–13].

The Kozeny–Carman equation stems from the Hagen–Poiseuille relationship (an exact analytical solution for viscous flow in a capillary tube) and gives satisfactory predictions for unconsolidated porous media. Based on the Kozeny–Carman equation, the Darcian permeability k_1 defined in Eqs. (1) and (2) is written as

$$k_1 = \frac{\varepsilon^3}{K_k S_o^2 (1-\varepsilon)^2} \tag{11}$$

where ε is the volumetric void fraction (porosity) of the bed, K_k the Kozeny parameter, and S_o the specific surface area, that is, the surface area of solid phase exposed to the fluid per unit volume of solid phase. S_o is usually measured by the Brunauer–Emmett–Teller (BET) method. The value of the Kozeny parameter is a function of the structure of the medium, and similar channel shapes will have the same value of K_k over a range of values of ε and S_o. For spherical particles, K_k is equal to 4.8 and for "irregular particles" it is around 5.0.

The permeability constant k_1 is related to an equivalent spherical particle diameter d_p through S_o, since:

$$S_o = \frac{6}{d_p}. \tag{12}$$

Ergun [14], working with experimental results gathered in both viscous and inertial regimes, developed the following relationships to describe k_1 and k_2 for packed columns made of spheres, cylinders, tablets, nodules, round sand, and crushed materials (glass, coke, coal, etc.):

$$k_1 = \frac{\varepsilon^3 d_p^2}{150 \ (1-\varepsilon)^2}, \tag{13}$$

$$k_2 = \frac{\varepsilon^3 d_p}{1.75 \ (1-\varepsilon)}, \tag{14}$$

where d_p is the equivalent particle diameter given by Eq. (12).

Macdonald et al. [7] modified the constants of the Ergun equation by fitting a larger number of experimental data and considering that, besides Eq. (12), the equivalent diameter for nonspherical particles could be given by:

$$d_p = \phi d_v \tag{15}$$

where d_v is the average volumetric diameter and ϕ is the particle sphericity. The relationships found for k_1 and k_2 are:

$$k_1 = \frac{d_p^2 \ \varepsilon^3}{180 \ (1-\varepsilon)^2}, \tag{16}$$

$$k_2 = \frac{d_p \ \varepsilon^3}{1.8 \ (1-\varepsilon)} \quad \text{for smooth particles,} \tag{17}$$

$$k_2 = \frac{d_p \ \varepsilon^3}{4.0 \ (1-\varepsilon)} \quad \text{for roughest particles.} \tag{18}$$

For particles with intermediary roughness, the numerical value in Eqs. (17) and (18) would lie between 1.8 and 4.0. Pressure-drop predictions across beds of coarser granular particles ($d_p > 50$ to 100 μm) from these equations should give an accuracy of about ± 50 % and hold over a wide porosity range ($0.36 < \varepsilon < 0.92$) [7].

Despite criticisms, Ergun-like relationships are widely used at present to predict fluid flow through consolidated and unconsolidated porous media in several fields, including groundwater hydrology, petroleum engineering, water purification, industrial filtration, and concrete infiltration. Many works have tried to establish similar correlations, usually of empirical type and applicable to a single class of porous materials; for instance, for a bimodal grain size distribution [15], for a bed of particles which are themselves very porous [16], for consolidated porous media that generally exhibit high tortuosity and for those having large or bimodal pore or grain size distribution [17, 18], and for highly porous anisotropic consolidated media [19]. Lin and Kellett [20] described several permeability equations for powder compacts, and Innocentini et al. [21] related k_1 and k_2 constants to pore size and porosity for consolidated refractory ceramics. Comprehensive theoretical and empirical reviews on fluid flow through granular media are also available in the literature [22–25].

4.2.4.2
Fibrous Media

Fluid-dynamic applications of fibrous media are essentially related to filtration processes of liquid or gaseous suspensions and more recently to heat-transfer devices. Fibrous media can be classified into woven, nonwoven (felted), and sintered media, and are shaped as candles, coupons, bags, or cartridges. For room-temperature applications, fibers are made from paper, cotton, wool, or synthetic materials (polyamide, polyester, polyacrylonitrile, polyethylene). For filtration at higher temperatures, fibers are made of special polymers (Nomex, Teflon), metals or metal alloys (copper, bronze, stainless steel), and mostly of ceramics (alumina, mullite, zirconia, silica). In this case, the fiber diameter usually varies between 2 and 20 μm, and porosity of the medium varies according to the fabrication method: $0.35 < \varepsilon < 0.5$ for woven, and $0.6 < \varepsilon < 0.95$ for felted media.

Typical fluid velocity in aerosol filtration is lower than 10 cm s^{-1}, and given the low gas density, inertial flow resistance has been often omitted from the analysis. For this reason, mostly Darcian permeability (k_1) data are found in the literature, varying from 10^{-15} to 10^{-10} m^2 [26, 27].

Seville et al. [28] related the permeability to the fiber structure of ceramic elements of very high porosity ($\varepsilon \approx 0.8$–0.95) and specific surface area [$S_o \approx (0.8\text{–}1.5) \times 10^6$ m^2 m^{-3}] through the Kozeny–Carman equation (Eq. (12)), in which the Kozeny constant K_k was experimentally found to be around 6.0. Innocentini [27] worked with air flow through a similar commercial fibrous medium ($\varepsilon \approx 0.8$) and determined average experimental values for k_1 and k_2 as 1.72×10^{-11} m^2 and 2.04×10^{-6} m, respectively.

MacGregor [29] modified the Kozeny–Carman equation for a textile assembly to model the flow of dyes through textile yarn packages, providing a method to predict the Darcian permeability based on fiber diameter d_f and fabric porosity ε:

$$k_1 = \frac{d_f^2 \varepsilon^3}{16K_k \ (1-\varepsilon)^2}. \tag{19}$$

K_k is around 5.5 [30], but this value can vary widely with the porosity of the medium [26].

Another frequently applied semi-empirical equation for Darcian air flow through random fiber media was proposed by Davies [31]:

$$k_1 = \frac{d_f^2}{64(1-\varepsilon)^{1.5}[1+56(1-\varepsilon)^3]}. \tag{20}$$

This equation is valid for the range $0.6 < \varepsilon < 1.0$, with $Re_f = \rho v_s d_f / \mu < 1$.

A correlation for fiber mats with porosities greater than 0.98, based on measurements on wool, cotton, rayon, glass, and steel-wool pads, in which the fiber size varied from 0.8 to 40 μm is given by [31]:

$$k_1 = \frac{d_f^2}{70(1-\varepsilon)^{1.5}[1+52(1-\varepsilon)^{1.5}]}. \tag{21}$$

Other permeability equations are described in the literature for felted and woven filtration fabrics [26, 32], and for deformable porous media [33, 34]. Theoretically derived permeability functions were discussed in refs. [35–38], while the flow patterns through fibrous media were recently reviewed in Ref. [39].

4.2.4.3 Cellular Media

Honeycombs

Honeycombs are cellular materials with a regular arrangement of parallel identical cells (channels) employed in a variety of fluid-flow applications: automotive emission-control catalysis, ozone-abatement catalysis, woodstove combustion, catalytic combustion, heat-exchange devices, molten-metal filtration, ultrafiltration, and diesel particulate filtration [40].

The honeycomb elements are pressed or extruded as cylinders and blocks (for gas flow) or tiles and disks (for liquid flow), which can be assembled in larger units to match the flow-area requirements. They are fabricated from metals (aluminum, stainless steel, special alloys) and sintered porous ceramics (cordierite, alumina, mullite, silicon carbide, zeolite, zirconia, spinel) to meet heat-transfer requirements or to withstand aggressive environments. The main processing variables are cell shape (triangular, square, hexagonal, or circular), wall thickness (0.2–1.5 mm), cell density (4–224 cells per cm^2), and percentage of open frontal area (30 to over 90 %). Porosity of the ceramic walls varies from less than 0.3 to over 0.5, with mean pore sizes of 4–50 μm.

Depending on the application, ceramic honeycombs can operate in different flow configurations. Those used as particulate traps are commonly referred as wall-flow, Z-flow or dead end filters, because the flow path is achieved by sealing half of the channels at the upstream end, in an alternating checkerboard manner, and the other half at the downstream end. Thus, particle-laden gas entering the upstream channels is constrained to flow through the porous walls, which act as surface filters, and exit through the downstream cells. On the other hand, honeycombs used in catalytic conversion or in molten-metal filtration operate in a parallel flow mode, with all cells open at both ends.

The pressure drop through clean *wall-flow honeycombs* can be modeled on the basis of three different contributions: the pressure drop across the porous walls ΔP_d, the friction pressure drop along the passageways ΔP_f, and the dynamic head pressure drop ΔP_h due to changes in flow area within the filtering element. Under optimal filtering conditions, ΔP_f and ΔP_h are minimized and the total pressure drop is mainly due to ΔP_d [41].

The flow through the porous walls is typically a case of flow through media treated in Sections 4.2.4.1 and 4.2.4.2. Since the operational pressure drop is low and the fluid velocity is normally below 20 cm s^{-1}, ΔP_d has been used in the literature with Darcy's law, in which k_1 is experimentally determined or predicted by Eqs. (13) or (16) for walls made of sintered particle grains, or by Eqs. (19)–(21) for bonded fibrous media. The Kozeny–Carman equation also can be used, provided that there is an appropriate value for the Kozeny constant K_k. It must be borne in mind that the superficial velocity used in ΔP_d calculations is given by the volumetric flow rate arriving at the honeycomb element divided by the total internal face area of channels, and L is the thickness of an individual channel wall. Experimental k_1 values were determined for clean wall ceramics made of cordierite ($k_1 = 4.3 \times 10^{-13}$ m^2) and SiC ($k_1 = 2.1 \times 10^{-12}$ m^2) [42].

Relations for ΔP_f and ΔP_h in square cells of clean honeycomb elements were proposed as [41]:

$$\Delta P_f = 8 c_f \rho v_s^2 \left(\frac{L_f}{W}\right)^3 \tag{22}$$

$$\Delta P_h = 16 \rho v_s^2 \left(\frac{L_f}{W}\right)^2 \tag{23}$$

where L_f is the filter or channel length, W the channel width, and c_f a friction coefficient dependent of the wall roughness and the Reynolds number.

Reliable evaluation of the total pressure drop through clean *parallel-flow honeycombs* must consider the flow resistance offered by the frontal solid area (friction and border effects) and along the channels (friction and turbulence). Simplifications can be made for some particular cases of honeycomb geometry, which lead to well-known theories of flow in channels, as described in the literature [4] and given in Table 3.

Table 3 Simplified models to estimate pressure drop through parallel-flow honeycombs.

Case	Honeycomb geometrical features	Simplified modeling	Typical situation
1	medium to long channels with high open frontal area	flow through set of parallel tubes	catalytic combustion
2	short channels with thick walls	flow through perforated plates	molten-metal filtration
3	short channels with thin walls	flow through screens	reduction of turbulence in wind tunnels

For the case of honeycombs with medium or long channels, flow resistance is essentially due to viscous friction and turbulence inside the parallel channels. If border (entrance and exit) effects are disregarded, then pressure drop is estimated through the Darcy–Weisbach equation [4]:

$$\Delta P_{ch} = \frac{f_d \rho L_f v_s^2}{2 d_h} \tag{24}$$

where f_d is the Darcy friction factor and d_h is the hydraulic diameter, defined for a single channel as four times its cross-sectional area exposed to flow divided by its sectional wetted perimeter. For circular channels, d_h is the cell diameter, and for square channels it is the cell width.

The fluid velocity v_s in Eq. (24) is obtained by dividing volumetric flow rate by the total frontal open area exposed to flow. Alternatively, the Fanning friction factor f_f can be used, considering that $f_d = 4\ f_f$. Friction factors can then be determined through empirical correlations [4, 43], or charts [44] for laminar and turbulent regimes, providing that the channel Reynolds number ($Re_{ch} = d_h v_s \rho/\mu$) and the roughness of cell walls are known.

The critical Reynolds number Re_c for transition from laminar to turbulent flow in noncircular channels varies with channel shape. In rectangular ducts, $1900 < Re_c < 2800$, and in triangular ducts, $1600 < Re_c < 1800$ [4].

At the channel entrance, a certain distance is required for flow to adjust from upstream conditions to a fully developed flow pattern. This distance depends on the Reynolds number Re_{ch} and on the upstream flow conditions. For a uniform velocity profile at the channel entrance, the computed length in laminar flow required for the centerline velocity to reach 99 % of its fully developed value is:

$$\frac{L_e}{d_h} = 0.370 \exp\left[-0.148 Re_{ch}\right] + 0.0550 Re_{ch} + 0.260. \tag{25}$$

In turbulent flow, the entrance length L_e is about:

$$\frac{L_e}{d_h} = 40. \tag{26}$$

The frictional losses at channel entrance are larger than those of similar length for fully developed flow. At the channel exit, the velocity profile also undergoes rearrangement, but the exit length is much shorter than the entrance length. At low Re_{ch}, it is about one channel radius. At $Re_{ch} > 100$, the exit length is essentially zero.

If the honeycomb channels are shorter than L_e, then Eq. (24) should no longer be applied. In this case, the pressure drop can be roughly estimated on the basis of the theory of flow through perforated plates (thick cell walls) or through screens (thin cell walls) as:

$$\Delta P_s = \left(\frac{1}{Y^2 C^2}\right)\left(\frac{1-\varphi^2}{\varphi^2}\right)\frac{\rho v_s^2}{2} \tag{27}$$

where, v_s is the superficial velocity based upon the total frontal area of the plate or screen of orifice/opening d_h, and φ the fraction of frontal area open for flow. The discharge (orifice) coefficient C is given as a function of the plate or screen Reynolds number $Re_s = d_h v_s \rho/(\varphi\mu)$, the hole pitch (center-to-center distance), and the thickness-to-diameter ratio. Y is the expansion factor for gases (for liquids, $Y = 1$). Charts and equations for C and Y are given in the literature [4, 45]. Studies on the fluid dynamics of different types of ceramic honeycombs are also available [40, 46–54].

Ceramic Foams

Foams are cellular structures composed of a three-dimensional packing of hollow polyhedra (cells) with edges randomly oriented in space. Depending on the application, foams can be tailored from different materials (polymers, metals, and ceramics) and with either open or closed cells, displaying different permeabilities. Most open-cell ceramic foams for fluid-flow applications are produced by two techniques which are well-described in the literature: ceramic replication of an organic substrate and foaming of ceramic slurries [55–57].

Ceramic replicas, commonly referred as *reticulated ceramics*, are produced from different raw materials (alumina, zirconia, silicon carbide, silica), display porosities varying from 70 % to greater than 90 %, with open tetrakaidecahedral cells of size from less than 100 μm to greater than 5 mm. Their geometrical features are determined by the polymeric sponge matrix, which is quoted in pores per linear inch (commercially available from 3 to more than 100 ppi). Depending on the application, finished elements can be shaped as disks, tubes, rods, rings, or in other custom-designed configurations.

Reticulated ceramics are primarily used in molten-metal purification, but their range of applications has recently spread to many other solid–fluid contact processes, such as catalytic combustion, hot-gas cleaning, gas combustion, and heat transfer. Contrary to honeycomb monoliths, reticulated ceramics have extensive pore tortuosity and flow patterns with high degrees of lateral mixing, which favor particle–filter contact and enhance trapping and conversion efficiencies in catalytic and filtration processes.

Foamed ceramics have porosities between 40 and 90 %, with closed or open nearly spherical cells, and pore sizes ranging from less than 10 μm to about 2 mm. They can be produced from several ceramic oxides (for filtration applications) or from high-purity materials such as hydroxyapatite and calcium phosphate for a variety of applications in the biomedical field, including materials for bone repair [58], carriers for controlled drug-delivery systems, and matrices for tissue engineering [59]. They are also used in ion-exchange processes and as filters for contaminated water [56].

Despite the importance of ceramic foams for fluid-flow applications, only in recent years has their permeability been addressed consistently in the literature. Hence, experimental data relating the permeability constants k_1 and k_2 to foam structure are still too scarce for a comprehensive analysis.

Permeability is expected to change according to several structural features of ceramic foams (pore count, porosity, cell size, strut thickness) that reflect different aspects of fluid–solid interaction.

The most common way to label a ceramic replica has been through its nominal pore count, and for this reason this quantity has been often related to permeability. In general, an increase in pore count results in a decrease in permeability, as it indirectly infers an increase in the number of cell boundaries (struts) per unit length and thus in the frictional area.

In Fig. 2, data gathered from the literature show that permeability constants of ceramic replicas vary by about three orders of magnitude over the commercial range of pore count (8–100 ppi). Interestingly, in this pore-count interval no defined trend is observed for porosity, which fluctuates between 0.75 and 0.95. Scattering in permeability values is relatively high for similar pore counts and can be explained on the basis of the following factors: 1) the nominal pore count is not a precise measure and may vary according to organic-foam manufacturer, 2) measuring techniques for structural parameters vary from author to author, 3) the compressibility effect on

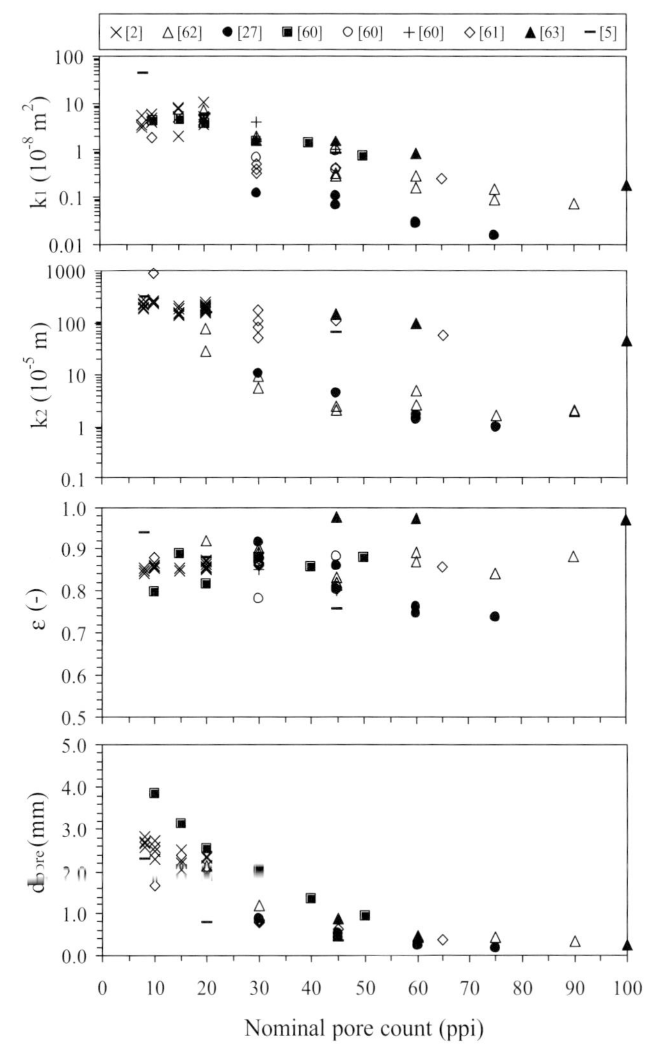

Fig. 2 Variation of properties with nominal pore count for reticulated ceramics.

permeability modeling is not considered for many studies with air flow, 4) in some studies, Darcy's law is applied without verification of its validity, and 5) the nominal pore count does not take into account variables of ceramic processing, such as thickness of the strut coating, blockage of cells, or shrinkage during sintering.

When the polymer matrix and ceramic composition of replicated foams are kept constant, other properties can be modified by varying the amount of slurry that impregnates the struts, generally via control of suspension viscosity and/or by squeezing the organic foam through preset rollers to remove the excess slurry. Although thickening of the struts may increase foam strength, this approach also leads to a pronounced decrease in permeability because of a reduction in porosity and blockage of cells. This can be seen in Fig. 3 for $SiC–Al_2O_3$ ceramic replicas produced with the same type of polymer matrix (10 ppi polyurethane foam) but coated with different slurry contents. The variables were percentage of foam compression and number of impregnation/compression cycles applied to each sample.

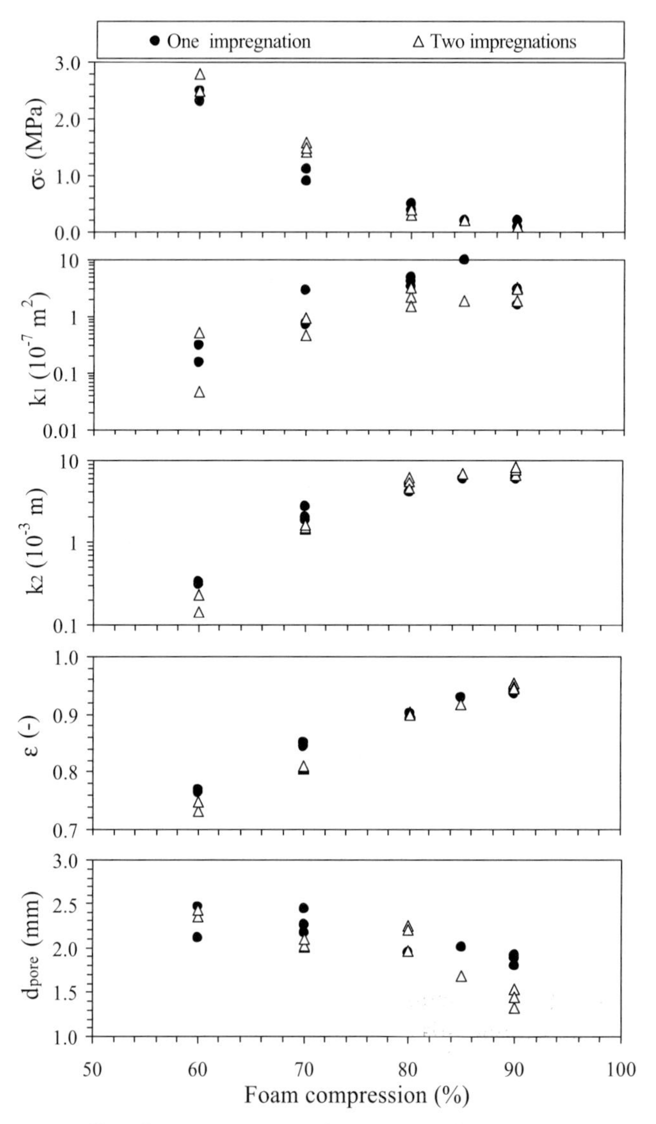

Fig. 3 Effect of ceramic coating on the properties of 10 ppi $SiC–Al_2O_3$ replicas (data from Salvini et al. [64]).

Although no influence of the number of impregnation cycles was observed on foam properties, reduction in foam compression from 90 % to 60 % increased the crushing strength about 20-fold (σ_c from 0.12 to 2.5 MPa), followed by a significant reduction in porosity (0.95 to 0.75) and permeability parameters (ca. 90 % for both k_1 and k_2). The apparent increase in mean pore diameter from 1.4 to 2.4 mm was in fact due to clogging of small pores and a shift of average cell size to higher values.

As seen in Figs. 4 and 5, the expected correlation between permeability and porosity, pore size, and strength depends significantly on processing method, as this determines the cellular structure. For replicas of different nominal pore counts, cells retain an approximately constant ratio between strut dimensions (width and length). Thus, porosity and all properties that depend on strut dimension ratio are only slightly affected. In this case, k_1 and k_2 are influenced mostly by changes in frictional area and tortuosity that result from cell-size variation. On the other hand, if the cell size of the polymer matrix is kept constant, permeability and mechanical strength are well correlated to porosity, which can be gradually varied according to the amount of slurry impregnating the struts.

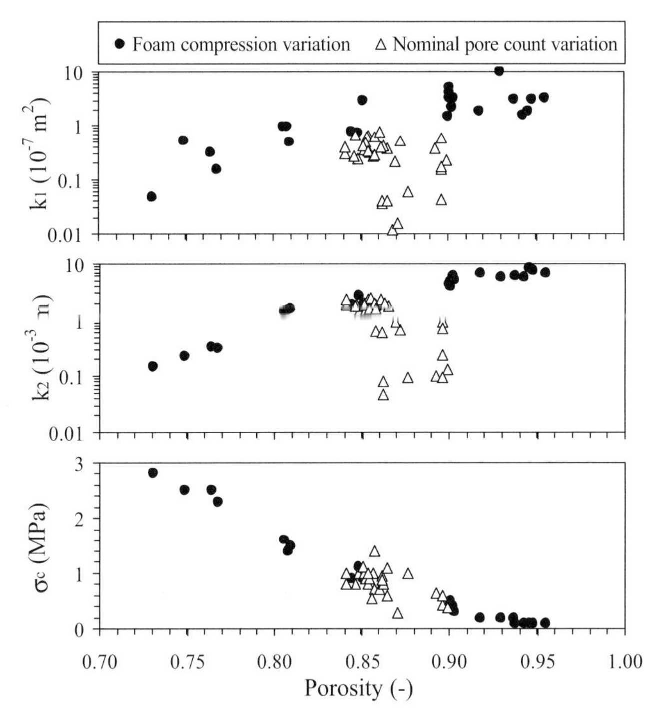

Fig. 4 Porosity dependence of properties for reticulated ceramics for two methods to produce the cellular structure (data from Salvini et al. [64]).

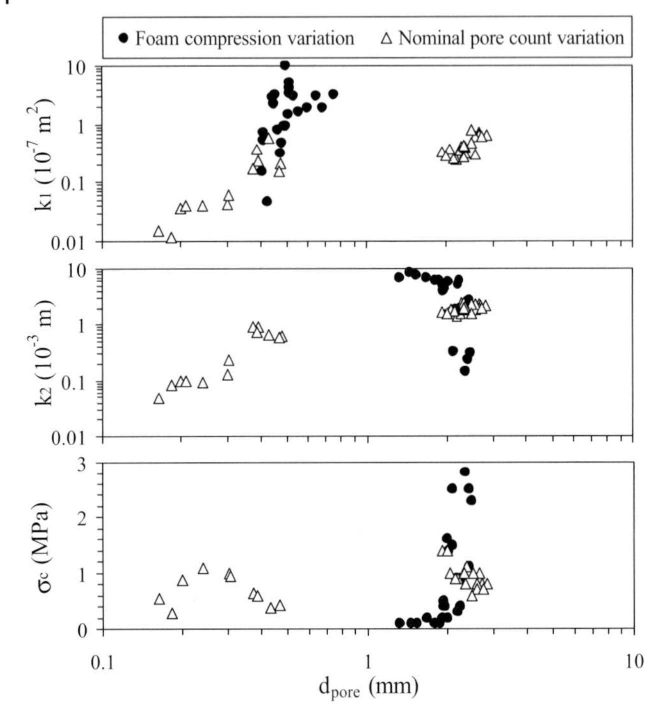

Fig. 5 Pore size dependence of properties for reticulated ceramics gor two methods to produce the cellular structure (data from Salvini et al. [64]).

Most models used in prediction of permeability parameters for ceramic foams are derived from correlations originally developed for granular beds, and this often leads to inaccuracies [5]. The major difficulty in derivation of such models is identifying a characteristic length that represents the cellular media realistically and thus replaces the particle size d_p in the Ergun-like equations described in Section 4.2.4.1. The most obvious approach involves use of an equivalent pore size d_c, usually identified as an equivalent circle diameter obtained, for instance, by image analysis of foam specimens. Philipse and Schram [62] proposed that d_p could be directly replaced by d_c in the Ergun equation and found experimentally that the k_1/k_2 ratio is roughly proportional to d_c in 20–90 ppi alumina–silica foams.

Another approach suggests that d_c could be represented by the cylindrical form of the hydraulic diameter, related to d_p by

$$d_p = 1.5\frac{(1-\varepsilon)}{\varepsilon}\ d_c. \tag{28}$$

Permeability constants k_1 and k_2 based on Ergun equations (13) and (14) can then be rewritten in terms of d_c respectively as:

$$k_1 = \frac{2.25}{150}\varepsilon d_c^2, \tag{29}$$

$$k_2 = \frac{1.5}{1.75}\varepsilon^2 d_c. \tag{30}$$

The numerical values in Eqs. (29) and (30) are based on granular beds and may be replaced by values that can be experimentally determined for cellular materials. Such an approach was investigated with $SiC–Al_2O_3$ ceramic replicas from 30 to 75 ppi [65]. Nevertheless, results showed that the accuracy of Eqs. (29) and (30) depends remarkably on the thickness of the foam specimens used for pore size evaluation.

A more sophisticated correlation for predicting permeability of foams has been proposed for metallic compositions in which the cellular medium is modeled as cubic unit cells, leading to the following correlations [63]:

$$k_1 = \frac{\varepsilon^2 d_c^2}{36\tau(\tau-1)}, \tag{31}$$

$$k_2 = \frac{\varepsilon^2(3-\tau)d_c}{2.05\tau(\tau-1)}, \tag{32}$$

where τ is the tortuosity of the porous matrix, that is, the mean distance covered by the fluid over the thickness of the crossed porous medium. The tortuosity can be measured in electric resistivity experiments or estimated in terms of the medium porosity ε by:

$$\frac{1}{\tau} = \frac{3}{4\varepsilon} + \frac{\sqrt{9-8\varepsilon}}{2\varepsilon}\cos\left\{\frac{4\pi}{3} + \frac{1}{3}\cos^{-1}\left[\frac{8\varepsilon^2 - 36\varepsilon + 27}{(9-8\varepsilon)^{1.5}}\right]\right\}. \tag{33}$$

In another recent model, proposed by Richardson et al. [61], the pressure drop values for 10 to 65 ppi reticulated foams made of Al_2O_3 and of ZrO_2 were associated to Ergun equations, resulting in the following expressions for k_1 and k_2:

$$k_1 = \frac{1}{\alpha S_o^2}\frac{\varepsilon^3}{(1-\varepsilon)^2}, \tag{34}$$

$$k_2 = \frac{1}{\beta S_o}\frac{\varepsilon^3}{(1-\varepsilon)}, \tag{35}$$

where the α and β are empirical parameters respectively given by

$$\alpha = 973 d_c^{0.743}(1-\varepsilon)^{-0.0982}, \tag{36}$$

$$\beta = 368 d_c^{-0.7523}(1-\varepsilon)^{0.07158}. \tag{37}$$

Three different models were examined for calculation of the specific surface area S_o, including one that is based on comprehensive work by Gibson and Ashby [66], in which the foam is treated as a regular packing of tetrakaidekahedra. Surprisingly, it has been found that the "hydraulic diameter model" proposed by Kozeny for packed beds and based on the pore diameter d_c, is sufficient to fit experimental data:

$$S_o = \frac{4\varepsilon}{d_c(1-\varepsilon)}. \quad (38)$$

Moreira et al. [5] compared the performance of all previous models to predict the pressure drop through SiC–Al_2O_3 replicas with 8 to 45 ppi. The foam structure was experimentally evaluated in terms of total and effective porosity, pore size, tortuosity, and specific surface area. Permeability parameters k_1 and k_2 were obtained from water- and airflow experiments. These authors found that Eqs. (34)–(38) using experimental values of S_o and ε exhibited the smallest deviations among all models, although these were still very high (56–109 %). Empirical expressions for k_1 and k_2 were then proposed based on the Ergun equation, fitting their data within a 12 % deviation:

$$k_1 = \frac{\varepsilon^3 d_c^{0.264}}{1.36x10^8(1-\varepsilon)^2}, \quad (39)$$

$$k_2 = \frac{\varepsilon^3 d_c^{-0.24}}{1.8x10^4(1-\varepsilon)}. \quad (40)$$

Compared to ceramic replicas, cellular ceramics produced by foaming of suspensions have seldom been characterized for experimental measurement and modeling of permeability. The few works available report data related to gel casting of foams, a technique that favors the generation of a porous structure with nearly spherical neighboring cells interconnected by circular windows. In these foams, a direct dependence of foam porosity on cell size and window size exists, and depending on processing variables, such properties can be suitably and more widely altered than those produced by the replication technique. Accordingly, k_1 and k_2 permeability

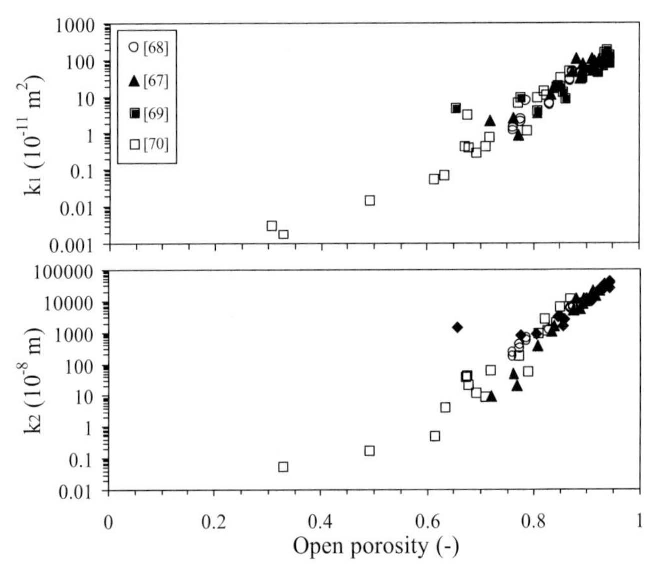

Fig. 6 Variation of permeability constants with porosity for gel-cast foams.

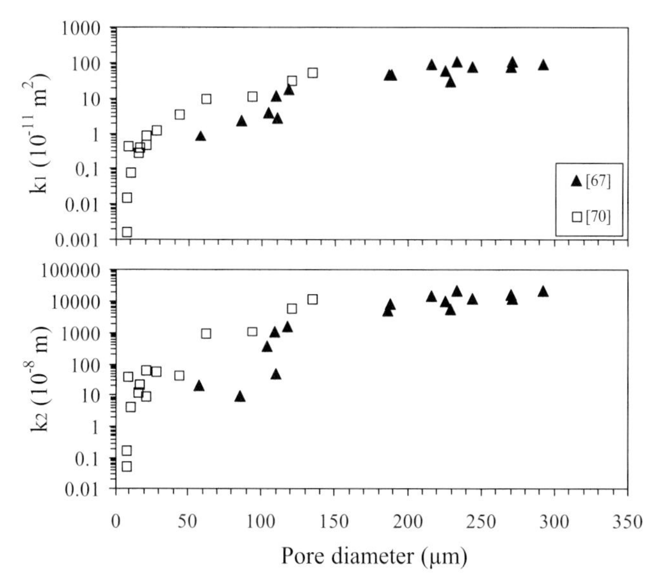

Fig. 7 Variation of permeability constants with pore size for gel-cast foams.

parameters for gel-cast foams are significantly wider in range than those of other ceramic materials. The available studies also point to a much stronger dependence of k_1 and k_2 on porosity and pore size, as shown in Figs. 6 and 7 for data gathered from the literature [68–71].

4.2.5
Viscous and Inertial Flow Regimes in Porous Media

A primary concern that exists when a porous medium is used in filtration, fluid mixing, or any other fluid-dynamic application is the determination of the energy (pressure) needed to achieve a required flow rate, and sometimes more importantly, to predict the dependence between pressure and flow rate. Under a *viscous flow regime,* the ΔP–v_s dependence will be linear, as stated by Darcy's law, and any increase in pressure will result in a proportional increase in flow rate. Under an *inertial flow regime,* the ΔP–v_s dependence will be quadratic, as stated by Forchheimer's equation, and to double the velocity will require between two and four times the energy input.

Despite its obvious importance, classification of flow regime for porous media is controversial, as it depends on the definition of the Reynolds number *Re*. This parameter was originally defined to characterize fluid flow through straight circular tubes and corresponds to the ratio between inertial and viscous forces [44]:

$$Re = \frac{\rho\, vd}{\mu} \tag{41}$$

where d is the pipe diameter, and v the average fluid velocity. For $Re < 2000$ the flow regime is said to be laminar, and for $Re > 2200$ it is characterized as turbulent.

Several authors [35, 71, 72] have proposed variations on the Reynolds number to take into account microscopic structural features which would enable characterization of the flow regime in porous media. Commonly, the tube diameter d in Eq. (41) has been replaced by a characteristic length of the pore structure δ, which can be considered as the equivalent particle diameter or the hydraulic pore diameter for granular beds [14, 35, 73], the strut diameter or the cell size for cellular materials [62, 71, 72, 74, 75], or the fiber length for fibrous media [31]. Similarly, the superficial velocity v_s defined earlier has sometimes been replaced by the *pore*, or *interstitial velocity* v_i. The lack of a standardized definition for these parameters has resulted in conflicting values for the transition between viscous and inertial regimes in the literature.

According to one of the approaches [73], an interstitial Reynolds number Re_i is defined as:

$$Re_i = \frac{d_c \rho\, v_s}{\varepsilon \mu} \tag{42}$$

where d_c is the equivalent pore diameter and ε is the porosity of the medium.

For $Re_i < 1$, the energy is dissipated into the porous medium only as a result of viscous friction between the fluid layers near the pore walls. The pressure drop in this case is linearly proportional to the fluid velocity, and Darcy's law is valid.

The key assumption that underlies Darcy's law is that the Reynolds number is considered to be sufficiently small that fluid inertia effectively plays no role in the dynamics of flow. In practice, Darcy's law shows good agreement with experiment for a wide range of flow, but it breaks down at $Re_i \approx 1$–10 due to the onset of inertial forces within the laminar regime, and not due to the onset of turbulence, as commonly thought [32, 35].

For $Re_i > 1$–10, inertial effects initiate in the laminar regime, with the emergence of disturbances in the streamlines owing to the curvature of flow channels. This additional flow resistance would explain the nonlinear relationship between pressure drop and fluid velocity represented by the term $\rho v_s^2/k_2$ in Forchheimer's equation, which holds until $Re_i \approx 150$.

For $150 < Re_i < 300$, an unsteady flow regime occurs, and for $Re_i > 300$ a highly chaotic flow regime is observed, which might be associated with true turbulence, as occurs in straight tubes.

The transition from laminar to turbulent flow in reticulated rigid foams was investigated by Seguin et al. [71, 72], based on the pore Reynolds number ($Re_p = \rho v_s d_c/\mu$). Some representative data of their study are shown in Table 4.

Table 4 Onset of turbulence in reticulated rigid foams. Adapted from Seguin et al. [71, 72].

	Nominal pore count			
	10 ppi	**20 ppi**	**45 ppi**	**100 ppi**
Re_i transition (Eq. 42)	390	470	>257	>82
% inertial effect on pressure drop at transition	83	85		
Re transition (Eq. 41)	4350	8050	>11 000	>11 000

Interestingly, these authors verified that laminar flow in tubes filled with reticulated foams extends far above the limit of $Re \approx$ 2000–2200 established for empty tubes. The "*laminarizing effect*", increases with decreasing pore size. A similar effect has been observed in the literature for ceramic foams with 4, 8, and 12 pores per centimeter [74–75].

The difficulty in evaluating the pore Reynolds number, due to the complexity in establishing a representative characteristic pore length δ, motivated some authors to develop alternative methods for determining the flow regime and thus verifing the applicability range for Darcy's law.

A criterion based on the velocity v_s, fluid properties μ and ρ, and features of the porous structure L, k_1, and k_2 was proposed by Ruth and Ma [76] to determine the relative contribution of viscous and inertial resistances and to decide whether Darcy's law or Forchheimer's equation is the best permeability equation for a given application.

For this purpose, the dimensionless Forchheimer number Fo was defined by these authors as:

$$Fo = \frac{\rho\, v_s}{\mu}\left(\frac{k_1}{k_2}\right). \qquad (43)$$

The Fo parameter represents the ratio between kinetic and viscous forces that contribute to fluid pressure drop. Since the ratio k_1/k_2 is expressed as length, Fo can be understood as an analogue of the pore Reynolds number, with $\delta = k_1/k_2$. Fo is related to the linearity in the pressure drop curve in the same way that Re_i is related to the laminarity of flow. Based on the Fo parameter, Forchheimer's equation (5) can be rewritten as:

$$\frac{\Delta P}{L} = \frac{\mu}{k_1} v_s\,(1 + Fo). \qquad (44)$$

The individual contribution of viscous and inertial resistances on the total pressure drop can now be estimated through:

$$\frac{\Delta P_{viscous}}{\Delta P_{total}} = \frac{1}{1+Fo}, \qquad (45)$$

$$\frac{\Delta P_{inertial}}{\Delta P} = \frac{Fo}{1+Fo}. \qquad (46)$$

Figure 8 shows the contributions of inertial and viscous resistances to the total pressure drop based on the Forchheimer number. For $Fo \ll 1$, inertial effects are negligible ($\Delta P_{total} \approx \Delta P_{viscous}$) and Eq. (10) reduces to Darcy's law. On the other hand, when $Fo \gg 1$, viscous effects can be disregarded ($\Delta P_{total} \approx \Delta P_{inertial}$) and the pressure drop can be reasonably estimated through:

$$\frac{\Delta P}{L} = \frac{\rho}{k_2} v_s^2 \tag{47}$$

For any other intermediary flow condition both terms are relevant and the complete Forchheimer's equation should be taken as primary permeability model. For compressible fluids, such as gases and vapors, the term ΔP in Eqs. (44)–(47) should be replaced by $(P_i^2 - P_o^2)/2\,PL$.

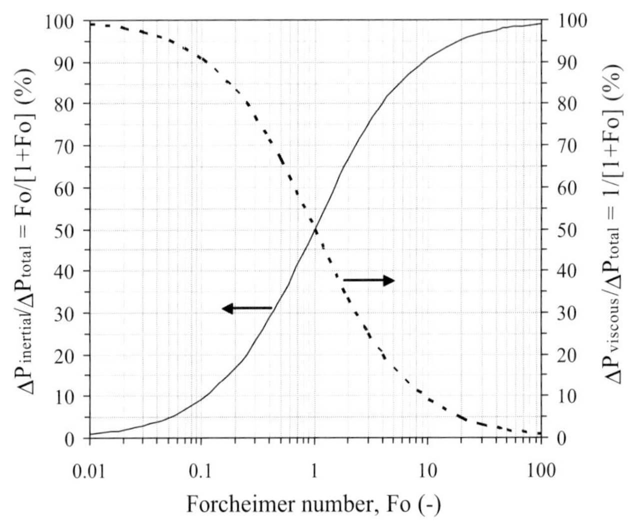

Fig. 8 Percentage of viscous and inertial pressure drops according to the Forcheimer number.

Despite the previous discussion, Darcy's law has been still preferred to Forchheimer's as the permeability equation in several engineering applications involving porous materials. This is either due to the proposed linearity between ΔP and v_s, which makes the inclusion of the law in mathematical modeling easier, or because the fluid velocity chosen for the desired application is low enough to disregard inertial effects. The abundance of studies dealing with Darcy's law has unfortunately spread its use in applications involving porous ceramics, frequently without previous analysis of its validity. The examples that follow illustrate the practical importance of choosing the appropriate permeability equation and the applicability of the Forchheimer number to help in the choice.

Example 1. Consider a 20 ppi SiC ceramic foam filter with 5.9 cm diameter, 2.14 cm thickness, and permeability constants $k_1 = 2.36 \times 10^{-8}$ m^2 and $k_2 = 1.85 \times 10^{-3}$ m. Determine the column head required to achieve a vertical fluid velocity of 0.10 m s^{-1} through the clean filter where the fluid is a molten metal ($\rho_{mm} = 8000$ kg/m^3, $\mu_{mm} = 2.0 \times 10^{-3}$ Pa·s). Compare the adequacy of Darcy's law and Forchheimer's equation.

Resolution:

a) Darcy's law for incompressible fluids: $\dfrac{\Delta P}{L} = \dfrac{\mu_{mm}}{k_1} v_s$,

$$\frac{\Delta P}{0.0214\text{m}} = \frac{2.00 \times 10^{-3}\,\text{Pa.s}}{2.36 \times 10^{-8}\,\text{m}^2}(0.10\text{ m/s}) \rightarrow \Delta P = 181.3\text{ Pa.}$$

The column head ΔH_{metal} is given by $\Delta H = \dfrac{\Delta P}{\rho_{mm} g}$

$$\rightarrow \Delta H_{metal} = \frac{181.3\text{Pa}}{(8000\text{kg/m}^3)(9.8\text{m/s}^2)} \rightarrow \Delta H_{metal} = 2.31\text{ mm.}$$

b) Forchheimer's equation for incompressible fluids:

$$\frac{\Delta P}{L} = \frac{\mu_{mm}}{k_1} v_s + \frac{\rho_{mm}}{k_2} v_s^2,$$

$$\frac{\Delta P}{0.0214\,m} = \frac{2.0 \times 10^{-3}\,\text{Pa.s}}{2.36 \times 10^{-8}\,\text{m}^2}(0.10\text{m/s}) + \frac{8000\text{ kg/m}^3}{1.85 \times 10^{-3}\,\text{m}}(0.10\text{ m/s})^2 \rightarrow \Delta P = 1106.8\text{ Pa.}$$

$$\Delta H_{metal} = \frac{1106.8\text{ Pa}}{(8000\text{ kg/m}^3)(9.8\text{ m/s}^2)} \rightarrow \Delta H_{metal} = 14.1\text{ mm.}$$

Table 5 Summary of data for Example 1.

Fluid	Darcy's law		Forchheimer's equation		*Fo*	$\frac{\Delta P_{viscous}}{\Delta P_{total}}$(%)	$\frac{\Delta P_{inertial}}{\Delta P_{total}}$(%)
	ΔP/Pa	ΔH/mm	ΔP/Pa	ΔH/mm			
Molten metal	181.3	2.3	1106.8	14.1	5.10	16.4	83.6

As seen in Table 5, Darcy's law predicted a pressure drop that is only 16.4 % of that given by Forchheimer's equation. In fact, if this column head of 2.3 mm were used, the actual velocity through the filter would be 3.6 cm s^{-1} and not 10 cm s^{-1} as desired. Under this flow condition (Fo = 5.1), the inertial resistance contributes more than 83 % of the total pressure drop, mainly because of the high fluid density. Thus, Darcy's law does not hold for this application, and Forchheimer's equation should be used to predict the flow behavior of metal through the foam.

Interestingly, if the fluid were air at 25 °C ($\mu = 1.83 \times 10^{-5}$ Pa·s, $\rho = 1.18$ kg m^{-3}) flowing at the same velocity, then $Fo = 0.08$ and viscous effects would represent 98 % of the total pressure drop, validating Darcy's law. However, even if air-flow tests were used to assess permeability data, as commonly found in certain foam catalogues, Darcian-like information of the type "v_s for a given ΔP" should be avoided, since it does not allow reliable prediction of inertial pressure drop for liquid-flow applications such as in foundries and steelmaking plants.

Example 2. A set of rigid fibrous alumina filters is to be used in a incineration facility for toxic waste to clean the exhaust gas stream at a temperature of 800 °C and absolute inlet pressure of 1 atm (1.013×10^5 Pa). Determine the operational filtration velocity through a single clean candle ($L = 1.63$ cm, $k_1 = 6.27 \times 10^{-11}$ m^2, $k_2 = 4.49 \times 10^{-6}$ m) if the maximum permissible pressure drop P_i–P_o is 150 mmH_2O (1500 Pa). Consider that fluid properties and velocity are based on the filter entrance ($\rho_{gas} = 0.33$ kg m^{-3}, $\mu_{gas} = 5.71 \times 10^{-5}$ Pa·s).

Resolution:

$$P_o = P_i - 1500 \text{ Pa} = (1.013 \times 10^5 - 1500) \text{ Pa} \rightarrow P_o = 0.998 \times 10^5 \text{ Pa}$$

a) Darcy's law for compressible fluids: $\dfrac{P_i^2 - P_o^2}{2PL} = \dfrac{\mu_{gas}}{k_1} v_s$ (in this case, $P = P_i$),

$$\text{Then } \frac{(1.013\times10^5 \text{ Pa})^2 - (0.998\times10^5 \text{ Pa})^2}{2\times1.013\times10^5 \text{ Pa}\times0.0163 \text{ m}} = \frac{5.71\times10^{-5} \text{ Pa.s}}{6.27\times10^{-11} \text{ m}^2} v_s$$

$\rightarrow v_s = 10.04$ cm s^{-1}.

b) Forchheimer's equation for compressible fluids: $\dfrac{P_i^2 - P_o^2}{2P_i L} = \dfrac{\mu_{gas}}{k_1} v_s + \dfrac{\rho_{gas}}{k_2} v_s^2$,

$$\frac{(1.013\times10^5 \text{ Pa})^2 - (0.998\times10^5 \text{ Pa})^2}{2\times1.013\times10^5 \text{ Pa}\times0.0163 \text{ m}} = \frac{5.71\times10^{-5} \text{ Pa.s}}{6.27\times10^{-11} \text{ m}^2} v_s + \frac{0.33 \text{kg/m}^3}{4.49\times10^{-6} \text{ m}} v_s^2$$

$\rightarrow v_s = 9.98$ cm s^{-1}.

Table 6 Summary of data for Example 2.

Fluid	$\Delta P = P_i$-P_o (Pa)	Darcy's law v_s/cm s^{-1}	Forchheimer's equation v_s/cm s^{-1}	*Fo*	$\frac{\Delta P_{viscous}}{\Delta P_{total}}$ (%)	$\frac{\Delta P_{inertial}}{\Delta P_{total}}$ (%)
Air	1500	10.04	9.98	0.008	99.2	0.8

Under this flow condition (*Fo* = 0.008), there is almost no contribution of inertial effects (0.8 %) to filter pressure drop, and both permeability equations converged to the same velocity value (Table 6). Thus, even when considering occasional fluctuations in v_s, P, T, or k_1 and k_2 values (due to pore clogging), Darcy's law could be safely used to model the filter air flow in such an incineration facility.

From these two examples, it becomes clear that the concept of high and low velocity in flow through a porous medium is not absolute and should not be used as the only criterion to validate Darcy's law. Although both ceramic foam and fibrous filter operated at the same fluid velocity, in the former case flow was mostly inertial, whereas in the latter it was purely viscous.

The use of *Fo* to determine the validity of each permeability equation therefore implies that both constants k_1 and k_2 must be available for analysis, which is anyway easier to accomplish experimentally than evaluating a characteristic length δ to use in the pore Reynolds number Re_i.

Recently, a substantial volume of permeability data based on Forchheimer's equation for highly porous media has become available in the literature. A compilation of representative experimental k_1 and k_2 values for several porous materials is shown in Fig. 9. Table 7 classifies these materials according to their location in Fig. 9.

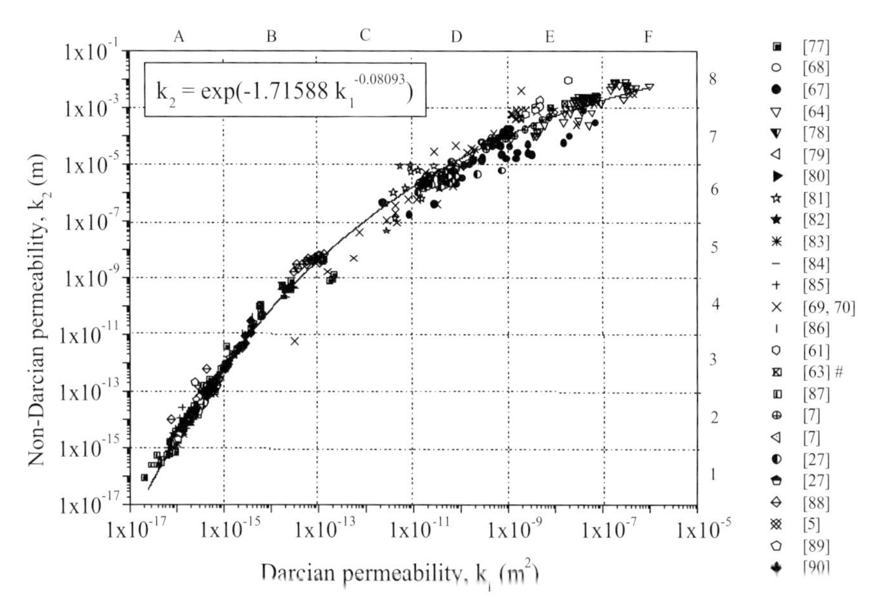

Fig. 9 Permeability data gathered from the literature for several porous materials. # Data for metallic foams.

Table 7 Classification of porous materials based on their location in Fig. 9.

Material	Location in Fig. 9
Green refractory castables, mortars	A1–A3
Fired castables, bricks	B3–B5
Starch-containing ceramics	B3–B4
Gel-cast foams	B3–B4–C4–C5–D6–E7
Fibrous filters	C6–D6
Honeycomb (wall-flow) filters	C5–C6
Granular beds (spherical particles)	D6–D7
Metallic reticulated foams (45–100 ppi)	E7
Ceramic reticulated foams (8–90 ppi)	E6–E7–F8

Despite all the different techniques, fluids, and flow conditions used to assess permeability data, it is remarkable that a clear correlation between both permeability parameters could link porous media of totally different structural features. The best fit ($r^2 = 0.9823$) considering the whole data set with 640 points was:

$$k_2 = \exp\left(-\frac{1.71588}{k_1^{0.08093}}\right) \tag{48}$$

where units of k_1 and k_2 are respectively m^2 and m.

Extensive literature data involving mainly air-flow applications could not be included in the map of Fig. 9 because only k_1 values were available. In such cases, however, given the apparent success of the proposed fitting, Eq. (48) could be used to find approximate values for the non-Darcian permeability k_2 and thus estimate the inertial contribution to flow resistance for a desired application.

4.2.6 Summary

A three-dimensional structure that combines a high degree of porosity and a controlled size range of interconnected pores makes cellular ceramics some of the most permeable rigid porous media currently used in engineering applications. Nevertheless, reliable evaluation of permeability parameters for these materials is recent and has faced different challenges.

The difficulty of defining a fluid-dynamic characteristic length for the cellular geometry has been the major obstacle for reliable permeability modeling. On the other hand, recent equations to predict permeability constants based on the cellular structure have not been extensively tested, and the influences of parameters such as pore count, pore size, porosity, and strut thickness are not yet well established.

Attempts to fit fluid flow data according to Darcy's law have also caused inaccuracies, since inertial effects generally prevail in pressure drops. The best approach currently available to assess permeability of cellular ceramics is still by fitting experimental data to Forchheimer's equation, in which both Darcian (k_1) and non-Darcian (k_2) permeability constants are obtained. The combined analysis of fluid and flow conditions and the use of Forchheimer's number *Fo* will then determine the most representative permeability parameter for a given application.

References

1 Innocentini, M.D.M., Pardo, A.R.F., Salvini, V.R. and Pandolfelli, V.C., *Am. Ceram. Soc. Bull.*, **1999**, *78*[11] 64–68.

2 Innocentini, M.D.M., Salvini, V.R., Coury, J.R. and Pandolfelli, V.C., *Am. Ceram. Soc. Bull.*, **1999**, *78*[9], 78–84.

3 Innocentini, M.D.M., Pardo, A.R.F. and Pandolfelli, V.C., *J. Am. Ceram. Soc.*, **2000**, *83*[6], 1536–1538.

4 Perry, R. and Green, D., *Perry's Chemical Engineers' Handbook*, 7th Ed., McGraw-Hill, New York, **1999**.

5 Moreira, E.A., Innocentini, M.D.M. and Coury, J.R., *J. Eur. Ceram. Soc.*, **2004**, *24*[10–11], 3209–3218.

6 Innocentini, M.D.M., Pardo, A.R.F., Salvini, V.R. and Pandolfelli, V.C., *J. Am. Ceram. Soc.*, **1999**, *82*[7], 1945–1948.

7 Macdonald, I.F., El-Sayed, M.S. and Dullien, F.A., *Ind. Eng. Chem. Fundam.*, **1979**, *18*[3], 199.

8 Kozeny, J., *Sitzungsber. Akad Wiss. Wien*, **1927**, *136*, 271–306.

9 Carman, P.C., *Trans. IChemE*, **1937**, *15*, 150–166.

10 Carman, P.C., *The Flow of Fluids through Porous Materials*, Academic Press, New York, **1956**.

11 Blake, F.E., *Trans. Am. Inst. Chem. Eng.*, **1922**, *14*, 415.

12 Burke, S.P. and Plummer, W.B., *Ind. Eng. Chem.*, **1928**, *20*, 1196.

13 Chilton, T.H. and Colburn, A.P. *Ind. Eng. Chem.*, **1931**, *23*, 913.

14 Ergun, S., *Chem. Eng. Progr.*, **1952**, *48*[2], 89–94.

15 Macdonald, M.J., Chu, C.F., Guilloit, P.P. and Ng, K.M., *AIChE J.*, **1991**, *37*[10], 1583–1587.

16 Pfeiffer, J.F., Chen, J.C. and Hsu, J.T., *AIChE J.*, **1996**, *42*[4], 932–939.

17 Garcia-Bengochea, I., Masce, A., Lovell, C.W. and Altschaeffl, A.G., *J. Geotech. Eng. Div.* **1979**, *105* (GT7-July), 839–856.

18 Bao, Y. and Evans, J.R.G., *J. Eur. Ceram. Soc.*, **1991**, 8, 81–93.

19 Mauran, S., Rigaud, L. and Coudevylle, O., *Transp. Porous Media*, **2001**, *43*, 355–376.

20 Lin, C.Y. and Kellett, B.J., *J. Am. Ceram. Soc.*, **1998**, *81*[8], 2093–108.

21 Innocentini, M.D.M., Pardo, A.R.F., Salvini, V.R. and V.C. Pandolfelli, *Cerâmica*, **2002**, *48*[305], 5–10 (in Portuguese).

22 Bear, J. *Dynamics of Fluids in Porous Media*, Dover Publishers, New York, **1988**.

23 Du Plessis, J.P. and Maslyah, J.H., *Transp. Porous Media*, **1991**, 6, 207–221.

24 Dullien, F.A.L. *Porous Media: Fluid Transport and Pore Structure*, 2nd ed., Academic Press, San Diego, **1992**.

25 Liu, S., Afacan, A. and Masliyah, J., *Chem. Eng. Sci.*, **1994**, *49*[21], 3565–3586.

26 Dullien, F.A.L., *Industrial Gas Cleaning*, Academic Press, San Diego, **1988**.

27 Innocentini, M.D.M., *Gas Filtration at High Temperatures*, Doctorate Thesis, Federal University of Sao Carlos, Brazil, **1997** (in Portuguese).

28 Seville, J.P.K., Clift, R., Withers, C.J. and Keidel, W., *Filtr. Sep.*, **1989**, July/August, 265–271.

29 McGregor, R., *J. Soc. Dyers Colourists*, **1965**, *81*, Oct., 429–438.

30 Fowler, J.L. and Hertel, K.L., *J. Appl. Phys.*, **1940**, *11*, 496–502.

31 Davies, C.N., *Proc. Inst. Mech. Eng.*, **1952**, *B1*, 185–198.

32 Wakeman, K.J. and Tarleton, E.S., *Filtration Equipment Selection Modelling and Process Simulation*, Elsevier Advanced Technology Oxford, **1999**.

33 Kyan, C.P., *Ind. Eng. Chem. Fund.*, **1970**, 9[4], 596–603.

34 Kamst, G.F., Bruinsma, O.S.L. and de Graauw, J., *AIChE J.* **1997**, *43*[3], 673–680.

35 Scheidegger, A.E., *The Physics of Flow through Porous Media*, University of Toronto Press, 3rd Ed., **1974**.

36 Jackson, G.W. and James, D.F., *Can. J. Chem. Eng.*, **1986**, 64, 364–374.

37 Brown, R.C., *Air Filtration: An Integrated Approach to the Theory and Application of Fibrous Filters*, Pergamon Press, Oxford, **1993**.

38 Koponen, A., Kandhai, D., Hellén, E., Alava, M., Hoekstra, A., Kataja, M., Niskanen, K., Sloot, P. and Timonen, J., *Phys. Rev. Lett.*, **1998**, *80*[4], 716–719.

39 Brown, R.C., Airflow through Filters – Beyond Single-Fiber Theory, in *Advances in Aerosol Filtration*, Ed.: K.V. Spurny, Lewis Publishers, Boca Raton, FL, **1998**, 153–172.

40 Lachman, I.M. and Mcnally, R.N., *Chem. Eng. Progr.*, **1985**, Jan., 29–31.

41 Ahluwalia, R.K. and Geyer, H.K., *Trans. ASME*, **1996**, *118*, July, 526–533.

42 Stobbe, P., Petersen, H.G., Hoj, J.W. and Sorenson, S.C., *SAE Papers*, **1988**, *932495*, 2151–2165.

43 Heck, R.M., Gulati, S. and Farrauto, R.J., *Chem. Eng. J.*, **2001**, 82, 149–156.

44 Bird, R.B., Steward, W.E. and Lightfoot, E.N., *Transport Phenomena*, Wiley, New York, **1960**.

45 Perry, R. and Green, D., *Perry's Chemical Engineers' Handbook*, 6th Ed., McGraw-Hill, New York, **1984**.

46 Abrams, R.F. and Goldsmith, R.L., Compact Ceramic Membrane Gas Filter, in *Gas Cleaning at High Temperatures*, Eds: Clift, R. and Seville, J.P.K., Blackie A&P, London, **1993**, pp. 346–362.

47 Akitsu, Y., Masaki, H. and Kyo, O., Ceramic Honeycombs for Hot Gas Cleaning, in *Gas Cleaning at High Temperatures*, Eds.: Clift, R. and Seville, J.P.K., Blackie A&P, London, **1993**, pp. 321–345.

48 Cilibert, D.F. and Lippert, T.E., Ceramic Cross Flow Filters for Hot Gas Cleaning, in *IChemE Symp. Ser.*, **1993**, *99*, 193–213.

49 Bishop, B. and Raskin, N.R., High Temperature Gas Cleaning Using Honeycomb Barrier Filter on a Coal-Fired Circulating Fluidised Bed Combustor, in *High Temperature Gas Cleaning*, Ed.: Schmidt, E., 1st Ed., Institut für Mechanische Verfahrenstechnik und Mechanik, Karlsruhe, **1996**, pp. 94–105.

50 Kwetkus, B.A. and Egli, W., The Ceramic Monolithic Filter Module: Filtration Properties and DeNOx Potential, in *High Temperature Gas Cleaning*, Ed.: E. Schmidt, E., 1st Ed., Institut für Mechanische Verfahrenstechnik und Mechanik, Karlsruhe, **1996**, pp. 278–290.

51 Dolecek, P., *J. Membr. Sci.*, **1995**, *100*, 111–119.

52 Nassehi, V., *Chem. Eng. Sci.*, **1998**, *53*[6], 1253–1265.

53 Sie S.T. and Calis, H.P., Parallel Passage and Lateral Flow Reactors, in *Structured Catalysts and Reactors*, Eds.: Cybulski, A. and Moulijn, J.A., Marcel Dekker, New York, **1998**, pp. 323–354

54 Versaevel, P., Colas, H., Rigaudeau, C., Noirot, R., Koltsakis, G.C. and Stamatelos, A.M., *SAE Paper*, **2000**, 2000-01-477, 1–10.

55 Saggio-Woyansky, J. and Scott, C.E., *Am. Ceram. Soc. Bull.*, **1992**, *71*[11], 1675–1682.

56 Sepulveda, P., *Am. Ceram. Soc. Bull.*, **1997**, *76*[10], 61–65.

57 Green, D.J. and Colombo, P., *MRS Bull.*, **2003**, April, 296–300.

58 Sepulveda, P., Binner, J.P.G., Rogero, S.O., Higa, O.Z. and Bressiani, J.C., *J. Biomed. Mater. Res.*, **2000**, *50*[1] 27–34.

59 Netz, D.J.A., Sepulveda, P., Pandolfelli, V.C., Spadaro, A.C.C., Alencastre, J.B., Bentley, M.V.L.B. and Marchetti, J.M., *Int. J. Pharm.*, **2001**, *213*, 117–125.

60 Acosta, F.A., Castillejos, A.H., Almanza, J.M. and Flores, A., *Metallurg. Mater. Trans. B*, **1995**, *26*, 159–171.

61 Richardson. J.T., Peng, Y. and Remue, D., *Appl. Catal. A*, **2000**, *204*, 19.

62 Philipse, A.P. and Schram, H.L., *J. Am. Ceram. Soc.*, **1991**, *74*[4], 728–732.

63 Du Plessis, P., Montillet, A., Comiti, J. and Legrand, J., *Chem. Eng. Sci.*, **1994**, *49*[21], 3545.

64 Salvini, V.R., Innocentini, M.D.M. and Pandolfelli, V.C., *Am. Ceram. Soc. Bull.*, **2000**, *79*[5] 49–54.

65 Innocentini, M.D.M., Salvini, V.R., Macedo, A. and Pandolfelli, V.C., *Mater. Res.*, **1999**, 2[4], 283–289.

66 Gibson, L.J. and Ashby, M.F., *Cellular Solids: Structure and Properties*, Pergamon Press, Oxford, **1988**.

67 Innocentini, M.D.M., Sepulveda, P., Salvini, V.R., Coury, J.R. and Pandolfelli, V.C., *J. Am. Ceram. Soc.*, **1998**, *81*[12], 3349–3352.

68 Sepulveda, P., Ortega, F., Innocentini, M.D.M. and Pandolfelli, V.C., *J. Am. Ceram. Soc.*, **2000**, *83*[12], 3021–3024.

69 Ortega, F., Sepúlveda, P., Innocentini, M.D.M. and Pandolfelli, V.C., *Am. Ceram. Soc. Bull.*, **2001**, *80*[4], 37–42.

70 Ortega, F.S., Innocentini, M.D.M., Valenzuela, F.A. and Pandolfelli, V.C., *Cerâmica*, **2002**, *48*[306], 79–85 (in Portuguese).

71 Seguin, D., Montillet, A. and Comiti, J., *Chem. Eng. Sci.*, **1998**, *53*[21], 3751–3761.

72 Seguin, D., Montillet, A., Comiti, J. and Huet, F., *Chem. Eng. Sci.*, **1998**, *53*[22], 3897–3909.

73 Dybbs, A. and Edwards, R.V. in *Fundamentals of Transport Phenomena in Porous Media*, Eds.: Bear, J. and Corapcioglu, Y., Martinus Nijhoff, Leiden, Boston, **1984**, p. 199.

74 Hall, M.J. and Hiatt, J.P., *Phys. Fluids*, **1994**, *6*[2], 469–479.
75 Hall, M.J. and Hiatt, J.P., *Exp. Fluids*, **1996**, *20*, 433–440.
76 Ruth, D. and Ma, H., *Transp. Porous Media*, **1992**, *7*, 255–264.
77 Innocentini, M.D.M., Pardo, A.R.F., Menegazzo, B.A., Bittencourt, L.R.M., Rettore, R.P. and Pandolfelli, V.C., *J. Am. Ceram. Soc.*, **2002**, *85*[6], 1517–1521.
78 Innocentini, M.D.M., Pileggi, R.G., Studart, A.R. and Pandolfelli, V.C., *Am. Ceram. Soc. Bull.*, **2001**, *80*[5], 31–36.
79 Innocentini, M.D.M., Salomão, R., Ribeiro, C., Cardoso, F.A., Bittencourt, L.R.M., Rettore, R.P. and Pandolfelli, V.C., *Am. Ceram. Soc. Bull.*, **2002**, *81*[7], 34–37.
80 Innocentini, M.D.M., Pileggi, R.G., Ramal Jr., F.T. and Pandolfelli, V.C., *Am. Ceram. Soc. Bull.*, **2003**, *82*[7], 1–6.
81 Vasques, R.A., *Processing of Porous Vitreous Substrates*, MSc. Dissertation, University of São Paulo, Brazil **2001** (in Portuguese).
82 Studart, A.R., Ortega, F., Innocentini, M.D.M. and Pandolfelli, V.C., *Am. Ceram. Soc. Bull.*, **2002**, *81*[2], 41–47.
83 Innocentini, M.D.M., Yamamoto, J., Ribeiro, C., Pileggi, R.G., Rizzi Jr., A.C., Bittencourt, L.R.M., Rettore, R.P. and Pandolfelli, V.C., *Cerâmica*, **2001**, *47*[304], 212–218 (in Portuguese).
84 Gerotto, M.V., Studart, A.R., Innocentini, M.D.M., Cabo, S.S. and Pandolfelli, V.C., *Am. Ceram. Soc. Bull.*, **2002**, *81*[9] 40–47.
85 Innocentini, M.D.M., Nascimento, L.A., Rizzi Jr., A.C., Paiva, A.E.M., Menegazzo, B.A., Bittencourt, L.R.M. and Pandolfelli, V.C., *Am. Ceram. Soc. Bull.*, **2003**, *82*[8], 1–5.
86 Innocentini, M.D.M., Studart, A.R., Pileggi, R.G. and Pandolfelli, V.C., *Cerâmica*, **2001**, *47*[301], 28–33 (in Portuguese).
87 Innocentini, M.D.M., Internal Report – In Confidence, Group of Microstructure and Engineering of Materials, Department of Materials Engineering, Federal University of Sao Carlos, Brazil, 2000.
88 Innocentini, M.D.M., Internal Report – In Confidence, Group of Microstructure and Engineering of Materials, Department of Materials Engineering, Federal University of Sao Carlos, Brazil, 2001.
89 Cardoso, F.A., Innocentini, M.D.M., Akiyoshi, M.M. and Pandolfelli, V.C., *J. Eur. Ceram. Soc.* **2003**, *24*[7], 2073–2078.
90 Salomão, R., Cardoso, F.A., Innocentini, M.D.M. and Pandolfelli, V.C., *Am. Ceram. Soc. Bull.*, **2003**, *82*[4], 51–56.

4.3
Thermal Properties

Thomas Fend, Dimosthenis Trimis, Robert Pitz-Paal, Bernhard Hoffschmidt, and Oliver Reutter

4.3.1
Introduction

Many cellular ceramic materials are used in high-temperature applications such as insulating materials, heat exchangers, solar receivers, and porous burners. Consequently, knowledge of their thermal properties is essential for the design engineer. Some quantities, such as the heat capacity, may be easily derived from the solid and gaseous components of the cellular ceramic; for others, such as the thermal conductivity, this derivation is complex and temperature-dependent. In addition, similar to the mechanical properties, the thermal properties strongly depend on the chemical composition and the manufacturing process. This emphasizes the need for measurements. As the literature provides little data on the thermal properties of cellular ceramics, this chapter describes the most important measurement techniques for heat conduction and heat transfer and provides some data. Knowledge of these thermal properties is also essential when solving heat-transfer and flow problems numerically with heterogeneous models.

4.3.2
Thermal Conductivity

Fourier's law describes heat transport in a medium and includes a definition of the thermal conductivity λ [W m^{-1} K^{-1}]. According to this law, heat transport is proportional to thermal conductivity and temperature gradient:

$$\dot{q} = -\lambda \nabla T \tag{1}$$

where $\dot{q}$ [W m^{-2}] denotes the heat flux density (heat per unit time and unit area), and T [K] the temperature.

In one dimension this equation has the form:

$$\dot{q} = -\lambda \partial T / \partial x, \tag{2}$$

which can be integrated for a homogeneous one-dimensional case:

Cellular Ceramics: Structure, Manufacturing, Properties and Applications.
Michael Scheffler, Paolo Colombo (Eds.)

ISBN: 3-527-31320-6

$$\dot{Q} = -\lambda A \frac{T_2 - T_1}{l} \tag{3}$$

where one can imagine $\dot{Q}$ [W] as the heat per unit time transferred through a rod of length l [m] and a cross sectional area A [m^2] linking two reservoirs at constant temperatures T_1 and T_2.

In this stationary case at each point of the rod the temperature is constant. For the transient case the heat flux in media is described by the following equation:

$$\rho c_P \frac{\partial T}{\partial t} = \lambda \nabla^2 T \tag{4}$$

or frequently written in the form

$$\frac{\partial T}{\partial t} = \frac{\lambda}{\rho c_P} \nabla^2 T \tag{5}$$

defining – because of its similarity to the equation of diffusion – the quantity thermal diffusivity κ [m^2/s]

$$\kappa = \frac{\lambda}{\rho c_P}. \tag{6}$$

These definitions are made for condensed or gaseous media, in which heat transport can be imagined as coupled vibrations of lattices or collisions of molecules or electrons. Cellular ceramics generally consist of a solid framework which is filled with a fluid. The fluid may flow through the solid framework in the case of open porosity. Consequently, thermal conductivity in a cellular solid is not a homogenous material property. It is a combination of several different mechanisms taking place at the pore-size level. Besides heat conduction through the solid walls or struts, there may be convective heat transfer in open or closed cells, heat conduction in the fluid, and radiative heat transfer. If all mechanisms are regarded separately, calculations of heat-transfer problems become quite complicated. Consequently, an "effective thermal conductivity" λ_{eff} has been introduced to simplify the calculations. The cellular material is considered as a "black box" and "local thermal equilibrium" is assumed. Then, the solution of the equations mentioned is still possible:

$$\dot{q} = -\lambda_{\text{eff}} \nabla T. \tag{7}$$

An enormous number of studies have been published on the theoretical prediction of the effective thermal conductivity of cellular materials. Good reviews of these studies are given by Kaviany [1] and Hsu [2] for porous media in general. Tsotsas and Martin [3] reviewed theoretical approaches and experimental methods to determine the effective thermal conductivity of packed beds. Taylor [4], Mottram and Taylor [5], and Hale [6] give very general and comprehensive reviews on the physical properties of composites which can also be applied to cellular ceramics. Collishaw [7] gives a more specific review on thermal conductivity of cellular materials. Ceramic foams have not been in the focus of these approaches; therefore, experimental data on the thermal conductivity of these materials are rare. Some experimental investigations have been performed by the authors and are reported below.

To roughly quantify the contributions of the single mechanisms, one has to consider that the thermal conductivity of ceramic materials is high compared to air, the most frequently used fluid. Thermal conductivity data of several (dense) ceramic materials and for air are given in Table 1. Consequently, thermal conduction by the air can be neglected in most cases of cellular ceramics as long as the porosity is not too high.

Table 1 Approximate thermal conductivities λ of some dense ceramic materials at room temperature compared to the heat conductivity of air.

Material	λ/W m^{-1} K^{-1}
Cordierite	1–3
Zirconium oxide (ZrO_2)	2.5–3
Sintered silicon carbide (S-SiC)	100–140
Clay-bound silicon carbide (CB-SiC)	10
Aluminium oxide (Al_2O_3)	25
Air (20 °C, 10^5 Pa)	0.02568
Air (100 °C, 10^5 Pa)	0.03139

The contribution of natural convection to heat transfer in closed cells can be estimated by calculating the Grashof number [8]

$$Gr = \frac{g\beta\Delta T_C l^3 \rho^2}{\mu^2} \tag{8}$$

which becomes 1000 or greater if convection heat transfer is of importance (g = acceleration due to gravity 9.81 m s^{-2}, β = volume coefficient of expansion for the gas, ΔT_C = temperature difference across one cell, l = cell size, ρ = density of the gas, μ = dynamic viscosity of the gas). The equation includes the cell size l as a main parameter of influence. For a particular fluid, a critical cell size can be calculated. For all cell sizes larger than this critical cell size, natural convection is of importance. By using data appropriate to air at 10^5 Pa (T = 300 K, $\beta = 1/T$, ΔT_C = 10 K, ρ = 1 kg m^{-3}, $\mu = 2 \times 10^{-5}$ N s m^{-2}), Gibson and Ashby calculate a critical cell size of 10 mm, which is larger than in most common cellular ceramics; thus, the contribution of natural convection to heat transfer can also be neglected [8].

This leaves three contributions, which are denoted as λ_R, λ_S, λ_C, representing radiative heat transport, heat transport through the solid, and convective heat transport by means of a flow through the open cells of the material (forced convection), respectively:

$$\lambda_{eff} = \lambda_R + \lambda_S + \lambda_C. \tag{9}$$

Note that *forced convection* only occurs if a flow of the fluid arises due to external forces or internal chemical reactions. Examples of the first case are exhaust catalysts and heat exchangers. Examples of the second case are porous burners and regeneration of diesel particle filters by burning off soot particles in the pores of the filter.

Closer examination of the contribution of forced convection λ_C reveals firstly that it is strongly dependent on the main direction of flow, that is, in the presence of flow through the cellular ceramic the effective thermal conductivity becomes anisotropic. Secondly, the contribution of forced convection is dependent on the geometry of the cells. In a foam, mixing of the fluid in directions perpendicular to the main flow direction leads to additional heat transfer. Imagining a tube filled with a cellular material, as illustrated in Fig. 1 and assuming cylindrical symmetry two effective conductivities describe the heat-transport properties of the material.

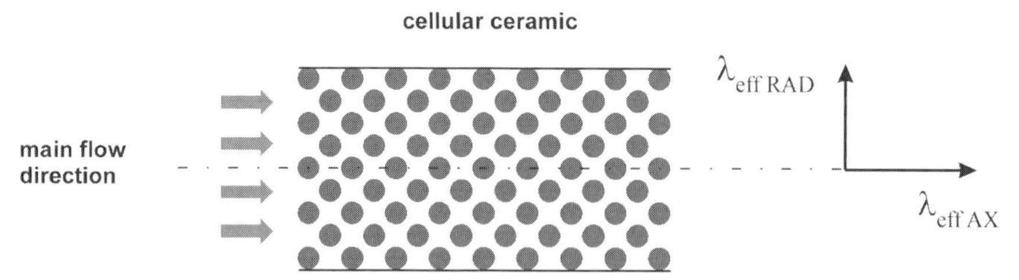

Figure 1 Anisotropic heat-transport properties of a cellular material inserted in a tube and subjected to flow.

This consideration leads to the two equations

$$\lambda_{\text{eff AX}} = \lambda_R + \lambda_S + \lambda_{\text{C AX}}, \tag{10}$$

$$\lambda_{\text{eff RAD}} = \lambda_R + \lambda_S + \lambda_{\text{C RAD}}. \tag{11}$$

The contributions λ_R and λ_S remain isotropic if an isotropic material is assumed. They are frequently denoted as $\lambda_{\text{eff},0}$, the effective thermal conductivity without flow

$$\lambda_{\text{eff},0} = \lambda_R + \lambda_S. \tag{12}$$

4.3.2.1
Experimental Methods to Determine the Effective Thermal Conductivity without Flow

Besides the approaches mentioned to predict the effective thermal conductivity of cellular ceramics theoretically, there are also several experimental methods to determine this quantity. For the case without forced convection, which excludes flow through the open pores of the material, methods can be employed which are also well known for dense ceramic materials. In the case of porous materials, in contrast to dense materials, sample dimensions should be chosen that are large compared to the pore dimensions. This is frequently not the case for the well known laser-flash method, for which sample dimensions of approximately 10 mm are required. It is the case for the transient plane-source technique and the hot-plate method, which are both presented here in more detail.

The *transient plane-source technique,* also known as the hot-disk method, uses a thin nickel double spiral as a heat source, placed between two identical material samples. Simultaneously it is used as a sensor for temperature measurement (Fig. 2). Depending on pore size and sample dimensions, sensor sizes of 10 to 50 mm are available. A defined quantity of electrical power is delivered in a defined time period to the hot disk, and the curve of the increasing temperature is recorded. The thermal conductivity of the sample material now influences the slope of the curve. A numerical fit of the temperature curve yields thermal conductivity λ and thermal diffusivity κ of the sample material. Specific heat capacity c_p is then given with the density ρ by

$$c_p = \frac{\lambda}{\rho\kappa}. \tag{13}$$

A detailed description of the method is given by Gustaffson [9].

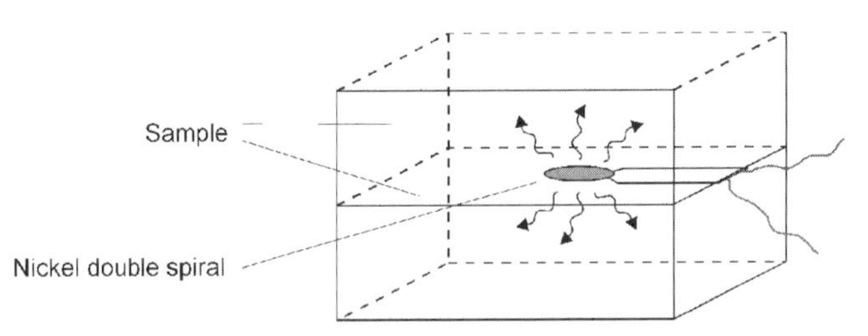

Figure 2 Transient plane-source technique to measure thermal conductivity of porous materials.

In the *hot-plate method,* a temperature gradient inside the test body generates a one-dimensional heat flow in the area of measurement. Then, the effective thermal conductivity without flow $\lambda_{\text{eff},0}$ [W m^{-1} K^{-1}] can be calculated from the heat flow density $\dot{q}$ [W m^{-2}], the sample thickness s [m], and the measured temperature difference ΔT [K]:

$$\lambda_{\text{eff},0}(T) = \frac{\dot{q} \cdot s}{\Delta T}. \tag{14}$$

Temperatures at the boundaries of the samples can be varied, so that the variation of the thermal conductivity as a function of temperature (in this case the average temperature) can be determined. For this purpose, the temperature difference should be kept small.

The measurement includes heat flow through the material as well as radiative heat transport. The measurement principle is illustrated in Fig. 3. Typical sample dimensions are a diameter of 200 mm and a thickness of 70 mm. The minimum sample thickness has to be large enough to suppress direct radiation from the hot to the cold plate and therefore depends on the extinction coefficient of the sample.

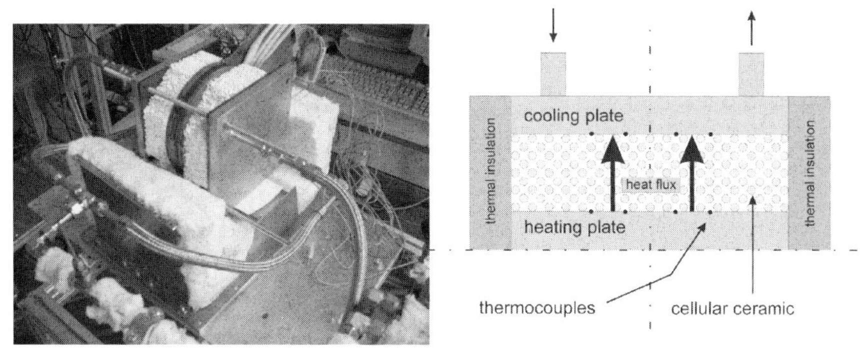

Figure 3 Photograph and sketch of an installation to measure thermal conductivity without flow by the hot-plate method (apparatus built at the Institute of Fluid Dynamics, Friedrich Alexander University of Erlangen-Nuremberg).

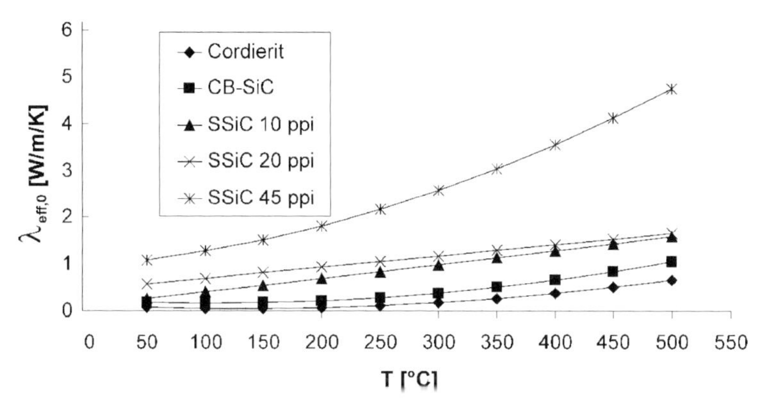

Figure 4 Thermal conductivities without flow of various ceramic foam materials with porosity of 0.76 determined by the hot-plate method [10].

Figure 4 shows thermal conductivity data of various ceramic foam materials [10]. As expected, thermal conductivity increases significantly with temperature. The reason for this is the increasing contribution of radiation. The radiative contribution can be quantified by considering that the heat flow density transferred by radiation from a body of temperature T_1 to a body of temperature T_0 increases in accordance with the Stefan–Boltzmann equation

$$\dot{q}_{\mathrm{R}}^{0} = \varepsilon\sigma(T_1^4 - T_0^4), \tag{15}$$

where σ denotes the Stefan–Boltzmann constant, and ε the emissivity of the cell material ranging between 1 for an ideal absorber and 0 for an ideal reflector. Now, imagining a cellular ceramic between these two bodies, radiation will then be attenuated in the material following an exponential law with the extinction coefficient k [m^{-1}] and the thickness of the material x [m]

$$\dot{q}_{\mathrm{R}} = \dot{q}_{\mathrm{R}}^{0} \mathrm{e}^{-kx}. \tag{16}$$

Recalling Eq. (2) and approximating

$$\partial T / \partial x \approx (T_1 - T_0)/x \tag{17}$$

and

$$(T_1^4 - T_0^4) \approx 4(T_1 - T_0) \cdot \overline{T}^3 \tag{18}$$

with the average temperature

$$\overline{T} = (T_1 + T_0)/2 \tag{19}$$

yields

$$\lambda_{\mathrm{R}} = 4x\varepsilon\sigma\overline{T}^3 \cdot \mathrm{e}^{-kx} \tag{20}$$

for the radiative contribution to the effective thermal conductivity. However, the contribution of heat conduction through the solid decreases with increasing temperature for ceramic materials. These two dependences explain the course of the curves in Fig. 4, which were all fitted with a second-order polynomial function.

4.3.2.2
Method to Determine the Effective Thermal Conductivity with Flow

In the above discussion on effective thermal conductivity with flow through the open cells of a ceramic material (see Fig. 1) the anisotropy of this quantity due to forced convection was mentioned. In the literature, this effect is often called *dispersion*. Assuming a steady flow in a tube filled with a cellular body one can imagine that heat transport is enhanced by mixing of the fluid in directions perpendicular to the tube axis. These dispersion effects depend on geometrical properties of the cells and on the flow conditions in such a way that they play an increasing role with increasing flow velocity. In a complex cell geometry, as in ceramic foams, it is difficult to derive the influence of the cell geometry on dispersion from fluid mechanics on the pore-size level. Therefore, there is a need to characterize this quantity as an effective, integrated property. For cellular ceramics a simple model derived from a packed bed leads to the following equations describing the influence of flow on the effective thermal conductivity [3, 11]. This procedure is legitimate as long as the macroscopic scales needed to describe the physical phenomenon are large in comparison to the microscopic scales describing the structure [3].

$$\lambda_{\mathrm{effAX}} = \lambda_{\mathrm{eff,0}} + \frac{\rho U_0 c_{p\mathrm{F}} d}{K_{\mathrm{AX}}} \tag{21}$$

and

$$\lambda_{\text{effRAD}} = \lambda_{\text{eff},0} + \frac{\rho U_0 c_{pF} d}{K_{\text{RAD}}} \tag{22}$$

Here ρ, c_{pF}, and U_0 denote density, specific heat, and velocity of the fluid, respectively, and d a characteristic length, which in this case is the diameter of a single cell. For the assumed cylindrical symmetry Pan et al. and Decker et al. have published an iterative experimental and numerical procedure to calculate the axial and radial mixing coefficients K_{RAD} and K_{AX}. This procedure can be applied to cellular materials in general [12, 13]. In their experiment heated air enters a tube filled with ceramic foam. The walls of the tube are kept at constant cold temperature. After some time, a stationary temperature distribution develops inside the foam. This temperature distribution is measured and compared with numerical calculations. These are carried out with varying mixing coefficients K_{RAD} and K_{AX} until a satisfying fit has been achieved.

As an example, results from a test are presented which was carried out on foam materials presented in Fig. 4 in the discussion on nonflow thermal conductivity. Figure 5 shows the effective thermal conductivities in radial direction as a function of temperature. Various flow velocities are compared. The curve denoted with zero velocity is identical to the effective thermal conductivity without flow. Because of the increasing contribution of radiative heat transfer it increases from about 0.1 W $\text{m}^{-1}\,\text{K}^{-1}$ at room temperature to 0.6 W $\text{m}^{-1}\,\text{K}^{-1}$ at 500 °C. For comparison, the thermal conductivity of a noncellular "dense" cordierite ceramic is about 1–3 W $\text{m}^{-1}\,\text{K}^{-1}$. With increasing flow velocity the effective thermal conductivity increases significantly.

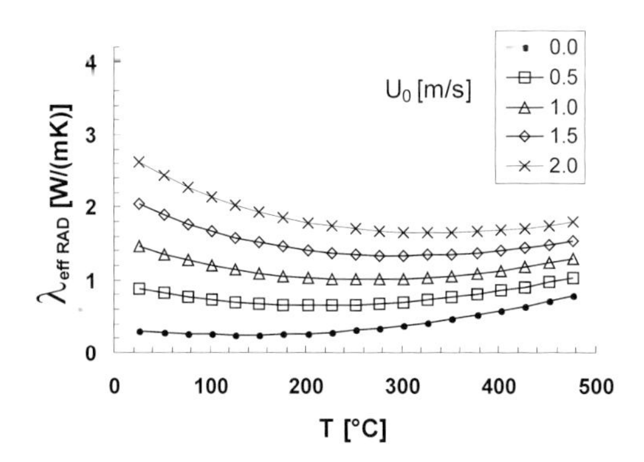

Figure 5 Effective thermal conductivities of a 20 ppi cordierite foam material perpendicular to the main flow direction for different flow velocities as a function of temperature.

Results for some other ceramic foam materials are presented in Tab. 2. From the mixing coefficients presented here effective thermal conductivities can be calculated from the fluid properties and flow conditions by using Eq. (1).

Table 2 Radial and axial mixing coefficients of various ceramic foam materials.

Material	n_{ppi}*	Porosity	K_{AX}	K_{RAD}
Cordierite	20	0.76/0.81	9.2	2.6
CB-SiC	10/20/45	0.76/0.81	8.63	6.96
S-SiC	10	0.76	0.55	3.07
S-SiC	20	0.76	13.22	1.6
S-SiC	45	0.76	18.52	0.90

* Cell density of ceramic foams (pores per inch).

4.3.3
Specific Heat Capacity

The specific heat capacity c_p [J kg^{-1} K^{-1}] characterizes the capability of a material to store energy in the form of heat. In contrast to the thermal conductivity, the specific heat of a cellular ceramic c_{pCC} consisting of the solid material s and the fluid f can be directly derived from the specific heat capacity of the components c_{pS} and c_{pF} as the sum of their weight fractions [8]:

$$c_{pCC} = c_{pF} \cdot \frac{m_f}{m_{cc}} + c_{pS} \cdot \frac{m_s}{m_{cc}}. \tag{23}$$

In most cases the mass fraction of the gas is so small that the specific heat capacity of the cellular ceramic is equal to that of the solid of which it is made. In the case of known chemical compositions, c_{pS} can be estimated with the well-known rule of Dulong and Petit [14]

$$c_{pS} = \frac{fk}{2\mu_{AM} m_H}, \tag{24}$$

which involves Boltzmann's constant k, the mass number of the atoms involved μ_{AM}, and the atomic mass unit m_H; f denotes the number of degrees of freedom of each particle. For most solids $f \approx 6$. The rule yields a good approximation and can be seen as an upper limit for "high temperatures". For most solids this limit is already reached at room temperature. For pure silicon carbide (SiC) and pure cordierite ($2\,MgO{\cdot}Al_2O_3{\cdot}SiO_2$) the rule yields values of 1244 and 1151 J kg^{-1} K^{-1}, respectively, which are close to those derived experimentally.

4.3.4
Thermal Shock

Thermal shock resistance is a quantity of interest to nearly every designer employing cellular or noncellular ceramics in high-temperature applications. Thermal shock occurs when changes in temperature lead to thermal stress due to the different amounts of thermal expansion in a piece of material. Due to the poor fracture toughness K_{IC} of most ceramic materials, thermal shock resistance of cellular ceramics is low compared to metallic materials. Thermal shock resistance of a material increases with increasing strength σ_C, and decreases with increasing elastic modulus E and thermal expansion α_l. Additionally, if heat flow is regarded, the thermal conductivity λ influences the thermal shock behavior. Experiments in which heated foams were quenched in water, oil, or cold air jets have shown the dependence of shock resistance on cell size, density, and material [15–19]. The thermal shock resistance rises with increasing cell size and is only weakly dependent on density. Thermal shock leads to the propagation of pre-existing cracks. The main source of stress is the temperature difference across the bulk and not across the individual struts, as could be shown by Orenstein, Green, and Vedula [15–17]. When cyclic thermal shocks are applied to ceramic foams there is more damage if higher temperatures or faster cooling rates are used [18]. By special processing, Colombo et al. were able to obtain materials with much higher shock resistances [19].

Usually, two quantities R_1 and R_2 characterize the thermal shock behavior. They are frequently called "hard" and "soft" thermal shock parameters (ν denotes Poisson's ratio):

$$R_1 = \frac{\sigma_C(1-\nu)}{\alpha_l E} \approx \Delta T_{MAX}, \tag{25}$$

$$R_2 = \frac{\sigma_C(1-\nu)\lambda}{\alpha_l E}. \tag{26}$$

The hard parameter R_1 gives an approximate maximum allowable temperature difference for a rapid superficial thermal shock. The soft parameter includes thermal conductivity and considers a long-term effect of the temperature gradient on the material. Heat flux into the material is taken into account. Table 3 includes the thermal shock parameters of selected dense ceramic materials. To describe the thermal shock behavior of ceramic foams the parameters ΔT_{10} describing the tempera-

Table 3 Quantities influencing the thermal shock behavior for a number of dense ceramic materials [20, 21].

Material	σ_C/MPa	ν	$10^6\,\alpha$/K^{-1}	λ/W m^{-1} K^{-1}	E/GPa	$R_1/10^2$ K	$R_2/10^3$ W m^{-1}
Aluminum oxide	200	0.25	8	25	360	0.52	1.3
Aluminum titanate	40	0.22	2.6	2	20	6.0	1.2
Cordierite	145	0.25	2.0	3	80	6.8	2.0
HIP* silicon nitride	650	0.26	3.4	32	290	4.9	16
Mullite	215	0.28	5	4.8	220	1.4	0.68
Silicon carbide (SiC)	450	0.17	4.6	70	410	2.0	14
Recrystallized SiC	100	0.17	4.8	27	280	0.62	1.7
Sintered SiC	300	0.19	4.5	140	350	1.5	22
Siliconized SiC	595	0.19	4.7	100	400	2.6	26
Zirconium oxide	800	0.3	9	3	200	3.1	0.93
Zirconium oxide, HIP	1600	0.3	9	3	200	6.2	1.9

* HIP = hot isostatically pressed.

ture difference for which the material experienced a 10 % reduction in Young's modulus [15] and the damage parameter D_E [17] were introduced. The damage parameter is defined as

$$D_E = 1 - \frac{E}{E_0}, \tag{27}$$

where E is the elastic modulus after thermal shock, and E_0 is the elastic modulus in the as-received state.

4.3.5 Volumetric Convective Heat Transfer

In some applications cellular ceramics are employed to heat or to cool a fluid which flows through the open pores of the material. Such applications are heat exchangers, solar receivers, electric air heaters with a cellular ceramic as a heating element, and air-cooling systems for gas turbines or combustion chambers. In all cases good convective heat transfer from the solid to the fluid or vice versa is needed. In some cases the problem of a fluid flowing through a cellular body is handled with *homogenous approaches,* considering fluid and solid as one medium with averaged effective quantities. Whenever a *heterogeneous approach* is used, the temperatures of fluid and solid are different. In this case the *convective heat transfer coefficient* α [W m^{-2} K^{-1}] describes the capability of a surface to transfer heat to a fluid which flows along the surface. It is defined by the equation which describes the typical volumetric heat-transfer problem:

$$\dot{q}_V = \alpha A_v (T_S - T_F) . \tag{28}$$

where $\dot{q}_v$ [W m^{-3}] denotes the volumetric heat flow, A_v [m^2 m^{-3}] the specific surface area of the cellular ceramic, and T_S [K] and T_F [K] the wall and fluid temperatures, respectively. Unlike many two-dimensional heat-transfer problems here the quantities $\dot{q}_v$ and A_v are volumetric properties.

Investigations of the convective heat transfer of open-celled foams revealed significant deviations (up to 100 %) from semi-empirical correlations for packed beds [22]. Therefore, for a detailed calculation of any heat-transfer problem in a cellular ceramic, it is necessary to determine the heat-transfer coefficient experimentally.

One way of measuring the heat-transfer coefficient is described in the following. It is an experiment based on the method by Younis and Viskanta [22]. The experimental setup is shown in Fig. 6. An air stream of alternating temperature flows through the porous sample. The air temperature can be regarded as a temperature wave $T(t,x)$, where x denotes the direction of air flow. This temperature wave is generated by an electrically driven heating element. The amplitude and the frequency of the temperature wave can be controlled. The tube through which the air flows is isolated, and the air temperature is measured at the inlet and outlet of the air flow through the sample. The sample induces a phase shift $\Delta\phi$ and an attenuation

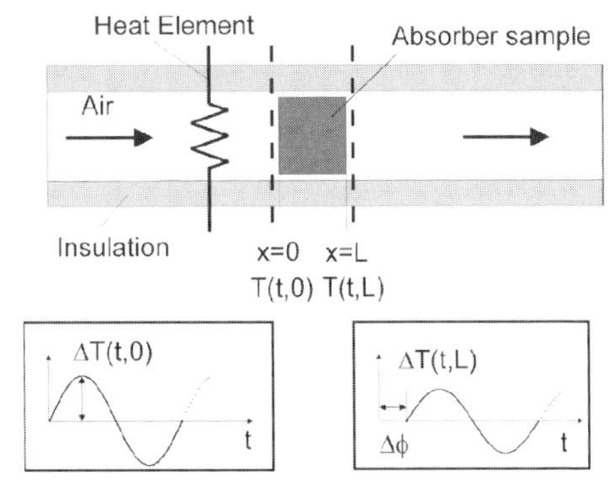

Figure 6 Experimental setup for measuring convective heat-transfer properties of cellular materials.

$$F(\omega) = \frac{\max(T(t,L))}{\max(T(t,0))} \tag{29}$$

of the temperature wave. From these, the product of the specific surface area A_v and the heat-transfer coefficient α can be calculated [23].

With the assumptions of a uniform temperature across any cross section of the tube $T = T(t,x)$, a constant heat transfer coefficient α and a negligible heat conductivity of the air, the energy transport equations of the air and the wall are:

$$(1 - Po)c_{pS}\rho_S \frac{\partial T_S}{\partial t} = \alpha A_v(T_F - T_S) + \lambda(1 - Po)\frac{\partial^2 T_S}{\partial x^2}, \tag{30}$$

$$\rho_F Po \frac{U_0}{Po} c_{pF} \frac{\partial T_F}{\partial x} + \rho_F Po \cdot c_{pF} \frac{\partial T_F}{\partial t} = A_v(T_S - T_F). \tag{31}$$

where the subscript F denotes the fluid and the subscript S the solid (ceramic), Po the porosity of the cellular ceramic, and c_{pF} and c_{pS} the corresponding specific heat capacities.

These equations are coupled differential equations. It could be shown by numerical simulations [23] that the analytical solution of this problem, which can achieved by neglecting the thermal conductivity λ of the wall, is valid in many cases with high porosity and thus low thermal conductivity. The general analytical solution is a dampened wave function. So by measuring the phase shift or the dampening the product of the heat-transfer coefficient and the specific area A_v can be calculated. As for most heat-transfer problems only the knowledge of the product A_v is essential, this is not a major drawback of this measurement technique.

As an example, results of an investigation on ceramic foam materials are presented [24]. The foam materials consist of cordierite, clay-bound silicon carbide (CB-SiC), and sintered silicon carbide (S-SiC). The results are presented as volumetric heat transfer coefficient αA_v as a function of fluid velocity. Figure 7 shows the results for various 20 ppi materials with a porosity of 0.76. The coefficients increase with increasing fluid velocity. The differences between materials of the same cell geometry but different material are small. In Fig. 8 the volumetric heat-transfer coefficients of CB-SiC foams with various cell diameters are compared. Heat transfer increases significantly with increasing ppi value, mainly due to the larger heat-transfer surface of materials with smaller cells. In the next section, the relationship between cell size and heat-transfer surface is quantified in more detail.

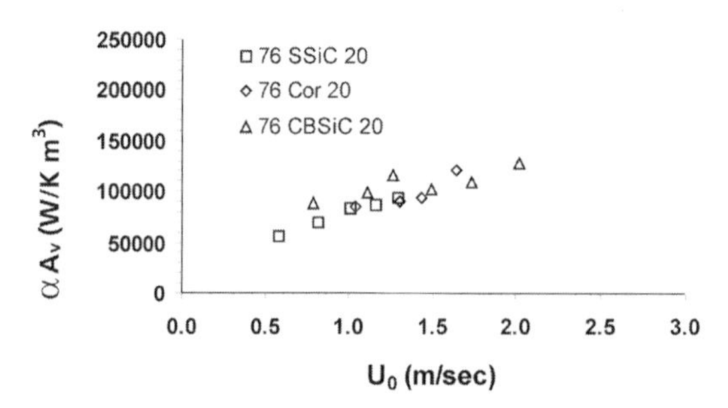

Figure 7 Volumetric heat transfer coefficients of various 20 ppi ceramic foam materials (porosity 76 %).

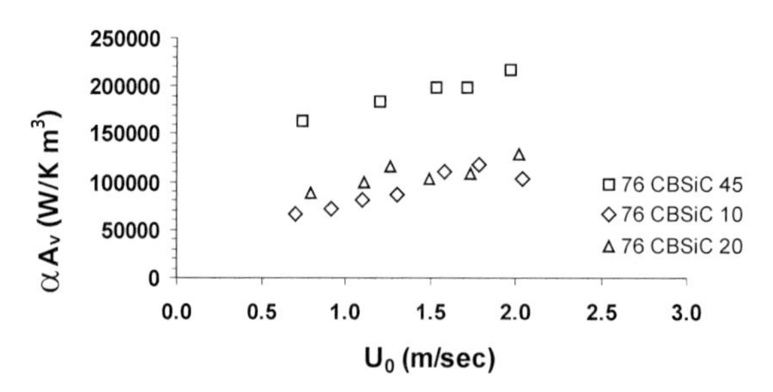

Figure 8 Volumetric heat-transfer coefficients of clay-bound silicon carbide foams with various cell diameters 10, 20, 45 ppi; porosity: 76 %).

4.3.5.1 Nusselt/Reynold Correlations and Comparison with Theoretical Data

To describe the heat-transfer properties of foams independent of their cell dimensions, similarity theory can be employed. Here, heat transfer is described by the Nusselt number *Nu*:

$$Nu = \alpha \cdot \frac{d_h}{\lambda} \tag{32}$$

where d_h is the hydraulic diameter, and λ the thermal conductivity of the fluid. For a regular channel geometry, for example, rectangular channels, the specific area A_v can be easily calculated geometrically. For these kinds of materials many *Nu–Re–Pr* correlations can be found in publications such as Kays and London [25].

To calculate Nusselt/Reynolds correlations from the volumetric heat-transfer data, assumptions and calculations concerning the pore level fluid velocity U_{PORE}, the hydraulic diameter d_h, and the specific surface area A_V [$m^2\ m^{-3}$] must be made. The calculation of the specific surface was performed with:

$$A_V = 35.7 \cdot n_{ppi}^{1.1461} . \tag{33}$$

This empirical equation is the result of a numerical investigation of various micrographs [26]. The fluid velocity in the pore $U_{PORE} = U_0/P_0$ is calculated from the porosity P_0 and the fluid velocity outside the cellular body U_0.

To calculate the hydraulic diameter d_h in a ceramic foam, various approaches have been published. Buck used a simple model in which the foam is considered to be a mesh made of cylindrical wires [27]. With this model, the length s of a characteristic cylindrical element and the hydraulic diameter d_h can be calculated:

$$s = \frac{0.0254}{1.5 \cdot n_{ppi}} \tag{34}$$

$$d_H = \sqrt{(1 - P_0) \cdot \frac{4s^2}{3\pi}} . \tag{35}$$

A further possibility to calculate d_h arises if the pore of the foam is considered to be a polyhedron made of solid struts, as was done by Gibson and Ashby [8]. In this case the cell of the foam consists of 24 vertices, linked by 6 faces consisting of 4 edges and 8 faces consisting of 8 edges, which results in a total of 36 edges (Fig. 9). Heat transfer can be regarded as a flow around the struts.

With the help of this model the diameter l [m] of the struts can be calculated:

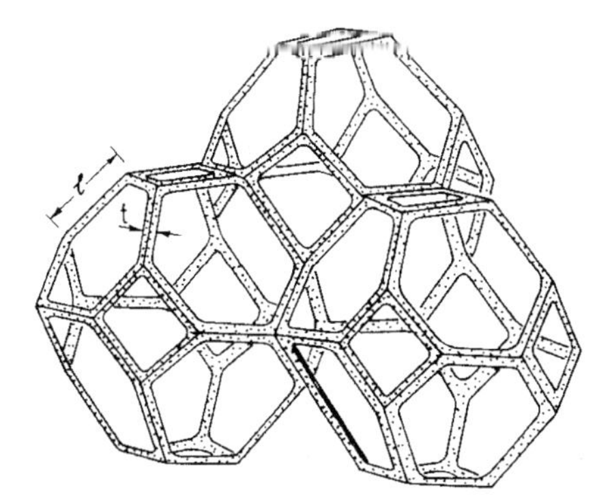

Figure 9 Tetrakaidecahedron as a model for foams investigated (after Gibson and Ashby [8]).

$$l = \frac{0.0196}{n_{ppi}} \tag{36}$$

$$\frac{t}{l} = \sqrt{(1 - P_0) \frac{1}{1.06}} . \tag{37}$$

Thus, the geometry of the pore can be derived from the macroscopic quantities porosity and cell density. As examples these quantities were calculated for some ceramic foams (Table 4). Additionally, the quantities derived from Buck's model and the reciprocal specific surface area, which also can be regarded as a characteristic length, are listed.

Table 4 Macroscopic and pore-level geometric data of various ceramic foams.

Material	n_{ppi}	Porosity/%	A_V/m^{-1}	d_h/mm, Buck	t, d_h/mm, Gibson–Ashby	A_v^{-1}/mm	l/mm, Gibson–Ashby	a	b
Cordierite	20	76	1100	0.27	0.47	0.9	0.98	0.09	0.74
CB-SiC	10	76	500	0.54	0.93	2.0	1.96	0.52	0.55
CB-SiC	20	76	1100	0.27	0.47	0.9	0.98	0.45	0.34
CB-SiC	45	76	2500	0.12	0.21	0.4	0.44	0.24	0.29
S-SiC	10	76	500	0.54	0.93	2.0	1.96	0.37	0.65
S-SiC	20	76	1100	0.27	0.47	0.9	0.98	0.13	0.63
S-SiC	45	76	2500	0.12	0.21	0.4	0.44	0.14	0.15
Cordierite	20	81	1100	0.24	0.41	0.9	0.98	0.03	0.96
CB-SiC	20	81	1100	0.24	0.41	0.9	0.98	0.31	0.40
S-SiC	10	81	500	0.48	0.83	2.0	1.96	0.95	0.35
S-SiC	20	81	1100	0.24	0.41	0.9	0.98	0.04	0.93
S-SiC	45	81	2500	0.11	0.18	0.4	0.44	0.04	0.71

For the Nusselt/Reynold plots in the further course of the text, the strut diameter t as calculated by Gibson and Ashby was chosen as hydraulic diameter d_h.

From the point of view of similarity theory the Nusselt number should be independent of the cell dimensions. However, the Nusselt/Reynold plots in Fig. 10 show that for the materials investigated this is not the case. The materials with larger cell dimensions (10 ppi) show significantly higher Nusselt numbers. In contrast, no significant differences are observed between the 20 and 45 ppi materials.

To derive Nusselt/Reynold relations, the experimental data were fitted with a potential function

$$Nu = aRe^{b}. \quad (38)$$

The parameters a and b, derived from heat-transfer measurements, are included in Table 4 for all materials investigated.

By means of the parameters derived from the measurements the Nusselt number of a foam can be calculated for a range of Reynold numbers of 5–150. An approach to explicitly determine the functional dependence of the Nusselt number on the cell density can now be performed by means of a modified fit. Three materials with a porosity of 0.76 were taken to perform a fit $Nu = aRe^b$ for cell densities of 10, 20, and

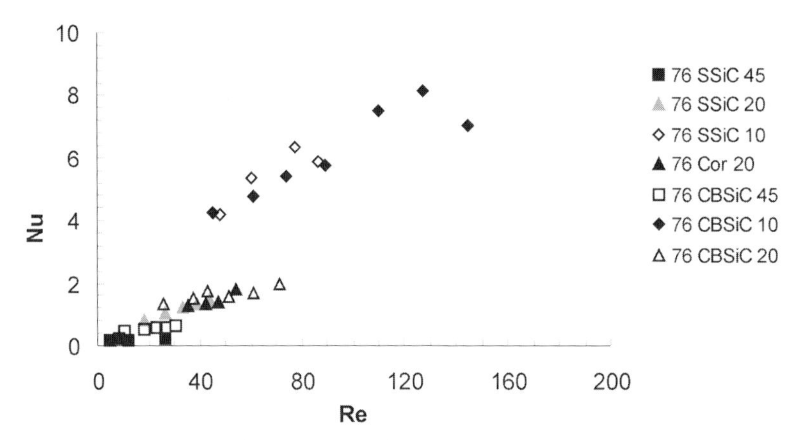

Figure 10 Nusselt/Reynold relationship of various ceramic foam materials (10, 20, 45 ppi; porosity 0.76)

45 ppi. The results, including the derived parameters, are shown in Fig 11. For the parameter a, the following function including the cell density n_{ppi} can be derived:

$$a = 4.8 \cdot n_{\text{ppi}}^{-1.1} \quad (39).$$

With this information a more general Nusselt/Reynold equation can be formulated:

$$Nu = 4.8 \cdot n_{\text{ppi}}^{-1.1} \cdot Re^{0.62} \quad (40).$$

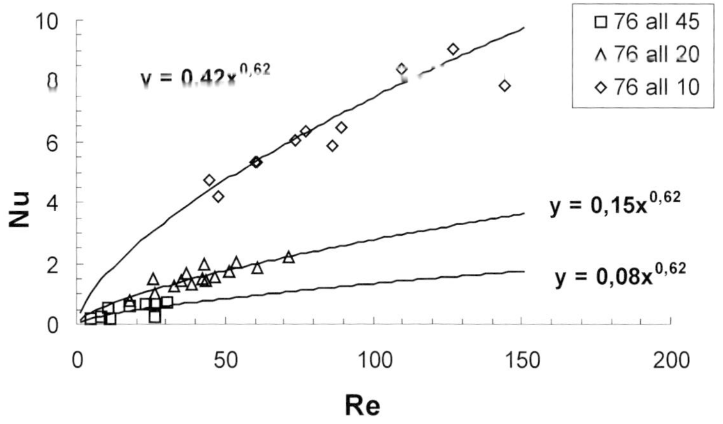

Figure 11 Potential fit of the results of various materials with a cell density of 10, 20, and 45 ppi (porosity 76%).

From the literature some information can be taken on heat-transfer properties of packed beds. For these porous media, experimentally derived Nusselt numbers are significantly lower than theoretically predicted values [28]. It was stated that the reason for this deviation is that a significant volume of the packed bed does not play a role in fluid flow due to inhomogeneities of the cell size. Smaller cell sizes induce higher pressure losses and consequently lower or vanishing flow velocities. The fact that in particular materials with smaller cell sizes show lower Nusselt numbers is an argument that in ceramic foams similar mechanisms take place.

In the following a comparison is presented between the data experimentally acquired for the foam materials mentioned above and a theoretical model describing heat transfer in ceramic foam materials. Before employing a certain model, the heat-transfer mechanism must be considered. If the flow through the cellular body is considered as a flow through a network of struts similar to the sketch in Fig. 9, the mechanism of heat transfer can be regarded as cross-flow over a network of cylinders. For this simple case the following relations can be found [29]:

$$Nu = 0.3 + \sqrt{Nu_{\text{lam}}^2 + Nu_{\text{turb}}^2}, \tag{41}$$

$$Nu_{\text{lam}} = 0.664\sqrt{Re} \cdot \sqrt[3]{Pr}, \tag{42}$$

$$Nu_{\text{turb}} = \frac{0.037 Re^{0.8} Pr}{1+2.443 Re^{-0.1}(Pr^{2/3}-1)}. \tag{43}$$

The equations are valid for a range of Reynold numbers of $10 < Re < 10^7$ and for a range of Prandtl numbers of $0.6 < Pr < 1000$. The experiments were carried out at temperatures between 40 and 60 °C and at flow speeds of 0.6–3 m s^{-1}. Prandtl and Reynold numbers were $Pr \approx 0.7$ and $5 < Re < 160$, that is, within the range of validity of Gnielinski's equations.

The comparison in Fig. 12 shows that the values derived from the model of Gnielinski are in good accordance with the values experimentally derived for 10 ppi foams. For higher cell densities (20 and 45 ppi) the experimental data deviate significantly from the model. This leads to the assumption that, similar to packed beds, flow vanishes in parts of the volume of the material. This effect is considered in Eq. (2).

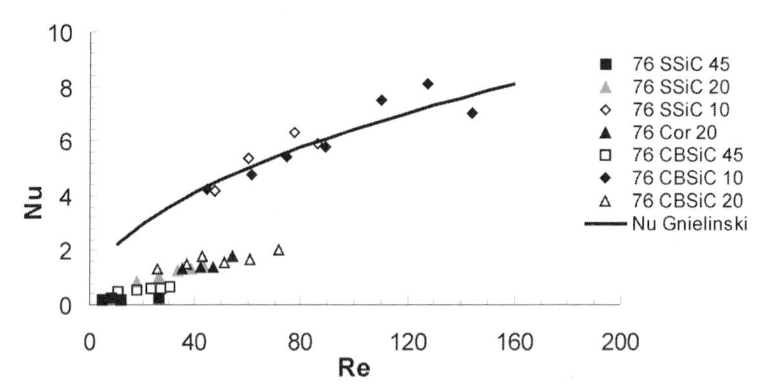

Figure 12 Comparison of experimentally derived Nusselt/ Reynold correlations with theoretical data for cross-flow over a cylinder (after Gnielinski).

4.3.6 Summary

This chapter could only provide a superficial glance at the thermal properties of cellular ceramics, which are needed when heat transfer problems are dealt with.

In the case of closed porosity the cellular ceramic material can simply be considered as a homogenous material with an effective thermal conductivity which may be determined with the methods described above. This quantity is significantly lower for cellular ceramics than for dense materials. Additionally, and in contrast to most dense materials, it increases with increasing temperature due to thermal radiation through the cells.

In the case of cellular ceramic materials with open porosity, which allows fluids to flow through the cells, heat-transfer mechanisms become a little more complex. In this case thermal conductivity increases due to forced convection and mixing of the fluid within the material. Furthermore, it becomes anisotropic since it depends on the main direction of flow. Data is provided in the chapter for a set of open-cell ceramic foam materials.

Futhermore, in the case of open-cell ceramic foams fluid-to-solid convective heat transfer must be dealt with as an additional physical quantity which is needed to solve heat-transfer problems. As is known from simpler heat-transfer problems, convective heat transfer increases with increasing flow velocity and is described with the theory of similarity by Nusselt numbers. An approach has been presented decribing the convective heat transfer in ceramic foams by a Nusselt/Reynold equation containing the macroscopic quantity cell density.

References

1 Kaviany, M. *Principles of Heat Transfer in Porous Media*, 2nd Ed., Springer-Verlag New York, Heidelberg, 1995.

2 Hsu, C.-T. Heat Conduction in Porous Media in *Handbook of Porous Media*, Vafai, K. (Ed.), Marcel Dekker, New York, 2000.

3 Tsotsas, E., Martin, H. *Chem. Eng. Process.* **1987**, *22*, 19–37.

4 Taylor, R. Thermophysical Properties in *International Encyclopedia of Composites*, Lee, S.M. (Ed.), Vol. 5,VCH, New York, 1991, pp. 530–548.

5 Mottram, J.T., Taylor, R. Thermal Transport Properties in *International Encyclopedia of Composites*, Lee, S.M. (Ed.),Vol. 5, VCH, New York, 1991, pp. 476–496.

6 Hale, D.K. *J. Mater. Sci.* **1976**, *11*, 2105–2141.

7 Collishaw, P.G., Evans, J.R.G. *J. Mater. Sci.* **1994**, *29*, 2261–2273.

8 Gibson, L.J., Ashby, M.F. *Cellular Solids.* Cambridge University Press, Cambridge, 1997.

9 Gustaffson S.E. *Rev. Sci. Instrum.* **1991**, *62*, 797

10 Becker, M., Decker, S., Durst, F., Fend, Th., Hoffschmidt, B., Nemoda, S., Reutter, O., Stamatov, V., Steven, M., Trimis, D. *Thermisch beaufschlagte Porenkörper und deren Durchströmungs- und Wärmeübertragungseigenschaften*, DFG Projekt DU 101/55-1, Final Report, 2003.

11 Schlünder, E.-U., Tsotsas, E. *Wärmeübertragung in Festbetten, durchmischten Schüttgütern und Wirbelschichten*, Georg Thieme Verlag, Stuttgart, 1988.

12 Pan, H.L., Pickenäcker, O., Pickenäcker, K., Trimis, D., Weber, T. *Experimental Determination of Effective Heat Conductivities in Highly Porous Media*, 5th European Conference on

Industrial Furnaces and Boilers, Porto, April 11–14, 2000.

13 Decker, S., Mößbauer, S., Nemoda, S., Trimis, D. Zapf, T. *Detailed Experimental Characterization and Numerical Modelling of Heat and Mass Transport Properties of Highly Porous Media for Solar Receivers and Porous Burners,* Sixth International Conference on Technologies and Combustion for a Clean Environment (Clean Air VI), Vol. 2, Paper 22.2, Porto, Portugal, 9–12 July, 2001.

14 Ashcroft, N.W., Mermin, N.D. *Solid State Physics,* Holt-Saunders International Editions, Tokio, 1981.

15 Orenstein, R..M., Green, D.J. *J. Am. Ceram. Soc.* **1992**, *75*, 1899–1905.

16 Vedula, V.R., Green, D.J., Hellmann, J.R., Segall, A.E. *J. Mater. Sci.* **1998**, *33*, 5427–5432.

17 Vedula, V.R., Green, D.J., Hellmann, J.R *J. Am. Ceram. Soc.* **1999**, *82*, 649–656.

18 Vedula, V.R. ,Green, D.J., Hellmann, J.R *J. Eur. Ceram. Soc.* **1998**, *18*, 2073–2080.

19 Colombo, P., Hellmann, J.R, Shellemann, D.L. *J. Am. Ceram. Soc.* **2002**, *85*, 2306–2312.

20 Morell, R. *An Introduction for the Engineer and Designer Handbook of Properties of Technical and Engineering Ceramics,* Part 1, Her Majesty's Stationery Office, London, 1985.

21 Schäfer-Sindlinger, A., Vogt, C.D. *Filtermaterialien für die additivgestützte und katalytische Dieselpartikelreduktion,* MTZ 3/2003, Jahrgang 64.

22 Viskanta, R., Younis, L.B. *Int. J. Heat Mass Transfer* **1993**, *36*, 1425–1434.

23 Hoffschmidt B. *Vergleichende Bewertung verschiedener Konzepte volumetrischer Strahlungsempfänger,* Deutsches Zentrum für Luft- und Raumfahrt, Köln, 1997.

24 Fend, Th., Hoffschmidt, Pitz-Paal, R., Reutter, O., Rietbrock, P *Energy* **2004**, *29*, 823–833.

25 Kays, W.M., London, A.L. *Compact Heat Exchangers,* 3rd Ed., McGraw-Hill Book Company, New York, 1984.

26 Adler, J. unpublished data of the Fraunhofer Institute of Ceramic Technologies and Sintered Materials, Winterbergstraße 28, 01277 Dresden, Germany, joerg.adler@ikts.fhg.de.

27 Buck, R. *Massenstrom-Instabilitäten bei volumetrischen Receiver-Reaktoren,* VDI Verlag, Düsseldorf, 2000.

28 Martin, H. *Chem. Eng. Sci.* **1978**, *33*, 913–919.

29 V. Gnielisnski, Forced Convection in Ducts in *Heat Exchanger Design Handbook,* Hewitt, G.F. (Ed.), Begell House, New York, 2002.

4.4
Electrical Properties

Hans-Peter Martin and Joerg Adler

4.4.1
Introduction and Fundamentals

Cellular ceramic solids can be of various types:

- Foams
- Honeycombs
- Biomimetic ceramic structures
- Connected fibers

The common feature of all the above cellular materials is a combination of solid and gaseous elements which are structured by more or less defined regular geometric shapes and positions.

Cellular materials differ from the "conventional" materials by the combination of a solid phase with closed or open regularly structured voids, tubes, or any other type of inhomogenities, such as pores. These inhomogenities change the electrical properties of cellular materials drastically if they differ from the ceramic. The most common inhomogenities are pores in a great variety of shapes and sizes. The general effects of porosity, such as dilution of one material, and the specific effects of porosity, which include effects determined by shape and distribution, are interlinked. This is of interest if electrical insulation or dielectric effects are aimed for an application. As long as no extreme temperature or high voltage is applied, the gaseous phase is just a passive component of the cellular material for electrical conduction or electrical capacity. The influence of the gaseous phase on the conduction mechanism is more complicated, while the influence on the dielectrical properties almost completely depends on the volume fraction of the gaseous phase.

The solid phase determines whether the material is electrically conducting, semiconducting, or insulating. It holds the electrical carriers and governs the electrical phenomena of the cellular materials in the vast majority of cases. The only exception could be a cellular material which is filled by an electrically conductive component which partly or completely substitutes the gaseous phase.

An obvious idea is that the electrical properties of the solid phase are identical with those of the same bulk material. However, manufacture of foam ceramics and conventional bulk ceramics can differ. Figure 1 presents the microstructure of conventional sintered silicon carbide and an SiC microstrut of a foam ceramic. There

Cellular Ceramics: Structure, Manufacturing, Properties and Applications.
Michael Scheffler, Paolo Colombo (Eds.)

ISBN: 3-527-31320-6

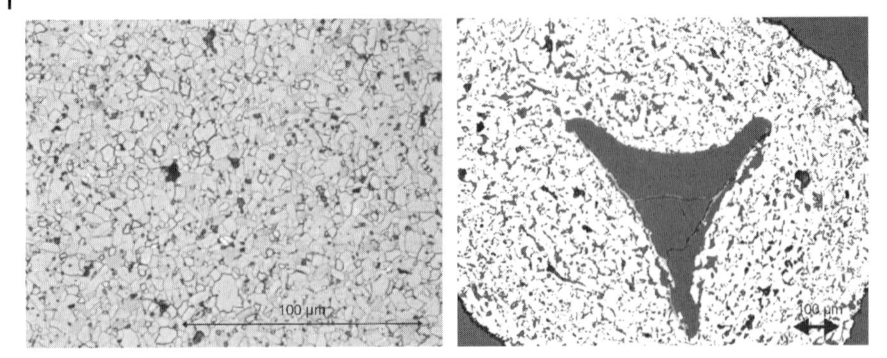

Figure 1 Micrographs of dense S-SiC (left) and S-SiC ligament from foam (right).

are significant differences in materials structure between the conventional and the ceramic foam material.

On the macroscopic scale of products made of cellular ceramics the electrical properties are influenced by the way in which the solid network is joined, the shape and geometry of the solid material, and interactions of the solid with the gaseous phase. A more detailed discussion is given below.

Since electrical applications have not been the driving force for the development of cellular materials, the electrical properties have not yet been investigated to the same extent as properties like permeability, mechanical strength, and thermal properties [1]. Nevertheless, several publications mention electrical data or report experiment results for cellular materials [1–7]. Moreover, there are links to mechanical and thermal behavior which are influenced by porosity or cellular effects. Changes in mechanical properties like Young's modulus and thermal conductivity are related to the electrical properties. This opens the opportunity to use models or considerations which were made for other properties. But care is needed in applying these models to electrical properties, since the physical character of the conduction or dielectric processes might require alteration or even exclude certain models of other physical phenomena. Chapters of this book on mechanical (4.1) and thermal properties (4.3) discuss a number of models which can be applied to electrical properties, for example, Minimum Solid Area (MSA) models. Furthermore, some aspects of porous materials which are related to cellular materials or even identical to them are discussed in [8]. In addition to the consideration of mechanical properties in relation to the porosity, electrical properties are also discussed in this book. It is stated that MSA models are consistent for the relation of porosity and electrical conductivity of different ceramic materials [8]. One should always be aware of the influence of pore size, pore shape, and other pore characteristics which could be significant for conduction. Even dielectric properties could be influenced by the character of the pores, particularly if the base materials are piezoelectric or ferroelectric [8].

The specific issue of cellular ceramics is the existence of large-scale "pores" such as tubular pores of honeycombs or macroscopic voids of foams. The "conventional" pores of ceramics are built up by ceramic grains. The maximum size of these pores depends on the grain size of the ceramic powder. The cellular pores are built up by macroscopic ceramic elements like walls of honeycombs or struts of foams, which consist of a number of grains. These pores can be manufactured in any size which is achievable by the applied technology. These ceramic elements may contain microscopic pores themselves. Ultimately, the porous effect is more complex than that known for "conventional" porous ceramics, and this must be considered in modeling cellular materials.

The application of cellular ceramics for electrical functions offers a number of unique features, such as tailoring of electrical resistance by the kind of cellular structure, anisotropic behavior of electrical conduction, combination of various types of cellular structures, modification of the electrical performance by using the properties of the gaseous phase, excellent transfer of energy from the cellular product into the surroundings (electrical heaters), and combination of excellent electrical conductivity with moderate thermal insulation.

Table 1 Electrical properties of bulk ceramic materials.

Material	Specific electrical resistance ρ_{el}/Ω m at 20 °C	Relative dielectrical constant ε_r	Electrical breakdown strength/kV mm^{-1}	Ref.
Alumina	10^{10} to 10^{13}	9	15	9
Zirconium dioxide	10^{6}	22	–	9
Magnesium oxide	10^{8}	10	–	9
Titanium dioxide	10^{6}	8	8	9
Porcelain	10^{7}	6–7	20	9
Steatite	10^{6}	6	15–20	9
Mullite	10^{7}	8	15–17	9
Cordierite	10^{6}	5	10	9
Aluminum nitride	10^{9}	20	20	9
Silicon nitride	$> 10^{11}$ *	8–12	20	9
Perovskites	10^{6}	350–3000	2	9
Silicon carbide (reaction bonded)	10^{-4} to 10^{-1} **	–	–	
Silicon carbide (sintered)	10^{-1} to 10^{4} **	–	–	
Silicon carbide (recrystallized)	10^{-1} to 10^{2} **	–	–	
Silicon carbide (LPS, dense)	10^{-1} to 10^{1} **	–	–	
Silicon carbide (LPS, porous)	10^{-4} to 10^{-3} **	–	–	
Molybdenum silicide ($MoSi_2$)	10^{-10} **	–	–	

* Ref. [10]. ** Experimental data from the authors.

The data in Table 1 are intended to give a basic idea about the range of a few electrical properties of ceramic materials. Unfortunately, almost no proper data are available for porous ceramic materials, so that the given data can not be applied directly for cellular materials.

A very important parameter is the specific electrical resistance. It is a materials constant which changes with temperature. It is the reciprocal of the specific electrical conductivity, which depends on the mobility, charge, and number of the electrical carriers per unit volume. Since the mobility and number of free electrical carriers depend on temperature, the specific electrical resistivity is also temperature-dependent. The specific resistivity can be obtained by measuring the electrical resistivity, the cross-sectional dimensions, and the length of a material sample, and finally calculated for practical use by Eq. (1)

$$\rho_{el} = R \cdot \frac{A}{l} \tag{1}$$

where ρ_{el} is the specific electrical resistance, R is the electrical resistance of the sample, A the cross-sectional area, and l the sample length.

The specific electrical resistivity data of Table 1 show that most of the ceramic materials are insulators ($\rho_{el} > 10^8\ \Omega$ m). This is due to the dominant ionic and covalent binding between the ions or atoms of ceramic compounds. In most cases no free electrical carriers are available for electrical transport through the ceramic material. Some of the materials become electrically conductive at high temperatures. For instance, the specific electrical resistivity of zirconium dioxide drops from $10^6\ \Omega$ m at 20 °C to about $10^{-2}\ \Omega$m at 1000 °C because of the activation of oxygen ions, which become able to move by diffusion through the material.

Some ceramic materials are semiconductive, for example, B_4C and SiC. The conductivity of SiC is due to the much smaller band gap between valence band and conduction band of the electrons. It reaches about 2.2 eV for β-SiC [11]. The resistance of these semiconducting materials decreases with increasing temperature because more free carriers (electrons and electron holes) are mobilized by thermal energy. Silicon carbide materials can have quite different electrical resistance depending on the type of manufacture. Even the same SiC material type, for example, sintered SiC (S-SiC), can differ by several orders of magnitude in resistivity just by slight variation of the sintering aid, sintering temperature, or other technological parameters. The porosity of cellular materials changes the parameters A, l, and ρ_{el} of Eq. (1), and this results in considerable changes in electrical resistance.

The insulators (specific electrical resistivity > $10^8\ \Omega$m) are known as dielectric materials. When an electric field is applied to these materials polarization of electrons or ions occurs inside the material. This effect leads to an increase in the capacitance of a capacitor on filling it with the dielectric material. The dielectric constant ε_r gives the ratio of the capacitance between a capacitor filled with material C_m and one filled with vacuum C_0 (Eq. (2)):

$$\varepsilon_r = \frac{c_m}{c_0}. \tag{2}$$

In most cases the cellular effect on the dielectrical constant is the dilution of the material by air-filled pores or cellular elements. Additional effects which may change the character of the ceramic material are particularly observed for ferroelectric or piezoelectric materials.

The complexity of the solid–gaseous interaction is illustrated by porous liquid-phase sintered SiC (LPS-SiC), which demonstrates the influence of a porous structure on the electrical resistance (Fig. 2).

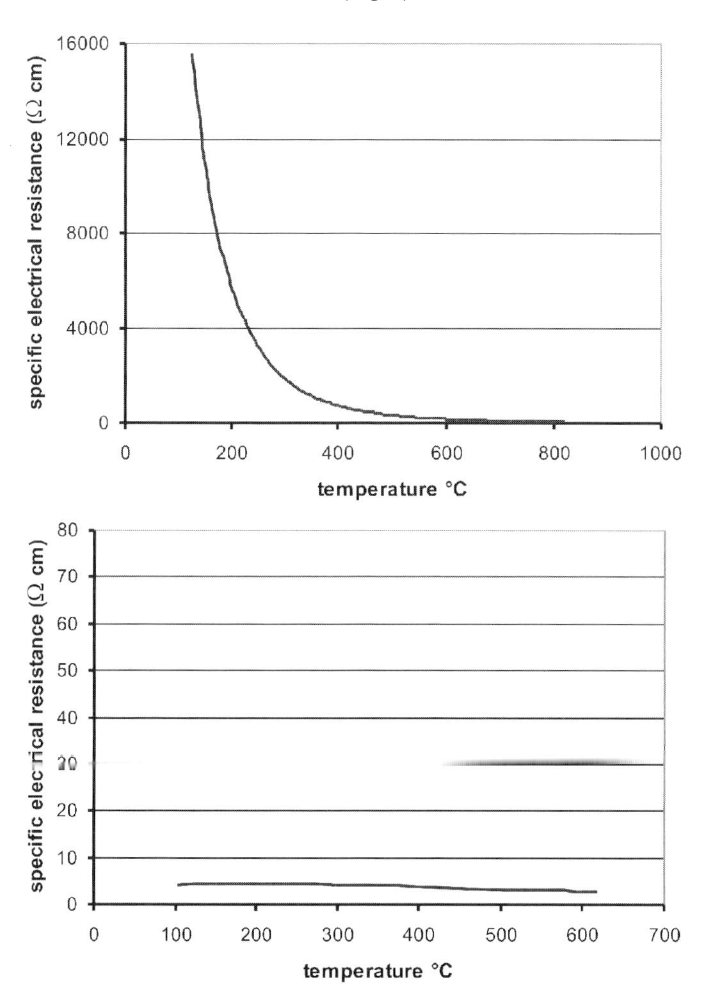

Figure 2 *R–T* characteristics of dense (top) and porous (bottom) LPS-SiC.

The electrical resistance can be modified by porosity in various manners:

- Increase of electrical resistance due to smaller cross-sectional area.
- Alteration of electrical resistivity due to alteration of the ceramic grains.
- Alteration of electrical resistivity due to formation of specific grain boundaries.
- Alteration of electrical resistivity due to interaction of ceramic grains with pores.

The first effect is a trivial one which needs no further discussion. The other effects can be caused by a number of specific reasons which can not be discussed in general. The electrical resistance of porous LPS-SiC (pore volume of 40 %) serves as an example to illuminate the above mentioned effects.

The manufacture of this type of material has been described elsewhere [12]. It is based on the well-known technology of LPS-SiC manufacture. High porosity is obtained by selecting suitable grain size distributions. The oxidic sintering additives form a liquid phase which is able to reprecipitate silicon carbide and thus promote the sintering process. The oxides themselves solidify as isolated aggregates at the triple junctions of the SiC grains without forming an interconnected layer around the silicon carbide grains. The SiC grains are doped by the oxide additives, so that the outer sphere of the SiC grains is highly conductive for electrical carriers, as was already discussed for dense LPS-SiC materials [13]. The porous silicon carbide is sintered according to the same mechanism as dense LPS-SiC, but its silicon carbide grains are completely accessible to the additives because of the high porosity. This shows the alteration of the SiC grains due to the porosity. Furthermore, the formation of intergrown SiC boundaries and segregations of oxide-dominated triple points predetermines the favorable path of electrical conduction and the final resistivity of the material demonstrating the effect of alteration of grain boundaries on the electrical resistivity. The outcome is a silicon carbide material with extraordinary electrical conductivity at room temperature and almost no change in electrical resistivity with increasing temperature. The interaction of the sintering additives and the ceramic grains is promoted by the pores during sintering, and this results in a level of electrical conductivity of a polycrystalline SiC material that has never before been achieved [14]. However, if the porous SiC material is heated under oxidizing atmosphere the resistance increases drastically at temperatures above 500 °C. The large surface area and accessibility of the SiC grains to the atmosphere promote oxidation along the grain boundaries. This interaction of the pores with the material again changes the electrical resistance drastically.

4.4.2 Specific Aspects of Electrical Properties of Cellular Solids

Whether a cellular ceramic is electrically conductive or insulating is determined by the solid ceramic in the same way as for other ceramics. The specific aspects of cellular ceramics are derived from the arrangement of the cellular structure. Furthermore, the existence and character of porosity inside the solid ceramic elements of the cellular structure strongly change the resulting electrical properties. Since the electrical properties are a product of the interaction between solid and gaseous phase, honeycombs, biomimetic ceramic structures, foams, and connected fibers must be discussed in different ways.

4.4.2.1
Honeycombs

Honeycomb materials are produced by extruding a plastic ceramic mass (Chapter 2.2). They mostly consist of the ceramic material and large gas-filled tubes. Tube diameters as small as 200 μm have been reported [5, 15] in addition to larger sized honeycombs which have been well established for a number of years. The interaction with the gas phase and the ceramic material is limited to the surface as long as the ceramic material is dense. Therefore, consideration of the electrical properties could be performed in a similar manner to the mechanical properties (Chapter 4.1). The MSA model, which gives the relation of the porous to the dense material, is given in that chapter already [$E/E_0 = (1-P)^n$, $n = 1$]. The straight tubular pores of honeycomb structures result in the upper limit of conductivity for porous or cellular structures being reached.

In most cases the cellular effect of a honeycomb ceramic on the electrical resistance is a simple geometrical one. The cross section is reduced by the lack of conductive material, so that the electrical resistance is increased in comparison to bulk material. An illustration is given in Fig. 3. The cross-sectional area of bulk material can be calculated by Eq. (3)

$$A_f = a \cdot b \tag{3}$$

and the cross-sectional area of honeycomb material by Eq. (4)

$$A_h = A_f - x \cdot (c_1 \cdot c_2) \tag{4}$$

where x is the number of tubular pores of the honeycomb, c_1 the horizontal width of a tubular pore, and c_2 the vertical width of a tubular pore.

The resistance can be calculated by Eq. (1) using the honeycomb cross-sectional area according to Eq. (4) by Eq. (5)

$$R_h = \rho_{el} \cdot \frac{l}{A_h} \tag{5}$$

where R_h is the electrical resistance of the honeycomb sample, ρ_{el} the specific electrical resistivity of the ceramic material, l the length of the honeycomb sample, and A_h the cross-sectional area of the honeycomb sample.

The cross section of honeycomb materials differs according to the direction of current flow. Figure 3 shows the cross section viewed parallel to the axis of extrudation. Current flow perpendicular to the extrudation axis is even more limited by a more reduced cross section. Consequently, the resistance of honeycomb materials depends on the direction of current flow. As long as the wall thickness of the honeycomb is thick enough that the specific surface area is in the range of that of bulk material, the honeycomb products can be considered like bulk ceramic material with regard to their electrical properties.

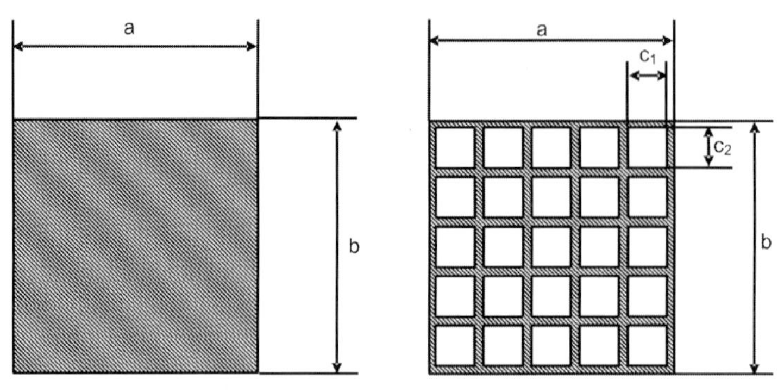

Figure 3 Cross section of bulk (left) and honeycomb (right) materials.

The effect of the honeycomb structure on the dielectric constant is very similar to the effect on the electrical resistance. The mass of the honeycomb material is reduced by the extent shown in Fig. 3. This leads to a proportional reduction in ε_r because of the substitution of ceramic material by air. In contrast to the electrical resistance, the dielectric behavior does not depend on the direction of the applied electrical field.

4.4.2.2
Biomimetic Ceramic Structures

Biomimetic ceramic materials have gained considerable attention in recent years. Their features can vary over a wide range depending on the biological structures which were used to produce the ceramic material [16–18]. Well-described types of biomimetic ceramics are wood-derived carbon materials and silicon carbide (Chapter 2.5) [19,20]. To our knowledge there is no extensive work published on investigations of biomimetic ceramic materials and their electrical properties. Therefore, just few general statements and predictions are given here.

Silicon carbide materials derived from different wood species could behave in various manners with regard to their electrical properties:

The biomimetic structure has a simple geometric effect on the cross-sectional area, as was described for honeycomb materials. The prediction becomes more difficult because the reduction of the cross section area is not constant along the axis of the products, as is illustrated in Fig. 4.

If the wood-derived ceramic is produced by liquid-silicon infiltration all pores can be filled by silicon. The material is no longer a real cellular material in this case. The electrical situation is completely changed because the former voids become conductive. If silicon infiltration is performed by gas-phase reaction the pores can stay open after reaction and sintering, and a honeycomblike material is obtained. The electrical conductivity decreases with increasing amount of pores.

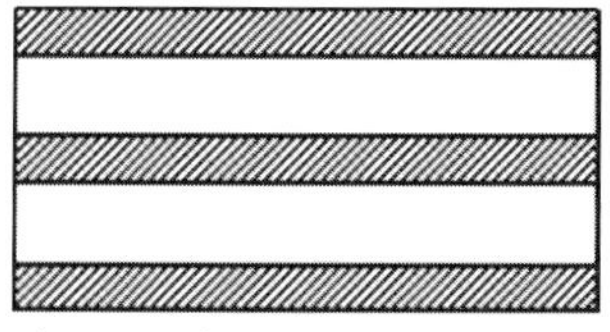

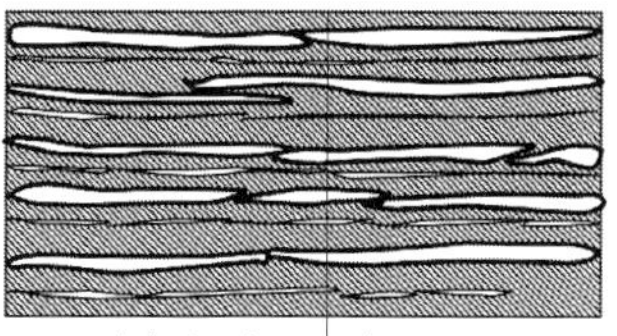

Figure 4 Principal structure of honeycomb (left) and wood-derived ceramic (right).

The biological template which was used to manufacture the biomimetic ceramic determines the resulting cellular structure. If wood is used as template for silicon-infiltrated silicon carbide or carbon materials the amount of free silicon and free carbon and the structure of the ceramic (pore size, pore shape, silicon distribution, etc.) are predetermined [18, 19]. The difference between wood-derived Si-SiC and conventional Si-SiC is a hierarchic arrangement of the carbon residues or of the free silicon. This can lead to favorable electrical paths along the conductive silicon or carbon structures. A tendency to decreasing conductivity can be expected if the pore volume of the biotemplate increases. The shape, structure, and arrangement of the template-derived pores govern the degree of modification of conduction. If dielectric materials like alumina or mullite are used for material production the dielectric constant decreases with increasing pore volume, but the dependence on the kind of pores is much less pronounced than for conductivity phenomena.

During sintering and application the silicon carbide material may undergo an interaction with the gaseous phase [20]. Since the wall thickness is in the range of several micrometers a strong interaction with the pores is enabled. The occurring gases depend on the sintering technology and the used raw materials. Doping of the silicon carbide, oxidation of the silicon carbide surface, or even a nitridation could alter the final product. This offers the opportunity to increase or decrease the electrical conductivity or dielectric constant of the ceramic material, depending on the kind of interaction and modification.

4.4.2.3
Ceramic Foams

Foams can be produced by different techniques and from a number of materials (see Part 2 of this book). Alumina, zirconia, glass, carbon, and silicon carbide are the major material types from which ceramic or ceramiclike foams have been made up to now [21–23]. Foams can be open-cell or closed-cell structures. In each case the solid phase is an interconnected matrix and allows transport of electrical carriers. A comparison of electrical properties of different foam types always requires a careful consideration of specific properties like foam density, ligament density, foam material type, and specific structure and composition of the material. Therefore, a straightforward comparison of different foams with regard to their electrical properties is rather vague. General parameters are given below for the estimation of the electrical resistance of ceramic foams.

The bulk electrical properties of the ligaments allow an estimation of the electrical trend of the foam materials. Hence, glass, alumina, and other oxides can be regarded as electrical insulators. Silicon carbide, carbon, and other electrical conducting or semiconducting foam materials can be considered to be conductive foams, too. The better the conductivity of the bulk material the better the conductivity of the foam material, as expected.

Table 2 Comparison of resistivity of bulk and foam ceramic materials.

Foam material	Bulk resistivity/Ωcm	Foam resistivity/Ωcm	Ref.
$La_{0.84}Sr_{0.16}Co_{0.02}MnO_3$	0.025	0.01 (13 vol% solids)	24
Vitreous carbon	0.001	0.2 (9 vol% solids)	25
Silicon carbide	10×10^4	$24 \cdot 10^4$ (12 vol% solids)	25

As discussed in Chapter 1.1 the electrical conductivity can be calculated if the conductivity of the solid is known (see Eqs. (11) and (15) of Chapter 1.1 and comment on page 10).

Foams can be regarded as porous ceramic materials with a three-dimensional array of hollow polygons [21]. Interaction with voids can alter the surface of the material and hence the electrical resistivity. Interaction with the voids determines the electrical resistivity more decisively for thin ligaments than for thicker ligaments. Internal porosity of the ligaments can itself alter the electrical properties and can promote the interaction of the ligament material with the voids, so that the thickness of the ligaments no longer limits the interaction. The actual effect depends on the type of interaction of the gaseous phase with the ligament material, and an increase or a decrease in electrical resistivity can result.

An illustration of this effect is given by foams with dense ligament material and porous ligament material. Densification of SiC foam struts can be achieved by silicon infiltration of reticulated foams (LigaFill). A recrystallized silicon carbide foam retains residual porosity in the struts. If these foams are heated in air up to temperatures of about 600 °C, the electrical resistance changes in the case of the porous-ligament material, while it remains the same for the dense-ligament foam (Table 3).

Table 3 Alteration of electrical resistance by interaction of SiC foam elements (40 × 40 × 25 mm) with air at 550 °C in dependence on the porosity of the strut material and duration.

Annealing time/h	Resistance/Ω of R-SiC foam (porous struts)	Resistance/Ω of Si-SiC foam (dense struts)
0	1.3	0.08
1	10.1	0.08
2	332	0.08

Besides the length of the electrical path the cross-sectional area is an important parameter for the electrical resistance of all materials. It determines the electrical resistance of a foam material and depends on foam density, ligament density, and foam type. Foam density and the ligament density determine how much solid phase exists in the overall cross section. For instance, 90 vol% of voids will result in approximately 10% of conductive solid phase in the cross section. However the cross-sectional area is not a constant quantity along the electrical path. It changes because of the nature of the foam structure, as illustrated by Fig. 5.

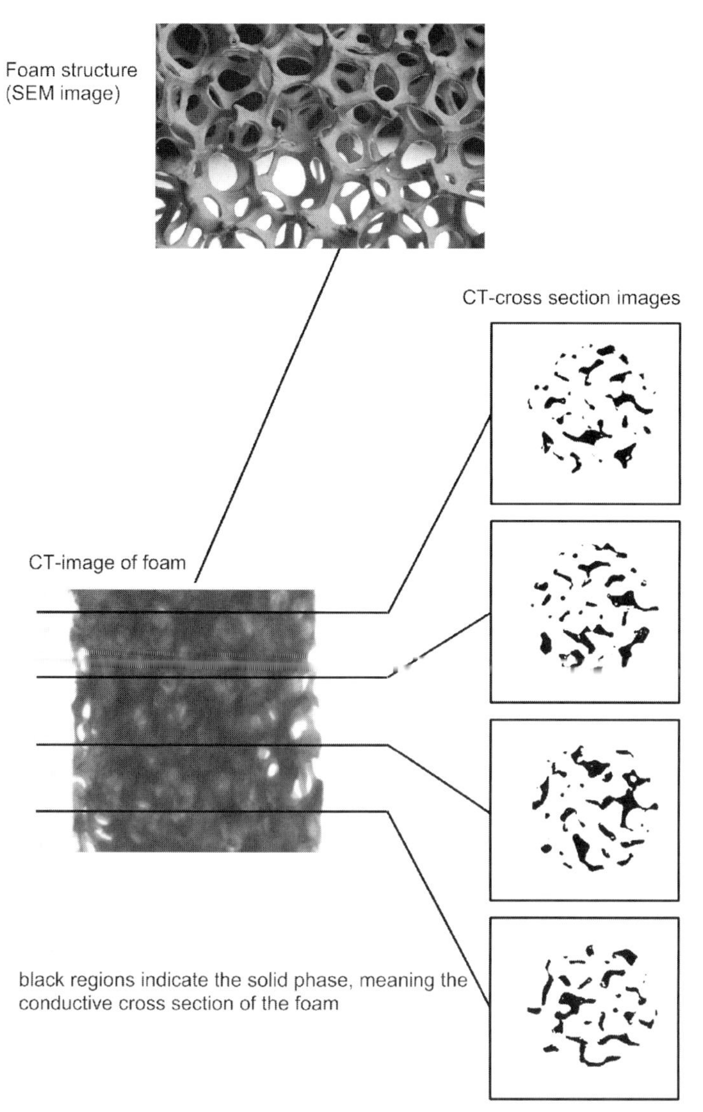

Figure 5 Illustration of the alteration of location and amount of conductive cross section of foam ceramics by computer tomography (CT), see also Chapter 3.1.

Looking at the variable cross sections it becomes obvious that a fixed factor like that used for Eq. (11) of Chapter 1.1 does not give exact conductivity values, since the effect of various cross sections is not regarded. A prediction of the actual electrical resistance needs a mathematical model which is able to take the various modifications of cross-sectional area, density of ligaments, and distribution along the electrical path into account. In Chapter 4.1 some models are given for Young moduli of foam structures which can be applied to electrical conduction as well. Additional models based on MSA, mixing, percolation, and effective-media theories have been published by a number of authors with regard to electrical conductivity [2, 8, 26].

The general effective-media equation (McLachlan) combines effective media and percolation theory, so that it is applicable for a wide range of different composite structures. Since foams are composites consisting of voids and interconnected solid phase their electrical conductivity might be described by the McLachlan equation [26, 27]:

$$(v_i) \cdot \frac{\sigma_i^{\frac{1}{s}} - \sigma_M^{\frac{1}{s}}}{\sigma_i^{\frac{1}{s}} + (\frac{1}{\Phi_c} - 1) \cdot \sigma_M^{\frac{1}{s}}} + (v_c) \cdot \frac{\sigma_c^{\frac{1}{t}} - \sigma_M^{\frac{1}{t}}}{\sigma_c^{\frac{1}{t}} + (\frac{1}{\Phi_c} - 1) \cdot \sigma_M^{\frac{1}{t}}} = 0 \tag{6}$$

where v is the volume fraction, Φ_c the critical volume fraction of conducting phase, σ the electrical conductivity, s the equation parameter 1 (insulating phase), t the equation parameter 2 (conducting phase), i the the index for insulating phase, c the index for the conducting phase, and M the index for the composite mixture.

The specific features of conductive foam ceramic allow a simplification of Eq. (6). The conductivity of the voids is almost zero in comparison to a semiconducting or conducting ceramic phase. The term of the insulating phase is reduced to Eq. (7):

$$(v_i) \cdot \frac{0 - \sigma_M^{\frac{1}{s}}}{0 + (\frac{1}{\Phi_c} - 1) \cdot \sigma_M^{\frac{1}{s}}} + (v_i) \cdot \frac{-\sigma_M^{\frac{1}{s}}}{(\frac{1}{\Phi_c} - 1) \cdot \sigma_M^{\frac{1}{s}}} = \frac{-v_i}{\frac{1}{\Phi_c} - 1} . \tag{7}$$

The modified Eq. (6) for foam ceramic is given by Eq. (8):

$$(v_c) \cdot \frac{\sigma_c^{\frac{1}{t}} - \sigma_M^{\frac{1}{t}}}{\sigma_c^{\frac{1}{t}} + (\frac{1}{\Phi_c} - 1) \cdot \sigma_M^{\frac{1}{t}}} = \frac{-v_i}{\frac{1}{\Phi_c} - 1} . \tag{8}$$

If the conductivity of the foam is to be predicted, Eq. (8) can be transformed into Eq. (9):

$$\sigma_M = \sigma_c \cdot \left[\frac{\frac{1}{\Phi_c} - 1 - \frac{v_i}{v_c}}{(\frac{1}{\Phi_c} - 1) \cdot (\frac{v_i}{v_c} + 1)} \right]^t . \tag{9}$$

Equation (9) requires the knowledge of the volume fractions of solid and void phase, the conductivity of the solid phase, and the parameters Φ_c and t. The parameter t needs to be fitted to the actual foam structure. Published values of t are in the range of 0.8–4.5 [27]. Φ_c is fairly low for foam structures in which percolation occurs as soon as any foam is consistent (< 0.05).

Figure 6 illustrates the relation of equation parameters σ_c and ν_i to the electrical conductivity of foam materials. They are the most important parameters for the alteration of electrical conductivity. The ceramic material which is used for the foam determines σ_c. The actual conductivity is determined by the volume content of voids, which is represented by ν_i. This parameter strongly alters the conductivity of foam, as it is shown by Fig. 6 (bottom). However, in contrast to conventional composites the conductivity changes moderately in almost linear relation to the content of

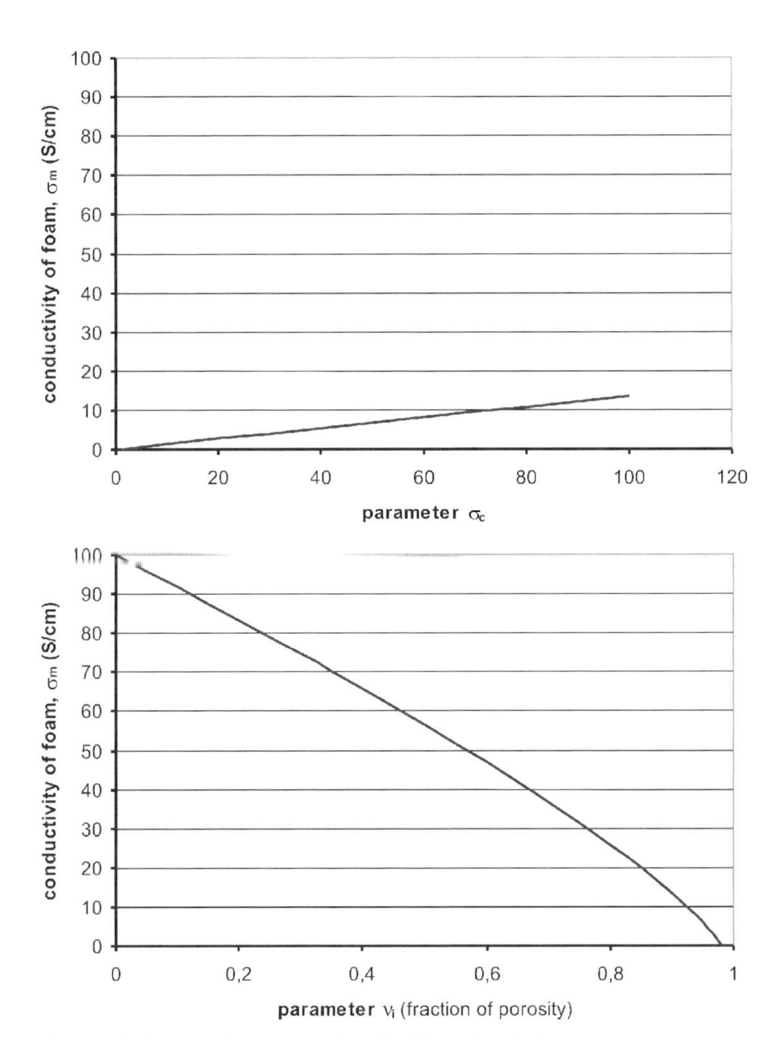

Figure 6 Influence of parameters from Eq. (9) on electrical conductivity of foam material. Top: parameter σ_c, ($\Phi_c = 0.02$; $t = 0.8$; $\nu_i = 0.9$, $\nu_c = 0.1$); Bottom: parameter ν_i ($t = 0.8$, $\Phi_c = 0.02$, ν_i= 0.9, ν_c= 0.1, and $\sigma_c = 100\ \mathrm{S\ cm^{-1}}$).

insulating phase (voids), since the percolation threshold parameter Φ_c is fairly low (about 0.02) for foam materials. The parameter t can vary in the range between 0.8 and 4.5, which does affect the foam conductivity but to a much lower degree than ν_i.

Ceramic foams have been manufactured and investigated with regard to their electrical conductivity [28]. The foams were obtained by direct foaming of silicon oxycarbide precursors. The use of the precursor alone leads to low conductivity in the range of 10^{-5} to 10^{-4} S m^{-1}. The same precursor was also investigated with various filler materials. The conductivity was increased to 30 S m^{-1} by applying fillers of about 30 wt% SiC, C, and $MoSi_2$. The use of Cu acetate and oxide can increase the electrical conductivity to about 4500 S m^{-1} even with relatively low Cu content. The direct electrical heating of precursor-derived foams is demonstrated in this paper up to 1200 °C.

The dielectric constant of insulating cellular materials is decreased by dilution of the solid volume by the gas-filled cells. This was already shown for porous ceramics, where the dielectric constant follows the rule of mixtures [8]. Exceptions exist with materials of high dielectric constant [8]. Pore-charging effects can alter the linear dependence of the dielectric constant on porosity. Additional effects of the voids on the solid phase can alter the material, and hence ε_r. The cellular structure does not influence the dielectrical properties with the same complexity as it does for the electrical conductivity. Changes in the dielectric constant of lead zirconate titanate (PZT) were reported in Ref. [29]. The foam material reaches dielectric constants in the range of 50–100, while the dense PZT material shows values above 2300.

4.4.2.4 Ceramic Fibers

Ceramic fibers have became widely used as a consequence of developments of the last few decades which have made them commercially available. Particularly electrically insulating fibers like alumina and silicate fibers have been widely used for thermal insulation and furnace construction. Electrical insulation has not been the main focus for these materials, but it is usually considered an advantage because there are no problems with touching electrical heating elements. Kamimura et al. [7] discuss the development of a Si_3N_4 fiber material which may be useful as highly temperature resistant material for electrical insulation of cables and electrical connections. More interesting here is the comparison given of electrical resistance with other fiber products made of alumina, quartz, and silicon carbide. As one would expect, woven or mat products made of the above materials show fairly high electrical resistivity at room temperature. Even the silicon carbide mat has an electrical resistivity of 10^6 Ω m. Except for the quartz fiber product all fiber tapes show almost constant electrical resistivity with increasing temperature. This behavior demonstrates that the electrical resistivity is not really dominated by the bulk ceramic material but by the resistance of the fiber-surface contacts in the mat. This is a specific effect of this class of cellular materials. The silicon carbide fiber product should show a drastic decrease in electrical resistivity if the semiconductive mechanisms becomes effective at high temperature. However, the electrical resistivity does not change. A slight increase could even be noted due to the growing thickness of the oxide layer around the silicon carbide fibers.

If a continuous fiber strand is used the effect of oxide layers does not influence electrical conduction as long as the ratio of the oxide volume is small compared to that of the conductive fiber. In this case the electrical resistance is derived from the total cross-sectional areas of the single fibers. The resistance or conductivity can be considered reciprocal to honeycomb materials, since there are long continuous fibers (tubes) and insulating space between them. Continuous silicon carbide fibers can differ in microstructure to conventional silicon carbide materials, since they are either precursor-derived [30, 31] or CVD-derived. CVD fibers can be controlled rather precisely with regard to their chemical composition and microstructure. Depending on the preferences of the application electrical properties can be tailored.

If fibers are interconnected by sintering, their structure resembles foam ceramics. But the cross-sectional area and the electrical properties are more homogeneous. An illustration of fiber contact is given in Fig. 7 (top) and shown for a Tyranno fiber mat in Fig. 7 (bottom).

Precursor-derived SiC fibers have been commercially available for more than 20 years and have been improved considerably since electron-beam curing was used for manufacture [30]. These precursor-derived fibers have rather high electrical resistance since they either include oxygen or exhibit an amorphous structure. Properties of cellular materials like tapes and networks will correspond to the fiber properties.

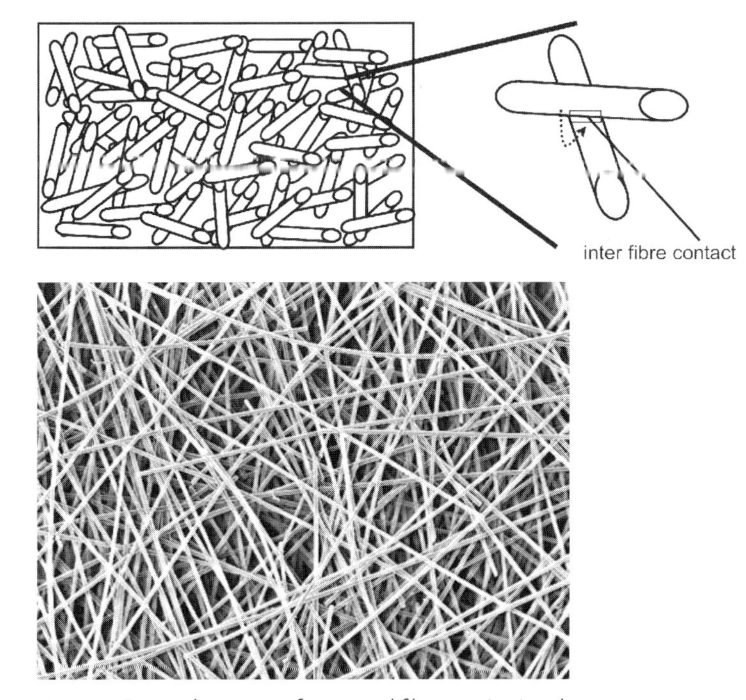

Figure 7 Principal structure of connected-fiber mat (top) and SEM image of Tyranno fiber mat (bottom).

The direction of the fiber axis, the length of the fibers, and the kind of fiber interconnections need to be considered to estimate or calculate the electrical resistance of fibrous materials. The general tendency is that a higher packing density of fibers increases the electrical conductivity and the dielectric constant of the connected fiber material. While the dielectric constant depends almost completely on the volume content of fiber material and the kind of the ceramic, the electrical conductivity is governed by further parameters such as length of fibers, interconnection of single fibers, and the fiber arrangement. Since the fibers are of uniform diameter and length the creation of a mathematical description is easier than for foam materials.

4.4.3 Electrical Applications of Cellular Ceramics

4.4.3.1 Foam Ceramic Heaters

Foam ceramics are interesting materials for electrical heating. Silicon carbide foams are available from different producers. Because of the semiconductivity of silicon carbide it can be directly heated by applying electrical power. The specific advantages by using foam ceramics as heater material are shown by Table 4.

Silicon carbide foam materials can be produced by a number of technological procedures:

- Clay bonding (CB-SiC), electrically insulating.
- Reaction sintering or silicon infiltration (Si-SiC), about 10^{-4} Ω m
- Recrystallization (R-SiC), about 10^{-1} Ω m
- Solid-phase sintering (S-SiC), about 10^{-2} Ω m
- Liquid-phase sintering (LPS-SiC), electrical properties not yet investigated

Table 4 Some advantages of ceramic foam heating elements.

Advantage	Application requirements
Low specific weight	- low-weight construction - small heat capacity - rapid heating and cooling
Large surface	- excellent heat transfer - intensive interaction with the environment
High specific strength	- use as structural element - good resistance to loading
Permeability	- heating of gases and liquids - quick cooling
No catastrophic failure	- high safety of heating device - avoidance of catastrophic breakdown - cost-effective maintenance
One-part heater	- one-step mounting of heater - low production cost

All manufacturing procedures can be used for foam production. The electrical properties are rather different for the different SiC material types with the same trends as observed for the related bulk materials (Table 1).

SiC foam ceramics for direct electrical heating have been developed and produced by Fraunhofer IKTS [32–34]. Cylindrical foam bodies made of Si-SiC (LigaFill) are used for demonstration of some principal heating features. The foam was manufactured according to the replication technique [31, 33]. Because of the good wetting of liquid silicon the small pores of the ligaments are filled with silicon during infiltration (Fig. 8). The resulting Si-SiC material contains about 20–40 wt% of free silicon. The foam cells are still open and unfilled.

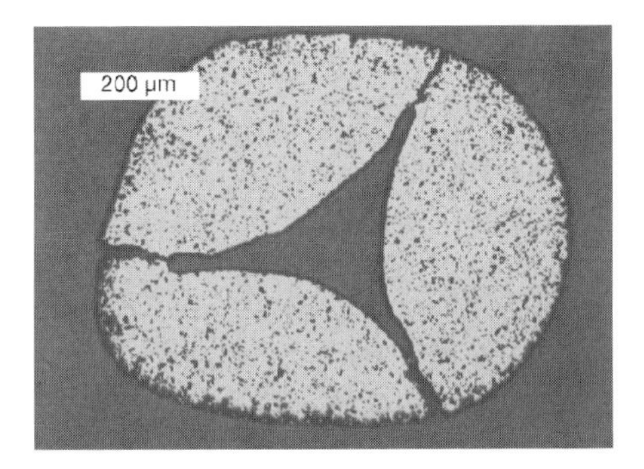

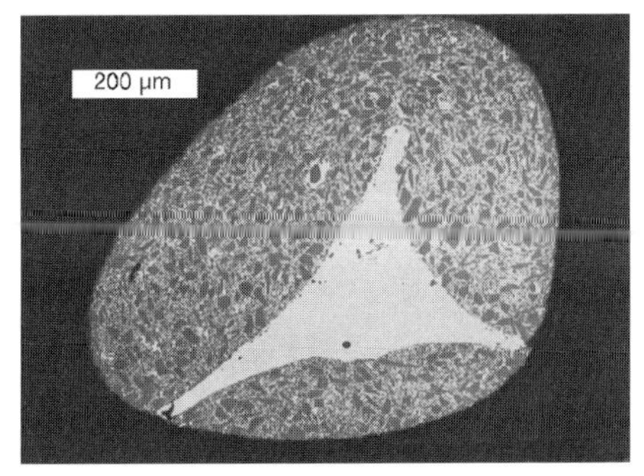

Figure 8 Micrographs of unfilled (top) and Si-filled (bottom) ligament of SiC foam.

Silicon is a better electrical conductor than silicon carbide. The rather high content of elemental silicon in combination with effects which have been discussed above means that the final LigaFill foam material exhibits outstanding electrical conductivity with an electrical resistance of less than 0.01 Ωcm [34]. Typical of bulk Si-SiC materials reach specific electrical resistance of around 0.03 Ωcm according to our experience. The LigaFill structure consists of Si-SiC coating and internal bulk silicon ligaments because of filling of the out-burned polymer ligaments.

The size of the produced foam ceramic heaters is 30 mm in diameter and a 60 mm in length. The cell size was varied from 10 to 60 ppi and the relative density was about 15%. The side ends of the cylinders were closed by disks of dense Si-SiC material. Graphite electrodes were mounted onto the Si-SiC ends to provide good electrical contact to the ceramic foam.

The foam could be heated up to 900 °C within 4 min by applying about 15 V/50 A. The inside and the middle of the sample are the hottest regions because of the temperature coefficient caused by cooling effects in the outer region of the foam. Table 4 gives some electrical data for direct electrical heating of the described samples.

Table 5 Heating behavior of Si-SiC foam ceramic in dependence on ppi and silicon content.

Cell size/ppi	Mass/g	Si/wt%	Voltage/V	Power/W	R/Ω (20 °C)	R/Ω (900 °C)	$T/$°C
10	30	25	16	800	0.25	0.32	900
30	30	30	15	850	0.23	0.27	900
60	30	35	14	880	0.21	0.21	900

The observations show that the temperature does not have a drastic effect on the electrical resistance of the ceramic foam. Given the characteristics for bulk Si-SiC materials, it conforms to the expectation of a slight positive temperature coefficient of the electrical resistance.

The size of cells affects the electrical resistance indirectly by causing different infiltration behavior of the ligaments and changing the composition of the materials. The small-diameter ligaments of fine foam materials are infiltrated by silicon more efficiently than large-diameter ligaments because of the resulting capillary force, which promotes the infiltration process. Consequently the ratio of silicon to silicon carbide is greater for fine foams, which results in lower electrical resistance.

A comparison of the heating procedure of foam ceramics and bulk ceramics was performed in addition to the above experiments. An Si-SiC sample with the same effective cross section and hence similar electrical resistance was chosen. Heating was quite different compared to the foam material. A significant decrease in the original resistance is observed with temperature increase. When the sample reached about 550 °C it cracked due to high energy load, so that a further heating up to 900 °C was impossible.

This experiment illuminates the outstanding performance of foam ceramics with regard to heating rate (ca. 200 K min^{-1} was tested) and thermal loading. Particularly the high ratio of surface to mass results in the above advantages, which are directly related to the cellular structure of the foam ceramics.

4.4.3.2
Electrically Conductive Honeycombs

The manufacture of electrical conductive honeycomb materials from a porcelain ceramic and coal tar pitch is reported by Alcaniz-Monge et al. [6]. The honeycomb ceramic is impregnated by the molten carbon precursor and then pyrolyzed up to 1073 or 1273 K under nitrogen atmosphere. A layer of carbon is formed which enables electrical conduction. The conductivity can be controlled by the temperature of carbonization, as demonstrated by the experimental results [1073 K (800 °C) –

$30\,(\Omega m)^{-1}$ and 1273 K (1000 °C) – $160\,(\Omega m)^{-1}$]. The aim of this development is the manufacture of electrodes for electrochemical applications.

A similar approach is reported in Ref. [4]. Cordierite honeycombs are impregnated with carbonaceous precursors. The electrical conductivity of the impregnated honeycomb reaches $0.31–110\,(\Omega m)^{-1}$, depending on precursor and pyrolysis temperature.

Electrical conductivity can be achieved by using conductive ceramic materials like silicon carbide. This is demonstrated in Refs. [35–37], which describe silicon carbide honeycomb materials which are used as soot filters for diesel engines. Regeneration can be performed by direct electrical heating of the filter. The electrical assembly and the resistance are shown in Ref. [37]. Heating ignites the settled soot particles at about 650 °C and regenerates the filter. Regeneration takes approximately 4 min and consumes 1.5 kW (24 V).

4.4.4 Summary

The electrical properties of cellular ceramic materials have not been the main focus of the development so far. Nevertheless, there are some specific effects which are unique to these materials.

The electrical behavior of cellular materials is determined by the type of ceramic material, the amount of pores/cells, the specific pore features, and the general cellular structure. The dielectric properties mainly depend on the volume fraction of pores and cells and the type of ceramic material. This causes exclusively a decrease in the dielectric constant in comparison to dense materials. The electrical conduction of cellular materials can be changed by the alteration of cross-sectional area, shape and size of cells and pores, surface composition, grain boundary character, and/or electronic properties of the material itself. Although this behavior is more complex the main trend leads to a decrease of the electrical conductivity of the cellular material in comparison to a dense material of the same kind. Because of the character of the used manufacturing procedures cellular ceramic material are different to the well-known bulk ceramic materials with regard to electrical properties.

Models of electric properties are related to models of other properties like Young's modulus or thermal conduction, which are based on more extensive work than those of electrical phenomena. Since the manufacture of cellular materials often differs from conventional ceramics (foam, connected fibers, etc.) the material properties have to be considered specifically for modeling.

Honeycombs, foams, and biomimetic materials can be used for numerous specific applications which include electrical features. Some examples were given in this chapter.

Acknowledgement

We acknowledge R.W. Rice for fruitful and critical discussion during the reviewing process, which was a valuable contribution to this chapter.

References

1 Gibson, L.J., Ashby, M.F. (Eds.), *Cellular Solids*, Cambridge University Press, 1999, pp. 283–309.
2 Kovacik, J., Simancik, F., *Scripta Mater.* **1998**, *39*, 239–246.
3 Tur, T.M., Pak, V.N., *Russ. J. Appl. Chem.*, **1999**, *72*, 1907–1910.
4 Valdes-Solis, T., Marban, G., Fuertes, A.B., *Microporous Mesoporous Mater.*, **2001**, *43*, 113–126.
5 Bardhan, P., *Curr. Opin. Solid State Mater. Sci.* *2*, **1997**, 577–583.
6 Alcaniz-Monge, J.A., Cazorla-Amoros, D., Linares-Solano, A., Morallon, E., Vazquez, J.L., *Carbon*, **1998**, *36*, 1003–1009.
7 Kaminmura, S., Seguchi, T., Okamura, K., *Radiat. Phys. Chem.*, **1999**, *54*, 575–581.
8 Rice, R.W., *Porosity of Ceramics*, Marcel Dekker, New York, 1998
9 www.keramverband.de/brevier/eigenschaften/daten.html
10 G. Petzow, M. Herrmann, Silicon Nitride Ceramics, *High Performance Non-Oxide Ceramics II, Structure and Bonding*, Vol. 102, Jansen, M. (Ed.), Springer Verlag, Berlin, Heidelberg, 2002, pp. 47–167
11 Kirschstein, G., Koschel, D. (Eds.), *Gmelin Handbook of Inorganic Chemistry, Si (B2); SiC part1*, Springer Verlag, Berlin, Heidelberg, New York, Tokyo, 1984.
12 Adler, J., Ihle, J., DE-Patent Appl., Akz. 10207860, 2002.
13 Martin, H.-P., Adler, J., "Design of Electrical and Thermal Properties for Liquid Phase Sintered Silicon Carbide, Materials Week 2001, Symposium K5 Multifunctional Ceramics, München 2001.
14 J. Ihle, J. Adler, M. Herrmann, H.-P. Martin, DE-Patent Appl., Akz. 10348819.7, 2003.
15 Carty, W.M., Lednor, P.W., *Curr. Opin. Solid State Mater. Sci.*, **1996**, *1*, 88–95
16 Gibson, L.J., Cellular Solids, *MRS Bull.*, **2003**, 270–271.
17 Sieber, H., Kaindl, A., Schwarze,D., Werner, J.-P., Greil, P., *cfi-Berichte DKG*, **2000**, *77*, 21–23.
18 Sieber, H., Hoffmann, Ch., Kaindl, A., Greil, P., *Adv. Eng. Mater.*, **2000**, *2*, 105–109.
19 Greil, P., Lifka, Th., Kaindl, A., *J. Eur. Ceram. Soc.*, **1998**, *18*, 1961–1973.
20 Vogli, E., Mukerji, J., Hoffman, Ch., Kldany, R., Sieber, H., Greil, P., *J. Am. Ceram. Soc.*, **2001**, *84*, 1236–1240.
21 Montanaro, L., Jorand, Y., Fantozzi, G., Negro, A., *J. Eur. Ceram. Soc.*, **1998**, *18*, 1339–1350.
22 Colombo, P., *Key Eng. Mater.*, **2002**, *206–213*, 1913–1918.
23 Green, D.J., Colombo, P., *MRS Bull.* **2003**, *4*, 296–300.
24 Will, J., Gauckler, L.J., Ceramic Foams as Current Collectors in Solid Oxide Fuel Cells (SOFC): Electrical Conductivity and Mechanical Behaviour, Proc. Fifth Intern. Symp. on Solid Oxide Fuel Cells (SOFC-V), Aachen, Germany, June 2–5, 1997, pp. 757–764.
25 Data sheet of ERG Materials and Aerospace Corporation, Oakland (CA), www.ergaerospace.com.
26 McLachlan, D.S., Blaszkiewicz, M., Newnham, R.E., *J. Am. Ceram. Soc.*, **1990**, *73*, 2187–2203
27 Runyan, J., Gerhardt, R.A., *J. Am. Ceram. Soc.*, **2001**, *84*, 1490–1496.
28 Colombo, P., Gambaryan-Roisman, T., Scheffler, M., Buhler, P., Greil, P., *J. Am. Ceram. Soc.*, **2001**, *84*, 2265–2268.
29 Creedon, M.J., Schulze, W.A., *Ferroelectrics*, **1994**, *153*, 333–339.
30 Yajima, S., Hayashi,J., Omori, M., Okamura, K., *Nature*, **1976**, *261*, 683–685
31 Takeda, M., Imai, Y., Ichikawa, H., Ishikawa, T., Kasai, N., Seguchi, T., Okamura, K., *Ceram. Eng. Sci. Proc.*, **1992**, *13* (7–8), 209–217.
32 J. Adler, M. Teichgräber, G. Standke, DE Patent 19621638, 2002.
33 Adler, J., Standke, G., *Keram. Z.*, **2003**, *55*, 694–703.
34 Adler, J., Standke, G., *Keram. Z.*, **2003**, *55*, 786–792.
35 Schäfer, W., Best, W., Schumacher, U., DE Patent 19809976 A1, 1999
36 Maier, H.R., Schumacher, U., Best, W., Schäfer, W., Europ. Patent 0796830 A1, 1996
37 Maier, H.R., Best, W., Schumacher, U., Schäfer, W., *cfi/Ber. DKG*, **1998**, *75*, 25–29

4.5
Acoustic Properties

Iain D. J. Dupère, Tian J. Lu, and Ann P. Dowling

4.5.1
Introduction

Many open-celled cellular ceramics have small cell sizes making them well suited for absorbing incident sound waves. Consequently, knowledge of the acoustic behavior of cellular ceramics is extremely important for various applications. The acoustic properties can be modeled in a number of different ways or indeed measured experimentally. A brief overview of the models and experimental techniques is given in this chapter. In particular, the acoustic behavior depends on the cell structure and is relatively independent of the material used. For example, porous metals behave in a very similar way to porous ceramics with the same cell structure. While there is not a huge amount of published data on the measured characteristics of cellular ceramics, we give an example of the acoustic behavior of a porous metal showing both a theoretical model and experimental measurements.

To consider the effect that cellular ceramics have on the propagation of acoustic waves, we begin by considering the propagation of acoustic waves in inviscid conditions. We then consider how this is modified by cellular ceramics. Readers familiar with inviscid acoustic propagation are directed to Section 4.5.6.

4.5.2
Acoustic Propagation

4.5.2.1
Linearized Equations of Motion

The amplitudes of acoustic waves are typically extremely small, enabling us to linearize the flow parameters ϕ to give

$$\phi = \overline{\phi} + \phi'$$

where ϕ represents any of the usual flow parameters such as pressure p, density ρ, temperature T, or velocity $\vec{u}$. The no mean flow mass conservation (continuity equation) and inviscid momentum equations then become

Cellular Ceramics: Structure, Manufacturing, Properties and Applications.
Michael Scheffler, Paolo Colombo (Eds.)

ISBN: 3-527-31320-6

$$\frac{\partial \rho}{\partial t} + \bar{\rho}\vec{\nabla} \cdot \vec{u}' = 0, \qquad \frac{\partial u}{\partial t} = -\frac{\vec{\nabla} p'}{\bar{\rho}}.$$

These equations give us the starting point for inviscid sound propagation in the absence of mean flow. In later sections we describe how the equations are changed by the presence of a cellular material.

4.5.2.2
Wave Equation

If we take the time derivative of the continuity equation we obtain

$$\frac{\partial^2 \rho'}{\partial t^2} + \bar{\rho}\frac{\partial(\vec{\nabla} \cdot \vec{u}')}{\partial t} = 0$$

Since this involves the time derivative of the velocity, it can be combined with the divergence of the inviscid momentum equation to give

$$\frac{\partial^2 \rho'}{\partial t^2} - \nabla^2 p' = 0$$

For perfect gases, the density and pressure fluctuations are linearly related, so we choose to write

$$p' = \bar{c}^2 \rho'$$

where $\bar{c}$ is a constant which will be determined later. Substituting for ρ' we obtain the following second-order differential equation:

$$\left(\nabla^2 p'\right) - \left(\frac{1}{\bar{c}^2}\right)\left(\frac{\partial^2 p'}{\partial t^2}\right) = 0.$$

Since the thermodynamic fluctuations p', ρ' and T' are linearly related and since the velocity must have the same time and spatial dependence, general low-amplitude waves satisfy the same differential equation:

$$\left(\nabla^2 \phi^2\right) - \left(\frac{1}{\bar{c}^2}\right)\left(\frac{\partial^2 \phi^2}{\partial t^2}\right) = 0,$$

i.e.

$$\left(\frac{\partial^2 \phi'}{\partial x^2}\right) + \left(\frac{\partial^2 \phi'}{\partial y^2}\right) + \left(\frac{\partial^2 \phi'}{\partial z^2}\right) - \left(\frac{1}{\bar{c}^2}\right)\left(\frac{\partial^2 \phi'}{\partial t^2}\right) = 0$$

where ϕ' is the acoustic variable. This equation is known as the wave equation because its solutions are waves. In an inviscid pipe at low frequency, the sounds waves will be planar, that is, one-dimensional. This equation tells us that one-dimensional waves are described by

$$p' = \hat{p}_1 e^{i\omega(t - x/\bar{c})} + \hat{p}_2 e^{i\omega(t + x/\bar{c})}$$

where $\hat{p}_1$ and $\hat{p}_2$ are the amplitudes of the forward and backward traveling waves, respectively, and ω is the angular frequency. We note that the constant $\bar{c}$, introduced earlier, is the propagation speed. These pressure waves would have associated velocity and density waves:

$$u' = \hat{u}_1 e^{i\omega(t-x/\bar{c})} + \hat{u}_2 e^{i\omega(t+x/\bar{c})},$$

$$\rho' = \hat{\rho}_1 e^{i\omega(t-x/\bar{c})} + \hat{\rho}_2 e^{i\omega(t+x/\bar{c})}.$$

The above equations represent one-dimensional acoustic waves propagating in the x-direction. However, similar relations exist when the waves propagate in two dimensions or radially.

4.5.2.3
Relationships between Acoustic Parameters under Inviscid Conditions

The velocity fluctuations are related to the pressure fluctuations via the inviscid momentum equation, which for one-dimensional waves in the positive x-direction gives

$$p' = \bar{\rho}\bar{c}u'$$

while for a wave propagating in the negative x-direction we have:

$$p' = -\bar{\rho}\bar{c}u'.$$

Thus.

$$p' = \bar{\rho}\bar{c}\hat{u}_1 e^{i\omega(t-x/\bar{c})} - \bar{\rho}\bar{c}\hat{u}_2 e^{i\omega(t+x/\bar{c})}$$

The density fluctuations are related to the linearized continuity equation:

$$\frac{\partial \rho}{\partial t} + \bar{\rho}\frac{\partial u'}{\partial x} = 0,$$

$$\rho' = \pm\frac{\bar{\rho}}{\bar{c}^2}u' = \frac{1}{\bar{c}^2}p' \quad \text{as before.}$$

In free space, the speed of propagation $\bar{c}$ is sufficiently high that there is insufficient time for heat flow giving isentropic flow, and thus the pressure fluctuation is related to the density fluctuation via

$$p' = \gamma R\bar{T}\rho', \quad \bar{c} = \sqrt{\gamma R\bar{T}}$$

where γ is the ratio of specific heat capacities (c_p/c_v) and R is the gas constant.

The above relations are strictly valid only when the effects of viscosity can be neglected. In practice the effects of viscosity on the propagation of acoustic waves is usually negligible in most situations, and thus they are commonly used to describe the propagation of acoustic waves. However, cellular materials are an exception to this because the pore sizes are usually so small that viscous effects must be taken into account. This modifies the momentum equation, which should now include the effect of viscous drag, and thus the relationship between the pressure and the velocity is altered in two ways. First, the waves propagate more slowly, which changes the effective speed of sound in the material. Second, it introduces some damping which converts the energy in the acoustic wave to heat. To allow for this, acoustic properties of the cellular ceramics must be considered. This can be done either theoretically or empirically.

4.5.2.4
Acoustic Energy

Acoustic waves carry acoustic energy. In the absence of a mean flow, the acoustic intensity, I_{ac}, that is, the rate at which this acoustic energy crosses a unit area, is given by:

$$I_{ac} = p'u'$$

Unsurprisingly, the sign of this quantity depends upon the direction of propagation. This quantity is proportional to p'^2 (because u' is itself proportional to p'). In the presence of a mean flow of low Mach number M, the expression for the acoustic intensity is modified [1] to give

$$I_{ac} = p'u'(1+M)^2$$

for waves propagating in the same direction as the mean flow, and

$$I_{ac} = p'u'(1-M)^2$$

for waves propagating in the opposite direction to the mean flow.

4.5.3
Acoustic Properties

4.5.3.1
Acoustic Impedance and Admittance

The *acoustic impedance* Z_{ac} of a fluid is defined as:

$$Z_{ac} = \frac{p'}{u'} = R_{ac} + iX_{ac}.$$

Z_{ac} is a complex quantity and is analogous to electrical impedance with p' analogous to electrical potential and u' analogous to electrical current. The real part describes the acoustic resistance and is related to the propagation speed of the acoustic waves. In air with no viscous effects it is equal to $\bar{\rho}\bar{c}$. The imaginary part describes the acoustic reactance. This is related to a net damping or generation of acoustic energy and is thus zero in air with no viscous effects. The reflection and transmission of sound across a boundary between two different fluids is directly related to their impedances, since the pressure and mass flow rate must be conserved across the boundary and since the pressure is related to the velocity through the impedance.

For a material the acoustic impedance can be defined in a number of different ways. Figure 1 shows a block of material in a pipe with an incident wave I propagating from left to right towards the material, and a reflected wave *R*. Waves *A* and *B* are formed on the other side of the block of material by a combination of transmission and reflection (including reflection from the closed end of the pipe). The acoustic pressure on the incident face is p_1, while the acoustic pressure on the transmission side is p_2. The acoustic particle velocity on the incident side is u_1. One useful acoustic property of such a material is the surface impedance, Z_{surf}, defined as:

$$Z_{\text{surf}} = \frac{p_1}{u_1}.$$

Z_{surf} depends upon the properties of the material, the thickness, and the location. It clearly has the advantage of describing the complete situation. However, it is not a function of the material alone. An alternative would be to consider the impedance of the material itself Z_{mat}, defined as:

$$Z_{\text{mat}} = \frac{p_1 - p_2}{u_1}.$$

Unlike Z_{surf}, Z_{mat} is independent of the external acoustic environment, which must be described separately. Z_{mat} is a property of the material as a whole and depends on both the fundamental properties of the material and its dimensions. The fundamental properties might be described using a material property impedance Z_{prop}, defined as:

$$Z_{\text{prop}} = \left(\frac{\bar{c}}{\omega}\right)\left(\frac{\frac{\partial p_1}{\partial x}}{u_1}\right)$$

where $\partial p_1/\partial x$ is the acoustic pressure gradient at the incident face within the material. The factor $\bar{c}/\omega$ is included to keep the dimensions the same.

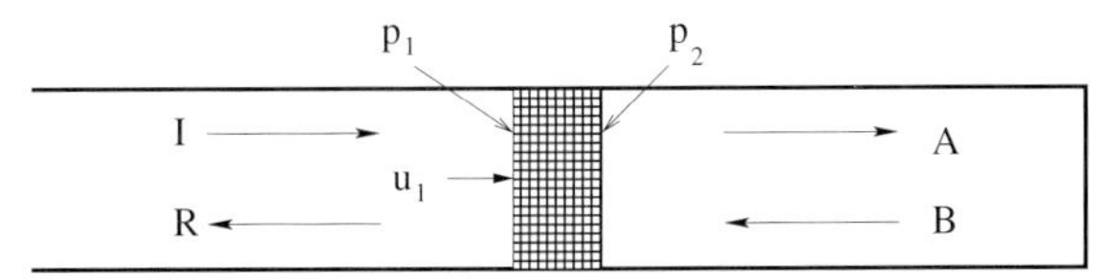

Fig. 1 The acoustic impedance of a material.

All three of these definitions for the acoustic impedance are, in general, functions of frequency.

The *acoustic admittance* is favored by some authors and is simply the reciprocal of the acoustic impedance.

4.5.3.2
Acoustic Wavenumber

The acoustic wavenumber k is defined as:

$$k = \frac{\omega}{c}$$

where c is the propagation speed. The wavenumber describes the propagation of the sound waves. If the propagation speed c and hence k is allowed to be complex, then the waves will decay or grow with distance.

4.5.3.3
Reflection Coefficient, Transmission Coefficient, and Transmission Loss

In many situations, it is useful to quantify the reflection from an end or material. This is usually done by using a reflection coefficient R_{refl}, defined (for the situation described in Fig. 1) as:

$$R_{\text{refl}} = \frac{R}{I}.$$

R_{refl} is a complex quantity. The waves in this definition can be referenced to any fixed point, but this is usually chosen to be the face of the material or the end in question. For example, for a closed end with waves referenced from the end, $R_{\text{refl}} = 1$, whereas for an open end R_{refl} is typically -1. For a material, R_{refl} is a function of frequency.

Transmission of acoustic waves can be described either by the transmission coefficient T_{trans}, or with a transmission loss T_{loss}. The transmission coefficient is defined, for the situation shown in Fig. 2, as:

$$T_{\text{trans}} = \frac{T}{I}$$

Like the reflection coefficient, the waves can be referenced to any location, but are usually referenced to the centre of the material. It, too, is a function of frequency. The transmission loss T_{loss} is defined as:

$$T_{\text{loss}} = 20 \log_{10}\left(\frac{T}{I}\right)$$

and is measured in decibels (dB).

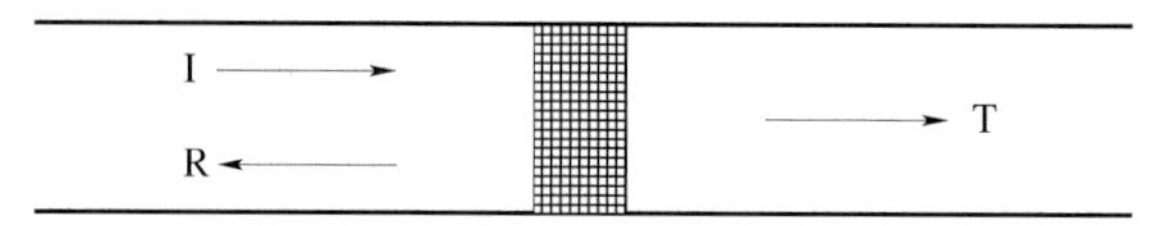

Fig. 2 Transmission.

4.5.3.4
Absorption Coefficient

Although the reflection and transmission coefficients provide useful information, it should be noted that small transmission or small reflection can be obtained without damping any acoustic energy at all. It is therefore also useful to quantify the absorption of the acoustic energy. This is usually done with the absorption coefficient Δ which is defined as the percentage of acoustic incident acoustic energy which is absorbed. For the situation shown in Fig. 2, which effectively describes an infinite pipe, this gives a unique definition:

$$\Delta = \frac{|I|^2 - |R|^2 - |T|^2}{|I|^2} = 1 - \frac{|R|^2 + |T|^2}{|I|^2}.$$

In practice, however, there is usually a second incident wave such as wave B in Fig. 1. In this situation, there are two possible ways to define the absorption coefficient: the absorption coefficient for the material alone and the absorption coefficient for the system as a whole.

For the material alone, there are two incident waves, I and B, and two waves propagating away from the material, R and A. The absorption coefficient is then defined as:

$$\Delta = 1 - \frac{|R|^2 + |A|^2}{|I|^2 + |B|^2}.$$

For the system as a whole, waves A and B are internal and so the absorption coefficient is defined as:

$$\Delta = 1 - \frac{|R|^2}{|I|^2} = 1 - |R_{refl}|^2.$$

4.5.4
Experimental Techniques

Acoustic properties of materials are usually measured by a moving-microphone technique or the two- (or four-) microphone technique. The term "impedance tube" is used variously to describe either the moving-microphone method or both methods. To avoid this confusion, we use the term "moving microphone". In both techniques the acoustical waves in the pipe are deduced by measuring the sound field at two points.

4.5.4.1
Moving-Microphone Technique

The moving-microphone technique is illustrated in Fig. 3. Here the material sample is mounted in a rigid tube. The tube is sealed at one end with a solid end leaving a cavity of length l behind the sample. On the other side of the sample the tube is excited over a range of frequencies by a loudspeaker. A moveable microphone is also placed in the tube on this side. It is moved along the tube and measurements of the amplitude and position (x_1 in Fig. 3) are taken at the points of the first maximum and first minimum in pressure amplitude. If the sample does not absorb, then a standing wave will be set up with zero amplitude at the first minimum because the forward and backward traveling waves have the same amplitudes. When the sample absorbs, however, the forward and backward traveling waves will have different magnitudes. Measuring the ratio between the amplitude at the first minimum and that at the first maximum provides enough information to decompose the sound field into the forward and backward sound waves. By measuring the distance to the first minimum, the phase between these waves can also be determined. These waves can then be used to find the surface impedance. Since the surface impedance depends on the size of the cavity l measurements must be taken with two different cavity lengths to completely describe the acoustic properties of the material. It is important that the sound waves are plane waves, so different diameter tubes should be used for different frequency ranges.

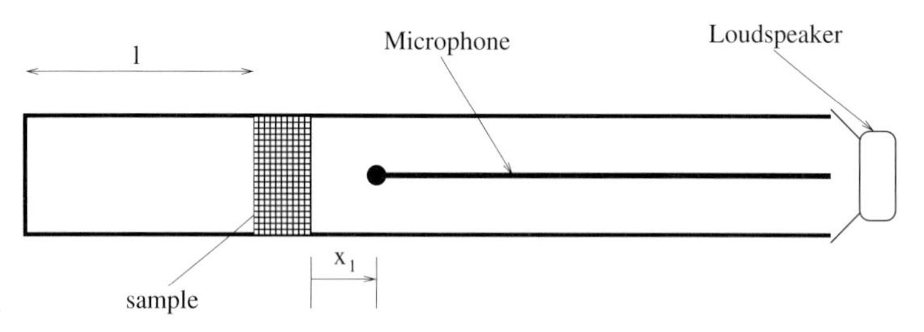

Fig. 3 Moving-microphone technique.

4.5.4.2
Two- and Four-Microphone Techniques

The two-microphone technique [2] is illustrated in Fig. 4. This is similar to the moving-microphone technique, except that here two microphones are used and located at fixed positions. Simultaneous measurements are then taken of the amplitude of each microphone and their relative phases. This also provides enough information to deduce the sound field and hence the surface impedance. As for the moving-microphone method, the material impedance can be calculated if measurements are made with two cavity lengths. However, another possibility is to take measurements on either side of the sample by using four microphones. Since this enables the user to calculate the waves on either side of the sample, this negates the need to take measurements with two cavity lengths. The material impedance Z_{mat}, which does not depend upon this length, can then be calculated directly. This is called the four-microphone technique by some authors, but since it essentially involves applying the two-microphone technique on either side of the sample, others refer to it as the two-microphone technique. As for the moving-microphone method, the sound waves must be planar, so different diameter tubes should be used for high-frequency measurements than for low-frequency measurements. An additional constraint for the two-microphone technique is that the microphones should be less than half a wavelength apart to prevent the matrix from being singular.

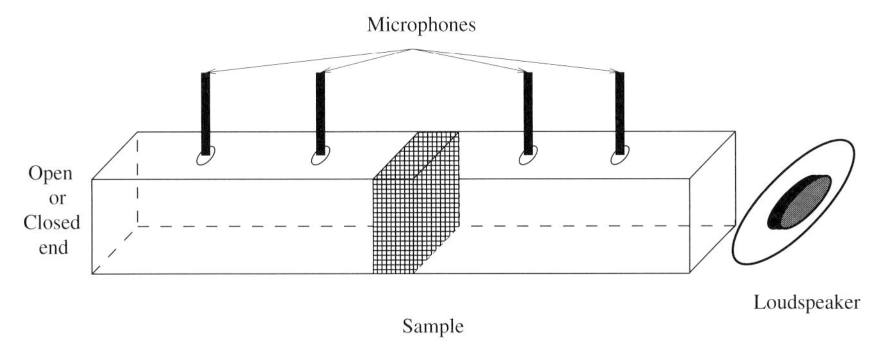

Fig. 4 Two-microphone technique.

The two-microphone technique has the disadvantage that the microphones must be calibrated relative to each other to ensure accurate measurements of the phase difference. However, it has the considerable advantage that the positions of the microphones are fixed and so can be measured more accurately.

4.5.5
Empirical Models

The characteristic impedance and the wavenumber of a material are convenient ways to characterize materials such as cellular ceramics. Delany and Bazley [3] showed empirically that these parameters are functions of the dimensionless parameter $\rho\omega/\sigma$, where σ is the static flow resistance. The static flow resistance is often obtained from Darcy's experimental law:

$$\bar{u} = C\frac{\partial \bar{p}}{\partial x}.$$

They suggest that these parameters be expressed as power-law functions derived empirically. Stinson and Champoux [4] consider the problem of rectangular and triangular pores and compare the theoretical predictions based upon Biot's method with the empirical correlations of Delany and Bazley [3] with reasonable agreement. Design charts based on this approach are provided by Bies and Hansen [5]. Experimental results for the acoustic properties of a semi-open metal foam are given by Lu et al. [6].

4.5.6
Theoretical Models

In cellular materials, the effects of viscosity become important, and so the relationship between the pressure fluctuation p' and the particle velocity fluctuation u' is altered. This is true both when the propagation is along the channel formed between the material elements and when it is normal to them. This alters the acoustic behavior considerably by introducing damping, changing the mean propagation speed, and changing the surface impedance. Some authors have modeled this by introducing effective physical parameters, such as a complex density, while others have gone back to the equations of motion including the viscous term, which they then solve. These techniques are reviewed in this section.

4.5.6.1
Viscous Attenuation in Channels (Rayleigh's Model)

A simple model of a cellular ceramic, favored by many authors, is to assume that the ceramic creates small channels through which the sound waves can propagate.

The effect of viscosity on the propagation of sound was considered in detail by Lord Rayleigh [7], who considered the problem of plane sound waves propagating in free space above an infinite plane surface. By solving the one-dimensional Navier–-Stokes equation for a harmonic wave with zero tangential velocity on the plane surface, he concluded that the coefficient of decay α is given by:

$$\alpha = \frac{8\pi^2\mu}{3\lambda^2\bar{\rho}\bar{c}}$$

where μ is the coefficient of viscosity and λ is the acoustic wavelength in meters.

Thus the sound must propagate over vast distances in free space above a plane for the viscous attenuation to be nonnegligible. A similar conclusion was reached by Lighthill [8–10]. When the propagation is through very narrow tubes, such as in some porous materials, the effect of the viscosity can be much more significant. Rayleigh extended the analysis for propagation over an infinite plane to consider the case of a narrow rigid tube in which the thickness of the boundary layer was nevertheless considered small in comparison with the diameter of the tube, so that the waves could still be treated as one-dimensional. This results in a modified wave equation for a narrow circular tube of the form [7]:

$$\frac{\partial^2 X}{\partial t^2}\left(1+\frac{2}{r}\sqrt{\frac{\nu}{2\omega}}\right)+\frac{2}{r}\sqrt{\omega\nu}\frac{\partial X}{\partial t}-\bar{c}^2\frac{\partial^2 X}{\partial x^2}=0$$

where X is the local volume such that the velocity is $\frac{1}{\pi r^2}\frac{\partial X}{\partial t}$; r the radius of the circular tube, ν the kinematic viscosity, and ω the angular frequency. The local volume X is used here simply for convenience, since it results in a wavelike equation. From this expression it is seen that the viscosity reduces the speed of the waves to give an approximate value of:

$$\bar{c}\left(1-\frac{1}{r}\sqrt{\frac{\nu}{2\omega}}\right)$$

which was also derived by Helmholtz [11]. This approach forms the basis of many theoretical analyses of porous materials. By far the most common configuration considered is propagation along hollow cylindrical tubes, which is considered variously by Arnott [12], Biot [13,14], Lambert [15–19], Stinson and Champoux [4], Selamet et al. [20], and Hovem and Ingram [21]. The technique is, however, general, and can be used to model the behavior of sound waves propagating along the outside of material elements in a cellular material.

4.5.6.2
Acoustic Damping by an Array of Elements Perpendicular to the Propagation Direction

Real porous materials contain elements which are not only parallel to the direction of propagation, but also elements which are perpendicular to the propagation direction. Since in most cases the magnitude of the acoustic velocity is very small, and the dimensions of the material elements are also generally very small, the Reynolds number associated with these cross-elements is on the order of unity, as pointed out by Attenborough [22]. In this regime the flow is very viscous and was first considered by Stokes [23], who derived a solution in the limit that the inertial terms in the momentum equation are negligible. For a sphere, an exact solution exists for which the drag coefficient C_D is related to the the Reynolds number Re by $C_D = 24/Re$, which is valid for steady flow with Reynolds numbers less than about 10. The drag observed due to a combination of spheres is discussed by Datta and Dea [24]. Urick [25] uses the Stokes–Lamb model as the basis for a model of the attenuation of sound by a spherical particle. He shows that attenuation derived by Lamb for the dissipation of sound by fog [26] can be attributed to a combination of viscous loss described by Stokes flow and scattering by the spherical object itself [22]. This model was extended by Umnova and Attenborough [27], who consider the effect of neighboring particles in acoustic propagation through suspensions by including crude outer boundary conditions to describe the interaction between the particles. For flow perpendicular to the axis of a cylinder, Stokes noted that no solution exists which satisfies all the boundary conditions. However, an approximate solution was obtained by Lamb [28] by partially including the inertial terms. In the limit of low Reynolds number this reduces to $C_D = 15.98/Re$. This was extended by Cheung et al. [28] to consider the interaction caused by two closely spaced cylinders, and by Deo [29] to describe the effect of multiple cylinders. While some authors have modeled the acoustic behavior of the elements by introducing a time-dependent drag using the formulas derived by Stokes and Lamb, it is also possible (see Sect. 4.5.6.5) to extend Stokes' work by including the unsteady term, $\partial\vec{u}'/\partial t$, still ignoring the inertial term. When this term is included a solution exists for both propagation past a sphere and past a cylinder without having to include the inertial term at all.

4.5.6.3
Generalized Models

A general model was given by Attenborough [22], who described the flow via the modified one-dimensional momentum and mass continuity equations:

$$\bar{\rho}\frac{\rho_{\text{eff}}}{\eta}\frac{\partial \vec{u}}{\partial t} + r_{\text{eff}}\,\vec{u} = -\vec{\nabla}p$$

and

$$\eta\frac{\partial \rho}{\partial t} + \vec{\nabla}\cdot(\rho\vec{u}) = 0$$

where ρ_{eff} is the effective density, η the porosity, r_{eff} the effective resistivity, and u is taken to be the average particle velocity. The porosity η used here is a geometrical factor and should include only nonsolid volume for which there is a path for the fluid to pass through the solid. The effective resistivity and density r_{eff} and ρ_{eff} are phenomenological parameters. Heat conduction from the walls can also be included by using the relationship:

$$\bar{c}^2\rho' = \gamma_{\text{eff}}\,p'$$

where γ_{eff} is the effective ratio of specific heats, which describes thermal conduction and has a value of 1 for adiabatic propagation and 1.4 for isothermal propagation. In practice, a value between these two extremes should be taken.

4.5.6.4
Complex Viscosity and Complex Density Models

For the simple case of a narrow cylindrical pore, Rayleigh's approach has a simple form. Beginning with the axisymmetric Navier–Stokes equation [30]:

$$\frac{\partial u'}{\partial t'} = -\frac{1}{\bar{\rho}}\frac{\partial p}{\partial z} + \nu\nabla^2 u'$$

which for harmonic waves $u' = \hat{u}e^{-i\omega t}$ gives:

$$i\omega u' = -\frac{1}{\bar{\rho}}\frac{\partial p}{\partial z} + \nu\left[\frac{\partial^2 u'}{\partial z^2} + \frac{1}{r}\frac{\partial}{\partial r}\left(r\frac{\partial u'}{\partial r}\right)\right].$$

The second term on the right-hand side is typically small and, when neglected, gives the solution (when the boundary conditions of zero velocity on the pore walls and zero shear stress at $r = 0$ are applied) [22]:

$$u' = \frac{-1}{i\omega\bar{\rho}}\left(\frac{\partial p}{\partial z}\right)\left[1 - \frac{J_0((i\omega/\nu)^{1/2}r)}{J_0((i\omega/\nu)^{1/2}r_0)}\right]$$

where r is the local radius, r_0 is the radius of the cylindrical pore, and J_0 is the zeroth order Bessel function of the first kind. This solution can been expressed more simply by defining a complex viscosity μ_{eff}, which is defined by comparing the wall drag due to the unsteady velocity profile given by the above equation with wall drag for steady Poiseuille flow. The complex viscosity is then given by:

$$\frac{\mu_{\text{eff}}}{\mu} = -\frac{1}{4}\left(\frac{r_0(\omega/\nu)^{1/2}(1+i)/\sqrt{2}F}{1-(2\sqrt{2}/r_0(\omega/\nu)^{1/2}(1+i)))F}\right)$$

where

$$F = \frac{J_1[r_0(\omega/\nu)^{1/2}(1+i)]}{J_0[r_0(\omega/\nu)^{1/2}(1+i/\sqrt{2}]}$$

and J_1 is the first-order Bessel function of the first kind. This approach was extended by Biot [13,14] for sound propagation along pores having arbitrary cross sections by including a dynamic shape factor.

Alternatively, the same result can be expressed by using the velocity averaged over the cross section $< u' >$ and writing a simplified momentum equation in terms of a complex density ρ_{eff}:

$$\frac{\partial p}{\partial z} = -\rho_{\text{eff}}\frac{\partial < u' >}{\partial t}.$$

ρ_{eff} is then given by:

$$\frac{\rho_{\text{eff}}}{\rho} = 1 - \frac{\sqrt{2}}{(r_0(\omega/\nu)^{1/2}(1+i))F}$$

This approach is favored by Wilson [31] and Brennan [32].

4.5.6.5
Direct Models

Rather than describing the acoustic behavior in terms of modified physical parameters, such as complex density, it is also possible to take a more direct approach in which the waves are solved directly in the limit of very low Reynolds number. This involves two terms: propagation parallel to the axes of the material elements, and perpendicular to the axes. This can be seen by considering the situation illustrated in Fig. 5 which shows both parallel and normal propagation.

For propagation parallel to the axes (i.e., in direction 2), the momentum equation implies:

$$r^2\frac{\partial^2 u_z}{\partial r^2} + r\frac{\partial u_z}{\partial r} + \frac{\partial^2 u_z}{\partial \theta^2} - \left(\frac{i\omega}{\nu}\right) r^2 u_z = \frac{r^2}{\mu}\frac{\partial p}{\partial z}.$$

This equation can be solved for a variety of different boundary conditions in terms of Kelvin functions. For the situation illustrated in Fig. 5, the boundary conditions are:

$$\frac{\partial u_z}{\partial 0} = 0 \quad \text{on } \theta_n = 0$$

$$u_2 \text{ and derivatives continuous} \quad \text{on } \theta_n = \pm\frac{\pi}{N}$$

$$u_z(r,\theta) = 0 \quad \text{on } r = r_0$$

$$\frac{\partial u_z}{\partial x_n} = 0 \quad \text{on } x = x_n$$

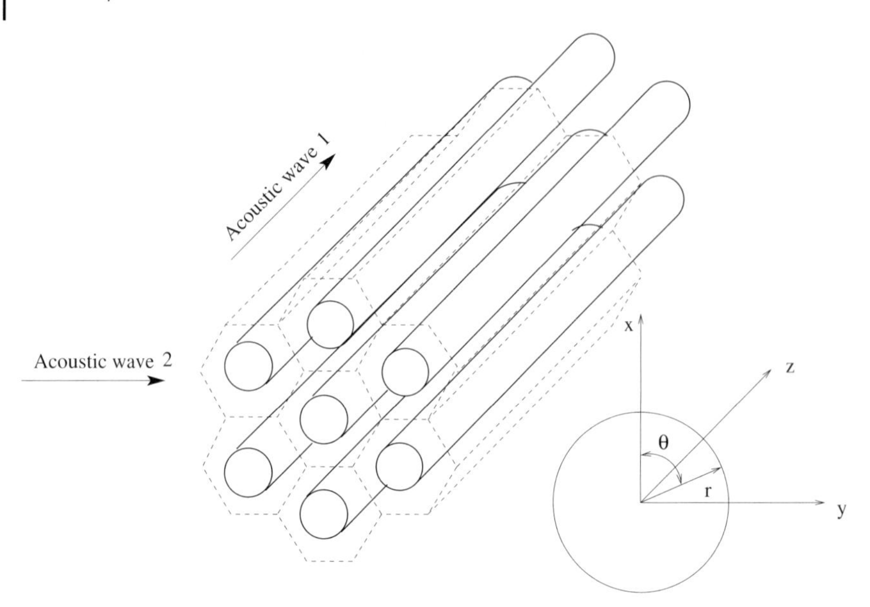

Fig. 5 Array of cylinders.

where N is the number of sides of the periodic boundary (shown as 6 in Fig. 5), r_0 the radius of the material element, $R = (\sqrt{\omega/\nu})r$ the nondimensional radius, θ_n a local angle chosen for each face (i.e., θ_n takes values between $-\pi/N$ and π/N and is zero at the center of each face), and x_n the distance normal to the periodic boundary (i.e. $x = x_n$ on the boundary and x_n is normal to the boundary). The general solution which satisfies the first three boundary conditions is:

$$u_z(R, \theta_n) = \sum_{k=-\infty}^{\infty} \cos(2kN\theta_k)(A_k Ke_{2Nk}(R) + B_k Be_{2Nk}(R)) - \frac{1}{\rho i\omega}\frac{\partial p}{\partial z}$$

where A_k and B_k are coefficients and Be_{2Nk} and Ke_{2Nk} are Kelvin functions of order $2Nk$. The remaining coefficients can be determined by applying the final boundary condition and using colocation along the periodic boundary shown.

When the propagation is normal to the axes, the flow field can be solved by following Stokes' procedure but including the inertial term. Taking the curl of the momentum equation we have:

$$\frac{\partial \vec{\omega}}{\partial t} = \nu \nabla^2 \vec{\omega}$$

where $\vec{\omega}$ is the unsteady vorticity. This has a boundary condition of zero vorticity on the periodic boundaries and must result from a velocity field with no flow on the cylinder surfaces. The pressure field is similarly solved by taking the divergence of the momentum equation to give:

$$\nabla^2\left(\frac{p-p_0}{\mu}\right)=0$$

Combining these equations; the drag on the acoustic wave is given by:

$$\text{Drag/unit length}=(\rho i\omega)2\pi r_1^2\left[\frac{f_1(R_0)-R_0f_1'(R_0)}{f_1(R_0)+R_0f_1'(R_0)}\right]D_1e^{i\omega t}$$

where:

$$f_1(R)=Ke_1(R)-\left(\frac{Ke_1(X_2)}{Be_1(X_2)}\right)Be_1(R).$$

These two models can be combined to give a model for a wave of arbitrary incidence. Since the acoustic properties are primarily dependent on the geometry, rather than the material, the acoustic properties of cellular ceramics are similar to the acoustic properties of cellular metal foams. As can be seen from Fig. 6 [33] (which compares prediction of this model and experiment for a 100 ppi FeCrAlY metal foam with a rigid backing and a 0.8 m cavity), the absorption, defined here in terms of the reflection coefficient, can be extremely high.

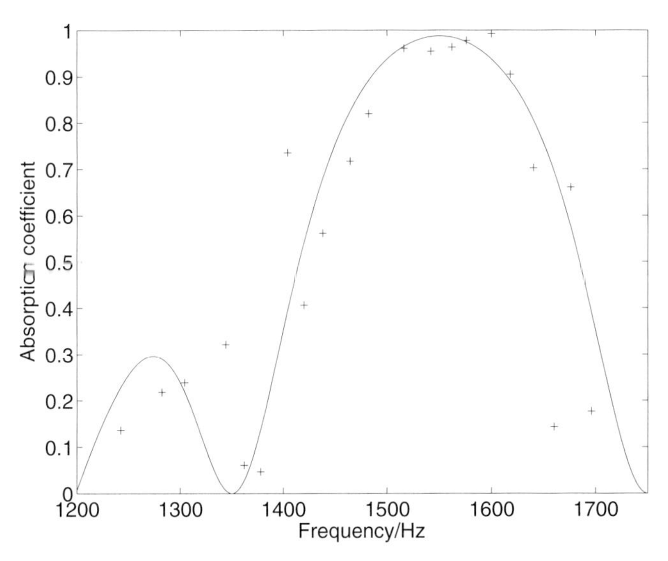

Fig. 6 Comparison between theory (solid line) and experiment (+) for the absorption coefficient for a metal foam block with rigid backing and a 0.8 m cavity at high frequency.

4.5.6.6
Biot's Model

The problem of waves propagating through an elastic porous isotropic solid was considered by Biot [13,14], both with and without viscous dissipation of the waves propagating through the fluid. With no viscous dissipation, Biot considered an elemental cube which is assumed large in comparison with the pore size, but small in comparison with the wavelength of the elastic waves in the fluid–solid aggregate. The equations of motion for the solid and the fluid are coupled by assuming that the system is conservative (i.e., the oscillation is about a state of minimum potential energy). The solution is then found in terms of four elastic constants which are determined empirically. Viscous dissipation of low-frequency sound waves is dealt with [13] by assuming Poiseuille-type flow. A similar approach was used by Morse and Ingard [34]. This assumption breaks down when the frequency of oscillation is sufficiently high that the size of the acoustic boundary layer is small in comparison with the pore size. To quantify this, Biot assumes that the assumption breaks down when pore diameter $2r$ is equal to the quarter wavelength of a viscous boundary layer of an acoustic wave above an infinite plane as considered by Rayleigh [7] and Lighthill [10]. This gives an upper frequency limit of $\pi\nu/16\,r$.

With Poiseuille flow in the pores, the stresses are recalculated with the new viscous shear stress giving rise to a wave which decays exponentially as it propagates through the porous material.

Higher frequency sound waves were treated by Biot in a later paper [14]. Here the limit on the frequency is less restrictive and set by the assumption that the wavelength is much larger than the characteristic pore size. The frequency-dependent dissipation is then calculated using a complex viscosity approach with a dynamic shape factor as described above. Biot's model has been used by a number of authors to characterize the acoustic properties of a variety of porous materials, including saturated sand (Hovem and Ingram [21]), rectangular and triangular pores (Stinson and Champoux [4]) and catalytic converters (Selamet et al [20]). Stinson and Champoux [4] also show that the shape factor used to account for the arbitrary shape should be frequency-dependent.

4.5.6.7
Lambert's Model

Biot's model was extended by Lambert [15–19] to include thermal effects. In the absence of heat flow the acoustic waves propagate isentropically and the temperature fluctuations are given by the equation:

$$\frac{T'}{\overline{T}} = (\gamma - 1)\frac{\rho'}{\overline{\rho}}$$

$$\Rightarrow \frac{\partial T'}{\partial t} = (\gamma - 1)\frac{\overline{T}}{\overline{\rho}}\frac{\partial \rho}{\partial t}$$

where T' is the temperature fluctuation, γ is the ratio of specific heats, and $\overline{T}$ is the mean temperature in the pore. Lambert assumes that the temperature fluctuations are related to the density fluctuations via a modified form of this equation:

$$\frac{\partial T'}{\partial t} + \omega_T T' = (\gamma - 1)\frac{\overline{T}}{\overline{\rho}}\frac{\partial \rho}{\partial t}$$

where ω_T is the reciprocal of the thermal time constant describing the conduction within the pore. Lambert then predicts ω_T using the first law of thermodynamics and Fourier's law for the heat conduction within a cell:

$$Q_T = N \int \left(-\lambda_T \frac{\partial T'}{\partial n} \right) d_s = \text{Vol} \left(c_v \frac{\partial T'}{\partial t} - \frac{\overline{p}}{\overline{\rho}^2} \frac{\partial \rho}{\partial t} \right)$$

where Q_T is the heat flux from a unit cell, N the average number of material elements per cell, λ_T is the thermal conductivity of the fluid in the pore, $\frac{\partial T'}{\partial n}$ the temperature gradient normal to the material elements, Vol the volume of the unit cell, and c_v the specific heat capacity at constant volume for the fluid. The temperature gradient is then estimated by introducing a thermal length characteristic, L_T, and approximating $\frac{\partial T'}{\partial n}$ by T'/L_T. The thermal length characteristic L_T is related to the Nusselt number by $Nu = d/L_T$, where d is a characteristic length scale for the material element. The integral then depends only on the geometry of the material elements and can be evaluated to give:

$$Q_T = c_v \text{Vol} \left(\frac{\partial T'}{\partial t} - \frac{\overline{p}}{c_v \overline{\rho}^2} \frac{\partial \rho}{\partial t} \right)$$

where I is the result of the integral with $\frac{\partial T'}{\partial n}$ removed. Substituting for $\overline{p} = \overline{\rho} R \overline{T}$ we have:

$$Q_T = \frac{T'}{L_T} I(\text{geometry}) = c_v \text{Vol} \left(\frac{\partial T'}{\partial t} - (\gamma - 1) \frac{\overline{T}}{\overline{\rho}} \frac{\partial \rho}{\partial t} \right).$$

Hence:

$$\omega_T = -\frac{I(\text{geometry})}{L_T c_v \text{Vol}} = -N_u \frac{I(\text{geometry})}{d c_v \text{Vol}}.$$

A similar approach was used by Arnott et al [12] to predict the properties of stacks of arbitrarily shaped pores.

4.5.7 Acoustic Applications of Cellular Ceramics

The good sound-absorbing properties exhibited by cellular ceramics make them a good choice for a number of different applications. On highways, cellular ceramics are currently being used to acoustically insulate road tunnels and as noise barriers. Cellular glass materials have also been used in the building industry for sound absorption. Since ceramics have very high melting points, cellular ceramics may also be suitable in the future as acoustic liners inside combustion chambers, particularly in low-NO_x combustors, where combustion oscillations are often a problem. A specific example of this type of application is presented in Chapters 5.5 and 5.6.

4.5.8 Summary

This chapter discussed the propagation of acoustic waves both in free space and through cellular materials. While it is traditional to ignore direct viscous effects, the small cell sizes which are present in cellular materials give rise to significant differences in the propagation characteristics. In particular, the propagation speed is reduced and the drag induced by the viscous effects also gives rise to attenuation of the acoustic waves.

A number of different ways in which these effects can be quantified, such as by defining an acoustic impedance, have been discussed. These can be measured empirically, and methods employed to measure the acoustic properties are discussed. We included a comparison between a theoretical model and experimental measurements for a porous metal.

It is also possible to model the acoustic properties empirically or theoretically. A number of different models exist and are described in Sections 4.5.5 and 4.5.6.

Finally, some examples of applications are given.

References

1 Morfey. C.L. *J. Sound Vibr.* **1971**, *14*(2): 159–170.
2 Seybert, A.F. and Ross, D.F. *J. Acoust. Soc. Am.* **1977**, *61*: 1362–1370
3 Delany, M.E. and Bazley, E.N. *Appl. Acoust.* **1970**, *3*: 105–116.
4 Stinson, M.R. and Champoux, Y. *J. Acoust. Soc. Am.* **1992**, 91(2): 685–695,
5 Bies, D.A. and Hansen, C.H. *Appl. Acoust.* **1980**, *13*: 357–391.
6 Lu, T.J., Chen, F. and He, D.. *J. Acoust. Soc. Am.* **2000**, *108*(4): 1697–1709.
7 Lord Rayleigh. *The Theory of Sound.* Dover Publications, New York, 1945.
8 Lighthill, M.J. *Proc. R. Soc. London* **1952**, *A211*: 564–587.
9 Lighthill, M.J. *Proc. R. Soc. London* **1954**, *A222*: 1–32.
10 Lighthill, M.J. *Waves in Fluids.* Cambridge University Press, 1978.
11 von Helmholtz, H. *Lehre von den Tonempfindungen.* Vieweg Verlag, Wiesbaden, 1877.
12 Arnott, W.P. Bass, H.E. and Raspet, R. *J. Acoust. Soc. Am.* **1991**, *90*(6): 3228–3237.
13 M.A. Biot. *J. Acoust. Soc. Am.* **1955**, *28*(2): 168–178.
14 Biot, M.A. *J. Acoust. Soc. Am.* **1955**, *28*(2): 179–191.
15 Lambert, R.F. *J. Acoust. Soc. Am.* **1982**, *72*(3): 879–887.
16 Lambert, R.F. *J. Acoust. Soc. Am.* **1983**, *73*(4): 1131–1138.
17 Lambert, R.F. *J. Acoust. Soc. Am.* **1983**, *73*(4): 1139–1146.
18 Lambert, R.F. *J. Acoust. Soc. Am.* **1985**, *77*(3): 1246–1247.
19 Lambert, R.F. *J. Acoust. Soc. Am.* **1990**, *88*(4): 1950–1959.
20 Selamet, A., Easwaran, V. Novak, J.M. and Kach, R.A. *J. Acoust. Soc. Am.* **1998**, *103*: 935–943.
21 Hovem, J.M. and Ingram, G.D. *J. Acoust. Soc. Am.* **1979**, *66*(6): 1807–1812.
22 Attenborough, K. *Phys. Rep.* **1982**, *82*(3): 179–227.
23 Stokes, G. *Trans. Cambridge Philos. Soc.* **1851**, *9*: 106–108.
24 Datta, S. and Deo, S. *Proc. Ind. Acad. Sci.* **2002**, *112*(3): 463–475.
25 Urick, R.J. *J. Acoust. Soc. Am.* **1948**, *20*: 283–290.
26 Lamb, H. *Hydrodynamics.* Cambridge University Press, 6th ed., **1932**.
27 Umnova, O., Attenborough, K. and Li, K.M. *J. Acoust. Soc. Am.* **2000**, *107*(6): 3113–3119.

28 Cheung, A.K.W., Tan, B.T., Hourigan, K. and Thompson, M.C. 14th Australasian Fluid Mechanics Conference, Adelaide University, Adelaide, Australia, 10–14 December 2001

29 Deo, S. *Sadhana* **2004**, *29*(4): 381–387.

30 Batchelor, G.K. *An Introduction to Fluid Dynamics.* Cambridge University Press, 1967.

31 Wilson, D.K. *J. Acoust. Soc. Am.* **1993**, *94*(2): 1136–1145.

32 Brennan, M.J. and To, W.M. *Appl. Acoust.* **2001**, *62*: 793–811.

33 Dupère, I.D.J., Lu, T.J. and Dowling, A.P. ASME International Mechanical Engineering, Congress and RD&D Expo, 13–19 November 2004, Anaheim, California, USA, (IMECE2004-60618), 2004.

34 Morse P.M. and Ingard, K.U. *Theoretical Acoustics.* McGraw-Hill, Singapore, 1968.

Part 5
Applications

Cellular Ceramics: Structure, Manufacturing, Properties and Applications.
Michael Scheffler, Paolo Colombo (Eds.)

ISBN: 3-527-31320-6

5.1
Liquid Metal Filtration

Rudolph A. Olson III and Luiz C. B. Martins

5.1.1
Introduction

In most molten-metal processing operations, the acts of melting, transporting, and alloying the metal in preparation for casting into desired shapes introduces undesirable nonmetallic inclusions into the melt. Molten metals are highly reactive and tend to interact with gases and refractories during processing steps; undesirable phases are absorbed either as liquid or solid. Examples of the introduction of such inclusions are molten aluminum reacting with atmospheric oxygen to form solid aluminum oxide, or molten cast iron reacting with atmospheric oxygen to form liquid slag phases.

Once metal is cast, these inclusions can result in defects that render the product unusable. In some cases, the inclusions are too small or too few to be detected in the as-cast part, and the defects are not detected until much further along in the process in machining and forming steps. Many modern advanced manufacturing systems rely on complete elimination of defects in the production path to facilitate and enhance productivity. Thus, to efficiently cast defect-free metal products, it is necessary to remove these nonmetallic inclusions from the melt, and one of the most effective methods of performing this task is the use of ceramic foam filters.

Ceramic foam filters (CFFs) were introduced in 1974 for filtration of molten aluminum used in the production of wrought aluminum alloys [1]. Commercial application in the aluminum industry started in 1976. Filtering of single-part mold castings also started with aluminum in 1977, followed by cast iron in 1983. Today, filtration through cellular ceramics plays a major role in processing several metals. CFFs are used for more than 50 % of the wrought aluminum cast in the world today, representing a total of about 650 000 filters per year. Filter sizes in this application range from 18 to 66 cm square, and the standard thickness is 5 cm. The production of cast iron parts is the second largest use. More than 50 % of cast iron parts produced today are filtered with CFFs. This usage represents a total of 400 000 000 filters per year with sizes ranging from 35 mm square to 150 × 300 mm; thickness typically ranges from 13 to 32 mm. Other metal products, such as aluminum castings, steel castings, copper alloys, and high-temperature superalloys, are filtered routinely with CFFs.

Cellular Ceramics: Structure, Manufacturing, Properties and Applications.
Michael Scheffler, Paolo Colombo (Eds.)

ISBN: 3-527-31320-6

Most CFFs used in molten-metal filtration are manufactured by the foam replication technique. In the process, polyurethane foam is coated with ceramic slurry and the resulting part is dried and fired. During firing, the polyurethane foam within the ceramic coating vaporizes and exits the structure, leaving behind the porous ceramic foam. The cycle time and temperature necessary to generate a bonded filter (for example, silica-bonded silicon carbide or phosphate-bonded alumina) is generally 1–2 h and about 1200 °C, whereas those for sintered ceramic filters (for example, zirconia-toughened alumina or partially stabilized zirconia) is generally 1–2 d and in excess of 1500 °C. The foam replication manufacturing process is capable of generating large volumes of filters with acceptable properties at relatively low cost.

5.1.2
Theory of Molten-Metal Filtration

Three filtration mechanisms operate alone or in combination in a filtration application. These are sieving, cake formation, and deep-bed filtration. Fig 1 shows a schematic of these three filtration modes. In most applications, deep-bed filtration is the dominant mechanism due to the large surface energies present in these systems and the small size of the inclusions that are removed. When an inclusion interacts with the filter wall, the strength of adhesion must be great enough to resist the force of continuous flowing metal and prevent its being swept away and reintroduced to the metal stream. The adhesion strength is correlated with the sum of the interfacial energies between inclusion/filter, metal/filter, and metal/inclusion. The change in free energy for separation of an inclusion from the melt to the filter wall is given as:

$$\Delta G = \gamma_{\mathrm{if}} - \gamma_{\mathrm{mf}} - \gamma_{\mathrm{mi}} \tag{1}$$

where ΔG is the Gibbs free energy, and γ the interfacial energy [2–3]. For the inclusion to remain fixed at the filter wall, the free energy must be sufficiently less than zero. This can be enhanced by reaction between the inclusion and the filter material, or by the metal remaining nonwetting to the filter and inclusion materials. An example of this scenario is provided in Fig. 2, where the metal has withdrawn at the interface because it does not wet the filter or the inclusion.

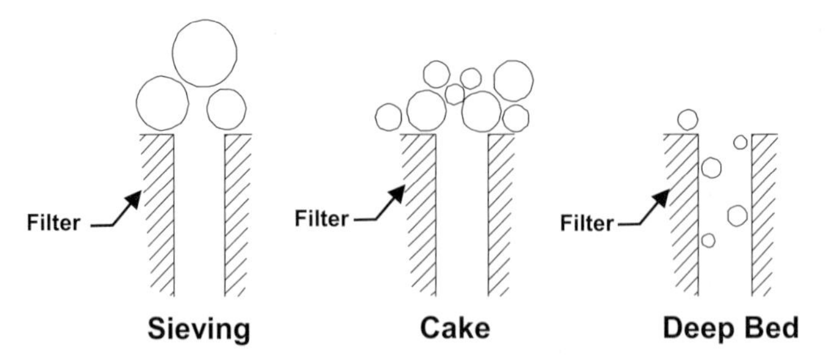

Fig. 1 Three possible modes of filtration.

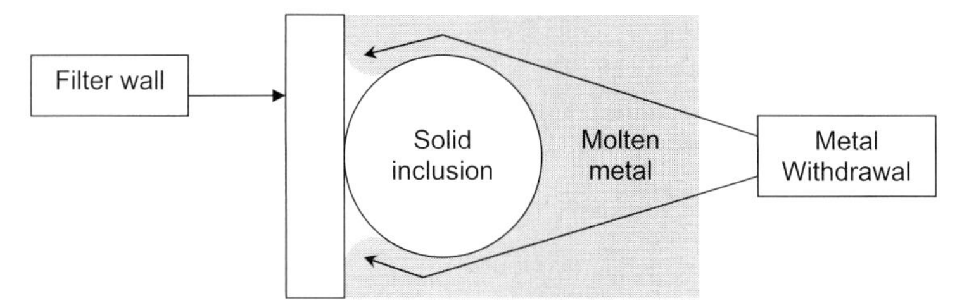

Fig. 2 Schematic of a solid inclusion particle at the interface between the filter and the molten metal.

Smith, Aubrey, and Miller [4] presented a model for filtration of liquid aluminum for wrought alloy application by CFFs. This model assumes that CFFs function as deep-bed filters, whereby the majority of the retained inclusion particles are smaller than the filter pore size and therefore are retained through the depth of the filter structure. Due to the high surface area and tortuous path inherent to CFFs, inclusions typically have a short transport distance to the filter wall. Figure 3 better illustrates deep-bed capture of inclusion particles through the depth of a CFF, with a concentration that is highest at the filter inlet surface and decreases towards the outlet surface. Figure 4 is a scanning electron micrograph of inclusions captured near the entrance of a CFF.

Filter performance models published in the literature are based on trajectory flow modeling, interception-collector theory, and probabilistic modeling. These models predict removal efficiency based on the characteristics of the filter media (cell size, density, thickness), metal flow rate (melt velocity), and the characteristics of the inclusion system (size, density, wetting characteristics) [5–8]. Smith, Aubrey, and Miller [4] constructed an interception-collision filtration model based on a model developed by Grandfield et al. [8]. The analysis assumed that gravity collection and

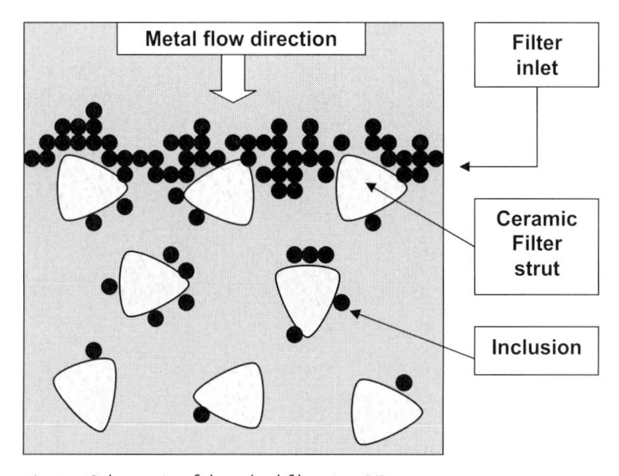

Fig. 3 Schematic of deep-bed filtration [4].

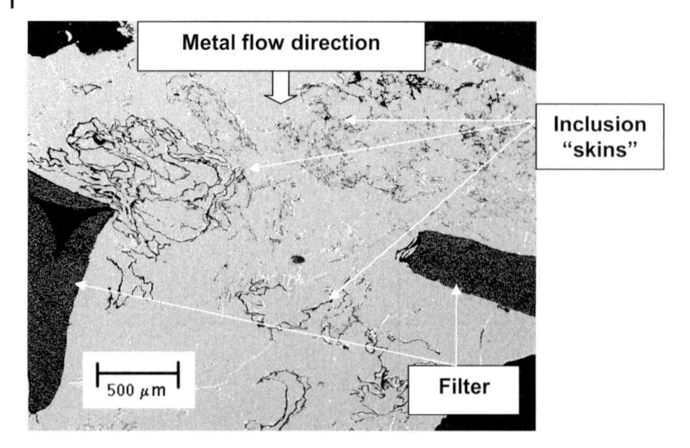

Fig. 4 Scanning electron micrograph showing typical inclusion retention at the entrance to a CFF used to filter aluminum. Note the large number of alumina skins.

direct particle collision with the CFF are the two most important parameters controlling removal efficiency. The total filter collection efficiency η_t is defined as:

$$\eta_t = \eta_c + \eta_g \tag{2}$$

where η_c is the direct particle collision collection efficiency, and η_g the gravity collection efficiency.

The collection efficiency for direct particle collision was defined as:

$$\eta_c = 3\, R_p / R_c \tag{3}$$

where R_p is the radius of the inclusion particle, and R_c the radius of the filter particle.

The gravity collection efficiency was defined as:

$$\eta_g = U_p/(U_p + U_\infty) \tag{4}$$

where the Stokes settling velocity of the inclusion particle U_p was defined as:

$$U_p = 2\, g(\rho_p - \rho_m)(R_p)^2 / 9\, \eta \tag{5}$$

where g is the acceleration due to gravity, η the melt viscosity, ρ_p the particle density, and ρ_m the melt density, and the metal velocity through the filter U_∞ as:

$$U_\infty = M\varepsilon / \rho_m A \tag{6}$$

where M is the mass metal flow per unit time, A the filter area, and ε the filter-bed porosity.

For deep bed filtration, the particle concentration C_z through depth z is calculated according to:

$$C_z = C_o \exp(-3\,\eta_t(1-\varepsilon)z/4\,\varepsilon R_c) \quad (7)$$

where C_o is the incoming concentration of inclusions.

Manipulation of the above equation allows the removal efficiency to be calculated as a function of filter pore size, melt velocity, filter thickness, and inclusion-particle size. Figures 5 and 6 are examples of how the model can be used to predict filtration

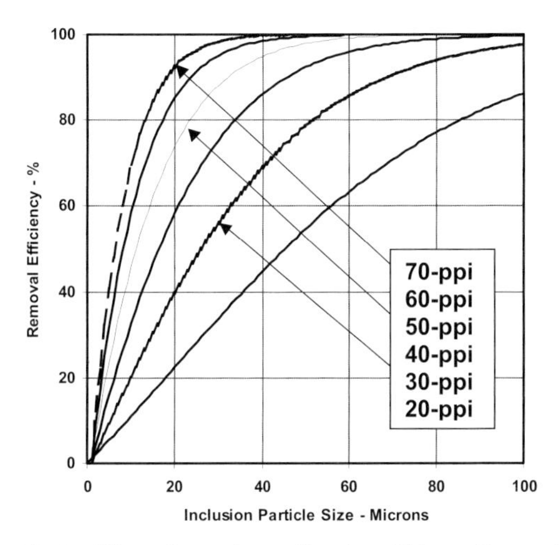

Fig. 5 Effect of pore size on filtration efficiency. Removal efficiency decreases as ppi decreases (used with permission from TMS, Warrendale, PA, USA, www.tms.org).

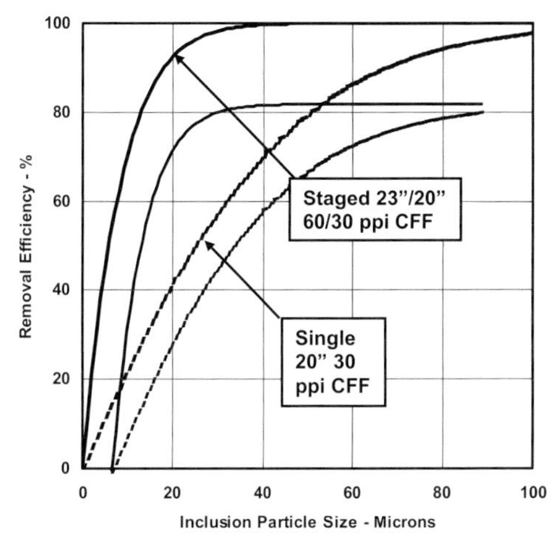

Fig. 6 Gain in removal efficiency from a second-stage fine-pore filter.

efficiency as a function of real-life variables [4, 9]. Figure 6 represents the efficiency of a staged filtration system, discussed in Section 5.1.3. Smith, Aubrey, and Miller [4] also numerically validated the model. Thus, removal efficiency tends to increase as filter pore size decreases, melt velocity decreases, filter thickness increases, and inclusion-particle size increases. The flipside to this scenario is that as filter pore size is reduced and thickness increased, pressure drop across the filter also increases, and as the melt velocity decreases, production rate also decreases; lower production rate and excessive pressure drop across the filter are both undesirable, so a balance must be struck.

5.1.3
Commercial Applications

5.1.3.1
Aluminum

Molten aluminum must be properly treated before filtration is performed in order to obtain acceptable properties in the final product. Waite [10] provides an excellent description of these processing steps. Once filtered, the aluminum is cast and wrought into final shape. "Wrought" aluminum signifies that products are rolled or extruded, which typically requires very clean metal. For these products, the required extent of filtration is dictated by the quality requirements of the final product. Common inclusions in molten aluminum that are removed by filtration are solids such as oxide skins, magnesium aluminate spinels, and borides, and liquids such as magnesium chloride salts and salt–oxide agglomerates.

Aluminum foils are produced commercially down to 5 μm in thickness; extremely clean metal is required. Metal used for aluminum beverage cans has one of the highest quality requirements; inclusion content is typically in parts per billion. Requirements for quality of aluminum extrusions result from processing needs such as increased extrusion speed or process yield, but can also depend on the sensitivity of final product requirements, such as copier drums and tubing.

Phosphate and/or silica are commonly used to bond the alumina aggregate to form the filter body. The filter composition is typically 80–90 % alumina with the balance phosphate and/or silica binder. The binder allows the filter to be manufactured with sufficient properties at high volume, relatively low temperature, near-zero part shrinkage, and low cost.

Selection of the filter pore size for a process is critical. The unit of measure for pore size is pores per inch (ppi). The pore size of CFFs for wrought aluminum varies from 20 to 70 ppi. Pore size selection is a function of incoming metal quality and final product requirements. Filters with 20–40 ppi are used for common applications, and with 50–70 ppi for high-end products. Pore size has a tremendous influence on both the flow rate of aluminum through the filter and the head required to prime the filter. Filtration of wrought aluminum alloys is performed at flow rates from 5 to 1200 kg min^{-1}. Filter use is semicontinuous; a furnace load of between 10 000 and 100 000 kg is cast through a disposable filter. The specific flow rate in mass flow per filtration area varies from about 0.1 to 0.2 kg min^{-1} cm^{-2}. When aluminum is first introduced to the filter, a thin skin of alumina forms at the interface between the metal and the filter. A head pressure is required to break through this skin and allow metal to prime the filter; the finer the pore size, the greater the head required for priming. In this application, it is imperative that the pore size across the filter is relatively uniform. If not, when the filter is primed and metal begins to flow, it will take the path of least resistance and only part of the filter will be primed.

Filtration requires the use of custom equipment designed to support the filter in the path of the molten metal and also provide means to preheat the filter before the molten metal is allowed to contact it. Molten aluminum is typically filtered at about 750–800 °C, and bonded alumina filters, although cost-effective, are not designed to withstand considerable thermal shock. Thus, carefully controlled preheating techniques are required to ensure the filter does not fail by thermal shock when molten metal is introduced. Preheating is especially critical for large filters of 60 cm square or more.

The patented staged filtration system is designed to use two filters in series [11]. These systems are commonly used in the production of wrought aluminum. Pictures and schematics of such systems are displayed in Figs. 7–9. The 60 cm filter has coarser pore size and removes much of the larger-sized inclusions from the melt, whereas the 50 cm filter is of finer pore size and removes the smaller inclusions that escape capture in the 60 cm filter. This system enables a plant to run relatively unclean metal yet generate a high-quality end product.

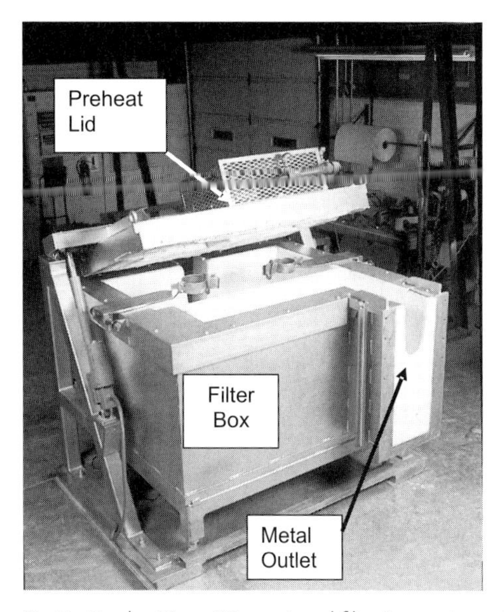

Fig. 7 Duplex 50 cm/60 cm staged filtration system with gas preheat lid. The filter is preheated from the top. The system uses 60 cm filters in series with underlying 50 cm filters. The 60 cm filters are coarser (used with permission from TMS, Warrendale, PA, USA, www.tms.org).

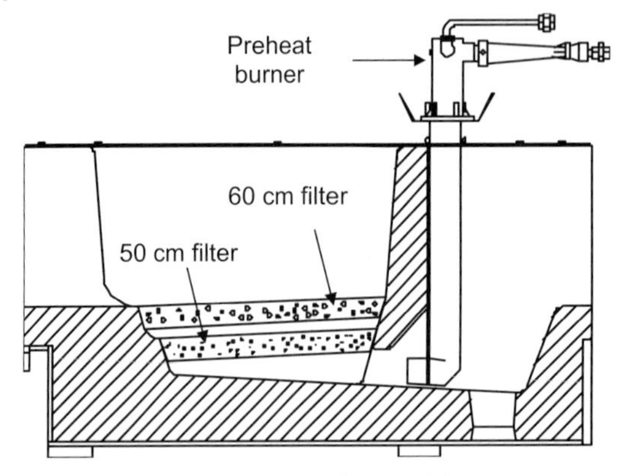

Fig. 8 Schematic illustration of a staged filtration bowl. Preheating of the filter is performed from the bottom by using a high-velocity burner (used with permission from TMS, Warrendale, PA, USA, www.tms.org).

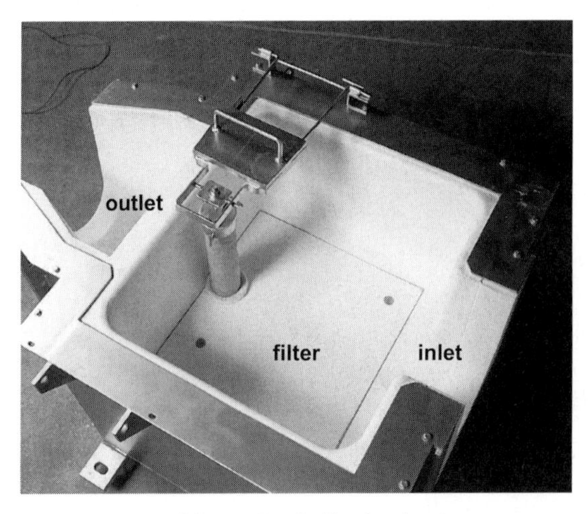

Fig. 9 Top view of the inside of a filter bowl with reaction-bonded alumina filter inside. The two dots on the corners of the filter disappear when the filter is preheated to the appropriate temperature (used with permission from TMS, Warrendale, PA, USA, www.tms.org).

5.1.3.2
Iron Foundry

Cast iron parts are used in a large number of applications, including automotive and mechanical assemblies. Iron can be easily cast into intricate shapes, producing a part highly similar to the final desired geometry, greatly reducing additional conversion.

Filtration is needed to meet the quality requirements established for complex-shaped parts manufactured in large-scale assembly lines such as engine blocks, crankshafts, and disk-brake rotors. These parts are individually cast in separate sand molds. Total cast weight per mold typically varies from 2 to 200 kg, and a mold may contain one or more parts. Filters are placed in the feeding system within the mold. Schematics of two feed systems are shown in Fig. 10. Silica or clay-bonded silicon carbide filters are used, ranging in size from 35 to 185 mm square, typically with pore size from 3 to 20 ppi.

The filter size and pore size selected for a process depend mainly on the type of iron being cast. Gray iron generally contains a lesser amount of inclusions; therefore a relatively high filter loading of 0.35–0.5 MPa and relatively fine pore size from 10 to 20 ppi can be used; filter loading is simply the mass of metal that can be poured through unit filter area. Ductile iron contains more inclusions, so filter loading must be lower and pore size coarser. Filter loads typically vary from 0.21 to 0.28 MPa and pore sizes from 3 and 10 ppi. Filter size is dictated by the need to keep the filter to choke area ratio preferably higher than 6. The choke is the smallest passage in the liquid-metal feeding system and controls metal flow velocity. The rule above prevents the filter from becoming a flow restriction and modifying mold feeding, which can greatly affect product yield and quality.

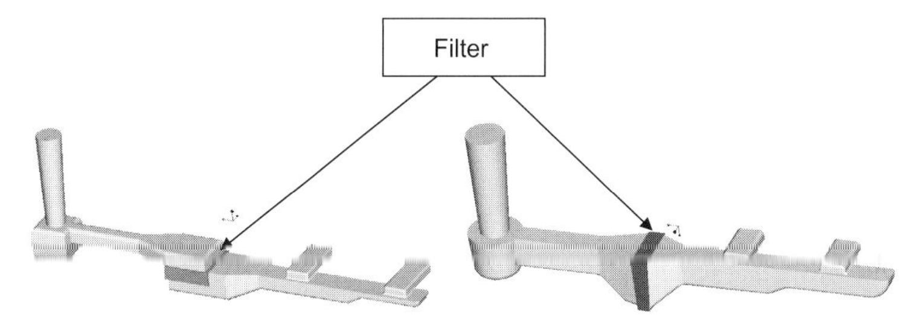

Fig. 10 Typical liquid-metal feeding canals with filter placed either horizontally or vertically.

Liquid inclusions are quite common in cast iron. Silicon, iron, and manganese from the melt can be oxidized to form SiO_2–FeO–MnO slag. A slag composed of these oxides can melt at temperatures as low as 1170 °C [12]. Capture of liquid slag by a silica-bonded silicon carbide ceramic foam filter is illustrated in Fig. 11 [13].

Filtering molten iron is a difficult task; the filter encounters a rapid temperature swing followed by very high mechanical load at high temperature. The casting time is only as much as 45 s, but the temperature of the molten iron is in excess of 1400 °C. The filter is not preheated when molten iron is introduced, so the temperature climbs very quickly from 20 to over 1400 °C. The filter is then quickly loaded with inclusions, and the head pressure above the filter continues to rise, generating significant loads.

Silica-bonded silicon carbide is an excellent material for use in iron filtration. The filter cannot be preheated, so it must have outstanding thermal shock resistance, which is afforded by the high thermal conductivity of silicon carbide. The silica

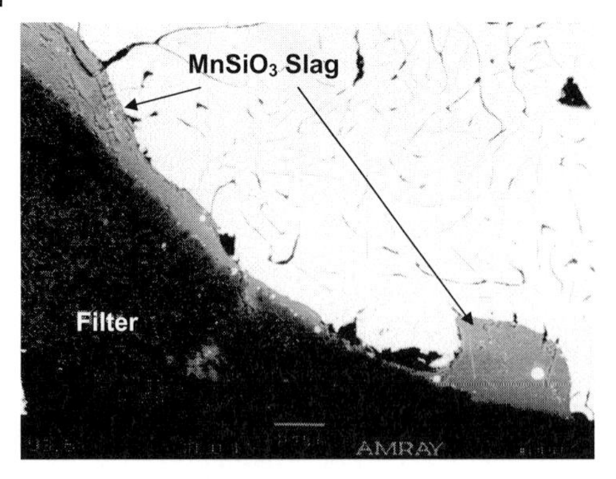

Fig. 11 Scanning electron micrograph in backscatter mode showing liquid manganese silicate slag wetted on a silica-bonded silicon carbide filter (used with permission of the American Foundry Society, Schaumburg, IL, USA, www.afsinc.org).

matrix and silicon carbide aggregate create a strong high-temperature bond when the composition is well designed. Filters formulated according to Hoffman and Olson [14] have modulus of rupture values greater than 0.7 MPa when ramped quickly to 1428 °C and tested at 45 s, which is a relatively long casting time. The composition, as well as the rheological characteristics of the slurry, is crucial to obtaining sufficient high-temperature strength.

Another important feature of ceramic foam filters in foundry applications is flow modification of the molten iron as it enters the casting. Turbulence in the molten metal stream as it enters the casting must be avoided at all costs, as it tends to promote formation of entrained gas bubbles and oxide inclusions. Water modeling has shown that ceramic foam greatly reduces the turbulence in a liquid stream as compared to screening-type filters [3]. Real-time X-ray images of molten metal flowing into a mold have shown that ceramic foam filters considerably reduce the turbulence of the melt and inhibit formation of these defects [15].

5.1.3.3
Steel

Steel is commonly used for complex mechanical parts such as pump and valve bodies. Steel is cast in sand molds in the nearly the same process as discussed above for iron. Steel sand castings are one of the most demanding filtration applications, even more so than iron, due to casting temperatures in excess of 1560 °C and low metal fluidity, as well as the practical limitation that the filter cannot be preheated. For these reasons, the choice of filter compositions is limited to partially stabilized zirconia. Magnesium is commonly used as the stabilizing agent.

The extent of zirconia stabilization in the filter is critical to its performance. If the extent of stabilization is too great, the filter can fail catastrophically due to thermal-shock-related phenomena. If the extent of stabilization is insufficient, the filter can

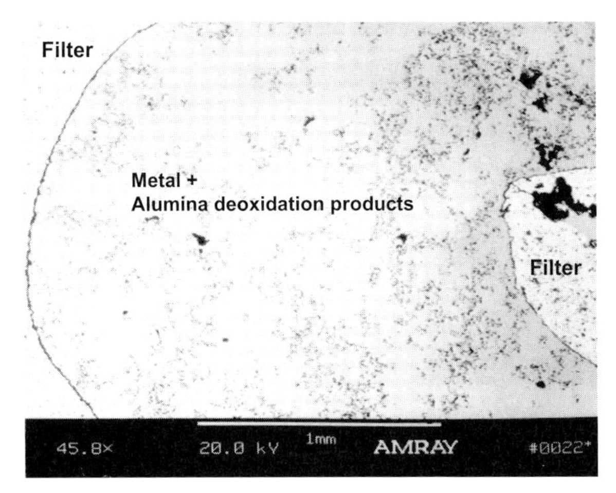

Fig. 12 Photomicrograph showing Al_2O_3 deoxidation inclusions bridging across the filter pore opening. Alloy type: CN-7M. Filter type: partially stabilized zirconia (used with permission of the American Foundry Society, Schaumburg, IL, USA, www.afsinc.org)

fail due to poor high-temperature strength. The composition and its manufacture are optimally designed to prevent failure by these two mechanisms.

Filtration of steel for castings is crucial because the presence of surface or subsurface oxide macro-inclusions can significantly impair their machinability, mechanical properties, and pressure-tightness. Oxide macro-inclusions are primarily the result of reoxidation during pouring. These inclusions must be removed in the gating system prior to entering the mold cavity.

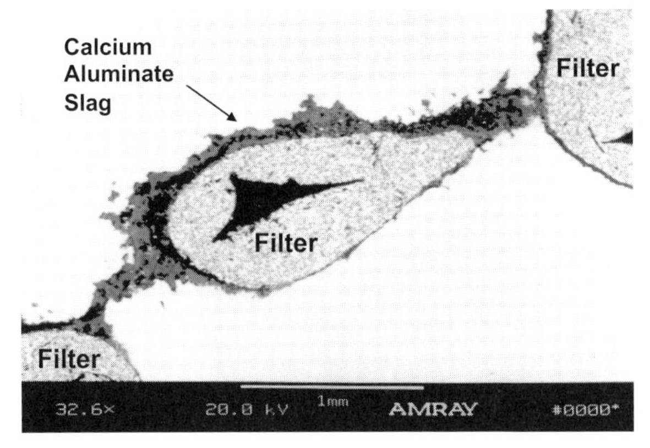

Fig. 13 Backscattered electron image showing liquid calcium aluminates which have wetted the filter structure. Energy-dispersive analytical X-ray analysis indicated the inclusion to be a mixture of $3\,CaO \cdot Al_2O_3$ and $12\,CaO \cdot 7\,Al_2O_3$. Deoxidation: Al and Ca wire injection (used with permission of the American Foundry Society, Schaumburg, IL, USA, www.afsinc.org).

Total cast weight per mold varies from 2 to 2000 kg, and a mold may contain one or more parts. Filter sizes ranges from 5 to 20 cm square with thickness of 2 or 3.8 cm. Common filter pore sizes are 3–5 and 6–10 ppi. Filter dimensioning is a strong function of the type of steel being cast. Aluminum-deoxidized steels are filtered at 0.17 MPa; silicon-deoxidized steels and stainless steels can be poured at up to 0.7 MPa.

Both solid and liquid inclusions are present in molten steel. Alumina particles form when aluminum is use to deoxidize the steel. These fine particles can easily agglomerate and be retained by the filter, as shown in Fig. 12, and liquid oxide slags are common, such as the calcium aluminate slag shown in Fig. 13 [16]. The large surface energies existing at these elevated temperatures allow the filter to collect both types of inclusions, as shown in the figures.

5.1.4 Summary

Ceramic foam filters provide an efficient and effective means of removing undesirable inclusions from molten metal. They must be refractory, corrosion-resistant, demonstrate sufficient thermal-shock resistance and high-temperature strength for the application, and be cost-competitive. In all molten-metal processing applications, the porosity must be sufficiently continuous and of consistent pore size from filter to filter to ensure the repeatability of filter performance in molten-metal processing operations, and the pore size must be appropriately selected to achieve optimum processing conditions and acceptable product. The deep-bed filtration mechanism is the most important, and capture of inclusions is based on interfacial energies between the molten metal, filter body, and inclusion type. Research continues in this area, and new CFF designs continue to be introduced to the marketplace in all areas of molten-metal filtration.

References

1 Dore, J.E., Yarwood, J.C., Preuss, R.K. US Patent 3,962,081, 1976.
2 Cramb, A.W., Jimbo, I. *Iron & Steelmaker* **1989**, *16*[6], 43–55.
3 Schmahl, J.R., Aubrey, L.S., Martins, L.C.B *Proc. 4th Int. Conf. Molten Aluminum Process.*, AFS, Orlando, FL, 1995.
4 Smith, D.D., Aubrey, L.S., Miller, W.C. *Light Metals*, TMS, San Antonio, TX, 1998, pp. 893–915.
5 Mutharasan, R., Apelian, D., Romanowski, C. *Light Metals*, TMS, Chicago, IL, 1981, pp. 735–750.
6 Gauckler, L.J., Waeber, M.M., Conti, C., Jacob-Duliere, M. *Light Metals*, TMS, New York, NY, 1985, pp. 1261–1283.
7 Netter, P., Conti, C. *Light Metals*, TMS, New Orleans, LA, 1986, pp. 847–860.
8 Grandfield, J.F., Irwin, D.W., Brumale, S., Simensen, C.J. *Light Metals*, TMS, Warrendale, PA, 1990, pp. 737–746.
9 L.S. Aubrey, D.D. Smith, Aluminium Cast House Technology, 6th Australian Asian Pacific Conference, P.R. Whiteley and J.F. Grandfield (Eds.), TMS, Sydney, Australia, 1999, pp. 133–157.
10 Waite, P., *Light Metals*, TMS, Seattle, WA, 2002, pp. 841–848.
11 Aubrey, L.A., Oliver, C.L., MacPhail, B.T. US Patent 5,673,902, 1997.
12 W.R. Maddocks, *Iron Steel Inst. (London) Carnegie Schol. Mem.* **1935**, *24*, 64; Figure 690 in *Phase Diagrams for Ceramists, Vol. I*, The American Ceramic Society, Columbus, OH, 1985.
13 Hoffman, W.I., Olson III, R.A. US Patent 6,663,776, 2003.
14 Schmahl, J.R., Aubrey, L.S. *AFS Trans.* **1993**, *101*, 1–12.
15 Foseco *Foundry Practice* **2003**, *238*[3], 18–26.
16 Aubrey, L.S., Schmahl, J.R., Cummings, M.A. *AFS Trans.* **1993**, *101*, 59–69.

5.2
Gas (Particulate) Filtration

Debora Fino and Guido Saracco

5.2.1
Introduction

Filtration through porous media can take place according to two main mechanisms: superficial or "cake" filtration and interstitial or "deep" filtration. In superficial filtration, a dust cake is formed over the porous filter during the process, eventually becoming the real filter media. Quite high efficiency of particle separation can be reached by the latter means, often exceeding 95 %. Conversely, deep filters lock dust inside their porous matrix by separation mechanisms such as inertial impaction, interception, and Brownian diffusion [1]. A lower filtration efficiency is generally reached (50–60 %) at equal pressure drop.

In both cases, cellular ceramics have met significant interest in the R&D community in the last decade for gas filtration applications. Figure 1 shows two examples of how cellular ceramics can serve as superficial or deep filtration devices.

Wall-flow monoliths (Fig. 1, left) having a honeycomb structure with adjacent channels blocked at opposite ends are typical surface filters: the filtered particles accumulate over the walls of the channels through which the dirty gas stream is fed. Foam filters (Fig. 1, right) trap particles inside their cellular structure (typically over their struts), as their pore size is generally much larger (> 200 μm) as opposed to their wall-flow counterparts (ca. 10 μm) [3].

A characteristic of wall-flow filters is their need for periodic cleaning or "regeneration", since the accumulation of trapped particulate causes an unacceptable increase in backpressure [4]. Conversely, foam filters are often referred to as "nonblockable" traps, since particulate blow off takes place after a certain mass holdup is reached (dependent on the dust and the nature and geometry of the filter), and thus prevents the pressure drop from rising beyond a given value. However, this is generally not acceptable owing to the very low filtration efficiency it entails [5].

This chapter address the application opportunities of the above-mentioned media in the field of gas filtration. Special attention is paid to their coupling with catalysts [6–9] to constitute *multifunctional reactors* [10] (i.e., catalytic filters) capable of carrying out, in addition to filtration, a catalytic reaction for the abatement of gaseous pollutants or particulate. Ceramic foams, used commercially for the filtration of molten metals (Chapter 5.1), are attracting increasing attention as catalyst supports

Cellular Ceramics: Structure, Manufacturing, Properties and Applications.
Michael Scheffler, Paolo Colombo (Eds.)

ISBN: 3-527-31320-6

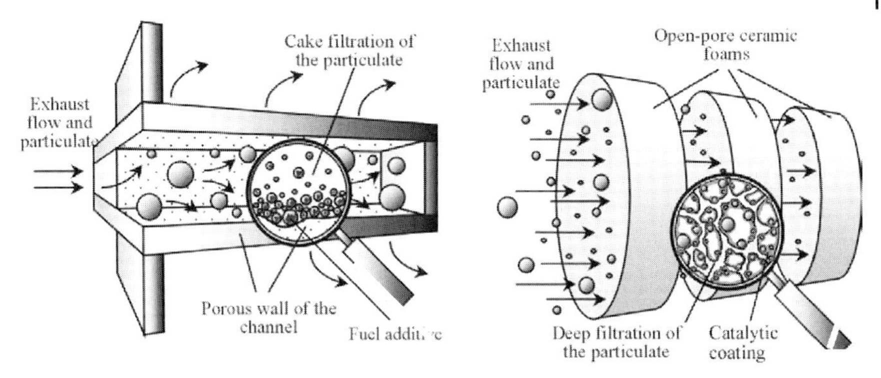

Fig. 1 Particulate traps for hot gas filtration. Left: wall-flow honeycomb monolith (superficial filtration); Right: ceramic foam trap (deep filtration). (Reprinted from [2], Copyright 2000, with permission from Elsevier Science Ltd.)

due to their high thermal stability, high porosity, and increased tortuosity relative to honeycombs. This application requires the development of complicated catalyst deposition routes inside the porous matrix [11].

5.2.2
Properties of (Catalytic) Cellular Filters

The potential techno-economical advantages of catalytic filters are those typical of multifunctional reactors: 1) Fewer process units (substitution of at least two process stages with a single reactor in which all operations of interest are carried out simultaneously); 2) Cost reduction (a likely, though not inevitable consequence of point 1) should be a decrease in investment costs); 3) Energy saving (catalytic filters may allow a more efficient management of energy and heat recovery in the tail-end treatment of flue gases from large boilers [12]); 4) Space saving (perhaps the most obvious advantage of coupling two operations into a single unit).

However, to take full advantage of these opportunities catalytic filters should have: 1) Good thermochemical and thermomechanical stability; 2) High particulate-separation efficiency: particulate should not markedly penetrate the filter structure since this would lead to pore obstruction and/or to catalyst deactivation; 3) High catalytic activity to attain nearly complete catalytic abatement at superficial velocities that are as high as possible (e.g., those employed industrially for dust filtration are 10–80 $m^3\ m^{-2}\ h^{-1}$ (standard temperature and pressure, STP) or even higher); 4) Low pressure drop (a certain increase in head compared with the virgin filters must be expected owing to the presence of the catalyst); 5) Sufficiently low cost to compete with conventional, well-established technologies. Requirement 3), in particular, should not be underestimated. For any degree of catalytic activity and filter thickness, a superficial velocity will indeed exist for which nearly complete catalytic abatement of pollutants can be achieved along with particulate filtration. However, it is mandatory to render this combination feasible at the superficial velocities typically adopted in industrial filtration. Otherwise, if the superficial velocity must be kept low to guarantee a sufficient residence time in the catalytic filter matrix for reaction purposes, the number of filter units would have to be larger than required for simple filtration when a given flow rate is to be treated. The investment costs will then increase and thus hamper the economic potential of the catalytic filters.

This chapter discusses how, and to what extent, the above properties can be exploited and the limitations overcome. Several flue gases (e.g., from coal-fired boilers, incinerators, diesel engines) are characterized by high loads of both particulate (e.g., fly ashes, soot) and gaseous pollutants (NO_x, SO_2, CO, volatile organic compounds, etc.), which must be removed for environmental protection purposes. In this context, several possible applications of catalytic filters based on cellular materials can be envisaged, some of which have already been successfully tested not only on the laboratory scale but also on the industrial scale.

5.2.3 Applications

5.2.3.1 Diesel Particulate Abatement

The high efficiency of diesel engines, their low operating costs, high durability, and reliability have provided them with a leading role in the heavy-duty vehicle market. This wide occurrence entails careful evaluation of the related environmental effects. Among the emitted pollutants, diesel particulate raises serious health concerns due to its carcinogenity [13], owing to the presence of polyaromatic hydrocarbons (PAH) and nitro-PAH in its so called soluble organic fraction (SOF), as well as to its size falling in the lung-damaging range (10–200 nm). In the field of particulate emission control, attention has mainly been paid so far to improvements in engine design [14], modification of fuel formulation, and use of alternative nonfossil fuels such as natural gas, alcohols, or esters [15], as well as the use of filtering and nonfiltering aftertreatment devices. The EURO IV regulations imposing particulate emissions lower than 0.025 g km^{-1} for passenger cars will force car manufacturers to adopt new solutions from 2005 onwards. These innovative aftertreatment systems will likely be based on catalytic filters.

Some commercial systems using wall-flow filters have already been launched. The *PSA (Peugeot-Citröen Societé d'Automobiles) system* (Fig. 2) is based on the following key components [16]:

- SiC wall-flow monolith: selected for its superior filtration efficiency and physical properties (high-temperature and thermal-shock resistance [17]).
- Active regeneration strategy: when trap regeneration is needed owing to high pressure drop detected by a sensor, fuel post-injection, after the main injection, enabled by the intelligent use of second-generation common-rail diesel

engines, induces an increase in exhaust-gas temperature and provides unburnt hydrocarbons to the preoxidizer.

- Preoxidizer: a catalytic converter that burns the above-mentioned unburnt hydrocarbons, thereby enhancing further the exhaust-gas temperature and igniting the trapped particulate.
- Ce fuel additive: this fuel additive leads to formation of CeO_2 particles well embedded in the structure of the diesel particulate and thus in very good contact with the soot, which lowers the ignition temperature by catalytic means to the benefit of savings in post-injected fuel [18].

This system is currently running on more than 500 000 cars with no apparent problems. However, it has some drawbacks [16]:

- CeO_2 deposits: the oxide derived from the additive remains in the traps and thus requires periodic cleaning or significant trap oversizing.
- High investment costs owing to the presence of many components (additives, additive storage tank and dosing pump, preoxidizer, pressure and temperature sensors, control electronics, common-rail diesel system).
- High operating costs: post-injected fuel does not contribute to driving power and thus corresponds to a cost (fuel penalty). A trade off with the fuel penalty caused by the increased trap pressure drop due to particulate loading (Fig. 3) leads to the determination of an optimal gap between subsequent regenerations of 300–400 km, which in any case entails an overall fuel penalty of about 4 %.

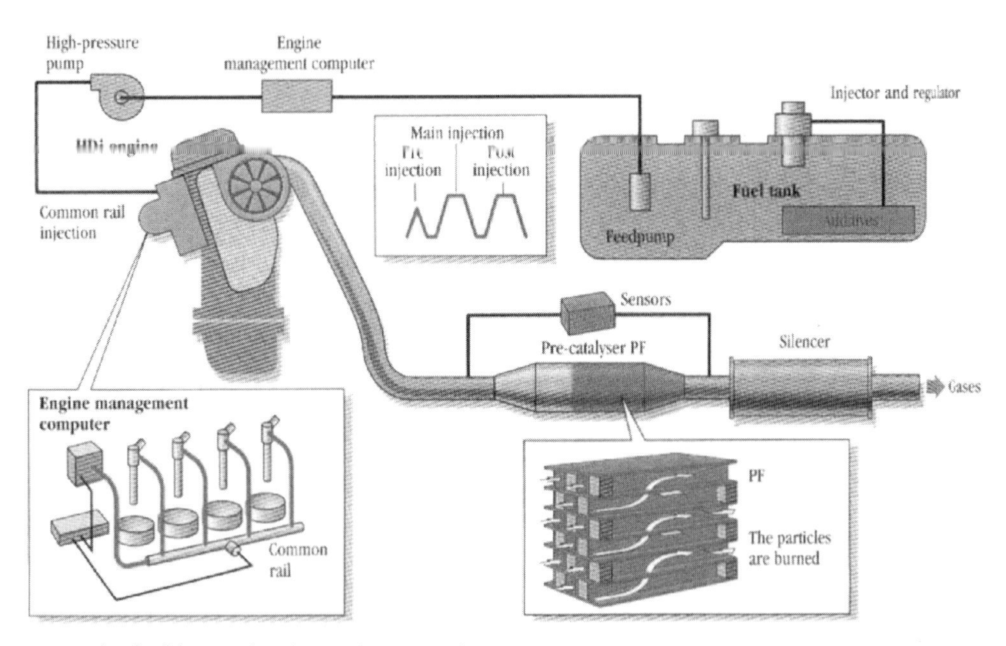

Fig. 2 Sketch of the PSA diesel particulate removal system. (Reprinted from [43], Copyright 2003, with permission from Elsevier Science Ltd.)

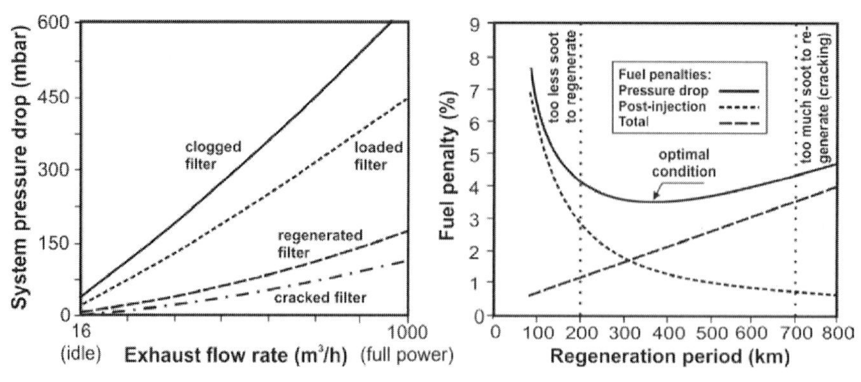

Fig. 3 Pressure drop and optimization of regeneration period for the PSA system (courtesy of Peugeot Citroen Societes des Automobiles, La Garenne Culombe, France).

Another patented and commercially available system is the *continuously regenerating trap* (CRT) system by Johnson Matthey. It exploits the pronounced oxidative activity of NO_2 towards carbonaceous particulate [19] and consists of a wall-flow trap with an upstream flow-through diesel oxidation catalyst, known as a preoxidizer (Fig. 4).

The preoxidizer converts about 90 % of the hydocarbons (HC) and CO present in the exhaust gas and promotes the reduction of at least 3 % of the nitrogen oxides. The most interesting feature of the CRT system, however, is its ability to promote continuous trap regeneration provided its operating temperature is kept in the range

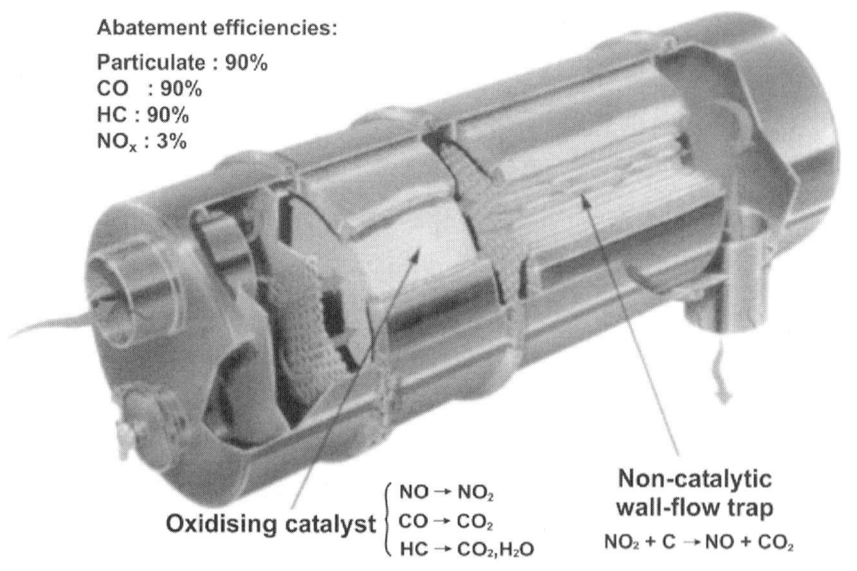

Fig. 4 The Johnson Matthey CRT system (adapted from www.jmcsd.com).

200–450 °C. Above 200 °C preoxidizer activity is sufficient to burn out the HC and CO as well as to convert NO to NO_2, which can rapidly react with diesel particulate, leading to its combustion and to formation of NO. Above 450 °C, thermodynamics start to disfavor NO_2 formation. The use of continuous regeneration means that extreme temperature gradients within the trap are avoided, which prolongs the trap life expectancy. A satisfactory performance of over 600 000 km has been reported [20].

The major drawback of the CRT system lies in the sensitivity of the preoxidizer to the presence of sulfur compounds [20], which has hampered significant introduction in the market. However, low-enough fuel sulfur contents (50 ppm) will be made mandatory all over Europe in 2005, and this may lead to extensive introduction of this technology. However, another weak point of the CRT system is its dependence on the presence of NO_x, as it is uncertain whether future diesel engines will produce sufficiently high NO_x-to-soot ratios to allow satisfactory operation.

One of the most attractive ways of catalyzing the combustion of the soot accumulated on the filter is the use of a catalytic coating on the filter itself, that is, a *catalytically coated trap* (CCT). On the basis of their chemical composition, the supported catalysts for the combustion of diesel particulate can be divided into two classes: noble metals and base metal oxides, frequently combined with alkali metal compounds. Moreover, perovskite-type oxides have been recently studied [21] for the simultaneous removal of nitrogen oxides and diesel particulate.

The most common precious metals used are Pt, Pd, and Rh [22]. The noble metal is intended to initiate oxidation of the easily combustible hydrocarbons adsorbed on the soot surface. This provides a local heat release which initiates oxidation of the relatively less reactive carbonaceous particulate. Moreover, Pt activates oxygen effectively and this enhances the rate of soot oxidation and minimizes formation of partially oxidized compounds, such as CO. Precious metals are often added in very small quantities to catalysts based on transition metals (e.g., copper, vanadium) to improve their low-temperature activity compared to those of the transition metals alone [23].

The oxides and/or salts of several base metals (e.g., vanadium, copper, molybdenum, manganese, cobalt, chromium, iron), known to catalyze graphite oxidation [24], have been used as soot-combustion catalysts. In general, based on the catalytic oxidation of graphite, only those oxides which are capable of being reduced by carbon to the metallic or lower oxidation state are likely to be active catalysts in soot oxidation. For instance, studies by McKee [24] on the copper-catalyzed oxidation of graphite showed that copper(II) salts form copper(II) oxide, which interacts with graphite and is reduced completely to metallic copper, which finally be reoxidized. The net result of this oxidation–reduction cycle is transfer of oxygen atoms from the gas phase to the graphite. The reduction of copper oxides to copper metal by carbon is favored by the comparatively low free energy of formation of these oxides, whereas the oxides of aluminum and zinc are stable in the presence of carbon and are therefore catalytically inactive.

These results can be extended to soot combustion: Van Doorn et al. [25] found that very stable oxides, such as those of aluminum and silicium, exhibit no activity at all, whereas Ahlström and Odenbrand [23] explained the high activity of V_2O_5 in terms of the capability of vanadium of existing in several oxidation states with small energy differences.

The combination of two of the above active metals [26], such as, for example, the use of copper vanadates, is also very frequent in the literature on the catalysis of soot combustion: in this case, copper cations can change their oxidation state easily between +II and +I, and vanadium can readily change its oxidation state between +V and +IV.

Alkali metal compounds, alone or added to transition metal compounds [27], proved to be active in the catalysis of soot combustion. The catalytic effect of alkali metals in carbon oxidation was explained in the early 1950s [28] by the fact that alkali metal atoms on the carbon surface act as sites for chemisorption of oxygen, which weakens the C–C surface bonds and promotes the desorption of gaseous oxidation products at low temperatures. Furthermore, the gasification activity of alkali metal compounds was found to be dependent on the nature of the alkali metal [29] and on the anion with which the alkali metal ion is associated. In the use of alkali metals in combination with transition-metal compounds, a synergistic effect can be observed [30]. Mross [31] reports on the addition of alkali metals to oxidic compounds: in this case, the electron-donating effect of the alkali metal ions increases the reactivity of the oxygen in the M=O bond (where M is a transition metal ion in the catalyst) and therefore catalyst doping with small quantities of alkali metal increases the reaction rate in the partial oxidation of hydrocarbons. Another possible explanation of the synergistic effect of alkali metal compounds is the capability of the alkali metal to lower the melting point of the catalytically active species and the viscosity of the resulting liquids [31]. Saracco et al. [32] showed that the addition of potassium chloride to several copper and potassium vanadates promotes the formation of eutectics with a melting point lower than those of pure vanadates. According to these authors, the role of the alkali metal compound (in this case potassium chloride) is that of bringing into the liquid phase the active components (e.g. vanadates), which, once the carbon surface has been wetted, catalyze its oxidation through redox processes.

The contact between soot and catalyst under operating conditions seems to be the most critical parameter controlling the reaction rate; this contact mainly depends on the outer surface of the catalyst layer and on the dispersion of the soot on the catalyst surface itself.

This problem was first underlined by Inui and Otowa [33], who compared the effect of support structure on soot conversion. They deposited the same amount of a copper-based catalyst on three different supports (a honeycomb monolith, a ceramic foam, and alumina pellets), uniformly loaded with the same quantity of a benzene-generated soot. The ceramic foam showed better performance than the honeycomb or the pellets. This was explained in terms of better contact between soot and catalyst, since ceramic foams expose a larger catalyst surface compared with the other two counterparts. Inui and Otowa obtained further confirmation of the influence of contact on soot combustibility from tests employing two different methods for establishing this contact: the "dry method" (placing the deposited catalyst in an air stream containing the soot) and the "wet method" (dipping the catalyst into soot-containing benzene, followed by benzene evaporation). They observed that the soot loaded on the supports by the wet method was more easily oxidized than that loaded by the dry method under the same reaction conditions, because in the former case soot was in closer contact with the catalyst. On the basis of these results, the authors concluded that the activity of a catalyst can be improved chiefly by improving its contact with soot, possibly approaching the contact conditions achieved with fuel additive based catalysts.

With the above background, catalytic foam traps were developed recently in the framework of the activities of an EU project (CATATRAP) by depositing through tailored techniques Cs–V-based catalysts on the pore walls of zirconia-toughened alumina foams developed by Centro Ricerche FIAT and Saint Gobain. On the basis of previous experience by other researchers [27, 34], the catalysts employed in this project enabled mobility of active species either by liquid-phase formation ($CsVO_3$ + KCl [1, 35, 36]) or by oxygen spillover ($Cs_4V_2O_7$ [37]), which facilitated improved contact between catalyst and carbon. Similar catalysts and catalytic foams were also developed in Italy [38, 39] and the Netherlands [40–42].

Such catalytic systems cannot be employed with wall-flow filters, as their pores are so fine that the liquid catalyst phase would be rapidly sucked into them, plugging the trap. For this reason, perovskite-type catalysts, for example, $LaCrO_3$, were lined on the walls of the inlet channels of cordierite and SiC wall-flow traps via an ad-hoc-developed combustion synthesis technique [43–45] (Fig. 5).

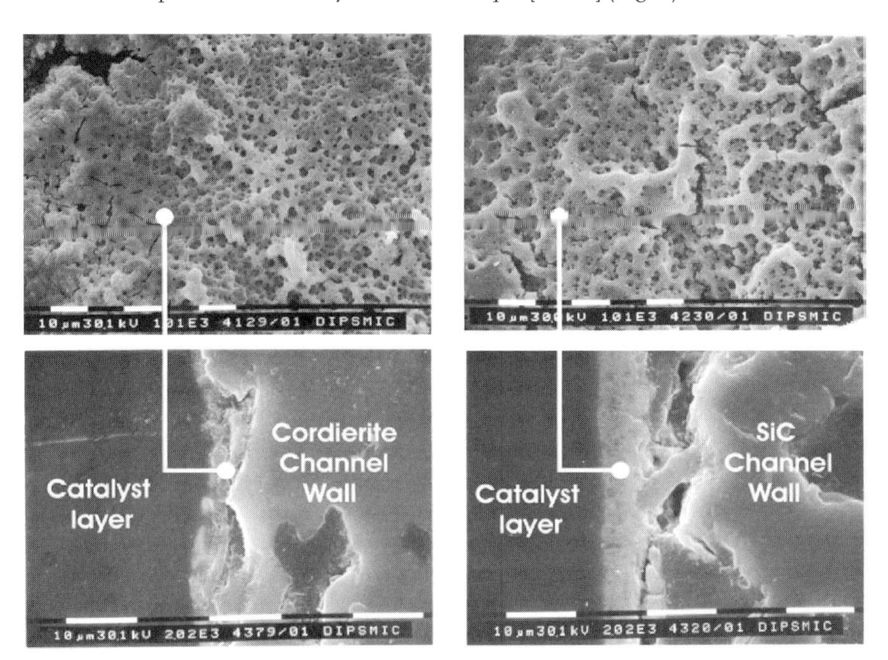

Fig. 5 Front and cross-section SEM views of layers of $LaCr_{0.9}O_3$ catalyst deposited on the inlet channel walls of ceramic traps. Left: cordierite channel wall; Right: SiC channel wall. (Reprinted from [45], Copyright 2003, with permission from Korean Institute of Chemical Engineers.)

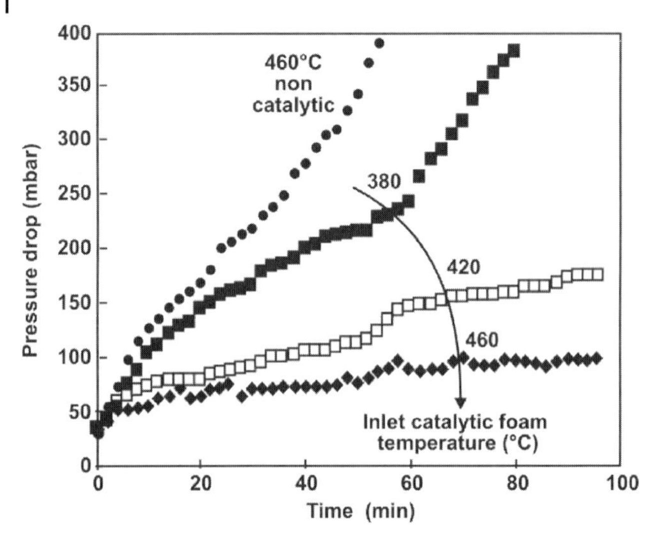

Fig. 6 Performance of catalytic foam traps during stationary runs. Catalyst: $CsVO_3$+KCl (30 wt % in the foam); foam trap: zirconia-toughened alumina; porosity: 80 %; 65 ppi (manufacturer: ACF-Selee Corporation, Hendersonville, North Carolina); superficial velocity: 0.7 m s^{-1}; soot feed concentration: 0.106 g m^{-3} (STP).

The performance of the developed traps was tested in the filtration of soot-laden streams generated by substoichiometric combustion of acetylene or by a diesel engine placed on a bench.

The most ambitious goal of deep-filtration catalytic foams is self-regeneration at normal diesel engine exhaust gas temperatures (180–350 °C) without the use of any active system to increase the exhaust gas temperature, such as those employed in the PSA system. Basically, this would mean that the catalyst should be capable of burning out the soot as soon as it is filtered, so that the soot holdup in the trap and its pressure drop are kept constant (balance-point operation). Figure 6 shows how this condition can be reached with the $CsVO_3$ + KCl catalyst only at rather high temperatures (above 420 °C) that are seldom reached in the exhaust stream. Better results could be achieved with spillover catalysts such as $Cs_4V_2O_7$ [43]. However, lower self-regeneration temperatures than 350 °C could not be reached, although this is 100 °C lower than what is achievable with fuel additives [16]. Some concerns remain regarding the stability of these catalysts owing to their long-term evaporation or sensitivity to water ($Cs_4V_2O_7$ shows some water solubility).

These issues, together with the average filtration efficiency of ceramic foams (50 % at acceptable pressure drops), hamper the application of CATATRAP technology, even though EURO IV legislation limits were achieved in standard driving cycles.

The solid nature of the perovskite catalysts employed for wall-flow filter activation (Fig. 5) means poorer catalyst-to-carbon contact conditions, despite the fact that the combustion synthesis technique produces a highly corrugated catalyst layer. As for noncatalytic traps, regeneration is not viable at normal diesel exhaust temperatures,

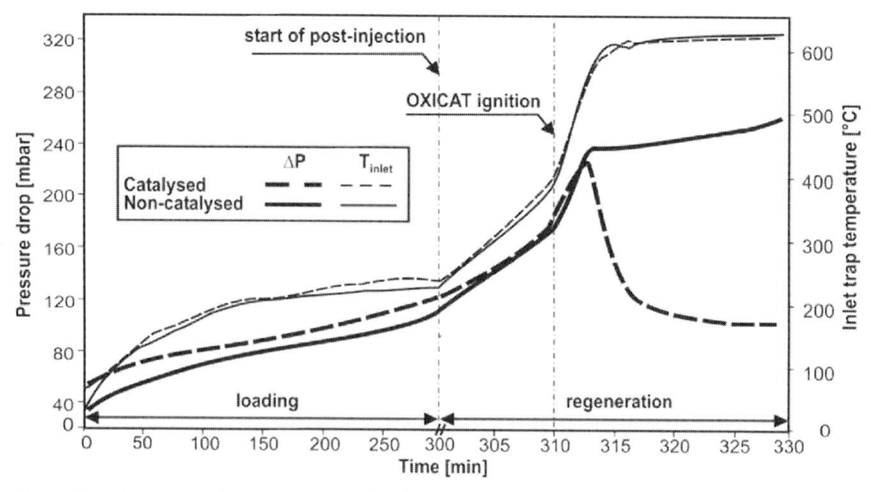

Fig. 7 Trap loading and regeneration cycles for a catalytic ($LaCrO_3$) and a noncatalytic wall-flow trap (Cordierite by Corning). (Reprinted from [45], Copyright 2003, with permission from Korean Institute of Chemical Engineers.)

and active measures must be adopted. Runs were performed on an engine bench provided with a fuel post-injection system and a preoxidizer like that employed in the PSA system. After a prolonged trap-loading period (final soot content: $10\,g\,L^{-1}$), regeneration was induced by the rapid temperature rise caused by fuel post-injection/combustion. Figure 7 shows how the presence of the catalyst over the trap enables much faster regeneration compared to a noncatalytic trap and has the benefit of savings in post-injected fuel and a lower fuel penalty for the particulate-abatement system.

This feature, together with the absence of fuel additives and the related storage and dosing systems, the high filtration efficiency of the wall-flow monoliths, and the good catalyst stability, makes this technology quite attractive for car manufacturers, who are believed to be adopting this technology soon.

In the context of diesel particulate removal from mobile sources two pioneering studies are noteworthy. Ciambelli et al. are currently developing a catalytic foam trap whose regeneration is induced by microwaves [46]. To optimize energy consumption and the related fuel penalty, the foam material must not absorb microwaves (the authors used alumina foams), whereas the catalyst composition must be tuned to maximize microwave absorption and redox properties. In this way, heat would be released exactly where needed (i.e., at the catalyst–soot locations) and thereby maximize overall system efficiency. Conversely, Setiabudi et al. [42] developed an aftertreatment device combining a catalytic foam trap ahead of a wall-flow catalytic filter. The foam was activated with a Pt catalyst whose role was mainly that of enabling NO to NO_2 oxidation to promote the CRT effect. Besides this, the foam can act as a sort of particulate agglomerator by favoring contact among soot particles, to the benefit of their filterability. The combination of both filters allowed good particulate collection and a faster regeneration. The combination of two filters may increase the pressure drop somewhat.

The use of catalytic foam filters has also been recently proposed and tested for the treatment of flue gases from stationary sources such as large boilers based on combustion of fuel, wood, peat, and coal on ferry boats, incinerators, or simply large diesel engines for power generation [43]. A specific filter baghouse reactor was conceived as shown in Fig. 8, on the grounds of the fact that foams can be produced in a tubular shape. A quite active $Cs_4V_2O_7$ catalyst was deposited into zirconia-toughened alumina foam traps (pore size: 50 ppi, equivalent to about 400 μm, thickness 17 mm), whose abatement performance was evaluated in a specific pilot plant based on a diesel engine. Good abatement efficiency (about 50 %), coupled with low pressure drop across the trap (< 10^4 Pa), were obtained for superficial velocities (2 m s^{-1}) and temperatures (about 400 °C) of industrial interest. In this context, important catalyst-related issues are:

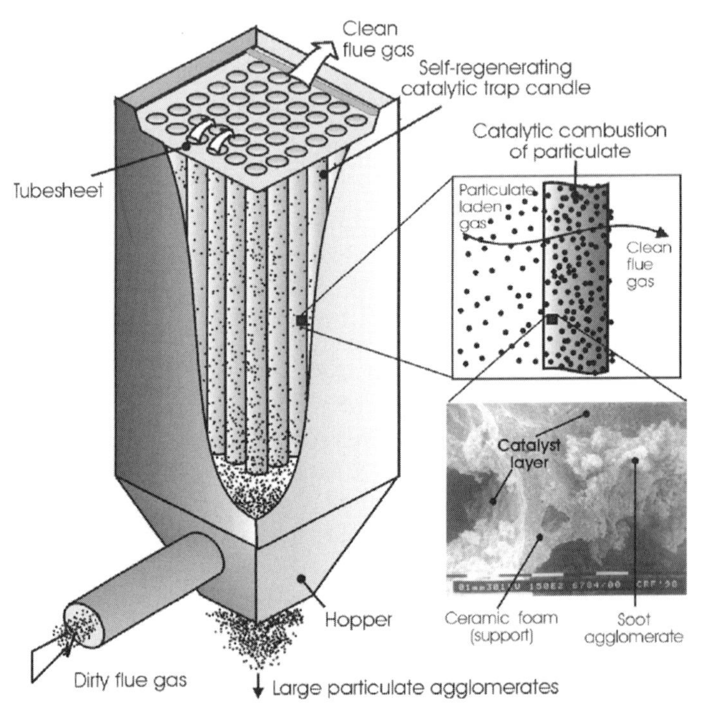

Fig. 8 Schematic view of a catalytic-trap baghouse and of the trap microstructure.

- Catalyst loading must be optimized to balance the need for high activity (filtration and catalytic conversion must be achieved in the limited space and time constraints imposed by the system) and the increase in pressure drop induced by the presence of the catalyst in the porous matrix. A catalyst deposition route based on filter impregnation with an aqueous suspension of the catalyst powders, followed by microwave drying and calcination, was adopted to achieve an optimal catalyst loading (14 wt %).

- Catalyst thermochemical stability: the highly active pyrovanadate ($Cs_4V_2O_7$) catalyst was found to be somewhat sensitive to water vapor. Some catalyst deactivation was indeed noticed in the long term at the front side of the filter. This is clearly visible in Fig. 9, where the orange color of the front side of the regenerated foam is attributed to the localized formation of the less active metavanadate $CsVO_3$. Nevertheless, the activity remained high enough to guarantee the above-mentioned efficiency, but more stable catalyst compositions are probably needed to achieve long-term viability of this technology.

Fig. 9 View of a foam containing $Cs_4V_2O_7$ catalyst after hydrothermal ageing. Left part: after regeneration by carbon combustion in oven at 450 °C; Right part: before regeneration. Inlet side in front of view. (Reprinted from [13], Copyright 2002, with permission from Elsevier Science Ltd.)

In addition, Fig. 10 clearly shows the superior performance of such a catalytic trap in comparison to a noncatalytic one. The presence of the catalyst allows the rapid oxidation of a significant fraction of the trapped soot and thereby reduces its holdup in the foam and consequently the pressure drop.

Figure 10 allows some conclusions to be drawn on modeling issues, described later in more detail. After measurement of key kinetic parameters for the catalytic combustion of carbon particulate (i.e., activation energy, reaction order in oxygen) and a deep characterization of the permeation properties of the filter, a mathematical model was validated by using experimental data obtained with catalytic and noncatalytic traps [1]. The agreement of the model with the experimental data was in both cases good, which is particularly promising for design purposes.

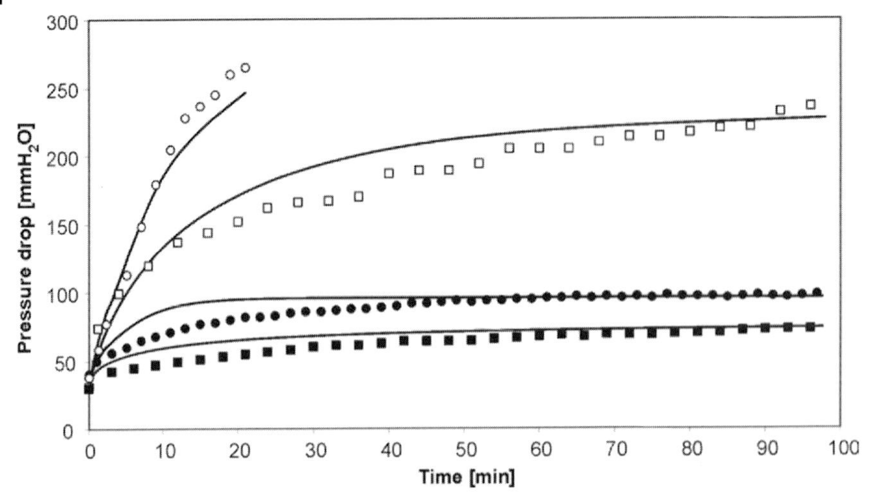

Fig. 10 Experimental and calculated (——) pressure drops across a noncatalytic (empty symbols) and catalytic (full symbols) foam trap as a function of soot feed concentration: squares: 0.085 g m^{-3} (STP); circles: 0.135 g m^{-3} (STP). Operating conditions: superficial velocity: 2 m s^{-1}; foam temperature: 440 °C. (Reprinted from [43], Copyright 2002, with permission from Elsevier Science Ltd.)

5.2.3.2
Abatement of Gaseous Pollutants and Fly-Ash

In this application, the porous body of the filter is used to host a catalyst capable of promoting the abatement of gaseous pollutants from large stationary sources (e.g., waste incinerators, pressurized fluidized-bed coal combustors, large diesel engines, boilers, biomass gasifiers, etc.). A list of commercially available inorganic filters potentially suitable to host catalysts for the above application is provided in Table 1.

This concept dates back to the late 1980s, when Babcock & Wilcox filed several patents concerning the so called SO_x-NO_x-Rox Box process [47], in which SO_2 and NO_2 removal (the former by adsorption on lime, the latter by catalytic reduction with ammonia) is accomplished by using V–Ti-coated catalytic fiber filters. Earlier studies of the authors at Politecnico di Torino concerned the development of ceramic catalytic filter candles based on sintered alumina grains for the simultaneous removal of fly ashes and nitrogen oxides [48, 49], as well as fly ashes and volatile organic compounds [50].

No cellular materials were employed in the above cases. The only viable mechanism to remove fly ashes is indeed cake filtration in bag filters, followed by periodic cleaning by reverse jet pulse. As opposed to diesel particulates, fly ashes cannot be removed by catalytic combustion and there is simply no benefit in having them trapped inside the filter where the catalyst is present. Penetration of fly ashes into the filter matrix might even lead to catalyst deactivation and must be prevented. This can be achieved with ceramic filters based on fibers or made of sintered grains. The latter

Table 1 Commercially available inorganic filters for high-temperature applications.

Filter type	Producers	Main applications
Rigid ceramic sintered filters	Cerel, Universal Porosics, Industrial Filters and Pumps, NOTOX, Schumacher, US Filters, Ibiden, etc.	coal gasification, fluidized-bed coal combustion, waste incineration, etc.
Pulp-type $SiO_2 \cdot Al_2O_3$ fiber candle filters	BWF, Cerel, etc.	separation of metal dust, fluidized-bed coal combustion, waste incineration
Ceramic woven fabric filters	3M, Tech-in-Tex	catalyst recovery, coal-fired boilers, metal smelting, soot filtration
Ceramic cross-flow filters	Coors	applications up to 1500 °C
Ceramic cordierite monoliths	Corning, Ceramem, NGK insulators, etc.	coal gasification, fluidized-bed coal combustion, waste incineration, soot filtration, etc.
Ceramic (SiC, ZTA, ZTM) foam filters	Selee Corp., Saint Gobain, Ecoceramcs, etc.	hot-metal filtration, diesel particulate removal, etc.
Sintered porous metal powder filters	Pall, Mott, Newmet, Krebsöge, Fuji, etc.	catalyst and precious metal recovery
Sintered stainless steel semirigid fiber filters	Bekaert, Memtec, etc.	catalyst and metal dust recovery, soot filtration, etc.

filters are generally made of a thin front layer with reduced pore size to achieve surface dust filtration and an inner structure with large pores to reduce pressure drop, provide sufficient mechanical strength, and to host the catalyst.

A new catalytic filter concept employing cellular materials has, however, been conceived recently with the perspective of applying it to the treatment of flue gases from waste incineration. In this context [51], Goretex fabric filter bags are widely adopted to remove both fly ashes and carbon particles suitably dispersed in the flue gases to remove dioxins and heavy metal traces. To provide the filters with the additional function of NO_x and VOC (for example dioxins) removal by catalytic redox processes, a catalytically active ceramic foam candle can be inserted in each filter bag, according to the scheme reported in Fig. 11 [52]. As the Goretex filter bags generally work at 200–210 °C, an innovative catalyst, active in such a temperature range for the selective catalytic reduction (SCR) of NO_x with ammonia and simultaneous combustion of VOCs, must be developed and applied over the foam structure to obtain the above-mentioned multifunctional operation.

Currently adopted SCR units use V_2O_5–TiO_2 catalysts which operate at 320–400 °C in honeycomb catalytic converters fed with space velocities of 20 000 h^{-1} [53]. A specific catalyst has thus been developed at Politecnico di Torino, based on a mechanical mixture of TiO_2 supported V_2O_5–WO_3 [54] and CeO_2-supported MnO_x catalysts recently proposed by other authors [55].

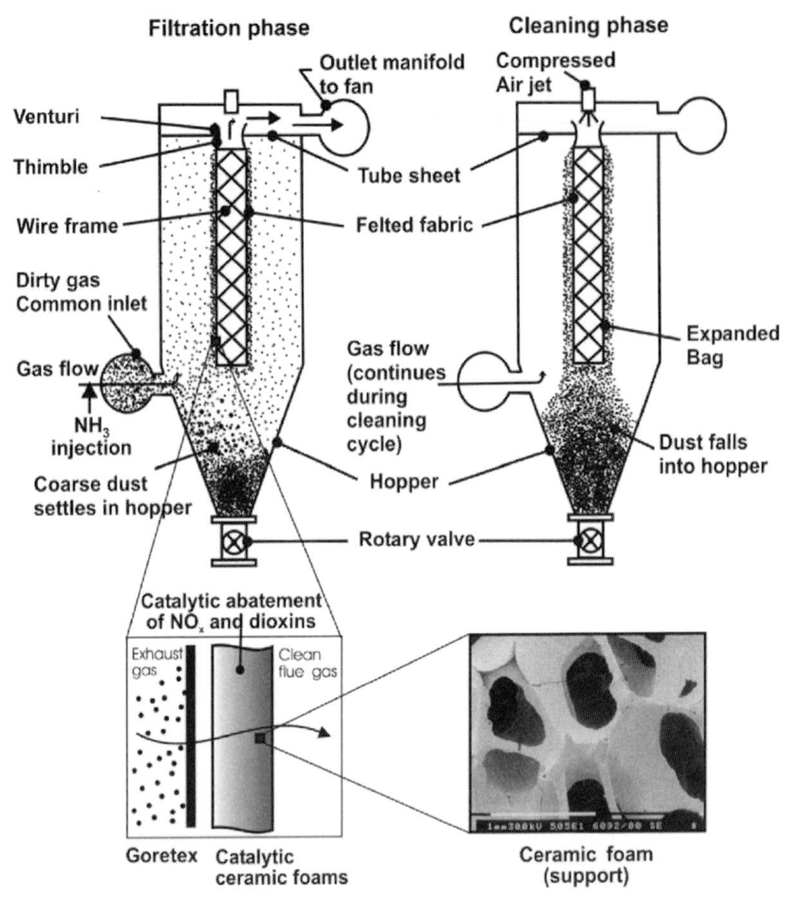

Fig. 11 Schematic view of a multifunctional filter for the simultaneous abatement of fly ashes, nitrogen oxides, and dioxins. (Reprinted from [52], Copyright 2004, with permission from Elsevier Science Ltd.)

Such catalysts were lined inside a support structure [zirconia-toughened alumina (ZTA) foam disk; 50 ppi (equivalent to a pore size of 4×10^{-4} m); thickness: 10 mm; manufacturer: Saint Gobain] by special operating procedures (impregnation in ultrasonic bath followed by microwave drying and final calcination at 400 °C). This procedure was tailored to achieve a catalyst distribution as even as possible throughout the disk body. Moreover, it enabled good adhesion to the porous walls of the structured support, and a rough surface of the catalyst layer with only a moderate increase in pressure drop across the porous medium. Subsequent impregnation steps, with intermediate drying stages, were carried out to reach a catalyst loading of 18 wt%.

The obtained catalytic foam was then tested on a synthetic gas mixture representing real incinerator flue gases (1000 ppm NO, 1/1 NO/NH_3 molar ratio, 10 vol% O_2,

200 ppm benzene; balance He) to check its pollutant-abatement performance at superficial velocities of industrial interest. On the basis of the work in Ref. [56], the selection of benzene as a representative for PAHs is conservative, since its catalytic combustion is initiated at higher temperatures than those of the much more harmful PAHs and chlorinated aromatics mentioned above.

Figure 12 shows how the prepared filter can achieve NO conversion as high as about 90 % in the temperature range 200–250 °C. Reduced conversion at higher temperatures is due to occurrence of the undesired oxidation of NH_3 to NO.

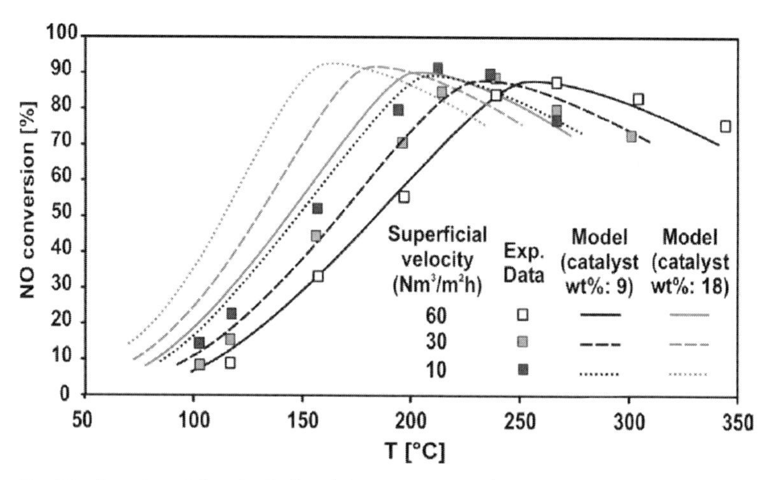

Fig. 12 Experimental and calculated data concerning the NO conversion performance of catalytic foams (catalyst: MnO_x-CeO_2+V_2O_5-WO_3·TiO_2) at different superficial feed velocities. (Reprinted from [52], Copyright 2004, with permission from Elsevier Science Ltd.)

A similar conversion efficiency was also obtained for benzene (80 % at ca. 210 °C).

A model developed by Saracco and Specchia [49] was employed to fit the experimental data after suitable evaluation of the reaction kinetics of the catalyst and the permeation properties of the catalytic foam. Figure 12 shows how good agreement between experimental data and model calculations could only be reached by assuming that the catalyst loading of the filter was only half of the true value.

After careful structural analysis of the catalytic foams under the scanning electron microscope (SEM), a likely explanation for the fact that the observed abatement activities were lower than expected was found in the tendency of the deposited catalyst to occasionally clog the pores of the foam (see Fig. 13). Pore clogging would lead to portions of the porous matrix in which the catalyst is present but cannot be adequately reached by the reactants, and its activity thus remains unexploited. From this viewpoint and on the grounds of the earlier described model predictions, it can be assumed that only half of the catalyst present can actually exert its activity. Hence, additional developmental work must be spent to further improve the catalyst deposition route.

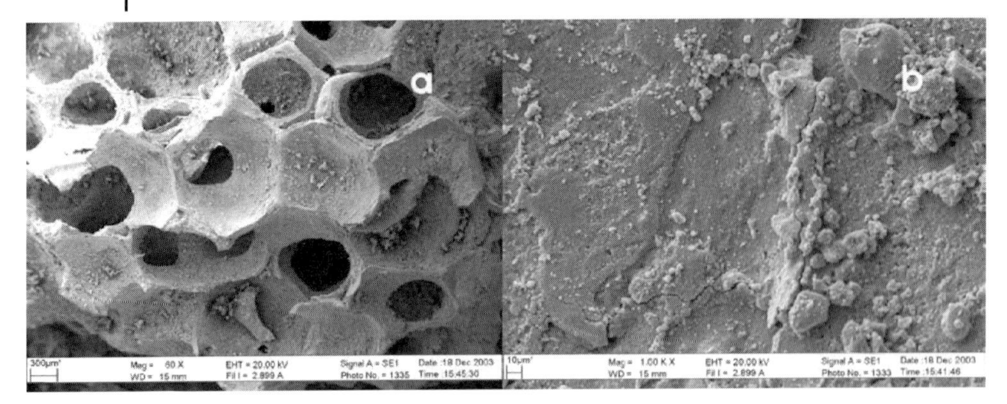

Fig. 13 SEM images of a catalytic foam filter (18 wt % $MnO_x{\cdot}CeO_2$). Left: view of some pores showing tendency for pore plugging by the catalyst; Right close-up view of a catalyzed pore wall. (Reprint from [52], Copyright 2004, with permission from Elsevier Science Ltd.)

Honeycomb cellular materials for the low-temperature reduction of NO_x with ammonia were recently synthesised by Valdés-Solìs et al. [57, 58]. The carbon substrate (Fig. 14) was obtained by coating the ceramic cellular monolith with a polymer solution (carbon precursor). Phenolic (resol and novolac) or furan resins and polysaccharides (sucrose and dextrose) were used as carbon precursors. The coated material was cured, carbonized (700–1000 °C in N_2), and activated (steam treatment at 700 °C leading to a specific surface area of about 800 $m^2\ g^{-1}$) prior to impregnation with the active phase. Manganese oxide was selected as the catalyst and deposited over the support. The produced structured catalysts showed a rather good, though improvable, NO_x conversion (in the range 34–73 %) at 150 °C for a space velocity of about 4000 h^{-1}. Gasification of the support was negligible under the above-mentioned conditions.

Fig. 14 SEM image of a carbon–ceramic cellular monolith (14 wt % carbon). (Reprinted from [57], Copyright 2001, with permission from Elsevier Science Ltd.)

5.2.4 Modeling

The modeling of mass transfer and reaction in catalytic filters and traps can be compared, in a first approximation, with the twin problem concerning honeycomb catalysts. The pores of the filters are counterparts of the channels of the monolith, whereas the catalyst layer deposited on the pore walls of the filter are related to the walls separating the honeycomb channels, which are in general exclusively made of catalytic material or catalyst-lined ceramic structures [57, 58].

Owing to the comparatively small size of the pores (100–500 μm, as opposed to a size of a few millimeters on average for the honeycomb channels) and the thinness of the catalyst layer (a few micrometers, compared some tenths of a millimeter for the catalytic wall of the honeycomb channels), both internal and external mass-transfer limitations to pollutant conversion in catalytic filters can frequently be neglected. Hence, the achievable conversion per unit catalyst mass is maximized, which is favorable owing to the need to convert the noxious gases within short space and time constraints. The above-mentioned advantage is, however, compensated, at least in part, by the higher pressure drop entailed by the filter. However, this last feature is a rather obvious consequence of the combination of particulate filtration and catalytic reaction.

The most frequently used approach to modeling catalytic filters is of the *pseudohomogeneous* type. The porous matrix is considered as ideally homogeneous by lumping the prevalent constituting parameters (porosity, tortuosity, permeability, catalyst concentration, etc.) into average values per unit volume. This allows material, component, and energy balances throughout the porous structure to be described by means of a set of partial differential equations (see, e.g., Ref. [59] for a typical set of these equations).

One of the most accurate modeling studies on catalytic filters, performed by Saracco and Specchia [59, 60], involved a γ-Al_2O_3-coated α-Al_2O_3 granular filter on which a model reaction (2-propanol dehydration, catalyzed by the γ-Al_2O_3 itself) was performed. The major conclusion drawn by the authors was that a certain degree of catalyst bypass occurs in filters prepared with a single catalyst deposition step. In other words, they concluded that a single catalyst deposition step induces a uneven distribution of catalyst in the filter, which means that some pores have a higher catalyst loading than others. This forced them to adopt for these filters a pseudohomogeneous model based on a bimodal pore size distribution to account for the fact that the less catalytically active pores are also more permeable, which implies lower overall pollutants conversion throughout the filter. However, a second deposition cycle seemed to remediate the above uneven distribution, and thus enabled the use of a simpler model based on a monomodal pore size distribution.

However, on the basis of a similar modeling approach for NO_x-abatement filters [49] (see also at the end of Section 5.2.3.2), it was also emphasized that high loadings of catalyst may result in a similar unevenness problem as long as pore plugging prevents some catalyst from being reached by the permeating gases.

All the above modeling studies showed therefore the importance of having a proper knowledge of the pore texture and catalyst distribution within the catalytic filter, since these parameters can seriously affect its performance. This suggests, in line with Ref. [61], the need for proper characterization of the porous structure of catalytic filters (pore connectivity, pore size distribution, presence of dead-end pores, etc.), since each of these features may play a primary role in reactor performance. On the basis of such characterization, valuable information could be drawn for selecting or optimizing preparation routes. This is an even more stringent requirement for cellular structures for diesel particulate removal where, in addition to the catalytic function, filtration mechanisms must taken into account.

Several EU projects (GROWTH GRD2-2001-50038: "Simulation Tool for Dynamic Flow Analysis in Foam Filters" – STYFF-DEXA; GROWTH 99 GRD1-1999-10588: "System Level Optimization and Control Tools for Diesel Exhaust Aftertreatment" – SYLOC-DEXA; GROWTH 99 GRD1-1999-10451: "Advanced Regeneration Technologies for Diesel Exhaust Particulate Aftertreatment" – ART-DEXA) have been or are being carried out to develop and apply an experimentally validated simulation tool that allows accurate characterization of foam materials applications and to enable the systematic optimization of cellular structures with high separation efficiency for diesel exhaust aftertreatment.

In such efforts, pseudohomogeneous models [1, 5] have been developed that include, in addition to mass and heat balances, proper expressions for the filtration mechanisms of inertial impaction, interception, and Brownian diffusion, as well as an additional mechanism for the blow-off of trapped particulate in the flowing gases. The model could properly predict the experimental behavior of the trap, as already discussed (Fig. 10), with the use of a single fitting parameter related to the re-entrainment mechanism.

This performance could perhaps be further improved by a completely different approach based on the Lattice Boltzmann (LB) method which is more suitable to achieve a real understanding of the physical and chemical phenomena occurring inside the foam matrix and how they interact.

Pseudohomogeneous approaches, typical of conventional computational fluid dynamics (CFD) methods based on adaptive unstructured computational grids, become rather time-consuming due to the necessary high grid resolution near complex-shaped surfaces, as found for example inside a filter material. An efficient alternative for the analysis of this kind of flow is generally considered to be the LB method [62]. With this method boundary conditions on complex, rigid surfaces can be satisfied on a fixed equidistant grid (lattice) with second-order accuracy. Essential advantages lie in the simple underlying algorithm, which facilitates parallel implementation, intrinsic stability, and capability to deal with arbitrarily shaped geometrical boundaries. This method is therefore ideally suited for studying flow phenomena in detail. The underlying geometrical model for given materials can be generated by computer tomographic methods [3, 63, 64] or lower order correlation functions [65], as shown in Fig. 15. A typical computed velocity pattern inside a porous ceramic foam is shown in Fig. 16.

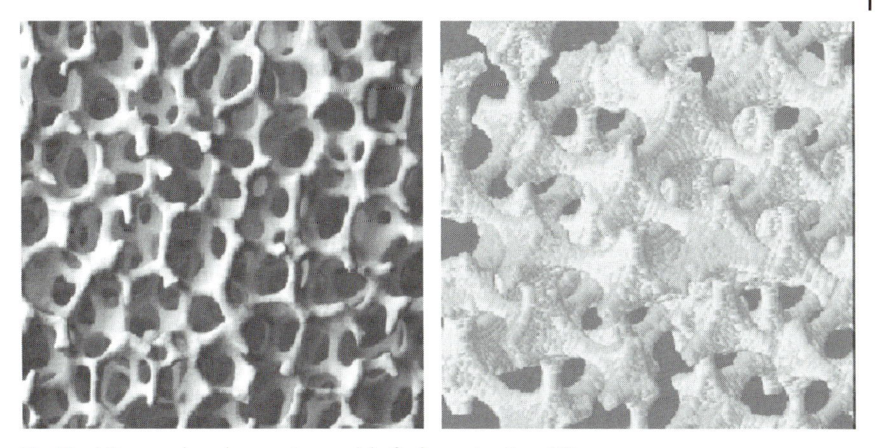

Fig. 15 Micrograph and computer model of a foam structure [62].

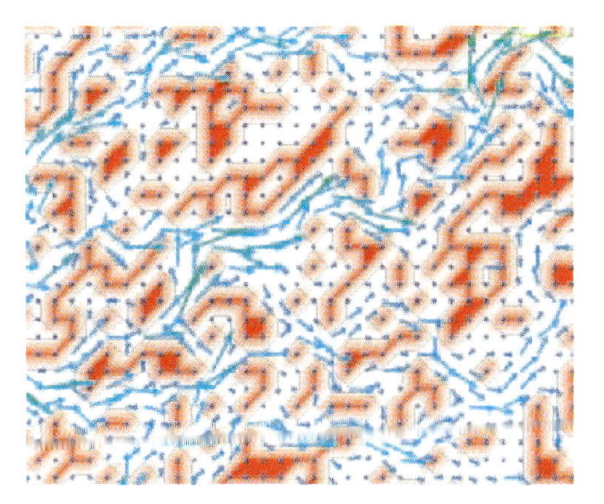

Fig. 16 Simulated flow pattern inside a ceramic foam (Lattice Boltzmann method) [62].

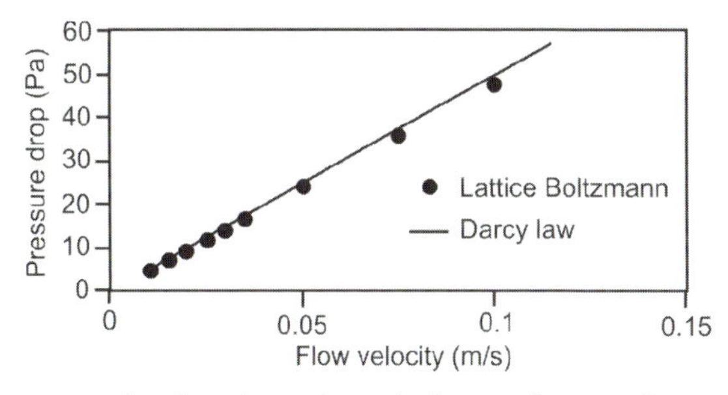

Fig. 17 Flow through a package of spheres as function of pressure drop (porosity: 0.46; permeability: $8 \times 10^{-8}\ m^{-2}$) [62].

Furthermore, the mean pressure-drop coefficients of porous structures can be estimated by integration over the complete flow domain. Usually, these coefficients are estimated by experiments or by empirical correlation functions such as Darcy's law [65]. A comparison of pressure drop coefficients predicted by LB calculations and Darcy's law is depicted in Fig. 17.

5.2.5 Summary

The basic properties and the potential of cellular materials in the field of gas filtration have been outlined, with particular emphasis on catalytic filters for simultaneous abatement of dust (including soot) and gaseous pollutants. On the basis of the discussed literature, and of the growing interest of producers of high-temperature filters, it seems reasonable to predict that penetration into the market will probably be gained in a few years. In this context, the leading application will definitely be diesel particulate removal from vehicle exhaust gases, for which the basic cost and durability requirements seems to have been reached, at least for some prototype systems.

As far as large-scale industrial applications are concerned (bag-house filters for incinerators, boilers, etc.), the extent of market penetration will largely depend on the long-term durability of the catalytic filters and on the initial investment costs of these rather innovative products. The work already done on catalytic filters demonstrated that it is possible to achieve nearly complete catalytic conversion of gaseous pollutants (e.g., NO_x) at superficial feed velocities of industrial interest [10–60 m^3 m^{-2} h^{-1} (STP)], which makes coupling of filtration and catalytic abatement convenient. The key issue is now to assess how the catalytic filters will resist long-term exposure to relatively harsh environments containing potential poisons for the catalyst itself (e.g., fly ash, sulfur, chlorinated compounds, steam, etc.).

Furthermore, preparation techniques should also be improved, always bearing process economics in mind, so that catalytic layers are thoroughly attached to the pore walls of the filters and exhibit good resistance to the mechanical stresses which arise from thermal fatigue and the jet-pulse cleaning technique. A key challenge is to increase the catalyst loading of the cellular material without causing occasional pore plugging and consequent high pressure drop. This future research lies in the hands of materials scientists.

Finally, new models are needed for the microstructural design (i.e., definition of optimal pore size, type of pore texture, optimized catalytic coating thickness; etc.) of catalytic and noncatalytic cellular filters. Such models, based for instance on the Lattice Boltzmann approach, should help achieve a more realistic representation of the basic mechanism governing flow through porous media and filtration of particles.

References

1 Ambrogio, M., Saracco, G., Specchia, V., *Chem. Eng. Sci.*, **2001**, *56*, 1613–1621.

2 Saracco, G., Russo, N., Ambrogio, M., Badini, C., Specchia, V., *Catal. Today*, **2000**, *60*, 33–41.

3 Boretto, G., Amato, I., Merlo, A.M., Characterization of Ceramic Foam by Computed Tomography with High Spatial Resolution. World Ceramic Congress and Forum on New Materials Proceedings, 9th CIMTEC, Florence, June 14–19, 1998.

4 Konstandopoulos, A.G., Kostoglou, M. *Combust. Flame* **2000**, *121*, 488–500.

5 Ambrogio, M., Saracco, G., Specchia, V., van Gulijk, C., Makkee, M., Moulijn, J.A., *Sep. Purif. Technol.*, **2002**, *27*, 195–209.

6 Swars, H., German Patent 3,619,360, **1987**.

7 Cai, J., Chen, D., Wan, H. *J. Chim. Ceram Soc.*, **1994**, *22*, 458–469.

8 Pestryakov, A.N., Fyodorov, A.A., Gaisinovic, M.S., Shoruv, V.P., Fyodorova, I.V., Gubaydulina, T.A. *React. Kinet. Catal. Lett.*, **1995**, *54*, 167–172.

9 Banhart, J., *Progr. Mater. Sci.*, **2001**, *46*, 559–632.

10 Dautzember, F.M., Mukherjee, M., Process Intensification using Multifunctional Reactors, *Chem. Eng. Sci.*, **2001**, *56*, 251–267.

11 Carty, W.M., Lednor, P.W., *Curr. Opin. Solid State Mater. Sci.*, **1996**, *1*, 88–95

12 Saracco, G., Specchia, V., Catalytic Filters for Flue-Gas Cleaning, in *Structured Catalysts and Reactors*, (Eds.: Cybulski, A., and Moulijn, J.A.), Marcel Dekker, New York, **1998**, pp. 417–434.

13 Mauderly, J.L. in *Environmental Toxicants: Human Exposures And Their Health Effects*, (Ed.: Lippmann, M.), 2nd edition, Wiley, New York, **1999**, pp. 193–241.

14 Zelenka, P., Kriegler, W., Herzog, P.L., Cartellieri, W.P., *NO_x Control Strategies for Diesel Engines*, SAE Paper 900602, **1990**.

15 Knothe, G., Bagby, M.O., Ryan Iii, T.W., Callahan, T.J., Wheeler, H.G., *Vegetable Oils as Alternative Diesel Fuels: Degradation of Pure Triglycerides During the Precombustion Phase in a Reactor Simulating a Diesel Engine*, SAE Paper 920194, **1992**.

16 Savat, O., Marez, P., Belot, G., *Passenger Car Serial Application of a Particulate Filter System on a Common-Rail Direct-Injection Diesel Engine*, SAE Paper 2000-01-0473, **2000**.

17 Evans, A.G., Hutchinson, J.W., Fleck N.A., Ashby, M.F., Wadley H.N.G., *Prog. Mater. Sci.*, **2001**, *46*, 309–327.

18 Lepperhoff, G., Lüders, H., Barthe, P., Lemaire, J., *Quasi-Continuous Particle Trap Regeneration by Cerium-Additives*, SAE Technical Paper 950369, **1995**.

19 Cooper B.J., Thoss J.E., *Role of NO in Diesel Particulate Emission Control*, SAE Paper 890404, **1989**.

20 Allanson, R., Cooper, B.J., Thoss, J.E., Uusimäki, A., Walker, A.P, Warren, J.P., *European Experience of High Mileage Durability of Continuously Regenerating Diesel Particulate Filter Technology*, SAE Paper 2000-01-0480, **2000**.

21 Teraoka Y., Nakano K., Shangguan W. F., Kagawa S., *Catal. Today*, **1996**, *27*, 107.

22 Marinangeli, R.E., Homeier, E.H. and Molinaro, F.S. in *Catalysis and Automotive Pollution Control* (Eds.: Crucq, A., Frennet, A.), Elsevier Science Publishers, Amsterdam, **1987**, p. 457.

23 Ahlström, A.F., Odenbrand, C.U.I., *Appl. Catal.*, **1990**, *60*, 157.

24 Mc Kee, D.W., *Carbon*, **1970**, *8*, 131.

25 Van Doorn, J., Varloud, J., Meriaudeau, P., Perrichon, V., Chevrier, M., Gautier, G., *Appl. Catal. B*, **1992**, *1*, 117

26 Setzer, C., Schütz, W., Schüth, F. in *New Frontiers in Catalysis* (Eds.: L. Guczi et al.), Proceedings of the 10th International Congress on Catalysis, Budapest, July 1992, Elsevier Science Publishers, Amsterdam, **1993**, p. 2629.

27 Watabe, Y.,Yrako, K., Miyajima, T., Yoshimoto, T., Murakami, Y., SAE Paper 830082, **1983**.

28 Sato, H., Akamatsu, H., *Fuel*, **1954**, *33*, 195.

29 Mc Kee, D.W., Chatterji, D., *Carbon*, **1975**, *13*, 381.

30 Neeft, J.P.A., PhD Thesis, Delft University, **1995**.

31 Mross, W.D., *Catal. Rev. Sci. Eng.*, **1983**, *25*, 591.

32 Saracco, G., Serra, V., Badini, C., Specchia, V., *Appl. Catal. B*, **1996**, *11*, 329.

33 Inui, T., Otowa, T., *Appl. Catal.*, **1985**, *14*, 83.

34 Pattas, K.N., Samaras, Z.C., Patsatzis, N., Michalopulou, C., Zogou, O., Stamatelos, A., Barkis, M., SAE Paper 900109, **1990**.

35 Fino, D., Saracco, G., Specchia, V., *Chem. Eng. Sci.*, **2002**, *57*, 4955–4962.

36 Fino, D., Russo, N., Badini, C., Saracco, G., Specchia, V., *AIChE J.*, **2003**, *49*, 2173–2180.

37 Fino, D., Saracco, G., Specchia, V. *Ind. Ceram.*, **2002**, *22*, 37–43.

38 Ciambelli, P., Palma, V., Russo, P., Vaccaro, S., *Catal. Today*, **2002**, *73*, 363–370.

39 Ciambelli, P., Palma, V., Russo, P., Vaccaro, S., *Catal. Today*, **2002**, *75*, 471–478.

40 van Setten, B.A.A.L., Bremmer, J., Jelles, S.J., Makkee, M., Moulijn, J.A., *Catal. Today*, **1999**, *53*, 613–621.

41 van Setten, B.A.A.L., Spitters, C.G.M., Bremmer, J., Mulders, A.M.M., Makkee, M., Moulijn, J.A., *Appl. Catal. B*, **2003**, *42*, 337–347.

42 Setiabudi, A., Makkee, M., Moulijn, J.A., *Appl. Catal. B*, **2003**, *42*, 35–45.

43 Fino, D., Fino, P., Saracco, G., Specchia, V., *Chem. Eng. Sci.*, **2003**, *58*, 951–958.

44 Fino, D., Russo, N., Saracco, G., Specchia, V., *J. Catal.*, **2003**, *217*, 367–375.

45 Fino, D., Fino, P., Saracco, G., Specchia, V., *Korean J. Chem. Eng.*, **2003**, *20*, 445–450.

46 Palma, V., Russo, P., D'Amore, H., Ciambelli, P., *Top. Catal.*, **2004**, *30/31*, 261–264.

47 Doyle, J.B., Prish, E.A., Downs, W. (Babcock and Wilcox Co.), US Patent 4,793,981, **1988**.

48 Saracco, G., Specchia, S., Specchia, V., *Chem. Eng. Sci.*, **1996**, *51*, 5289–5297.

49 Saracco, G., Specchia, V., *Appl.Therm. Eng.*, **1998**, *18*, 1025–1035.

50 Saracco, G., Specchia, V., *Chem. Eng. Sci.*, **2000**, *55*, 897–908.

51 Bonte, J.L., Fritsky, K. J., Plinke, M.A., Wilken, M., *Waste Manag.*, **2002**, *22*, 421–426.

52 Fino, D., Russo, N., Saracco, G., Specchia, V. *Chem. Eng. Sci..*, **2004**, *59*, 5329–5336.

53 Forzatti, P., *Appl. Catal. A*, **2001**, *222*, 221–236.

54 Koebel, M., Madia, G., Elsener, M., *Catal. Today*, **2002**, *73*, 239–247.

55 Qi, G., Yang, R.T., *J. Catal.*, **2003**, *217*, 434–441.

56 Weber, R., Sakurai, T., Hagenmaier, H., *Appl. Catal. B*, **1999**, *20*, 249–256.

57 Valdés-Solìs, T., Marbán, G., Fuertes, A.B., *Catal. Today* **2001**, *69*, 259–264.

58 Valdés-Solìs, T., Marbán, G., Fuertes, A.B., *Microporous Mesoporous Mater.* **2001**, *43*, 113–126.

59 Saracco, G., Specchia, V., *Ind. Eng. Chem. Res.* **1995**, *34*, 1480–1487.

60 Saracco, G., Specchia, V., *Chem. Eng. Sci.* **1995**, *50*, 3385–3394.

61 Mc Greavy, C., Draper, L., Kam, E.K.T., *Chem. Eng. Sci.* **1995**, *49*, 5413–5422.

62 Wassermayr, C., Brandstätter, W., Prenninger, P. *THIESEL 2002 Conference on Thermo- and Fluid-Dynamic Processes in Diesel*, **2002**, CD-ROM Proceedings.

63 Bernsdorf, J., Brenner, G., Durst, F., *Comput. Phys. Commun.*, **2000**, *129*, 247–255.

64 Hall, M. J., Bracchini, M., *J. Am. Ceram. Soc.*, **1997**, *80*, 1298–1305.

65 Singh, M., Mohanty, K. K., *Chem. Eng. Sci.*, **2000**, *55*, 5393–5403.

5.3
Kiln Furnitures

Andy Norris and Rudolph A. Olson III

5.3.1
Introduction

A definition for kiln furniture might be movable articles, similar to chairs and tables, which are necessary and useful for firing materials at high temperature. At the simplest level, when a ceramic or metal component is fired, it must sit on something, and this something is usually kiln furniture.

When a ceramic or powdered metal component is manufactured, it typically undergoes a multistep heating process called a thermal cycle, an example of which is shown in Fig. 1. Initially, the green component (prior to firing) is typically heated to an intermediate temperature to allow organic binder or chemically bound water to exhaust from the component at an acceptable rate. Next, the component is heated to relatively high temperature to sinter or chemically bond the material together. Finally, the component is returned to room temperature, possibly with another hold at some other temperature for further processing; this last step might be annealing or exposure to a gas such as oxygen, argon, or hydrogen, to control redox conditions in the final component. The component will often shrink during the cycle, by as much as 30–40 %. The final microstructure, geometry, composition, properties, and performance of the component are a function of the firing process.

One of the main reasons kiln furniture is used to support ware in a kiln is that manufacturers are trying to meet product design specifications and reduce variation in their processes. The goal is to manufacture the component to fall within certain quality-, property-, and performance-based tolerances. Well-designed kiln furniture facilitates this because it provides an inert, nonstick, thermally stable, flat (or supportive) substrate on which a component can sit during its thermal cycle. If the furniture provides these functions without breaking or deforming during thermal cycling, then its continued use will tend to reduce variability in the characteristics of the ware, and the final product will more closely meet the design specifications of the product. For these reasons, kiln furniture is an essential part of many ceramic and metal powder manufacturing processes, but with this vital association come some limitations.

Traditional kiln furniture tends to be dense, which is attractive from the standpoint of mechanical strength, but its high mass is unfavorable with respect to energy consumption, weight, ergonomics, and thermal shock resistance in relatively fast

Cellular Ceramics: Structure, Manufacturing, Properties and Applications.
Michael Scheffler, Paolo Colombo (Eds.)

ISBN: 3-527-31320-6

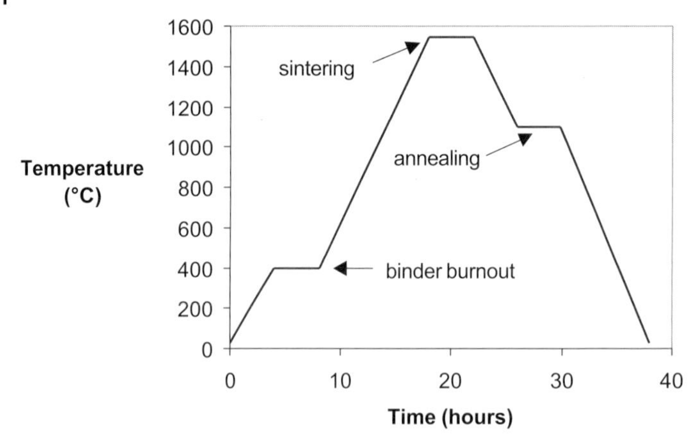

Fig. 1 Example of a thermal cycle for processing a green ceramic component.

thermal cycles. Dense kiln furniture also tends to have relatively high silica content to lower thermal expansion and enhance thermal shock resistance [1]. This may be unacceptable in situations where silica is a contaminant to the ware. Silica also migrates quickly under reducing conditions, as solid SiO_2 converts to vaporous SiO when reduced. Making the kiln furniture highly porous can lessen these limitations. This type of product is usually termed *cellular ceramic* or *ceramic foam*.

The pore size used for kiln furniture generally ranges from about 5 μm to 5 mm, depending on the application, and the volume fraction of porosity tends to be between 50 and 85 %. Just about any pore size or density can be manufactured within this range by using fugitive pore formers, but these processes typically have trouble manufacturing material suitable for use as kiln furniture at large pore size and high volume fraction of porosity. The pore size of reticulated foam manufactured by the foam replication process ranges from about 200 μm to 5 mm and the volume fraction of porosity can be very high (85 %). Processes that generate foam through the use of stable air bubbles or self-expanding slurries containing gas-generating materials can make foam with finer pore size averaging between about 50 μm and 1 mm, but can have difficulty making foam pores in the 1–5 mm range. Casting foam in molds by gelation or coagulation processes produces foams with an average pore size in the range of 50–300 μm. Freeze casting of ceramic slurries generates 5–50 μm sized pores through the formation of ice crystals and the associated 9 % expansion. Vacuum-formed fiber compacts also tend to have pores in this size range. In all of these processes, the pore structure within the foam tends to be interconnected, which is different from structures such as brick containing bubbled alumina, in which the pore size may average hundreds of micrometers, but the pores are virtually disconnected.

The characteristics of lower density, good refractoriness, thermal stability, and improved thermal shock resistance can give cellular ceramics enhanced performance over traditional dense kiln furniture. In addition, the enhanced thermal shock resistance allows furniture to be made essentially silica-free, so reactivity with ware tends to be lower. With these benefits comes one obvious trade-off: lower mechanical strength. For instance, the modulus of rupture (MOR) of reticulated ceramic kiln furniture tends to be 3–7 MPa, whereas the MOR for dense furniture tends to be 30–70 MPa or more. Thus, cellular ceramic furniture may not be able to support a particularly heavy load across a span, or display enough resiliency to function as a pusher plate. Due to its lower strength, the maximum load cellular kiln furniture can support without creep (deformation under mechanical load at high temperature) is generally lower than that of dense furniture. Taken as a whole, these drawbacks must be weighed against the benefits when evaluating cellular ceramics for application as kiln furniture. Examples where cellular kiln furniture performs extremely well are stackable 7 × 20 × 1 cm setters for firing oxygen sensors, 20 × 25 × 1.3 cm plates for firing powder metal components, stackable 10 × 10 × 1 cm two-rail setters for firing dielectric components, and stackable 10 × 15 × 20 cm saggers for firing pelletized ceramic media. There are many niches in this traditional market segment where the performance of a cellular ceramic adds value to an operation. These performance benefits as applied to kiln furniture are the main topic of this chapter.

5.3.2 Application of Ceramic Foam to Kiln Furniture

Advantages of ceramic foam for use as kiln furniture can include 1) longer life, 2) better uniformity of atmosphere surrounding the fired ware, 3) reduction of frictional forces generated during ware shrinkage, 4) chemical inertness such that it does not react with atmosphere or fired ware, and 5) cost benefits.

5.3.2.1 Longer Life

Rapid advances in the areas of ceramics and powder metallurgy have resulted in increased complexity and decreased size and mass of the manufactured components. The same level of attention has not been applied to the kiln furniture on which the components are fired. Relatively large and heavy furniture is still commonly used to fire very small, low-mass components. The refractory nature of the furniture also means it tends to have a high coefficient of thermal expansion (CTE) and low thermal conductivity, which translate to poor thermal shock resistance. In many cases the kiln furniture, and not the component, limits the thermal cycle, which is an intolerable fact given the constant pressures of productivity improvement in the current manufacturing environment.

Cellular kiln furniture tends to have longer life than traditional dense furniture when an aggressive firing profile is employed, as is commonly used in the manufacture of electrical or powder metal components. There is a continuous desire to speed the thermal cycle for these products to enhance productivity, reduce cycle time, and lower cost. A standard cycle for such products might be 2–3 h to 1150 °C with hold, then cooling over 1–1.5 h to about 150 °C, followed by immediate removal from the kiln. The fast cooling rate is especially harsh on ceramic kiln furniture, as it generates tensile stresses in the body, and it is well known that ceramics are generally an order of magnitude weaker in tension than compression. Figure 2 demonstrates how tensile stresses develop in kiln furniture during cooling. The CTE of most materials is positive, that is, they expand as the temperatures increases, and contract as it decreases. When kiln furniture is cooled, the exterior cools faster than the interior, which causes the exterior to contract faster than the interior, and induces tensile stress on the surface. Under similar thermal conditions, materials with higher CTE will tend to generate larger tensile stresses than those with much lower ones (e.g., alumina and zirconia both have CTEs of about 9×10^{-6} mm mm °C^{-1}, whereas that of mullite is about 4.5×10^{-6} and that of cordierite less than 2×10^{-6}). Ultimately, the most important question is performance-based: how long will the kiln furniture last?

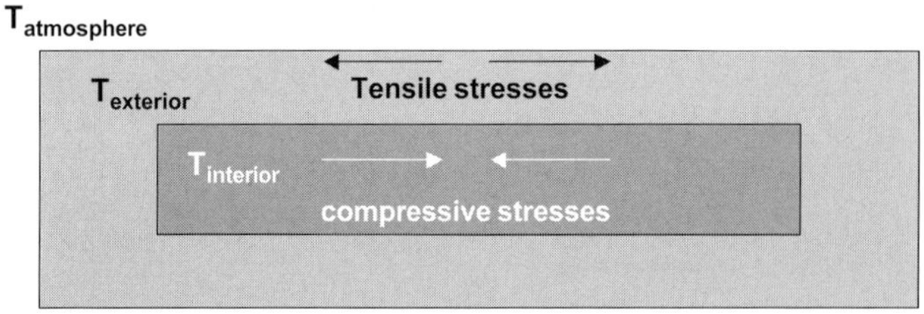

Fig. 2 Schematic cross section of a cooling piece of kiln furniture with a plate configuration. The temperature $T_{atmosphere}$ represents the kiln atmosphere. $T_{atmosphere} < T_{exterior} < T_{interior}$. In this representation, tensile stresses develop in the exterior, whereas the interior is in compression.

Ceramic foam furniture is weaker than dense ceramic furniture, but this can be used to advantage to gain longer lifetimes in applications having aggressive firing cycles. Hasselman [2–4] developed several thermal shock resistance parameters that characterize the behavior of ceramics subjected to thermal stress. These parameters represent two main criteria: 1) resistance to crack initiation, and 2) resistance to crack growth, which is correlated with minimization of the release of stored elastic strain energy.

If cracks are to be completely avoided in the furniture and not allowed to initiate, the following parameter must be considered [3]:

$$R[^\circ\mathrm{C}] = \frac{\sigma_y (1-\nu)}{\alpha E} \tag{1}$$

where R is the maximum temperature gradient a component can withstand before cracks are initiated (the higher the value, the greater the resistance), E is Young's modulus (Pa), ν Poisson's ratio, σ_y yield strength (Pa), and α coefficient of thermal expansion (mm mm °C^{-1}). The units of α can be understood from its definition (Eq. (2))

$$\alpha = \frac{\Delta L}{L(\Delta T)}. \tag{2}$$

To completely avoid crack initiation with every thermal cycle, the furniture must have a composition with sufficiently high yield strength, low Young's modulus, and low thermal expansion while remaining sufficiently inert and retaining its refractoriness. In this case, ceramic foam is not very different from dense ceramic, as discussed below.

At densities between about 10 and 25 %, the yield strength of reticulated ceramic foam follows the relationship

$$\sigma_y/\sigma_s = C(\rho/\rho_s)^n \tag{3}$$

where ρ is the density of the foam (g cm^{-3}), C a constant, and ρ_s and σ_s are the density and yield strength of the solid ceramic material, respectively. The value of n is typically around 2, regardless of whether testing is performed in compression or bending. A similar relationship is found for the dependence of Young's modulus on density:

$$E/E_s = C(\rho/\rho_s)^n \tag{4}$$

where E_s is the Young's modulus of the solid ceramic material. Experimental evidence from Gibson and Ashby [5], as well as a rigorous finite-element modeling approach by Roberts and Garboczi [6], has shown that n is also about 2 in this relationship. Because E and σ_y display nearly the same dependence on density, according to Eq. (1), these parameters do not significantly influence the resistance to crack initiation with decreasing density from dense solid to porous foam. Note that Poisson's ratio is relatively constant with change in density [5, 6], so it would also have minimal influence on R. The one parameter in Eq. (1) that is indirectly influenced by density is α, because the thermal gradient across the foam is dependent on thermal conductivity, which is influenced by the presence of porosity. More detail on thermal shock resistance as influenced by thermal conductivity is provided in Refs. [2–4] and Chapter 4.3 of this book.

Once cracks or flaws have been nucleated in an article of furniture, the dominant parameter becomes resistance to crack propagation, and it is then desired to reduce the amount of stored elastic strain energy released upon failure. The following equation can be used to estimate this behavior [3]:

$$R'''[\mathrm{Pa}^{-1}] = \frac{E}{(\sigma_y)^2(1-\nu)}. \tag{5}$$

R''' is also considered a resistance parameter, as it is inversely related to the extent of damage incurred at failure; the larger the R''' value, the less energy is released on failure and the lesser the extent of crack propagation, so the resistance to crack propagation is increased. In this equation, the yield strength and Young's modulus are inverted from Eq. (1), and the yield strength is now a squared term. Because E and

σ_y display the same dependence with respect to density, and yield strength is now squared, a reduction in density from dense solid to porous foam will have a beneficial impact on the resistance to crack propagation. Because cellular ceramic furniture is inherently weaker than dense furniture, it tends to perform better when flaws are present. A general experiment demonstrates this behavior.

Several reticulated ceramic foam samples with dimensions of 20 × 40 × 80 mm and pore size of 40 pores per inch (ppi) were subjected to repeated thermal cycling under extreme conditions. The three compositions tested – stabilized zirconia/alumina composite (YZA), zirconia-toughened alumina (ZTA), and partially magnesia stabilized zirconia (Mg-PSZ) – are routinely used as kiln furniture in various applications. The samples were cycled repeatedly between 200 and 1150 °C. A high-velocity gas burner was used for heating, and compressed air for cooling. A picture of the apparatus is provided in Fig. 3. The elapsed time to traverse this temperature range was only about 15 s, as shown in Fig. 4. The yield strength of thermally shocked samples was tested using an Instron model 4206. Samples were tested in three-point bending mode and the MOR was calculated according to the following equation:

$$\mathrm{MOR[kPa]} = \frac{3 \cdot \mathrm{load} \cdot \mathrm{span}}{2 \cdot \mathrm{width} \cdot (\mathrm{height})^2}. \tag{6}$$

The retained strength as a function of cycling is shown in Fig. 5. Table 1 lists the number of thermal cycles and the number of samples per test, as well as values for MOR, ceramic foam density, standard deviations, and a projection to 10 000 cycles based on the curve fits in Fig. 5. A substantial amount of strength is lost in the first

Fig. 3 Thermal shock rig used to thermally cycle ceramic foam. The high-velocity gas burner is at the back, and the cooling port injects compressed air at an angle from the side. Four ceramic foam samples with approximate dimensions of 20 × 40 × 80 mm reside in the stack.

cycle, but beyond that, damage is only incremental with successive cycling and appears to be predictable. The loss in strength in the first cycle would be dependent on Eq. (1), whereas subsequent strength loss would be dependent on Eq. (5). Note that the ZTA had the greatest initial MOR, but lost the most strength in the first cycle. The experiment was repeated with dense refractory samples of dimensions 80 × 40 × 10 mm having a high-alumina composition (90 % Al_2O_3/10 % SiO_2), but these samples did not survive one cycle without breaking completely in half.

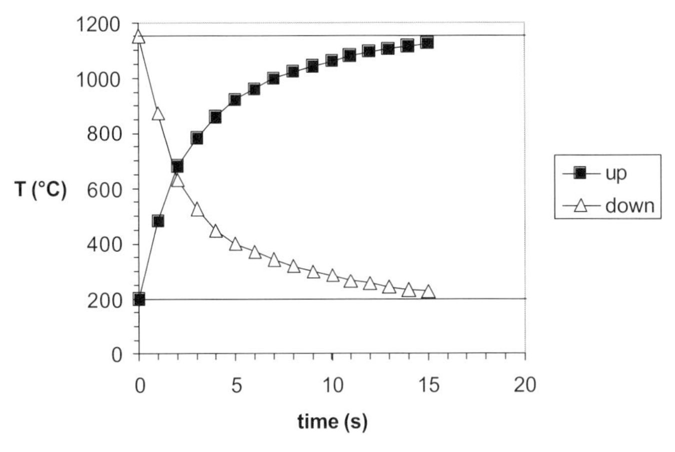

Fig. 4 Thermal cycle of ceramic foams subjected to thermal shock testing.

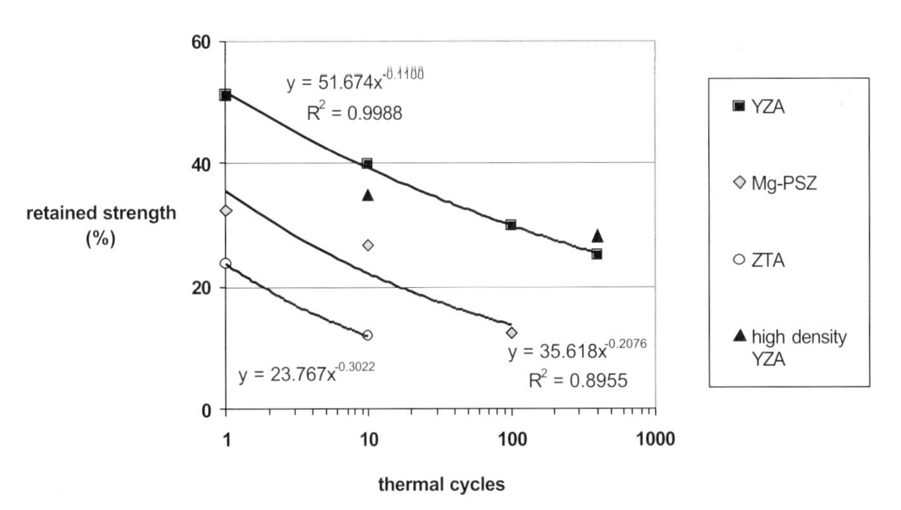

Fig. 5 Representation of thermal shock behavior of ceramic foam as retained strength versus number of cycles. The data for high-density YZA represent samples having density of 20–21 %; four samples were tested at 10 cycles, and six samples at 400 cycles.

Projection of the strength to 10 000 cycles suggests that ZTA and Mg-PSZ would be too weak to handle, yet YZA is four times stronger than the others and would still have some integrity. The experiment was also repeated on ten samples of YZA with a higher density of 20–21 %. As shown in Fig. 5, the data still fall on the same curve as those for 14 % dense material. This is in agreement with Vedula et al. [7–9], who stated that thermal shock resistance of ceramic foam is not a strong function of density. The projected strength of the higher density YZA at 10 000 cycles is over 700 kPa.

Under these extreme conditions, ceramic foam performs relatively well, and even the most aggressive powder-metal or electrical-component thermal cycles cannot match the severity of that used in this experiment. The better performance of YZA is due to the microstructural and material-dependent properties of the composite (of important note is that YZA is entirely composed of highly inert, refractory materials and does not contain silica to reduce its CTE). When properly designed and implemented, the lifetimes of cellular kiln furniture in these applications tend to be on the order of months to years.

Table 1 Modulus of rupture (MOR) for thermally shocked ceramic foams.

Ceramic	Cycles	MOR/kPa	SD*/kPa	Density/%	SD*/%	Retained strength/%	No. of samples
YZA	0	1219	269	14.0	0.61	100	18
	1	624	256	13.9	2.55	51	16
	10	487	189	14.2	0.51	40	16
	100	364	80	14.4	0.66	30	16
	400	308	199	14.4	0.56	25	6
	10 000**	222**				18.2**	
Mg-PSZ	0	1075	112	14.6	0.46	100	18
	1	348	219	14.9	0.76	32	7
	10	287	183	14.7	0.49	27	16
	100	134	95	15.8	1.00	12	8
	10 000**	57**				5.3**	
ZTA	0	2647	451	14.5	0.55	100	18
	1	629	301	14.5	0.26	24	4
	10	314	359	14.2	0.53	12	16
	10 000**	39**				1.5**	

* SD = standard deviation. ** Projected.

5.3.2.2
More Uniform Atmosphere Surrounding the Fired Ware

The open structure of cellular ceramic provides better access to the bottom of the part in contact with the setter, allowing the gas in the atmosphere to interact more uniformly with the bottom as well as the sides and top. For example, certain titanate-based electrical ceramics must be fired in an oxidizing atmosphere to finely tune desired electrical characteristics such as dielectric constant and loss tangent. In some cases, easily reduced dopants such as Sn or Zn are used in the component to achieve desired effects. Firing on a dense surface may sometimes result in reducing conditions at the bottom of the component in contact with the setter, and the desired electrical characteristics are not achieved because the redox chemistry of the component is compromised. Promotion of gas flow to the bottom of the component helps to prevent reducing conditions. Typical kiln furniture for this particular application is made of calcia-stabilized zirconia. A similar situation arises in the manufacture of translucent alumina components. These are also fired on reticulated calcia-stabilized zirconia foam plates in an oxidizing atmosphere to more effectively burn out organic binder prior to sintering in hydrogen.

5.3.2.3
Reduction of Frictional Forces during Shrinkage

The sintering of powder metal components is another process where the properties of cellular kiln furniture are utilized. Metal injection molded (MIM) parts experience more extensive binder burn out and have greater shrinkage than more traditional parts (> 20 %). At firing temperatures of 1100 °C and below, powder metal parts are typically sintered on a metal mesh belt or on dense cordierite plates, but at higher temperatures, the refractoriness, chemical inertness, and thermal shock resistance of cellular ceramics is required. The furniture must be stable in hydrogen atmosphere and must not react with the metal; hence, silica-bearing furniture is usually avoided.

In such cases of high shrinkage, frictional forces between the component and the furniture must be kept low to prevent the part from failing. When the part is shrinking at high temperature, the organic binder is gone, and the component is simply held together by interparticle contacts, so it does not have considerable strength. Such interparticle contacts between the component and the furniture create friction as the part shrinks. If the frictional forces exceed the strength of the component at any time in the cycle, the part tears or cracks. The porosity at the surface of the furniture minimizes the point contacts between the component and the furniture, inhibiting reaction between the two and reducing frictional forces during shrinking.

5.3.2.4
Chemical Inertness

As extremes are approached in temperature and dimensions for a particular application, recrystallized silicon carbide is often preferred for strength and low mass, yet even in these cases, a cellular plate may be used as a barrier between the component and the SiC kiln plate. For example, slip-cast ceramic sputtering targets are sintered on relatively thin, porous plates of zirconia-toughened alumina stacked on thin SiC plates to maximize kiln loading; the SiC plates provide the necessary support and the ceramic foam protects the part from silica contamination from the SiC plate.

In some cases, the ware contains elements such as zinc or manganese that tend to be very mobile at high temperature and have an undesirable propensity to migrate from the ware. These elements are commonly used in electrical components to finely tune certain properties, and when they are leached out, the properties degrade. The component may also be composed of a very reactive ceramic, such as iron oxide used to make ferrite components. To combat this effect, the composition of the cellular furniture is designed to more closely match that of the ware to reduce the chemical potential between the two. For example, one cellular ceramic manufacturer markets porous zinc oxide kiln furniture for firing of zinc oxide electronic components for use in computers and cell phones. The kiln furniture must retain the other primary properties that allow it to function and, depending on the composition, it can be easier to manufacture such custom formulations in the form of a cellular ceramic than in dense form.

5.3.2.5
Cost Benefits

Cellular kiln furniture can also have cost benefits. The 50–85 % lower mass compared to dense kiln furniture translates to much less material to heat in each thermal cycle and lower energy costs. Lighter loads also translate to ergonomic benefits, which can save money through greater productivity and fewer injuries, or may enable the use of robots to load kilns.

Alumina is fairly cheap, so dense alumina kiln furniture is relatively economical, but it can react detrimentally with certain materials. Firing on zirconia can minimize reactions, but in dense form the cost is typically prohibitive because it is many times more expensive than alumina ($ 0.70/kg versus $2.50 kg for fine powder). The combination of a zirconia cellular ceramic set onto a dense alumina-based setter provides a sacrificial barrier in a cost effective fashion, combining the high strength of the dense alumina setter with the low weight of the cellular zirconia plate. Compositions and uses of reticulated ceramic kiln furniture are listed in Table 2.

Table 2 Compositions and uses of atypical reticulate ceramic kiln furniture.

Material	Typical use temperature/°C	Thermal shock resistance	Typical use
Al_2O_3, 92%	1480	very good	titanates, alumina
Al_2O_3, 99.5%	1550	good	titanates, powdered metals
Al_2O_3, zirconia-toughened	1480	good	powdered metals, electrical components
Cordierite	1260	excellent	firing silver electrodes
YZA	1550	very good	powdered metals, electrical components
Zirconia, fully stabilized (Ca)	1450	good	titanates
Zirconia, partially stabilized (Mg)	1650	good	titanates, zirconia

5.3.3 Manufacture of Kiln Furniture

5.3.3.1 Foam Replication Process

The foam replication process [10] (see also Chapter 2.1) manufactures cellular ceramics to a selected pore size and geometric shape, and thus makes design work economical and flexible. Molds or forming dies, which tend to be expensive, are not used. Design and construction of molds and dies are also time-consuming. Dimensional adjustments with foam are fast and easy. In the process, polyurethane foam of desired pore size and geometry is first selected and shaped; this material functions as the precursor structure to the ceramic foam. The final pore size of the ceramic is determined by the pore size of the polyurethane foam and the firing shrinkage of the ceramic. The pore size range for replication-type cellular ceramic is about 200–5000 μm. Dimensional extremes are also material dependent, ranging up to about 60 mm for the thickness and 500 mm for the length of planar structures. Minimum dimensions are set by the pore size of the chosen precursor and the need to maintain sufficient mechanical strength. Once the polyurethane foam is prepared, it is coated with slurry of appropriate composition and rheology, dried, and fired.

Between drying and firing, the foam can be further processed:

- The open porosity on selected faces can be sealed by different coating techniques.
- The green part can be machined into intricate shapes.
- Multiple parts can be assembled and bonded to form a unique whole.

Integrated posts or rails can be incorporated into the furniture for stacking purposes. Some applications require very large components with dimensions greater than 1 m. These cannot be manufactured in one piece due to limitations in the manufacturing process, but can be produced by using modular design, where components are interlocked by bevels or shiplap joints and bonded after firing. Figures 6 and 7 show examples of shapes created by the foam replication process for use in specialty kiln applications; these represent only a small portion of the potential variations.

The porosity in reticulated kiln furniture provides several advantages, but also presents one disadvantage in some cases. When the component being fired has fine geometric detail, undergoes very high shrinkage, and/or tends to slump at high temperatures during sintering, it can get caught in the porosity at the surface (ware grabbing) or slump slightly into the face of the furniture, and this creates an imprint of the foam on the part. Reducing the pore size can help to inhibit these defects, but as the pore size is decreased, grinding the reticulated foam to a smooth finish becomes increasingly difficult. The struts themselves are much smaller and weaker as a result, and cannot hold up to the stresses generated in the grinding process, so they tend to break instead of being polished. In addition, the thermal shock resis-

Fig. 6 Examples of reticulated ceramic foam products.

Fig. 7 A ceramic foam sagger created by bonding interlocking pieces. The modular design allows the sagger to be stacked.

Fig. 8 Side view of skinned ceramic foam kiln furniture having dimensions 19.5 × 19.5 × 1.5 cm. The pore size is 15 ppi. The ceramic skin is about 1–2 mm thick.

tance of the foam is reduced as the pore size is decreased, as shown by Vedula et al. [7–9], and this is an undesirable characteristic. For these cases, experimental reticulated kiln furniture is being tested. A novel concept involves applying a continuous, thin "skin" of dense ceramic on one or more sides of a ceramic foam article, as shown in Fig. 8 [11]. This solves some of the problems described above without entirely losing the low density of ceramic foam. The foam can be skinned on two sides, creating a sandwich and providing a flat surface on the bottom for sliding or additional strength. The skin can also be a different composition from the foam [12].

5.3.3.2
Foams Manufactured by using Fugitive Pore Formers

Cellular ceramics made by using fugitive pore formers can achieve much smaller pore sizes than those from reticulated foam, typically between 5 and 200 μm. These products are generally pressed, cast, or extruded. Micromass is a commercial product of this type having 100 μm pores connected by 10 μm necks [13–15]. A polished cross section is shown in Fig. 9. The standard commercial composition is zirconia-toughened alumina, but other compositions have been developed for special applications. The pores are formed by using polymeric spheres. The porosity is about 70 %, and as a result the product can be easily machined with relatively good precision. For a 17 × 28 × 1 cm plate, a flatness of 500 μm is achievable, which is measurably better than is achieved when grinding reticulated foam. The fine porosity inhibits ware grabbing and imprinting, and the good machinability and flatness help to meet tight tolerances in the fired ware. Strength is well retained on cycling, as demonstrated by data in Fig. 10, which compares the MOR values of Micromass plates that were cycled in a commercial production facility for electrical components for several months with those of new uncycled plates.

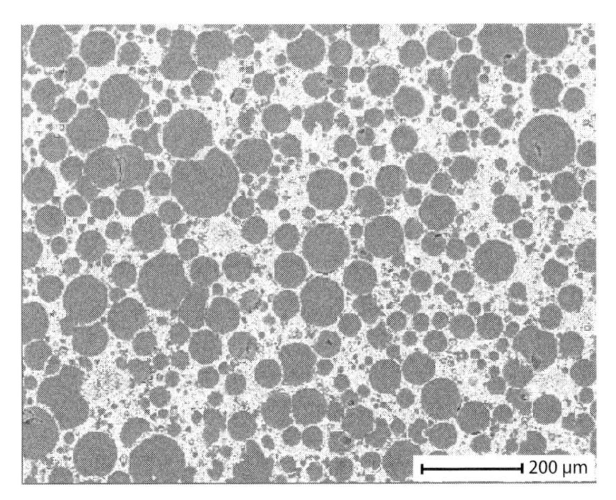

Fig. 9 Scanning electron micrograph of a polished cross section of zirconia-toughened alumina (Micromass).

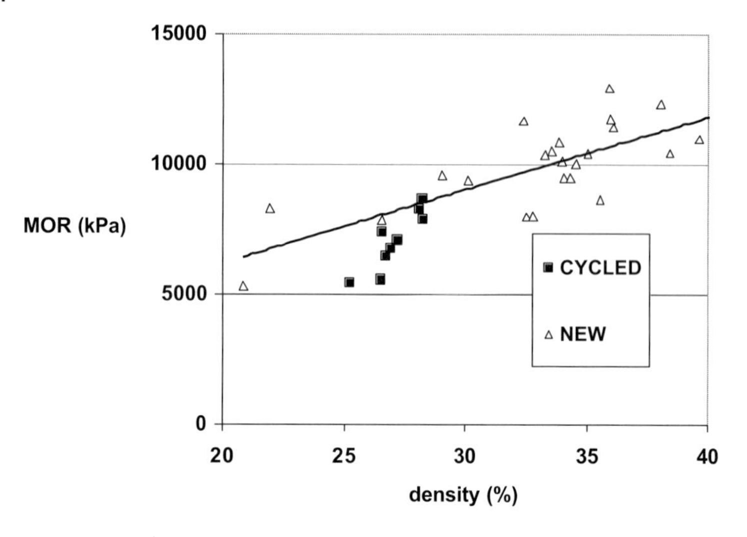

Fig. 10 Comparison of the MOR of Micromass kiln furniture plates cycled for several months at a production facility for electrical components with an aggressive thermal cycle versus those for new uncycled plates over a broad density range. The curve represents a power-law fit to the NEW data.

5.3.4 Summary

Performance benefits of cellular ceramic kiln furniture can include:

- Enhanced lifetime when aggressive thermal cycles are employed.
- Improved uniformity of the atmosphere surrounding the fired ware, which prevents redox defects in the region where the part touches the furniture.
- Reduction of frictional forces generated during ware shrinkage, and inhibition of defects caused by these forces.
- Low reactivity, as cements and silica-bearing materials are not needed to form cellular ceramics or reduce CTE, and points of contact between the ware and the furniture are reduced.
- Lower cost due to lighter weight, as less energy is required to heat the furniture.
- Ergonomic benefits, which can lead to greater productivity and fewer injuries.
- Geometry is easily customized to meet the complex needs of the marketplace.

References

1 Edwards, C.L., Keyzer, K.E., *Ind. Heating,* October **2000**.
2 Hasselman, D.P.H, *J. Am. Ceram. Soc.* **1969**, *52* [11], 600–604.
3 Hasselman, D.P.H, *Bull. Am. Ceram. Soc.* **1970**, *49* [12], 1033–1037.
4 Lee, W.L., Rainforth, W.M., *Ceramic Microstructures – Property Control by Processing,* Chapman and Hall, London, UK, 1994, pp. 109–112.
5 Gibson, L.J., Ashby, M.F., *Cellular Solids – Structure and Properties,* 2nd ed., Cambridge University Press, Cambridge, UK, 1997, p. 186.
6 Roberts, A.P., Garboczi, E.J., *J. Mech. Phys. Solids* **2002**, *50* [1] 33–55.
7 Schwartzwalder, K., Somers, H., Somers, A.V., US Patent 3,090,094, 1963.
8 Vedula, V.R., Green, D.J., Hellman, J.R., *J. Am. Ceram. Soc.* **1999**, *82* [3], 649–656.
9 Vedula, V.R., Green, D.J., Hellman, J.R., *J. Eur. Ceram. Soc.* **1998**, *18,* 2073–2080.
10 Vedula, V.R., Green, D.J., Hellman, J.R., *J. Am. Cer. Soc.* **1992**, *75* [7], 1899–1905.
11 Olson III, R.A., Bowen, G., Redden, M., Heamon, M., Patent Pending, 2004.
12 Morris, J.R., US Patent 4,568,595, 1984.
13 Butcher, K.R., Pickrell, G.R., *Mater. Res. Soc. Symp. Proc.* **1999**, *549,* 9–15.
14 Mathews, S., Pickrell, G., *Am. Ceram. Soc. Bull.* **1999**, *78* [2], 77–78.
15 Pickrell, G.R., Butcher, K.R., Lin, C., US Patent 6,773,825, 2004.

5.4
Heterogeneously Catalyzed Processes with Porous Cellular Ceramic Monoliths

Franziska Scheffler, Peter Claus, Sabine Schimpf, Martin Lucas, and Michael Scheffler

5.4.1
Introduction

Catalytic processes are the driving force in industrial chemistry. More than 80 % of all base chemicals and products have encountered a catalyst (homogeneous or heterogeneous) during their production or conversion. Most of these processes are heterogeneously catalyzed, which means that the catalyst is in a different physical condition (solid) than the starting material/product (liquid and/or gas phase), and the majority of processes are operated in a small temperature window within the range 100–600 °C. Only a few processes are carried out at higher temperatures.

In classical heterogeneous processes the catalyst typically consists of an active component, a binder, and several additives. Often starting from a paste, these mixtures are shaped, for instance, by extrusion or tablet pressing. The resulting products are spheres, cylinders, or tablets with lengths and diameters of several millimeters. These catalyst bodies are used in fixed bed reactors, often with dimensions of several or some tens of cubic meters, activated, and act for a certain time as a catalyst. Reviews on catalytic processes and how to select and make a catalyst can be found in [1–3]. Beside the well-established classical processes operated with *shaped bodies* novel processes were established in the automotive and chemical industries over the past 30 years making use of *ceramic monolith catalysts.* A well-known example is the three-way catalyst (TWC) used in automotive exhaust gas cleaning since 1974. This development was triggered by fouling of shaped catalysts in this process, caused by incomplete burning of fuel [4], and the power loss due to the resulting pressure drop over the fixed catalyst bed. Stacked corrugated sheets of cordierite, coated with alumina and impregnated with hexachloroplatinic acid, were used first in 1961, similar to the honeycombs which are used nowadays [5]. Figure 1 shows some characteristics of both types of ceramic monoliths (honeycombs and ceramic foams) compared to common fixed-bed catalysts.

From the viewpoint of catalysis, ceramic monoliths have a low geometrical surface area (about 1–5 m^2 per liter of support volume). Therefore, to apply these structures in heterogeneously catalyzed processes, for instance, for the production of chemicals or as three-way catalytic converters for the transformation of pollutants into harmless gases before releasing them into the environment, they must be coated

Cellular Ceramics: Structure, Manufacturing, Properties and Applications.
Michael Scheffler, Paolo Colombo (Eds.)

ISBN: 3-527-31320-6

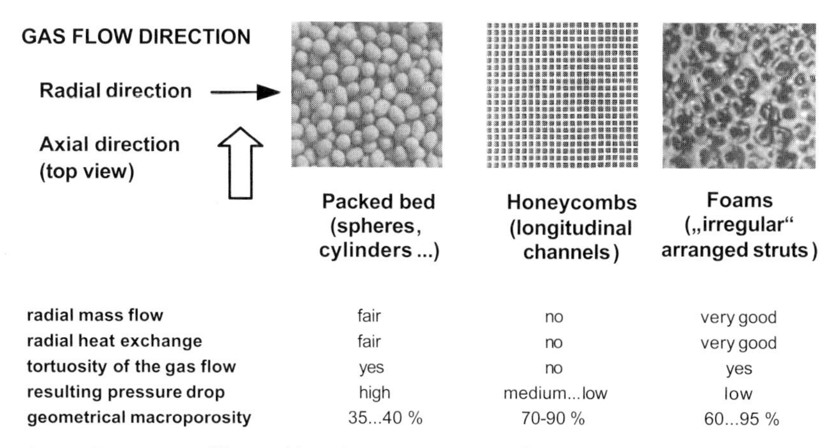

	Packed bed	Honeycombs	Foams
radial mass flow	fair	no	very good
radial heat exchange	fair	no	very good
tortuosity of the gas flow	yes	no	yes
resulting pressure drop	high	medium...low	low
geometrical macroporosity	35...40 %	70-90 %	60...95 %

Fig. 1 Comparison of flow and heat-dynamic properties of shaped catalysts, honeycombs, and foams.

with a thin active layer, often of an inorganic oxide. This oxide layer (often applied as a so-called wash coat) contains the catalytically active material, for example, precious metals incorporated into the porous oxide structure, which is responsible for the catalytic transformation of the starting material. However, the high surface area of the wash coat components are not only needed for a high dispersion of the incorporated metals, but they also can take part in the catalytic reaction, for example, by strong metal–support interactions, by triggering the catalytic (electronic) properties of the metal component.

This chapter describes established and novel developments in heterogeneous catalysis using ceramic monoliths (honeycombs and foams). Exemplary methods of modifying the monolith surface, increasing the surface area, and loading with active components will be discussed, and an overview of the use of ceramic honeycombs and foams in industrial chemistry, automotive applications, and current research will be given. This chapter, however, can provide only a small insight into heterogeneously catalyzed processes. For detailed information, some references ranging from basic principles to common applications in heterogeneous catalysis are included to allow the reader a more profound study of this matter.

5.4.2
Making Catalysts from Ceramic Monoliths

Most of the available porous ceramic materials (for manufacture, see Part 2, especially Chapters 2.1, 2.2, and 2.6) do not themselves show sufficient catalytic performance for technically relevant processes. To exploit the fluid-dynamic and structural advantages of porous ceramics for heterogeneously catalyzed processes, generation of the desired type of catalytically active sites on the ceramic surface is necessary.

Common industrial catalysts often consist of several catalytically active components (e.g. metals, metal cations, simple binary or complex multicomponent metal

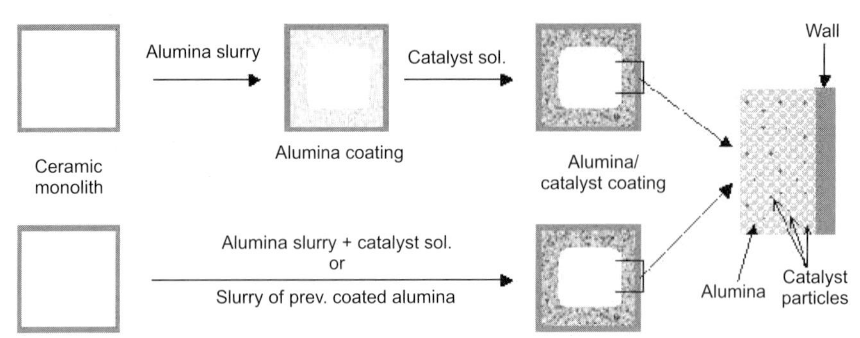

Fig. 2 Top: Initial application of bare secondary support (alumina) on walls of monolith followed by application of catalytically active precursor on the porous layer of secondary support. Bottom: Application of catalytic secondary support (alumina) simultaneously (or previously) loaded with active precursor onto walls of monolith. Right: Primary support (monolith) and porous secondary support (alumina) loaded with active component(s). Adopted from Ref. [6].

oxides) [1], mainly supported on silica, alumina, carbon, or titania. Monolithic elements made of these support materials can be produced by extrusion of the support material with some additional binder, but the mechanical strength is significantly lower than that, for instance, of cordierite. Therefore, to combine the advantages of both materials, for example, alumina is deposited on cordierite, and afterwards (or simultaneously) the catalytically active component(s) are added. These procedures are schematically demonstrated in Fig. 2.

To find a suitable coating technique for a particular application, one must take into account 1) which is the most appropriate type of support (regarding the application conditions), 2) what are the necessary catalytically active compounds, 3) is the specific surface area of the support large enough for the process, and 4) what are the demands on the properties (layer thickness, bonding strength, thermal and chemical resistance).

5.4.2.1
Enlargement of Surface Area and Preparation for Catalyst Loading

Cordierite as a classical support in automotive exhaust gas cleaning has a relatively low BET surface area of about 0.7 $m^2\ g^{-1}$, compared to conventional catalysts with about 100–1000 $m^2\ g^{-1}$. For the majority of catalytic processes a larger surface area is necessary than is provided by the support. Surface area can be augmented prior to or simultaneously with loading with catalytically active components. Typical methods are *wash coating* and *dip coating* with slurries containing high surface area ceramic powder. An overview of different methods of coating monoliths with catalyst support material is given in Ref. [6], which focuses on monoliths for catalytic combustion, and Ref. [7], with focus on monoliths for multiphase reactions. The most convenient methods (coating with colloidal solutions, sol–gel coating, slurry coating, and polymerization coating of carbon) are discussed for coating monoliths with alumina, silica, and carbon. An example of surface area enlargement of an α-alumina foam with a pore size of 30 ppi with alumina is given in Ref. [8]: A slurry consisting of 36 wt% boehmite, $Al(NO_3)_3 \cdot 9\,H_2O$ as binder, calculated amounts of $La(NO_3)_3 \cdot 9\,H_2O$ to give La_2O_3, which prevents the γ-Al_2O_3 to α-Al_2O_3 transition during calcination, and small amounts of starch and glycerol as viscosity modifier was used. The predried ceramic α-Al_2O_3 foam disks were immersed in the slurry, and excess slurry was removed by draining after immersion. The coated disks were dried and calcined at 800 °C. Immersion and calcination were repeated several times until a desired surface area was achieved. After repeated processing a wash-coat loading of up to 13.5 wt% was achieved, and the BET surface area was increased from 2 $m^2\,g^{-1}$ (uncoated foam) to 12 $m^2\,g^{-1}$ at this degree of loading. This same method can be applied to honeycombs and foams.

5.4.2.2
Loading with Catalytically Active Components and Activation

Loading with catalytically active component can be performed with a number of methods and chemical compounds bearing the desired component. The most common in the field of supported catalyst preparation are wash or slurry coating, impregnation, adsorption, ion exchange and deposition or precipitation. After loading with active component thermal treatment for solvent removal, calcination, and consolidation are necessary. In the case of transition metals or precious metals as the active component, chemical reduction to form the dispersed metal from loaded metal oxides or metal ions, often carried out in situ prior to the start of the catalytic process, is a further step. Among the numerous publications dealing with specific methods of catalyst preparation some economical aspects and aspects of catalyst development are described in Ref. [9], and some general aspects of catalyst preparation and a classification of methods are given in Ref. [10]. The principles described for honeycomb monoliths therein are also applicable for ceramic foam monoliths. A scheme illustrating the different steps and routes from monolithic ceramics (extrusion and foam formation as well) to the final monolithic catalyst is given in Fig. 3.

A large group of heterogeneously catalyzed reactions proceeds on metal cations or well-dispersed metal nanoclusters or on a surface that provides an interaction between a metal and a second surface-active component such as a metal oxide or a zeolite structure. The most common way to distribute the desired metal ions across the support surface is impregnation with an aqueous solution of a suitable soluble salt. Nitrates or simple salts of organic acids (e.g., acetate, oxalate) are preferred, since these anions are completely removable by thermal treatment. The impregnation conditions (concentration, solid/liquid ratio, temperature, and pH value) can vary over a broad range and depend on the metal type and the desired amount.

A specific example of the coating of a ceramic foam with active component and its activation is given in [11]: Pt was deposited by impregnation of an α-Al_2O_3 ceramic foam with a saturated solution of hexachloroplatinic acid (H_2PtCl_6) in water. After

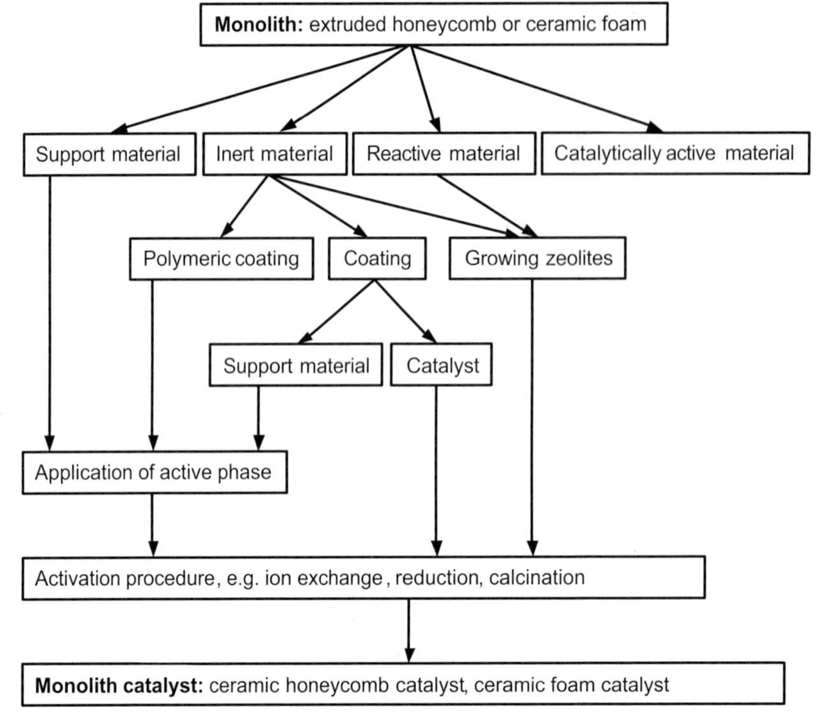

Fig. 3 Path leading to a monolithic catalyst. Adopted from Ref. [7] and modified.

impregnation the monolith was dried in nitrogen, calcined in air, and reduced in hydrogen. The Pt amount was 3–6 wt %. To fabricate a Pt–Au bimetallic catalyst the Pt/α-Al_2O_3 catalyst was impregnated thrice with a warm, saturated solution of gold trichloride ($AuCl_3$) in water, and the calcination and reduction steps were repeated as described above.

5.4.2.3
Zeolite Coating: A Combination of High Surface Area and Catalytic Activity

Zeolites are a large group of crystalline, porous materials with a broad diversity of porosity (type of pore system and pore size created by the specific order of the crystallographic arrangement) and chemical composition [12, 13]. Some special zeolites are used in a great number of catalytic and sorption processes. Due to their well-defined pore sizes zeolites can restrict accessibility to the educts (educt selectivity), products (product selectivity), or transition states (transition state selectivity) during a reaction. In combination with their adjustable acidity and their capability as metal (cluster) hosts, the catalytic properties of those catalysts can be tailored over a wide range [14–17].

Since zeolites are powders with a particle size in the micrometer range after hydrothermal synthesis, they must be shaped before use in technical applications. Instead of conversion to shaped bodies they can also be attached to the surface of ceramic honeycombs or of ceramic foams.

The wash coating procedures (dip, wash, or slurry coating [18]) make use of binders such as alumina, silica, or layered silicates. Other additives are organic agents such as cellulose or starch to improve adhesion between the zeolite-loaded slurry and the support surface during the coating procedure. Depending on the coating procedure and the texture of the support the solid/liquid ratio of such slurries can vary between 1/1 and 1/100. Considering the viscosity of the slurries a minimum cell size of the ceramic foams or a minimum channel size of the ceramic honeycomb is required, and pore filling can be accelerated by vacuum treatment. To achieve a homogeneous layer either the amount and composition of the slurry is adjusted such that it completely soaks the support, or the excess slurry can be drained. The remaining solvent can be removed, for example, by microwave drying [19] or thermal treatment at slightly elevated temperatures, and this is often followed by a calcination/activation step.

Whereas the slurry and dip-coating methods either require the addition of binder [20] or are very limited in terms of layer thickness [21], in situ crystallization of zeolites on ceramic surfaces can be used to prepare binder-free zeolite layers with good adherence to the ceramic surface and a broad variability of zeolite structure, chemical composition, particle size, and controlled layer thickness. The first work to synthesis zeolite on porous ceramics can be traced back to attempts to synthesize dense layers of zeolite A or silicalite on α-Al_2O_3 for membrane applications [22]. Further work was carried out on commercially available ceramic honeycombs and foams made of silicon carbide, α- and γ-alumina, mullite, cordierite, and zirconia, or alternatively with polymer-derived microcellular SiOC foams. These ceramics have a high chemical resistance with respect to zeolite crystallization conditions, such as a high alkalinity up to pH 14, temperatures up to 200 °C, and autogenous water pressure at this temperature. The advantage of the use of "inert" materials is that the synthesis procedures can easily be covered by recipes from conventional zeolite powder syntheses, for instance, as described in Refs. [23–26].

Whereas the above-mentioned materials were considered to be inert supports under common conditions of zeolite synthesis, other porous ceramics act not only as supports but also as a source of framework builders (Si^{IV} and Al^{III}) of the zeolite. Examples are ceramic foams fabricated by a self-foaming procedure of specific silicone resins (see Chapter 2.1.) with a tailored amount of SiO_2, SiC, Si metal, or γ-Al_2O_3. Such tailored materials allow partial transformation of the porous ceramic into the desired zeolite type by hydrothermal treatment [27, 28]. The ceramic foam acts as substrate and as silicon source under synthesis conditions (175 °C, tetrapropylammonium hydroxide in sealed, Teflon-lined stainless steel autoclaves under autogenous pressure). Figure 4 shows an SEM image of a zeolite silicalite-1 layer on the surface of a polymer-derived ceramic foam (pyrolyzed in argon at 1000 °C) after 72 h of zeolite synthesis time.

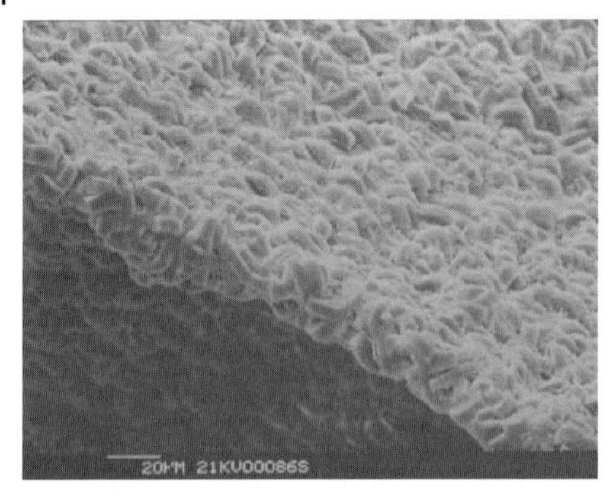

Fig. 4 SEM images of dense zeolite layer on polymer-derived ceramic foam.

A specific group of porous materials – biotemplated SiSiC ceramics derived, for instance, from rattan wood by carbonization and infiltration of $Si_{(l)}$ or $SiO_{(g)}$ (see Chapter 2.5.) – was also shown to be useful as reactive support for partial transformation into zeolite coating [29]. The SiC which is formed by the rattan carbon template is inert, while the excess silicon of the composite material is reactive and can be transformed into a zeolitic layer under zeolite synthesis conditions. By this route using natural materials as template, very complex, hierarchically ordered structures are available. In the case of rattan, an anisotropic unidirectional channel system with channel sizes from 250 to 400 μm is available, and the zeolite coating showed a layer thickness of at least 10–20 μm. Figure 5 (left) shows the channel opening of a rattan-templated SiSiC ceramic after zeolite crystallization by partial transformation, and Fig. 5 (right) an open cylindrical channel along the axial direction.

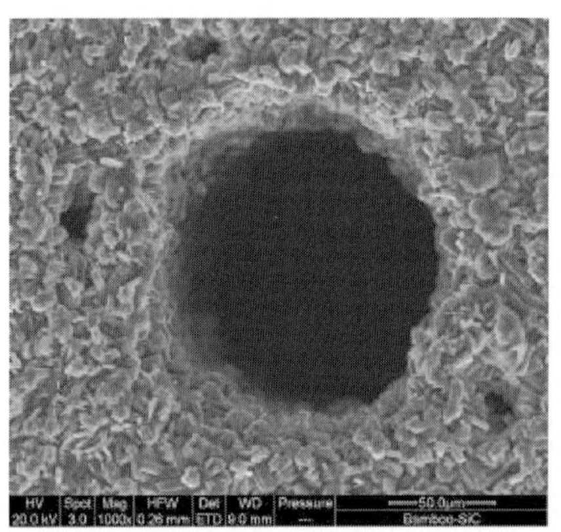

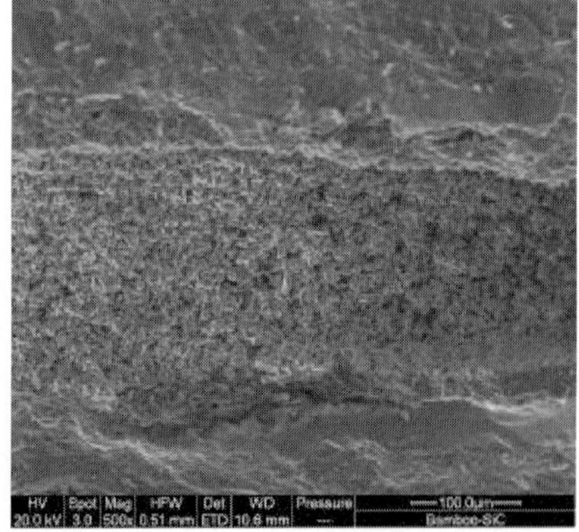

Fig. 5 Channel opening of a rattan-templated SiSiC ceramic after zeolite crystallization by partial transformation (left) and open cylindrical channel along the axial direction (right).

5.4.3
Some Catalytic Processes with Honeycomb Catalysts

Monolith catalysts were long confined to applications in gas–solid processes, mainly in the area of exhaust gas purification. The largest application is in the automotive industry for cleaning exhaust gases. A short and general overview of the applications of monoliths for gas-phase catalytic reactions covering ceramic honeycombs, ceramic foams, and metallic honeycombs is given in Ref. [30]. These applications are:

- Three-way catalysts
- Diesel catalysts
- Ozone abatement in aircraft and photocopiers
- Natural-gas engines
- Catalysts for small engines like motorcycles, chain saws, and lawn mowers
- Catalysts for the destruction of NO_x
- Destruction of volatile organic compounds (VOCs) from restaurants
- Catalytic combustion

In recent years, monolithic catalysts were also used for multiphase reactions, in which gas and liquid phase pass the reactor simultaneously.

The following sections deal with catalytic applications of honeycomb catalysts, starting with the three-way catalysts for automobile exhaust cleaning, diesel engines, and NO_x removal from industrial flue gas, followed by other applications in catalysis, in both gas-phase and multiphase reactions, mainly research applications.

5.4.3.1
Automotive Catalysts

Based on the number of the monoliths the cleaning of automotive exhaust gas is the most common use of honeycomb catalysts. The exhaust gases emitted by combustion engines contain, besides carbon dioxide (CO_2) and water, nitrogen oxides (NO_x), carbon monoxide (CO), unburned hydrocarbons (HC), sulfur dioxide (SO_2), and oxygen in variable amounts depending on the manner of driving and type of engine. The main pollutants which should be removed are carbon monoxide, hydrocarbons, and nitrogen oxides. Catalytic purification of exhaust gases has been known since the 1930s [31]. A patent on the production of structured ceramic articles by Corning [32], submitted in September 1958, refers to the use of honeycomb monolithic structures as catalyst supports in emission control. Starting from 1970, due to air pollution control acts, first total oxidation catalysts [33, 34], the predecessors of the gasoline three-way catalysts (TWC), were introduced as exhaust gas cleaning catalysts [35, 36]. As catalyst support, activated alumina was used in the form of beads and honeycomb monolith. Active metal was in both cases platinum and palladium [37]. These oxidation catalysts merely removed CO and hydrocarbons from the exhaust gas stream, and NO_x passed unhindered. The breakthrough for the solution of this problem was achieved in 1976 by Volvo with the introduction of the TWC with lambda probe.

In the TWC the catalyst support is applied by a wash-coating procedure to the monolith structure, which consists of cordierite, metal, or, for applications at high temperatures, of SiC.

The wash coat is applied in several layers of different composition [35]. For catalytic exhaust gas cleaning, cell densities between 400 and 1200 cpsi (cells per square inch) are of interest [38].

The first generation oxidation catalysts were based on a combination of Pt and Pd [37] supported on temperature-stabilized γ-alumina. The next generation of automotive catalysts additionally reduce NO_x to nitrogen. As additional metal component rhodium was introduced. Possible metal combinations are Pt/Rh [39, 40], Pd/Rh [35], and Pt/Pd/Rh [41]. The disadvantage of rhodium is the formation of inactive aluminate phases with the alumina support above 800 °C [37]. In the mid-1990s, the development of Pd-only TWC catalysts [42–44] appeared desirable because platinum (24 €/g) and rhodium (21 €/g) are much more expensive than palladium (8 €/g).

Cerium oxide as an additive to the wash coat expands the window of optimal pollutant conversion because of its oxygen-storage capacity [45]. The insertion of ZrO_2 in the CeO_2 crystal lattice prevents the sintering of ceria at temperatures above 1000 °C [46–48]. Phase transformation of the high surface area γ-alumina into the low surface area α-alumina at temperatures above 1000 °C can be prevented by a number of different additives [49]. The main principle is to exchange the hydrogen in the AlOH groups of the γ-alumina for other elements (e.g. lanthanum). The

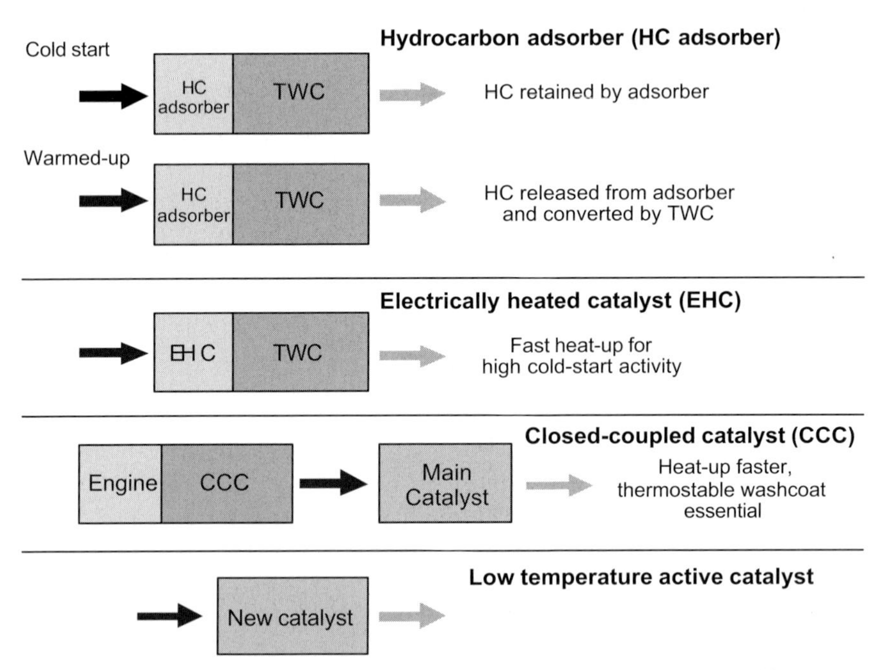

Fig. 6 Some strategies for the abatement of engine start-up emissions [51]. (Reprinted from Catalysis Today, 77, J. Kaspar, P. Fornasiero and N. Hickey, Automotive catalytic converters: current status and some perspectives, 419–449, Copyright 2002, with permission from Elsevier.)

AlOH groups sinter at higher temperatures by loss of water and form Al-O-Al bonds. Ce–Zr mixed oxides also improve the thermal stability of the alumina [50] and promote noble metal dispersion [51].

Of major importance for the total balance of the exhaust gas catalyst is pollutant conversion at low temperatures, especially when the engine is cold-started. In the first two minutes of cold start the engine produces around 60–80 % of the emitted hydrocarbons [35]. Figure 6 describes schematically some strategies to prevent emission of pollutants at lower engine temperatures:

- Hydrocarbon adsorber in front of the TWC: the zeolite wash coat [52] traps hydrocarbons at lower exhaust gas temperatures and desorbs the hydrocarbons when the TWC reaches its working temperature.
- A closed coupled catalyst, mounted near the engine, which reaches the working point faster. The maximum temperature attained is about 1050 °C.
- An electrically heated metal monolith: by applying a electrical voltage in the start up phase of the motor, the catalyst heats up immediately and achieves acceptable hydrocarbon conversion rates three times faster [35] than the conventional TWC.

For high fuel efficiency, it is necessary to operate under lean conditions. However, under these conditions the TWC catalyst is not effective in reducing NO_x. One solution is the use of *NO_x storage–reduction (NSR) catalysts* that trap NO_x under lean conditions on, for instance, barium oxide (Fig. 7a) [53, 54].

Platinum oxidizes in a first step the NO under lean (oxygen-rich) conditions to NO_2, which migrates to the barium oxide and forms barium nitrate. After a certain period, when the storage capacity is depleted, the motor management switches to rich conditions for a short time (Fig. 7b) and the stored barium nitrate re-forms NO_2. The NO_2 is reduced by the platinum catalyst component to nitrogen. The reducing agent is CO or hydrocarbons. As a disadvantage, barium oxide forms bar-

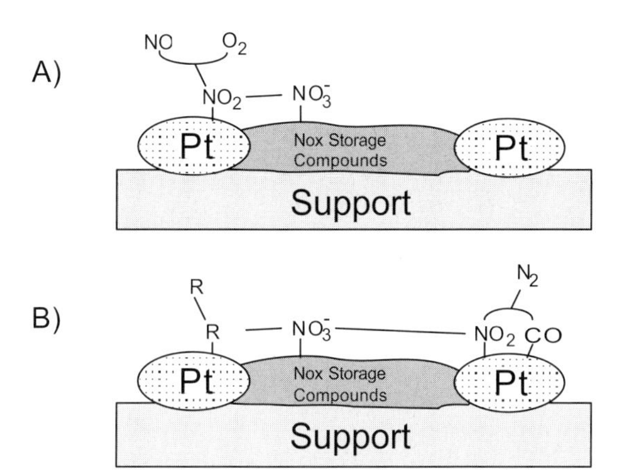

Fig. 7 NO_x trapping under lean conditions (A) and NO_x conversion under rich conditions (B) on NSR catalyst (NO_x storage compound, e.g., barium oxide). After Refs. [53, 54].

ium–sulfur compounds in the presence of sulfur oxides. Such compounds are very stable up to high temperatures and reduce the NO_x storage capacity of the catalyst. Therefore, reduction of the sulfur content in the fuel is essential for the function of the NO_x storage system.

5.4.3.2
Diesel Engine Catalysts

With progress in the emission control of gasoline-fueled vehicles, the comparatively small pollutant emissions of diesel vehicles moved in the last few years into the center of attention. In addition, the number of diesel-engined cars increased due to their lower fuel consumption. Diesel engines operate with excess oxygen under lean conditions. Solutions for this working area of the catalysts were already discussed in the preceding section. Additionally, diesel engines emit soot, soluble organic fractions (SOF), and SO_2. The cleaning of the exhaust gases of diesel engines can be divided into three tasks: 1) total oxidation of HCs, SOF, CO, aldehydes, and SO_2; 2) trapping of particles and removal of NO_x, and 3) reduction of NO_x to nitrogen (DENOXing). The removal of sulfur trioxide is necessary due to its capability to form sulphate, which deactivates the precious metals or attacks the support material [55]. Due to the lower temperature of diesel exhaust gas, the oxidation catalyst must be highly active at temperatures below 250 °C [56]. Common catalysts are composed of $Pt/CeO_2/Al_2O_3$ with a tailored porous system that allows adsorption at lower temperatures for removal of SOF below the light-off temperature [55]. At present great efforts are dedicated to the development of catalytically active diesel soot traps based on honeycomb or foam monoliths to remove soot and NO_x simultaneously [57–69]. A more detailed description of particulate diesel traps, materials used for their preparation, and working principle can be found in Chapter 5.2 of this book.

After conversion of CO, HC, SOF, and soot, removal of nitrous oxide under oxygen excess, especially for systems with continuous regeneration trap (CRT), is necessary. Thus, typical lean-burn diesel engines require procedures for reduction of nitrogen oxides in the presence of oxygen. Technically important DENOX processes are selective catalytic reduction (SCR) with ammonia (NH_3-SCR) [70, 71] or hydrocarbons (HC-SCR) [72] and the NO_x storage–reduction catalysts (NSR) [53] (see Section 5.4.3.1).

NH_3-SCR is driven by the reaction of NO and NH_3 to form nitrogen and water when NO is the major NO_x component (Eq. (1)):

$$4\,NO + 4\,NH_3 + O_2 \rightarrow 4\,N_2 + 6\,H_2O. \tag{1}$$

In the case of equimolar amounts of NO and NO_2 in the exhaust gas, the reaction can be described by Eq. (2) [73, 74]:

$$4\,NH_3 + 2\,NO + 2\,NO_2 \rightarrow 4\,N_2 + 6\,H_2O. \tag{2}$$

At temperatures below 200 °C the formation of ammonium nitrate is a serious problem for NH_3-SCR. During its decomposition at higher temperatures ammonium nitrate can form the stable N_2O, or on initial ignition explosive decomposition can occur [70]. Generally, honeycomb monoliths covered with $TiO_2/WO_3/V_2O_5$ wash coat have been used for SCR of NO_x with ammonia. For safety reasons ammonia is produced on-board diesel trucks by hydrolysis of urea in a preheated catalyst zone (Fig. 8).

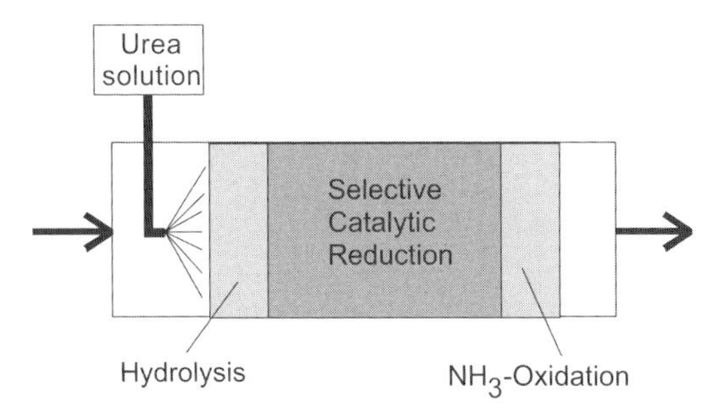

Fig. 8 Scheme of NH_3-SCR [51] with urea injection. (Reprinted from Catalysis Today, 77, J. Kaspar, P. Fornasiero and N. Hickey, Automotive catalytic converters: current status and some perspectives, 419–449, Copyright 2002, with permission from Elsevier.)

The HC-SCR process is similar to NH_3-SCR, except that hydrocarbons are employed as reducing agents, commonly in combination with copper-loaded zeolite catalysts (Cu-ZSM-5) [72, 75].

5.4.3.3 Catalytic Combustion for Gas Turbines

The catalytic reduction of NO_x to nitrogen in power plant exhaust gas based on the SCR method is described in Ref. [76]. A relatively new application for monolith reactors is catalytic combustion in gas turbines. In conventional gas turbines fuel is burnt by flame combustion at around 1800 °C. Because of materials limitations it is necessary to cool the hot gases with bypassed compressed air having temperatures of 1100–1500 °C. Due to the high flame temperature the combustion leads to large amount of NO_x in the exhaust gas. For this reason catalytic combustion at lower temperature is desirable [77]. The catalyst should withstand continuously temperatures up to 1500 °C and thermal shocks of 1000 K s^{-1}. The existing solutions therefore consist of hybrid combustion systems [78] in which the heterogeneously catalyzed conversion has the function of preheating the fuel and air for the downstream homogeneous combustion.

5.4.3.4 Applications of Honeycomb Catalysts for Other Gas Phase Reactions

Apart from the large and well-established area of exhaust gas purification, only a few other examples for uses of monolithic catalysts in gas-phase reactions have become established. Corning developed a novel extrusion method to make bulk transition metal oxide honeycomb catalysts, for instance, extrusion of iron oxide honeycomb catalysts for the dehydrogenation of ethylbenzene to styrene, a process with a worldwide capacity 20×10^6 t/a. In industry styrene is synthesised mostly in radial-flow fixed-bed reactors. The overall economics could be improved with parallel-channel honeycomb catalysts and axial flow reactors, which provide low pressure drop while making more efficient use of reactor volume, with better heat- and mass-transfer characteristics compared to a conventional radial packed bed. Schematics of both reactors are given in Fig. 9. Addiego et al. found styrene selectivity of greater than 90 % and ethylbenzene conversions in excess of 60 % in bench-scale testing under conventional conditions without apparent deactivation or loss of mechanical integrity [79].

Schanke et al. used a monolith reactor for Fischer–Tropsch synthesis to produce hydrocarbons and other aliphatic compounds, such as methane, synthetic gasoline, waxes, and alcohols [80]. The starting material is a mixture of hydrogen and carbon monoxide (synthesis gas). The monolithic catalyst consists of alumina, silica, zeolite, or titania, and a conventional Fischer–Tropsch catalyst with a precious metal precursor.

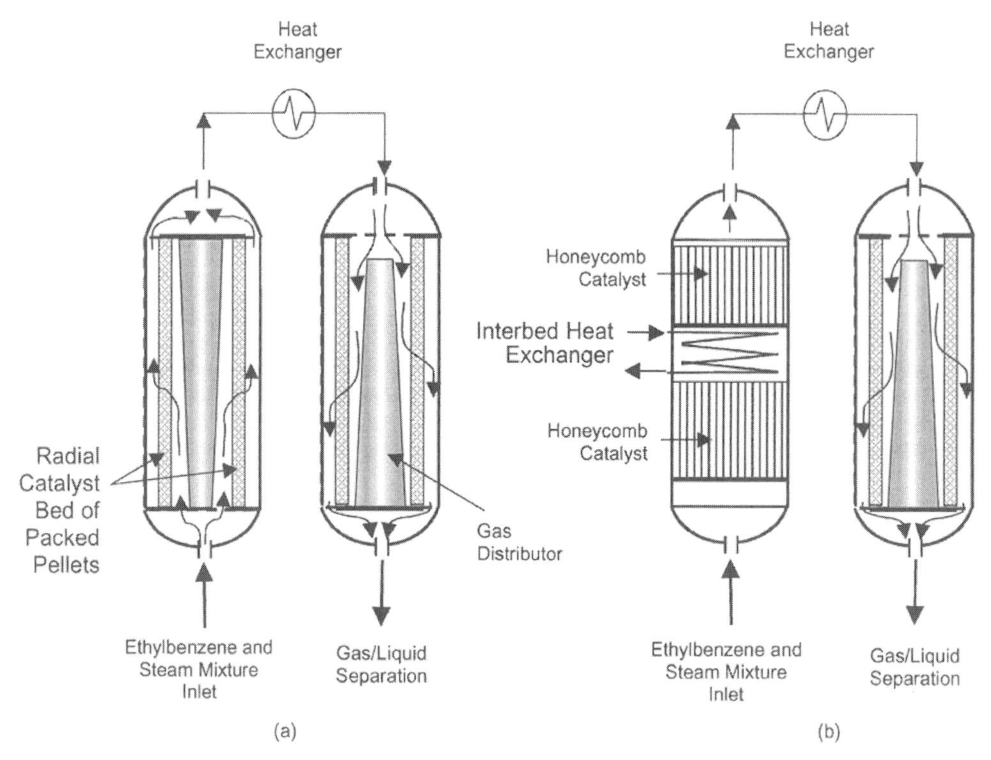

Fig. 9 a) First-stage radial-flow reactor with cylindrical packed catalyst bed. b) Honeycomb catalyst reactor with interbed heat exchanger. In both configurations, effluent is sent to a second-stage reactor to dehydrogenate remaining ethylbenzene [79]. (Reprinted from Catalysis Today, 69, William P. Addiego, Wei Liu and Thorsten Boger, Iron oxide-based honeycomb catalysts for the dehydrogenation of ethylbenzene to styrene, 25–31, Copyright 2001, with permission from Elsevier.)

The strongly exothermic nature of Fischer–Tropsch synthesis requires effective heat transfer for successful reactor operation. The authors also compared different materials of monolithic structures (cordierite, γ-alumina, and steel) in Fischer–Tropsch synthesis [81]. Each was compared to corresponding powder catalysts. Cordierite monoliths were as active and selective to C_{5+} (pentane and higher hydrocarbons) as comparable powder catalysts, steel monoliths were found to have lower activity and C_{5+} selectivity than comparable powder catalysts and cordierite monoliths, and alumina monoliths gave comparable selectivities to powder catalysts but lower activity.

Nicolau et al. (Celanese International Corporation) synthesized vinyl acetate by vapor-phase reaction of ethylene, oxygen, and acetic acid using a ceramic honeycomb as reactor. The catalysts consist of Pd and Au deposited on silica-coated cordierite or mullite monoliths. Catalyst space–time yields for vinyl acetate production of 434 g L^{-1} h^{-1} at 192 °C have been reported [82, 83].

For the numerical simulation of monolithic catalysts, several proposals have been made in recent years. A newly developed computational tool is DETCHEM^{MONOLITH} for the transient two- and three-dimensional simulation of catalytic combustion monoliths, which is especially useful for spatially structured monolithic catalysts in which the timescales of variation of the gas phase are much smaller than those of the thermal changes in the monolithic structure, for example, for catalytic combustion, conversion of natural gas, and automotive catalytic converters. The developed computer code for the first time offers the possibility of performing transient 2D and 3D monolith calculations with such detailed models for transport and chemistry in the individual channels. As an example, the application of the tool to the hydrogen-assisted catalytic combustion of methane in a platinum-coated honeycomb monolith is demonstrated in Ref. [84].

5.4.3.5
Honeycomb Catalysts for Gas/Liquid-Phase Reactions

Since the introduction of the monolith multiphase reactor several investigations have been carried out to increase the utilization of monolithic structures as catalyst support in multiphase reactions. Current research is mainly focused on hydrogenation reactions. These multiphase applications are mostly operated in cocurrent mode [85]. In monoliths, different flow regimes can occur. Most interesting are Taylor (bubble or slug flow) and film flow. Figure 10 demonstrates bubble or slug flow in a capillary. In this type of flow gas bubbles and liquid slugs move with constant velocity through the monolith channels, and the gas is separated from the catalyst only by a thin liquid film. For this flow regime a higher mass transfer of gas to the catalyst is expected compared to that of a trickle-bed reactor. The better plug-flow behavior (gas and liquid have sharp residence-time distribution) is expected to result in higher selectivities towards the desired product in conversions with unwanted consecutive reactions [86]. A more detailed description of fluid flow through cellular ceramics can be found in Chapter 4.2 of this book.

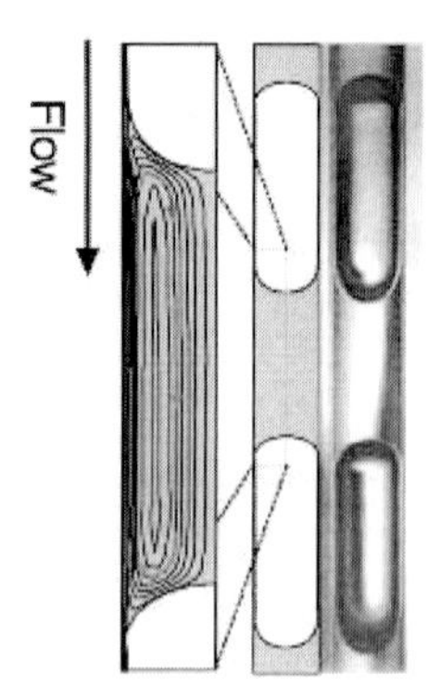

Fig. 10 Taylor (bubble or slug) flow in a capillary. The CFD picture (left) shows the liquid circulation patterns [86]. (Reprinted from Chemical Engineering Science, 56, T.A. Nijhuis, M.T. Kreutzer, A.C.J. Romijn, F. Kapteijn and J.A. Moulijn, Monolithic catalysts as efficient three-phase reactors, 823–829, Copyright 2001, with permission from Elsevier.)

Regarding multiphase reactions advantages of monolithic reactors are:

- Low pressure drop.
- Low transport resistance and short diffusion paths to/from the active site for reactants and products, which decreases the possibility of side/consecutive reactions and hence increases selectivity.
- Easy reactor scale-up.
- No catalyst separation problems.

Despite these advantages, only a few structured reactors are used in chemical industry for two reasons. The structured reactors contain less catalyst than a fixed bed and hence require a more stable catalyst than fixed bed or batch process. In three-phase processes extremely good mixing is required to bring together the reactants in the gas phase and those in the liquid phase at the active site of the catalyst to produce good productivity and selectivity. This is the justification for all the work devoted to understanding the hydrodynamics of structured reactors.

Presently, there is only one *large-scale industrial application* of monolithic catalysts in a multiphase process [38]. Akzo developed a monolith-based process for H_2O_2 production in which the hydrogenation step of the anthraquinone (AQ) autoxidation process is performed under relatively mild conditions (a few bar, 40–70 °C) [87, 88]. Instead of conventional monolith coated with a porous wash coat, reinforced amorphous silica was used throughout the wall [89]. The Pd active phase was deposited by electroless deposition, which involves chemical reduction of a Pd-containing solution [87]. The main reason for choosing a monolithic catalyst was to avoid the transport of fine catalyst particles with the liquid to the oxidation reactor, which results in the decomposition of hydrogen peroxide [7]. The concept of using a monolithic catalyst on a large scale was successful. Anthraquinones dissolved in an ordinary organic solvent could be hydrogenated with extremely high selectivity and high productivity. The idea of using a monolith for this reaction was put forward by researchers at the Chalmers University of Technology, Göteborg and was picked up by Eka Chemicals. The first large-scale reactor was brought on stream in the early 1990s.

Today the new type of reactor is used in plants with a total annual capacity of 200 000 t [87].

Kapteijn, Moulijn et al. built a *pilot-scale setup* and investigated two model reactions [86]: the hydrogenation of α-methylstyrene as a (hydrogen) mass-transfer-limited reaction and the hydrogenation of benzaldehyde to benzyl alcohol as a conversion with an unwanted consecutive reaction. In comparison with a trickle-bed reactor higher productivities for the mass-transfer-limited reaction and higher selectivities for the hydrogenation of benzaldehyde were found. For the former reaction the overall mass transfer of 0.5–1.5 s^{-1} indicates that excellent mass transfer can be achieved for gas–liquid–solid reactions in monolithic reactors [90].

One of the first examples of *research on monolithic catalysts* was a monolithic reactor investigated by the Moulijn group in cooperation with DSM Research, Netherlands, in the selective hydrogenation of unsaturated hydrocarbons. A mixture of styrene and 1-octene in toluene, representative for hydrocarbon mixtures subjected to hydrotreating, was hydrogenated in a cordierite monolith with alumina wash coat impregnated with Pd [91]. Operating the monolithic reactor in the Taylor flow regime gave considerably higher reaction rates than those reported in the literature so far. Nevertheless, external mass-transfer limitation was present under the conditions investigated. The fact that reaction rates could be strongly enhanced by operating at higher liquid loadings indicated that nonuniform or incomplete catalyst wetting may partially control conversion. Proper design of the gas–liquid distributor was found to be the critical factor.

In *comparisons of monolith and slurry reactors*, the former has the added advantage of allowing easy catalyst separation. Comparative analysis of monolithic reactors and slurry reactors were done by Moulijn et al. [92], Winterbottom et al. [93], and Hatziantoniou et al. [94]. The Moulijn group investigated the liquid-phase hydrogenation of 3-hydroxypropanal to 1,3-propanediol. Mathematical modeling of both reactors over a broad range of operating conditions showed that the monolithic reactor performed better in terms of both productivity and selectivity [92]. The hydrogenation of 2-butyne-1,4-diol to *cis*-2-butene-1,4-diol was additionally compared with a stirred tank reactor (STR) [93]. The selectivity to *cis*-2-butene-1,4-diol, an important intermediate in the production of Endosulfan (insecticide) and vitamins A and B_6, was significantly higher for the monolith and the slurry reactor (98.0–99.3 %) than for the STR (90–95 %) at conversions of 2-butyne-1,4-diol approaching 100 %. The hydrogenation of mixtures of nitrobenzene and *m*-nitrotoluene was investigated by Hatziantoniou et al. [94]. The activity of the catalyst was so high that the mass-transfer steps were rate-determining, and mass transfer of hydrogen directly from the gas plugs to the channel wall was found to be the dominant transport step. The decreased selectivity of aniline formation found in the monolithic Pd catalyst was explained by the influence of film-transport resistance near the channel wall. The catalytic hydrogenation of nitrobenzoic acid to aminobenzoic acid was used also for a quantitative study of influences of operating conditions on the observed reaction rate in a *single-channel monolith reactor* operated in Taylor flow regime [95]. An improvement in a process for hydrogenating nitroaromatics by employing a monolithic catalyst system was claimed by Machado et al. of Air Products and Chemicals

[96]. The same authors also claimed a gas–liquid reaction process including a liquid-motive ejector as a gas–liquid distributor, and monolith catalyst. The invention relates to process for carrying out gas–liquid reactions such as those employed in the hydrogenation or oxidation of organic compounds [97, 98].

Other examples for *hydrogenation reactions* are the hydrogenation of dimethyl succinate over cordierite-based monolithic copper and nickel catalysts prepared by the wash coat technique, studied by Cybulski et al. [99], and the Friedel–Crafts acylation of aromatics using zeolite-coated (BEA and FAU) monoliths [100]. The viability of using a monolithic catalyst support system was also demonstrated for a condensation/polymerization reaction. For the production of siloxane fluids tripotassium phosphate was used as catalyst, and the experiments were performed in a single-channel flow reactor with 15 mm i.d. and 500 mm catalyst-coated length [101].

A *multifunctional reactor* integrating reaction and a separation process based on a monolithic system was developed by Moulijn et al. Internally finned monolith coated with solid acids (zeolite BEA or the ion-exchange resin Nafion, both prepared by dip coating with 5 wt% with gas and liquid flowing countercurrently) were used in the esterification of 1-octanol with hexanoic acid, with removal of the side product water from the liquid reaction mixture by means of reactive stripping [102, 103]. This special type of monolith, which is optimally suited for stripping operations, is shown in Fig. 11. The fins stabilize the flow and provide additional surface area for the catalyst.

The reactive-stripping operation in a monolithic reactor results in a significantly better performance of the catalyst, since not only is the inhibiting effect of water reduced, but also conversions beyond equilibrium conversion are possible. This type of monolithic reactor was originally developed by the same authors for deep-HDS (hydrodesulfurization), in which the H_2S produced has a strong inhibiting effect on the catalyst activity, and countercurrent operation is therefore highly attractive.

A *rotating monolith reactor* was used for catalytic propane dehydrogenation by Stitt et al. [104]. The reaction is equilibrium-limited, strongly endothermic, and normally carried out at high temperatures. Catalyst deactivation due to the deposition of carbonaceous species on the surface (catalyst time on line for a given cycle is on the order of 10–10 000 min) is conventionally countered by subjecting the catalyst to periodic regeneration. The authors developed a rotating monolith reactor, in which a cylindrical block of honeycomb monolith rotates past various feed zones subjecting the catalyst successively to propane and regenerating gas. The exothermic nature of the regeneration reactions is used at least in part to provide heat to the endothermic dehydrogenation reaction via regenerative heat transfer facilitated by the movement of the solid monolith. The catalyst exhibits very high activity and selectivity in the period shortly after regeneration. Process modeling shows the design to be feasible in terms of matching the heats of reactions and achieving high conversions, but questions were raised over its practicability from the viewpoints of mechanical design and process stability. A rotating monolith was also used by Moulijn et al. to achieve alternate contact of the monolith with gas and liquid phase in the bottom of the reactor and gas phase in the top of the reactor [105]. Another alternative to mixing of the gas and liquid phase inside the reactor is saturating the liquid with gas,

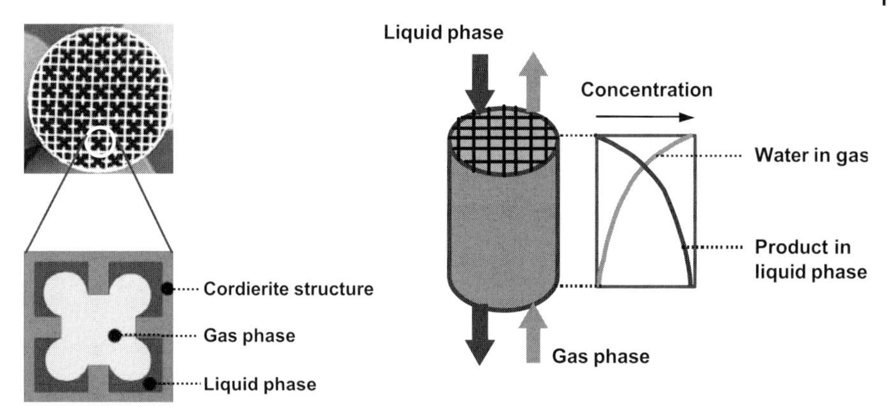

Fig. 11 Internally finned monolith; photographic view of a cross section (top left), schematic view with representation of the gas and liquid in a channel [102] (bottom left), and countercurrent stripping configuration, including the concentration profiles [103] (right). (Left: Reprinted from Chemical Engineering Science, 57, T.A. Nijhuis, A.E.W. Beers, F. Kapteijn, J.A. Moulijn, Water removal by reactive stripping for a solid-acid catalyzed esterification in a monolithic reactor, 1627–1632, Copyright 2002, with permission from Elsevier. Right: Reprinted from Catalysis Today, 66, A.E.W. Beers, R.A. Spruijt, T.A. Nijhuis, F. Kapteijn and J.A. Moulijn, Esterification in a structured catalytic reactor with counter-current water removal, 175–181, Copyright 2001, with permission from Elsevier.)

followed by passing the saturated liquid through the structured reactor (circulation). This concept was realised by Vaarkamp et al. and was tested for the hydrogenation of tetralin and of 1-hexene [106].

A *four phase system*, as applied industrially for the partial oxidation of benzene, was studied by Moulijn et al. [107], namely, the hydrogenation of α-methylstyrene (AMS) with $Ni–Al_2O_3$ with addition of water.

A relatively new type of monolith is based on carbon. Nevertheless, *carbon-based monoliths* are rarely a topic in the scientific literature. The combined favorable properties of carbon and monolithic structures, for instance, chemical stability, adjustable porosity, high void fraction, and large geometric surface area, create a support with great potential in adsorption processes and in catalytic processes such as residential water purification, controlling emissions of volatile organic compounds, storage of natural gas for gas-powered vehicles or equipment, indoor air purification, industrial respirators, automotive cabin-air filters, ventless hoods, chemical separations, NO_x and SO_x control, and exhaust traps for automotive cold-start emissions [38, 103]. Applications in the field of adsorption as well as an overview of the preparation of carbon-based monolith are given in [108] for both coated- and integral-type structures. The latter were produced by extrusion, and the former by coating a ceramic monolith with a molten carbon precursor by a chemical vapor deposition (CVD) method using cyclohexene or, more frequently, by a liquid polymer such as a resole or furan-type resin. The manufacture of foams solely from carbon is described in Chapter 2.6 of this book. Several applications of carbon-based monoliths in multiphase reactions were investigated: oxidation of cyclohexanol to adipic acid [104], hydrogenation of cinnamaldehyde to cinnamyl alcohol [110, 111], and selective oxidation of cyclohexanone [112, 113]. More details of preparation and characterization can be found in Refs. [114, 115]. Another application could be found in patents and patent applications published by Nordquist et al. of Air products and Chemicals concerning carbon coated monolith catalysts for hydrogenation in a monolithic reactor under conditions of immiscible liquid phases [116–118].

The following *conclusions* could be drawn. Regarding multiphase reactions, advantages of monolithic reactors could be proved, but they are still seldom used because of lack of experience, especially in larger scale processes. Compared to conventional agitated slurry reactors, the productivity of a monolithic reactor is usually much higher, while the energy costs for mixing/circulation are much lower than for a mechanically agitated slurry reactor. Heat removal from the monolithic reactor is simpler since no operation with a slurry is needed. Operation of a monolithic reactor is safer because 1) a smaller amount of hazardous materials is handled in the reaction zone, 2) the reactants can be separated immediately from the catalyst in the case of a process interrupt, and 3) filtration of a catalyst is avoided. Slurry reactors are superior to monolithic reactors when a catalyst is rapidly deactivated: replacement of the used catalyst with a fresh one is simpler and faster in this reactor [92].

5.4.3.6
Other Research Applications of Honeycomb Catalysts

Ceramic monoliths, as special type of micro/minireactors, can be applied for high-throughput screening of heterogeneous catalysts [119–120]. If a monolith material is chosen that is impermeable for gases and fluids (e.g., Cordierite 410, Inocermic GmbH, Hermsdorf, Germany), each channel of the monolith represents a single fixed-bed reactor and can contain a different catalyst material, so that up to 200 catalyst compositions can be tested in parallel (channel diameter 2.6 mm (72 cpsi), channel length 75 mm) [120]. The monolith was coated uniformly with the catalyst support material (Al_2O_3, SiO_2, TiO_2, etc.) by a wash coat procedure. It could be shown that it was possible to prepare in a reproducible manner catalytically active coatings on the wall of single channels of the monoliths by a channel-by-channel procedure. Thus, these multichannel reactors, coupled with mass spectrometry or gas chromatography, could be used for fast screening of heterogeneous catalysts. Because temporally and spatially resolved sampling is needed and complex gaseous mixtures must be analyzed, a special 3D positioning system, which allows measurements at high temperatures, is needed as the central element of the equipment. The high efficiency and reliability of the channel-by-channel preparation and the developed screening method was demonstrated for the total oxidation of hydrocarbons and carbon monoxide in the presence of further components such as O_2, H_2O, CO_2, NO, SO_2, and inert gases over precious metal catalysts.

5.4.4
Catalytic Processes with Ceramic Foam Catalysts

While ceramic honeycomb structures with their unidirectional channel systems are well established in mobile and industrial catalytic processes (see Section 5.4.3), ceramic foam supports have been the subject of an increasing number of investigations in the last ten years. A reason for the "late discovery" of ceramic foams for catalytic applications might be seen in the unsatisfactory mechanical properties at the time when the first honeycomb structures where used in heterogeneous catalysis.

From the viewpoint of catalytic performance ceramic foams can be superior in some of the applications using honeycomb monoliths or packed-bed catalysts. While in some special cases in honeycomb structures plug flow can cause a temperature profile within a single channel, and in packed catalyst beds hot spots with an offset of up to 300 K can occur, in rigid, open-cell foam structures excellent mixing in the radial direction occurs as a consequence of the flow tortuosity, and mass and heat transfer in radial directions is excellent. Compared to catalyst pellets with an equivalent surface area, heat transfer in ceramic foams can be as much as five times higher [121]. The use of ceramic foam structures in heterogeneous catalysis was discussed with respect to advantages for the following categories of chemical processes in Ref. [122]:

- Heat-transfer-limited processes: many processes with heat-transfer limitation are carried out in long, small-diameter reactor tubes in order to achieve the desired temperature by heat exchange through the reactor walls. A significant pressure drop is the consequence. Larger voids between the pellets, realized by the use of ceramic foams, can reduce pressure drop, enhance radial heat transfer, and increase the effectiveness factor, if other parameters such as the catalytically active surface and the residence times are comparable.
- Pore-diffusion limited processes: processes with pore-diffusion limitation are mostly carried out with small-diameter catalyst pellets which result in a high pressure drop. By using ceramic foams with a coated surface the reactants have direct access to the catalyst without or with significantly decreased inner diffusion. The consequence is a reduced pressure drop and an improved effectiveness factor.
- Processes with selectivity-control problems: when an intermediate product is desired, short residence times realized by high flow rates are necessary. This operational mode always causes a high pressure drop, and ceramic foams are the ideal candidates to overcome such problems.

The pressure drop can be reduced significantly by the use of ceramic foams rather than catalyst pellets. In fact, catalyst pellets can provide a significantly higher surface area. By modification of the monoliths by several coating procedures the surface area of foams and honeycombs can be increased drastically without changing the cellular performance of a monolith. Coating procedures, however, lead to an increase in surface roughness, which affects the pressure drop. The pressure drop increased with increasing surface area [8]. Thin homogeneous layers of catalytically active substances can be fabricated by a tailored coating (for coating methods of honeycombs and ceramic foams, see Section 5.4.2). These coatings have the advantage of a large specific surface area without the disadvantage of outer diffusion through shaped catalyst parts.

The following section deals with catalytic applications of ceramic foams, mostly in laboratory-scale experiments. The processes described herein are of very heterogeneous nature with respect to the reaction mechanisms and process parameters such as temperature, pressure, and space velocity. However, a more practical classification is given on the basis of (potential) applications of ceramic foam catalysts.

5.4.4.1
Improvement of Technical Processes for Base Chemicals Production

The promising properties of ceramic foam catalysts were investigated in several types of catalytic processes, whereby a marked improvement in terms of higher selectivity and/or yield could be expected in fast reactions or reactions with a pronounced heat development such as 1) oxidative dehydrogenation of alkanes, 2) partial oxidation of ethylene, and 3) steam and dry reforming of methane. The important requirements in such chemical processes are short contact times and good temperature control.

The investigation of the substitution of the commonly used catalyst (e.g., V_2O_5/SiO_2) in the oxidative dehydrogenation of ethane or other alkanes by ceramic monoliths exemplifies the potential of these materials. The oxidative dehydrogenation of ethane was carried out on a 45–80 ppi alumina foam coated with 1–5 wt % Pt, Rh, or Pd. With the Pt-loaded monolith a conversion of about 80 % with a C_2H_4 selectivity of greater than 70 % can be achieved, comparable to those of currently used catalysts. A remarkable improvement is that the contact time in the ceramic foam catalyst is reduced to 5–10 ms, which would allow the construction of a reactor 100 to 1000 times smaller in size than currently used for the same olefin yield [123].

Table 1 list more results from investigations regarding the application of ceramic monoliths in industrially important reactions, whereby the desired improvement can be in selectivity, yield, energy requirements, special product composition, or reducing working temperature or reactor size.

In a special and effective construction the combination of catalytic combustion and reforming is realized. To increase heat conduction, NiCr metal foam is used as cellular structure and alumina-coated by plasma spraying prior to impregnation with the catalyst component [137]. The reactor consists of alumina-cladded nickel catalyst foam metal core inserted into a metal tube, which itself is inserted into an LaCo-impregnated core-cladding foam. The inner foam operates as reformer, while the outer foam is used for heating the inner foam by methane combustion in air. With a methane conversion in the range 50–63 % the feasibility of this combined catalytic combustion/reforming apparatus was demonstrated, but no relations of the data to technical processes are given. A detailed description of ceramic coating by plasma spraying with alumina is given in Ref. [138]. Due to their higher ductility as compared to ceramic foams, the ceramic-coated metal foams could have several advantages in mobile applications.

Table 1 Investigations on the application of ceramic monoliths in industrially important reactions.

Chemical process	Experimental setup	Aims/results	Refs.
Dehydrogenation of ethane and other alkanes	45–80 ppi alumina foam, 1–5 % Pt, Rh, or Pd	Pt on ceramic foams improves olefin yield and selectivity, shorter contact times	123, 124
Ethane dehydrogenation	Cr_2O_3 on ZrO_2 foam, Pt-modified	prevention of deactivation	125
Methane dehydrogenation	Al_2O_3 foam, 1–20 % Pt or Rh	higher product selectivity	126
Oxidative dehydrogenation of propane and butane	Pt/ZrO_2 foam monolith	good activity, higher selectivity	11, 127
Dehydrogenation of C_5 and C_6 alkanes	alumina foam with Pt	mechanistic studies	128
Ethane dehydrogenation	20, 45, and 80 ppi alumina foam	influence of the cell size	129
Partial oxidation of liquid fuels	alumina foam with 5 % Rh	use of higher alkanes for H_2 production	130–132
Methane reforming ($CH_4 + H_2O \rightarrow 3\,H_2 + CO$)	α-Al_2O_3-foam pellets with NiO, TiO_2	lower working temperature, higher effectiveness factor	122, 133
Dry methane reforming ($CH_4 + CO_2 \rightarrow 2\,H_2 + 2\,CO$)	γ-Al_2O_3 foam with alumina wash coat, Rh and PtRe impregnation	14 times higher turnover numbers	134
Methane conversion	Al_2O_3 and SiC foams, Ru	using solar energy for process heating	135 and Chapter 5.7
HCN production and Ostwald process	honeycombs, extruded metallic monoliths, and ceramic foams	catalytic performance: ceramic foams > honeycombs > metallic monoliths	136

5.4.4.2
Hydrogen Liberation from Liquid Precursors/Hydrogen Cleaning for Fuel Cell Applications

In mobile fuel-cell systems in situ hydrogen generation from liquid sources at ambient pressure and temperature is preferred, and methanol and ethanol have been suggested for catalytic on-board hydrogen production. For long-term operation of the catalyst, however, high-purity hydrogen with a CO content of less than 50 ppm is required [139], and high-quality hydrogen for solid-polymer fuel cells (SPFCs) can be produced by steam reforming of methanol [140]. This endothermic process (Eq. (3)) can be operated at 150–280 °C over wash coat catalyst (e.g., alumina-supported platinum). The thermal energy to operate the process is provided by combustion of part of the methanol feed (Eq. (4)).

$$CH_3OH + H_2O \rightarrow 3\,H_2 + CO_2 \quad (3)$$

$$2\,CH_3OH + 3\,O_2 \rightarrow 2\,CO_2 + 4\,H_2O \quad (4)$$

It was demonstrated that 20–25 ppi aluminum foam and a heat-exchanger structure, both with a catalytic wash coat, provided about 100 % CH_3OH conversion at 260 °C. In comparison to a packed-bed reformer (3 mm pellet diameter) the amount of catalyst could be reduced by factor of four in both structures, and the optimal temperature control led to greater than 90 % methanol conversion over a period of 450 h. Better performance, attributed to the significantly higher heat transport and faster thermal response at startup, was found, and the structural advantage of the foam was clearly demonstrated.

Steam reforming of ethanol (Eq. (5)) for the same application was investigated on catalyst pellets, ceramic foams, and honeycombs with Ru as catalytically active component [141]. The active-metal loading on the cordierite monolith (400 channels per square inch) and the alumina–zirconia foam (1/16-inch γ-alumina extrudate, 50 ppi) was carried out by wash coating with a $Ru(NO)(NO_3)_3/\gamma$-alumina slurry followed by in situ reduction, typically at 750 °C under hydrogen. The catalysis experiments were carried out in the temperature range from 400 to 1000 °C. The temperature was controlled by combustion of part of the ethanol (Eq. (6)).

$$C_2H_5OH + 3\,H_2O \rightarrow 2\,CO_2 + 6\,H_2 \quad (5)$$

$$C_2H_5OH + 0.61\,O_2 + 1.78\,H_2O \rightarrow 2CO_2 + 4.78\,H_2 \quad (6)$$

With ethanol conversion greater than 95 % and H_2 selectivity of 90–97 % the impregnated foams showed the best results at medium space velocities, (lowest yield of CH_4 and CH_3OH, which are undesired byproducts; high CO_2 selectivity, less CO).

Reticulated SiC foams (18 mm diameter, 45 mm in length), which were washcoated with γ-Al_2O_3 to increase the specific surface area and impregnated by different procedures using several Ru salts, showed excellent performance in CO removal by oxidation in the temperature range from 100 to 160 °C. A CO content in the feed gas (75 % H_2, 18 % CO, 4 % H_2O, 4000 ppm CO, 6000 ppm CO_2, 2 % N_2) of less than 30 ppm was achieved after conversion [139].

5.4.4.3
Automotive and Indoor Exhaust Gas Cleaning

Significant reduction of exhaust emissions of internal combustion engines can be achieved inter alia by reducing emissions of hydrocarbons (HC), CO, and NO_x during the cold start phase. A promising method is *microwave-enhanced conversion* of the pollutants. A microwave absorbing ceramic foam of the composition SiO_2–NiO–SiC–Fe_2O_3 was developed and covered with a Pd-doped slurry of CeO_2–ZrO_2–Al_2O_3 [142]. This catalyst was heated by microwave radiation during operation and compared to a conventionally heated ceramic foam of the same type with a simulated exhaust gas mixture. The exhaust gas inlet temperature was varied. Prior to the measurements the catalyst was reduced under hydrogen and aged under working conditions. The microwave-heated catalyst achieved 50 % conversion at an exhaust-gas inlet temperature of 60 °C, while the conventionally heated catalyst reached the same conversion rate only at 320 °C. At 110 °C the conversion was 100 % HC, 95 % NO_x, and 92 % CO for the microwave-heated catalyst. Conventional heating required a temperature of 450 °C. Moreover, the catalyst showed an almost constant pollutant conversion over a broad range of space velocity from 20 000 to 60 000 h^{-1}.

Diesel-soot converters operate on the basis of a supported molten-salt catalyst consisting of eutectic mixtures of Cs_2O, V_2O_5, MoO_3, and Cs_2SO_4 [143–145]. An alternative system consists of a Pt-loaded SiC foam, partly combined with a soot-loaded foam or a SiC membrane wall-flow reactor to assist diesel-soot combustion by NO_2 reduction [146]. More details of these systems are given in Chapter 5.2.

In a *photocatalytic oxidation reactor* for air cleaning by catalytic combustion of volatile organic compounds (VOCs) the catalyst is activated by UV light. The most common, most powerful oxidizing agent is TiO_2, and activation is carried out by excitation at a wavelength of 300–400 nm. The mechanism may be more complex, but a set of simplified equations for the activation of the semiconductor oxide, the formation of the reactive hydroxyl radicals ($OH^{\bullet}$) and the total oxidation of VOCs are described according to Ref. [147] in Eqs. (7)–(10):

$$TiO_2 + h\nu \rightarrow h^+ + e \qquad \text{(activation)} \qquad (7)$$

$$OH^- + h^+ \rightarrow OH \qquad \text{(oxidation)} \qquad (8)$$

$$O_{2(ads)} + e^- \rightarrow O_{2(ads)} \qquad \text{(reduction)} \qquad (9)$$

$$OH^{\bullet} + \text{pollutant} + O_2 \rightarrow \text{products } (CO_2, H_2O, \text{etc.}) \qquad (10)$$

Requirements for this setup are a large-volume reactor, a low pressure drop, and an efficient contact among the photons, the catalyst, and the gaseous reactants. Several types of reactor design were developed, with membranes, honeycombs, or even the inner wall of a quartz glass tube as the catalyst support [147]. Novel developments make use of ceramic monoliths. The kind of monolith – honeycomb or foam – affects the radiation field of the UV source, and model calculations must be performed for optimal reactor design. First simulations of the UV absorption process in cellular ceramics (foams and honeycombs) for a tailored reactor are described in Ref. [148], and two different reactor designs were realized in Ref. [149]. One type of reactor consists of an alternating arrangement of UV lamps and titania-coated foam monolith plates. The monoliths are arranged perpendicular to the flow direction, and lamps are mounted at the inlet and outlet and between the plates (Fig. 12).

In the annular-type reactor a UV lamp is inserted in a hollow cylinder of ceramic foam. First catalytic experiments demonstrated 80 % conversion of isopropanol as a model compound.

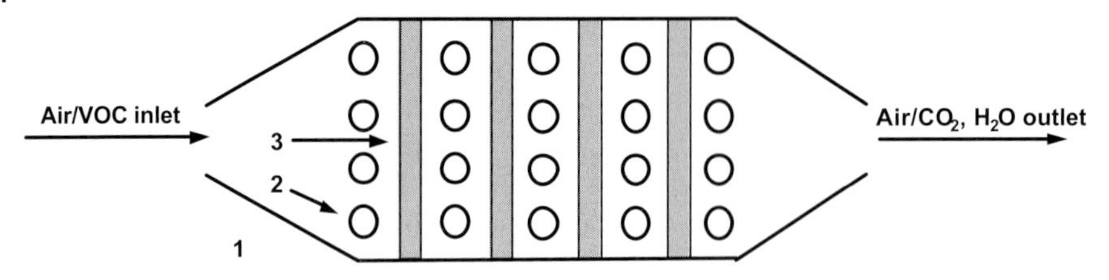

Fig. 12 Schematic of a PCO reactor [149]. 1) Reactor cabinet, 2) UV lamps, 3) TiO_2/ceramic foam.

Catalyst systems for *ozone decomposition*, which are necessary for deozonisation of, for example, airplane passenger cabins and photocopiers, work over a temperature range of 2–177 °C, typically 20–50 °C in most technical applications. The degree of decomposition is between 69 and 100 %, but greater than 90 % in most systems. A review on ozone decomposition including several catalyst systems, decomposition mechanism, reactivity of ozone-decomposition catalysts, and reactor designs/support materials is given in Ref. [150]. Ceramic cordierite and reticulated carbon foams (both 20 ppi) were used as monolithic catalyst supports [151]. The foams were coated with a slurry consisting of activated carbon (AC) and metal oxide or a system of AC and two metal oxides. The highest decomposition rate was achieved over $AC/MnO_2/Fe_2O_3$ supported on activated carbon foam, which achieved a value of greater than 90 % when gum arabicum was used as disperser in the slurry.

Selective catalytic reduction of NO_x was investigated on zeolite-coated structured catalyst packings [152]. An α-alumina foam (45 ppi) and a cordierite monolith honeycomb (300 csi with a square channel cross section of 1×1 mm) were used as supports. The zeolite coating was generated by in situ supported crystallization [23] followed by ion exchange. The catalyst-loaded alumina foams were arranged to form a parallel-passage reactor (PPR, Fig. 13). Catalytic NO_x reduction as shown in Eq. (1) was carried out in the temperature range 200–350 °C in the PPR and at 125–370 °C

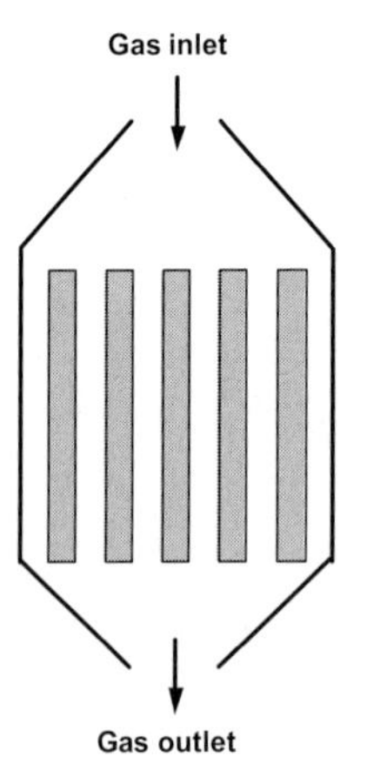

Fig. 13 Schematic of the parallel passage reactor [152].

in the monolith reactor. An optimum operation temperature for the PPR was found to be 300 °C, and an NO conversion of 80 % was found at a gas hourly space velocity (GHSV) of 3000 h^{-1} for the PPR and at 18 000 h^{-1} for the monolith reactor. However, the efficiency of the PPR was 17 % in this experiment, compared to the honeycomb monolith with about 95 %. The general workability of this novel reactor type employing a zeolite coating grown in situ was demonstrated. Due to the low efficiency factor it was concluded that the design of the PPR was not optimal and a lateral arrangement and optimized plate thickness was suggested for ongoing work. The low pressure drop in this system is of paramount importance.

5.4.4.4 Catalytic Combustion in Porous Burners

A further application of ceramic foams with a catalytically active surface was studied to improve the performance of porous burners. When a catalytically active coating is applied in porous burners, the combustion temperature of the burner fuels can be decreased, which results in reduced CO, HC, and NO_x emissions. Coating a ceramic foam (60 ppi, mullite, 12 mm in thickness and 120 mm in diameter) with $LaMnO_3$ lead to a significant decrease in burner temperature: complete methane combustion was found at about 800 °C with no coating, while a $LaMnO_3$ catalyst on the foam surface lead to a temperature of 700 °C over a nonprecoated and to less than 650 °C on an La_2O_3-precoated foam [153]. Further details on porous burners can be found in Chapter 5.5.

In Sections 5.4.3 and 5.4.4 the use of ceramic monoliths in several catalytic applications was demonstrated. Common ceramics such as cordierite, mullite, γ- and α-alumina, and silicon carbide were applied, but also special developments with tailored specific properties, such as microwave-absorbing of metal oxides have been reported. Optimal radial and axial heat and mass flows allow the operational temperature range to be narrowed to the needs of a specific reaction. The range of service temperature is broad, from room temperature up to 1400 °C, and the reaction conditions vary in a similarly broad range. This makes a careful selection of material, surface treatment and further modification of the ceramic monoliths (honeycombs and foams) necessary.

5.4.5 Summary

Ceramic honeycomb structures are a well-established class of catalyst supports. Uses range from mobile applications such as automotive and diesel exhaust gas catalysis, to indoor gas cleaning and production-site exhaust gas cleaning, and use in industrial chemical processes. A large number of applications are still the subject of research and development with major potential for application.

Open-cell ceramic foams have considerable potential for use as catalyst supports in heterogeneously catalyzed processes in the chemical industry and environment protection. In comparison to some of the well-established classical processes, ceramic foam catalysts showed superior catalytic performance, albeit so far on the laboratory scale. Due to the low pressure drop and the short contact times this class of supported catalysts has the potential for reduction of reactor size, which is associated with significant cost and energy savings in the chemical industry.

Development of novel materials, for example, the fabrication of electrically conductive monoliths with self-heating properties, and an improvement of their mechanical properties make both classes of ceramic support promising candidates for energy saving and for present and future applications in heterogeneous catalysis, where multifunctionality is a demand.

Acknowledgement

The authors gratefully acknowledge the Deutsche Forschungsgemeinschaft (DFG), the Bundesministerium für Bildung und Forschung (BMBF), the Fonds der Chemischen Industrie (FCI) and the Alexander-von-Humboldt Foundation (AvH), for financial support.

References

1 Knözinger, H. in: *Ullmann's Encyclopedia of Industrial Chemistry*, Electronic Version, Chapter "Heterogeneous Catalysis and Solid Catalysts", Wiley-VCH, Weinheim 2003.
2 Ertl, G., Knözinger, H., Weitkampp, J. (Eds.), *Handbook of Heterogeneous Catalysis*, Wiley-VCH, Weinheim 1997.
3 Gates, B.C. (Ed.), *Catalytic Chemistry*, John Wiley & Sons, New York 1991.
4 Cole, E.L., Knowles, E.C., US Patent 3 155 627, 1964.
5 Smith, G.R., US Patent 3 088 271, 1963.
6 Geus, J.W., van Giezen, J.C., *Catal. Today* **1999**, *47*, 169–180.
7 Nijhuis, T.A., Beers, A.E.W., Vergunst, T., Hoek, I., Kapteijn, F., Moulijn, J.A., *Catal. Rev.* **2001**, *43*, 345–380.
8 Richardson, J.T., Peng, Y., Remue, D., *Appl. Catal. A* **2000**, *204*, 19–32.
9 Gallei, E., Schwab, E., *Catalysis Today* **1999**, *51*, 535–546.
10 Campanati, M., Fornasari, G., Vaccari, A., *Catal. Today* **2003**, *77*, 299–314.
11 Huff, M., Schmidt, L.D., *J. Catal.* **1995**, *155*, 82–94.
12 Baerlocher, Ch., Meier W.M., Olson, D.H., *Atlas of Zeolite Framework Types*, 5th ed., Elsevier, Amsterdam 2001.
13 http://www.iza-stucture.org.
14 Rabo, J.A., Schoonover, M.W., *Appl. Catal. A* **2001**, *222*, 261–275.
15 Kripylo, P., Wendlandt, K.-P., Vogt, F., *Heterogene Katalyse in der chemischen Technik*, Deutscher Verlag für Grundstoffindustrie, Leipzig, Stuttgart 1993.
16 Corma, A., *J. Catal.* **2003**, *216*, 298–312.
17 Karge, H.G., Weitkamp, J. (Eds.), *Molecular Sieves, Science and Technology*, Vols. 1–4, Springer-Verlag, Heidelberg 1998, 1999, 2002, 2004.
18 Boix, A.V., Zamaro, J.M., Lombardo, E.A., Miró, E.E., *Appl. Catal. B* **2003**, *46*, 121–132.
19 Saracco, G., Russo, N., Ambrogio, M., Badini, C., Specchia, V., *Catal. Today* **2000**, *60*, 33–41.
20 X. Xu, J.A. Moulijn, in: B. Delmon, P.A. Jacobs, R. Maggi, J.A. Martens, P. Grange, G. Poncelet (Eds.), *Stud. Surf. Sci. Catal.* **1998**, *118*, 845–854.
21 Buciuman, F.-C., Kraushaar-Czarnetzki, B., *Catal. Today* **2001**, *69*, 337–342.
22 Masuda, T., Hara, H., Kouno, M., Kinoshita, H., Hashimoto, K., *Microporous Mater.* **1995**, *3*, 565–571.
23 Seijger, G.B.F., Oudshoorn, O.L., van Kooten, W.E.J., Jansen, J.C., van Bekkum, H.,

van Den Bleek, C.M., Calis, H.P.A., *Microporous Mesoporous Mater.* **2000**, *39*, 195–204.

24 Clet, G., Peters, J.A., van Bekkum, H., *Langmuir* **2000**, *16*, 3993–4000.

25 Katsuki, H., Furuta, S., Komarneni, S., *J. Am. Ceram. Soc.* **2000**, *83*, 1093–1097.

26 Zampieri, A., Colombo, P., Mabande, G.T.P., Selvam, T., Schwieger, W., Scheffler, F., *Adv. Mater.* **2004**, *16*, 819–823.

27 Scheffler, M., Zeschky, J., Zampieri, A., Herrmann, R., Schwieger, W., Scheffler, F., Greil, P., *Ceram. Trans.* **2003**, *154*, 49–59.

28 Scheffler, F., Zampieri, A., Schwieger, W., Zeschky, J., Scheffler, M., Greil P., *Adv. Appl. Ceram.* **2005**, *104*, 43–48.

29 Zampieri, A., Sieber, H., Selvam, T., Mabande, G.T.P., Schwieger, W., Scheffler, F., Scheffler, M., Greil., P., *Adv. Mater.* **2005**, *17*, 344–349.

30 Heck, R.M., Gulati, S., Farrauto, R.J., *Chem. Eng. J.* **2001**, *82*, 149–156.

31 Cottrell, F.G., US Patent 2 121 733, 1938.

32 Hollenbach, R.Z., US Patent 3 112 184, 1963.

33 Houdry, E.J., US Patent 2 946 651, 1960.

34 Henderson, D.S., Briggs, W.S., Stover, W.A., Thomas, A.H., US Patent 3 295 919.

35 Heck, R.M., Farrauto, R.J., *Appl. Catal. A* **2001**, *221*, 443–457.

36 Gandhi, H.S., Grahem, G.W., McCabe, R.W., *J. Catal.* **2003**, *216*, 433–442.

37 Heck, R.M., Farrauto, R.J. (Eds.), *Catalytic Air Pollution Control*, Van Nostrand Reinhold, New York 1995.

38 Williams, J.L., *Catal. Today* **2001**, *69*, 3–9.

39 Hu, Z., Allen, F.M., Wan, C.Z., Heck, R.M., Steger, J.J., Lakis, R.E., Lyman, C.E., *J. Catal.* **1998**, *174*, 13–21.

40 Martin, L., Arranz, J.L., Prieto, O., Trillano, R., Holdago, M.J., Galán, M.A., Rives, V., *Appl. Catal. B* **2003**, *44*, 43–52.

41 Gonzales-Velasco, J.R., Botas, J.A., Ferret, R., Gonzáles-Marcos, M.P., Marc, J.-L., Gutiérrez-Ortiz, M.A., *Catal. Today* **2000**, *59*, 395–402.

42 Kobayashi, T., Yamada, T., Kayano, K., *Appl. Catal. B* **2001**, *30*, 287–292.

43 Kim, D.H., Woo, S.I., Noh, J., Yang, O., *Appl. Catal. A* **2001**, *207*, 69–77.

44 Noh, J., Yang, O., Kim, D.H., Woo, S.I., *Catal. Today* **1999**, *53*, 575–582.

45 Miki, T., Ogawa, T., Haneda, M., Kakuta, N., Ueno, A., Tateishi, S., Matsuura, S., Sato, M., *J. Phys. Chem.* **1990**, *94*, 6464–6467.

46 Martinez-Arias, A., Fernadez-Garcia, M., Hungria, A.B., Iglesias-Juez, A., Duncan, K., Smith, R., Anderson, J.A., Conesa, J.C., Soria, J., *J. Catal.* **2001**, *204*, 238–248.

47 Vlaic, G., Di Monte, R., Fornasiero, P., Fonda, E., Kašpar, J., Graziani, M., *J. Catal.* **1999**, *182*, 378–389.

48 Vlaic, G., Fornasiero, P., Geremica, S., Kašpar, J., Graziani, M., *J. Catal.* **1997**, *168*, 386–392.

49 Johnson, M.F.L., *J. Catal.* **1990**, *123*, 245–252.

50 Monte, R.D., Fornasiero, P., Kašpar, J., Rumori, P., Gubitosa, G. Graziani, M., *Appl. Catal. B* **2000**, *24*, 157–167.

51 Kašpar, J., Fornasiero, P., Hickey, N., *Catal. Today* **2003**, *77*, 419–449.

52 Lafyatis, D.S., Ansell, G.P., Bennet, S.C., Frost, J.C., Millington, P.J., Rajaram, R.R., Walker, A.P., Ballinger, T.H., *Appl. Catal. B* **1998**, *18*, 123–135.

53 Fridell, E., Soglundh, M., Westerberg, B., Johansson, S., Smedler, G., *J. Catal.* **1999**, *183*, 196–209.

54 Shinjoh, H., Takahashi, N., Yokota, K., Sugiura, M., *Appl. Catal B* **1998**, *15*, 189–201.

55 Farrauto, R.J., Voss, K.E., *Appl. Catal. B* **1996**, *10*, 29–51.

56 Clerc, J.C., *Appl. Catal. B* **1996**, *10*, 99–115.

57 Cooper, B.J., Jung, H.J., Thoss, J.E., US Patent 4 902 487, **1990**.

58 Summers, J.C., Van Houtte, St., Psaras, D., *Appl. Catal. B* **1996**, *10*, 139–156.

59 Jelles, S.J., Makkee, M., Moulijn, J.A., *Top. Catal.* **2001**, *16/17*, 269–273.

60 Fino, D., Fino, P., Saracco, G., Specchia, V., *Chem. Eng. Sci.* **2003**, *58*, 951–958.

61 Ciambelli, P., Palma, V., Russo, P., Vaccaio, S., *Catal. Today* **2002**, *73*, 363–370.

62 Oi-Uchisaw, J., Wang, S., Nanba, T., Ohi, A., Obuchi, A., *Appl. Catal. B* **2003**, *44*, 207–215.

63 Neri, G., Rizzo, G., Galvagno, S., Donato, A., Musolino, M.G., Pietropaolo, R., *Appl. Catal. B* **2003**, *42*, 381–391.

64 Fino, D., Russo, N., Saracco, G., Specchia, V., *J. Catal.* **2003**, *217*, 367–375.

65 Kureti, S., Weisweiler, W., Hizbullah, K., *Appl. Catal. B* **2003**, *43*, 281–291.

66 Ciambelli, P., Palma, V., Russo, P., Vaccaio, S. *J. Mol. Catal. A* **2003**, *204–205*, 673–681.

67 Caldeira Leit Leocadio, I., Braun, S., Schmal, M., *J. Catal.* **2004**, *223*, 114–121.

68 Milt, V.G., Pissarello, M.L., Miró, E.E., Quercini, C.A., *Appl.Catal. B* **2003**, *41*, 397–414.
69 Fino, D., Fino, P., Saracco, G., Specchia, V., *Appl.Catal. B* **2003**, *43*, 243–259.
70 Koebel, M., Madia, G., Elsner, M., *Catal. Today* **2002**, *73*, 239–247.
71 Koebel, M., Elsener, M.. Kleemann, M., *Catal. Today* **2000**, *59*, 335–345.
72 Shelef, M., *Chem. Rev.* **1995**, *95*, 209–225.
73 Kato, A., Matsuda, S., Nakajima, F., Kuroda, H., Narita, T., *J. Phys. Chem.* **1981**, *85*, 4099–4102.
74 Tuenter, G., Leeuwen, W., Snepvangers, L., *Ind. Eng. Chem. Prod. Res. Dev.* **1986**, *25*, 633–636.
75 Ritscher, J.S., Sandner, M.R., US Patent 4 297 328, 1981.
76 Forzatti, P., *Appl. Catal. A* **2001**, *222*, 221–236.
77 Forzatti, P., *Catal.Today* **2003**, *83*, 3–18.
78 Thevenin, P.O., Menon, P.G., Järas, S.G., *Cattech* **2003**, *7*, 10–22.
79 Addiego, W.P., Liu, W., Boger, T., *Catal. Today* **2001**, *69*, 25–31.
80 Schanke, D., Bergene, E., Holmen, A. US Patent 6 211 255, **2001**.
81 Hilmen, A.-M., Bergene, E., Lindvåg, O.A., Schanke, D., Eri, S., Holmen, A., *Catal. Today* **2001**, *69*, 227–232.
82 Nicolau, I., Colling, P.M., Johnson, L.R., US Patent 5 705 679, **1998**.
83 Nicolau, I., Colling, P.M., Johnson, L.R., US Patent 5 854 171, **1998**.
84 Tischer, S., Correa, C., Deutschmann, O., *Catal. Today* **2001**, *69*, 57–62.
85 Heibel, A.K., Heiszwolf, J.J., Kapteijn, F., Moulijn, J.A., *Catal. Today* **2001**, *69*, 153–163.
86 Nijhuis, T.A., Kreutzer, M.T., Romijn, C.J., Kapteijn F., Moulijn, J.A., *Chem. Eng. Sci.* **2001**, *56*, 823–829.
87 Edvinsson Albers, R., Nyström, M., Silverström, M., Sellin, A., Dellve, A.-C., Andersson, U., Herrmann, W., Berglin, Th., *Catal. Today* **2001**, *69*, 247–252.
88 Berglin, Th., Herrmann, W., US Patent 4 552 748, **1985**.
89 Berglin, Th., Herrmann, W., EP Patent 102 934, **1994**.
90 Kreutzer, M.T., Du, P., Heiszwolf, J.J., Kapteijn, F., Moulijn, J.A., *Chem. Eng. Sci.* **2001**, *56*, 6015–6023.
91 Smits, H.A., Stankiewicz, A., Glasz, W.Ch., Fogl, T.H.A., Moulijn, J.A., *Chem. Eng. Sci.* **1996**, *51*, 3019–3025.
92 Cybulski, A., Stankiewicz, A., Edvinsson Albers, R.K., Moulijn, J.A., *Chem. Eng. Sci.* **1999**, *54*, 2361–2358.
93 Winterbottom, J.M., Marwan, H., Stitt, E.H., Natividad, R., *Catal. Today* **2003**, *79/80*, 391–399.
94 Hatziantoniou, V., Andersson, B., Schöön, N.-H., *Ind. Eng. Chem. Process Des. Dev.* **1986**, *25*, 964–970.
95 Bercic, G., *Catal. Today* **2001**, *69*, 147–152.
96 Machado, R.M., Parrillo, D.J., Boehme, R.P., Broekhuis, R.R., US Patent 6 005 143, 1999.
97 Machado, R.M., Broekhuis, R.R., US Patent 6 506 361, 2003.
98 Broekhuis, R.R., Machado, R.M., Nordquist, A.F., *Catal. Today* **2001**, *69*, 87–93.
99 Cybulski, A., Chrzaszcz, J., Twigg, A.V., *Catal. Today* **2001**, *69*, 241–245.
100 Beers, A.E.W., Nijhuis, T.A., Kapteijn, F., Moulijn J.A., *Microporous Mesoporous Mater.* **2001**, *48*, 279–284.
101 Awdry, S., Kolaczkowski, S.T., *Catal. Today* **2001**, *69*, 275–281.
102 Nijhuis, T.A., Beers, A.E.W., Kapteijn, F., Moulijn, J.A., *Chem. Eng. Sci.* **2002**, *57*, 1627–1632.
103 Beers, A.E.W., Spruit, R.A., Nijhuis, T.A., Kapteijn, F., Moulijn, J.A., *Catal. Today* **2001**, *66*, 175–181.
104 Stitt, E.H., Jackson, S.D., Shipley, D.G., King, F., *Catal. Today* **2001**, *69*, 217–226.
105 Edvinsson, R.K., Moulijn, J.A., WO Patent 98/30323, 1998 (to DSM).
106 Vaarkamp, M., Dijkstra, W., Reesink, B.H., *Catal. Today* **2001**, *69*, 131–135.
107 Wolffenbuttel, B.M.A., Nijhuis, T.A., Stankiewicz, A., Moulijn, J.A., *Catal. Today*, **2001**, *69*, 265–273.
108 Vergunst, T., Linders, M.J.G., Kapteijn, F., Moulijn, J.A., *Catal. Rev.* **2001**, *43*, 291–314.
109 Kapteijn, F., Heizwolf, J.J., Nijhuis, T.A., Moulijn, J.A. *Cattech* **1999**, *3*, 24–41.
110 Vergunst, Th., Kapteijn, F., Moulijn, J.A., *Catal. Today* **2001**, *66*, 381–387.
111 Vergunst, Th., Ph.D. thesis, DUP Science, Delft, The Netherlands, **1999**.
112 Crezee, E., PhD thesis, DUP Science, Delft, The Netherlands, **2003**.
113 Crezee, E., Barendregt, A., Kapteijn, F., Moulijn, J.A., *Catal. Today* **2001**, *43*, 283–290.

114 García-Borejé, E., Kapteijn, F., Moulijn, J.A., *Catal. Today* **2001**, *69*, 357–363.
115 Vergunst, Th., Kapteijn, F., Moulijn, J.A., *Carbon* **2002**, *40*, 1891–1902.
116 Nordquist, A.F., Wilhelm, F.C., Waller, F.J., Machado, R.M., US Patent 20030036477, 2003.
117 Nordquist, A.F., Wilhelm, F.C., Waller, F.J., Machado, R.M., US Patent 6 479 704, 2002.
118 Nordquist, A.F., Wilhelm, F.C., Waller, F.J., Machado, R.M., US Patent 20030027718, 2003.
119 Claus, P., Hönicke, D., Zech, T., *Catal. Today* **2001**, *67*, 319–339.
120 M. Lucas, P. Claus: High Throughput Screening in Monolith Reactors for Total Oxidation Reactions. *Appl. Catal.*, Special Issue "Combinatorial Catalysis" (Guest Ed.: Maier, W.F.), **2003**, 35–43.
121 Richardson, J.T., Remue, D., Hung, J.-K., *Appl. Catal. A* **2003**, *250*, 319–329.
122 Twigg, M.V., Richardson, J.T., *Trans. IChemE* **2002**, *80*, 183–189.
123 Huff, M., Schmidt, L.D., *J. Catal.* **1994**, *149*, 127–141.
124 Huff, M., Schmidt, L.D., *J. Phys. Chem.* **1993**, *97*, 11 815–11 822.
125 Flick, D.W., Huff, M.C., *Appl. Catal. A* **1993**, *187*, 13–24.
126 Hickman, D.A., Schmidt, L.D., *Science* **1993**, *259*, 343–346.
127 Liebmann, L.S., Schmidt, L.D,. *Appl. Catal. A* **1999**, *179*, 93–106.
128 Dietz III, A.G., Carlsson, A.F., Schmidt, L.D. *J. Catal.* **1998**, *176*, 459–473.
129 Bodke, A.S., Bharadwaj, S.S., Schmidt, L.D. *J. Catal.* **1998**, *179*, 138–149.
130 Krummenacher, J.J., West, K.N., Schmidt, L.D. *J. Catal.* **2003**, *215*, 332–343.
131 O'Connor, R.P., Klein, E.J., Henning, D., Schmidt, L.D., *Appl. Catal. A* **2003**, *238*, 29–40.
132 Schmidt, L.D., Klein, E.J., Leclerc, C.A., Krummenacher, J.J., West, K.N., *Chem. Eng. Sci.* **2003**, *58*, 1037–1041.
133 Jager, B., *Stud. Surf. Sci. Catal.* **1998**, *119*, 25.
134 Richardson, J.T., Garrait, M., Hung, J.-K., *Appl. Catal. A* **2003**, *255*, 69–82.
135 Wörner, A. Tamme, R., *Catal. Today* **1998**, *46*, 165–174.
136 Hickman, D.A., Huff, M., Schmidt, L.D., *Ind. Eng. Chem. Res.* **1993**, *32*, 809–817.
137 Ismagilov, Z.R., Pushkarev, V.V., Podyacheva, O.Yu., Koryabkina, N.A., Veringa, H., *Chem. Eng. J.* **2001**, *82*, 355–360.
138 Ismagilov, Z.R., Podyacheva, O.Yu., Solonenko, O.P., Pushkarev, V.V., Kuz'min, V.I., Ushakov, V.A., Rudina, N.A., *Catal. Today* **1999**, *51*, 411–417.
139 Wörner, A., Friedrich, C., Tamme, R., *Appl. Catal. A* **2003**, *245*, 1–14.
140 de Wild, P.J., Verhaak, M.J.F.M., *Catal. Today* **2000**, *60*, 3–10.
141 Liguras, D.K., Goundani, K., Verykios, X.E., *Int. J. Hydrogen Energy*, **2004**, *29*, 419–427.
142 Zhenming, Y., Jinsong, Z., Xiaoming, C., Qiang, L., Zhijun, X., Zhimin, Z., *Appl. Catal. B* **2001**, *34*, 129–135.
143 van Setten, B.A.A.L., Bremmer, J., Jelles, S.J., Makkee, M., Moulijn, J.A., *Catal. Today* **1999**, *53*, 613–621.
144 Saracco, G., Badini, C., Specchia, V., *Chem. Eng. Sci.* **1999**, *54*, 3035–3041.
145 van Setten, B.A.A.L., Spitters, C.G.M., Bremmer, J., Mulders, A.M.M., Makkee, M., Moulijn, J.A., *Appl. Catal. B* **2003**, *42*, 337–347.
146 Setiabudi, A., Makkee, M., Moulijn, J.A., *Appl. Catal. B* **2003**, *42*, 35–45.
147 Zhao, J., Yang, X., *Build. Environ.* **2003**, *38*, 643–654.
148 Changrani, R., Raupp, G.B., *AIChE J.* **1999**, *45*, 1085–1094.
149 Raupp, G.B., Alexiadis, A., Hossain, Md. M., Changrani, R., *Catal. Today* **2001**, *69*, 41–49.
150 Dehandapini, B., Oyama, S.T., *Appl. Catal. B* **1997**, *11*, 129–166.
151 Heisig, C., Zhang, W., Oyama, S.T., *Appl. Catal. B* **1997**, *14*, 117–129.
152 Seijger, G.B.F., Oudshoorn, O.L., Boekhorst, A., van Bekkum, H., van den Bleek, C.M., Calis, H.P.A., *Chem. Eng. Sci.* **2001**, *56*, 849–857.
153 Cerri, I., Saracco, G., Specchia, V., *Catal. Today* **2000**, *60*, 21–32.

5.5
Porous Burners

Dimosthenis Trimis, Olaf Pickenäcker, and Klemens Wawrzinek

5.5.1
Introduction

Combustion processes dominate the entire energy and transport sectors. The different burner types mainly depend on the fuel type (gaseous, liquid, or solid) and the application requirements (operating conditions, size, and emission limits). Although all possible fuel types are presently being used, there are clear signs that natural gas is becoming of increasing importance, especially in fields where low-emission thermal energy is needed. This is the case in domestic appliances such as household heating and hot-water supply systems, where low cost is another important factor that burner must have to be of interest to the market. In industrial processes such as drying, coating, and preheating, more stringent emission requirements and increased process efficiency are the driving forces for further developments in burner technology.

Two types of burner emissions are of major importance: 1) emissions due to incomplete combustion such as carbon monoxide and unburned hydrocarbons, and 2) hazardous combustion byproducts such as nitrogen oxides and sulfur oxides (greenhouse gases and/or toxic). Concerning emissions due to incomplete combustion, several measures can be applied, such as premixed combustion instead of diffusion-type flame, longer residence time in the reaction region, better insulation of the combustion chamber, higher combustion chamber temperature, more homogeneous mixing of the reactants, etc. Concerning combustion byproducts, the amount of sulfur oxides is related only to the sulfur content of the applied fuel (which is highest for solid fuels and lowest for natural gas), while the amount of nitrogen oxides, which are the most important nowadays in environmental protection, basically depends on the combustion temperature and the residence time at high temperatures. Thus, all efforts dealing with the abatement of nitrogen oxide emissions focus on reduction of the combustion temperature while taking into account that a spatially homogeneous minimum combustion temperature is compulsory to avoid incomplete combustion.

Several concepts for low-NO_x burners emerged in the last two decades, especially for burners fired with natural gas. A reduction in combustion temperature is common to all concepts, but it is realized in different ways: exhaust gas recirculation, internal recirculation regions, staged combustion with heat decoupling, and direct

Cellular Ceramics: Structure, Manufacturing, Properties and Applications.
Michael Scheffler, Paolo Colombo (Eds.)

ISBN: 3-527-31320-6

heat extraction from the combustion region by interaction and heat exchange with a porous solid matrix are the major strategies applied to reduce the combustion temperature homogeneously (without allowing disturbances from hot spots and cold region) so that NO_x emissions are reduced, without paying the penalty of increased emissions due to incomplete combustion.

A significant decrease in NO_x emissions was achieved with the development of radiant porous surface burners. Such burners produce a flame sheet very near to the burner support, which operates at a high temperature and radiates a part of the heat released by combustion to the appliance. Thus, the combustion temperature and hence the NO_x emissions decreases.

A further improvement was achieved by the development of volumetric porous radiant burners, which completely trap the flame inside a porous structure and have the same benefits as radiant surface burners even at significantly higher heat loads. They also show a higher turn-down ratio than conventional burners.

In recent years ceramic porous structures have been increasingly used as burner supports, mainly for natural gas combustion in domestic and industrial appliances. Cellular ceramic structures offer an interesting alternative to metallic burner supports due to higher operating temperatures, more efficient heat extraction through the radiating porous surface and/or volume, and potentially better performance concerning flashback safety. The burner supports can be categorized as those for catalytic combustion, for surface combustion, or for volumetric porous combustion (Fig. 1). In most cases flat plate geometries or cylinders are used. The ceramic supports can be sintered or fiber-based, while the materials range from mullites and alumina to silicon carbide fibers.

In this chapter an overview of the different burner types using porous ceramic burner supports is given, and the major properties affecting the combustion process

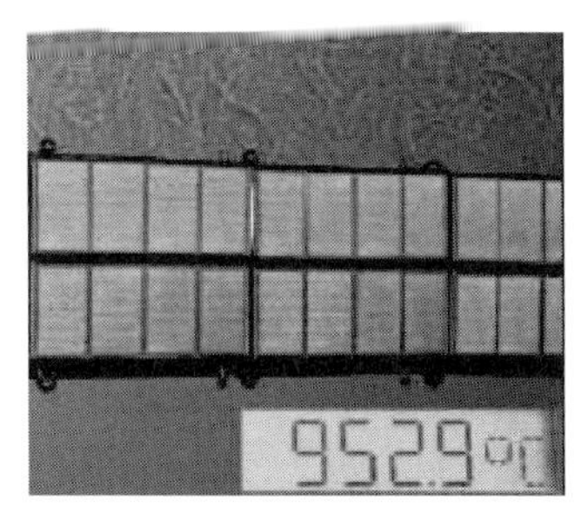

Surface Radiant Burners
Middle wave length
Thermal loads ca. 120-600 kW/m²
(up to 3000 kW/m² in blue flame mode)
Radiating temperature ca. 900-1100 °C

Catalytic radiators
Long wave length
Thermal load ca. 50 kW/m²
Radiating temperature ca. 600 °C

Volumetric Porous Burners
Short wave length
Thermal loads ca. 200-4000 kW/m²
(up to 8000 kW/m² with air preheating)
Radiating temperature ca. 1100-1500 °C

Fig. 1 Different types of porous burners operated with gas [1].

and flame stabilization are discussed. The case of volumetric porous burners stabilized by flame quenching is discussed in more detail. Also an overview on the different structures, their characteristic properties, and the suitable applications is given.

5.5.2
Flame Stabilization of Premixed Combustion Processes in Porous Burners

Flame propagation of premixed fuel/oxidant mixtures in porous inert media depends on the structure and physical properties of the solid matrix and on the properties of the combustible gas. The resulting flame propagation modes can be classified into different regimes, of which some important parameters are given in Table 1. This chapter discusses only regimes in which no pressure gradient occurs in the reaction zone, that is, the low-velocity regime (LVR) and the high-velocity regime (HVR). The solid matrix strongly influences the reaction conditions of combustion. Heat transport in the burner is dominated by the properties of the solid material, especially under real operating conditions, where the porous matrix reaches high temperatures. Ceramic spheres and other packing material can be used, but cellular ceramics are the most interesting structures for practical applications. In Ref. [2] an overview on applicable porous media for porous burners is given.

Table 1 Flame propagation regimes in porous media [3–6].

Regime	Speed of combustion wave/m s^{-1}	Mechanism of flame propagation
Low-velocity regime (LVR)	0–10^{-4}	heat conduction and interphase heat exchange
High-velocity regime (HVR)	0.1–10	high convection
Rapid-combustion regime (RCR)	10–100	convection with low pressure gradient
Sound-velocity regime (SVR)	100–300	convection with significant pressure gradient
Low-velocity detonation (LVD)	500–1000	self-ignition with shock wave
Normal detonation (ND)	1500–2000	detonation with momentum and heat loss

Flame stabilization of premixed combustion in porous media differs significantly from that of free premixed flames. Stabilization depends mainly on the heat-transport properties of the solid matrix. Heat transport in a porous medium is often described by an effective thermal conductivity which comprises radiation and heat conduction of both solid and gas phases and additionally gas convection and dispersive mechanisms (see Chapter 4.3). The effective heat transport inside of a porous medium is 2–3 orders of magnitude higher than in free flames and can be considered the dominant parameter for flame propagation in most cases. Compared with free premixed flames, the higher heat transport leads to 10–30 times faster flame

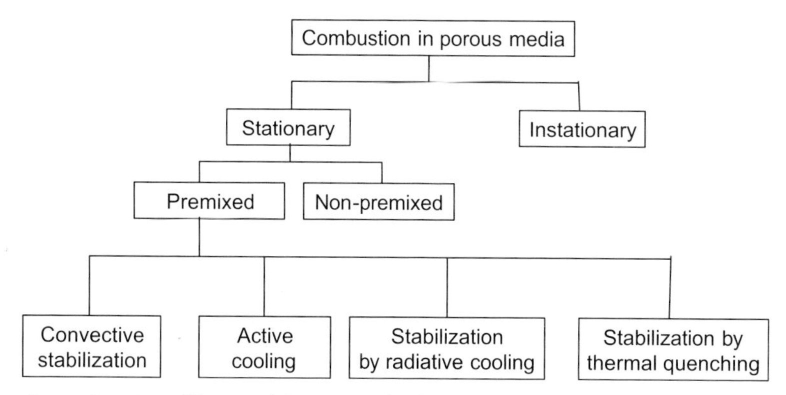

Fig. 2 Overview of flame stabilization methods in porous media.

propagation [7, 8], which hampers flame stabilization in porous media. For solving the problem of flame stabilization in porous media, different approaches have been developed, which are summarized in Fig. 2.

In the LVR and HVR regimes most relevant to burner applications, flame propagation or extinction can be described by a modified Péclet number *Pe* formulated with the laminar flame speed S_L instead of a flow velocity

$$Pe = \frac{S_L d_{p,\text{eff}} \rho_f c_{p,f}}{\lambda_f} = \frac{S_L d_{p,\text{eff}}}{a_f} \tag{1}$$

where $d_{p,\text{eff}}$ is an equivalent porous cavity diameter, $c_{p,f}$ the specific heat capacity, ρ_f the density, λ_f the thermal conductivity, and a_f the thermal diffusivity of the gas mixture. The modified Péclet number describes the ratio between the heat release due to combustion in a pore and the heat removal on the walls of a pore. It also describes the ratio of the characteristic porous cavity diameter to the laminar flame front thickness. This ratio must exceed a critical value for flame propagation in a cold porous medium. According to Ref. [3] flame propagation in porous media is possible if:

$$Pe \geq 65 \pm 45. \tag{2}$$

If $Pe < 65$ then the flame structures extinguish (quenching), since heat is transferred to the porous matrix at a higher rate than it is produced. Flame propagation in porous media in the LVR and HVR regimes (Table 1) is mostly dominated by convection but the high conductivity and the radiation properties of the solid matrix also influence the flame speed, especially at higher temperatures. The quantity $d_{\text{p,eff}}$ represents an equivalent length scale of the pore size for heat transport. Whether or not flame propagation will occur in a porous inert medium can be decided by the choice of the pore size of the solid matrix. Thus, a critical pore size exists above which flame propagation and below which flame quenching occurs. Nevertheless, flame propagation may also occur in subcritical cavities if the temperature of the matrix is high enough that reactions are not quenched by the low wall temperature.

Thus, combustion processes inside subcritical porous media need a special starting procedure to preheat the porous matrix to the operating temperature. This kind of operation is often called filtration combustion.

5.5.2.1
Flame Stabilization by Unsteady Operation

Combustion in subcritical porous media (low characteristic porous cavity size) will only take place if the temperature of the matrix is high enough to ignite the mixture. Flame propagation is driven mainly by the thermal conductivity of the solid material and interphase heat exchange, while the heat transport in the gas phase is negligible. The so-called combustion wave travels very slowly through the porous matrix with a speed of 10^{-5} to 10^{-4} m s^{-1} [9–11], either in or against the flow direction. In the former case superadiabatic and in the latter subadiabatic combustion temperatures occur (in comparison to the adiabatic flame temperature). The direction of the wave is mainly dependent on the heat capacities of the solid and gas phases and on interphase heat exchange. For a standing wave only one operational point is possible for a certain gas mixture. However, the very low speed of the combustion wave allows operation with varying gas velocities by changing the flow direction when the combustion wave reaches the end of the reactor. Figure 3 shows the temperature profiles in a reactor with a subcritical porous matrix. In this case the combustion wave travels in the flow direction. The heat of combustion heats the porous matrix downstream of the combustion wave.

When the flow direction is reversed, the hot porous matrix preheats the fresh gas mixture, and superadiabatic combustion occurs [10, 11]. With this principle gas mixtures with very low heat content can be burned. The flame speeds in such reactors are about 2–4 times higher than the laminar flame speed [12, 13].

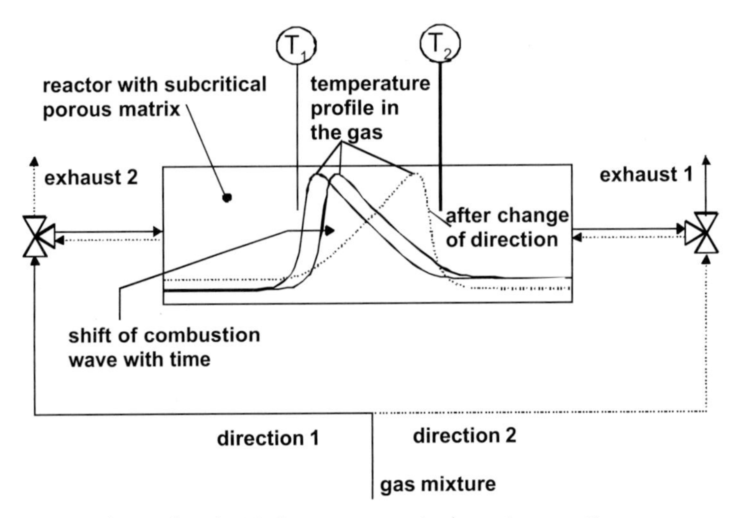

Fig. 3 Scheme of a subcritical porous reactor in alternating operation.

5.5.2.2
Flame Stabilization under Steady Operation by Convection and Cooling

In porous media the same flame propagation mechanisms act as in free flames, but the higher effective thermal conductivity must be considered. The effective thermal conductivity and thermal diffusivity are 2–3 orders of magnitude higher than in a gas. Following the simplified theory of flame propagation [14] the flame speed S is proportional to the square root of the temperature diffusivity:

$$S \sim \sqrt{\frac{a_f}{\tau}} \tag{3}$$

where a_f is the thermal diffusivity, and τ a characteristic timescale of the reaction. From this it results that the flame speed in porous media is 10–30 times higher than the laminar flame speed. One possibility to stabilize combustion in porous media with supercritical pore sizes (flame propagation is possible) is to induce a change in flow speed by a stepwise or continuous change in the cross sectional area (convective stabilization). For methane/air mixtures in porous media the required flow speed to avoid flashback lies in the range of 4–12 m s^{-1}, which is about 10–30 times higher than in free flames. The exact velocity depends on the heat-transport properties of the applied porous structures and the operating conditions. This type of flame stabilization was already used in the early 1900s [15–17]. The major drawback of this flame-stabilization principle is that the power modulation range is rather small, and for low powers with corresponding low flow velocities, flame flashback may occur.

Another possibility for convective stabilization is active control of the air and/or the fuel flow rate based on temperature measurements in the porous matrix. By changing the air and/or the fuel flow rate, the flow velocity and the flame speed changes and the flame front can be stabilized by active control in a desired position within the porous matrix [18]. This stabilization principle can be used for subcritical and for supercritical pore sizes with respect to the flame-propagation criterion according to Eq. (2), but the operational range is small, and for each configuration a single operating condition is found by active control.

Another possibility for flame stabilization is to cool the reaction zone, for example, by embedded water-cooled tubes in the main reaction zone, which was also already realized in the early 1900s [19]. In Fig. 4 (left) the principle of flame stabilization by cooling of the reaction zone is schematically shown. Combustion takes place inside the pores, and the heat of reaction is conducted to the embedded water tubes.

Cooling of the reaction zone can also alternatively be realized by intense radiation from it. This principle is used nowadays in surface radiant burners, which are widely used in domestic and industrial appliances. As shown in Fig. 4 (right) the combustion region must be located close to the surface of the porous matrix in order to extract heat by radiation. However, the combustion process may be completed outside of the porous matrix, depending on the operating conditions. The porous surface extracts heat from the flame and radiates it to the environment. Effective heat transport by radiation only occurs at high temperatures above the ignition limits. This means that for radiation-cooled burners the porous matrix must be subcriti-

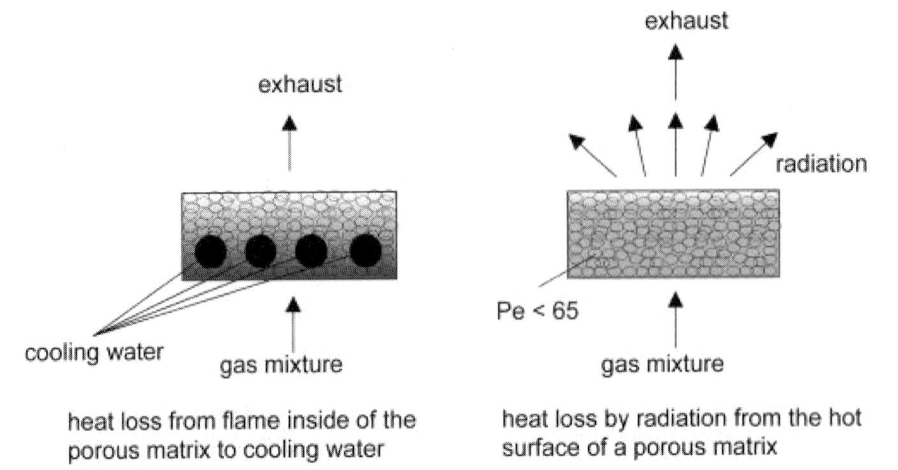

Fig. 4 Flame stabilization in porous media by cooling.

cal with respect the modified Péclet number (Eq. (1)), because otherwise the flame could move into the porous matrix, which eventually would lead to flashback. A significant drawback of this stabilization principle is that by increasing the heat load and correspondingly the flow rate of the incoming gas/air mixture, the flame location moves completely outside of the porous matrix and the burner operates in the so called blue-flame mode with significantly increased emissions. Thus, the beneficial combustion process at least partly inside of the porous matrix is only possible for a small range of power modulation (see Section 5.5.4).

5.5.2.3
Flame Stabilization under Steady Operation by Thermal Quenching

Research and development on stationary combustion completely trapped in the cavities of porous inert media started in the early decades of the 1900s. Bone [20] designed the first boilers and muffle heaters, while Lucke [21] built radiant room heaters, crucible furnaces, and cooking stoves, operating with town gas. However, even earlier pioneers such as Welch [22], Mitchell [23], Ruby [24], and Schnabel [25] developed the first appliances for different fields, most probably independently of each other. The porous media used in the early stages were packings of aluminosilicate pebbles or spheres. Also in the 1930s [26] and up to the present time [18, 27–30], there was limited but continuous research and development activity in porous-medium combustion. Major drawbacks in porous-medium combustion were the deficiencies of the flame stabilization concepts for a stationary combustion process completely trapped inside a porous medium.

A novel combustion technique based on combustion in porous media was developed recently [2, 31–35]. The major novelty of this work is the combustion stabilization principle based on thermal quenching, which allows extremely stable operation of the premixed combustion process in the porous matrix for a wide range of power modulation. The flame stabilization layer is located inside the porous matrix and is well defined by the matrix design. The combustion process is stabilized by a sudden change of the pore size, corresponding to a change in the Péclet number in the combustion reactor. In region A of the porous burner, the porous-body properties are chosen in such a way that flame propagation is not possible by reducing the equivalent porous cavity diameter (subcritical Péclet number). Region A functions as a preheating region and flame trap at the same time. In combustion region C, the pores are large enough that flame propagation is possible (supercritical Péclet number). At the interface between regions A and C the ignition temperature is reached. In Fig. 5 the schematic setup of such a porous burner with the preheating region A and the combustion region C is shown. Depending on the actual application, additional regions may be directly combined with the described basic design of the porous burner, for example, heat exchangers, insulation, premixing chambers, and so on.

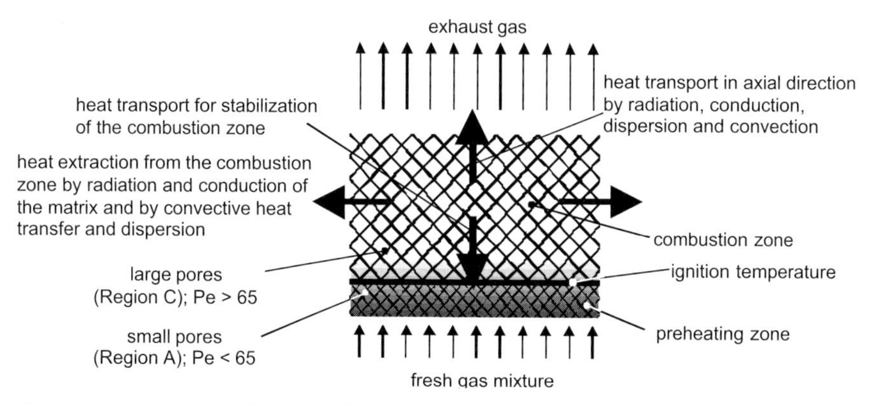

Fig. 5 Heat transport and flame stabilization in a two-layer porous burner [31].

Under stationary conditions heat transfer from the combustion region C to the flame trap in region A may heat the latter to temperatures above the ignition temperature. To prevent flashback the amount of heat which is transported against the flow direction must not be higher than the amount which is carried convectively by the fresh gas mixture into the combustion region, because otherwise a combustion wave can develop starting from the hot interface between region A and C and possibly travel against the flow direction. This can be achieved by means of a region A of low effective thermal conductivity, which allows only a small amount of heat transport against the flow direction. In contrast, region C should have a high effective thermal conductivity, because this allows operation at high flow rates without the danger of blowoff. In this respect, besides the design criteria based on the modified Péclet number and the corresponding equivalent porous cavity diameters, the choice of the heat-transfer properties of regions A and C is crucial for the power modulation range of such porous burners.

The principle of flame stabilization by changing the Péclet number can be used advantageously in many application fields. However, in the Péclet number criterion mainly heat-transport processes are considered and, if diffusive mass transport becomes dominant, incorrect critical pore diameters may be predicted.

5.5.2.4 Diffusive Mass-Transport Effects on Flame Stabilization

The influence of diffusive mass transport on the stabilization of a flame in a porous medium is not well known. In Refs. [3, 36] flame instabilities were observed for gas mixtures containing hydrogen. These instabilities were linked to the increased influence of diffusive mass transport, which decreased the dominance of the heat transport in the flame stabilization. The ratio between diffusive mass transport and heat transport can be described by the Lewis number Le_c of a component c of the gas mixture:

$$Le_c = \frac{\lambda_f}{D_c \rho_f c_{P,f}} = \frac{a_f}{D_c} \tag{4}$$

which is the ratio between the temperature diffusivity a_f and the diffusion coefficient of the component D_c. According to Refs. [3, 37] the flame structure changes for Lewis numbers smaller than unity. For many widely used gases (e.g. methane/air mixtures), Le is close to unity and therefore its influence is often neglected. However, mixtures with Lewis numbers far below unity require the consideration of mass diffusion. For example hydrogen/air and hydrogen/chlorine mixtures have Lewis numbers of about 0.4. For such low values one can expect a strong influence of diffusion on flame stabilization in porous media. This influence is neglected in the previously described Péclet number criterion, because only the laminar flame speed is considered, which does not account for deformations of the flame front. This means that the constant value of the critical Péclet number is not sufficient for the design of a porous burner flame trap at low Lewis numbers.

Recent flame propagation experiments in porous media for mixtures with different Lewis numbers show the dependence of the critical Péclet number on the Lewis number [38]. The flame propagation limits were experimentally determined in packed beds of spheres. For each packing the critical Péclet number was calculated for the lean and rich flame propagation limits with the laminar flame speed at the corresponding equivalence ratio. The average of the critical Péclet numbers of different packings was plotted against the Lewis number (Fig. 6). For Lewis numbers close to and larger than unity the criterion of Ref. [3] of $Pe_{\text{crit}} \approx 65$ could be confirmed, but for mixtures with lower Lewis number the critical Péclet number was found to lie far below this limit.

The results of this experimental study show that combustion gas mixtures with a high diffusivity compared with the thermal diffusivity of the mixture, that is, Lewis numbers less than unity, have smaller critical pore diameters or critical Péclet numbers than mixtures with Lewis number equal to or larger than unity. The Lewis number must be calculated for the reaction component that is lacking for a stoichiometric mixture. The results of Fig. 6 can be used for better design of flame traps for highly diffusive mixtures.

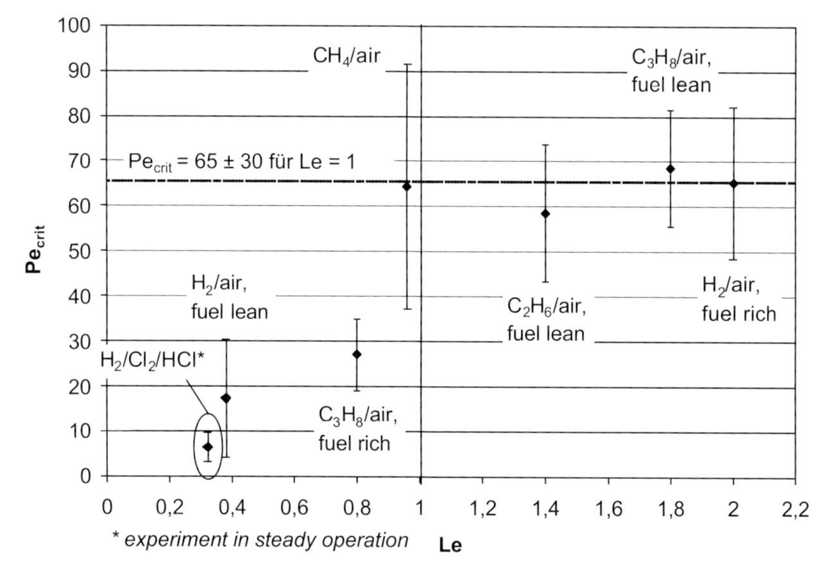

Fig. 6 Critical Péclet number versus Lewis number [38].

5.5.3
Catalytic Radiant Surface Burners

Most catalytic surface burners operate with nonpremixed (i.e., diffusion) combustion. The gas flows through the flat burner support, which is coated with a catalyst (in most cases noble metals such as Pt and Pd distributed on a wash coat layer, e.g., γ-alumina) on the combustion side. Ceramic burner supports may have the form of a porous plate or a flexible fiber mat. The oxygen for the combustion process is provided by the surrounding air and is transported to the catalytic combustion region by diffusion processes. The surface power load is very limited due to these diffusion processes, and typical values are in the range up to 50 kW m^{-2}. The free radiating surface at such low heat loads leads to a very low surface and combustion temperatures in the range up to 600 °C. Combustion at such low temperatures is only possible in the presence of a catalyst. According to Wien's displacement law [39] the radiation maximum moves to shorter wavelength the higher the temperature of the emitting surface is. Thus, catalytic gas IR burners have a relatively long wavelength of the emitted radiation. Due to the relatively low temperatures and the flashback-safe diffusion mode of operation, the material requirements for the burner support are relatively low, and the main focus of such developments lies on the catalyst technology.

Premixed catalytic radiant surface burners have not reached wide commercial usage yet and are mainly the subject of research activities. In most cases premixed catalytic supported combustion [40] is performed (in contrast to full catalytic combustion) at surface temperatures of about 1100 °C.

5.5.4
Radiant Surface Burners

Radiant surface burners operate by stabilizing a premixed flame near or partially inside a noncombustible porous burner support. The enthalpy of combustion released in the gas phase heats the porous matrix, which then emits thermal radiation to a heat load. In small-scale applications radiant surface burners have already shown performance gains over conventional open-flame burners in the form of higher efficiencies, lower NO_x emissions, and more uniform heating.

In most cases flat plate geometries with many small holes or with a volumetric foamlike porous structure are used as burner support. A flat flame sheet is produced on the surface of the ceramic plate, and depending on its position relative to the ceramic plate, it heats the ceramic plate, which radiates heat to the appliance.

The ceramic supports may be sinter- or fiber-based, while the materials range from mullites and alumina up to silicon carbide fibers. Important properties of the structures are their effective thermal conductivity, optical thickness, and emissivity, which affect the flashback behavior and the radiation properties. Especially the effective conductivity (which describes the effective heat transport inside the porous structure due to conduction, dispersion, and radiation, see Chapter 4.3) is of special importance with regard to flashback safety. In general, flashback safety is increased if the heat load is increased and the ceramic burner support is cooled more effectively by convection of the fresh air/gas mixture. Ceramic supports with high effective conductivity transport a high heat flux through the burner support to the back side where the upstream air/gas mixture is located. As a result the temperature of the ceramic support increases on the back side and may ignite the incoming air/gas mixture (increased tendency for flashback). At the same time the temperature of the radiating surface decreases due to increased heat losses to the upstream direction. Thus, ceramic burner supports with very low effective conductivity are advantageous for this type of application and result in increased safety, a higher turn-down ratio, and higher radiative operating temperatures. The lowest effective conductivity is achieved for fiber-based structures (ca. $0.1\ W\ m^{-1}\ K^{-1}$), while the highest values are reached with sintered plates having low microporosity (ca. $2\ W\ m^{-1}\ K^{-1}$).

Most radiant surface burners operate in the radiant mode at heat loads between 100 and 600 $kW\ m^{-2}$ (for natural gas under atmospheric conditions and without air preheating). At higher heat loads the flame sheet lifts to a distance of several millimeters from the burner support and thus the burner support cools down. This mode of operation is also called the "blue-flame mode", in which no significant radiation is emitted. In the heat load range between the two modes a transition from the radiant mode to the blue-flame mode takes place. The maximum surface temperatures of the burner support are reached for heat loads of approximately 300–400 $kW\ m^{-2}$ (depending on the actual burner-support properties) and they are in the range of up to 1100 °C.

Radiant surface burners can be also operated at significantly higher heat loads up to 3000 $kW\ m^{-2}$ in the blue-flame mode. In this mode the lifted flame reaches significantly higher temperatures, while the temperature of the ceramic burner plate decreases. Emissions are correspondingly higher, since no heat extraction through radiation of the "cold" burner surface takes place. One interesting concept to improve the performance of such burners in the blue-flame mode is to use ceramic structures with a bimodal pore size distribution. This delivers a corresponding bimodal velocity distribution through the ceramic structure. Thus, portions of the flame in the positions of lower velocity remain at the burner surface, while those in positions with higher velocities are lifted from the surface. Thus, even at higher heat loads a fraction of the heat is still extracted by radiation.

5.5.5
Volumetric Porous Burners with Flame Stabilization by Thermal Quenching

Volumetric combustion in porous media offers exceptional advantages compared to conventional combustion technologies with free flames. Unlike conventional premixed combustion processes, combustion in porous media does not operate with free flames. Rather, a flameless combustion takes place in the three-dimensionally arranged cavities of a porous medium. The superior heat-transfer properties of the porous matrix are superposed on the heat transfer in the gas phase and coupled through the intensive heat exchange between the gas and the porous matrix. Radiation, conduction, and heat transfer due to dispersion imposed by the porous matrix are the dominant heat-transfer mechanisms. The relatively new technology of porous-medium burners is characterized by higher burning rates, increased flame stability with low noise emissions, and controllable, homogeneous combustion-zone temperatures which lead to a reduction in NO_x and CO emissions. Porous-medium burners also show a high power density that translates into compact designs and, depending on the flame stabilization principle used (Section 5.5.2.3), an extremely high power turndown ratio of up to 1:30. Additionally, complex combustion chamber geometries, adapted to the specific needs of the individual applications, which are not feasible with conventional state-of-the-art combustion techniques, are possible.

Figure 7 shows a photograph of a stabilized combustion process in a porous medium with the temperature distribution. The good heat-transfer properties keep the maximum temperature low, which also keeps NO_x emissions low. A very homogeneous temperature field is also achieved, and this keeps the carbon monoxide concentration in the waste gas low, too. A considerable part of the heat from the combustion region is also transported upstream, and hence higher flame speeds are up to 30 times higher than with laminar free flames. This translates into a greater power density and better space utilization.

Compared to more conventional combustion processes with free premixed flames, properly stabilized combustion processes in porous media lead to the following advantages, which result mostly from the very intense heat transport inside the porous structure and the stabilization principle:

- Wide, infinitely variable dynamic power range up to 1:30.
- High power density with heat loads per cross section up to 8 MW m^{-2} under atmospheric pressure for methane/air mixtures.

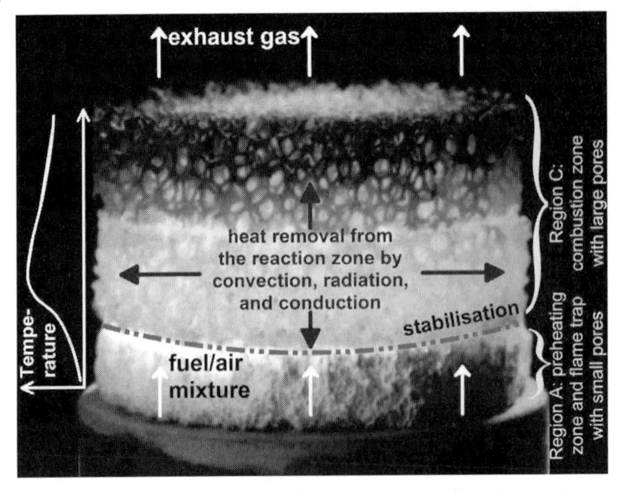

Fig. 7 Porous burner with flame stabilization by thermal quenching.

- Low CO and NO_x emissions over the complete dynamic power range.
- Stable combustion in an extended equivalence ratio range of $\Phi = 0.5$–2.5 for CH_4/air mixtures (excess-air ratios: $\lambda = 0.4$–2.0).
- High combustion stability due to the heat capacity of the porous material.
- Geometrical flexibility of the burner cross section, allowing good adaptation of heat-transfer to application demands.

Due to these outstanding properties, porous-burner technology is attractive for different fields of application.

5.5.5.1
Materials and Shapes for Porous-Medium Burners

A special feature of porous-burner technology is its dependency on special highly temperature resistant porous ceramic components. Therefore, a short overview of materials and shapes which are suitable for combustion in porous media is given. More details are reported in Ref. [2].

The most important materials and forms for porous burners are sintered silicon carbide (S-SiC) foams as well as silicon-infiltrated composite SiC-based foams, and mixerlike structures made of Al_2O_3 fibers, ZrO_2 foams, or C/SiC structures. Table 2 lists the most important data for these materials. The maximum use temperatures are rather theoretical values that are not always reached in practice. For some applications also iron–chromium–aluminum alloys and nickel-base alloys can be used. All of the mentioned materials are substantially different with regard to manufacturing and properties. Al_2O_3 and ZrO_2 materials can be used at temperatures above 1650 °C. Metals and SiC materials do not meet this qualification; however, they show outstanding characteristics with regard to thermal shock resistance, mechanical strength, and conductive heat transport. Wire meshes and mixerlike structures have low heat capacity and correspondingly good startup behavior and small pressure drop.

Table 2 Important properties of Al_2O_3, SiC, and ZrO_2.

Property	Unit	Al_2O_3	SiC	ZrO_2
Maximum use temperature in air	°C	1900	1650	1800
Thermal expansion coefficient α (20–1000 °C)	10^{-6} K^{-1}	8	4–5	10–13
Thermal conductivity λ at 20 °C	W m^{-1} K^{-1}	20–30	80–150	2–5
Thermal conductivity λ at 1000 °C	W m^{-1} K^{-1}	5–6	20–50	2–4
Specific thermal capacity	J g^{-1} K^{-1}	0.9–1	0.7–0.8	0.5–0.6
Thermal stress resistance parameter, hard shock, R ($\sigma/E\alpha$)	K	100	230	230
Thermal stress resistance parameter, mild thermal shock, R' ($R\lambda$)	10^{-3} W m^{-1}	3	23	1
Total emissivity at 2000 K	–	0.28	0.9	0.31

The overall performance of the porous body is always a combination of the base material itself and the actual porous structure. Therefore, for both the suitable materials and for the shapes of porous structures the basic properties are given below. However, the emissivity data are not very reliable and strongly depend on the actual surface structure.

Alumina structures (Fig. 8a) can be in principle used up to process temperatures in the range of 1900 °C, although the technical temperature limit of these structures is nowadays at about 1700 °C. Alumina-based materials show an intermediate thermal conductivity ranging from 5 W m^{-1} K^{-1} at 1000 °C to about 30 W m^{-1} K^{-1} at 20 °C. Alumina shows intermediate thermal expansion, intermediate resistance to thermal shock, and an overall emissivity at 2000 K of about 0.28.

High-quality silicon carbide materials (Fig. 8b) are characterized by a maximum usage temperature of about 1650 °C, high thermal conductivity in the range of 20 W m^{-1} K^{-1} at 1000 °C and 150 W m^{-1} K^{-1} at 20 °C, very low thermal expansion, and very good resistance to thermal shock. The overall emissivity at 2000 K is about 0.8–0.9.

Temperature-resistant metal alloys (Fig. 8d) can be used for temperatures lower than 1250 °C. Their properties include high thermal conductivity ranging from 10 W m^{-1} K^{-1} at 20 °C to about 28 W m^{-1} K^{-1} at 1000 °C, high thermal expansion, and an extremely good resistance to thermal shock. The emissivity of metals varies strongly with the surface finish and the surface itself (0.1–0.7).

Of all the presented materials, zirconia (Fig. 8c) shows the highest temperature resistance, up to 2300 °C. The thermal conductivity of solid zirconia is not significantly temperature dependent and in the range of 2–5 W m^{-1} K^{-1}. The emissivity at 2000 K is about 0.31. Porous zirconia-based mixer structures, as shown in Fig. 8c, combine the extremely high maximum temperature resistance of solid zirconia with a short start-up phase and an extremely good resistance to thermal shock, due to the high inner porosity of the ZrO_2 lamellas.

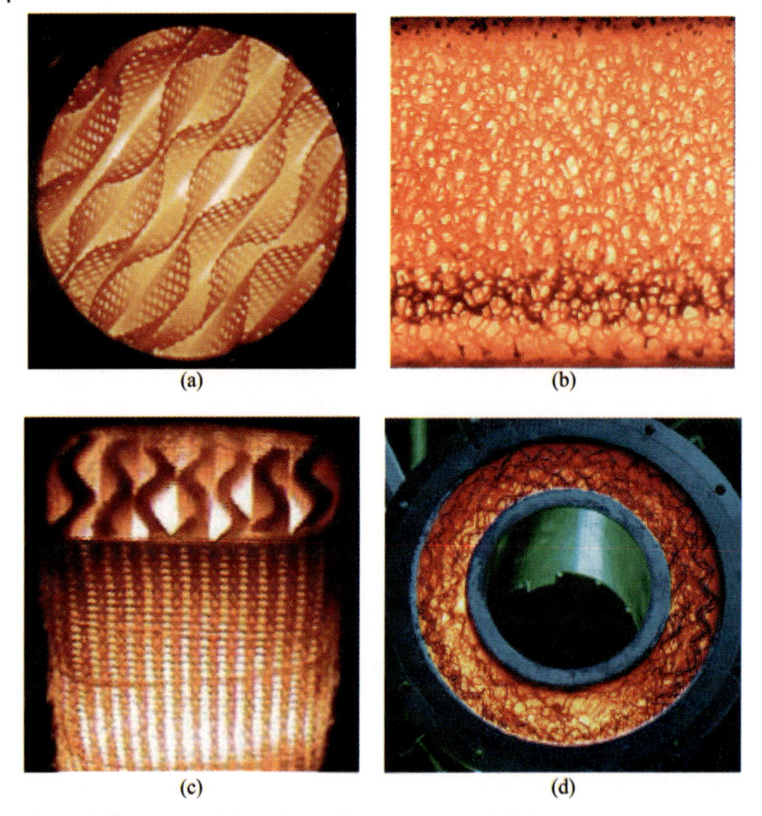

Fig. 8 Different materials in the combustion region. a) Al_2O_3 fiber-based lamellar structure; b) SiC foam; c) ZrO_2 foam-based lamellar structure; (d) Fe–Cr–Al alloy wire mesh.

Static mixer structures (Fig. 8a and 8c) are characterized by a low conductive heat transport, a short start-up phase, excellent radiative heat transport, excellent dispersion properties, and very low pressure drop.

Wire meshes (Fig. 8d) show poor conductive heat-transport and dispersion properties, due to their high porosity. On the other hand, they have a short start-up phase, excellent radiative heat-transport properties, and very low pressure drop.

Ceramic foams (Fig. 8b) are widely used for porous burners. Foam structures feature good conductive heat transport, a rather long start-up phase due to lower macroporosity, intermediate radiative heat-transport properties, intermediate dispersion properties, and relatively high pressure drop.

5.5.5.2
Applications of Volumetric Porous Burners

Due to its outstanding properties, such as the wide operating range with regard to thermal power, small dimensions, and low pollutant emissions, porous-burner technology can be advantageously applied to many different industrial branches. In the following paragraphs a few selected applications of porous-burner technology are presented.

Gas Infrared Heater for T > 1400 °C

For more than 50 years gas infrared burners have been available on the market. They are used for heating purposes, for warming, preheating, and drying purposes in nearly all branches of industry where convective heat transfer and material warming by contact heat is not possible. The function of conventional gas infrared heaters is based on the combustion of a premixed gas/air mixture with free flames on the surface of a ceramic or metal plate (see Section 5.5.4). This plate is mainly heated by conductive and radiative heat-transport mechanisms resulting in temperatures of about 800–1100 °C regardless of whether sintered ceramic perforated plates, metal fibers, or ceramic fibers are used as burner support and radiant surface. The energy supply and thus the gas and air supply must be increased to raise the temperature at the radiant surface, and this results in an increase in flow velocity. The flame, burning at or in the surface of the radiant area, loses contact to the surface in case of higher flow velocity. This means the gas infrared burner becomes a gas burner with open flame (blue-flame mode).

Considering the above-mentioned restrictions of conventional gas infrared heaters, volumetric porous burners offer exceptional advantages for industrial applications with a need for a high-temperature radiative heat source. The infrared porous-medium burner RADIMAX, which was developed on the basis of the previously described principles and with the assistance of numerical tools [41], can achieve temperatures up to 1550 °C, which result in higher radiation power density and efficiency. Thus, processes such as drying, coating, preheating, and so on can be realized in a more efficient way with significantly higher process speeds or reduced space requirements.

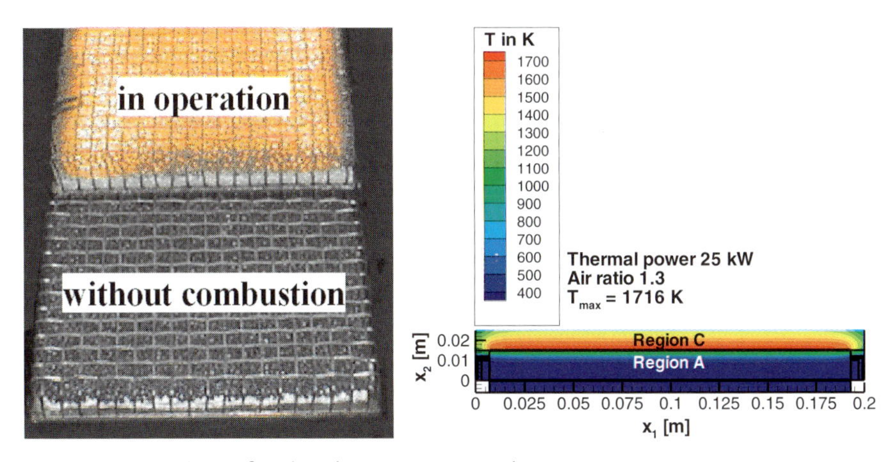

Fig. 9 Porous-medium infrared gas burner prototype and simulated temperature field [41].

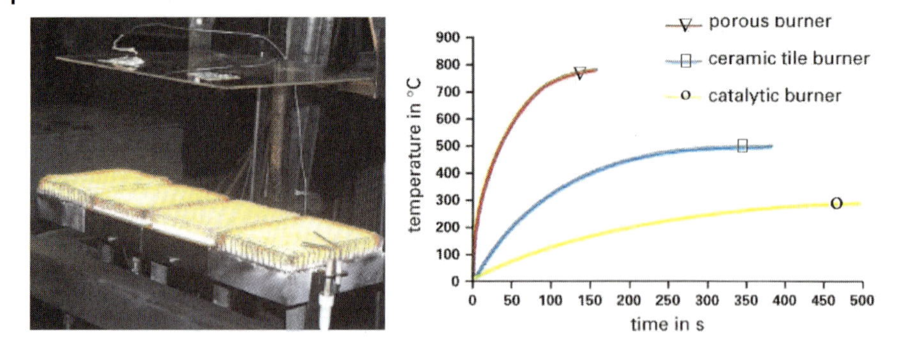

Fig. 10 Heating time of a steel plate with different infrared burners.

Figure 9 shows the radiant heater and the numerically simulated temperature distribution in the porous burner. It consists of a 20 mm thick alumina fiber plate with 1 mm pore diameter in the preheating region and a 15 mm, 10 ppi SiC foam in the combustion region. The numerically and experimentally determined maximum temperatures are more than 1400 °C, reaching up to 1550 °C for this burner geometry, in comparison to 1100 °C for conventional infrared radiant heaters. The thermal power density can be up to 1500 kW m^{-2}, in contrast to the radiant-mode limit of about 400 kW m^{-2} of conventional systems.

In Fig. 10 an experimental comparison of the time for *heating steel plate* with different radiant burners is shown. The experimental arrangement with four porous burners can be seen on the left, while the diagram on the right shows the superior performance of the porous burner in this application in comparison to conventional radiant heaters having the same radiative surface area. The conventional radiant heaters were operated at their respective maximum radiant output, while the porous burner was operated at 70 % of its maximal power.

Application in Domestic Gas Boilers

An important aspect for the development of new gas burners for household applications is the fact that the power requirement for the heating operation decreases more and more, due to the improved insulation of the buildings. This can be related to the trend of lowering energy costs by energy-saving buildings. However, for hot-water production the energy requirement remains unchanged, since high comfort is strongly requested in this field. Taking into account this discrepancy, burners for heating systems are required to have a wide power-modulation range. This also relates directly to the number of burner starts. As the highest emissions occur during the warm-up phase, a higher power modulation automatically leads to a decrease in waste-gas emissions [35]. Also in stationary operation, emissions of porous media burners are minimized in comparison to many domestic gas appliances working with free flames because it is possible to control the combustion temperature with the porous material in the combustion region. Table 3 shows for a 30 kW domestic gas boiler operating with a porous burner that the NO_x and CO emissions are clearly below the most stringent European emission standards.

Table 3 Emissions of a 30 kW porous-burner gas boiler and European standards

	German standard DIN 4702	Swiss standard	German standard "Blauer Engel"	Hamburg promoting program	Porous-burner excess-air ratio		
					1.25	1.3	1.4
NO_x (mg/kW·h)	200	80	60	20	33.6	30.3	22.4
CO (mg/kW·h)	100	60	50	15	5.2	10.3	10.3

Another aspect that makes the use of volumetric porous-medium combustion especially interesting for household applications is the compactness of porous-medium burner units. Power densities of 3500 kW m^{-2} (cf. 300–600 kW m^{-2} for conventional systems) can be realized with various gases and air ratios [42]. Because of the small burner sizes separate heating rooms are no longer necessary. Instead the porous burner units can be installed in small wall niches or even outside the house.

Figure 11 shows a sketch and a photograph of a complete 30 kW condensing gas boiler for domestic use based on porous-burner technology. It is about half as large as a conventional heating system for the same nominal power output and shows a power turndown ratio of 1:10 (3–30 kW).

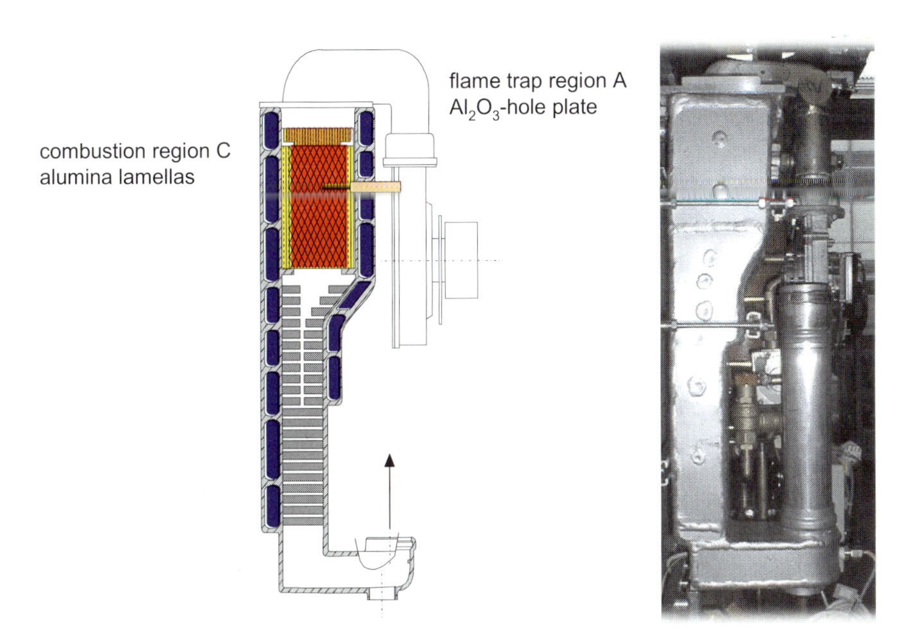

Fig. 11 Porous burner and integrated heat exchanger unit for household applications.

Oil Burner for Domestic Boilers Based on Volumetric Porous Burner

Oil burners have shown to date very poor power modulation especially at low power outputs, as usually needed in domestic appliances. Liquid biofuels like FAME (fatty acid methyl ester, also known as biodiesel) are also not compatible with conventional oil-burner technologies. Furthermore, the integration of oil burners in wall-hung systems requires further reductions in burner size. In the EC-funded BIOFLAM project [43] new liquid-fuel-fired condensing boilers were developed showing such major features as a power modulation of at least 10:1, ultralow CO and NO_x emissions over the entire power-modulation range, significantly greater compactness than conventional liquid-fuel-fired boilers, and compatibility with renewable liquid fuels like FAME. To reach these goals the cool-flame vaporization process [44] and the porous-medium burner were combined.

Figure 12 shows a schematic of the BIOFLAM unit. The liquid fuel is entering the vaporizer through a spray nozzle. A fan is used for pressurizing the combustion air, which is split between primary and secondary air. The primary air enters the vaporizer through an annular gap around the injection nozzle. The secondary air enters the burner section, is preheated and subsequently enters the mixing chamber and is mixed with the cool-flame products. The complete mixture enters the porous

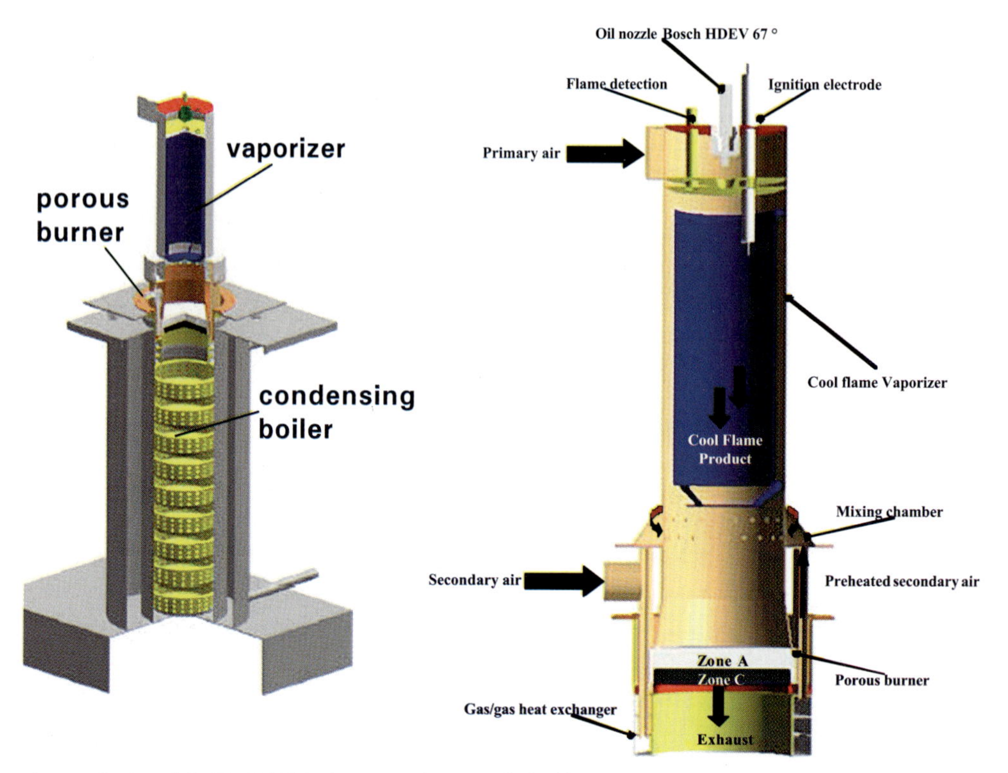

Fig. 12 Design of the BIOFLAM boiler unit with porous fuel-oil burner.

burner at a temperature of about 270 °C. The exhaust gases enter the condensing boiler after the porous burner.

Figure 13 shows the emission characteristics of the first prototypes over the entire power-modulation range. Emissions for nonstaged operation of the cool-flame vaporizer with preheated air at 350 °C are shown (1st version) for comparison purposes. Clearly, the staged vaporizer concept (2nd version) requires much less air preheating (only the secondary air is preheated at a much lower temperature of 200 °C) and results in significantly less nitrogen oxide emissions, due to the resulting reduced combustion temperatures, although a lower excess-air ratio was set. Details on this ongoing development can be found in Ref. [43].

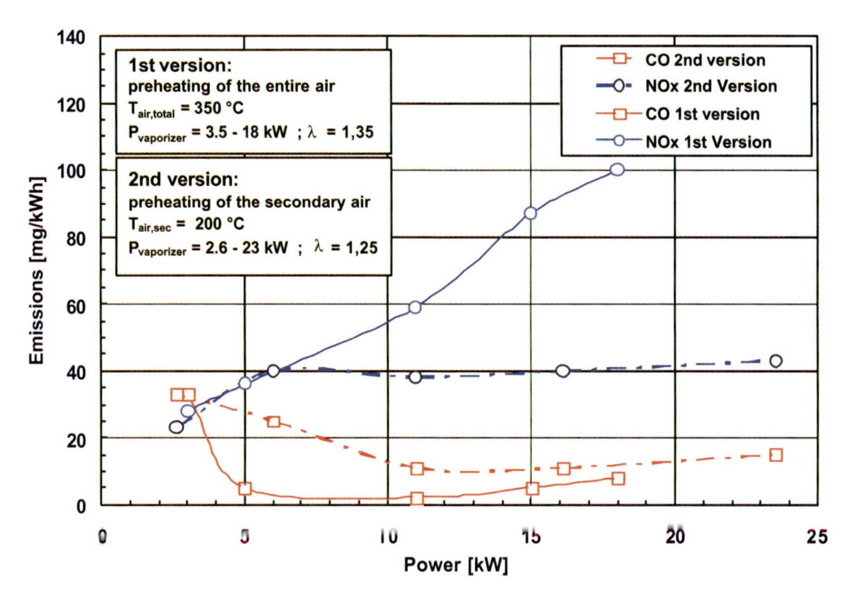

Fig. 13 Emission characteristic of the BIOFLAM burner.

The excellent emissions characteristic of the BIOFLAM boiler unit operating with a staged cool-flame vaporizer and a porous burner is comparable to the emission levels of low-emission gas burners, which represents a breakthrough for oil burners.

Porous Burner as Chemical Reactor for HCl Synthesis

The advantages of porous burners can also be used for different gas-phase chemical reactions, for example, the *synthesis* of *HCl* from H_2 and Cl_2. Conventional reactors for HCl synthesis from H_2 and Cl_2 operate with free-flame diffusion burners at high hydrogen excess ratios due to the flashback safety aspects of premixed combustion with such mixtures. The diffusion burner mode leads to extremely long reactors (up to 6 m) operating with large excess of hydrogen (up to 50 %) to achieve a complete conversion without any unconverted chlorine and thus meet the extreme purity demands of the HCl product.

For this application of volumetric porous burners, three main parameters differ greatly compared with CH_4/air combustion: The stoichiometric adiabatic temperature of an H_2/Cl_2 flame of 2400 °C is significantly higher than that of CH_4/air combustion. The maximum temperature and the chemical resistance of the solid material are limiting parameters of the porous burner. Additionally, the laminar flame speed of the stoichiometric H_2/Cl_2 system of 2.2 m s^{-1} is five times higher than that of the CH_4/air system [45]. Finally the Lewis number of H_2/Cl_2 with hydrogen excess, in the range of 0.3–0.4, is significantly lower than unity and results in a significantly lower critical Péclet number for flame stabilization. In Table 4 the major differences between methane/air and chlorine/hydrogen combustion are summarized.

Table 4 Differences between methane/air, methane/oxygen, and chlorine/hydrogen combustion [38].

	CH_4/air	CH_4/O_2	Recirculation rate HCl/Cl_2			
			0	1	2	3
S_L/cm s^{-1}	43	450	220	71	34	14
$T_{adiabatic}$/ °C	2000	2800	2200	1800	1450	1200
$T_{ignition}$/ °C	615	556	ca. 200			
Le	0.96	0.83	0.38	0.36	0.34	0.32
Pe_{crit}	65	30	< 10			
$d_{p,eff}$/mm at Pe_{crit}	3.6	0.19	0.2	0.4	0.8	1.8
$D_{spheres}$/mm at Pe_{crit}	9.04	0.82	0.6	1.2	2.3	5.1

By reducing the combustion temperature and the flame speed, the critical pore diameter for flame stabilization based on thermal quenching can be increased to realistic values for a large-scale industrial reactor despite the very low critical Péclet number of approximately 7 resulting from the low Lewis number of the H_2/Cl_2 mixture. HCl recirculation can be applied to reach these goals [46]. As indicated in Table 4, a recirculation ratio of 3 between HCl and Cl_2 must be applied to be able to realize a flame trap as a packed bed of ceramic spheres with a diameter of 5 mm showing an effective pore size of about 1.8 mm. Alternatively, other inert components such as H_2O can be added to the mixture of reactants to reach the same target.

In Fig. 14 a laboratory porous burner (top) built of graphite is shown. Alumina sphere packings were used as porous structure, due to their robustness and chemical stability against the aggressive radicals in this application. Stable operation was achieved as expected at an HCl recirculation rate of 3, leading to an extremely pure HCl product with only 5 % excess hydrogen and a reactor height of 20 cm. The lower part of Fig. 14 shows a pilot reactor for the production of 30 t d^{-1} of HCl having a cross section of 1 × 1 m and an overall height of 0.5 m. A conventional reactor for the same load and product quality requirements would have 2 m diameter and a height of 6 m and would operating at 30–50 % excess hydrogen. Details on this ongoing development can be found in Ref. [38].

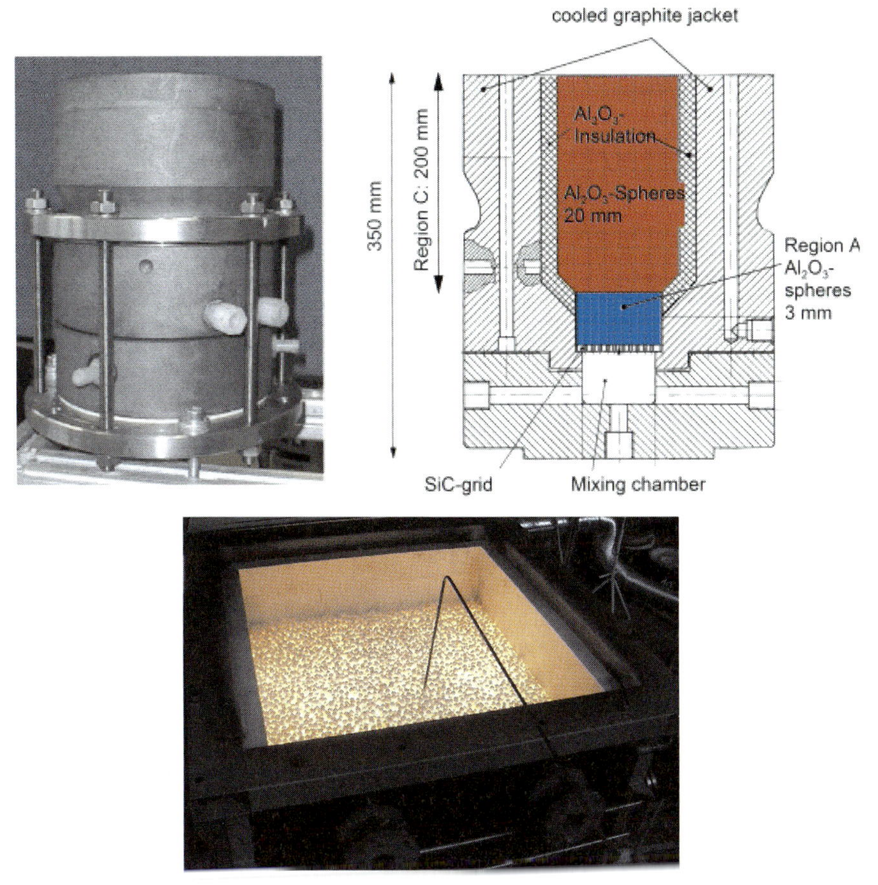

Fig. 14 Laboratory porous burner for HCl sythesis built from graphite housing with alumina packings and pilot chemical reactor for 30 t d^{-1} production of HCl.

Oxy-Fuel Radiant Porous Burner

Oxy-fuel burners are increasingly applied in high-temperature applications because of the primary high efficiency without recuperation (low investment costs) and the reduced NO_x emissions due to the very low nitrogen concentration in the furnace environment. However, due to the relatively low flow rates in comparison to conventional air–fuel burners and the higher combustion temperature, a significant part of the heat has to be transferred by radiation from the gas phase, which makes large gas volumes necessary. However, the extremely high combustion temperature of oxy-fuel burners does not allow the construction of oxy-fuel radiators in a conventional way.

The above considerations led to the design of an annular-gap-shaped combustion chamber, which is shown in Fig. 15. Only a very lean methane–air mixture enters

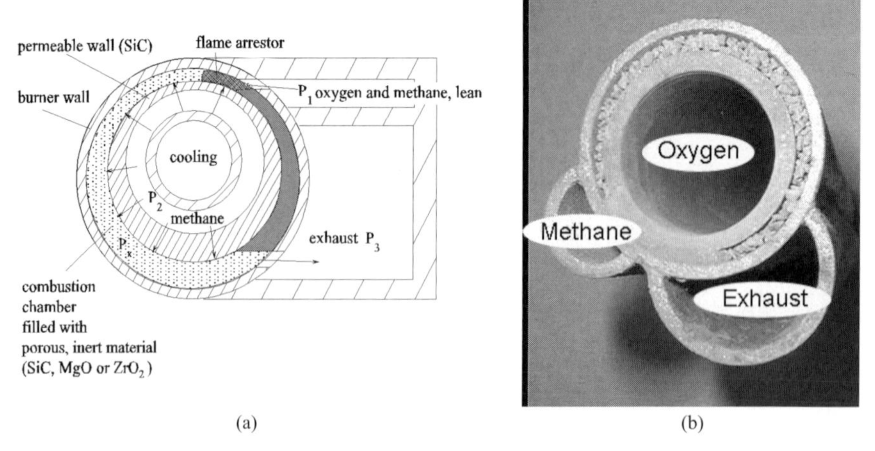

Fig. 15 a) Schematic diagram of the oxy-fuel radiant burner design and b) partly mounted oxy-fuel radiant burner with SiC porous body.

the actual combustion chamber in the tangential direction. The rest of the methane is pressed through the permeable inner tube, thus leading to a continuously staged combustion process. The annular gap is primarily fed with a lean methane–oxygen mixture (excess oxygen ratio of about 5, Fig. 15a). Alternatively oxygen staging can be applied, as shown in Fig. 15b.

Details of this development can be found in Ref. [47]. Development work on the improvement of the applied ceramic components is still going on for this application.

5.5.6 Summary

In the present chapter an overview of the different porous burners using cellular ceramics as burner supports was given, and the major properties affecting flame stabilization and the overall combustion process were discussed. It was indicated that the flame-stabilization concept may cause significant drawbacks in the operational characteristics. Conventional radiant surface burners, for example, are limited to a maximum surface temperature of about 1100 °C due to the applied flame stabilization concept.

The utilization of volumetric porous-medium combustion with flame stabilization based on thermal quenching is a good solution that overcomes most of the drawbacks of previous flame-stabilization concepts for combustion in porous media. However, volumetric porous burners require cellular ceramics which are stable in the long term at high temperatures in aggressive atmospheres. Such cellular ceramics are already reality but not yet widely available.

Selected application fields of porous-medium combustion technology were outlined. Taking advantage of its outstanding benefits like compact design, high power turndown ratio, and load-independent and minimal waste-gas emissions, porous-burner technology is predestined to find use in many different industrial branches.

In addition to the applications described in Section 5.5.5.2, porous burners have been successfully developed and tested in the following application fields:

- As a heat source for novel steam engines (see Ref. [48]).
- As a thermal partial oxidation reactor for hydrogen or synthesis-gas production (see Ref. [49]).
- As a reactor for the controlled destruction of fluorochlorohydrocarbons (see Ref. [38]).
- As preheating and off-gas burner for fuel-cell systems.
- As radiant burners in high temperature glass-melting furnaces.

The described basic principles and design criteria for flame stabilization allow the transfer of porous-burner technology to different applications with a wide range of significantly different requirements.

References

1 GoGas Goch GmbH & Co., Dortmund, Company Flyer: "Drying Techniques".

2 Pickenäcker, O., Pickenäcker, K., Wawrzinek, K., Trimis, D., Pritzkow, W.E.C., Müller, C., Goedtke, P., Papenburg, U., Adler, J., Standke, G., Heymer, H., Tauscher, W., Jansen, F., *Interceram*, **1999**, *48*, 326–330, 424–434

3 Babkin, V.S., Korzhavin, A.A., Bunev, V.A., *Combust. Flame*, **1991**, *87*, 182–190.

4 Pinaev, A.V., Lyamin, G.A., *Combust. Explos. Shock Waves USSR*, **1989**, *25*, 448–458, translated from *Fizika Goreniya i Vrzyva, 25*, 75–85.

5 Lyamin, G.A., Pinaev, A.V., *Combust. Explos. Shockwaves USSR*, **1986**, *22*, 553–558, translated from *Fizika Goreniya I Vrzyva, 22*, 64–70.

6 Lyamin, G.A., Pinaev, A.V., *Combust. Explos. Shockwaves USSR*, **1987**, *23*, 399–402, translated from *Fizika Goreniya I Vrzyva, 23*, 27–30.

7 De Soete, G.: *Stability and Propagation of Combustion Waves in Inert Porous Media*, Eleventh Symposium (International) on Combustion, The Combustion Institute, Pittsburgh, pp. 959–966, 1966–1967.

8 Korzhavin, A.A.; Bunev, V.A., Abdullin, R.Kh., Babkin, V.S., *Combust. Explos. Shock Waves*, **1982**, *18*, 628–631.

9 Bingue, J.P., Saveliev, V.A., Fridman, A. A., Kennedy, L.A., *Int. J. Hydrogen Energy*, **2002**, *27*, 643–649.

10 Hanamura, K., Echigo, R., Zhdanok, S.A., *Int. J. Heat Mass Transfer*, **1993**, *36*, 3201–3209.

11 Gavrilyuk, V.V., Dmitrienko, Y.M., Zhdanok, S.A., Minkina, V.G., Shabunya, S.I., Yadrevskaya, N.L., Yakimovich, A.D., *Theor. Found. Chem. Eng.*, **2001**, *35*, 589–596.

12 Tseng, C., *Int. J. Hydrogen Energy*, **2002**, *27*, 699–707.

13 Diamantis, D.J., Mastorakos, E., Goussis, D.A., *Combust. Theor. Modeling*, **2002**, *6*, 383–411.

14 Zeldovich,Y.B., Frank-Kamenetskii, D.A., *Zh. Fiz. Khim.*, **1938**, *12*, 100–120.

15 Lucke C.E., US Patent 755,376, 1901.

16 Lucke C.E., US Patent 755,377, 1901.

17 Bone W.A., Wilson J.W., McCourt, C.D., US Patent 1,015,261, 1910.

18 Martin J.M., Stilger J.D., Holst, M.R., US Patent 5,165,884, 1991.

19 Wedge, U., US Patent 1,225,381, 1915.

20 Bone, W.A., *J. Franklin Inst.*, **1912**, *173*, 101–131.
21 Lucke, C.E., *J. Ind. Eng. Chem.*, **1912**, *5*, 801–824.
22 Welch, W., British Patent 5293, 1890.
23 Mitchell, A., British Patent 7078, 1898.
24 Ruby, C.F., US Patent 737,279, 1902.
25 Schnabel, R., German Patent 218,998, 1908.
26 Hays, J.W., US Patent 2,095,065, 1933.
27 Korzhavin, A.A., Bunev, V.A., Abdullin, R.Kh., Babkin, V.S., *Combust. Explos. Shock Waves*, **1982**, *18*, 628–631.
28 Hsu P.F., Howell J.R., Matthews R.D., *ASME J. Heat Transfer*, **1993**, *115, 744–750.*
29 Ellzey, J.L., Goel, R., *Combust. Sci. Technol.*, **1995**, *107, 81–91.*
30 Fu, X., Viskanta, R., Gore, J.P., *J. Thermophys. Heat Transfer*, **1988**, *12, 164–171.*
31 Trimis, D., Durst, F., *Combust. Sci. Technol.*, **1996**, *121, 153–168.*
32 Durst, F., Kesting, A., Mößbauer, S., Pickenäcker, K., Pickenäcker, O., Trimis, D., *Gaswärme Int.*, **1997**, *46, 300–307.*
33 Brenner, G., Pickenäcker, K., Pickenäcker, O., Trimis, D., Wawrzinek, K., Weber, T., *Combust. Flame*, **2000**, *123, 201–213.*
34 Pickenäcker, O., Trimis, D., *J. Porous Media*, **2001**, *4, 197–213.*
35 Mößbauer, S., Pickenäcker, O., Pickenäcker, K., Trimis, D., *Clean Air*, **2002**, *3, 185–198.*
36 Kennedy, L.A., Fridman, A.A., Saveliev, A.V., *Fluid Mech. Res.*, **1995**, *22, 1–26.*
37 Barenblatt, G.I., Zeldovich, Ya.B., Istratov, A.G., *Prikl. Mekh. Tekhn. Fiz.*, **1962**, *4*, 21–26.
38 Wawrzinek, K.: *Untersuchungen zur Prozessführung und Stabilisierung exothermer Hochtemperaturprozesse in porösen Medien für Anwendungen in der chemischen Reaktionstechnik, PhD-Thesis, Friedrich-Alexander-University of Erlangen-Nurenberg, Technical Faculty, 2003.*
39 Modest, M.F., *Radiative Heat Transfer*, Academic Press, San Diego 2003, pp. 6–11.
40 Bröckerhoff, P., Emonts, B., *Gaswärme Int.*, **1997**, *46*, 243–251.
41 Kesting A., Pickenäcker K., Trimis D., Cerri I., Krieger R., Schneider H.: *Development of a Highly Efficient Gas Infrared Heater by Means of Combustion in Inert Porous Media, International Gas Research Conference (IGRC), Amsterdam, Netherlands, 2001.*
42 Durst, F., Pickenäcker, K., Trimis, D., *gwf Gas Erdgas*, **1997**, *138, 116–123.*
43 Brehmer, T., Heger, F., Lucka, K., von Schloss, J., Abu-Sharekh, Y., Trimis,D., Heeb, A., Köb, G., Hayashi, T., Pereira, J.C.F., Founti, M., Kolaitis, D., Molinari, M., Ortona, A., Michel, J.-B., Theurillat, P.: *BIOFLAM Project: Application of Liquid Biofuels in New Heating Technologies for Domestic Appliances Based on Cool Flame Vaporization and Porous Medium Combustion, 7th Int. Conference on Energy for a Clean Enviroment, Lisbon, 2003.*
44 Lucka, K., Köhne, H.: *Usage of Cold Flames for the Evaporation of Liquid Fuels, 5th Int. Conference on Technologies and Combustion for a Clean Enviroment, Lisbon, 1999, pp. 207–213.*
45 *Gmelins Handbuch der anorganischen Chemie, Chlor, Ergänzungsband, Teil B, System No. 6, Verlag Chemie, Weinheim/Bergstraße, 1968.*
46 Wawrzinek, K., Kesting, A., Künzel, J., Pickenäcker, K., Pickenäcker, O., Trimis, D., Franz, M., Härtel, G., *Catal. Today*, **2001**, *69, 393–397.*
47 Kesting, A., Pickenäcker, O. Trimis, D., Durst, F.: *Development of a Radiation Burner for Methane and Pure Oxygen Using the Porous Burner Technology*, 5th Int. Conference on Technologies and Combustion for a Clean Environment, Lisbon, 1999.
48 Buschmann, G., Haas, T., Hoetger, M., Mayr, B.: *IAV's Steam Engine – A Unique Approach to Fulfill Emission Levels from SULEV to ZEV, 2001-01-0366, SAE 2001 World Congress, Detroit, 2001.*
49 Al-Hamamre, Z., Trimis D., Wawrzinek, K.: *Thermal Partial Oxidation of Methane in Porous Burners for Hydrogen Production, 7th Int. Conference on Energy for a Clean Enviroment, Lisbon, 2003.*

5.6
Acoustic Transfer in Ceramic Surface Burners

Koen Schreel and Philip de Goey

5.6.1
Introduction

When using a burner of any type in a closed combustion system, noise problems can occur due to coupling between the oscillating velocity/pressure field associated with the sound and the resulting oscillating heat release by the combustion process. When using cellular ceramics as a burner material, the relevant question is how this coupling takes place. The interaction between porous materials and acoustics is treated in Chapter 4.5, but the material presented there covers the reflection/absorption of cellular materials. For the frequencies of interest for combustion applications (80–800 Hz) this (viscous) damping of the acoustic wave can be neglected. Instead, heat transfer between the flame and the porous material is the key process in coupling with the acoustic field. The use of cellular ceramics as burner materials is extensively treated in Chapter 5.5, which is recommended reading before studying this chapter. We also refer to Chapter 4.5 for an introduction to acoustics, and Chapter 4.3 for the thermal properties of cellular ceramics.

The interaction between surface-stabilized flames and acoustics belongs to the field of thermo-acoustics. Early examples of thermo-acoustic phenomena include the "singing flame" reported in 1777 by Higgins [1], the Sondhauss tube [2], and the Rijke tube [3], each showing that heat sources can produce sound when placed in a tube. Since then, these combustion-driven instabilities have been studied experimentally by numerous authors in various configurations, for example, Putnam et al. [4] and Schimmer et al. [5]. Lord Rayleigh [6] was the first to pose a theoretical criterion for acoustic instability in these devices. The so-called Rayleigh criterion states that the energy in the acoustic field increases when the following inequality holds:

$$\int_0^T p'(t)q'(t)dt > 0 \tag{1}$$

where p' and q' are the oscillating parts of the pressure and heat release, respectively, and the integration is over time. Putnam et al. [7], and Putnam [8] put this into a mathematical formulation. Raun et al. [16] presented an extensive overview. Still, the Rayleigh criterion is phenomenological, and more fundamental studies are still going on to provide the necessary information on the exact distortion of the acoustic field by the flame.

Cellular Ceramics: Structure, Manufacturing, Properties and Applications.
Michael Scheffler, Paolo Colombo (Eds.)

ISBN: 3-527-31320-6

When a surface burner is operated in the blue-flame mode, Bunsen-type flames are stabilized on the surface. In this case the interaction with an acoustic field is governed mainly by flame surface variations, which were studied experimentally by, for example, Durox et al. [10] and Ducruix et al. [11] and analytically by Fleiffil et al. [21]. In the last-cited work, the flow field is described by a Poiseuille flow and the profile is assumed to be undistorted by the flame. The motion of the flame is determined by using the G equation with constant burning velocity. Although the latest results [22, 23] show significant progress in understanding, there are still unresolved issues regarding the exact nature and origin of the unsteady velocity field.

Since the operation of a porous ceramic burner in the blue-flame mode leads to higher NO_x emissions, the practically more relevant case is the radiant mode, in which a flat flame is stabilized on the surface. Since the oscillating velocity field is one-dimenional, this is theoretically better understood. McIntosh et al. [12], McIntosh [13], Van Harten et al. [14], and Buckmaster [15] pioneered this area. Raun et al. [9], McIntosh [17], and, more recently, McIntosh [18] and McIntosh et al. [20] used the flame/acoustic transfer function model to investigate Rijke tube oscillations. These flames are anchored to a burner plate, and the acoustic field is calculated from the reacting flow equations that are approximated by low-Mach number and high activation energy asymptotics. From this analysis, the transfer function for the acoustic velocity arises by which the acoustic quantities outside the flame are coupled. All these analysis assume that heat transfer between the gas flow and the porous ceramic is ideal, that is, the local gas temperature is always equal to the local temperature in the ceramic.

The latest research at the Eindhoven University of Technology is also based on the acoustic interaction of flames stabilized on flat surface burners. A simple analytical model based on fluctuating heat generation due to the dynamics of flames stabilized on top of porous burners, derived by Rook et al. [24], show behavior similar to the much more complex model of McIntosh. Numerical results for a burner with ideal heat transfer were found to be in agreement with the analytical model. Experiments performed on a cooled brass perforated-plate burner confirmed these results. Local multidimensional flow and heat-transfer effects were studied by Rook et al. [25]. Schreel et al. [26] experimentally investigated the behavior of ceramic surface burners, which have a much higher surface temperature than cooled metal burners, and found reasonable agreement with the extended theory for burners with a large heat transfer. Recently, however, detailed measurements and numerical studies have shown that most realistic surface burners cannot be considered to have ideal heat transfer. If this is the case, the other material parameters, such as heat transfer αS, the effective conductivity $\lambda_s \tau_s$, the specific heat $c_{p,s}$, and radiative properties such as the emissivity ε also have an effect on the acoustic behavior [27].

In this chapter, the state-of-the-art knowledge of the interaction between acoustics and combustion on realistic porous ceramic burners is presented analytically, numerically, and experimentally. In the next section, a short introduction will be given on the network analysis of 1D acoustic systems and an extended version of acoustic transfer is introduced with respect to that given in Chapter 4.5. Furthermore, the relevant transport equations for the description of combustion are presented. In the remaining sections, an analytical description of acoustic transfer is given, followed by experimental work on several different porous-burner materials. The last section is devoted to a full numerical simulation of the transfer function, including an accurate description of volumetric heat transfer in the porous ceramic and other relevant ceramic material properties.

5.6.2
Acoustic Transfer

When analyzing the acoustic behavior of complete systems in which a burner is used, one has to realize that, although the combustion process can be considered as a source of acoustic energy, it is the complete system geometry that determines whether spontaneous acoustic oscillations will occur or not. A generally applied method to analyze complete systems is so-called network analysis [32, 33], in which each physical element of the system is represented by a transfer function. The overall transfer function of the system can then be analyzed for resonances.

If the wavelength of the sound of interest is much longer than the typical duct diameter, the system can (acoustically) be considered as one-dimensional, and a relatively simple expression of the transfer function can be given. The velocity can be written as (cf. Chapter 4.5)

$$u = \bar{u} + u' \tag{2}$$

where $\bar{u}$ is the average velocity and u the acoustic fluctuations. The acoustic velocities and pressures on both sides of an element can then be related via a transfer matrix. For a burner-stabilized flat flame, it can be shown that the transfer matrix has (in the low Mach number approximation) the following form

$$\begin{bmatrix} p'_{\mathrm{b}} \\ u'_{\mathrm{b}} \end{bmatrix} = \begin{bmatrix} 1 & 0 \\ 0 & T_{22} \end{bmatrix} \begin{bmatrix} p'_{\mathrm{u}} \\ u'_{\mathrm{u}} \end{bmatrix} \tag{3}$$

which means that the complete transfer matrix is determined by T_{22}, the coupling between the velocity fluctuations upstream of the burner and flame (u'_{u}) and downstream of the flame (u'_{b}). This matrix element can be determined by considering the detailed dynamics of the flame on top of the burner and the heat-transfer fluctuations towards the burner.

The 1D flame motion is governed by the set of N conservation equations for the species mass fractions Y_i

$$\phi\rho_g \frac{\partial}{\partial t} Y_i + \phi\rho_g u \frac{\partial}{\partial x} Y_i + \frac{\partial}{\partial x}\left(\phi\rho D_i \frac{\partial}{\partial Y_i}\right) = \phi\dot{\rho}_i, \quad i = 1,\ldots,N, \tag{4}$$

where $\dot{\rho}_i$ is the production rate of mass of species i in the flame by chemical reactions, combined with the continuity equation

$$\frac{\partial \rho_g \phi}{\partial t} + \frac{\partial}{\partial x}(\phi\rho_g u) = 0, \tag{5}$$

for the gas flow and the ideal gas law

$$p = \rho_g RT \sum_i (Y_i/M_i). \tag{6}$$

Note that the momentum conservation equation reduces to a constant pressure law $p = p_u$ for low Mach numbers. The energy conservation equation describing combustion on a ceramic surface burner must be split into two parts for the two continuous phases, yielding for the gas

$$\phi \rho_g c_{p,g} \frac{\partial T_g}{\partial t} + \phi \rho_g u c_{p,g} \frac{\partial T_g}{\partial x} - \frac{\partial}{\partial x}\left(\phi \lambda_g \frac{\partial T_s}{\partial x}\right) = \alpha S(T_s - T_g) - \phi \sum_{i=1}^{N} h_i \dot{\rho}_i \tag{7}$$

and for the porous material

$$(1-\phi)\rho_s c_{p,s} \frac{\partial T_s}{\partial t} - \frac{\partial}{\partial x}\left((1-\phi)\tau_s \lambda_s \frac{\partial T_g}{\partial x}\right) = -\alpha S(T_s - T_g) + \frac{\partial q_{rad}}{\partial x}. \tag{8}$$

In the above equations, the subscripts s and g indicate whether a quantity relates to the solid material or the gas, respectively. The properties of the solid material are reflected in the volumetric heat-transfer coefficient αS, the porosity ϕ, the heat capacity $\rho_s c_{p,s}$, and the product of tortuosity and thermal conductivity $(\tau_s \lambda_s)$. With regard to tortuosity, note that two tortuosities exist for a porous material. One describes the tortuosity of the "holes" and concerns the effective distance traveled by a flowing medium, and the other describes the solid and concerns the effective distance over which heat conduction takes place. In this context the latter is meant (see also Chapter 4.3).

5.6.3 Analytical Model

As an intuitive model for the interaction between the flame and the acoustic waves, a kinematic description can be applied [25] in which the heat transfer between the burner and the flame is considered to be perfect. This means that the gas temperature and the temperature of the porous solid are always the same, and Eqs. (7) and (8) can be combined into one equation for the temperature T of the single continuum (m)

$$\rho_m c_{p,m} \frac{\partial T}{\partial t} + \rho_m u c_{p,m} \frac{\partial T}{\partial x} - \frac{\partial}{\partial x}\left(\lambda_m \frac{\partial T}{\partial x}\right) = \frac{\partial q_{rad}}{\partial x} - \phi \sum_{i=1}^{N} h_i \dot{\rho}_i. \tag{9}$$

Due to the large heat capacity of the burner, the temperature inside the material is steady in this case, and the material parameters of the burner then do not influence the flame dynamics at all. The only parameter of interest is the burner surface temperature. A variation in the standoff distance of the flame (induced by the acoustic velocity fluctuations) then results in a displacement of the complete flame structure, except for the temperature in the burner, which remains steady. This leads to a feedback loop [24, 26], by which (convective) enthalpy waves emanating from the burner surface induce a fluctuating flame temperature and therefore also a fluctuating burning velocity of the flame. The flame motion can be amplified or damped by these enthalpy waves, depending on the frequency. The resulting transfer function reads:

$$T_{22} = \frac{T_{\text{surf}}}{T_{\text{u}}} + \left(\frac{\overline{T}_{\text{b}} - T_{\text{surf}}}{T_{\text{u}}}\right) A(\hat{\omega}) + \left(\frac{T_{\text{ad}} - \overline{T}_{\text{b}} + T_{\text{surf}} - T_{\text{u}}}{T_{\text{u}}}\right) \left(\frac{1 - A(\hat{\omega})}{\text{i}\hat{\omega}}\right) \frac{1}{2}\left(1 + \sqrt{1 + 4\text{i}\hat{\omega}}\right). \tag{10}$$

where the coefficient $A = s'_L/u'$ is the ratio between the fluctuating burning velocity and the acoustic velocity:

$$A(\hat{\omega}) = \left[1 + \frac{2}{Ze} \frac{\overline{T}_{\text{b}} - T_{\text{u}}}{T_{\text{ad}} - \overline{T}_{\text{b}}} \text{i}\hat{\omega} \exp\left(-\frac{\overline{\psi}_{\text{f}}}{\delta} \frac{1}{2}\left(1 - \sqrt{1 + 4\text{i}\hat{\omega}}\right)\right)\right]^{-1}. \tag{11}$$

In these equations, T_{u}, T_{surf}, T_{b}, T_{ad}, and T_{a} are the temperatures of the unburnt mixture, the surface, the flame, the adiabatic flame, and the activation temperature. Furthermore, $\hat{\omega}$ is the frequency scaled with the flame transit time δ/u, where δ is the flame thickness, and $\overline{\psi}_{\text{f}}$ the average flame standoff distance. For more details, the reader is refered to Ref. [24].

This transfer function shows a resonancelike behavior, as can be seen in Fig. 1. For low frequencies the magnitude of T_{22} is determined by the thermal expansion of the flame, then increases up to a resonance frequency, after which the magnitude falls to 1 for even higher frequencies. The resonance frequency is mainly a function of the standoff distance of the flame, and thus of the surface temperature of the burner.

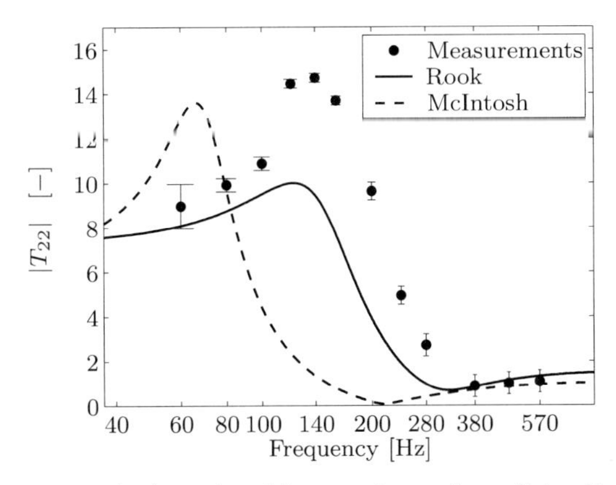

Fig. 1 Absolute value of the acoustic transfer coefficient T_{22} as a function of frequency for a perforated brass plate. The curves were calculated with two analytical models, both assuming perfect heat transfer.

For arbitrary Lewis numbers an analytical expression for the transfer function has been found by McIntosh et al. [20], based on a matched asymptotic analysis of the full equations. Also this model shows reasonable agreement with the measurements, but the elevated burner surface temperatures of radiant burners are not accomodated in the model of McIntosh. An analytical model for acoustic transfer when the heat transfer coefficient αS is finite is not available.

5.6.4
Acoustic Transfer Coefficient for Realistic Porous Ceramics

The analytical approach presented in the Section 5.6.3 gives good insight into the physics, but it remains to be seen whether the accuracy is high enough for the purpose of acoustic modeling of a complete combustion system. The effect of realistic values of αS has been studied by Schreel et al. [27], and it was found that the assumption of infinitely fast heat transfer is not adequate for most realistic radiant burners. The experimentally observed large variation in acoustic behavior, especially for higher frequencies, can not be explained by a variation in surface temperature alone. Thus, a more realistic heat-transfer model needs to be used.

In this section, the numerical and experimental results obtained by Schreel et al. [27] are presented. Results were obtained for four different ceramic materials, completed by the results for a perforated brass plate, which serves as a reference case exhibiting infinitely fast heat transfer. Material 1 consisted of sintered fibers with a diameter of 25 μm. The fibers had an inner core of silicate, an outer shell of silicon carbide, and were arranged randomly within a layer. Material 2 was the unsintered version of material 1, with additional holes perforated in the plate. Because the material is not sintered, the porosity is noticeably larger. Material 3 was a perforated ceramic plate consisting of foamed cordierite with a closed-cell structure. The porosity was thus entirely determined by the perforation pattern, which was hexagonal (pitch 1.95 mm) with a hole diameter of 1 mm. Material 4 was a ceramic foam made of cordierite. It had a reticulated structure with 24 pores per centimeter and an average hole size of 0.3 mm. All relevant material properties are presented in Table 1. The value of the volumetric heat-transfer coefficient is based on the application of known Nusselt relations for an approximate geometry of the solid (these are discussed with the presentation of the measurements), and is normalized with respect to the volume of the plate. The porosity was derived from a comparison between the measured density of the porous material and the literature value of the density of the bulk solid. This does not take into account a distinction between micro- and macroporosity, nor does it consider the occurrence of closed-cell structures. For the materials used here this seems to be adequate, but in general the more elaborate techniques described in Chapter 4.3 should be applied. For the heat capacity and conductivity literature values were used. The (apparent) emissivity was assumed to be 0.85 for all materials.

First, modeling results for T_{22} are presented for a variation of αS, ϕ, $\rho_s c_{p,s}$, $\tau_s \lambda_s$, and ε. Experimental results are presented afterwards, and compared to the modeling results based on the best known values of the material properties.

5.6.4.1
Numerical Results

The porous solid can adequately be modeled with a volume-averaged continuum approach as provided by Eqs. (4)–(6), (7), and (8). These equations with the appropriate boundary conditions can in principle be solved with a good time-dependent computational fluid dynamics (CFD) solver, but the problem is quite complex and not all codes are suited for the combination of porous flow and combustion. The results presented here were obtained with the software package Chem1D, developed in-house [37] at the Einhoven University of Technology. For the purpose of this work, the chemical model as proposed by Smooke [39] is used. This model is limited to lean CH_4/air flames, but it can accurately describe the flame dynamics and offers a high computational performance gain over more complex models like GRI [40]. The diffusion was modeled by using the EGlib library of Ern and Giovangigli [41], which incorporates complex transport processes including the Dufour and Soret effects.

The actual properties of the material can be described by the parameters in Eqs. (7) and (8), which are discussed above and tabulated in Table 1. The radiation term q_{rad} in Eq. (8) is in most cases described by the Rosseland model inside the porous material, with an extinction coefficient of 15 cm^{-1}. At the surface the radiative heat loss is described as $\varepsilon\sigma T^4$. An even better modeling of the radiative heat transfer near the surface could be obtained by using the discrete-ordinate method or a similar technique capable of dealing with optically thin media, but the results will show that radiative effects at the surface do not have a strong influence on the acoustic transfer function.

Table 1 Relevant properties of the burner materials.

Burner material	αS/W cm^{-3} K^{-1}	ϕ	$\rho_s c_{p,s}$/J cm^{-3} K^{-1}	$f_s \lambda_s$/W cm^{-1} K^{-1}	ε
1. Sintered ceramic fibers	7	0.926	3.56	0.05	0.85
2. Unsintered ceramic fibers	1	0.970	3.56	0.05	0.85
3. Perforated ceramic plate	1.24	0.240	0.86	0.02	0.85
4. Ceramic foam	0.5	0.820	3.90	0.02	0.85

When modeling a stationary flame, the temperature distribution inside the ceramic is clearly different for the gas and the solid (see Fig. 2 for an example). At the surface the solid temperature is significantly lower than the gas temperature due to the radiative heat loss at the surface. The solid, however, is a better heat conductor than the gas and at some position inside the flame holder the gas and solid temperature are equal. Below that point, the solid temperature is higher than the gas temperature.

From a variation of the parameters, their relative importance can be judged and from this the accuracy can be estimated with which they need to be known for modeling an arbitrary material. For all the calculations a fuel lean methane–air flame was used at a stoichiometric ratio of 0.8 and a gas mixture velocity of 17 cm s^{-1}.

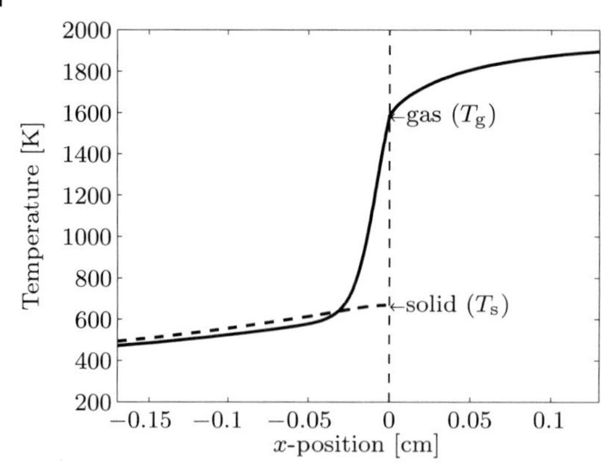

Fig. 2 Temperature profile of the gas and solid for a flame on a porous ceramic material. The vertical dashed line represents the boundary of the porous material.

The volumetric heat-transfer coefficient is theoretically the parameter which determines whether it can be assumed that a material exhibits perfect heat transfer. A series of transfer function calculations with αS in the range of 1–80 W cm^{-3} K^{-1} is presented in Fig. 3. For large values of αS a resonancelike shape is found for T_{22}, which is damped for lower values of αS. The cutoff frequency also decreases with lower values of αS. There is only influence on the low-frequency behavior of the transfer coefficient for very low values of αS. For values of αS larger than 40 W cm^{-3} K^{-1}, the transfer function becomes insensitive to variations in αS, that is, heat transfer becomes ideal. The transfer function is particularly sensitive to variations in the region 1–10 W cm^{-3} K^{-1}, which corresponds to the range of values encountered in the burner materials under consideration. Clearly, the approximation that heat transfer is ideal is not valid. This parameter should be given special attention when modeling an arbitrary material if no direct comparison with measurements of the transfer function is available. The measurement or modeling of the volumetric heat-transfer coefficient is not trivial [29, 42], although when great care is taken reliable results can be obtained [30]. In comparison, the analytical model gives good results for lower frequencies but fails for frequencies higher than 500 Hz.

A series of transfer functions with varying porosity ϕ is presented in Fig. 4. Also in this case the influence on the transfer function is significant, but now the magnitude of the transfer function is affected for the frequency range below the fall-off frequency. The fall-off frequency itself is hardly affected. Although the porosity is important, it is difficult to measure it accurately enough to be able to model an arbitrary material.

The influence of the thermal conductivity $\tau_s \lambda_s$ is almost negligible and is not presented in a figure. We found only significant influence on the transfer function for variations of an order of magnitude. This can be understood from Eq. (8). The term containing the thermal conductivity is proportional to $\partial T_s / \partial x$. As can be seen from Fig. 2, the slope of the temperature is quite moderate. Normally thermal conductivity

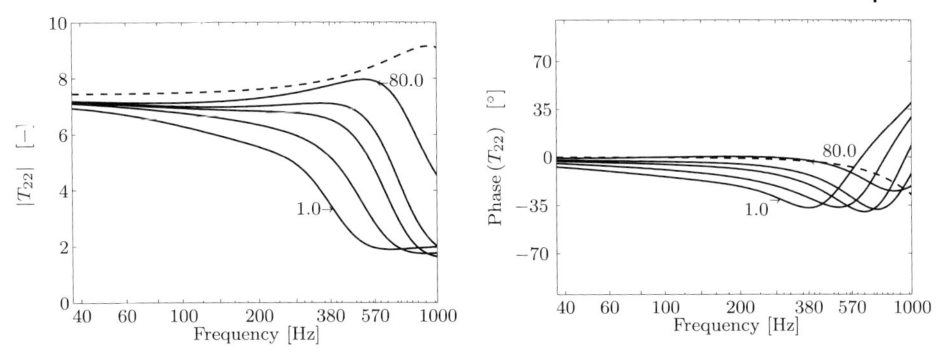

Fig. 3 Absolute value and phase of the acoustic transfer coefficient T_{22} as a function of frequency for burner material 1 with $\alpha S/\mathrm{W\,cm^{-3}\,K^{-1}}$ values of 1 (lower solid curve), 2, 4, 7, and 80 (upper solid curve). The dashed curve is the analytical model by Rook.

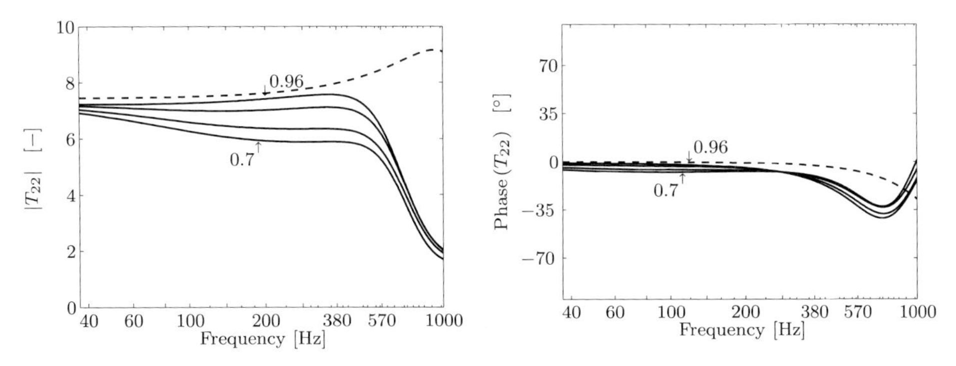

Fig. 4 Absolute value and phase of the acoustic transfer coefficient T_{22} as a function of frequency for burner material 1 with porosity ϕ of 0.7 (lower solid curve), 0.85, 0.926, and 0.96 (upper solid curve). The dashed curve is the analytical model by Rook.

is known with an accuracy better than a factor of 2, which means that no special attention is needed.

Even less significant is the heat capacity of the solid $c_{p,s}$. Again from Eqs. (7) and (8) it can be seen that, since the heat capacity of the solid is three orders of magnitude larger than that of the gas, the thermal inertia of the solid is too large to feel the fast fluctuations of the gas temperature.

Figure 5 presents results for varying emissivity. Some effects can be observed, but only for unrealistic values of ε. Based on this we chose a value of $\varepsilon = 0.85$ for all materials. Although most ceramic materials have an emissivity higher than 0.9, due to the porosity the apparent emissivity is somewhat lower. Even if an error is made of ± 0.05, this hardly has any influence on the accuracy of the modeling.

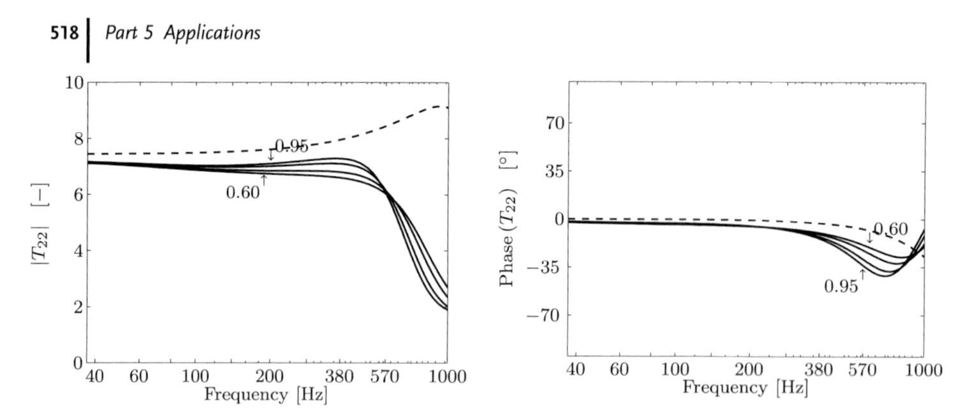

Fig. 5 Absolute value and phase of the acoustic transfer coefficient T_{22} as a function of frequency for burner material 1 with emissivity ε of 0.6 (lower solid curve), 0.7, 0.85, and 0.95 (upper solid curve). The dashed curve is the analytical model by Rook.

5.6.4.2
Measurements

The experimental determination of the transfer function element T_{22} involves accurate measurement of the velocity fluctuations directly upstream and downstream of the burner. A proven and reliable method to do this would be the four-microphone technique (see Section 4.5.4.2), if it were not for the too strong variation of the temperature in the flue gases. This can be accounted for by using additional microphones and assuming a certain function for the dependence of the velocity of sound on the distance from the flame, but this can be quite cumbersome and experimentally challenging due to the high temperatures involved. The approach used by Schreel et al. [26, 27] is to use the two-microphone technique [32, 35, 36] in the upstream (cold) part of the flow, and to directly measure the velocity immediately downstream of the flame by laser Doppler velocimetry (LDV). Details about the setup can be found in Ref. [26]. A short description is given here.

The setup consists essentially of an approximately 60 cm long tube with an inner diameter of 5 cm, placed vertically to minimize buoyancy effects. The gas inlet is located at the bottom of the tube, and the flame holder is approximately 7 cm from the exit. Some grids are fitted right after the entrance to settle the flow. In the lower part of the tube a hole is made in the side which is coupled through a flexible hose to a loudspeaker. The part of the tube downstream of the flame holder is water-cooled at nominally 50 °C to avoid condensation of water. The acoustic properties of the tube itself were studied, and it was found that a rather weak resonance occurs around 640 Hz, but this frequency region is not used for this research. In this case frequencies up to 800 Hz are applied, avoiding the 600–680 Hz band. Two calibrated pressure transducers are mounted in the wall of the tube upstream of the flame. Downstream of the flame optical access for the LDV measurement is provided by three small holes 4 mm above the flame holder. The LDV equipment consists of a 20 mW HeNe laser used in forward scatter mode and a counter-based signal analyzer (Disa 55L series). In principle one does not measure the transfer matrix element of the flame in this way, but the transfer matrix element of the flame combined with the flame holder. Test measurements showed however that the transfer matrix element of the flame holder itself is very close to unity for all materials at the frequencies of interest and can be neglected.

The measurement results are presented in Figs. 6–9. Correspondence with modeling is good if the correct values for αS are chosen. Burner material 1 has a relatively high value of αS. A sharp cutoff frequency can be observed at a quite high frequency. Burner material 2 is made from the same fibers, but much more loosely packed. The effect of the resulting lower value of αS clearly shows in the observed transfer function. Estimates based on a Nusselt relation for cylinders [27] yield values for the volumetric heat transfer coefficient αS of approximately 7 and $1\,\mathrm{W\,cm^{-3}\,K^{-1}}$ for materials 1 and 2, respectively.

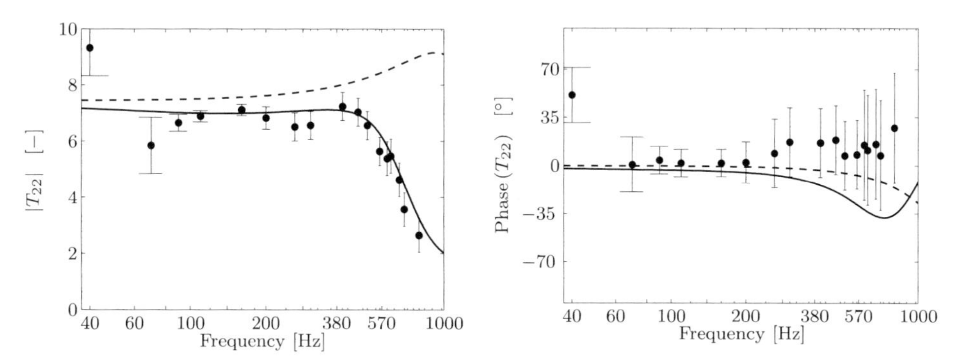

Fig. 6 Absolute value and phase of the acoustic transfer coefficient T_{22} as a function of frequency for burner material 1. The dashed curve is the analytical model by Rook.

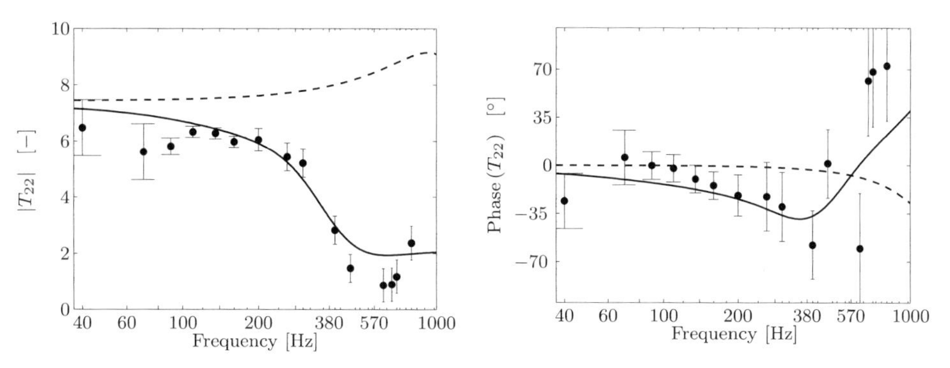

Fig. 7 Absolute value and phase of the acoustic transfer coefficient T_{22} as a function of frequency for burner material 2. The dashed curve is the analytical model by Rook.

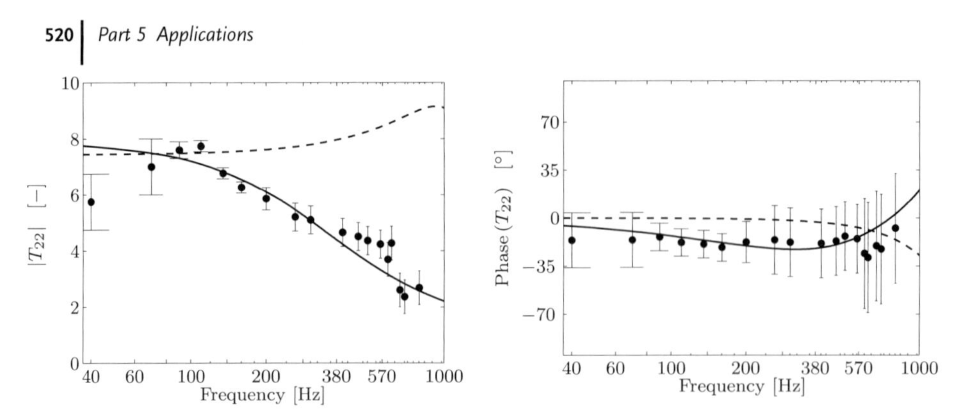

Fig. 8 Absolute value and phase of the acoustical transfer coefficient T_{22} as a function of frequency for burner material 3. The dashed curve is the analytical model by Rook.

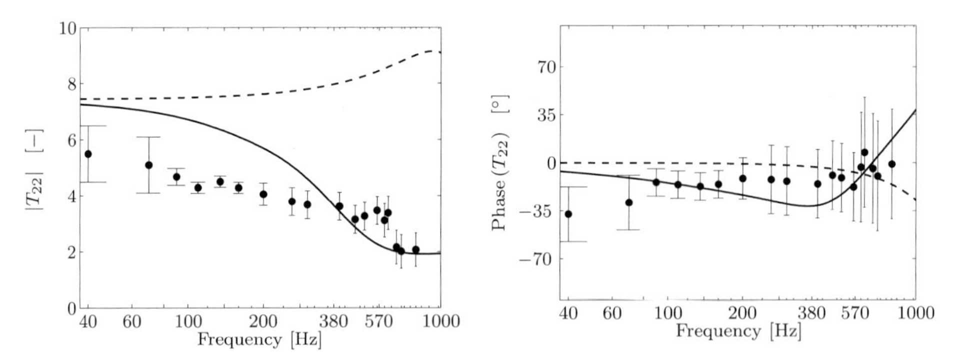

Fig. 9 The absolute value and phase of the acoustical transfer coefficient T_{22} as a function of frequency for burner material 4. The dashed curve is the analytical model by Rook.

For a perforated ceramic plate like material 3, the structure can be quite accurately modeled as a collection of cylindrical holes. This leads to the estimation that αS = 1.24 W cm^{-3} K^{-1} for material 3, which yields a good correspondence between the numerical model and the measurements.

For material 4 (ceramic foam), the αS value was estimated using an empirical correlation [31] leading to a value of 0.5 W cm^{-3} K^{-1}. Unfortunately, the numerical model shows flashback for values of αS below 1 W cm^{-3} K^{-1} (at which value the numerical curve is plotted), so no real comparison can be made, although an extrapolation of the curve of αS = 1 W cm^{-3} K^{-1} to lower values of αS seems to indicate a reasonable correspondence.

When judging the applicability of the analytical model, it can be concluded that for high values of αS and frequencies up to 500 Hz the results are good, but large discrepancies arise outside these limits.

5.6.5 Summary

Depending on the needed accuracy, the acoustic transfer function of a radiant surface burner can be obtained by using an analytical model, by measurement, or by numerical modeling. Based on experiments, it turns out that the acoustic transfer function is strongly affected by heat transfer between the flame and the flame holder. This heat transfer is mainly governed by the volumetric heat-transfer coefficient and to some extent by the porosity. It is shown that the heat transfer can be assumed to be ideal (i.e., the gas temperature and the solid temperature are identical) for values above $40\,W\,cm^{-3}\,K^{-1}$. In practice these values do not occur for ceramic materials. Other material properties like heat capacity, heat conductivity, and emissivity are of minor importance.

The correspondence between numerical modeling and experiments is good. The key is to determine the volumetric heat-transfer coefficient, since the modeling is so sensitive to its value. This can be done on the basis of Nusselt relations, but this is not very easy for difficult geometries. The direct measurement of the volumetric heat transfer coefficient has the preference, as can be done with the methods described in Chapter 4.3 or tby Fend et al. [30]. Modeling of an arbitrary material without accurate knowledge of the volumetric heat- transfer coefficient is not advisable.

References

1 Higgins B. *J. Nat. Phil. Chem. Arts* **1802**, *1*, 129.
2 Sondhauss C. *Pogg. Ann. Phys. Chem.* **1850**, *79*, 1.
3 Rijke P.L. *Phil. Mag.* **1859**, *17*, 319.
4 Putnam A.A., Dennis W.R. *Trans. ASME* **1953**, *75*, 15.
5 Schimmer H., Vortmeijer D. *Combust. Flame* **1977**, *28*, 17.
6 Lord Rayleigh J.W.S. *Nature* **1878**, *18*, 319.
7 Putnam A.A., Dennis W.R. *J. Acoust. Soc. Am.* **1956**, *28*, 246.
8 Putnam A.A. *Combustion-Driven Oscillations in Industry*, American Elsevier, New York, 1971.
9 Raun R.L. Beckstead M.W. *Combust. Flame* **1993**, *94*, 1–24.
10 Durox D., Baillot F., Searby G., Boyer L. *J. Fluid Mech.* **1997**, *350*, 295–310.
11 Ducruix S., Durox D., Candel S. 28*th Proc. Combust. Inst.* **2000**, *28*, 765.
12 McIntosh A.C. Clark J.F. *Combust. Sci. Tech.* **1984**, *38*, 161–196.
13 McIntosh A.C. *Combust. Sci. Tech.* **1986**, *49*, 143–167.
14 Van Harten A., Kapila A.K., Matkowsky B.J. *SIAM J. Appl. Math.* **1984**, *44*, 982.
15 Buckmaster J.D. *SIAM J. Appl. Math.* **1983**, *43*, 1335.
16 Raun R.L., Beckstead M.W., Finlinson J.C., Brooks K.P. *Prog. Energy Combust. Sci.* **1993**, *19*, 313–364.
17 McIntosh A.C. *Combust. Sci. Tech.*, **1987**, *54*, 217–236.
18 McIntosh A.C. *Combust. Sci. Tech.* **1990**, *69*, 147–152.
19 McIntosh A.C. *Combust. Sci. Tech.* **1993**, *91*, 329.
20 McIntosh A.C., Rylands, S. *Combust. Sci. Tech.* **1996**, *113/114*, 273.
21 Fleifil M., Annaswamy A.M., Ghoniem Z.A., Ghoneim A.F. *Combust. Flame* **1996**, *106*, 487–510.
22 Schuller, T., Durox, D., Candel, S., *Combust. Flame* **2003**, *134*, 21–34.
23 Preetham Lieuwen T. *AIAA J.* **2004**, 4035

24 Rook R., Groot G.R.A., Schreel K.R.A.M., Aptroot R., Parchen R., de Goey L.P.H. in *Proc. ECSBT2*, **2000**, Vol. I, pp. 149–159.

25 Rook R., de Goey L.P.H., Somers L.M.T., Schreel K.R.A.M., Parchen R. *Combust. Theor. Model.*, **2002**, *2*, 223–242.

26 Schreel K.R.A.M., Rook R., de Goey L.P.H., *Proc. Combust. Inst.* **2002**, *29*, 115.

27 Schreel K.R.A.M., van den Tillaart E., de Goey L.P.H., *Proc. Combust. Inst.* **2004**, *30*, 17111.

28 Lammers F.A, de Goey L.P.H. *Combust. Flame* **2003**, *133*, 47–61.

29 Howell J.R., Hall M.J., Ellzey J.L., *Prog. Energy Combust. Sci.* **1996**, *22*, 121–145.

30 Fend T., Hoffschmidt B., Pitz-Paal R., Reutter O., Rietbrock P., *Energy*, **2004**, *29*, 823–833.

31 Barra A.J., Diepvens G., Ellzey J.L., Henneke M.R., *Combust. Flame* **2003**, *134*, 369–379.

32 Munjal M.L., *Acoustics of Ducts and Mufflers*, John Wiley and Sons, New York, 1987.

33 Dowling A.P., Ffowcs Williams J.E., *Sound and Sources of Sound*, Ellis Horwood, Chichester, 1983 .

34 Bouma P.H., de Goey L.P.H., *Combust. Flame* **1999**, *119*, 133–143.

35 Seybert, A.F., Ross, D.F., *J. Acoust. Soc. America*, **1977**, *61*, 1362–1370.

36 Paschereit C.O., Gutmark E., Weisenstein W. *AIAA* **1998**, *38*, 1025.

37 Chem1D, a one-dimensional laminar flame code, Eindhoven University of Technology, The Netherlands, http://www.combustion.tue.nl/chem1d, version 2.2.

38 Somers L.M.T., PhD thesis, Eindhoven University of Technology, 1994, available at http://www.tue.nl/bib/.

39 Smooke M.D., Giovangigli V., *Mechanisms and Asymptotic Approximations for Methane-Air Flames*, Springer Verlag, Berlin, 1991.

40 Smith G.P., Golden D.M., Frenklach M., Moriarty N.W., Eiteneer B., Goldenberg M., Bowman C.T., Hanson R.K., Song S., Gardiner, Jr. W.C., Lissianski V.V., Qin Z., http://www.me.berkeley.edu/gri_mech.

41 A. Ern, V. Giovangigli, *Multicomponent Transport Algorithms*, Lecture Notes in Physics, m24, New Series Monographs, Springer Verlag, Berlin, 1994.

42 Fu X., Viskanta R., Gore J.P., *Exp. Therm. Fluid Sci.* **1998**, *17*, 285–293.

43 Bejan A., *Heat Transfer*, John Wiley and Sons, New York, 1993.

5.7
Solar Radiation Conversion

Thomas Fend, Robert Pitz-Paal, Bernhard Hoffschmidt, and Oliver Reutter

5.7.1
Introduction

This chapter covers the use of cellular ceramic materials as absorbers in volumetric solar receivers. A receiver is a central element of *solar tower technology*, which converts concentrated solar radiation into high-temperature heat. In a volumetric absorber cellular material is employed to absorb concentrated solar radiation and to transfer the energy to a fluid flowing through its open cells. This is explained more in detail in Section 5.7.2. The concentrated radiation is generated by a large number of controlled mirrors (heliostats), each of which redirects the solar radiation onto the receiver as a common target on the top of a tower (Fig. 1). There are different concepts to exploit the heat generated by a volumetric receiver.

In one concept ambient air is forced through the open pores of the material and is heated to about 700 °C. It is then used to generate steam for a conventional steam-turbine process (Fig. 2). This idea of an open volumetric air receiver was first presented in a study in 1985 [1]. Since then, the technology has been successfully proven in a number of projects [2–4]. Key components of a 2.5 MW_{th} facility were tested by the TSA (Technology Program Solar Air Receiver) consortium under the leadership of the German company Steinmüller on top of the CESA 1 tower at the

Fig. 1 Views of the Californian 10 MW test plant Solar Two (left) and the Spanish 1.2 MW test plant Cesa 1 (right).

Cellular Ceramics: Structure, Manufacturing, Properties and Applications.
Michael Scheffler, Paolo Colombo (Eds.)

ISBN: 3-527-31320-6

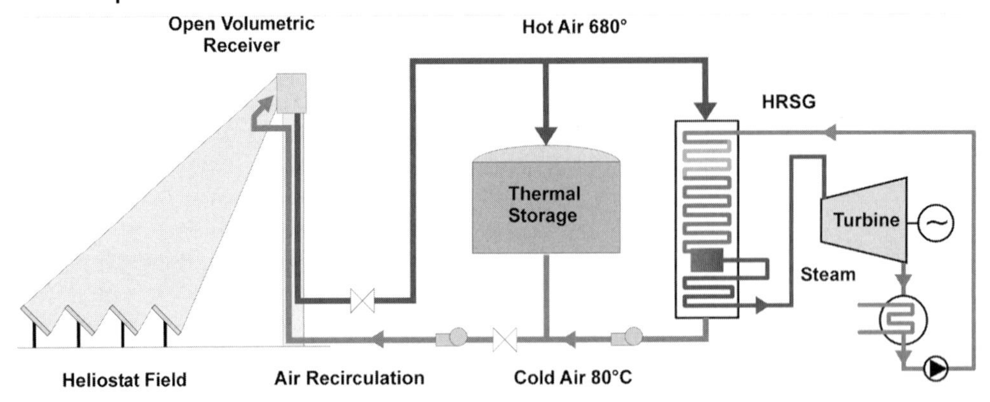

Fig. 2 Flow chart of a power plant using cellular ceramics as an open volumetric air receiver.

Plataforama Solar de Almería in Southern Spain in 1993. A ceramic 3 MW_{th} receiver was successfully tested by a European consortium in 2002 and 2003 within the SOLAIR project [5].

Another concept employs a closed, pressurized loop, in which the volumetric absorber is separated from the surroundings by a quartz glass window. Figure 3 shows a volumetric receiver module with a closed loop and a secondary concentrator. The receiver on top of the solar tower is composed of a set of these modules next to each other. The secondary concentrator boosts the radiation flux on the absorber. This high concentration can be used to integrate solar heat into a gas-turbine cycle by using pressurized air [6–8]. Use of a gas-turbine cycle promises higher efficiencies.

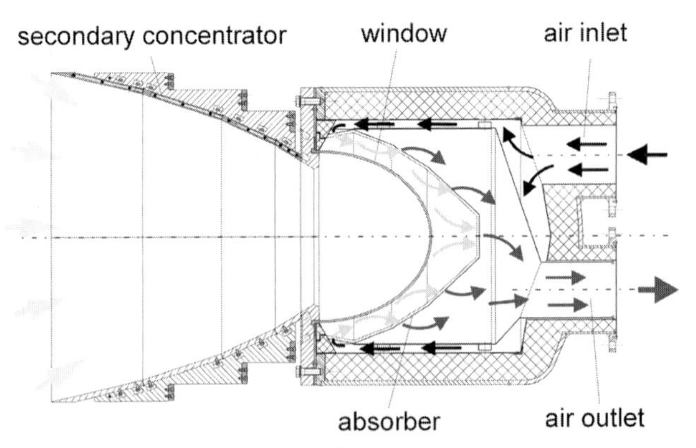

Fig. 3 Drawing of a pressurized volumetric receiver. The cellular ceramic used as an absorber heats the pressurized air. If a catalytically active absorber is used then the same receiver can be used for reforming processes.

In the same type of closed receiver, cellular ceramic materials have been used as an absorber/reactor for chemical reactions. Part of the solar energy transferred by the absorber to the fluid is stored in chemical energy. For example, solar reforming of methane was investigated by the Sandia National Laboratories (SNLA) and the German Aerospace Centre (DLR) in the CAESAR Receiver [9, 10], as well as by the Weizmann Instute of Science (WIS) and the DLR in the SCR (Solar Chemical Reactor-Receiver) [11–13]. The absorber is catalytically coated, and the inlet fluid is a mixture of methane and carbon dioxide.

5.7.2
The Volumetric Absorber Principle

The principle of the volumetric absorber is illustrated in Fig. 4. To demonstrate the advantages of the volumetric absorber the principle of a simple tubular absorber is shown for comparison. Because cold ambient air enters the material at the front of the volumetric absorber, where it is facing the radiation, the material temperature can be kept low. In ideal operation, the temperature distribution shown on the lower right-hand side of Fig. 4 should be realized. The low temperature at the front minimizes thermal radiation losses, which occur in accordance with the well-known Stefan–Boltzmann law $\dot{q} = \varepsilon\sigma T^4$. In the inner absorber volume the temperature increases and the temperature difference between fluid and solid vanishes. Usually, this is already the case after a couple of cell diameters, for example, in the case of an 80 ppi ceramic foam after 1–2 mm. In contrast to this increasing temperature distribution from the inlet to the outlet of the absorber module in case of an ideal volumetric absorber, the temperature distribution of a simple tubular absorber is disadvantageous. It is shown in the graph on the lower left-hand side of Fig. 4. Here the fluid which is to be heated flows inside a tube. The solar radiation heats the tube, which in turn heats the fluid. The significantly higher temperature at the outer tube surface leads to higher radiation losses. The temperature at the outer tube surface is limited by the temperature resistance of the material employed. To avoid destruction of the tube material, the intensity of the concentrated radiation must be kept low compared to volumetric absorbers. This makes it necessary to install larger absorber apertures to achieve similar amounts of total power.

Volumetric absorbers usually consist of materials with a high open porosity, which must withstand temperatures of 1000 °C and more. A high porosity is needed to allow the concentrated solar radiation to penetrate into the volume of the cellular material. This volume is called extinction volume. The structures of the porous material have to be small to achieve the large surface areas necessary to transfer heat from the material to the gaseous fluid flowing through the open pores of the material. Even though the extinction volume decreases with decreasing structure size, the increased surface area and the increased heat transfer due to smaller hydraulic diameters leads to the desire for structures that are as small as possible as long as the porosity can be kept high.

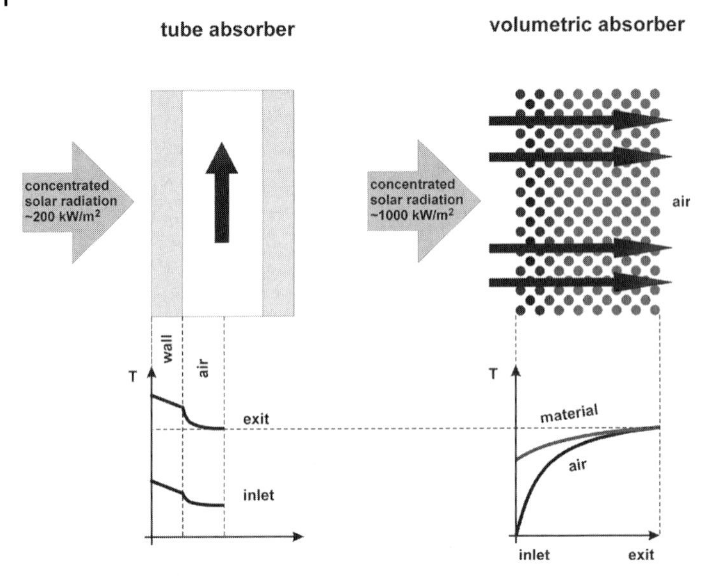

Fig. 4 Open volumetric absorber principle compared to a tubular absorber.

When volumetric absorbers are used as chemical reactors, solar energy is converted into heat and chemical energy. For example in the SCR Receiver for methane reforming [11–13] an aluminum oxide ceramic foam is used as absorber. A wash coat made of highly porous γ-aluminum oxide was applied to enlarge the active surface area, and rhodium was used as catalyst.

5.7.3
Optical, Thermodynamic, and Fluid-Mechanical Requirements of Cellular Ceramics for Solar Energy Conversion

The general requirements and the resulting material properties of a material to be used as a volumetric absorber are summarized in Table 1. To achieve high values of *optical absorption* in the 250–2500 nm wavelength range, that is, the solar spectrum, dark ceramics must be employed as solar absorbers or black coatings must be applied to the front surface of the cellular ceramic. Silicon carbide has been used for this purpose and reaches weighted solar absorption values of 0.9. Oxide ceramics usually have white, highly reflecting surfaces. Black coatings can be used to increase the solar absorption. For example Pyromark high-temperature paint was used for coating a cordierite cellular ceramic material, which then reached absorption values of 0.95 and more. In porous materials the radiation enters the extinction volume of the material, and the pores or channels then act as small cavities which improve absorption due to multiple reflections at the cell walls. The use of selective coatings or selective absorbers has been investigated, and theoretical predictions state that

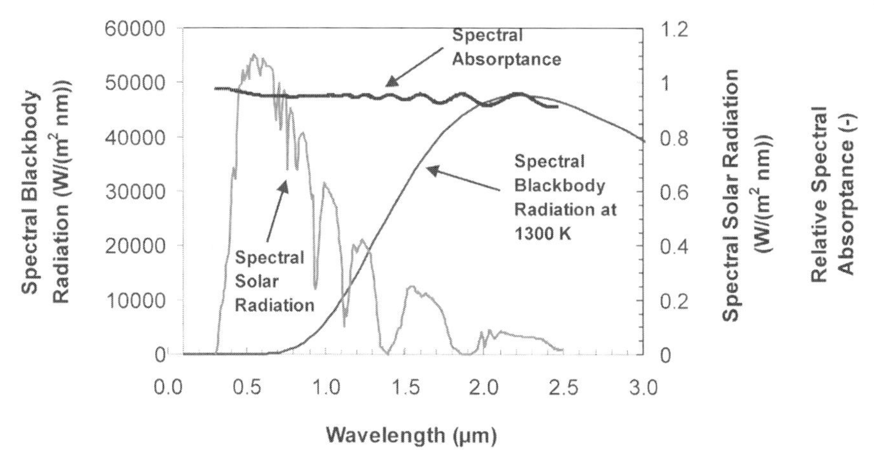

Fig. 5 Spectral absorptance of a silicon carbide short-fiber mesh material, spectral solar radiation, and spectral black-body radiation at 1300 K.

thermal radiation losses can be minimized [14]. However, due to the overlap of the spectral emission curve of a black-body radiator at 1300 K and the solar spectrum, the effect of selectivity is much smaller than for less concentrating technologies such as flat-plate collectors or parabolic trough systems [2]. Figure 5 shows the spectral solar irradiation (not concentrated), the spectral absorptance of a silicon carbide short-fiber mesh material, and the spectral black-body radiation at 1300 K.

Table 1 Optical, thermodynamic, and resulting material requirements of candidate absorber materials

Optical/thermodynamic requirements	Material requirements
High absorption	dark color
Optical extinction	high porosity
Heat-transfer surface	high cell density
High flux	temperature resistance
Radial heat transport	thermal conductivity
High permeability	3D structure

Experimentally the *absorptance* ε in the solar spectrum can be determined indirectly by measuring the solar weighted hemispherical reflectance $\rho_{2\pi S}$ and using the simple equation $\varepsilon = 1 - \rho_{2\pi S}$. This equation holds under the assumption that the transmittance is negligible, which it is for all feasible absorbers. It is calculated from the spectral hemispherical reflectance $\rho_{2\pi}(\lambda)$ according to a standard [15]. The quantity $\rho_{2\pi}(\lambda)$ is measured in the wavelength range $250 < \lambda < 2500$ nm with a UV-Vis-NIR spectrometer equipped with an integrating sphere. Then a standard solar spectrum is used [16] to obtain solar weighted hemispherical reflectance values from the following formula:

$$\rho_{2\pi S} = \frac{\sum_{i=0}^{450} \rho_{2\pi}(\lambda_i) \cdot E(\lambda_i)}{\sum_{i=0}^{450} E(\lambda_i)} \tag{1}$$

where $E(\lambda_i)$ denotes direct normal spectral irradiance at wavelength λ_i in 5 nm steps ($\lambda_0 = 250$, $\lambda_1 = 255$,..., $\lambda_{450} = 2500$ nm).

High porosity is needed to let the solar radiation penetrate into the volume of the material. As shown schematically in Fig. 6, the intensity of the radiation I is attenuated in the cellular material due to a multiple reflection/absorption process approximated by an exponential law

$$I = I_0 e^{-kx}. \tag{2}$$

Thus, a measure for this property is the optical extinction coefficient k.

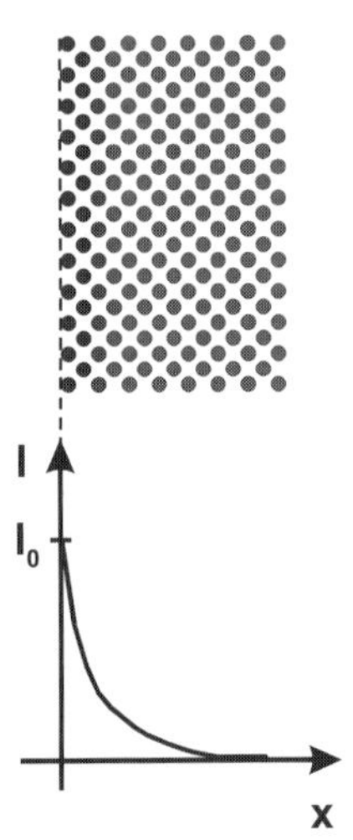

Fig. 6 Attenuation of the intensity of solar radiation in a cellular ceramic material used as a volumetric absorber.

Experimentally the *extinction coefficient* k can be measured indirectly by performing transmittance measurements on samples of cellular materials of different thicknesses, as illustrated in Fig. 7. The measured intensities $I(x_i)$ are then fitted with the above-mentioned exponential function to determine the extinction coefficient k. Alternatively, pictures of the cross section of the cellular material can be evaluated to find the penetration depth and thus the extinction coefficient [13].

In the case of regular geometries such as extruded monoliths, this coefficient can also be easily calculated from the channel diameters, wall thicknesses, and the aperture angle of the incoming radiation. In the case of fiber materials or foams, models must be used to describe the geometry of the material. For example, considering a fiber mesh as layers of parallel cylinders, as in Fig. 7, leads to the following equation if scattering is neglected:

$$k(\gamma) = -\frac{1}{d_H} \ln\left(1 + \frac{4}{\pi\cos\gamma}(P_0 - 1)\right). \tag{3}$$

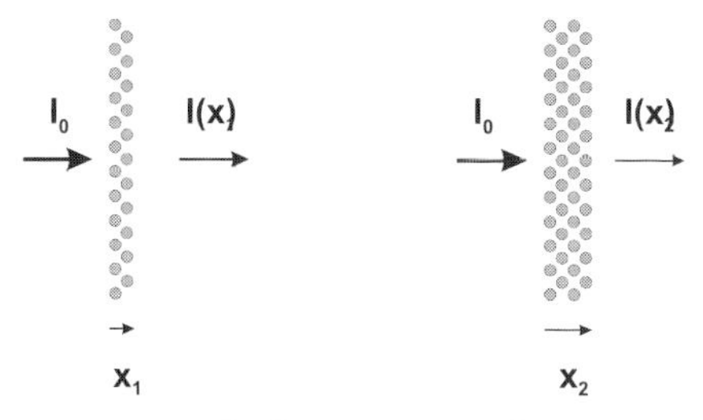

Fig. 7 Principle of measurement of extinction coefficients.

In the case of a fiber material the hydraulic diameter d_H is simply the fiber diameter. If the diameter increases at constant porosity this means that the distance between neighboring fibers increases and thereby k decreases.

The third important requirement of a cellular ceramic used as a volumetric solar absorber is the *heat-transfer surface area*, which is a more general requirement that is also important for high-temperature heat exchangers. The quantity describing this property is called specific surface area A_V [$m^2\ m^{-3}$]. This surface area is responsible for efficient solid-to-fluid heat transfer. High values of A_V are achieved by using small cells or small fiber diameters. Similar to the extinction coefficient, models can be used to estimate this quantity.

A comparison of the different structures can be found in Section 5.7.4, and values of thermophysical properties are listed in Table 2. In the case of ceramic foams, the following empirical equation yields the specific surface area as a function of the cell density n_{ppi}. It was derived from a numerical investigation of various micrographs [17]:

$$A_V = 35.7 \cdot n_{ppi}^{1.1461}. \tag{4}$$

Table 2 Geometric and thermophysical properties of selected cellular ceramics used as volumetric solar absorbers [19].

Material	ε	k/m^{-1}	A_V/m^2 m^{-3}	d_H/mm	K_1/K_2	P_0	λ/W m^{-1} K^{-1}
Fiber mesh	0.95	2000	8000	0.025	0.015×10^{-4}	0.95	0.08
SiC catalyst	0.96	140	1000	2	0.019×10^{-4}	0.51	50
SiSiC catalyst	0.90	140	1000	2	0.089×10^{-4}	0.51	11
SSiC foam, 10 ppi	ca. 0.94	191	500	1.7	4.7×10^{-4}	0.76	0.2
SSiC foam, 20 ppi	0.94	406	1000	0.8	0.6×10^{-4}	0.76	0.5
SSiC foam, 80 ppi	ca. 0.94	1620	5400	0.2	0.15×10^{-4}*	0.76	2.7

* Results from a 20 ppi/80 ppi sandwich foam sample.

In the case of fibrous materials A_V can be easily calculated from the porosity P_0 and the fiber diameter d_H. The porosity of the fibrous material can be calculated from the total volume V and the volume of the fibers V_{fiber} by

$$P_0 = \frac{V - V_{fiber}}{V}. \tag{5}$$

Assuming $l \gg d_H$ for the length of the (cylindrical) fibers, the surface area O_{fiber} and volume V_{fiber} of n fibers can be calculated by

$$O_{fiber} = l\pi d_H n, \tag{6}$$

$$V_{fiber} = l\pi \frac{d_H^2}{4} n. \tag{7}$$

Combining these equations and using the definition of the specific area $A_V = O_{fiber}/V$ yields the following simple equation:

$$A_V = \frac{4(1-P_0)}{d_H}. \tag{8}$$

Comparing Eqs. (3) and (8) shows that a higher value of A_V leads to a higher extinction coefficient and hence a lower penetration depth, because of the geometric similarity of a smaller pore structure of the material.

Up to now we showed how the material properties porosity P_0 and cell dimensions d_H influence the specific surface area and the extinction coefficient, but how do they affect the efficiency of a volumetric absorber? This can be shown by considering that P_0 and d_H also influence the heat-transfer properties of the cellular ceramic. The solid-to-fluid heat transfer of a cellular ceramic can be described by the equation

$$\dot{Q} = \alpha A_V V(T_S - T_F) \tag{9}$$

where α denotes the convective heat-transfer coefficient, T_S and T_F are solid and fluid temperature, respectively, and V is the total volume of the cellular body. An important result from the theory of volumetric absorbers is the following conclusion for the efficiency of a volumetric absorber η, which is defined as the power of the fluid mass flow at the outlet of the absorber divided by the power of the concentrated solar radiation penetrating through the aperture area of the absorber [18]. It states that the deeper the radiation penetrates into the absorber volume (small k) and the better the heat transfer (large αA_V) the better the volumetric effect of the absorber. This means that the front temperature of the absorber and hence the thermal radiation losses decrease. This leads to higher efficiencies if the other parameters like irradiance and mass flow are kept constant.

$$\frac{k}{\alpha A_V} \to 0 \quad \Rightarrow \quad \eta \to 1. \tag{10}$$

Equation (3) can be approximated numerically in the relevant range of porosity from 0.3 to 0.95 and an incident angle of radiation of 20° as the following function of the porosity

$$k \sim (1 - P_0)^{1.35}. \tag{11}$$

From Eq. (2) it directly follows

$$A_V \sim (1 - P_0). \tag{12}$$

Since the convective heat-transfer coefficient is independent of the porosity it is easy to conclude

$$\frac{k}{\alpha A_V} \sim (1 - P_0)^{0.35}, \tag{13}$$

$$P_0 \to 1 \Rightarrow \frac{k}{\alpha A_V} \to 0 \Rightarrow \eta \to 1. \tag{14}$$

For cell dimension d_H Eqs. (3) and (8) yield

$$k \sim \frac{1}{d_H} \text{ and } A_V \sim \frac{1}{d_H}. \tag{15}$$

The heat-transfer theory applied for cellular ceramics (see Chapter 4.3) yields

$$\alpha \sim \frac{1}{d_H^r} \text{ with } r < 1 \tag{16}$$

resulting in

$$\frac{k}{\alpha A_V} \sim d_H^r, \tag{17}$$

so that the following can be concluded for the cell dimension d_H:

$$d_H \to 0 \Rightarrow \frac{k}{\alpha A_V} \to 0 \Rightarrow \eta \to 1. \tag{18}$$

These calculations conclusively show that both a high porosity and a high cell density (corresponding to a low characteristic length d_H) are beneficial for the efficiency of a volumetric absorber.

Furthermore, materials that retain their *strength* and *corrosion resistance* at high temperatures are needed if high solar radiation fluxes on the order of 1000 kW m^{-2} are to be absorbed. Silicon carbide materials, which offer excellent high-temperature properties and thermal shock resilience, can be used. Material properties are discussed in more detail in Chapter 4.3.

Thermal conductivity is another property essential for a safe operation of a volumetric absorber. This and the *permeability* of cellular ceramics to air flow have a significant influence on the flow through the open pores of the material [13, 19]. It could be shown that permeability properties of the porous material especially influence the homogeneity of the air flow. This is discussed in detail in the Section 5.7.6. Permeability of a solid with open cells can be characterized by the Forchheimer equation describing the dependence of the pressure loss Δp divided by the length L of the absorber as a function of fluid flow velocity U_0:

$$\frac{\Delta p}{L} = \frac{\mu_{dyn}}{K_1} U_0 + \frac{\rho_F}{K_2} U_0^2 \tag{19}$$

where K_1 is the viscosity coefficient, K_2 the inertial coefficient, ρ_f the density of the fluid, and μ_{dyn} the dynamic viscosity of the fluid.

In this equation the linear term has a major contribution when frictional forces dominate the flow, for example, as in catalyst supports with straight channels. The quadratic term has a major contribution when inertial forces dominate the flow, for example, as in foams or sponges, in which the flow has to follow the tortuous paths of the linked cell walls.

Experimentally, *permeability* measurements can be carried out with a simple setup shown in Fig. 8. A fan generates an air flow through the porous sample. The pressure difference between the outlet and the inlet is measured. A mass flow and a temperature sensor allows the determination of the air flow velocity.

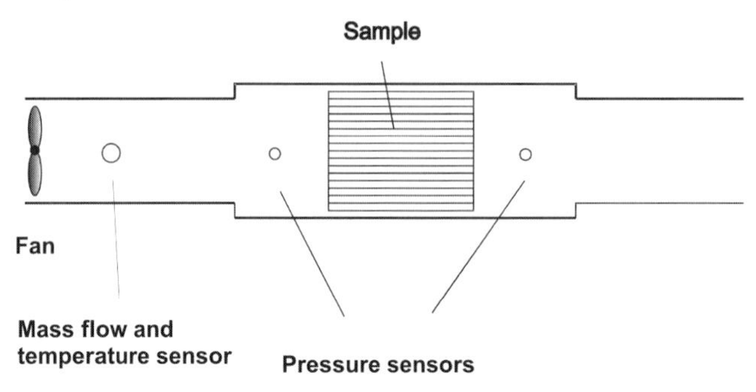

Fig. 8 Test setup used for pressure-loss measurements

5.7.4
Examples of Cellular Ceramics Used as Volumetric Absorbers

An overview of some of the physical properties of materials for volumetric absorbers is given in Table 2. Although the materials are at least partly some form of silicon carbide because of its favorable heat and heat shock resistance properties there is still a wide variety of construction parameters and hence physical properties.

5.7.4.1
Extruded Silicon Carbide Catalyst Supports

Silicon carbide extrusion technology has been widely employed for manufacturing diesel particle filters. Thus, this technology also offers the chance for cost-effective manufacturing of solar absorbers, because only a slight modification of the manufacturing process is necessary. After various laboratory-scale tests a 3 MW modular receiver system was manufactured and tested at the Plataforma Solar de Almería (PSA) in Southern Spain within the project SOLAIR, co-funded by the European Commission and a consortium of European companies and research institutions [5]. This receiver system contains absorber elements made of extruded silicon carbide ceramic and is known as Hitrec (high temperature receiver). The absorbers, manufactured by the Danish company Heliotech, have a rectangular honeycomb structure with parallel channels (Fig. 9). The channel width of 2 mm and wall thickness of 0.8 mm represent a compromise between material stability and optimal

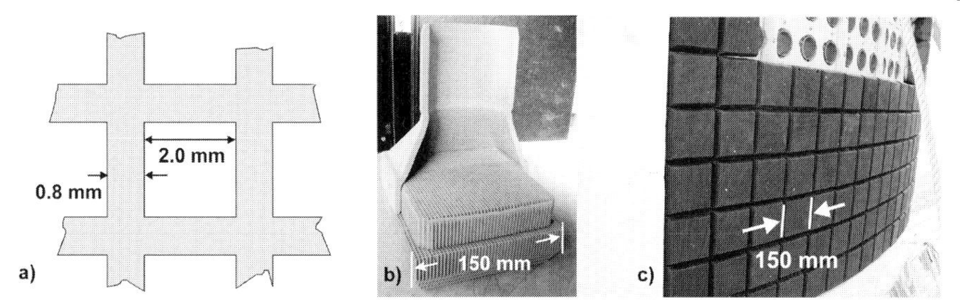

Fig. 9 a) Sketch of the channel geometry seen from the front, b) photograph of a single absorber module standing on its front side with the cup holding the parallel-channel absorber cut in half for a better view, c) photograph of the SOLAIR 3 MW receiver from the front during installation showing many installed absorber modules next to each other. In the upper part of the picture there are still some openings awaiting modules (photographs by P. Stobbe, Heliotech).

heat-transfer characteristics, which would require thinner walls. This absorber has excellent resistance to high temperatures up to 1600 °C. Air outlet temperatures of over 1000 °C have been reached.

5.7.4.2
Ceramic Foams

In general, ceramic foams are manufactured by replication of polyurethane foams. A variety of materials are available, and mechanical properties and temperature resistance are strongly dependent on sintering conditions and additives. Furthermore, several infiltration techniques allow further reduction of wall porosity. Ceramic

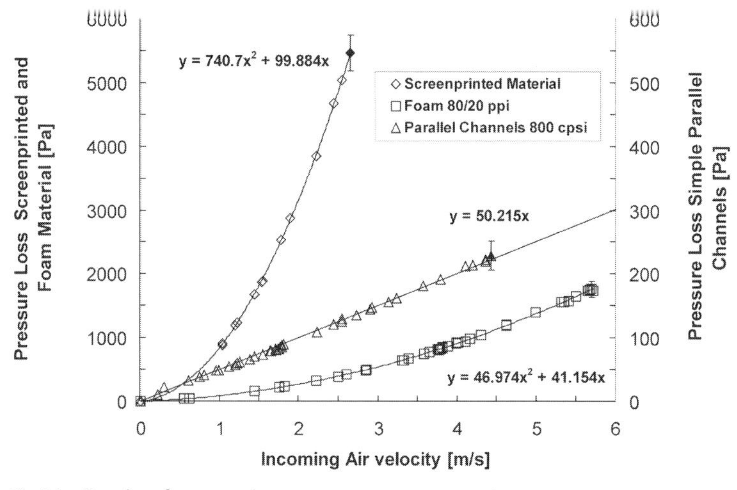

Fig. 10 Results of pressure-loss measurements comparing ceramic foam material, screen-printed material, and a simple parallel-channel catalyst support material.

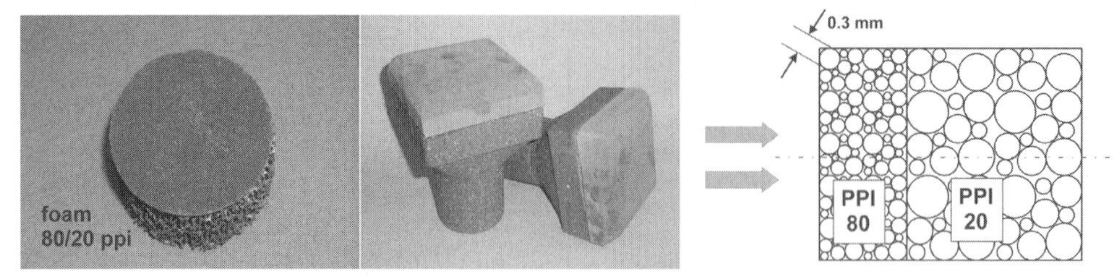

Fig. 11 Photograph and sketch of 80 ppi/20 ppi silicon carbide foam absorber samples manufactured for laboratory-scale (left, 70 mm diameter) and 200 kW (middle, ca. 125 mm, two absorbers inserted into cups) efficiency tests (photographs by P.M. Rietbrock, DLR).

foams were first tested as volumetric solar absorbers in a project investigating solar CO_2 steam reforming of methane [10, 11, 20]. Since then a number of materials have been tested on a laboratory scale [19]. Especially silicon carbide foams show excellent absorption and permeability properties. The permeability properties of various absorber materials can be seen in Fig. 10. To achieve a large specific surface area for a good solid-to-fluid heat transfer, foams with small cell dimensions must be employed. As an example, an 80 ppi foam material is shown in Fig. 11. It was tested in combination with a 20 ppi substrate, as shown on the right. The 80 ppi material serves as the absorbing and heat-transferring medium, and the 20 ppi material improves mechanical stability [21].

5.7.4.3
SiC Fiber Mesh

The SiC fiber mesh material shown in Fig. 12 originally was developed for burners. The material was tested as a solar absorber in the German project SOLPOR [22]. Samples were manufactured by Schott Glas in Mainz, Germany. The material is commercially available under the trade name Ceramat FN. It consists of silicon carbide fibers of 25 µm diameter glued together to form a layer 3.5 mm thick. The fibers are oriented in directions perpendicular to the direction of the air flow, which

Fig. 12 Silicon carbide fiber mesh material Ceramat FN manufactured by Schott Glas, Mainz, Germany (photograph by P.M. Rietbrock, DLR).

is beneficial for radial heat transport and good heat transfer properties. Due to its small cell dimensions the potential efficiency is rather high. Results of efficiency tests are presented in Section 5.7.5.

5.7.4.4
Screen-Printed Absorbers (Direct-Typing Process)

The direct-typing process is a new method to create cellular bodies with predetermined pore structures. Three-dimensional (3D) objects can be manufactured from silicon carbide by means of the direct-typing modified screen-printing method. The 3D cellular bodies are built from multiple thin layers in a way that offers many degrees of freedom for construction parameters. Further experiments by the process developer Bauer R&D suggest the possibility of integrating construction elements made of different materials within the monolithic structure.

In the above-mentioned SOLPOR project, four cylindrical cellular absorber test samples with a diameter of 100 mm and a length of 15 mm were manufactured by Bauer R&D for tests at the DLR laboratories [21]. The cell geometry of the samples was according to the drawings in Fig. 13. They consist of parallel channels with a

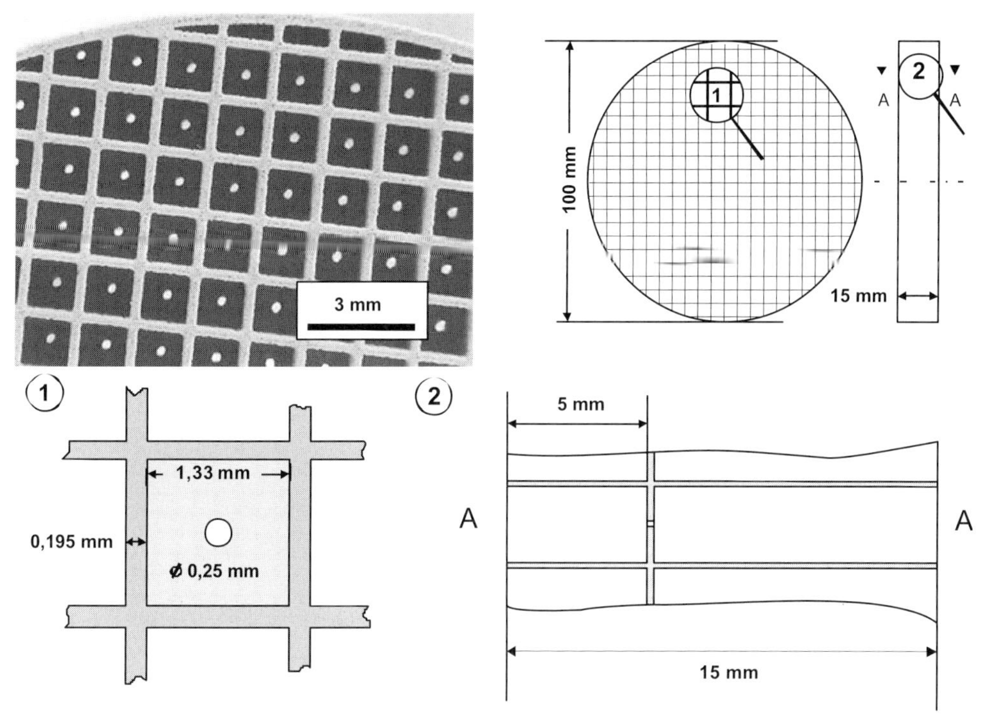

Fig. 13 Cellular silicon carbide absorber sample manufactured by Bauer R&D. Top left: photograph of a front detail; Top right: drawing of the absorber sample (top and lateral view); Bottom left: detailed drawing of a single channel (front view); Bottom right: drawing in section of a single channel (lateral view). Photograph by P.M. Rietbrock, DLR.

square cross section and an open cell width of 1330 μm. The wall thickness is 195 μm. In contrast to extruded cellular monoliths and typical catalyst supports, each channel has a plane of 270 μm thickness arranged perpendicular to the channel direction. Each plane has a small hole of 150 μm diameter which serves as an orifice in each individual channel. This additional plane changes the fluid flow from pure Darcy flow to a more turbulent flow, which is an important prerequisite for an application as a volumetric absorber.

5.7.4.5
Material Combinations

A front material having high specific surface area, excellent absorption, and high porosity to achieve volumetric absorption of the concentrated solar radiation (Schott Ceramat FN, see Section 5.7.4.3) has been combined with a material having beneficial thermal conductivity properties and a quadratic pressure loss characteristic (SiC catalyst support, see Section 5.7.4.1). This combination is called advanced Hitrec. For tests in concentrated solar radiation, the two materials have been glued into a ceramic tube without directly connecting them. Figure 14 shows the combined absorber element.

Fig. 14 Absorber element made out of an SiC fiber mesh and a SiC catalyst support (advanced Hitrec; photograph by P.M. Rietbrock, DLR).

5.7.5
Absorber Tests

Efficiency and performance tests can be carried out at test sites at various research centers, from laboratory-scale dimensions up to large-scale tests of 3 MW and more [23, 24]. Generally, an installation capable of generating concentrated solar radiation is needed. Two of a number of important installations of this kind in Europe are Cesa 1 and the SSPS tower at the Plataforma Solar de Almería (PSA) in Spain with total powers of 3000 and 200 kW, respectively, and the Solar Furnace at the German Aerospace Centre (DLR) in Germany with a total power of 25 kW [25].

In general, the efficiency η of a solar absorber is defined as the useful power of the heat and the reaction enthalpy of the fluid mass flow generated by the absorber P_{use}, divided by the power of the concentrated radiation penetrating through the aperture of the absorber (power on aperture, POA)

$$\eta = \frac{P_{use}}{POA}. \tag{20}$$

To assess an absorber it is useful to also measure the temperature distribution of the front surface of the absorber, so that any hot spots or other irregularities can be detected.

In the case of an open-air receiver there is no chemical reaction, and heat is the only useful power. The general principle of measurement is shown in Fig. 15 along with a photograph of the top of the SSPS tower, where a 200 kW receiver test bed is installed. A fan forces ambient air to flow through the sample followed by an air/water heat exchanger. The following quantities are monitored during the test: direct normal incidence (DNI, W m^{-2}) of the solar radiation, the material front (peak) temperature, the air temperature at the outlet of the absorber, the air temperature at the outlet of the heat exchanger, the total air mass flow, the water mass flow of the heat exchanger, and the water temperatures at the heat exchanger inlet and outlet. From these values the power of the air flow and the power of the water flow at the outlet of the heat exchanger can be calculated. In addition to these continually measured data, periodic measurements are made. First, the solar flux density penetrating through the aperture area (POA) is measured. This aperture area corresponds to the diameter of the absorber sample. Second, the temperature distribution on the absorber front is measured. These two measurements are carried out several times during the experiment. At these times efficiency is calculated by (power water + power air)/POA with the POA value at this time and averages of the other quantities. The POA values have to be taken as representative for the time of averaging. Optical flux measurements are usually performed with a camera/target method, and the temperature distribution is measured with an infrared camera. One camera/target method is described in more detail by Neumann and Groer [25].

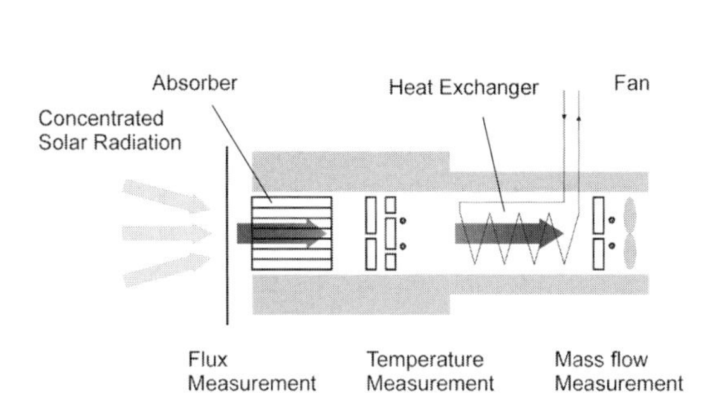

Fig. 15 Test bed used for measurement of open volumetric absorber efficiency in concentrated solar radiation (photograph by P.M. Rietbrock, DLR).

As examples from a wide variety of materials tested [21, 22], results of efficiency measurements on four cellular ceramic materials are briefly presented here. These tests were conducted under solar radiation fluxes of around 1 MW m^{-2}. With a 20 ppi silicon carbide foam, efficiencies of 80 % were achieved at air outlet temperatures of 700 °C (Fig. 16). As mentioned above, smaller cell dimensions lead to improved heat transfer in the extinction volume. Thus, an 80 ppi silicon carbide foam was used as a 2 mm outer layer on a body of a 20 ppi silicon carbide foam material. The heat transfer surface area A_V available is 5400 m^2 m^{-3} for the 80 ppi foam as opposed to 1100 m^2 m^{-3} for the 20 ppi foam. Consequently, the efficiency could be increased to values of more than 90 %. Because of its lower strength properties, the 80 ppi layer only serves as a functional material providing absorption and solid-to-fluid heat transfer.

Two further examples are shown in Fig. 17. An extruded silicon carbide catalyst carrier (Hitrec) consisting of parallel channels of 2 mm width and advanced Hitrec (see Section 5.7.4.5) with a front surface made of a fiber mesh material consisting of silicon carbide fibers of 25 μm diameter. Again the very different cell dimensions lead to differing values of the specific surface area A_V of 1000 m^2 m^{-3} (Hitrec) and 8000 m^2 m^{-3} (fiber mesh). Consequently, advanced Hitrec shows improved efficiency values of up to 95 % compared to 80–85 % for the Hitrec material. However, operating the fiber mesh material and the 80 ppi foam at high temperatures causes oxidation problems, which become more severe with decreasing cell dimensions. The SiC catalyst support material has been additionally proven on a 3 MW scale in the European project SOLAIR [5].

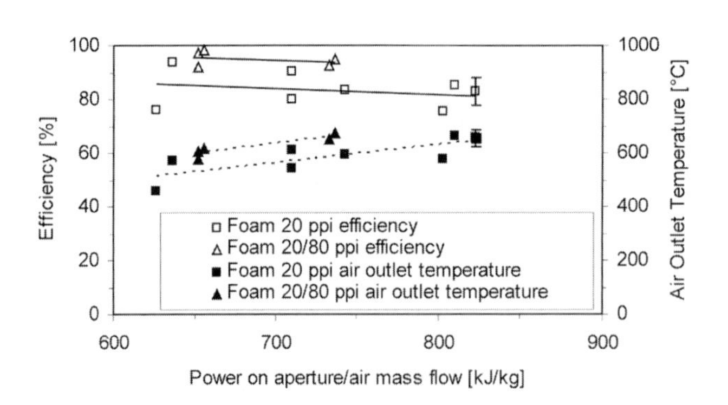

Fig. 16 Results of the efficiency measurements on a sandwich-foam material (80 ppi/20 ppi) compared to a simple foam material (20 ppi). Photographs by P.M. Rietbrock, DLR.

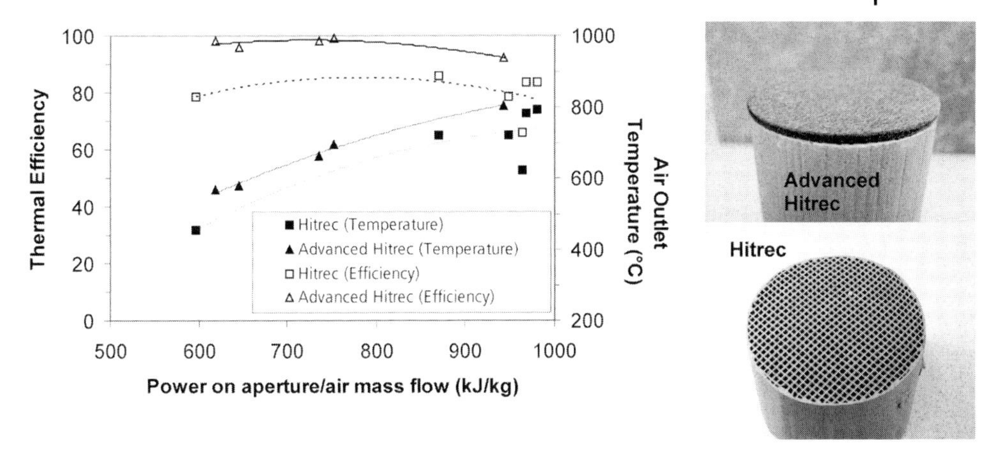

Fig. 17 Efficiency properties of a silicon carbide catalyst carrier material (Hitrec) compared to advanced Hitrec (photographs by P.M. Rietbrock, DLR).

5.7.6 Physical Restrictions of Volumetric Absorbers and Flow Phenomena in cellular ceramics

Much theoretical and experimental research has been conducted on the properties and physical restrictions of the open volumetric absorber principle [18, 24, 26, 27]. Several independent experiments in the field of solar applications concerning the flow through porous materials have indicated a relationship between absorber temperature and resistance to flow [10, 13]. Local high solar flux leads to a lower mass flow and high material temperatures. Local low solar flux leads to a high mass flow with a low material temperature (Fig. 18). This means that the local absorber temperature can exceed the upper operating temperature of the material and lead to its destruction although the average air outlet temperature is low. The main cause of this behavior is the temperature-dependent increase in the viscosity of the fluid.

Several theoretical approaches [27] and numerical simulations [13, 18, 19] lead to a fairly good agreement between calculations and experiments, and general tendencies could be shown. The most important influence on flow stability comes from the pressure loss characteristic of the porous media. A linear dependency of the pressure drop on the flow velocity (Darcy flow) can lead to instabilities; a purely quadratic dependency (Dupuit, Forchheimer) can not. The problem is that in solar applications usually there is a mixture of both linear and quadratic behavior. In the following it is shown with a simple model under which conditions instabilities can occur in flow through porous media.

In flow through a porous structure the mass flow density is determined by the pressure difference between the two sides of the structure. In extended structures with large void spaces on both sides of the structure, the pressure difference is nearly the same across the whole structure, because there is instant pressure equal-

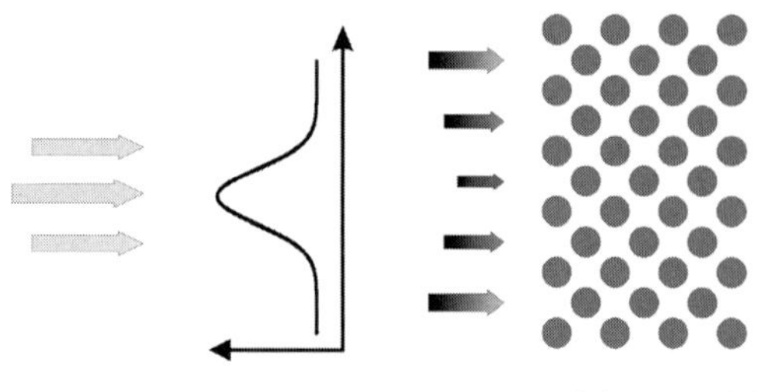

Fig. 18 Instabilities in the flow through porous materials used in solar applications.

ization on both sides. Instability occurs when a certain pressure drop can cause different mass flow densities. In the following a one-dimensional model described by Kribus [27] is extended to find general conditions for instabilities in volumetric absorbers.

The pressure drop across the sample can be described by Darcy's law in the Forchheimer extension:

$$-\frac{dp}{dx} = \frac{\mu_{dyn}}{K_1} U_0 + \frac{\rho_F}{K_2} U_0^2 \tag{21}$$

where p denotes the pressure, x the coordinate in the direction of the flow, K_1 and K_2 the viscosity and the inertial coefficient, μ_{dyn} the dynamic viscosity, ρ_F the density of the fluid and U_0 the velocity of the fluid.

If air is used as fluid, the ideal gas law can be taken as valid over a wide temperature range:

$$p = \rho RT \tag{22}$$

with the specific gas constant $R = 287\,\mathrm{J\,kg^{-1}\,K^{-1}}$ and the air temperature T. The dependence of the dynamic viscosity on air temperature at high temperatures can be approximated by Eq. (23) [27]. The same result can be obtained by numerically fitting literature data, as given in Ref. [28].

$$\mu_{dyn}(T) = \mu_{dyn}(T_0)\left(\frac{T}{T_0}\right)^{0.7}. \tag{23}$$

The mass flow density can be expressed as

$$\dot{m}_A = \rho U_0. \tag{24}$$

Inserting this into the above pressure drop equation gives

$$-pdp = \left(\frac{\mu_{dyn}(T_0)}{K_1 T_0^{0.7}} T^{0.7}\, \dot{m}_A RT + \frac{RT\dot{m}_A^2}{K_2}\right) dx. \tag{25}$$

This equation can be integrated by assuming a deep absorber for which the length is much larger than the hydraulic diameter of the structure and the temperature distribution along its length can be well approximated by the outlet air temperature. In practice the temperature rise of the fluid from inlet to outlet takes place in the first few millimeters, whereas absorber thicknesses are on the order of several centimeters for reasons of stability. This integration gives

$$\frac{p_0^2 - p_{out}^2}{2} = \left(R \frac{\mu_{dyn}(T_0)}{K_1 T_0^{0.7}} T_{out}^{1.7} \dot{m}_A + \frac{R T_{out} \dot{m}_A^2}{K_2} \right) L \tag{26}$$

where L denotes the length of the absorber, and T_0 and T_{out} are the inlet and outlet temperatures of the air, respectively. Now consider an absorber and its energy balance neglecting radial heat transfer

$$I_0 = \dot{m}_A C_{PF} (T_{out} - T_0) + \beta \sigma T_{out}^4 \tag{27}$$

where I_0 is the solar radiation flux incident on the surface of the absorber, C_{PF} the specific heat capacity of the air, σ the Stefan–Boltzmann constant, and β a correction factor which describes the thermal losses through radiation. If β is smaller than 1, then the surface temperature of the outside of the absorber is lower than the air outlet temperature ("volumetric operation") and the radiation losses are smaller. This is the case when the solar radiation can enter the inside of the porous structure and the incoming air cools the front surface well. If $\beta > 1$ the conditions are similar to those in a pipe absorber. The surface temperature is higher than the air outlet temperature ("nonvolumetric operation").

Rearranging the energy balance for the mass flow density and inserting it into the integrated pressure drop equation gives:

$$\frac{p_0^2 - p_{out}^2}{2} = RL \frac{\mu_{dyn}(T_0)}{K_1 T_0^{0.7}} T_{out}^{1.7} \frac{I_0 - \beta \sigma T_{out}^4}{C_{PF}(T_{out} - T_0)} + \frac{RLT_{out}}{K_2} \left(\frac{I_0 - \beta \sigma T_{out}^4}{C_{PF}(T_{out} - T_0)} \right)^2 . \tag{28}$$

For a better understanding the left-hand side can be seen as the product of the pressure drop and the mean pressure $\Delta p \cdot p_{mean}$. In the case of a structure having a linear pressure drop relation, that is, $K_2 = \infty$, the above equation becomes much simpler. A plot of the quadratic pressure difference versus temperature shows what happens in the case of instability (Fig. 19). If one or more temperatures and the corresponding mass flows are possible for the same pressure drop, there is an instability. In the graph lines of constant pressure drop are drawn, which intersect the curves with high solar flux at three points. Thus, for the same pressure drop different temperatures are possible. Parts of the absorber can have a low mass flow through them, and others a high mass flow. The low mass flow can lead to local overheating and thus to the destruction of the absorber.

Figure 19 shows the quadratic pressure drop as a function of the temperature for different solar fluxes and a material with purely linear pressure drop characteristic ($K_2 = \infty$). One can see that instability only exists above a certain solar flux.

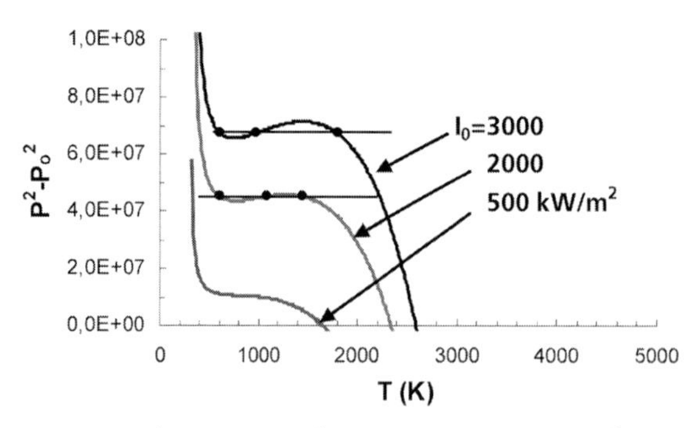

Fig. 19 Quadratic pressure drop versus air temperature for different solar fluxes.

Looking at instability mathematically an ambiguity can only occur if the curve of the pressure drop shows a zero point in the derivative of the outlet temperature which lies in the physically relevant range between the maximum of the inlet temperature T_0 and the maximum temperature T_{max}, which is reached at zero mass flow:

$$T_{max} = \sqrt[4]{\frac{I_0}{\sigma}}. \tag{29}$$

This zero point of the derivative of the pressure difference depends only on T_{out}, T_0, β, σ, and I_0. Setting the derivative to zero gives the following equation:

$$0 = 0.7 I_0 T_{out} + 5.7\beta\sigma T_0 T_{out}^4 - 4.7\sigma T_{out}^5 - 1.7 T_0 I_0. \tag{30}$$

This equation can only be satisfied if the solar flux is above a critical value; below this value, there is no solution. The value can be obtained by solving the equation for I_0 and finding the minimum of the obtained expression. The minimum occurs at $T_{out} = 2.95\ T_0$. The critical flux above which instabilities can occur is:

$$I_{0,crit} = 1694\beta\sigma T_0^4. \tag{31}$$

For example for $\beta = 1$ and $T_0 = 300$ K this means a critical flux of 778 kW m^{-2}.

This calculation is only valid for a porous medium with a purely linear pressure drop characteristic, that is, $K_2 = \infty$. From a physical point of view the introduction of a quadratic term in the pressure drop equation considers the forming force of the porous medium on the fluid flow, while in the purely linear equation of Hagen–Poiseuille only friction forces are taken into account. Looking at the curves of the quadratic pressure drop difference for the case of $K_2 \neq \infty$, the trend for the case of constant solar flux can be seen in Fig. 20. A change in the pressure-drop characteristic of the absorber has a significant influence on the curves. The lower the value of K_2 the less probable instabilities become. The values of K_1 and K_2 for the curves shown in the graph are of realistic magnitude. For example, the SiC parallel-channel catalyst support with a cell density of about 80 cpsi has $K_1 = 10^{-7}$ m^2 and $K_2 = 0.011$ m.

For foam ceramics the values of K_2 are even lower and can be less than 10^{-4} m. This means in practice that when a ceramic foam is used as a volumetric air absorber at a solar flux of 2000 suns instabilities are not expected. However, if other fluids are used, instabilities can occur at significantly lower flux densities. This is valid, for example, for solar methane-reforming receivers.

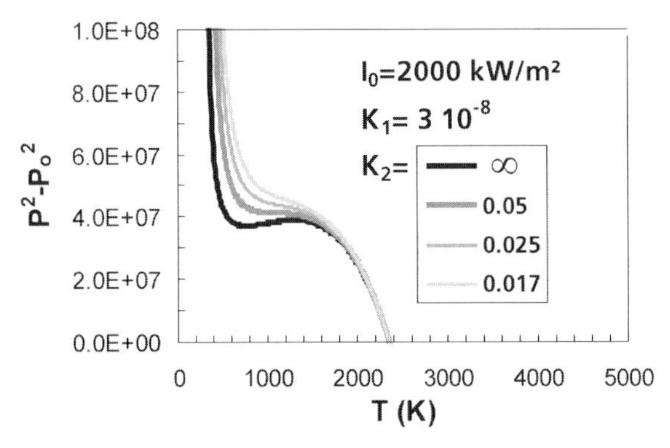

Fig. 20 Quadratic pressure drop difference as a function of the air outlet temperature for several values of the inertial coefficient.

The mathematical description of the more general case is shown in the following. Setting the derivative of the pressure drop difference to zero results in the equation:

$$\frac{K_1}{K_2} = \frac{c_{pr}\mu_{dyn}(T_0)}{T_0^{0.7}} \frac{T_{out}^{0.7}(T_{out} - T_0)\left(I_0(0.7T_{out} - 1.7T_0) + \beta\sigma T_{out}^4(-4.7T_{out} + 5.7T_0)\right)}{I_0^2(T_{out} + T_0) + I_0\sigma T_{out}^4(6T_{out} - 10T_0) + (\sigma T_{out}^4)^2(-7T_{out} + 9T_0)} \equiv f(I_0, T_{out}). \tag{32}$$

If the ratio of K_1 and K_2 is greater than $f(I_0, T_{out})$ there is no ambiguity in the pressure drop for different temperatures. The maximum was also found in the usual way by solving the system of equations by setting the partial equations to zero. The maximum of $f(I_0, T_{out})$ is found for $\beta = 1$ at $T_{out} = 1342$ K and $I_0 = 4.167$ MW m^{-2}. This gives a critical ratio of K_1 and K_2 at:

$$\frac{K_1}{K_2} = 1.94 \cdot 10^{-6}\,\text{m}. \tag{33}$$

Considering the given boundary conditions of a certain setup it is possible to find suitable material parameters which give stable operation of the solar absorber.

In the above-discussed simple model used to study the systematic dependency on the parameters under which instabilities occur, not all parameters are included. A further important parameter of flow stability is thermal conduction of the absorber. In the case of an inhomogeneous temperature distribution, heat flow can level out local hot regions. The better the thermal conduction in the absorber is, the less possible instabilities become. Nevertheless, even without instabilities a positive feedback between solar radiation and mass flow distribution is observed. This means that in locations with increased solar irradiation the local mass flow density is reduced.

5.7.6.1
Experimental Determination of Nonstable Flow

Flow instability can be investigated in a transient experiment, sketched in Fig. 21. Samples of cylindrical shape are heated to 800 °C in a tube heater at zero fluid velocity U_0. At a defined time t_1, U_0 at the entrance of the unblocked section is set to $U_0 \approx 1$ m s^{-1}. At a time $t_2 \approx t_1 + 0.5$ s fluid velocity in the blocked section is also set to $U_0 \approx 1$ m s^{-1}. Experimentally, this time shift is realized by a blocking mechanism which covers an area in the lower half for 0.5 s. By monitoring the front surface of the sample with an infrared camera, the temperature distribution during the cooling phase can be recorded.

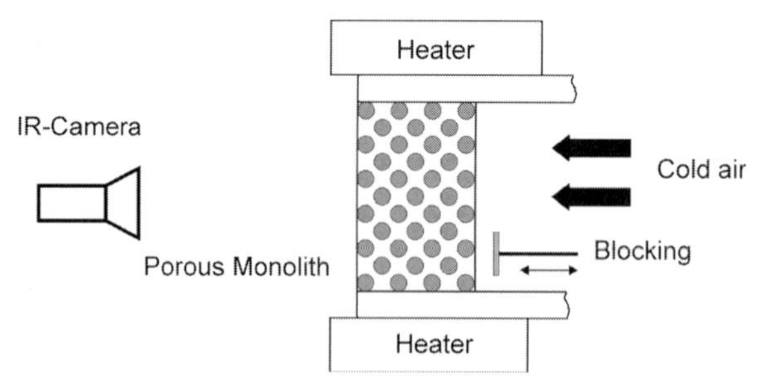

Fig. 21 Experimental setup for investigating the front-temperature distribution of a heated porous monolith cooled with ambient air.

As an example, the homogeneity of the temperature distribution during the cooling phase of a porous material through which cold air is flowing is compared for two materials: a cordierite ceramic foam (20 ppi) and a cordierite catalyst support (400 cpsi). Though having similar bulk thermophysical properties, the geometric structure of the material leads to different pressure-drop characteristics and thus to completely different flow properties.

Figure 22 shows the results of such an experiment. Slight temperature inhomogeneities observed at the front of the foam are rapidly compensated, whereas the catalyst support shows a permanent inhomogeneous temperature distribution. The evolving temperature distribution not only depends on the initial blocking but also on inhomogeneities in the initial temperature before the onset of fluid flow.

In conclusion it can be said that instabilities are a problem for volumetric absorbers, but they can be overcome by choosing an appropriate design. Instabilities are more likely to occur at higher concentrations, but this is no relief as volumetric

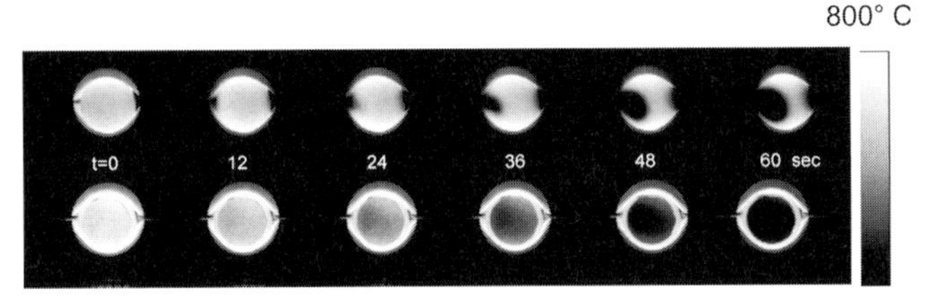

Fig. 22 Cooling behavior of air-cooled preheated porous materials: a cordierite catalyst supported (top) and a cordierite foam (bottom).

absorbers are designed for high solar concentrations. Care should be taken to choose a design of an absorber in which the pressure drop characteristic has a strongly quadratic behavior and radial heat exchange is high. In designing an volumetric receiver the feedback between temperature and mass flow must be taken into account.

5.7.7 Summary

Volumetric absorbers are the absorbers of choice for high-temperature applications utilizing highly concentrated solar radiation. As ceramics offer high temperature resistance they play a major role in the design of volumetric absorbers. Many different designs of porous structures can be manufactured from ceramics. When the ceramic is coated with a catalyst it can be used as a chemical absorber/reactor and thus allows thermochemical storage of solar energy. Many successful tests have shown the feasibility of ceramic absorbers in high-temperature solar applications. By considering the different physical properties needed for high efficiency together with construction parameters to avoid unstable flow, it is possible to design highly efficient absorbers with efficiencies above 80 %.

References

1 Fricker, H., *Bull. SEV/VSE*, 1985, *76*, 10–16.
2 Winter, C.J., Sizmann, R.L., Vant-Hull, L.L. (Eds.), *Solar Power Plants*, Springer-Verlag, Berlin, 1991.
3 Meinecke, W., Bohn, M., Becker, M., Gupta, B. (Eds), *Solar Energy Concentrating Systems*, C.F. Miller Verlag, Heidelberg, 1994, pp. 18–19, 68.
4 Chavez, J.M., Kolb, G.J., Meinecke, W., *Second Generation Central Receiver Technologies – A Status Report*, Becker, M., Klimas, P.C. (Eds.), Verlag C.F. Müller, Karlsruhe, 1993.
5 Hoffschmidt, B., Dibowski, G., Beuter, M., Fernandez, V., Téllez, F., Stobbe, P., *Test Results of a 3 MW Solar Open Volumetric Receiver*, Proceedings of the ISES Solar World Congress 2003 "Solar Energy for a Sustainable Future", June 14–19 2003, Göteborg, Sweden.

6 Buck, R., Bräuning, T., Denk, T., Pfänder, M., Schwarzbözl, P., Tellez, F. *J. Sol. Energy Eng.* 2002, *124* (1), 2–9.
7 Romero, M., Buck, R. , Pacheco, J.E.. *J. Sol. Energy Eng.* 2002, *124* (2), 98–108.
8 Kribus, A., Zaibel, R., Carrey, D., Segal, A., Karni, J., *Sol. Energy* 1997, *62*, 121–129.
9 Wörner, A., Tamme, R., *Catal. Today*, 1998, *46*, 2–3, 165–174.
10 Muir, J.F., Hogan, R.E., Skocypec, R.D., Buck, R., *Sol. Energy* 1994, *52*, 467–477.
11 Abele, M., Bauer, H., Buck, R., Tamme, R., Wörner, A., Design and Test Results of a Receiver-Reactor for Solar Methane Reforming in *Solar Engineering*, Davidson, J.H., Chavez, J. (Eds.), American Society of Mechanical Engineers, New York, 1996, pp 339–346.
12 Buck, R., Abele, M., Bauer, H., Seitz, A., Tamme, R., Development of a Volumetric Receiver-Reactor for Solar Methane Reforming in *Solar Engineering*, Klett, D.E., Hogan, R.E., Tanaka, T. (Eds.), American Society of Mechanical Engineers, New York, 1994, pp 73–78.
13 Buck, R., *Massenstrom-Instabilitäten bei volumetrischen Receiver-Reaktoren*, Fortschrittsberichte VDI Reihe 3 Nr. 648, VDI Verlag, Düsseldorf, 2000.
14 Cordes, S., Fiebig, M., Pitz-Paal, R., First Experimental Results from the Test of a Selective Volumetric Air Receiver, 6th International Symposium on Solar Thermal Concentrating Technologies 1992, Mojacar, Spain.
15 ASTM Standard E903-82 (1993) Standard Test Method for Solar Absorptance, Reflectance, and Transmittance of Materials Using Integrating Spheres in *Annual Book of ASTM Standards 1993*, Vol. 12.02, American Society for Testing and Materials, Philadelphia, PA., pp. 512–520.
16 ASTM Standard E891-87 (1992) Standard Tables for Terrestrial Direct Normal Solar Spectral Irradiance for Air Mass 1.5 in *Annual Book of ASTM Standards 1993*, Vol. 12.02, American Society for Testing and Materials, Philadelphia, PA., pp. 481–486.
17 Adler, J., unpublished data of the Fraunhofer Institute of Ceramic Technologies and Sintered Materials, Winterbergstraße 28, 01277 Dresden, Germany, joerg.adler@ikts.fhg.de.
18 Hoffschmidt, B., Vergleichende Bewertung verschiedener Konzepte volumetrischer Strahlungsempfänger, Deutsches Zentrum für Luft- und Raumfahrt Köln, 1997.
19 Pitz-Paal, R., Hoffschmidt, B., Böhmer, M., Becker, M., *Sol. Energy* 1997, *60*, 135–150.
20 Möller, S., Buck, R., Tamme, R., Epstein, M., Liebermann, D., Meri, M., Fisher, U., Rotstein, A., Sugarmen, C., Solar Production of Syngas for Electricity Generation: SOLASYS Project Test-Phase, Proc. 11th SolarPACES Int. Symposium on Concentrated Solar Power and Chemical Energy Technologies, September 4–6, 2002, Zurich, Switzerland.
21 Fend, Th., Reutter, O., Pitz-Paal, R., Hoffschmidt, B., Bauer, J., *Sol. Energy Mater. Sol. Cells* 2004, *84* (1/4), 291–304.
22 Fend, Th., Hoffschmidt, B., Pitz-Paal, R., Reutter, O., Rietbrock, P., *Energy* 2004, 29 (5/6), 823–833.
23 Haeger, M., *Phoebus Technology Program: Solar Air Receiver (TSA)*, PSA Tech. Report: PSA-TR 02/94, July 1994.
24 Palero, S., Romero, M., Estrada, C.A., Experimental Investigation in Small Scale Volumetric Solar Receivers to Study the Mass Flow Instability and its Comparison to Theoretical Models, Proceedings of the ISES Solar World Congress 2003 "Solar Energy for a Sustainable Future", June 14–19, 2003, Göteborg, Sweden
25 Neumann, A., Groer, U., *Sol. Energy* 1996, *58* (4–6), 181–190.
26 Garcia-Casal, X., Ajona, J.I., *Sol. Energy* 1999, 67 (4–6), 265–268.
27 Kribus, A, Ries, H, Spirkl, W., *J. Sol. Energy Eng.* 1996, *118*, 151–155.
28 Fricke, J., Borst, L., *Energie*, R. Oldenbourg Verlag, München, 1984, p. 243.

5.8 Biomedical Applications: Tissue Engineering

Julian R. Jones and Aldo R. Boccaccini

5.8.1 Introduction

Life expectancy is increasing as healthcare and technology improve, but not all body parts can maintain their function with the ageing process. Bone and cartilage are needed to support the ageing body even though the cells that produce them become less active with age, while the heart, kidneys, and liver must operate for much longer than ever before.

The most common bone disease is osteoporosis, which causes a loss of bone density and affects everyone as they age. Bone is a natural composite of collagen (polymer) and bone mineral (ceramic). Collagen is a triple helix of protein chains, which has high tensile and flexural strength and provides a framework for the bone. Bone mineral is a crystalline calcium phosphate ceramic [hydroxyapaptite, $Ca_{10}(PO_4)_6(OH)_2$] that contributes the stiffness and compressive strength of bone. Cortical bone is a dense structure with high mechanical strength and is also known as compact bone. Trabecular bone is a network of struts (trabeculae) enclosing large voids (macropores) with 55–70 % interconnected porosity, which is a supporting structure [1]. The struts of the trabecular bone, which are present in the ends of long bones such as the femur or within the confines of the cortical bone in short bones, are most affected by osteoporosis. The loss of bone density is caused by a reduction in number of osteogenic cells (osteoblasts) with age. The only treatment for severe cases of osteoporotic joints is total joint replacement.

Surgical procedures for the repair of tissue lost as a result of trauma or by the excision of diseased or cancerous tissue involve graft implants (transplants). Grafts can be taken from a donor site in the same patient (autograft), from another human donor (homograft), or from other living or nonliving species (xenografts). Grafts can be used to attain the ultimate goal of restoration of a tissue to its original state and function, but there are many limitations [2]. Autografts and homografts are restricted by limited material availability and complicated multistage surgery to the detriment of the harvest site. Homografts and xenografts run the risk of disease transmission. Therefore, there is a great demand for synthetic substitutes specially designed and manufactured to act as a scaffold for the regeneration of tissues to their natural state and function, which constitutes the fundamentals of the tissue engineering discipline [3].

Cellular Ceramics: Structure, Manufacturing, Properties and Applications.
Michael Scheffler, Paolo Colombo (Eds.)

ISBN: 3-527-31320-6

This chapter reviews how cellular ceramics are being considered in bone reconstruction and other orthopaedic applications with focus on the development of highly porous scaffolds for bone tissue regeneration. Where appropriate, reference to the application of cellular ceramics in other areas of tissue engineering and implant technology are made. The rest of the chapter is organized in the following manner: Section 5.8.2 includes a general consideration of the new scientific field of regenerative medicine and the opportunities for biomaterials, Section 5.8.3 covers the field of bioceramics, that is, ceramics used in medical applications in general, and presents the classification of bioceramics accepted in the current literature, Section 5.8.4 covers the field of porous scaffolds for tissue engineering, while Section 5.8.5 is devoted specifically and extensively to cellular structures of hydroxyapatite, calcium phosphate, and bioactive glass. Reference to other bioceramics used in cellular form in tissue engineering applications is also made in Section 5.8.5. Finally, Section 5.8.6 contains a summary of the chapter, identifying progresses made and needs for future developments, particularly in the area of bioactive porous scaffolds.

5.8.2 Regenerative Medicine and Biomaterials

Regenerative medicine is a general term used to describe techniques that are being developed to regenerate diseased or damaged tissues to their original state and function. Tissue engineering and tissue regeneration are two branches of this broad field. In some tissue engineering applications synthetic or natural porous scaffolds are used as templates for tissue growth [1, 3]. There is the potential for stem cells to be extracted from a patient and seeded on a (natural or synthetic) scaffold of the desired architecture in vitro, where they will be given the biological signals to proliferate and differentiate, and new tissue will grow, ready for implantation [4, 5]. Tissue regeneration techniques involve the direct implantation of an engineered scaffold (seeded with cells or not) into a defect to guide and stimulate tissue repair in situ. In both cases the scaffold should be resorbed (dissolve) as the tissue grows, leaving no trace of damage or surgery [3, 6]. In any application of natural or artificial scaffolds, choice and design of the material are important.

Any material that is implanted into the body should be biocompatible, that is, noncytotoxic (not toxic to cells). There are three classes of noncytotoxic materials: bioinert, bioactive, and bioresorbable [7].

No material is completely inert on implantation, but the only response to the implantation of *bioinert materials* is encapsulation of the implant by fibrous tissue (scar tissue). Examples of bioinert materials are medical-grade alumina, zirconia, stainless steels, and high-density polyethylene, materials that are used, for example, in total hip replacements.

Millions of orthopaedic prostheses made of bioinert materials have been implanted, an example of which is the Charnley total hip replacement, which is heralded as one of the most successful surgical inventions [7, 8]. However, long-term monitoring of 20 000 Charnley joints revealed that after 25 years of implantation

24 % required revision surgery [8]. The most common reason for failure was aseptic loosening of the femoral stem, where bone resorption occurred due to a mismatch in the Young's modulus of bone and the metal stem (stress shielding) [9].

Resorbable materials are those that dissolve in contact with body fluids and whose dissolution products can be secreted via the kidneys. The most common biomedical resorbable materials are polymers that degrade by chain scission such as poly(glycolic acid), PGA, poly(lactic acid), PLA, and their copolymers (PLGA). Some bioceramics are also resorbable in vivo, such as calcium phosphates [7], as discussed below.

Bioactive materials stimulate a biological response from the body such as bonding to living tissue [10]. There are two classes of bioactive materials intended for bone reconstruction and orthopaedic implants: class B bioactive materials bond to hard tissue (bone) and stimulate bone growth along the surface of the bioactive material (osteoconduction). Examples of class B bioactive materials are synthetic hydroxyapatite and tricalcium phosphate ceramics.

Class A bioactive materials not only bond to bone and are osteoconductive but they are also osteoproductive, that is, they stimulate the growth of new bone on the material away from the bone/implant interface and can bond to soft tissue such as gingiva (gum) and cartilage [10]. Examples of class A bioactive materials are bioactive glasses.

The mechanism of bone bonding to bioactive materials is thought to be due to the formation of a hydroxyapatite (HA) layer on the surface of the materials after immersion in body fluid. This layer is similar to the apatite layer in bone and therefore a strong bond can form. The layer forms quickest on the class A bioactive materials [10].

There is a wealth of specialized literature on bioceramics; some relevant books and review articles about the general use of porous bioceramics in biomedical applications are those in Refs. [7, 10–13].

5.8.3
Bioactive Ceramics for Tissue Engineering

Ceramic materials used for medical implants as well as in the repair and reconstruction of diseased or damaged parts of the muscoskeletal system and of other tissues, including dental applications, are called bioceramics [7, 11]. Many ceramic compositions have been tested for use in the body, but only a few have achieved human clinical application to date. Depending on the type of response in the body, bioceramics can be broadly classified as bioinert, bioactive, and resorbable, following the classification given above. Three factors influence the choice of ceramics and glasses as biomaterials: 1) physical and mechanical properties, 2) degradation of the material in the body, and 3) biocompatibility.

Bioactive ceramics are the bioceramics of choice for applications in regenerative medicine strategies and are the focus of the present chapter. The most widely used bioactive bioceramics are briefly described in this section, while the specific development of bioactive ceramics of cellular structure and their applications in bone tissue engineering and regeneration are considered in detail in subsequent sections.

Hydroxyapatite [HA, $Ca_{10}(PO_4)_6(OH)_2$] is the main mineral constituent of teeth and bones. *Synthetic hydroxyapatite* has been used in several clinical applications such as a filler for bone defects, bone spacers and plates, bone-graft substitutes, and as a coating of the metal femoral stem in total hip replacement [11, 13–15]. The aim of the HA coating is to improve the interface between metal and bone [9]. Due to its excellent biocompatibility and bioactivity, HA has attracted major interest and research efforts in the last 20 years and, as discussed in detail below, is one of the most extensively considered materials for production of porous structures for bone tissue regeneration and tissue engineering scaffolds.

Tricalcium phosphate (e.g., β-TCP, $Ca_3(PO_4)_2$, with Ca/P = 1.5) is an osteoconductive material that is also resorbable in the body. β-TCP is usually used in conjunction with synthetic HA to improve the resorbability of HA in applications such as the filling of bone defects left by cysts, sinus-floor augmentation, and bone cements. These bioresorbable ceramics became commercially available in the early 1980s for medical and dental applications [16]. *Biphasic calcium phosphate* consists of a mixture of HA and β-TCP. The bioactivity of this material can be controlled by manipulating the HA/β-TCP ratio. In this group of bioceramics, also calcium phosphate cements (CPCs) should be included, a term which encompasses a wide variety of formulations [17, 18]. Unlike sintered HA, CPCs can be actively remodelled in vivo [19].

Selected compositions of silicate glasses (e.g., 45S5 Bioglass) and glass ceramics (e.g., apatite-wollastonite) as well as some calcium phosphate glasses are *bioactive glasses* [7, 10]. The first bioactive silicate glass composition (46.1 SiO_2, 24.4 Na_2O, 26.9 CaO, 2.6 P_2O_5, in mol %), was reported in 1971 by Hench [20]. This bioactive glass, known as 45S5 Bioglass, is now used in the clinic as a treatment for periodontal disease (Perioglas) and as a bone filling material (Novabone) [10]. Bioglass implants have also been used to replace damaged middle ear bones, restoring hearing to thousands of patients [7, 10].

After immersion in body fluid bioactive glasses undergo a dissolution process that is instrumental in the formation of an apatite layer, which in the case of bioactive glasses is a carbonated hydroxyapatite layer (HCA). This not only means the glasses are resorbable in the body, but recent findings have shown that the dissolution products of bioactive glasses up-regulate seven families of genes that regulate osteogenesis and the production of growth factors [21, 22]. These findings may provide the reasons why certain compositions of bioactive glasses are specially effective in promoting bone growth and are class A bioactive materials (see Section 5.8.2) [10]. As discussed in detail below, cellular solids made from bioactive glasses (glass foams) are being increasingly considered as scaffolds in different tissue engineering strategies.

5.8.4
Scaffold Biomaterials for Tissue Engineering

Essentially, the above-mentioned materials and combinations thereof are the biomaterials available for scaffold design for tissue engineering applications. An ideal scaffold is one that mimics the extracellular matrix of the host tissue so that it can act as a template in three dimensions on which cells attach, multiply, migrate, and function. The criteria for an ideal scaffold for bone regeneration can be summarized as follows [3]:

- It acts as template for tissue growth in three dimensions, that is, it has an interconnected pore network with pores diameters in excess of about 100 μm for cell penetration, tissue ingrowth, vascularization, and nutrient delivery to the center of the regenerating tissue [23, 24].
- It is made from a material that is biocompatible and bioactive, that is, it bonds to the host tissue without formation of scar tissue.
- It exhibits a surface texture that promotes cell adhesion and adsorption of biological metabolites [25].
- It influences the genes in the bone generating cells to enable efficient cell differentiation and proliferation.
- It resorbs at the same rate as the tissue is repaired, with degradation products that are nontoxic and can easily be excreted by the body, for example, via the respiratory or urinary systems.
- It is made by a processing technique that can produce irregular shapes to match that of the defect in the bone of the patient and that can be adapted for mass production.
- It exhibits mechanical properties sufficient to be able to regenerate tissue in the particular application such as bone in load-bearing sites.
- It has the potential to be commercially producible to the required ISO or FDA standards.
- It can be sterilized and maintained as a sterile product to the patient.

The first criterion above refers to porosity and pore structure, which are key parameters determining the properties and the applicability of scaffolds for tissue engineering [1, 3, 26]. Figure 1 shows a summary of the different functions related to the pore structure in a tissue engineering scaffold [27]. In general, scaffold porosity, pore morphology, and pore orientation must be tailored to the particular tissue under consideration, and extensive evidence in the literature documents the critical influence of scaffold porosity and pore structure on the success of bone-tissue engineering approaches. For example, bone morphogenic protein (BMP)-induced osteogenesis has been shown to depend on pore size [28], porous structure [29], and overall scaffold geometry [30, 31] of hydroxyapatite-based scaffolds. The list above summarizes the ideal criteria for a versatile scaffold, but all the criteria may not have to be fulfilled for all applications. For example, for bone regeneration, where the scaffold is implanted directly into a bone defect, mechanical properties of the scaffold are critical, and the modulus of elasticity and mechanical strength of the porous material should match that of natural bone. If a scaffold with a modulus much lower than the host bone is implanted into a load-bearing site, the scaffold will fracture. If the modulus of the scaffold is much higher than that of bone the load will be transmitted through the scaffold instead of the bone (stress shielding), causing bone resorption rather than bone regeneration. On the other hand, for tissue engineering applications where, for example, osteoblast cells grow and proliferate onto the porous scaffold and new bone forms ex vivo, only the mechanical properties of the final tissue-engineered construct is critical.

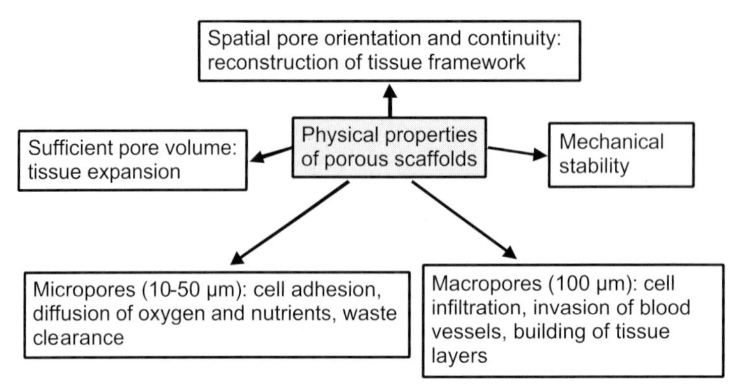

Fig. 1 Schematic diagram showing the different functions of a tissue engineering scaffold in dependence on its porosity and pore structure.

5.8.5
Cellular Bioceramics as Scaffolds in Tissue Engineering

HA, TCP, and bioactive glasses have all been used successfully in the clinic as bone filler materials in powder form [10–13, 32–34], but the challenge is to develop them into 3D porous scaffolds, that is, in a cellular structure with the properties listed above. Moreover such 3D porous structures, if correctly designed, could be used as carriers to encapsulate drugs, effectively forming drug-delivery systems [35–39]. The following sections focus on this challenge and review current developments based on HA, TCP, and bioactive glass scaffolds exhibiting cellular structure. Moreover, available studies on other bioceramics are also briefly discussed for completeness. In this chapter only inorganic (bioceramic) foams are covered. Porous composites and composite foams formed by combination of natural or synthetic biopolymers and inorganic bioactive phases (HA, TCP, bioactive glasses) will not be discussed as they fall outside the scope of the present volume. Review articles are available describing research into cellular polymer/ceramic composites for tissue engineering scaffolds [40, 41] and recent examples are presented in Refs. [42, 43]. Biomimetic strategies to develop HA layers on organic foams have also been developed [44–46].

5.8.5.1
HA and Other Calcium Phosphates

HA and other calcium phosphates such as TCP in porous form have been widely applied as bone substitute and bone filler [13–15, 47–50]. These bodies usually exhibit macroporosity (pore diameter > ca. 50 μm) and microporosity (pore diameter < 10 μm). As mentioned above, macropore diameters should be in excess of about 100 μm to allow bone ingrowth, and if bone is required to penetrate within the ceramic implant, macropores should be large enough that they are connected by pore windows with diameters in excess of 100 μm to allow bone ingrowth. However, the presence of micropores is significant as they may act as attachment sites for cells such as osteoblasts (bone-growing cells). Moreover, interconnected micropores are important because they allow HA bodies to be machinable, a requirement to produce complex shapes. It has been discussed that optimum pore sizes required for bone ingrowth differ between in vitro and in vivo environments, and an excellent discussion about the design strategies of 3D porous structures for bone tissue engineering scaffolds is provided in the chapter of Baksh and Davies in Ref. [1].

The classical way to produce porous HA and TCP ceramics with pore sizes greater than about 100 μm is sintering of powders with suitable porogenic additives, such as paraffin, naphthalene, hydrogen peroxide [13, 51], and other organic particles [28, 48, 52–55], which are burnt out at elevated temperature with evolution of gases and leave mainly spherical pores in the HA or TCP structure. Traditional methods based on ceramic-slip foaming [56, 57], salt leaching [58,59], emulsion [60], and dual-phase mixing of polymer and ceramic slurries [61] have also been investigated, as well as the use of naturally occurring porous calcium-based structures, such as in the hydrothermal conversion of coral or bone [62, 63].

Different forms of porous HA substrates produced by the above methods are available commercially, such as Endobon (Merck, Darmstadt, Germany), Bio-Ostetic (Berkeley Advanced Biomaterials, Inc., Berkeley, CA, USA) and Pro Osteon (Interpore International, Irvine, CA, USA). Endobon is manufactured from bovine cancellous bone, whereby the organic portion of the bone is removed and the bone mineral (still maintaining the trabecular structure) is hydrothermally treated to obtain hydroxyapatite. The advantage of this product is that it mimics the structure of trabecular bone well and has high strength, but there is concern over the possibility of disease transfer. Pro Osteon is made by the hydrothermal conversion of marine coralline excavations to hydroxyapatite. The porous structure resembles trabecular bone, but of course is a less accurate mimic than HA produced from natural bone itself. A resorbable form of Pro Osteon has recently been developed [64]. Bio-Ostetic is an alternative resorbable porous implant based on TCP and HA.

There is little long-term data on the survivability of these implants, although what exists is favorable. After two years of implantation of Pro Osteon 200 with rigid anterior plating, as a bone-replacement implant in the cervical spine, there has been no record of plate breakage, screw breakage, resorption of the implant, or pseudarthrosis [65].

These implants are limited to bone repair and have generally been designed to repair bone as bone replacements, rather than by bone regeneration.

Traditional fabrication methods are in general not suitable for production of highly porous foams of sufficient strength, as required for tissue engineering scaffolds. In fact, most commercial suppliers of porous HA and calcium phosphate materials for bone reconstruction or substitution recognize the need for more research focusing on developing new products with improved strength. Furthermore, pore interconnectivity obtained by traditional fabrication methods is low when compared to the total volume of the pores [52, 66]. As a consequence, in the last 15 years numerous novel fabrication procedures have been developed for production of improved cellular HA and TCP structures for scaffold applications, as described in subsequent paragraphs, and the subject is of continuous interest.

Template or Replication Techniques

Methods based on powder-filled polymer sponges, which are widely used for production of ceramic foams (see Chapter 2.1) have been used also for many years for production of porous HA and other calcium phosphate cellular structures. Recent reports dealing specifically with scaffolds for tissue engineering are available [67–72]. In these methods, sponges having the desired pore structure are infiltrated with a ceramic slurry and then dried and fired. The polymer substrate, usually polyurethane (PU), is removed by pyrolysis and the HA skeleton is sintered. Polyurethane systems free from silicones are favored to avoid contamination of the ceramic on pyrolysis [73]. Recently, HA foams with porosities in excess of 90 %, reticulated open-cell structure, and crushing strength of 0.2 MP have been produced by this method [73]. Sacrificial PU foams have been also used to produce glass-reinforced HA foams [74]. Addition of a phosphate glass to HA leads to increased mechanical strength of the foams, and methods have been developed to incorporate soluble bioactive glasses as reinforcement in HA [75–77]. The polymer-sponge method has been also proposed to manufacture macroporous calcium phosphate glass scaffolds of composition $CaO–CaF_2–P_2O_5–MgO–ZnO$ [70].

In related developments, HA foams with tailored pore structure have been produced from slurries of high solids content (60 wt %) by using commercially available HA powders and adding a dispersant [78]. Polyurethane foams and rapid-prototyping models (see also below) were used as substrates, which could be eliminated by heat treatment at 650 and 700 °C, respectively. Porosities in the range 50–96 % were produced by using different types of polymer substrates [78].

Several approaches are being investigated to improve the mechanical properties of HA foams produced by replication methods. It has been reported, for example, that combination of replicating techniques with gel casting of foams (see below) leads to HA cellular structures of improved fracture strength [67]. Compression strengths in the range 0.5–5 MPa were measured for these foams, depending on HA concentration in the starting suspension [67]. Another investigated approach to increase the mechanical strength of HA foams is coating of the porous structure with polymer or polymer/HA composite layers. In particular, the brittleness and low strength of HA foams obtained by the reticulate method have been improved by applying coatings based on poly(ε-caprolactone) (PCL)/HA composites [35]. The PCL coating can be used also to effectively entrap antibiotic drugs and thus the construct can be considered a drug delivery system and used to enhance bone ingrowth and regeneration in the treatment of bone defects.

Foaming of Ceramic Suspensions and Gel-Casting Processes

Approaches to build foamlike structures into ceramic suspensions are being investigated as a means to obtain cellular HA structures. The methods are based on the addition of foaming agents to ceramic suspensions and forming foam structures by agitation. However, the consolidation of the formed foam, that is, its transformation into a rigid network, is a critical step that is difficult to accomplish by using traditional shaping methods involving simple liquid removal [79, 80]. Novel setting mechanisms are being developed with this aim, which can make use of nonporous molds offering advantages in terms of shaping capability and shape complexity. In these techniques, which may be grouped under the heading "direct consolidation techniques" [80], the structure of the ceramic slips is consolidated without powder compaction or removal of liquid, thus preserving the homogeneity achieved in the slurry state. Starch consolidation, protein consolidation, and direct coagulation casting (DCC) are some of the methods proposed to achieve porous structures that mimic that of cortical and/or trabecular bone [79, 81–83]. Well dispersed aqueous suspensions of HA powders with solids loadings as high as 60 vol% in the presence of suitable amount of ammonium polycarbonate as dispersant are used. The consolidator agents considered include etherified potato starch modified by hydroxypropylation and cross-linking, albumin, chicken egg white, and mixtures of two polysaccharide powders, namely, agar and locust bean gum [80, 84].

To obtain complex shapes and pore structures resembling the different parts of bone, well-known methods such as slip casting, foaming, and use of fugitive porogenic additives should be combined with starch- and protein-consolidation techniques. It is claimed that smart combination of traditional and advanced techniques will enable the structure of the HA scaffold to be tailored according to the application requirements [79].

Gel-casting of foams is another recently developed method to produce macroporous ceramics [85–87]. This method yields compounds in various porosity fractions and controlled pore size that are noncytotoxic and have optimized strength and open spherical-cellular structure [86, 88]. Gel-casting involves dispersion of an aqueous suspension of HA powder with polyacrylate derivatives as dispersing agent. Acrylic monomers are also incorporated into the suspensions to promote gelation by in situ polymerization. Prior to this, the mixtures are foamed by agitation, aided by the addition of nonionic surfactants that reduce the surface tension of liquid–gas interfaces and serve as stabilizers of the foams. The gelation process of foamed suspensions is promoted by addition of initiator and catalyst for in situ polymerization of the added monomers by means of the redox system of ammonium persulfate and tetramethylethylenediamine. Before gelation, the mass is cast into molds of the required size and shape and dried. A subsequent sintering stage is required at 1350 °C for at least 2 h for densification of the HA matrix [86, 88].

An improved method, termed "slurry expansion", has been recently developed that leads to isotropic cellular HA structures with porosity close to 80 vol % and bimodal pore size distribution [89]. Figure 2 shows the obtained scaffold micro- (left) and macrostructures (right). Compressive strength and Young's modulus were in the ranges 6–9 MPa and 0.5–2.5 GPa, respectively [89].

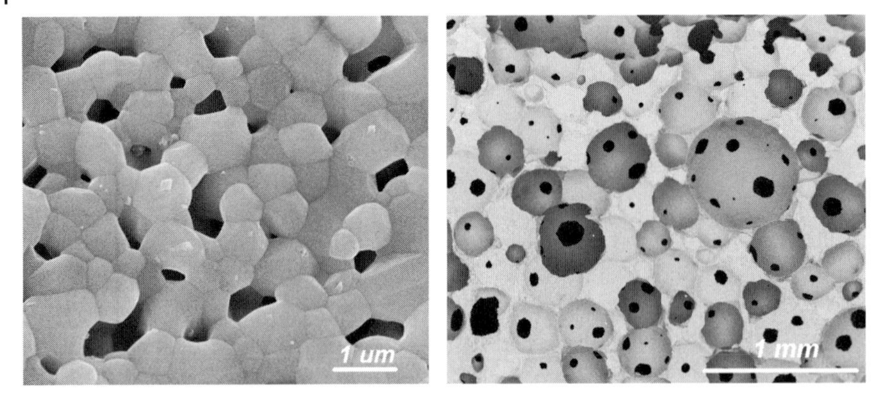

Fig. 2 SEM images showing microstructure (left) and macrostructure (right) of HA foam scaffolds prepared by the slurry-expansion method of Martinetti et al. [89]. (Photos courtesy of Dr. R. Martinetti, FIN-Ceramica, Faenza, Italy.)

Alternative routes to produce calcium phosphate scaffolds at low temperature, to avoid the formation of high-temperature apatite phases, have been developed [90]. This method has the objective of increasing the in vitro and in vivo solubility of the scaffold by yielding a calcium-deficient hydroxyapatite. α-TCP powder is used as starting material and is mixed with hydrogen peroxide as foaming agent. Consolidation is not obtained by sintering but through low-temperature setting reactions [90]. Similarly, highly porous carbonated apatite bodies have been produced recently at low temperatures from mixtures of poly(vinyl alcohol) fibers, sodium chloride, and nanocrystalline carbonated apatite powders [91].

Use of Naturally Occurring Cellular Structures

Naturally occurring porous structures are considered to fabricate HA scaffolds [92]. A frequently used structure is coral. Hydrothermal and solvothermal methods are used to transform natural coral into HA after removal of the organic component by, for example, immersion in sodium hypochlorite [93]. Pore size in typical coral formations is in the range 200 to 300 μm. The porosity is interconnected and the structure resembles that of trabecular bone.

Wood has been proposed as a suitable template to fabricate porous HA scaffolds [94]. Heating wood in a nonoxidizing atmosphere at temperatures above 600 °C results in decomposition of the polyaromatic constituents to form a carbon residue which can reproduce the original porous structure. A typical microstructure of a pyrolyzed specimen of *Quercus alba* is shown in Fig. 3 [94], which exhibits in principle adequate macropore size (> 100 μm) and pore interconnectivity required for bone-tissue engineering applications. This porous carbon substrate can be subsequently transformed into a calcium phosphate skeleton after impregnation with a suitable precursor and calcination [94].

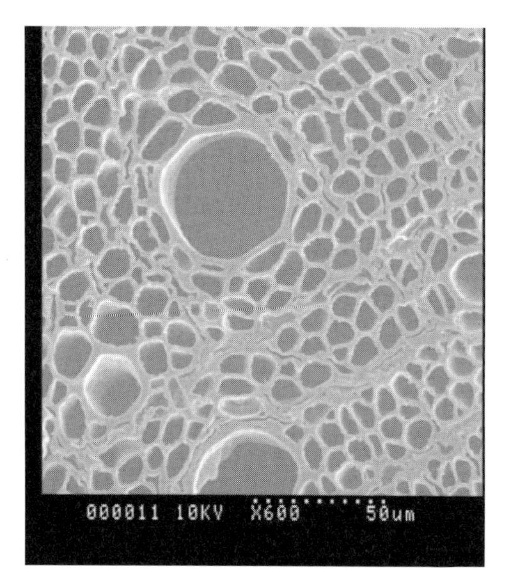

Fig. 3 Typical microstructure of a pyrolyzed specimen of *Quercus alba*, which exhibits adequate macropore size (> 100 µm) and pore interconnectivity for bone-tissue engineering. This porous carbon substrate can be subsequently transformed into a calcium phosphate skeleton after impregnation with a suitable precursor and calcination [94]. (Micrograph courtesy of Prof. K.A. Gross, Monash University, Australia.)

Solid Free-Form Fabrication

The complete design optimization of 3D macroporous structures for tissue engineering scaffolds is most likely to be satisfactorily achieved through solid free-form fabrication (SFF) and rapid prototyping (RP) techniques. SFF and RP refer to a variety of technologies capable of producing 3D physical models from 3D computer data sets or CAD files [95, 96]. These techniques, as commonly used to produce ceramic porous structures, are considered in detail in Chapter 2.3 of this book.

Few reports on the use of RP and SFF techniques for production of cellular HA and TCP constructs for tissue engineering or related biomedical applications have been published. Among existing systems, stereolithography (SLA) [97, 98], extrusion free-forming or fuse deposition modeling (FDM) [95, 99], negative mold inkjet (IJ) printing [100, 101], and selective laser sintering (SLS) [102] are possible choices. The typical flow diagram for HA scaffold design and manufacturing by indirect RP techniques, schematically shown in Fig. 4, involves three main steps: 1) mold microstructure design, 2) ceramic slurry development, and 3) binder burnout and sintering.

HA scaffolds with interconnecting square pores have been produced by using Unigraphics CAD software [98]. Scaffolds were fabricated by casting a ceramic slurry into molds fabricated by stereolithography. These HA scaffolds have been developed further, and extensive in vivo studies have been carried out in animal models [98]. HA scaffolds with macroporosity of controlled size and shape have been also fabri-

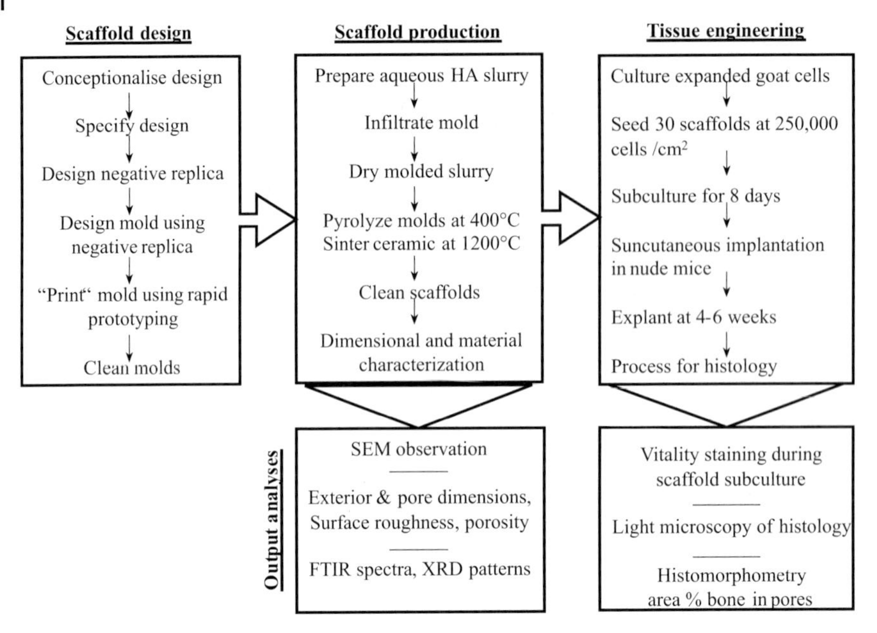

Fig. 4 Typical flow diagram for HA scaffold design and manufacturing by RP techniques, after Wilson et al. [96].

cated with ink-jet printing systems [64] and by the commercial TheriForm SFF process [103]. Manufacturing techniques utilizing commercially available RP technologies have been developed to produce standardized HA scaffolds with defined architectural parameters [96]. The system constructs models by sequentially depositing material in thin parallel planar layers. Each layer is milled to a specified thickness prior to deposition of the next layer. This layer-by-layer process continues until the entire mold is constructed. Molds are filled with an aqueous HA slurry, prepared conventionally [61] by using a vacuum-infiltration device. Pyrolysis of the mold and other organic components, as well as final sintering of the ceramic scaffolds, occurred by designed heat-treatment in air at a maximum temperature of 1250 °C [96]. A perspective view of a typical scaffold fabricated by this technique is shown in Fig. 5.

Extrusion free-forming is another promising technique used to produce hydroxyapatite scaffolds with highly controlled pore structure [99]. Figure 6 shows a typical microstructure of a latticework of HA produced by extrusion free-forming. The advantage of this method is that multinozzle or variable-nozzle operation is possible, which can produce scaffolds with large pores (150 μm) for blood-vessel infiltration with intersecting fine-porous structure which should provide accommodation for active osteoblast cells [99].

Combination of SFF techniques with conventional sponge-fabrication techniques have also been proposed [104]. RP systems have been also developed for fabrication of chitosan–HA [103] and polyetheretherketone–HA [102] composite scaffolds.

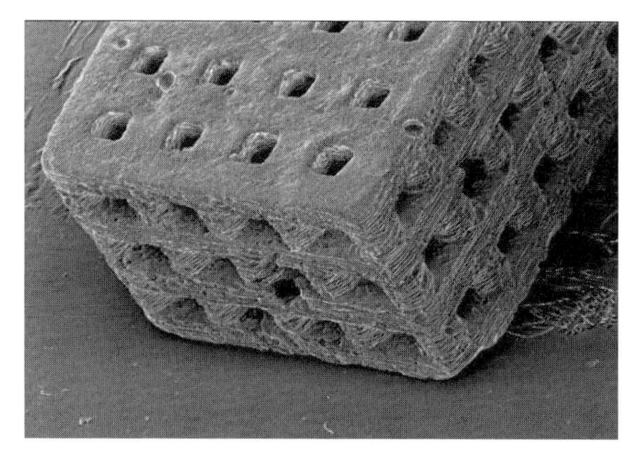

Fig. 5 Perspective view of a typical HA scaffold fabricated by the RP technique developed by Wilson et al. [96]. (Micrograph courtesy of Dr. C.E. Wilson, Twente University, The Netherlands.)

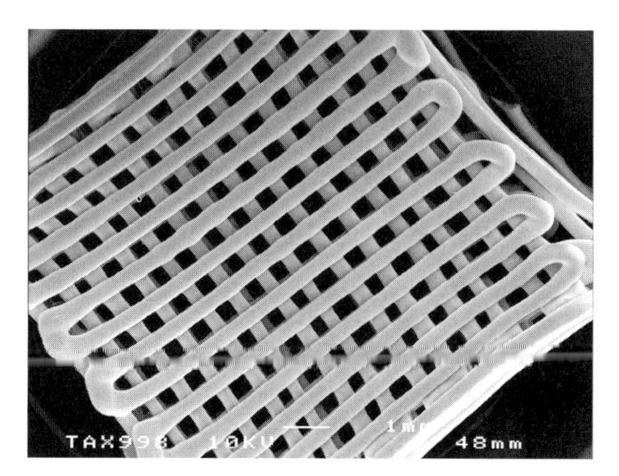

Fig. 6 Typical microstructure of a sintered latticework of HA produced by extrusion free-forming by the method developed by Gomes de Sousa and Evans [99] (Courtesy Ms. C. Gomes de Sousa, Queen Mary University of London, UK.)

However, the number of HA-based systems fabricated to date is relatively small, and data on scaffold performance are scarce, so that a conclusive comparison of the different RP and SFF methods in terms of optimal 3D design of porous structures, attained properties, and in vitro or in vivo behavior is not yet possible.

There is continuous research focused on the improvement of computational optimization procedures to be coupled with RP and SFF techniques. This should aid in creating scaffold structures such that both the porous ceramic scaffold and the eventual regenerated tissue will match host tissue stiffness, while at the same time meeting constraints on scaffold porosity, material, and fabrication method [97, 98].

5.8.5.2
Melt-Derived Bioactive Glasses

Although melt-derived bioactive glasses have had clinical success in particulate form, such as the commercially available 45S5 Bioglass composition [10], it has proven difficult to produce scaffolds with interconnected pore networks from these materials. Livingston et al. [33] produced a simple sintered scaffold by mixing 45S5 Bioglass powders, with particle size range of 38–75 μm, with 20.2 wt % of camphor ($C_{10}H_{16}O$) particles with a size range of 210–350 μm. The mixture was dry-pressed at 350 MPa and heat treated at 640 °C for 30 min. The camphor decomposed to leave porous Bioglass blocks. The pores were in the range of 200–300 μm in diameter, but total porosity was just 21 %, hence distances between pores were large, and interconnectivity was low.

Yuan et al. [34] produced similar scaffolds by foaming Bioglass 45S5 powder with a dilute H_2O_2 solution and sintering at 1000 °C for 2 h to produce a porous glass ceramic. Although pore diameters were in the desired range of 100–600 μm, the pores were irregular in shape with orientated channels running through the glass and large distances between channels. Interconnectivity was therefore low. The samples were implanted into the muscle tissue of dogs and were found to induce bone growth, containing osteoblasts and osteocytes, in soft tissue. Pathological calcification, that is the formation of nonhealthy bone that does not contain osteocytes, was also observed in pores near the surface of the material. Bone formed directly on the solid surface and on the surface of crystal layers that formed in the inner pores and in pores containing pathological calcification. Osteogenic cells were observed to aggregate near the material surface and secrete bone matrix, which then calcified to form bone. However, although the implants had a porosity of about 30 % only 3 % bone was formed.

The few, rather unsatisfactory, results available indicate that optimized highly porous bioactive glass scaffolds cannot be produced from melt-derived glass by conventional powder-technology methods. Thus the most favored bioactive glasses for produce porous scaffolds are those derived by sol–gel processing.

5.8.5.3
Sol–Gel-derived Bioactive Glasses

The sol–gel process is an alternative method to produce bioactive glasses [25]. An advantage of gel-derived bioactive glasses over their melt-derived counterparts is that they exhibit a mesoporous texture (pores diameters in the range 2–50 nm). This texture enhances bioactivity and resorbability of the glasses [105, 106]. A disadvantage of gel-derived glasses is that there is no clinical experience with them as yet; therefore, the route to clinical use would be longer than if melt-derived glasses were used.

The sol–gel process involves the production of a colloidal solution (sol) of Si-O groups by the hydrolysis of alkoxide precursors such as tetraethyl orthosilicate (TEOS) in excess water under acidic catalysis. Simultaneous polycondensation of

Si-O groups continues after hydrolysis is complete, beginning the formation of the silicate network. As the network connectivity increases, viscosity increases, and eventually a gel forms. The gel is then subjected to carefully controlled thermal processes of ageing (60 °C) to strengthen it, drying (130 °C) to remove the liquid by-product of the polycondensation reaction, and thermal stabilization/sintering (600–800 °C) to remove organics from the surface of the gel and to densify the structure. A chemically stable glass is then produced [107].

Sol–gel-derived bioactive glass scaffolds with cellular structure have been produced in several ways.

Many authors have cast silica-based sol–gel glasses around removable *templates* such as close-packed latex or polystyrene spheres [108–110] with similar techniques as used for HA slurries, except that the sol gels around the organic spheres, and the polymer is burnt out during stabilization of the gel. The pore network produced is very homogeneous and interconnected due to the close packing of the spheres. However pore size is limited by the difficulty in producing organic spheres with diameters on the order of 500 μm required to produce an ideal scaffold for tissue engineering applications.

Foaming agents are chemical additives that produce gas bubbles in a liquid. Gun et al. [111] added 6–10 vol % hydrogen peroxide to the base-catalyzed sol–gel process for methyl silicate materials. Decomposition of the hydrogen peroxide formed bubbles which remained throughout polymerization. Crack-free rods (20 cm length × 0.5 cm diameter) were prepared that did not shrink during ageing. This technique could be applied to bioactive gel glasses, but there were few pores greater than 10 μm in diameter.

Polymer foaming techniques have been applied to sol–gel systems. Viscosity was controlled to stabilize bubbles created by Freon 11 (CCl_3F) droplets dispersed in a sol to form porous silica with porosities of 55–90 % and corresponding mean cell diameters of 30–1000 μm [112]. Bubbles formed because Freon 11 has a boiling point of 23.8 °C, and incubation was carried out at temperatures above this boiling temperature. The foam structure was stabilized by addition of an anionic surfactant and a rapid polymerization reaction, which caused fast gelation. The viscosity was controlled by adjustment of pH with H_2SO_4. A surfactant is generally used to stabilize any bubbles formed in the liquid phase by reducing the surface tension of the gas–liquid interfaces [113]. The bending strength of the silica foams made by these methods range from 2 MPa at a porosity of 86 % to approximately 8 MPa at a porosity of 66 %.

Direct foaming of sol–gel-derived bioactive glasses has been performed by vigorous agitation, with the aid of surfactants, to produce scaffolds that fulfill many of the criteria of the ideal scaffold [114]. The surfactant lowers the surface tension of the sol and stabilizes bubbles created by air entrapment. A gelling agent (hydrofluoric acid, HF) is added to the sol to increase the gelation time to a few minutes. As gelation occurs the bubbles are stabilized permanently. The highly porous foam is then heat-treated in the same way as a normal sol–gel glass. Figure 7 shows an SEM image of a typical foam of 70S30C bioactive glass (70 mol % SiO_2, 30 mol % CaO).

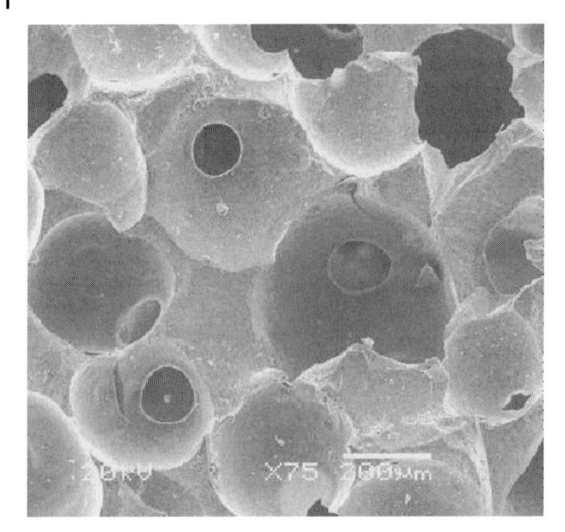

Fig. 7 SEM image of a typical sol–gel-derived bioactive glass foam of composition 70 mol % SiO_2, 30 mol % CaO.

The properties of the pore network can be controlled at each stage of the process, for example, by the type and concentration of catalyst and surfactant, the glass composition, and the process temperature [115, 116]. The scaffolds can be produced with interconnected macropores, porosities in the range of 60–90 %, and a modal interconnected pore diameter of up to 150 μm [115] with many interconnections in excess of 300 μm in diameter.

The composition and textural pore size can be used to tailor the rate of resorbability of the scaffold and therefore the rate of release of elements that affect the gene expression of osteoblasts [115]. Due to the nature of the sol–gel process the scaffolds can be produced in many shapes, which are determined simply by the shape of the casting mold, or scaffolds can be cut to a required shape.

Primary human osteoblast cells, harvested from the tops of femurs removed during total hip replacements, have been seeded on the foamed glasses in vitro. The cells attached, proliferated, and secreted bone extracellular matrix, which mineralized after 10 d of culture. Figure 8 shows an SEM image of primary human osteoblasts cultured on a 58S bioactive glass foam for 10 d [117]. The micrograph shows a mineralized bone nodule inside a macropore. A bone nodule is a group of cells that have laid down some extracellular bone matrix. This bone nodule has mineralized without the addition of mineralization supplements to the culture media, which indicates the great potential of the bioactive glass foam as osseous tissue scaffold.

The only ideal scaffold criterion not addressed is the matching of mechanical properties of the scaffolds to bone for in situ bone-regeneration applications. Foams optimized by sintering have compressive strength of about 2.5 MPa, which is similar to that of trabecular bone (2–10 MPa) [118], but the fracture toughness and tensile strength of the foams are much lower than the values measured for bone. The mechanical properties of these foams should be sufficient for tissue engineering

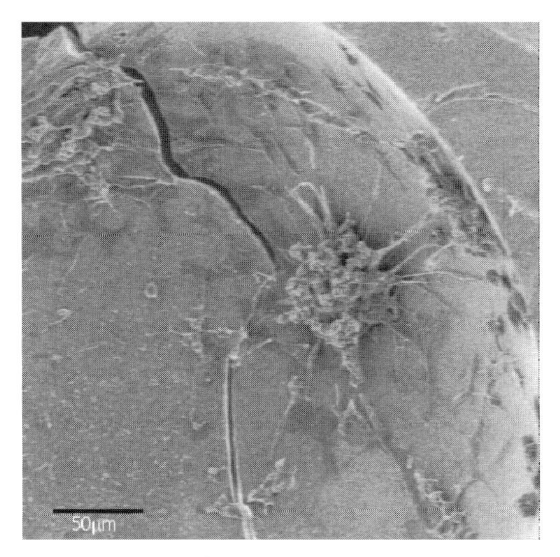

Fig. 8 SEM image of primary human osteoblasts cultured on a bioactive glass foam (58S) for 10 d. The micrograph shows a mineralized bone nodule inside a macropore [117]. (Micrograph courtesy of Dr. Julie Gough, University of Manchester, UK.)

applications, where bone would be grown on a scaffold in the laboratory before implantation. To improve the toughness of the scaffolds, however, the complete structure of bone should be mimicked more closely. The foamed glasses and ceramics mimick bone mineral, but bone is a natural composite of bone mineral and collagen. An organic phase should therefore be introduced into the scaffold to provide toughness for in situ bone regeneration applications. In recent developments, antimicrobial macroporous scaffolds based on sol–gel glasses doped with silver ions have been developed [119]. However, these foams, which combine bioactive and antimicrobial properties, are yet to be investigated further in terms of in vitro and in vivo behavior.

The interactions between cells and surfaces play a major biological role in cellular behavior. Cellular interactions with artificial surfaces are mediated through adsorbed proteins. A common strategy in tissue engineering is to modify the biomaterial surface selectively to interact with a cell through biomolecular recognition events. Adsorbed bioactive peptides can allow cell attachment on biomaterials, and permit three-dimensional structures modified with these peptides to preferentially induce tissue formation consistent with the cell type seeded, either on or within the device [120, 121]. To achieve full functionality, peptides must be adsorbed specifically. The surfaces of bioactive glass foam scaffolds can be modified with amine and mercaptan groups [122]. The modified scaffolds have been used as carriers for laminin, which was selected as a molecular model for the eventual tailoring of scaffold chemistry to engineer growth of arterial and lung tissues. The covalent bonds between the binding sites of the protein and the ligands on the scaffold surface did not denaturate the protein. The 3D architecture of the foams mimics the scale and interconnectivity of structures needed to grow a pulmonary artery and lung lobe in vitro. In vitro studies show that foams modified with chemical groups and coated with laminin maintained bioactivity and improved proliferation of mouse lung epithelial (MLE-12) cells compared to uncoated foams [123]. Sustained and controlled release from the scaffolds over a 30-day period was achieved as laminin release from the bioactive foams followed the dissolution rate of the material network [123].

Sol–gel-derived bioactive glasses can be processed to produce similar structures to the HA bone grafts used clinically. Although they have controllable resorbability and the potential to stimulate bone cells at the genetic level to encourage bone regeneration, sol–gel bioactive glass foams have not yet been submitted for FDA or EU (CE mark) approval. However, melt-derived bioactive glasses have FDA and EU approval and are commercially available as regenerative bone filler, so it is a matter of time before the sol–gel glass scaffolds are approved for implantation.

5.8.5.4
Other Bioceramics Exhibiting Cellular Structure

Research is also being carried out on the development of cellular structures from bioceramics other than calcium phosphates, hydroxyapatite, or bioactive glasses. For example, zirconia porous scaffolds coated with biphasic calcium phosphate (HA + TCP) have been developed [124]. Zirconia foams (90 % porosity, pore size 500–600 μm) exhibiting highly interconnected porosity were fabricated by a replication method using reticulated PU sponges. Slurry dipping was used to obtain calcium phosphate coatings of homogeneous thickness (ca. 30 μm) uniformly covering the pore walls of the zirconia foams. The adhesive strength of the coating layer was as high as 20–25 MPa. The biphasic calcium phosphate layer improved osteoconduction and stimulated the proliferation of bone cells [124].

Controlled porosity, bioinert alumina scaffolds with pore sizes in the range of 300–500 μm have been processed using a RP process based on fused deposition [95]. The technique allows production of scaffolds with designed volume fraction and pore size, which were used to evaluate mechanical properties and biological response and their interrelationship with porosity parameters [95]. The coating of porous alumina substrates with HA layers has been proposed to impart bioactive behavior to the scaffolds [125].

Alumina foams coated with HA or TCP have been prepared also by a PU-sponge method, followed by multiple slurry-dipping in aqueous HA or TCP suspensions and final heat treatment [126]. In these foams, the alumina structure provides the mechanical strength (compressive strength of 12 MPa for a porosity of 75 %), which is much higher than that of equivalent foams made totally of HA [127–129]. Good osteoconduction of HA-coated alumina scaffolds, equivalent to that of pure HA scaffolds, was proved by in vitro and in vivo studies [126].

Titania foams are gaining increasing attention for applications as tissue engineering scaffolds due to the high biocompatibility of TiO_2 [130, 131]. Although traditionally classified as a bioinert material [132], it has been recently shown that surface-modified porous titania structures can exhibit bioactive behavior [133–135]. Titania cellular structures have been produced by the replication method using fully reticulated polyester-based PU foams [130]. The aqueous ceramic suspension was prepared from titania powder with mean particle size of 0.3 μm in a concentration of 57 vol %. The heating schedule for burning out the polymers and sintering the foams comprised a first stage at 450 °C for 1 h and subsequent heating at 1150 °C. Pore sizes were 445 or 380 μm, depending on the PU foam used. Initial cell culture work with fibroblasts showed high adhesion of the cells to the foam structure [130], and further in vitro and in vivo are expected on these novel materials.

As a new generation of light, tough, and high-strength materials for medical implants and bone substitution, biomorphic SiC porous ceramics coated with bioactive glass have been developed [136]. Biomorphic SiC is fabricated by molten-Si infiltration of carbon templates obtained by controlled pyrolysis of wood. The structure exhibits high porosity (up to 70 %), high anisotropy, and resembles the structure of bone. The uniform and adherent bioactive glass coating is then applied by pulsed laser ablation. In vitro tests in simulated body fluid demonstrated the formation of HA layers after 72 h [136]. It is also claimed that biomaterials with high silicon content, such as SiC-based structures, do not present adverse physiological effects in the body, but little is known on the long term in vivo behavior of biomorphic SiC scaffolds, in particular considering the presence of impurities in the starting wood.

5.8.6
Properties of Some Selected Bioactive Ceramic Foams

Many techniques are being employed with different degree of success to produce cellular bioceramics tailored for applications as tissue engineering scaffolds, particularly for bone tissue and to a lesser extent for other (soft) tissues. Table 1 lists different characteristics of selected bioactive ceramic foams developed for bone-tissue engineering, including porosity characteristics, methods of fabrication, and typical mechanical properties. Mechanical properties of cortical bone are also shown for comparison. For HA-based scaffolds, innovative foaming of ceramic suspensions coupled with gel casting of foams have led to very satisfactory results in terms of porous structure and subsequent in vitro and in vivo behavior. Similarly, scaffolds produced by rapid prototyping and solid free-form fabrication methods exhibit highly ordered microstructures and can be readily manufactured to desired, complex shapes. It is intriguing that these methods have been developed so far mainly for HA scaffolds but not for bioactive glasses. Gel casting of foams produces bioceramic scaffolds with a structure very similar to that of trabecular bone. Both the gel-cast foamed HA and sol–gel-derived bioactive glass foams showed favorable results in both in vitro and in vivo tests for bone regeneration. In general, optimized bioactive glass or HA scaffolds have the potential to fulfil eight out of nine of the criteria for an ideal scaffold for tissue engineering applications (as presented in Section 5.8.2). The only criterion not fulfilled is related to mechanical properties. Although the scaffolds can exhibit compressive strengths similar to those of trabecular bone, while maintaining a pore network suitable for tissue engineering, they have low flexural strength. This may not affect tissue engineering applications, where tissue is grown in bioreactors in vitro, but the scaffolds are not yet suitable for direct implantation in load-bearing applications.

Table 1 Selected bioactive ceramic foams for bone tissue engineering: porosity characteristics, methods of fabrication and typical mechanical properties. Data for cortical bone are also shown for comparison.

Material	Porosity	Modal interconnected pore diameter	Method of fabrication	Compression strength (σ_{max}) [MPa]	Ref.
HA	> 90 %,	up to 1 mm	replication of open celled polymer foam	0.2	73
HA	76–80% (Fig. 2)	30–120 μm	gel-cast foams	4.4–7.4	89
HA	70–90%	>100 μm	combination of replicating techniques with gelcasting of foams	0.5–5	67
Bioactive glass (70S30C)	70–95%	100–140 μm	direct foaming of sol-gel derived bioactive glasses and subsequent sintering	0.5–2.5	119
HA	40%, controlled and isotropic (non-spherical)	380–450 μm	solid freeform fabrication	30	99
Trabecular bone	50–90%	>100 μm	n/a	2–10	10
Endobon®	60–88%	400–600 μm	thermally treated bovine HA	1–11	137

5.8.7 Summary

The topics covered in this chapter demonstrate that the development of cellular bioceramics for tissue engineering and regenerative medicine is a very active research field worldwide. This R&D sector is expected to continue to attract funding and efforts from a wide range of disciplines, including cell biology, materials science, biochemistry, pharmaceutical sciences, clinical medicine.

Acknowledgement

The authors acknowledge stimulating discussions and conversations with Professor Larry Hench and Professor Dame Julia Polak (both at Imperial College London). The authors wish to thank Lloyds Tercentenary Foundation and EPSRC (UK) for support.

References

1 Davies J.E. (Ed.), *Bone Engineering*; EM^2 incorporated, Toronto, 2000.
2 Jones J.R., Hench L.L. Biomedical Materials for the New Millennium: A Perspective on the Future. *J. Mater. Sci. Technol.* **2001**, *17*, 891–900.
3 Lanza R.P., Langer R., Vacanti J.P., *Principles of Tissue Engineering*, 2nd ed. Academic Press, San Diego, 2000.
4 Ohgushi H., Caplan A.I. *J. Biomed. Mater. Res.* **1999**, *48*, 913–927.
5 Takezawa, T. *Biomaterials* **2003**, *24*, 2267–2275.
6 Freyman T.M., Yannas I.V., Gibson L.J. *Prog. Mater. Sci.*, **2001**, *46*, 273–282.
7 Hench L.L., Wilson J. (Eds.), *Introduction to Bioceramics*. World Scientific, Singapore, 1993.
8 Berry D.J., Harmsen W.D., Cabanela M.E., Morrey M.F. *J. Bone Jt. Surg. Am.* **2002**, *84A* (2), 171–177.
9 Bradley J.G., Andrews C.M., Lee K., Scott C.A., Shaw D. *Key Eng. Mater.* **2000**, *192*, 1013–1020.
10 Hench L.L. *J. Am. Ceram. Soc.* **1998**, *81*, 1705–1728.
11 Ravaglioli A., Krajewsky A. (Eds.), *Bioceramics and the Human Body*, Elsevier Applied Science, London and New York, 1992.
12 Simske S.J., Ayers R.A., Bateman T.A. *Mater. Sci. Forum* **1997**, *250*, 151–182.
13 Suchanek W., Yoshimura M. *J. Mater. Res.* **1998**, *13*, 94–117.
14 Hing K.A., Best S.M., Tanner E., Bonfield W., Revell P. *J. Biomed. Mater. Res.* **2004**, *68A*, 187–200.
15 LeGeros R.Z. Biological and Synthetic Apatites. In *Hydroxyapatite and Related Materials*, Brown P.W., Constantz B. (Eds.), CRC Press, Boca Raton, 1994, pp. 3–28.
16 LeGeros R.Z., Lin S., Rohanizadeh R., Mijares D., Legeros J.P. *J. Mater. Sci. Mater. Med.* **2003**, *14*, 201–209.
17 Barralet J.E., Grover, L., Gaunt, T., Wright, A.J., Gibson I.R. *Biomaterials* **2002**, *23*, 3063–3072.
18 Georgescu G., Lacout J.L., Freche M. *Key Eng. Mater.* **2004**, *254–256*, 201–204.
19 Frankenburg E.P, Goldstein S.A., Bauer T.W., Harris S.A., Poser R.D. *J. Bone Jt. Surg. Am.* **1998**, *80A*, 1112–1124.
20 Hench, L.L., Splinter, R.J., Allen, W.C., Greenlee, T.K., *J. Biomed. Mater. Res.* **1971**, *74*, 1478–1570.
21 Xynos I.D., Hukkanen M.V.J., Batten J.J., Buttery L.D.K., Hench L.L., Polak J.M. *Calc. Tiss. Int.*, **2000**, *67*, 321–329.
22 Xynos I.D., Edgar A.J., Buttery, L.D.K., Hench, L.L., Polak, J.M. *Biochem. Biophys. Res. Comm.* **2000**, *276*, 461–465.
23 Lu J.X., Flautre B., Anselme K., Hardoiun P., Gallur A., Deschamps M., Thierry B. *J. Mater. Sci. Mater. Med.* **1999**, *10*, 111–120.
24 Okii N., Nishimura S., Kurisu K., Takeshima Y., Uozumi, T. *Neurol. Med.* **2001**, *41*, 100–104.
25 Hench L.L. *Curr. Opin. Solid State Mater. Sci.* **1997**, *2*, 604–610.
26 Liu D.-M., Dixit V. (Eds.), *Porous Materials for Tissue Engineering*, Trans. Tech. Publications, Zürich, Switzerland, 1997.
27 Hutmacher D.W. *Biomaterials* **2000**, *21*, 2529–2543.
28 Tsuruga E., Takita H., Itoh H., Wakisaka Y., Kuboki Y. *J. Biochem. (Tokyo)* **1997**, *121*, 317–324.
29 Kuboki Y., Takita H., Kobayashi D., Tsuruga E., Inoue M., Murata M., Nagai N., Dohi Y., Ohgushi H. *J. Biomed. Mater. Res.* **1998**, *39*, 190–199.

30 Jin Q.M., Takita H., Kohgo T., Atsumi K., Itoh H., Kuboki Y. *J. Biomed. Mater. Res.* **2000**, *52*, 491–499.
31 Ripamonti U., Ma S., Reddy A.H. *Matrix* **1992**, *12*, 202–212.
32 Gibson, I.R., Bonfield, W. *J. Biomed. Mater. Res.* **2002**, *59*, 697–708.
33 Livingston T., Ducheyne P., Garino J. *J. Biomed. Mater. Res.* **2002**, *62*, 1–13.
34 Yuan H., de Bruijn J.D., Zhang X., van Blitterswijk C.A., de Groot K. *J. Biomed. Mater. Res.* **2001**, *58*, 270–276.
35 Kim H.-W., Knowles J.C., Kim H.-E. *Biomaterials* **2004**, *25*, 1279–1287.
36 Krajewski A., Ravaglioli A., Roncari E., Pinsco P., Montanari L. *J. Mater. Sci. Mater. Med.* **2000**, *11*, 763–772.
37 Bajpai P.K., Benghuzzi H.A., *J. Biomed. Mater. Res.* **1998**, *22*, 1245–1251.
38 Jarcho M. *Clin. Orthop.* **1981**, *157*, 259–278.
39 Rose F.R., Cyster L.A., Grant D.A., Scotchford C.A., Howdle S.M., Shakesheff K.M. *Biomaterials* **2004**, *25*, 5507–5514.
40 Boccaccini A.R., Maquet V.M. *Compos. Sci. Technol.* **2003**, *63*, 2417–2429.
41 Laurencin C.T., Lu H.H., Polymer-Ceramic Composites for Bone-Tissue Engineering, in *Bone Engineering*, Davies J.E. (Ed.), EM^2 incorporated, Toronto, Canada, 2000, pp 462–472.
42 Maeda H., Kasuga T., Nogami M., Kagami H., Hata K., Ueda M., *Key Eng. Mater.* **2004**, *254–256*, 497–500.
43 Malafaya P.B., Reis R.L. *Key Eng. Mater.* **2003**, *240–242*, 39–42.
44 Lickorish D., Ramshaw J.A.M., Werkmeister J.A., Glattauer V., Howlett C.R. *J. Biomed. Mater. Res.* **2004**, *68A*, 19–27.
45 Kokubo T., Hata K., Nakamura T., Yamamuro T. *Bioceramics* **1991**, *4*, 113–120.
46 Miyaji F., Kim H.M., Handa S., Kokubo T., Nakamura T. *Biomaterials* **1999**, *20*, 913–919.
47 Van Blitterswijk C.A., Grote J.J., Kuijpers W., Daems W.T., de Groot K. *Biomaterials* **1986**, *7*, 137–146.
48 Liu D.M., *J. Mater. Sci. Mater. Med.* **1997**, *8*, 227–232.
49 Rosa A.L., Beloti M.M., Oliveira P.T., Van Noort R. *J. Mater. Sci. Mater. Med.* **2002**, *13*, 1071–1075.
50 Shimaoka H., Dohi Y., Ohgushi H., Ikeuchi M., Okamoto M., Kudo A., Kirita T., Yonemasu K. *J. Biomed. Mater. Res.* **2004**, *68A*, 168–176.
51 Navarro M., Del Valle S., Ginebra M.P., Martinez S., Planell J.A. *Key Eng. Mater.* **2004**, *254–256*, 945–948.
52 Bouler J.M., Trecant M., Delecrin J., Royer J., Passuti N., Daculsi G. *J. Biomed. Mater. Res.* **1996**, *32*, 603–609.
53 Koc N., Timucin M., Korkusuz F. *Key Eng. Mater.* **2004**, *254–256*, 949–952.
54 Koc N., Timucin M., Korkusuz F. *Ceram. Int.* **2004**, *30*, 205–211.
55 Albuquerque J.S.V., Nogueira R.E.F.Q., Pinheiro da Silva T.D., Lima D.O., Prado da Silva M.H. *Key Eng. Mater.* **2004**, *254–256*, 1021–1024.
56 Peelen J.G.J., Rejda B.V., De Groot K. *Ceramurgia Int.* **1978**, *4*, 71–74.
57 Tamai N., Myoui A., Tomita T., Nakase T., Tanaka J., Ochi T., Yoshikawa H. *J. Biomed. Mater. Res.* **2002**, *59*, 110–117.
58 Yoon I.J., Park T.G. *J. Biomed. Mater. Res.* **2001**, *55*, 401–408.
59 Harris L.D., Kim B.S., Mooney D.J. *J. Biomed. Mater. Res.* **1998**, *42*, 396–402.
60 Ambrosio A.M., Sahota J.S., Khan Y., Laurencin C.T. *J. Biomed. Mater. Res.* **2001**, *58*, 295–301.
61 Li S.H., de Wijn J.R., Layrolle P., de Groot K. *J. Biomed. Mater. Res.* **2002**, *61*, 109–120.
62 Roy D.M., Linnehan S.K. *Nature* **1974**, *247*, 220–222.
63 Dard M., Bauer A., Liebendorger A., Wahlig H., Dingeldein E. *Acta Odonto. Stom.* **1994**, *185*, 61–69.
64 Thalgott J.S., Fritts K., Giuffre J.M., Timlin M., *Spine* **1999**, *24* (13), 1295–1299.
65 Walsh W.R., Loefler A., Nicklin S., Arm D., Stanford R.E., Yu Y., Harris R., Gillies R.M. *Eur. Spine J.* **2004**, 13 (4), 359–366.
66 Charriere E., Lemaitre J., Zysset Ph. *Biomaterials* **2003**, *24*, 809–817.
67 Ramay H.R., Zhang M. *Biomaterials* **2003**, *24*, 3293–3302.
68 Zhang Y., Zhang M. *J. Biomed. Mater. Res.* **2002**, *61*, 1–8.
69 Fidancevska E., Milosevski M., Bossert J., Boccaccini A.R. *Ceram. Crist.* **2001**, *132*, 30–34.
70 Lee Y.–K., Park Y.S., Kim M.C., Kim K.M., Kim K.N., Choi S.H., Kim C.K., Jung H.S., You C.K., Legeros R.Z. *Key Eng. Mater.* **2004**, *254–256*, 1079–1082.

71 Kitamura M., Ohtsuki C., Ogata S., Kamitakahara M., Tanihara M. *Key Eng. Mater.* **2004**, *254–256*, 965–968.

72 Miao X., Hu Y., Liu J., Wong A.P. *Mater. Lett.* **2004**, *58*, 397–402.

73 Ebaretonbofa E., Evans J.R.G. *J. Porous Mater.* **2002**, *9*, 257–263.

74 Queiroz A.C., Teixeira S., Santos J.D., Monteiro F.J. *Key Eng. Mater.* **2004**, *254–256*, 997–1000.

75 Lemos A.F., Santos J.D., Ferreira J.M.F. *Key Eng. Mater.* **2004**, *254–256*, 1033–1036.

76 Santos J.D., Knowles J.C., Reis R.L., Monteiro F.J., Hastings G.W., *Biomaterials* **1994**, *15*, 5–11.

77 Lopes M.A., Knowles J.C., Santos J.D., Monteiro F.J., Olsen I. *J. Biomed. Mater. Res.* **1998**, *41*, 649–655.

78 Deisinger U., Stenzel F., Ziegler G. *Key Eng. Mater.* **2004**, *254–256*, 977–980.

79 Lemos A.R., Ferreira J.M.F. *Key Eng. Mater.* **2004**, *254–256*, 1037–1040.

80 Lemos A.F., Ferreira J.M.F. *Key Eng. Mater.* **2004**, *254–256*, 1041–1044.

81 Lyckfeldt O., Ferreira J.M.F., *J. Eur. Ceram. Soc.* **1998**, *18*, 131–135.

82 Lyckfeldt O., Brandt J., Lesca S., *J. Eur. Ceram. Soc.* **2000**, *20*, 2551–2556.

83 Olhero S.M., Tari G., Coimbra M.A., Ferreira J.M.F., *J. Eur. Ceram. Soc.* **2000**, *20*, 423–430.

84 Lemos A.F., Ferreira J.M.F. *Key Eng. Mater.* **2004**, *254–256*, 1045–1048.

85 Sepulveda P., Binner J.G.P., *J. Eur. Ceram. Soc.* **1999**, *19*, 2053–2060.

86 Sepulveda P., Bressiani A.H., Bressiani J.C., Meseguer L., Koenig Jr. B. *J. Biomed. Mater. Res.* **2002**, *62*, 587–592.

87 Sepulveda P. *Am. Ceram. Soc. Bull.* **1997**, *76* (10), 61–65.

88 Sepulveda P., Binner J.G.P., Rogero S.O., Higa O.Z., Bressiani J.C. *J. Biomed. Mater. Res.*, **2000**, *50*, 27–34.

89 Martinetti R., Dolcini L., Belpassi A., Quarto R., Mastrogiacomo M., Cancedda R., Labanti M. *Key Eng. Mater.* **2004**, *254–256*, 1095–1098.

90 Almirall A., Larrecq G., Delgado J.A., Martinez S., Planell J.A., Ginebra M.P. *Biomaterials* **2004**, *25*, 3671–3680.

91 Tadic D., Beckmann F., Schwarz K., Epple M., *Biomaterials* **2004**, *25*, 3335–3340.

92 Ben-Nissan B., *Curr. Opin. Solid State Mater. Sci.* **2003**, *7*, 283–288.

93 Kim S. R, Lee J.H., Kim Y.T., Jung S.J., Lee Y.J., Song H., Kim Y.H. *Key Eng. Mater.* **2004**, *254–256*, 969–972.

94 Rodriguez-Lorenzo L.M., Gross K.A. *Key Eng. Mater.* **2004**, *254–256*, 957–960.

95 Bose S., Darsell J., Kintner M., Hosick H., Bandyopadhyay A. *Mater. Sci. Eng. C* **2003**, *23*, 479–486.

96 Wilson C.E., de Bruijin J.D., van Blitterswijk, C.A., Verbout A.J., Dhert, W.J.A. *J. Biomed. Mater. Res.* **2004**, *68A*, 123–132.

97 Levy R.A., Chu T.M., Halloran J.W., Feinberg S.E., Hollister S. *Am. J. Neuroradiol.* **1997**, *18*, 1552–1525.

98 Chu T.-M.G., Orton D.G., Hollister S.J., Feinberg S.E., Halloran J.W. *Biomaterials* **2002**, *23*, 1283–1293.

99 Gomes de Sousa C., Evans J.R.G. *J. Am. Ceram. Soc.* **2003**, *86*, 517–519.

100 Adolfsson E. *Key Eng. Mater.* **2004**, *254–256*, 1025–1028.

101 Limpanuphap S., Derby B. *J. Mater. Sci. Mater. Med.* **2002**, *13*, 1163–1166.

102 Tan K.H., Chua C.K., Leong K.F., Cheah C.M., Cheang P., Abu Bakar M.S., Cha S.W. *Biomaterials* **2003**, *24*, 3115–3123.

103 Roy T.D., Simon J.L., Ricci J.L., Rekow E.D., Thompson V.P., Parsons J.R. *J. Biomed. Mater. Res.* **2003**, *67A*, 1228–1237.

104 Taboas J.M., Maddox R.D., Krebsbach P.H., Hollister S.J. *Biomaterials* **2003**, *24*, 181–194.

105 Li P., Zhang F. *J. Non-Cryst. Solids* **1990**, *119*, 112–116.

106 Sepulveda P., Jones J.R., Hench L.L. *J. Biomed. Mater. Res.* **2002**, *61* (2), 301–311.

107 Hench L.L., West J.K. *Chem. Rev.* **1990**, *90*, 33–72.

108 Lebeau B., Fowler C.E., Mann S., Farcet C., Charleux B., Sanchez C. *J. Mater. Chem.* **2000**, *10*, 2105–2108.

109 Khramov A.N., Collinson M.M. *Chem. Commun.* **2001**, *8*, 767–768.

110 Yan H., Zhang K., Blandford C.F., Francis L.F., Stein, A. *Chem. Mater.* **2001**, *13*, 1374–1382.

111 Gun J., Lev O., Regev O., Pevzner S., Kucernak J. *Sol-Gel Sci.* **1998**, *13*, 189–193.

112 Wu, M., Fujiu, T., Messing, G.L. *J. Non-Cryst. Solids* **1990**, *121*, 407-412.

113 Rosen M.J., *Surfactants and Interfacial Phenomena*, 2nd ed., John Wiley & Sons, New York, 1989, pp. 277–303.
114 Sepulveda P., Jones J.R., Hench L.L. *J. Biomed. Mater. Res.* **2002**, *59* (2), 340–348.
115 Jones J.R., Hench L.L. *J. Mater. Sci.* **2003**, *38*, 3783–3790.
116 Jones J.R., Hench L.L. *J. Biomed. Mater. Res. Appl. Biomater.* **2004**, *68B*, 36–44.
117 Gough J.E., University of Manchester, personal communication, 2004.
118 Jones J R, Ehrenfried L, Hench L L. *Key Eng. Mat.* **2004**, *254–256*, 981–984.
119 Saravanapavan P., Gough J.E., Jones J.R., Hench L.L., *Key Eng. Mater.* **2004**, *254–256*, 1087–1090.
120 Healy K.E. *Curr. Opin. Solid State Mater. Sci.* **1999**, *4*, 381– 387.
121 Mansur H.S., Vasconcelos W.L., Lenza R.F.S., Oréfice R.L., Reis, R., Lobato Z.P., *J. Non-Cryst. Solids* **2000**, *273*, 109–115.
122 Lenza R.F.S., Jones J.R., Vasconcelos W.L., Hench L.L. *J. Biomed. Mater. Res.* **2003**, *67A*, 121–129.
123 Tan A., Romanska H.M., Lenza R., Jones J., Hench L.L., Polak J.M., Bishop A.E. *Key Eng Mater* **2003**, *240–242*, 719–724.
124 Kim H.-W., Kim H.-E., Knowles J.C., *Key Eng. Mater.* **2004**, *254–256*, 1103–1106.
125 Bose S., Darsell J., Hosick H.L., Yang L., Sarkar D.K., Bandyopadhayay A., *J. Mater. Sci. Mater. Med.* **2002**, *13*, 23–28.
126 Jun Y.-K., Kim W.H., Kweon O.-K., Hong S.-H. *Biomaterials* **2003**, *24*, 3731–3739.
127 Chang B.S., Lee C.K., Hong K.S., Youn H.J., Ryu H.S., Chung S.S., Park K.W., *Biomaterials* **2000**, 21, 1291–1298.
128 Le Huec J.C., Schaeverbeke T., Clement D., Faber J., Rebeller A.L., *Biomaterials* **1995**, *16*, 113–118.
129 Hattinangadi A., Bandyopadhyay A. *J. Am. Ceram. Soc.* **2000**, *83*, 2730–2736.
130 Haugen H., Will J., Koehler A., Hopfner U., Aigner J., Wintermantel E. *J. Eur. Ceram. Soc.* **2004**, *24*, 661–668.
131 Polonchuk L., Elbel J.L.E., Wintermantel E., Eppenberger H.M. *Biomaterials* **2000**, *21*, 539–550.
132 Fredel M.C., Boccaccini A.R. *J. Mater. Sci.* **1996**, *31*, 4375–4380.
133 Jokinen M., Paetsi M., Rahiala H., Peltola T., Ritala M., Rosenholm J.B. *J. Biomed. Mater. Res.* **1998**, *42*, 295–302.
134 Nygren H., Eriksson C., Lausmaa J., *J. Lab. Clinical Med.* **1997**, *129* (1), 35–46.
135 Nygren H., Tengvall P., Lundstrom, I., *J. Biomed. Mater. Res.* **1997**, *34*, 487–492.
136 Gonzalez P., Borrajo J.P., Serra J., Liste S., Chiussi S., Leon B., Semmelmann K., de Carlos A., Varela-Feria F.M., Martinez-Fernandez J., de Arellano-Lopez A.R. *Key Eng. Mater.* **2004**, *254–256*, 1029–1032.
137 Hing K.A., Best S.M., Bonfield W. *J. Mater. Med. Mater. Med.* **1999**, 10, 135–145.

5.9 Interpenetrating Composites

Jon Binner

5.9.1 Introduction

The conventional approaches to producing composites consisting of two phases typically result in materials that consist of a discrete phase dispersed in an otherwise homogeneous matrix material. That is, while the matrix phase is continuous throughout the material, the dispersed phase is either not interconnected in any of the three dimensions or, if continuous fibers are used, connected in a single dimension. Such composites are defined as 0–3 and 1–3 respectively in Fig. 1 [1]. However, a change in connectivity between the phases, from 0–0 (zero connectivity between either of the phases present) through to 3–3 (both phases interpenetrating in all three dimensions) can result in substantially different properties. Consider as an example a mixture of electrically conducting and insulating phases. The two extremes types of connectivity, 0–0 and 3–3, and the intermediate 0–3, are likely to display isotropic behavior with the 0–0 composite being insulating and the 3–3 composite conducting, whilst the other seven levels of connectivity will show different degrees of anisotropy. Truly interpenetrating composites are those that are defined as 3–3.

Although 3–3 connectivity is fairly common in natural composites such as bone and wood, there have been relatively few attempts at creating it synthetically [2]. The difficulty with fabricating a truly interpenetrating network lies in achieving the required connectivity and spatial distribution of the phases, especially on a fine scale. However, the ability to fabricate by design such an interpenetrating microstructure raises the possibility of developing materials with truly multifunctional properties; each phase contributes its own characteristics to the macroscopic properties of the composite. For instance, one phase might provide high strength or wear resistance, while the other contributes a different property such as electrical conductivity. In addition, since many continuum properties change abruptly at the percolation threshold (the limit where connectivity is gained or lost), possibly the greatest potential for making materials capable of exhibiting novel behavior lies with those that have marginally interconnected microstructures [2]. This necessitates the ability to control the fraction and structure of both phases, preferably with a degree of independence. However, until more of these interpenetrating composites are produced and their properties systematically investigated, there is little other guidance as to how the interconnectivity might affect the overall properties.

Cellular Ceramics: Structure, Manufacturing, Properties and Applications.
Michael Scheffler, Paolo Colombo (Eds.)

ISBN: 3-527-31320-6

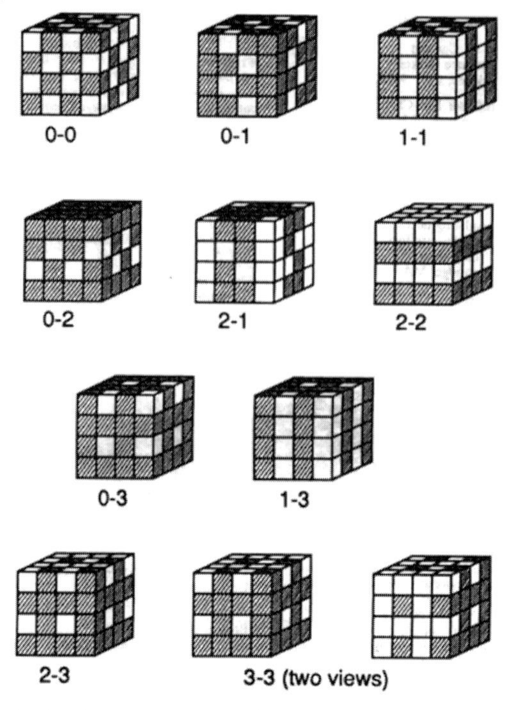

Fig. 1 The ten different levels of connectivity that can exist in composites, as defined by Newnham [1] (Newnham, R.E., Skinner, D.P., Cross, L.E., Mat. Res. Bull., 1978 13 525–536). Reprinted with permission from Elsevier.

One route to achieving tailorable 3–3 composites is infiltration of a second phase into porous materials that display complete pore interconnectivity. Provided the structure of the initial porous material can be precisely controlled in terms of the degree of porosity, the size and shape of the pores, the size of the windows between them, and the nature of the struts separating them, then there is the opportunity to design and fabricate interpenetrating composites with customized structures. Hence, the infiltration of cellular ceramics offers the potential for producing tailored ceramic-based interpenetrating composites with 3–3 interconnectivity.

5.9.2
Metal–Ceramic Interpenetrating Composites

A number of methods are available for fabricating cellular ceramics, and some of these yield structures that exhibit connected, open-celled porosity (see Part 2, and especially Chapters 2.1, 2.6, 2.7, and 2.8 of this book). Such structures can then be infiltrated with metals, polymers, or different ceramics by a number of techniques. Perhaps the majority of work involves introducing molten metal alloys by squeeze or die casting [3–8], the size of the porosity determining the degree of pressure required, though it must also be born in mind that the cellular ceramic preform must have the required strength to survive the process.

In this field, most work has involved the use of aluminum and aluminum alloys, for example, the work of Mattern et al. [6]. In this case, alumina cellular ceramics produced by using starch-based sacrificial pore-forming agents were infiltrated with the aluminum–silicon alloy AlSi12 by squeeze casting. A pressure of 100 MPa was found to be sufficient to infiltrate even the microporous walls of preforms sintered at low temperatures, and thus fully interpenetrating composites were achieved both on the macro- and microscale.

In principle, a wide range of metals and their alloys can be used provided the ceramic has sufficient thermal shock resistance to survive contact with the molten metal and does not react with it to form unwanted reaction products that deleteriously affect the properties of the resulting material. For example, Zeschky et al. [7] produced interpenetrating magnesium/ceramic composites by infiltrating polysiloxane-derived ceramic foam performs (82 % porous, 800 μm average cell diameter) with AZ 31 magnesium alloy by squeeze casting at 680 °C. Interfacial bonding was achieved by formation of spinel and cordierite in a reaction layer at the metal/ceramic interface. The dense metal–ceramic composites had significantly higher elastic modulus, yield strength, and creep resistance than the monolithic alloy at both room temperature and 135 °C. The results suggested that filler-loaded preceramic polymers have a high potential for optimization of lightweight cast metal components.

Di et al. produced Mg-alloy-based interpenetrating composites using "ecoceramics" as performs [8]. These are porous carbon materials that are obtained by impregnating wood-based waste materials with phenolic resin by using an ultrasonic impregnation system and then carbonized at a high temperature under vacuum. These cellular ceramic performs were then infiltrated with ZK60A Mg alloy in a vacuum high-pressure infiltration furnace to yield dense and homogenous interpenetrating composites with good mechanical properties and excellent damping capability.

As indicated above, 3–3 composites can display a range of useful properties, typically including increased strength, stiffness, hardness, wear and abrasion resistance, lower mass and thermal expansion coefficient, and better resistance to elevated temperatures and creep compared to the matrix metal while retaining adequate electrical and thermal conductivity. However, a disadvantage seems to be that they are inherently brittle as a result of the continuous nature of the ceramic phase. Whilst this limits their potential applications, one for aluminum-alloy-infiltrated ceramic foams is as brake disks for automobiles and trains.

This application in particular has been investigated during a research and development project involving Aachen University, Thyssen Guss AG, and SAB Wabco BSI in Germany [9]. The primary goal was to reduce the mass of the bogie of high-speed railway coaches (Fig. 2a). Since the brake disks and ancillary equipment comprise about 20 % of the total bogie mass, any reduction could have significant consequences. A range of different A356 aluminum-alloy-based metal-matrix composites were compared to the spheroidal-graphite iron disks currently used, and potential mass reductions of about 40 % were identified, while thermal conductivities that were more than five times higher reduced the problem of hot spots that could lead to thermal cracking. In addition, it is known that some conventional metal-matrix composites can offer very low wear rates, for example, aluminum alloy composites reinforced with silicon carbide particles are now being used for the brake disks of the Lotus Elise sports car [10]. Wear of the disks is minimal and their expected life exceeds 160 000km. However, although the infiltrated reticulated ceramics used in the work at Aachen University (Fig. 2b), showed a low rate of wear when tested under realistic loads, particularly for the higher ppi ceramics, they created unacceptably large wear for the brake pads, to the extent that further development of the existing pad material would be required [9].

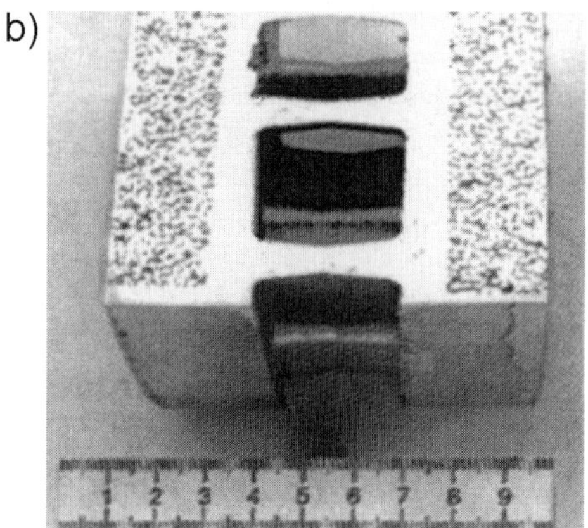

Fig. 2 Bogie for a high-speed railway carriage (a) and prototype ceramic foam reinforced brake disk (b), from Ref. [9] (Zeuner, T., Materials World, 1998, 17–19). Reprinted with permission from the Institute of Materials, Minerals and Mining.

5.9.3 Polymer–Ceramic Interpenetrating Composites

Similar approaches can also be applied to the infiltration of cellular ceramics with liquids that can be polymerized or cross-linked to result in interpenetrating polymer–ceramic composites. As for the metal–ceramic composites, the distribution of the two phases can be tailored by controlling the degree of porosity, pore size, and pore shape in the precursor ceramic foam. For example, anisotropy can be created by squeezing the ceramic foam while it is still flexible. Production also requires only three steps: 1) casting of the foam to the desired shape, 2) drying and sintering of the preform, and 3) infiltration of the preform with a suitable polymer.

Regarding the third step, evidence suggests that the use of (low) pressure to force the polymer into the ceramic preform results in a higher overall density, and hence superior properties, compared to sucking the polymer into the ceramic by using a vacuum [11]. The latter method can result in entrainment of air in the polymer.

Potential applications for polymer-infiltrated ceramic foams can be found in the building, automotive, and sensor industries [12]. For the first-named, the combination of their light weight, waterproofness, and capability for net-shape manufacture could result in a number of applications if manufacturing costs can be kept low enough. Possibilities include fire-resistant internal walls, cladding panels, and roofing tiles. In the automotive industry, while the inherently brittle nature of the continuous ceramic phase again prohibits structural applications, a potentially major use could be as fire-resistant underhood insulation.

However, the first application for interpenetrating polymer-ceramic composites will probably be in the field of sonar hydrophones. As long ago as 1978 it was shown that these type of composites can display a larger and more tailorable piezoelectric coefficient than dense piezoelectric ceramics, should have better acoustic coupling to water, and their buoyancy can be more easily adjusted than that of high-density lead zirconate titanate (PZT) ceramic. Furthermore, a more compliant material is more resistant towards mechanical shock and has a higher damping, which is desirable in a passive device [13].

Two approaches have been investigated with respect to replacing dense ceramics in hydrostatic applications. The first involved finding the desired properties in a single material, something that was more or less achieved with poly(vinylidene fluoride), PVF. However, whilst the hydrostatic voltage coefficient, compliance, and flexibility are desirably high and the density suitably low, a low piezoelectric strain coefficient means that the material is not of interest as an active device. In addition, while a high voltage sensitivity means it should be good for passive devices, when used as a hydrophone the material must be fixed to a curved surface that can flex in response to pressure changes. The difficulty lies in designing a sealed flexible mount for the polymer that will function when exposed to the high pressures that exist deep in the ocean and still retain sensitivity near the surface [13].

The second approach therefore involved the development of piezoelectric ceramic–polymer composites in which the properties of both materials are combined. Wenger et al. investigated the potential of such composites with 0–3 and 1–3 connectivity [14–17]. While they found them to be quite suitable for passive sensors, and they had the advantage of being flexible, composites with higher level connectivity offered superior responses.

Skinner et al. [13] used the replamine form process to produce 3–3 piezoelectric composites (Fig. 3) and investigated their properties. The process consisted of vacuum impregnating a coral skeleton with wax and then leaching away the skeleton with hydrochloric acid to leave a wax negative of the original coral. This was vacuum-infiltrated in turn with a PZT slip and then the wax was burned out at 300 °C to leave a replica of the original coral structure made of PZT. This was sintered and again infiltrated with a high-purity silicone rubber. The resultant rigid composite was connected with electrode layers and poled in an electric field; a flexible version of the composite was produced by crushing it to break the ceramic struts. While this lowered the permittivity by interrupting the electric flux, because the ceramic pieces were still held in position by the polymer matrix, stresses could still be transmitted. The net result was a composite with a high hydrostatic charge coefficient, while the permittivity was lowered and therefore the piezoelectric voltage coefficient was greatly enhanced. For the unbroken composites, a rigid polymer such as epoxy or polyester could be used, and a low-density, high-coupling resonator could be fabricated.

Fig. 3 Piezoelectric ceramic–polymer composite made by the replamine form process, from Ref. [13] (Skinner, D.P., Newnham, R.E. & Cross, L.E., Mat. Res. Bull. 1978 13 599–607). Reprinted with permission from Elsevier.

Shrout et al. [18] also investigated a PZT–polymer composite with 3–3 connectivity, was produced by mixing a PZT powder with poly(methyl methacrylate) spheres of 50–150 μm diameter in a 30:70 volume ratio. The mixture was pressed to form green bodies that were slowly heated to 400 °C to volatilize the spheres. The porous ceramic was then sintered and vacuum-infiltrated with silicone rubber or an epoxy resin. The resulting piezoelectric characteristics were found to be similar to those of the replamine form process described above, but the microstructure was more randomly orientated. The main advantage of this process lay in the fewer processing steps required.

While these two processing technologies were developed in the late 1970s, two more recent and sophisticated processes were patented in the late 1990s. General Electric Company in the USA developed a piezoelectric composite with anisotropic 3–3 connectivity [19–21]. A ceramic preform of interconnected lamellae was produced by freeze drying a ceramic slurry. The connectivity of the lamelli in the *z* direction was greater than in the other directions as a result of using an unidirectional heat flow. The preform was then sintered and vacuum infiltrated with a polymer or a low acoustic impedance glass or cement. Poling and addition of the electrodes produce the final composite material.

The second approach involved using the rapid prototyping concept of fused deposition [22, 23]. The object was built up in layers from a molten polymer filament that was extruded through a nozzle and deposited in the desired pattern by a computer-controlled platform that could be moved in the *x* and *y* directions by a suitable computer-aided design (CAD) file. The height of the part was generated by dropping the platform slightly in the *z* direction at the end of each layer. Two versions of the basic approach were used, a “direct” and an “indirect” route (Fig. 4). In the former, a polymer filled with ceramic powder was used to construct the desired green body directly; the polymer was subsequently burnt out by using the same techniques as in injection molding. In the indirect route, a polymer mold, created by the deposition process, was subsequently cured and then filled with PZT ceramic slurry. After drying the latter, the mold was removed by a heat treatment that resulted in the controlled combustion of the polymer. With both routes the ceramic green bodies produced were sintered prior to being encapsulated in the required polymer

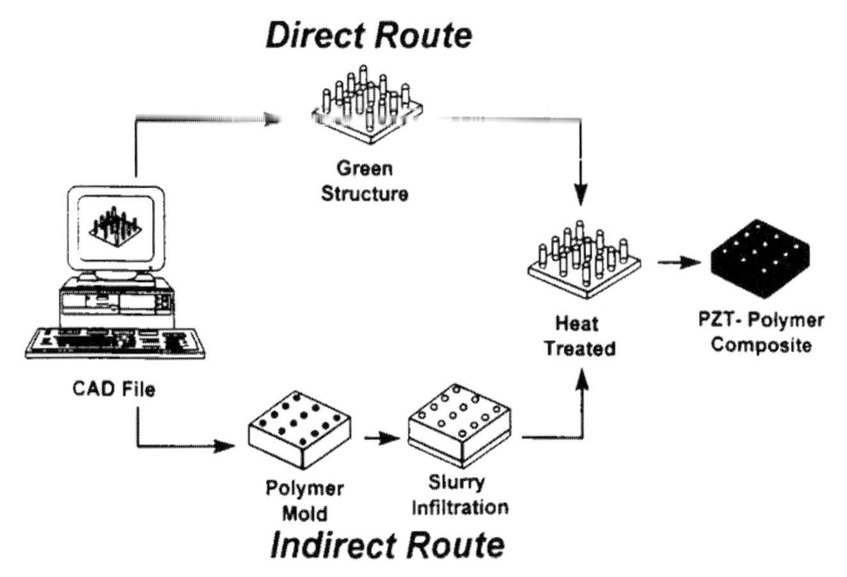

Fig. 4 Direct and indirect routes for the production of polymer–ceramic piezoelectric composites by rapid prototyping, after Ref. [23] (Bandyopadhyay, A., Panda, R.K., Janas, V.F., Agarwala, M.K., Danforth, S.C. and Safari, A., J. Am. Ceram. Soc. 1997 80 [6] 1366–1372). Reprinted with permission of The American Ceramic Society, www.ceramics.org. Copyright [1997]. All rights reserved.

to achieve composites that could have a range of different connectivities ranging from 1–3, shown in the figure, to 3–3. While this approach clearly yielded the greatest degree of sophistication it is by no means a simple route and a large number of process steps are required, increasing the cost and reducing the applicability for mass production.

5.9.4 Summary

Interpenetrating composites are now the subject of increasing research around the world, and new and more sophisticated routes to their fabrication are under development. They offer the promise of unusual combinations of properties and the ability to tailor properties to an extent that has not been available before. To this end, the infiltration of cellular ceramics with molten metals or polymers offers the chance for a simple and easy fabrication route to ceramic-based interpenetrating composites. In addition, since many continuum properties change abruptly at the percolation threshold, possibly the greatest potential for making materials capable of exhibiting novel behavior lies with those that have marginally interconnected microstructures. In this respect, the precise control over the cellular structure that is increasingly possible, as the other chapters in this book demonstrate, offers tremendous advantages.

Acknowledgements

The author would like to acknowledge the work of some of his former research students, Mr. Lars Monson and Mr. Suresh Talluri.

References

1 Newnham, R.E., Skinner, D.P., Cross, L.E., *Mater. Res. Bull.*, 1978 *13* 525–536.
2 Clarke, D.R., *J. Am. Ceram. Soc.* 1992 *75* [4] 739–759.
3 Howes, M.A.H., Advanced Composites, Conf. Proc. of the Am. Soc. for Metals, Dearborn, Michigan, USA, 1985, pp. 223–230.
4 Peng, H.X., Fan, Z., Evans, J.R.G., *Mater. Sci. Technol.*, 2000 *16* [7–8] 903–907.
5 Binner, J.G.P., *Ceram. Eng. Sci. Proc.*, 2003 *24* [3] 125–134.
6 Mattern, A., Huchler, B., Staudenecker, D., Oberacker, R., Nagel, A., Hoffmann, M.J., *J. Eur. Ceram. Soc.* 2004 *24* 3399–3408.
7 Zeschky, J., Lo, J.H., Scheffler, M., Hoeppel, H.-W., Arnold, M., Greil, P., *Z. Metallkd.* 2002 *93* [8] 812–818.
8 Di, Z., Xian-qing, X., Tong-xiang, F., Bing-he, S., Sakata, T., Mori, H., Okabe, T., *Mater. Sci. Eng.* 2003 *A351* 109–116.
9 Zeuner, T., *Mater. World*, 1998 6 [1] 17–19.
10 Hollins, M., *Metal Matrix Composites VI*, The Royal Society, London, UK, 1997.
11 Binner, J.G.P., Talluri, S., presented at Materials Congress, London, UK, April 5–7, 2004.
12 Monson, L., Diplomarbeit, Aachen University, 2001.
13 Skinner, D.P., Newnham, R.E., Cross, L.E., *Mater. Res. Bull.* 1978 *13* 599–607.

14 Wenger, M.P., Blanas, P., Shuford, R.J., Das-Gupta, D.K., *Polym. Eng. Sci.* 1996 *36* [24] 2945–2953.

15 Wenger, M.P., Almeida, P.L., Blanas, P., Shuford, R.J., Das-Gupta, D.K., *Polym. Eng. Sci.* 1999 *39* [3] 483–492.

16 Wenger, M.P., Blanas, P., Shuford, R.J., Das-Gupta, D.K., *Polym. Eng. Sci.* 1999 *39* [3] 508–518.

17 Wenger, M.P., Das-Gupta, D.K., *Polym. Eng. Sci.* 1999 *39* [7] 1176–1188.

18 Shrout, T.R., Schulze, W.A., Biggers, J.V., *Mater. Res. Bull.* 1979 *14* 1553–1559.

19 General Electric Company, US Pat. No. 5591372, 1997.

20 General Electric Company, US Pat. No 5660877, 1997.

21 General Electric Company, Europ Pat. No. EP 0764994, 1996.

22 Safari, A., Janas, V.F., Bandyopadhyay, A., Panda, R.K., Agarwala, M., Danforth, S.C., US Pat. No. 6004500, 1999.

23 Bandyopadhyay, A., Panda, R.K., Janas, V.F., Agarwala, M.K., Danforth, S.C., Safari, A., *J. Am. Ceram. Soc.* 1997 *80* [6] 1366–1372.

5.10
Porous Media in Internal Combustion Engines

Miroslaw Weclas

5.10.1
Introduction

Modern direct-injection (DI) combustion engines are characterized by low specific fuel consumption and high raw emission levels. Reduction of these emissions (especially NO_x and soot particles) requires complex post-treatment techniques and engine-management systems. Thus, future combustion engines require new concepts with extreme reduction of raw emissions from the primary combustion process with at least unchanged or reduced fuel consumption. Advantageously, this would even be associated with a reduced need for expensive post-treatment of exhaust gases in future vehicles.

Conventional diesel engines work with an injection pressure of up to 2000 bar and local combustion temperatures can reach up to 2400 K. A problem is the time required to obtain a homogeneous mixture before ignition, and moreover the diffusion-controlled combustion process leads to a flame front and temperature gradient during combustion. These processes, each of which is of very complex nature, lead to the trade-off of the diesel engine: the formation of soot particles and NO_x. Soot particles are formed in oxygen-lean regions and/or at peak temperatures of the combustion process, while NO_x is formed in oxygen-rich regions. It is not possible in conventional combustion systems to decrease the amount of one of the components without increasing the amount of the other.

Current technologies such as electronically controlled high-pressure injection systems, variable valve control, exhaust gas recirculation (EGR), and combinations thereof, however, do not automatically solve the problem of engine emissions under all operational conditions. It is thus necessary to establish new concepts for mixture formation and combustion that allow development of future "clean reciprocating engines".

Generally, a possible solution is a combustion system operating with a gaseous, homogeneous air/fuel (A/F) mixture (from very lean to nearly stoichiometric mixture compositions), characterized by a homogeneous combustion process in the cylinder. Such a system could also reduce the specific fuel consumption under conditions of near-zero combustion emissions. A possible route to achieve these conditions is the use of porous media in the combustion zone of a combustion engine. In this chapter the use of ceramic foams as porous media (PM) in the combustion

Cellular Ceramics: Structure, Manufacturing, Properties and Applications.
Michael Scheffler, Paolo Colombo (Eds.)

ISBN: 3-527-31320-6

zone is described and the advantages of this development are discussed. The emission sources and their origin as a complex interplay of various parameters is schematically shown in Fig. 1.

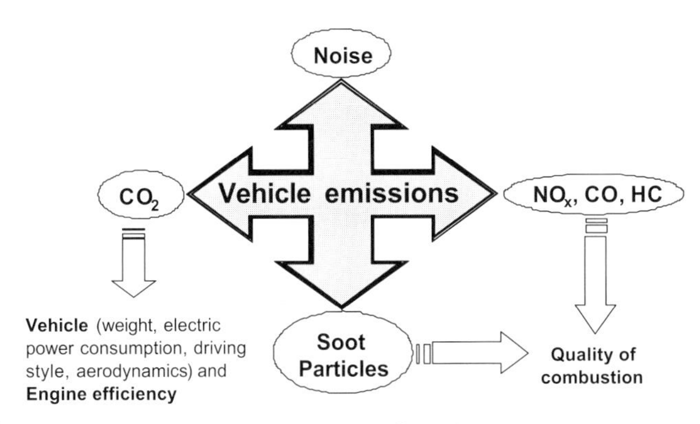

Fig. 1 Main components and sources of vehicle emissions.

5.10.2 Novel Engine Combustion Concepts with Homogeneous Combustion Processes

Novel concepts for future engine combustion systems must allow homogenization of the combustion process over a wide range of engine operational conditions caused by vehicle load, speed, traffic conditions, and driver habits. Some additional requirements for the future engine follow:

- Lowest possible specific fuel consumption
- Zero or near-zero combustion emissions
- Homogeneous combustion (from homogeneous ultralean to (nearly) stoichiometric mixtures)
- Higher power density
- High durability and low cost.

Already realised technologies are:

- Control of air supply (intelligent valve technology, variable intake system geometry, non-throttle operation)
- Variable fuel supply (high-pressure injection, electronically controlled injection strategies, nozzle design, water injection)
- Supercharging and downsizing, variable compression ratio
- EGR (cold and hot), and exhaust post-treatment (NO_x, HC, CO, and soot).

This chapter focuses on homogeneous combustion processes in internal combustion engines. Homogeneous combustion can be facilitated by the use of porous media, similar to porous-burner technology (see Chapter 5.5 of this book).

Homogeneous combustion in internal combustion (IC) engines is defined as a process in which a 3D ignition of the homogeneous charge is followed by simultaneous heat release in the entire combustion chamber volume with a homogenous temperature field (Fig. 2). Besides volumetric ignition, which plays a dominant role, further aspects must be taken into account: 1) Combustion temperature, especially for close-to-stochiometric A/F compositions. It is necessary to lower this temperature from the adiabatic level to allow near-zero NO_x emissions. 2) The service range of engine loads, that is, the extent to which the homogeneous combustion system can operate over a wide range of engine loads, from very lean to nearly stoichiometric A/F compositions. 3) A wide range of vehicle speed must also be considered. To satisfy the above conditions it is necessary to control the ignition timing under variable operational conditions, and to control the rate of heat release for different mixture compositions. Moreover, for low emissions it is necessary that the liquid fuel is completely vaporized prior to ignition.

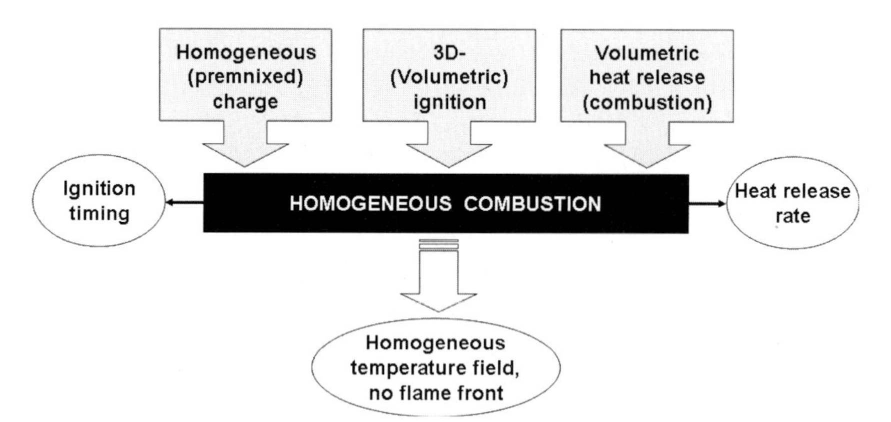

Fig. 2 Conditions and requirements for homogeneous combustion in engines.

According to the definition given above, three steps of the mixture formation and combustion can be selected that define the ability of a given system to operate as a homogeneous combustion system: 1) homogenization of cylinder charge, 2) ignition conditions, and 3) heat-release process in the temperature field.

A critical factor of a system operating with homogeneous combustion is the ignition process. Among the four different ignition strategies – local ignition by spark plug, thermal self-ignition by compression, controlled (low-temperature chemical) autoignition, and 3D thermal PM self-ignition – the last three can provide a homogeneous combustion process.

A critical point is the control of ignition timing and heat-release rate under variable engine loads. This is of special importance in so-called homogeneous-charge compression-ignition (HCCI) systems, which are currently under investigation [1–3]. The following aspects must be considered for further development of HCCI systems (in general for all homogeneous combustion systems) for a wide range of engine operational conditions: controlling ignition timing; controlling burn rate; extending the service range to high engine loads; minimizing HC and CO emissions; NO_x emissions at high loads; cold-start conditions; and engine transient response [4, 5].

5.10.3 Application of Porous-Medium Technology in IC Engines

Porous-medium (PM) technology is defined here as the use of the specific and unique features of highly porous media (cellular structures) for supporting mixture formation, ignition, and combustion processes in IC engines [5–7]. Most of these processes when performed in a PM volume have drastically different features from those observed in a free volume.

Generally, the most important parameters of PM for application to combustion technology are specific surface area; heat-transport properties (optical thickness, conductivity); heat capacity; accessibility to fluid flow and flame propagation; pore size; pore density; pore structure; thermal resistance; mechanical resistance and mechanical properties under heating/cooling conditions; surface properties; and electrical properties (for electrical heating of PM structures). The most important specifications of porous materials as applied to combustion processes in engines are listed in Table 1.

Table 1 Critical parameters for porous media applications in IC engines

Parameter	Range of values required/expected	Suitable materials
Specific surface area	large: must be adapted to particular application	all available materials
Heat-transport properties (optical thickness, conductivity)	excellent; especially important for PM engine concept	SiC, NiCrAl, ZrO_2
Heat capacity (as compared to gas)	large (a few hundreds and more): must be adapted to particular application; in case of PM engine defines the engine dynamic properties	all available materials; preferably SiC, NiCrAl, Al_2O_3
Permeability to flow and flame propagation	for gas or liquid flow more than 80 % porosity; preferably more than 90 % porosity with open pores (cells)	all available materials; preferably SiC, Al_2O_3
Pore size and size distribution	typical pore size from 1 to 5 mm. Pore density from 30 to 8 ppi; for flame propagation under pressure see Péclet number criterion: pore size greater than 1 mm	all available materials; preferably foams and random structures
Pore shape	in principle all available shapes are suitable (open cells)	all available materials; preferably foams and random structures with regular pore shapes
Strength (bending, compression, fatigue)	according to concept and place in engine	SiC, NiCrAl, ZrO_2

Table 1 Continued

Parameter	Range of values required/expected	Suitable materials
Thermal shock resistance	high, especially for PM engine concept	SiC, ZrO_2
Corrosion resistance	high, especially in the atmosphere of burned gases	all available ceramic materials and light metals
Electrical properties	for direct heating: high electrical resistance and homogeneous energy distribution (preferably voltage 12 V and current 10–80 A); attainable temperatures of 1500 K	preferably foams and other regular structures; good experience with SiC for homogeneous temperature field; relatively long heating duration (minutes)
Available maximum temperature	depends on the application; PM engine concept $T_{max} < 2000$ K; MDI concept $T_{max} < 1500$ K; two-stage combustion concept $T_{max} < 1800$ K	preferably SiC, Al_2O_3
Porous-medium mechanical stability	important under conditions of high temperature and pressure; very critical factor in the case of ceramic material mounted in the piston top; accelerations up to 500 g	metal foams
Porous medium, mounting in engine components	very critical factor, especially in the case of ceramic materials; possibly with a high-temperature ceramic adhesive; important is also that the porous reactor can be cold or very hot	metal foams
Variable geometry	important for all engine applications (adapting to available space and shape)	all available materials; preferably metal foams
Long-term stability	should be very high; however, almost unknown area for engine applications	must be realized for engine applications

A large specific surface area can be utilized for excellent interphase heat transfer in the PM volume, fuel (spray) distribution throughout the PM volume, and for very fast liquid vaporization in the PM volume. This is supported by excellent heat-transfer properties, especially by strong thermal radiation of the solid phase. Together with a high heat capacity and thermal resistance of the PM material, this kind of 3D structures can be used for realization of very clean and efficient homogeneous combustion in a PM volume. High porosity of porous structures (more than 80 %) leads very good properties for gas flow, spray, and flame throughout the PM volume. Direct electrical heating of porous reactors can be utilized for cold-start conditions and vaporization of liquid fuel, but also for afterburning and self-cleaning process. Especially for ceramic materials the method of inserting the PM reactor in engine components (e.g., engine head or piston top) may significantly limit the applicability of this technology to engines. Here further investigations and development of new materials are necessary. These requirements become critical if the porous medium is used directly for controlling the combustion process (as a so-called PM reactor) under high-pressure conditions. Structural features and properties of porous media, especially ceramic foams can be found in Chapters 2.1, 2.6, 3.1, and 4.3 of this book.

The following engine processes can be supported by the application of highly porous medium:

- *Energy recirculation in engine cycle* in the form of hot burned gases or combustion energy recirculation. This may significantly influence thermodynamic properties of the gas in the cylinder and can control the ignitability (activity) of the charge. This energy recirculation can be performed under different pressure and temperature conditions during the engine cycle. Additionally, this heat recuperation can be used for controlling the combustion temperature.
- *Fuel injection in the PM volume*: especially unique features of liquid-jet distribution and homogenization throughout the PM volume (effect of self-homogenization) [8] are very attractive for mixture formation in the PM volume (see Fig. 3).
- *Fuel vaporization in PM volume*: a combination of large heat capacity of the PM material, large specific surface area, and excellent heat transfer in the PM volume make the vaporization of liquid fuel very fast and complete.
- *Mixing and homogenization in the PM volume*: unique features of the flow properties inside 3D structures allow very effective mixing and homogenization in the PM volume.
- *3D thermal PM ignition* (if the PM temperature is at least equal to the ignition temperature under certain thermodynamic conditions and mixture composition): this new kind of ignition is especially effective if the PM volume automatically creates the combustion chamber volume [6, 9, 10].
- *Heat release in the PM volume under controlled combustion temperature*: there is only one kind of combustion, and this allows homogeneous combustion conditions almost independent of the engine load with the possibility of controlling the combustion temperature [6, 9, 10].

Four new concepts concerning applications of PM technology to mixture formation and combustion in IC engines can be considered:

- Combustion system with mixture formation and homogeneous combustion in the PM volume, so-called *PM engine concept* [9, 10].
- Mixture formation system, with heat recuperation, vaporization, and chemical recombination in PM volume, so-called *MDI concept* (mixture direct injection) [7].
- *Intelligent engine concept* based on the MDI system permitting homogeneous combustion conditions (in a free cylinder volume) over a wide range of engine operational conditions [7].
- Phased combustion system for conventional DI diesel engine, with temporal and spatial control of mixture composition by utilization of interaction between diesel jet and PM structure, so-called *two-stage combustion*.

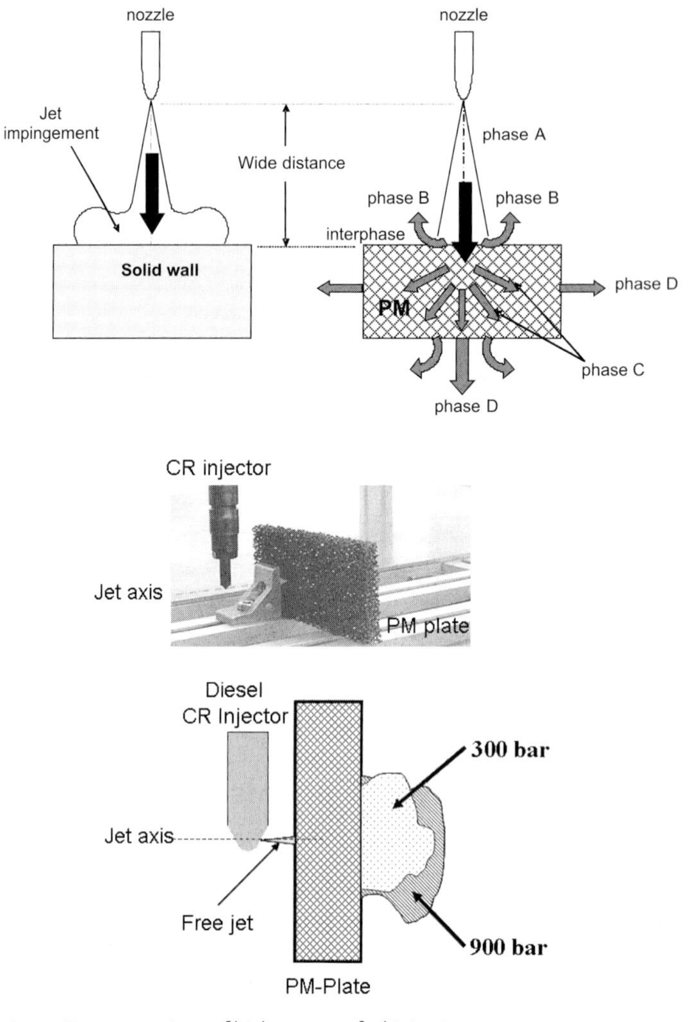

Fig. 3 Top: comparison of high-pressure fuel injection onto solid wall (left) and into porous medium (right). Common-rail diesel injection onto porous plate: experimental setup (middle) and jet dispersion after passing a ceramic foam plate (bottom).

Most of the PM technologies reported in the literature focus on internal heat recuperation in engines [11–22]. The main goal of such PM applications in internal combustion engines is to influence the thermal efficiency of the engine by internal heat recuperation. There are also concepts combining heat regeneration and catalytic reduction of toxic components, gaseous and particulate. Heat flux and energy recirculation in such engines have been described in detail in Ref. [11]. In this case the heat recuperator is attached to a rod and moves inside the cylinder, synchronized to the piston movement. The main advantage of such internal (in-cylinder) heat recuperation between burned gases and fresh air is high volumetric efficiency of the cylinder, which is necessary for high power density of the engine.

This kind of application of the porous medium to internal combustion engine deals with energy balance of the cycle. Engines with heat recuperation could realize much higher combustion temperatures which result in much higher NO_x emissions.

The presented engine concept could also be extended by application of catalytic porous insert offering afterburning of combustion products such as particles. A similar concept was recently analyzed by Hanamura and Nishio [12]. Also in this engine the maximum combustion temperature is higher than adiabatic owing to heat recuperation in a porous medium.

Another application of porous-medium technology to internal combustion engines covers exhaust post-treatment systems, especially catalytic converters and particle filters. More information on application of porous-medium technology, especially to diesel particle filters, can be found in the literature [23–28] (see also Chapter 5.2 of this book).

5.10.4
The PM Engine Concept: Internal Combustion Engine with Mixture Formation and Homogeneous Combustion in a PM Reactor

Here PM engine is defined as an internal combustion engine in which a homogeneous combustion process is realized in a porous-medium volume. The following individual processes of the PM engine can be realized in the volume of a porous medium: internal heat recuperation, fuel injection, fuel vaporization, mixing with air, homogenization of charge, 3D thermal self-ignition, and homogeneous combus-

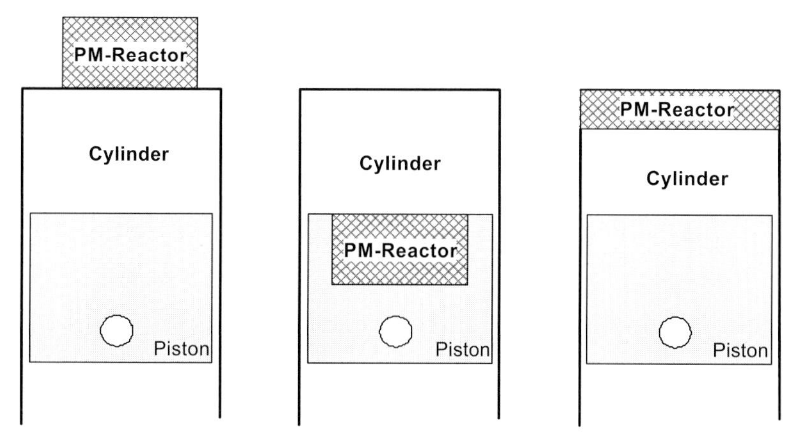

Fig. 4 Possible locations of PM reactor in PM engines: in head (left), in piston (middle), and in cylinder (right).

tion. The TDC (top dead center) compression volume is equal to the PM volume, and this limits the size of engine combustion chamber. Outside the PM volume no combustion occurs in the cylinder.

PM engines may be classified with respect to the timing of heat recuperation in engine as:

- Engines with periodic contact between PM and cylinder (so-called closed PM chamber).
- Engines with permanent contact between PM and cylinder (so-called open PM chamber).

Another classification criterion concerns the location of the PM reactor in the engine: engine head, cylinder, or piston (Fig. 4).

Another interesting feature of PM engines is their ability to operate with different liquid and gaseous fuels. Independent of the fuel, this engine is a 3D PM thermal self-ignition engine. Finally, the PM engine concept can be applied to both two- and four-stroke engines.

5.10.4.1 PM Engine with Closed PM Chamber

The operation of an engine with a closed PM chamber (periodic contact between working gas and PM heat recuperator) is shown in Fig. 5. At the end of the expansion stroke (Fig. 5e) the valve controlling timing of the PM chamber closes and fuel can be injected into the PM volume. This chamber is a low-pressure chamber and sufficient time is available for fuel supply and vaporization in the PM volume.

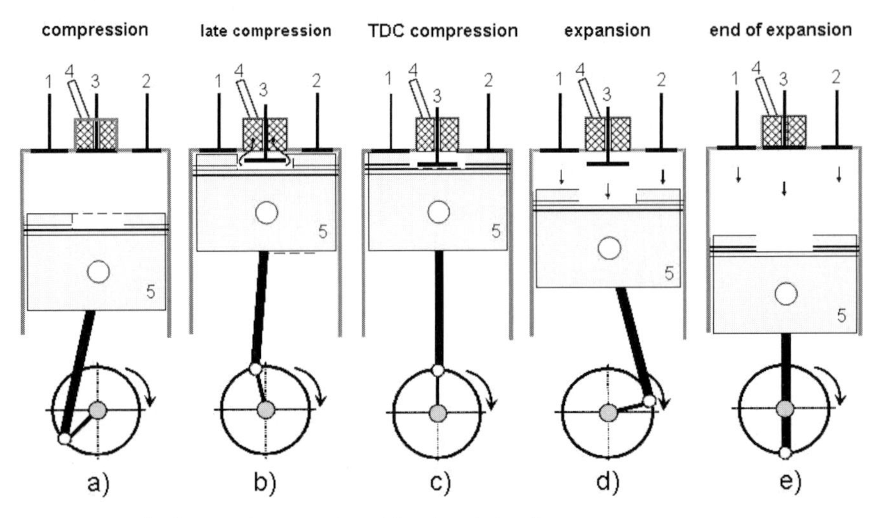

Fig. 5 Principle of PM engine operation with a closed chamber; 1) intake valve, 2) exhaust valve, 3) PM chamber valve, 4) fuel injector, 5) piston.

Simultaneously, other processes can be performed in the cylinder volume and continued through exhaust, intake, and compression strokes (Fig. 5a). Near the TDC of compression (Fig. 5b) the valve in the PM chamber opens and the compressed air flows from the cylinder to the hot PM containing fuel vapor. Very fast mixing of both gases occurs before mixture ignition in the whole PM volume (Fig. 5c). The resulting heat release process occurs simultaneously in the entire PM volume. Three necessary conditions for a homogeneous combustion are fulfilled here: homogenization of charge in the PM volume, 3D thermal self-ignition in the PM volume, and volumetric combustion with a homogeneous temperature field in the PM volume. Additionally, the PM acts as a heat capacitor and controls the combustion temperature.

5.10.4.2
PM Engine with Open PM Chamber

Another possible realization of the PM engine is a combustion system with a permanent contact between working gas and PM reactor. In this case it is assumed that the PM combustion chamber is mounted in the engine head, as shown in Fig. 6. During the intake stroke (Fig. 6a) the PM heat capacitor has a weak influence on the in-cylinder thermodynamic air conditions. Also during early compression stroke, only small amount of air contact the hot PM. This heat exchange process (nonadiabatic compression) increases with continuing compression timing (Fig. 6b), and at the TDC the whole combustion air is entrapped in the PM volume. Near the TDC of compression the fuel is injected into the PM volume (Fig. 6c) and very fast fuel vaporization and mixing with air occur in the 3D structure of the PM. A 3D thermal

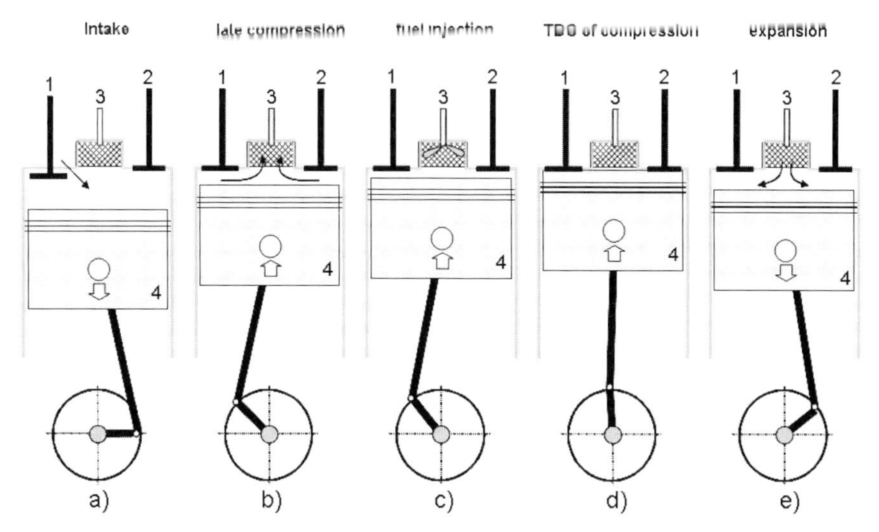

Fig. 6 Principle of PM engine operation with an open chamber; 1) intake valve, 2) exhaust valve, 3) fuel injector, 4) piston.

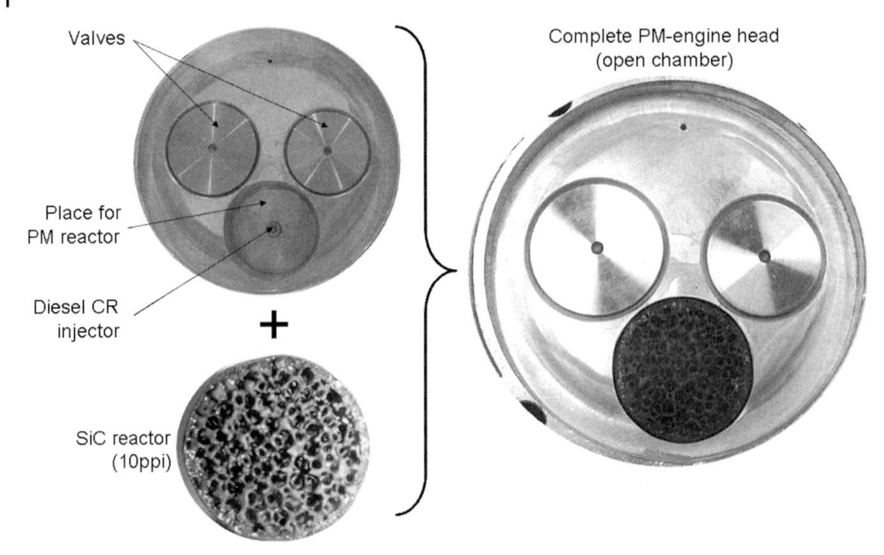

Fig. 7 View of the PM engine head with open chamber built on the basis of one-cylinder DI diesel engine

self-ignition of the resulting mixture follows in the PM volume together with a volumetric combustion characterized by a homogeneous temperature distribution in the PM combustion chamber (Fig. 6d). During the following expansion stroke the heat is transformed into mechanical work (Fig. 6e). Again, all necessary conditions for homogeneous combustion are fulfilled in the PM combustion volume. To describe the main thermodynamic properties of the PM engine cycle, a thermodynamic model was proposed, and a more detailed thermodynamic analysis of the PM engine concept can be found in Ref. [9]. An example of a PM engine head with open chamber and PM reactor mounted in the engine head is shown in Fig. 7. The porous-medium reactor made of SiC is mounted in the head of a single-cylinder direct injection diesel engine. The injection nozzle is positioned inside the porous medium reactor and permits very good homogenization throughout the PM volume. The reactor volume is approximately 45 cm^3 with a mean pore diameter of 3 mm. The outer surface of reactor consists of solid wall with thermal inslation.

5.10.5
An Update of the MDI Engine Concept: Intelligent Engine Concept with PM Chamber for Mixture Formation

The MDI (mixture direct injection) concept covers the mixture-formation and heat-recuperation system. This concept offers homogenization of the combustion process by performing fuel vaporization, its chemical recombination (low-temperature oxidation processes, e.g., cool and blue flames), and energy recirculation in a porous-medium chamber. The enthalpy of the burned gases is partly transferred to the porous medium and can later be supplied back to the cylinder. This energy is utilized for both vaporization of liquid fuel and for its chemical recombination in the PM volume [7, 20, 21].

A practical approach to the MDI system requires a porous-medium chamber to be mounted in proximity to the cylinder and equipped with a valve allowing contact between PM chamber and cylinder volume. The engine cycle described below models the real engine cycle, and timings for the PM chamber other than those presented here can be used. The MDI concept can be combined with conventional combustion modes, such as GDI (gasoline direct injection), HCCI (homogeneous-charge compression ignition), and with radical combustion (RC), and only control of the PM-chamber timing is necessary to select a combustion mode in the engine [7]. By applying the variable timing of the PM chamber, the MDI concept offers a

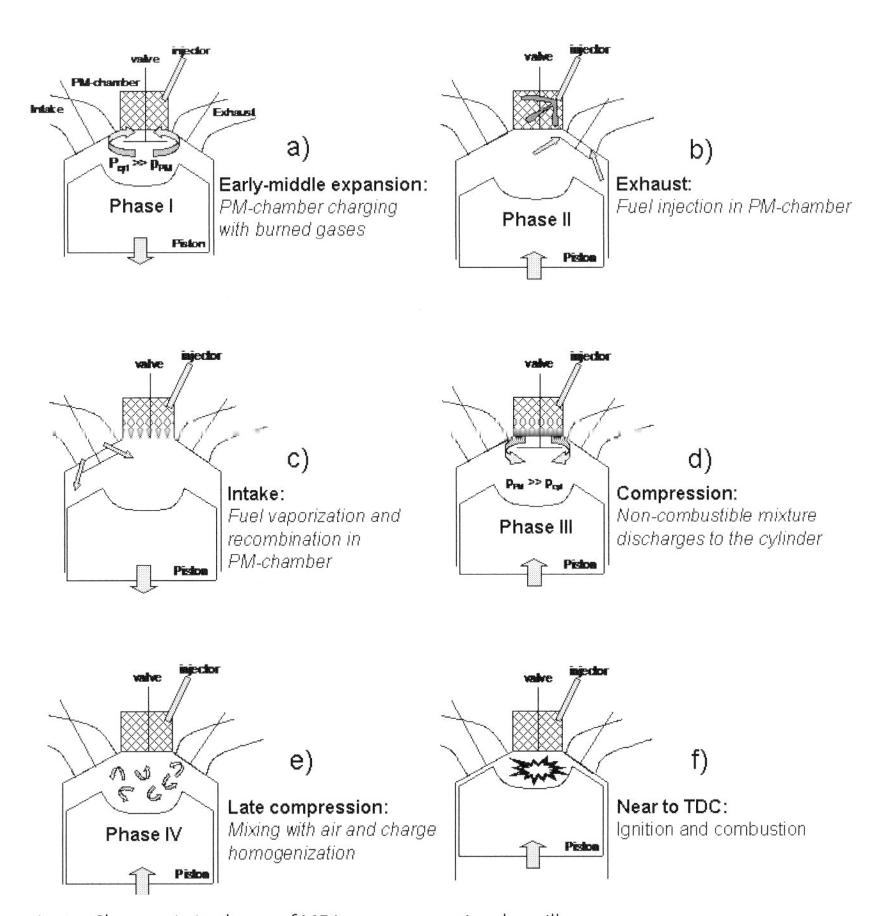

Fig. 8 Characteristic phases of MDI system operation, here illustrated as an updated concept in association with a porous medium in the engine head.

combination of individual combustion modes in one engine, as described below. Characteristic phases of the cycle with MDI mixture preparation are illustrated in Fig. 8. In phase I the PM chamber is charged with burned gases containing energy (Fig. 8a). In phase II the liquid fuel is injected into the PM chamber and fuel vaporization occurs (Fig. 8b). In phase III gaseous charge containing evaporated fuel, energy, and active radicals discharges from the PM chamber into the cylinder (noncombustible mixture; Fig. 8d), and in phase IV mixing with air in the cylinder and ignition of combustible mixture take place (Fig. 8e). The system considered in Fig. 7 consists of the cylinder with a moving piston and the PM chamber equipped with a poppet valve. This valve allows control of the PM-chamber timing. A detailed analysis of the MDI concept can be found in Ref. [7].

Combination of the MDI concept with other individual conventional combustion systems (GDI, HCCI, RC) provides several advantages, including extension of lean effective limit and improved ignitability of a homogeneous charge, reduction of temperature peaks under lean operation conditions (homogeneous charge), elimination of soot, better and faster homogenization of the charge in the cylinder, and control of the concentration of active radicals almost independent of the cylinder conditions.

For homogeneous combustion under variable engine load and speed, ignition conditions and charge reactivity are also required to be variable. This variability means variable ignition and combustion mode. The combination of these variable conditions allows not only realization of homogeneous combustion conditions but also control of ignition timing and heat-release rate. Both aspects define the practicability of the combustion system operating under homogeneous combustion conditions. Thus, the variable timing of the MDI concept allows control of the cylinder charge parameters, which is necessary for realization of homogenous combustion: TDC compression temperature; temperature history during the compression stroke; reactivity of the charge; homogeneity of the charge; and heat capacity of the charge. This variability defines the "intelligence" of the engine.

5.10.6
Two-Stage Combustion System for DI Diesel Engine

Another application of porous-medium technology to mixture formation and combustion in the DI diesel engine is the two-stage combustion concept. This concept offers control of mixture formation and combustion conditions in direct injection diesel engines by spatial splitting of the combustion process into three zones and two time phases, as shown in Fig. 9. Zone 1 is the volume of the inner part of the PM ring, zone 2 the volume in the PM ring itself, and zone 3 the free volume between outer part of the PM ring and the piston bowl.

Main features of the two-stage combustion system are:

The system operates under two characteristic conditions related to partial and full load (i.e., small amount of fuel injected under low pressure, and large amount of fuel injected under high pressure).

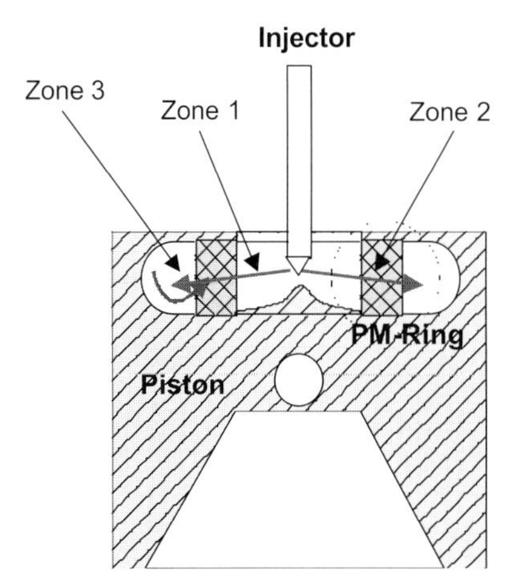

Fig. 9 Principle of PM ring in the piston bowl of a diesel engine for two-stage combustion system.

The PM divides the combustion chamber into three parts (zones 1 to 3) and significantly influences the fuel distribution, fuel vaporization, mixing with air, and generates turbulence during the gas flow through the PM ring.

The PM only takes part in the combustion process, but it significantly influences the temperature of the zone by accumulating part of the energy released, and improves self-ignition from the hot PM walls.

Generally, the system operates in such a way that the following mixture conditions are achievable: rich mixture (in a free volume), lean mixture (in a free volume), and any mixture composition in PM. Combustion of such a mixture permits low NO_x and low soot emissions.

Generally, two stages of the combustion process can be selected, independent of the operational conditions: early stage, from late compression until TDC (upward piston motion); late stage, from TDC until completion of combustion (downward piston motion).

Transition between the two stages of combustion is connected with a strong gas flow through the PM generating turbulence and significantly improving the mixing of gases. In both stages of combustion the porous medium (partly) controls the ignition and combustion process

Generally, the application of porous-medium technology to gas flow processes, reduces engine dependence on large-scale in-cylinder flow structures (e.g., swirl, tumble) and the turbulence generated in the cylinder. If the gas is pushed into or through the PM volume, strong heat transfer from the solid phase of the PM and gas is observed together with a spatial homogenization of the gas in the PM volume. One of most critical aspects in conventional combustion systems is injection of liquid fuel. If the liquid fuel is injected directly into the PM volume, fuel atomization and spray geometry are not critical. A self-homogenization process in the PM volume is observed and leads to spatial distribution of the liquid fuel throughout the

PM volume (Fig. 3). There are four characteristic phases of jet interaction with the porous medium [8]: phase A representing exiting from the nozzle and free jet formation; phase B representing jet interaction with the PM interface; phase C representing liquid distribution throughout the PM volume; and phase D representing liquid leaving the PM volume. Mixture formation and charge homogenization conditions in conventional engines are very complex and difficult to control. Significantly different conditions of mixture formation occur in porous media, especially where a 3D porous medium structure controls charge homogenization and fuel distribution in the PM volume. In this case, the conditions of mixture formation are almost independent of the engine operational conditions. The art of ignition and resulting combustion process in conventional engines depends on the fuel and injection conditions.

In a porous-medium reactor (e.g., PM engine system), independent of the engine operational conditions, 3D thermal PM self-ignition of the homogeneous charge is realized in the PM volume. The combustion process is characterized by a homogeneous and controlled temperature in the whole PM chamber volume, and no combustion occurs in the free volume of the cylinder. The maximum temperature is reduced by heat accumulation in the porous medium, giving rise to ultralow NO_x emissions independent of the engine operational conditions.

5.10.7 Summary

In this chapter novel concepts for combustion engines in diesel-fuelled vehicles were discussed. It was shown that porous media, especially those with a foam structure, can be used for a great variety of improvements in the combustion process. The key factor for NO_x abatement and elimination of soot emission is a homogeneous combustion of the air/fuel charge in the cylinder volume. This can be realized by homogeneous mixture formation, prevention of formation of a flame front having a temperature gradient, and 3D ignition in the entire combustion volume. All these processes can be controlled or influenced with the help of porous media/ceramic foams.

References

1 Kusaka, J., Yamamato, T., Daisho, Y., Simulating the Homogeneous Charge Compression Ignition Process Using a Detailed Kinetic Model for n-Heptane Mixtures, *Int. J. Engine Res.*, **2000**, *1* (3) 281–289.

2 Urushihara, T., Hiraya, K., Kakuhou, A., Itoh, T., Parametric Study of Gasoline HCCI with Various Compression Ratios, Intake Pressures and Temperatures, Proc. A New Generation of Engine Combustion Processes for the Future?, Ed. Duret, P., **2001**, pp. 77–84.

3 Coma, G., Gastaldi, P., Hardy, J.P., Maroteaux, D., HCCI Combustion: Dream or Reality? Proc. 13th Aachen Colloquium on Vehicle and Engine Technology, **2004**, Aachen, pp. 513–524.

4 Aceves, S.M., Flowers, D.L., Martinez-Frias, J., Smith, J.R., Dibble, R., Au, M., Girard, J.,

HCCI Combustion: Analysis and Experiments, SAE Technical Paper No. 2001-01-2077, **2001**.

5 Montorsi, L., Mauss, F., Bhave, A., Kraft, M., Analysis of a Turbocharged HCCI Engine using a Detailed Kinetic Mechanism, 2002 European GT-SUITE Users Conference, October 21, **2002**, Frankfurt, Germany.

6 Weclas, M., New Strategies for Homogeneous Combustion in I.C. Engines Based on the Porous Medium (PM)-Technology, ILASS Europe, June **2001**.

7 Weclas, M., Strategy for Intelligent Internal Combustion Engine with Homogeneous Combustion in Cylinder, Sonderdruck Schriftenreihe University of Applied Sciences in Nuernberg, **2004**, No. 26, pp. 1–14.

8 Weclas, M., Ates, B., Vlachovic, V., Basic Aspects of Interaction Between a High Velocity Diesel Jet and a Highly Porous Medium (PM), 9th Int. Conference on Liquid Atomization and Spray Systems ICLASS, **2003**.

9 Durst, F., Weclas, M., A New Type of Internal Combustion Engine Based on the Porous-Medium Combustion Technique, *J. Automobile Eng. IMechE D*, **2001**, *215*, 63–81.

10 Durst, F., Weclas, M., A New Concept of IC Engine with Homogeneous Combustion in Porous Medium (PM), 5th International Symposium on Diagnostics and Modelling of Combustion in Internal Combustion Engines, COMODIA, **2001**, Nagoya, Japan, Paper No. 2 27, pp. [illegible]

11 Park, C-W., Koviany, M., Evaporation-Combustion Affected by In-Cylinder, Reciprocating Porous Regenerator, *Trans. ASME*, **2002**, *124*, 184–194.

12 Hanamura, K., Nishio, S., A Feasibility Study of Reciprocating-Flow Super-Adiabatic Combustion Engine, The 6th ASME-JSME Thermal Engineering Joint Conference, **2003**, Paper No. TED-AJ03-547.

13 Leissner, H.F., US Pat. 1,260,408, **1918**.

14 Pfefferle, W.C., US Pat. 3,923,011, **1972**.

15 Bernecker, G., US Pat. 4,103,658, **1978**.

16 Firey, J.C., US Pat. 4,381,745, **1983**.

17 Siewert, R.M., US Pat. 4,480,613, **1984**.

18 Ferrenberg, A.J., US Pat. No. 4,790,284, **1988**.

19 Durst, F., Weclas, M., German Pat. 197 53 407, US Pat. 6,125,815, **1997**.

20 Weclas, M., German Pat. Appl. 198 13 891, **1998**.

21 Weclas, M, German Pat. Appl. 101 35 062.7, **2001**.

22 Durst, F., Weclas, M., Mößbauer, S., A New Concept of Porous Medium Combustion in I.C. Engines, International Symposium on Recent Trends in Heat and Mass Transfer, **2002**, Guwahati, India.

23 Mayer, A., Definition, Measurement and Filtration of Ultrafine Solid Particles Emitted by Diesel Engines, ATW-EMPA-Symposium, 19 April **2002**.

24 Particulate Traps for Heavy Duty Vehicles, Report of Swiss Agency for the Environment, Forests and Landscape (SAEFL), **2000**, Environmental Documentation No. 130.

25 Emission Control Retrofit of Diesel-Fueled Vehicles, Report of Manufacturers of Emission Controls Association, March **2000**.

26 Bovonsombat, P., Kang, B-S., Spurk, P., Klein, H., Ostgathe, K., Development of Current and Future Diesel After Treatment Systems, MECA/AECC Meeting, Bangkok, February 1, **2001**.

27 Schäfer-Sindlinger, A., Vogt, C.D., Hashimoto, S. Hamanaka, T., Matsubara, R., New Materials for Particulate Filters in Passenger Cars, *Auto Technol.*, **2003**, No. 5.

28 Jacob, E., D'Alfonso, N., Doering, A., Reisch, S., Rothe, D., Brueck, R., Treiber, P., PM-KAT: a Non-Blocking Solution to Reduce Carbon Particle Emissions of EuroIV Engines, **2002**, 23. Internationales Wiener Motorensymposium, 25–26 April 2002, Vol. 2: VDI Reihe 12, Nr. 490, VDI Verlag, Duesseldorf, 2002, pp. 196–216.

29 Weclas, M., Melling, A., Durst, F., Unsteady Intake Valve Gap Flows, SAE Technical Paper, No. 952477, **1995**.

30 Weclas, M., Melling, A., Durst, F., Flow Separation in the Inlet Valve Gap of Piston Engines, *Progr. Energy Combust. Sci.*, **1998**, *24* (3), 165–195.

31 Brenn, G., Durst, F., Trimis, D., Weclas, M., Methods and Tools for Advanced Fuel Spray Production and Investigation, *Atomiz. Sprays*, **1997**, *7*, 43–75.

5.11 Other Developments and Special Applications

Paolo Colombo and Edwin P. Stankiewicz

5.11.1 Introduction

Cellular ceramics are versatile, controllable, and tailorable materials suited to varied applications. Advanced cellular ceramics must be properly engineered and integrated within the system to fulfil the requirements of a given application. Component design includes selecting the proper cell size, pore size, amount of porosity, material, and fabrication technique to ensure optimal performance. The fabrication technique heavily influences the properties because it affects the morphology, the compositional purity, and the flaw population of the ceramic material that constitutes the cellular component. For some applications, high-purity, stoichiometric materials are necessary, for their enhanced strength and corrosion resistance relative to lower purity materials.

Over the years there have been substantial developments – most of them discussed in the previous chapters – in the field of cellular ceramics. These developments relate to fabrication and applications.

This final chapter briefly describes some recent advancements, not covered elsewhere in this book, with the additional primary intent of giving the reader a better sense of the ever-growing range of applications that can be fulfilled by these fascinating materials. Practical examples are presented of how tailoring the material of construction, pore size, porosity, foam structure, and fabrication technique results in unique practical and functional components with the crucial strength, flow characteristics, and materials properties required for optimum performance in a given application.

5.11.2 Improving the Mechanical Properties of Reticulated Ceramics

Successful application of cellular ceramics typically demands high mechanical properties, preferably at low apparent densities. Mechanical properties models, such as those proposed by Gibson and Ashby for foams and honeycombs (see Chapter 4.1 and Ref. [1]), require use of empirically derived coefficients. These coefficients depend on the specific foam material and fabrication method. Selection of the fabrication method, in turn, depends upon a judicious balance of performance require-

Cellular Ceramics: Structure, Manufacturing, Properties and Applications.
Michael Scheffler, Paolo Colombo (Eds.)

ISBN: 3-527-31320-6

ments, producer and user experience, and cost. Varied methods have been developed and used to improve mechanical properties, and each presents its own balance of benefits and limitations.

The production of ceramic foams by replication methods such as those introduced by Schwartzwalder and others (see Chapter 2.1) is still common in the industry, largely due to its simplicity and affordability. However, fabrication defects that significantly lower strength typically exist in the final product. Defects include longitudinal cracks that form during firing due to drying shrinkage and the gas evolution resulting from decomposition of the polymer foam substrate and to the mismatched thermal coefficients of polymer substrate and ceramic powder coating on heating. Increased strength would enhance structural efficiency and widen the applicability of these materials, particularly in advanced applications. Analysis of the fracture surface indicates that the origin of failure often develops in the strut corners near the apex of the triangular hole present in the hollow struts or is associated with macroscopic defects in the struts [1]. Furthermore, often due to cost constraints as well as for limiting shrinkage during firing, the ceramic material comprising the ligaments of the foam is often not fully sintered. This leads to porosity within the struts or cell walls. Thus, besides using a different fabrication procedure (e.g., direct blowing or gel casting, see Chapter 2.1), several approaches for improving the strength of ceramic foams have been proposed, including those discussed in the following.

5.11.2.1
Ceramic Foams by Reaction Bonding

Researchers at VITO (Mol, Belgium) are developing ceramic foams based on reaction-bonded aluminum oxide (RBAO) [2] that display bending strength far superior to that of conventional reticulated ceramics (see Fig. 1). The precursor Al/Al_2O_3

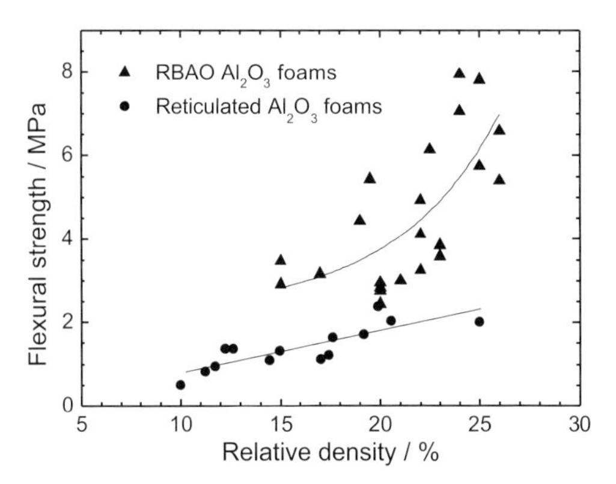

Fig. 1 Comparison of three-point bending strength of Al_2O_3 ceramic foams produced by the conventional Schwartzwalder process (reticulated) and by reaction bonding (RBAO). Data courtesy of J. Luyten (VITO).

mixture is intensively milled, and oxidation of the Al phase produces rather small grains with less glass phase at the grain boundaries [3]. Moreover, the precursor RBAO slurry wets the polyurethane sponge template better than conventional powder suspensions. This leads to better distribution of slurry on the template, with fewer cell blockings in comparison to poorly wetting slurries. Thus, the combination of a stronger coating and better wetting leads to fewer defects in the struts.

5.11.2.2
Overcoating of Conventional Reticulated Ceramics

By repeating the impregnating and drying steps several times, it is possible to increase the strut size and to increase the strength of a ceramic foam, while only moderately increasing its bulk density or reducing permeability [4]. Alternatively, it is possible to recoat the reticulated ceramic foam after the sintering step [5, 6]; by coating with a low-viscosity slurry, thicker struts are obtained, and the occurrence of large flaws is minimized by the recoating process. Indeed, collaborative research performed at EMPA (Duebendorf, Switzerland) and at the University of Bologna showed that recoating a presintered ceramic foam with a slurry of suitable rheological characteristics allows not only healing of the macroscopic flaws in the struts but also penetration into the structure and at least partial filling of the hollow struts (see Fig 2a, b). In this set of experiments, the recoating process led to an approximately 40 % increase in relative density with only about 3 % increase in strut thickness. This resulted in an order of magnitude increase in mechanical strength, from about 0.2 MPa to greater than 2 MPa for alumina foams with a relative density of 0.13 when sintered at 1550 °C. An effect of the temperature reached in the presintering step (during which the polymer sponge substrate is decomposed) and of the rheology of the second slurry was also observed. A lower temperature for the first sintering step (i.e., 1200 °C) and a lower viscosity slurry in the second infiltration resulted in higher strength after recoating, likely due to better access of the slurry into the sintered ceramic body. The diameter of the struts and their uniformity in thickness, both of which affect the mechanical properties, can be controlled in reticulated ceramics also by adopting recoating (in the green stage) and centrifuging steps, as demonstrated recently [7].

Another possibility is to coat low-cost, low-strength reticulated ceramic foams by chemical vapor deposition (CVD). Experiments performed at Ultramet (Pacoima, CA) demonstrated that this method can result in a very large increase in strength, as well as an improvement in corrosion resistance. For example, the crushing strength of a 10 ppi ZrO_2 foam increased from about 1.6 to about 5.0 MPa, after depositing a thin SiC layer.

An approach newly developed by Erbicol (Balerna, Switzerland) consists of immersing an SiC reticulated foam (presintered at a low temperature of 500–700 °C) into a carbon-containing precursor (e.g., phenolic resin or preceramic polymer), followed by pyrolysis at 1000 °C in inert atmosphere and infiltration with molten Si [8]. With this processing strategy, Si does not fill completely the struts but reacts with the carbon residue giving a SiC deposit that rounds off the tips of the triangular

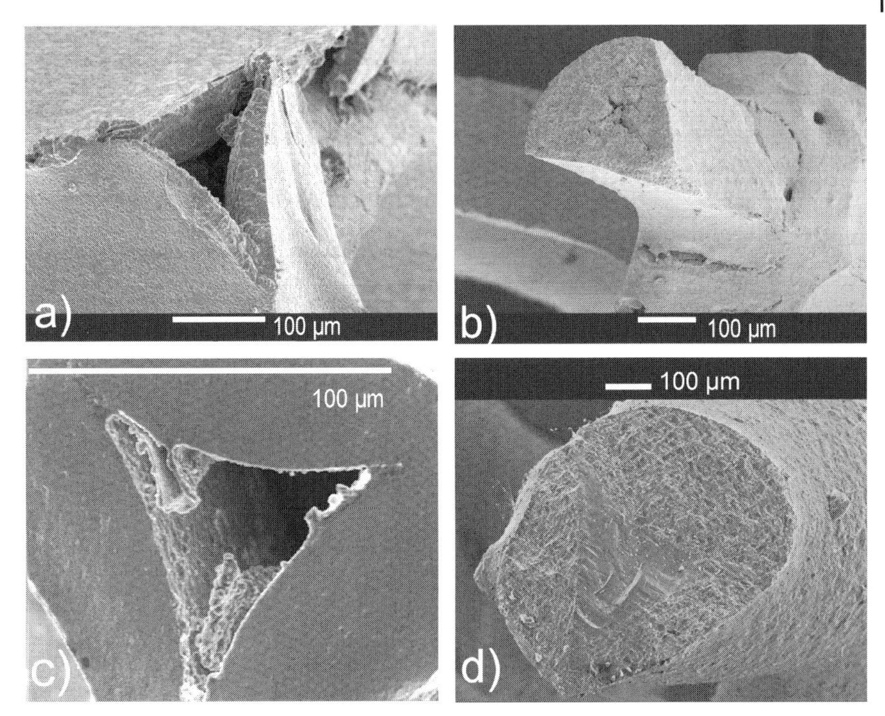

Fig. 2 Comparison between the fracture surface of a strut of a) a conventional reticulated foam, b) the same foam after recoating [images courtesy of C. Testa, U. Vogt (EMPA), and P. Colombo (University of Bologna)], c) strut in which the tips of the triangular void have been rounded off by a SiC coating [image courtesy of A. Ortona (ICIMSI)], and d) strut infiltrated by the LigaFill process [image courtesy of J. Adler (IKTS)].

voids within the strut (see Fig. 2c). This greatly reduces the formation of fatigue cracks during service, limits strut swelling due to the formation of silica from Si (only a limited amount of excess Si is present), and maintains a low thermal mass (low thermal capacity) and good permeability. The resultant ceramic foams are well suited for high-temperature applications, for instance, as porous-burner components, in which the material is subjected to an oxidative environment and severe thermal shock (especially in the power-off state in which the foam goes from 1500 °C to room temperature in a few seconds; see Chapter 5.5).

5.11.2.3
Infiltration of the Struts of Reticulated Ceramics

The possibility of filling the hollow struts of reticulated ceramics to increase their strength has been proposed in several patents; a sintered foam can be immersed in a suspension containing colloidal silica, alumina, or other refractory oxides [9], which because of their very small size can penetrate the voids of the structure and fill the microcracks. The colloidal silica sol is transformed at about 150 °C into silica gel retaining the refractory powder, before sintering the foam again at high temperature. With this process it is possible to modify the affinity of the foam surface for slag-type micro-inclusions in molten metals and thus improve the filtering efficiency. Alternatively, a sintered foam can be immersed in an aqueous sol of aluminum hydroxide or zirconium hydroxide, then refired [10].

An unique way to completely fill the struts of reticulated foams was developed at Fraunhofer IKTS (Dresden, Germany) using metal or glass melts [11]. The melts infiltrate readily into the struts if the melt has a sufficiently low viscosity and exhibits good wetting characteristics. More importantly, using a silicon melt allows complete infiltration in a single thermal process starting from a polyurethane foam coated with unsintered SiC powder and results in a high-strength SiC ceramic foam (see Fig. 2d). The process (trade name LigaFill) can be used for cell sizes ranging from 10 to 80 ppi and, because it consists of a single step, it is economically advantageous in comparison to competing technologies that require refiring. A compressive strength enhancement of more than 600 % compared to an unfilled sintered SiC foam has been measured, with maximum values reaching 24 MPa for relative densities of less than 30 %.

5.11.3 Microcellular Ceramic Foams

Conventional processing methods do not easily allow the fabrication of macroporous bodies in which the average cell size is less than 100 μm, because of the lack of convenient polyurethane foams with cell size above 100 ppi. Novel methodologies for the production of microcellular foams, in which the average cell size ranges from about 1 to less than 100 μm, are under development. A possible strategy is to infiltrate porous salt preforms with a molten preceramic polymer [12], while other approaches suggest using sacrificial fillers [mainly commercially available poly (methyl methacrylate), PMMA, microbeads or latex] either in combination with chemical precursors [13] or with preceramic polymers [14]. Another possibility is to dissolve CO_2 gas under pressure (5.5 MPa for 24 h at room temperature) in preceramic polymers and introduce a thermodynamic instability by rapidly dropping the pressure (at a rate of 2.9 MPa s^{-1}) [15]. A large shrinkage (ca. 30 % linear) occurs on firing, but it appears to be isotropic and thus limits the formation of cracks, and overall cost could become an issue for large components. Microcellular ceramic foams manufactured by the above processes have cell densities greater than 10^9 cells/cm^3 and cells smaller than 50 μm (Fig. 3). Because of the more uniform distribution of cell size (especially when sacrificial fillers are used), thinner struts, and the possibility of being fabricated with either open or closed cells, they have different properties than macrocellular materials and extend the range of applications available to ceramic foams.

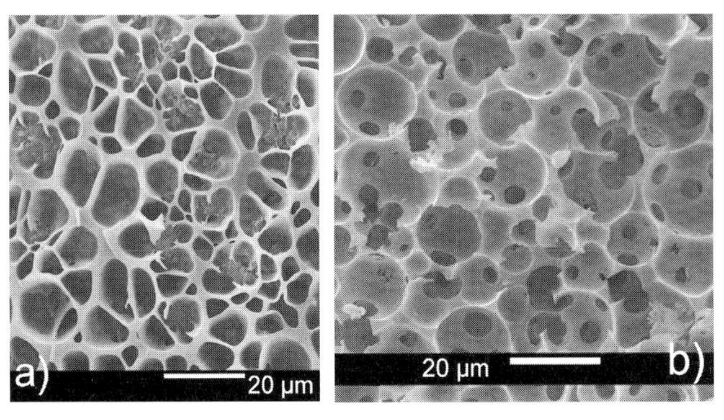

Fig. 3 Typical microstructure of a microcellular SiOC foam: a) closed-cell, from CO_2 processing [image courtesy of Y.-W. Kim (University of Seoul) and C. Wang, C. B. Park (University of Toronto)] and b) open-cell by burnout of PMMA microbeads [image courtesy of P. Colombo (University of Bologna)].

5.11.4
Porous Ceramics with Aligned Pores

Porous ceramics with aligned pores can be fabricated by several techniques (see Chapters 2.2 and 2.5) affording various degrees of porosity. With freeze-drying, a component (such as water or a solvent) is first allowed to solidify and grow unidirectionally, and then is eliminated by melting or directly by sublimation from the solid to the gas state under reduced pressure. The resulting structure is actually very complex and is comprised of continuously open, strongly textured channels with a flat shape and a size of about 10–40 μm, which afford high porosity (up to 70 vol %) [16–18].

Gas generation during an aqueous electrophoretic deposition (EPD) process, coupled with freeze-drying, was also used for producing thick (ca. 2 mm) porous ceramic layers containing many aligned continuous pores with a diameter of about 100 μm [19]. However, this process seems more suitable for the fabrication of membranes, rather than monolithic porous components.

Oriented continuous pores can also be fabricated by slurry coating of fugitive fibers, producing ceramic materials having pores with a diameter of 165 μm and 35 % open porosity. The pore size and the amount of porosity can be adjusted by varying the diameter of the sacrificial filaments and the solids concentration of the slurry, respectively [20].

Honeycomblike structures with oriented, tubular pores can be fabricated by using a gel-formation approach, in which ceramic particles are dispersed in a solution containing an inorganic polymer (alginate) that can be gelled [21]. Capillaries form in the direction of diffusion of the cross-linking agent (multivalent metal ions) through the slurry. The thickness of the components is limited to a few centimeters, and they have up to 80 vol % porosity and pore diameters ranging from 10 to 30 μm.

Drying should be carefully controlled to avoid crack formation in the ceramic components, and pronounced shrinkage occurs in the transition from wet gel to sintered body. Well-structured samples were obtained by optimizing the amount of powder in the slurry.

Highly porous cellular ceramic components with a thickness of about 1 cm and oriented cylindrical pores with a diameter up to about 2 mm were fabricated by forming hydrogen gas in an alumina sol–gel system by reaction of dispersed Al particles with H^+ ions. The large shrinkage and the difficulty of eliminating the solvent during drying limit the size of the components achievable with this method [22].

5.11.5 Porous Superconducting Ceramics

High-temperature superconducting ceramics have a host of possible applications, for which various shapes and sizes of the ceramic material are required. The simple extension of conventional ceramic shaping techniques to these materials is not possible because of the stringent requirement of achieving a specific microstructural texture affording superconducting properties suitable for the applications. Foams and interconnected porous structures of superconducting ceramics could provide solutions for some of the problems encountered in applications of bulk or film-type superconductors. For instance, porous components may find applications in resistive superconducting fault-current limiters (requiring efficient heat extraction from the superconducting components) or could be reinforced by infiltration with suitable phases. Ceramic superconductors with enhanced mechanical strength are useful for applications like flywheel energy storage and levitation devices or quasipermanent magnets for magnetic fields exceeding 17 T at 25 K.

A biaxial grain texture, the entire superconductor ideally being a large single grain, with fine and homogeneously distributed normal conducting particles, are the two essential microstructural features that make the superconducting oxide $YBa_2Cu_3O_{7-x}$ (123) suitable for practical applications. Such a microstructure has been successfully achieved by various melt-processing techniques [23]. However, intrinsic drawbacks of these techniques, such as large amounts of low-viscosity liquids and distortions due to large and anisotropic shrinkage, limit their straightforward extension to complex shaped bodies such as foams.

The manufacture of superconducting 123 foams and porous bodies with high critical current densities has become possible by an infiltration and growth process developed by Reddy and Schmitz [24, 25] at ACCESS Materials & Processes (Aachen, Germany); see Fig. 4. In this process, standard ceramic shape-processing techniques are coupled with methods to avoid the shrinkage and distortion that typically occur in melt processing during the peritectic conversion of the green body to a single superconducting grain. Reticulated foams produced by this method typically have strut thickness of a few hundred micrometers and pore sizes ranging from 10 to 100 ppi. 3D interconnected porous bulk structures with a relative density of about 0.35, a pore size of about 1 to 2 mm, and strut thickness in the millimeter range

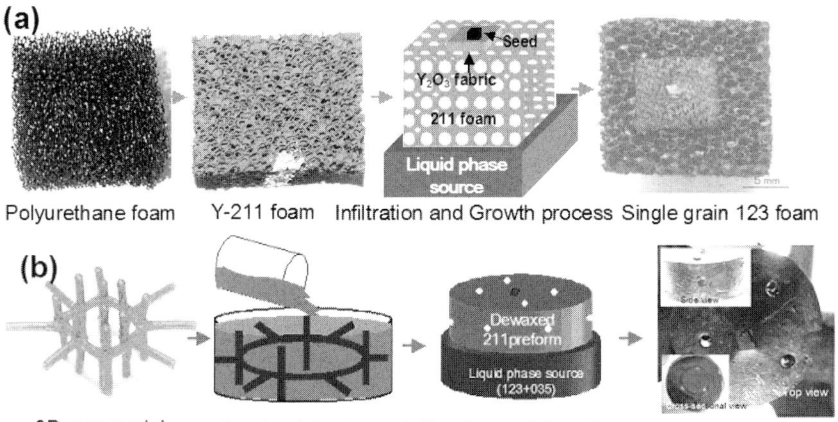

Fig. 4 Schematic diagrams detailing the fabrication by infiltration process of a) a single-grain superconducting $YBa_2Cu_3O_{7-x}$ (123) foam and b) a porous single-grain bulk material. The processing method involves two stages: in the first step a Y_2BaCuO_5 (211) porous structure is processed either as a replica of a commercial polyurethane foam or by creating a hollow replica of a wax model in 211 castings by standard ceramic processing. Subsequently this porous 211 body is converted to a single-grain 123 foam or porous bulk by infiltrating it with liquid phases from a liquid-phase source containing barium cuprates and copper oxides. Slow solidification through the peritectic temperature in the presence of a higher melting $NdBa_2Cu_3O_{7-x}$ ("Nd123") seed crystal having a similar crystal structure results in nucleation and growth of a single superconducting grain covering the entire skeleton of the 211 preform. Images courtesy of G.J. Schmitz (ACCESS, Materials & Processes) and E.S. Reddy (IRC in Superconductivity).

have also been produced by the approach described in Fig. 4b [25], or by using spherical wax balls of desired size [26].

The high surface area of the foams, which is controllable partly by selecting the pore size, also allows the investigation of fundamental aspects of superconductivity, such as the extent of surface pinning and hence the critical current densities.

5.11.6 Porous Yb_2O_3 Ceramic Emitter for Thermophotovoltaic Applications

Thermophotovoltaics (TPV) is a technique to convert radiation from a synthetic emitter to electricity by using commercial silicon photocells [27] (Fig. 5). Researchers at EMPA (Duebendorf, Switzerland) and PSI (Villigen, Switzerland) recently proposed the use of reticulated ceramics for TPV converters for residential heating systems. The main concept is to combine heat and power generation in one compact unit, providing electrically self-powered operation of the furnace. For gas-fired TPV systems, which would operate as a porous burner (see Chapter 5.5), Yb_2O_3 ceramic foams were chosen as the selective mantle emitter because of their favorable emission spectrum and because they allow a high mechanical and temperature stability

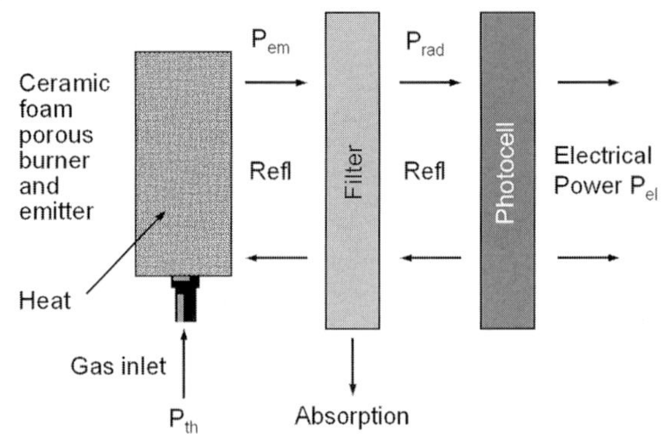

Fig. 5 The thermophotovoltaic principle with a ceramic foam-based mantle emitter. P_{th} is the thermal power of the burner which is supplied as combustion heat (proportional to the thermal power of the fuel), P_{em} the energy released by radiation from the emitter (ca. 10–20 % of the combustion heat), P_{rad} the energy available to the Si photocell after the filter, and P_{el} is the generated electric power. Image courtesy of U. Vogt (EMPA) and W. Durisch (PSI).

of the emitter structure to be obtained. Al_2O_3, Y_2O_3, CeO_2, or Zr_2O_3 was added as sintering additives, and processing was performed at 1550–1650 °C.

For this application, a specific pore size is required, since pore size and the porosity of the ceramic foam determine the gas flow resistance and the emission behavior. Experimentally, 20 ppi foams were found to allow easy ignition of the flame and to stabilize the flame in the ceramic emitter structure (see Chapter 5.5). To improve the mechanical stability of the foams, the presintered Yb_2O_3 foams can be impregnated with a ceramic slurry (recoating), followed by an additional sintering step. Alternatively, presintered Al_2O_3 foams can be recoated with a Yb_2O_3 slurry to give a defect-free emitting coating.

TPV systems with commercial silicon solar cells (Type SH2 from ASE GmbH, Alzenau, Germany) and the developed selective Yb_2O_3 foam emitter have been demonstrated to be suitable for application in residential heating systems. For self-powered operation of the heater, an electrical power of 120 W is sufficient, and safety requirements as well as cost effectiveness could be met. The Yb_2O_3 foam ceramic emitter has acceptable mechanical and thermal stability at 1400 °C. Thermal-shock tests up to 200 ignition cycles showed that the emitters have satisfactory thermal-cycling resistance, as required for TPV systems. The selectivity of the solar cell with respect to the emitter radiation is currently 10 %, and a successfully tested TPV prototype system had an overall electrical efficiency of $\eta_{sys} \approx 1$ % [28].

5.11.7 Ceramic Foams for Advanced Thermal Management Applications

Open-cell vitreous carbon foams with porosity ranging from 75 to 97 % can be used as lightweight, low thermal conductivity insulation. Heat is conducted through the foam by conduction through the solid ligaments, convection through any gas in the pores of the foam, and radiation (see Chapter 4.3). To minimize thermal conduction through the solid, the lowest mass or highest porosity foam is used, and the solid should have low intrinsic thermal conductivity. Small pore diameters reduce heat transport by convection. Radiative heat transfer can be reduced by filling the pores of the open-cell foam with an aerogel of low thermal conductivity. Figure 6 shows the thermal conductivity of a vitreous carbon foam, a carbon aerogel, and an aerogel-filled foam, as developed by Ultramet (Pacoima, CA). The data indicate that the filled reticulated vitreous carbon retains the excellent thermal insulation characteristics of low-density aerogel. Combining the carbon foam skeleton with the aerogel provides adequate structural performance. This material is being developed for lightweight insulation panels for aerospace applications.

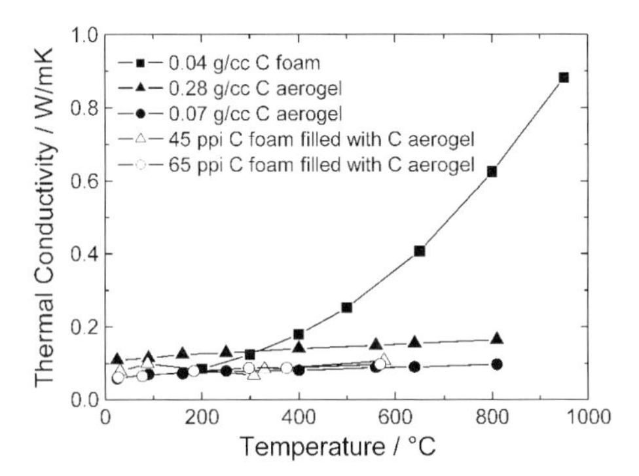

Fig. 6 Thermal conductivity of aerogel, foam, and aerogel-filled foam. Data courtesy of E. Stankiewicz (Ultramet).

Realization of a viable reusable launch vehicle (RLV) is a primary goal for NASA. Compared with the existing Space Shuttle orbiter, the next-generation vehicle needs lower operating costs, increased reliability and safety, and faster turnaround between flights. The options currently under consideration range from major upgrade programs to various new vehicle concepts. For the thermal protection system (TPS), the key goals are improvements in both durability and temperature capability and a reduction in maintenance cost. A research effort is underway at NASA Ames Research Center (Moffett Field, CA) to develop durable, oxidation-resistant, and reusable foams for both acreage and leading-edge TPS applications. These foam systems need low to moderate density and temperature capability comparable to those of carbon TPS systems (reusable at 1650 °C) with application of a suitable coating.

In a collaborative effort between researchers at The Pennsylvania State University and the University of Bologna, preliminary arc-jet testing was conducted on SiOC open-cell macrocellular (density ca. 0.73 g cm^{-3}, cell size ca. 700 μm, crushing strength ca. 5 MPa) and microcellular (density ca. 0.35 g cm^{-3}, cell size ca. 8 μm, crushing strength ca. 10 MPa) foams processed from preceramic polymers [29]. The tests were conducted in the NASA Ames 60 MW Interaction Heating Facility (IHF). Small diameter (2 cm) cylinders were tested for 120 s exposure to 1690 °C (see

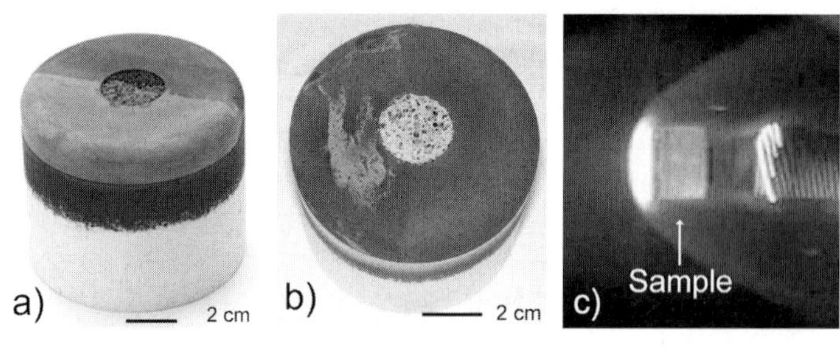

Fig. 7 a) The ceramic foam test specimen (macrocellular) is the black disk in the center of the holder (before testing). The bottom part was coated with a standard NASA tile coating. b) Same specimen after arc-jet testing (coated part is on the left-hand side; coating is damaged because it reached temperatures higher than those for which it was developed). c) The arc-jet nozzle exit and swing arm are visible to the left and right of the sample being tested (center). Images courtesy of M. Stackpoole (Eloret, NASA).

Fig. 7). Post-test phase analysis by X-ray diffraction confirmed that a phase change did not occur in the foam substrates during the test duration. Both types of samples showed minimal degradation (ablation or oxidation) to have occurred as a result of these test exposures.

Initial results are encouraging, but longer test durations are needed to better evaluate the long-term stability of these foam materials in the severe environments experienced in the arc-jet facility. Current work demonstrates that processing of preceramic polymer is a viable method to form refractory ceramic foams with tailorable properties at relatively low manufacturing temperatures (1200 °C), which display encouraging properties under simulated reentry conditions.

5.11.8
Ceramic Foams for Impact Applications

5.11.8.1
Hypervelocity Impact Shields for Spacecrafts and Satellites

Cellular materials are currently discussed as candidates for micrometeorite and debris protection systems (MDPS) for spacecraft, because of their unique combination of properties including low density, high stiffness, and energy absorption that can be tailored through the design of their microstructure. A typical MDPS for a manned mission is based on the principle of multiple shocks followed by radial expansion of the impacting projectile, induced by several thin layers (bumpers) placed between the external environment and the panel walls composing the spacecraft's shell [30]. The large amount of cell walls that a projectile impacting a cellular material would encounter along its path would induce a high number of consecutive shocks on the impactor. The protection effectiveness depends on several factors, among which are: 1) the ratio between the density of the projectile and of the cellular material; 2) the relative dimension of the projectile (fragments) with respect to the cell walls; 3) the number of walls (or cells) per unit length and 4) the relative dimension of the projectile (fragments) with respect to the cell volume. Factor 2 controls the intensity of a single shock; factor 3 the number of shocks to which the projectile is subjected along its path through the material; and factor 4 the expansion of the cloud of new fragments.

Researchers at CISAS Hypervelocity Impact Facility (Padova, Italy) and at University of Bologna tested macro- (cell size 200–800 mm) and microcellular (cell size 10–100 μm) ceramic foam cores (2 cm thick) sandwiched between two thin aluminum sheets (0.4 mm thick) by impacting them with Al projectiles (diameter 1–1.5 mm) launched at 4, 4.5, and 5 km s^{-1} using a two-stage light-gas gun [31]. The experimental data collected so far indicate that denser foams can withstand more energetic impacts with a given damage. Thicker struts or higher solid material density result in more effective shocks within the cellular structure. However, the total weight of the shield is a paramount concern for space applications, and a suitable compromise between performance and MDPS mass budget should be found. It is difficult to independently analyze the role of the cell size with respect to the penetration depth because cell size often influences the mechanical strength of the material, and compressive strength is a critical parameter in impact-related problems, especially during the final phase of penetration. The foams tested in these experiments were neither completely perforated nor broken into several macroscopic fragments when impacted [32].

Hypervelocity impact tests on ceramic foams showed that these materials, if well designed, can effectively arrest (in a restricted space) the debris cloud generated from the collision between the projectile and the external bumper (see Fig. 8). Their performance compares favorably with that of traditional Whipple shields (two aluminum sheets at a given distance, with no core material between them) when only very limited space is available for the MPDS. With respect to metal foam-based cores, ceramic foams can be used in high temperature environments (e.g., near-sun

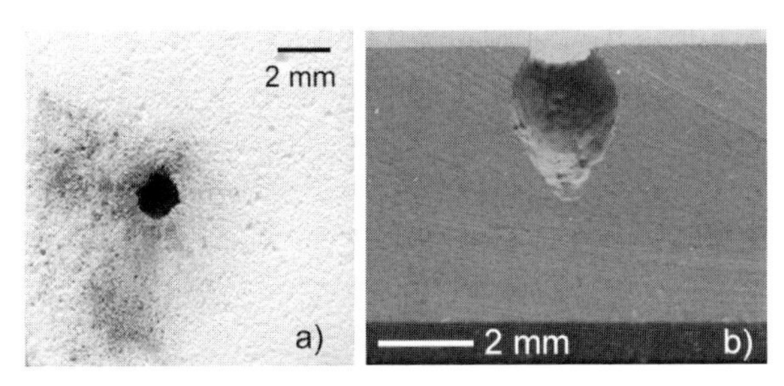

Fig. 8 Images of the impact crater caused by a 1 mm Al projectile at 4 km s^{-1}. a) Frontal view of a gel-cast Al_2O_3 foam (bulk density 0.731 ± 0.062 g cm^{-3}, compression strength 16.9 ± 4.5 MPa, average cell size 220 ± 34 μm – sample courtesy of Hi-Por Ceramics Ltd). b) Cross-section view of an SiOC microcellular foam (bulk density 0.359 ± 0.016 g cm^{-3}, compression strength 11.0 ± 1.9 MPa, average cell size 7.8 ± 1.5 μm). Images courtesy of A. Francesconi, D. Pavarin (CISAS), A. Arcaro and P. Colombo (University of Bologna).

orbiting probes) or as thermal protection shields for reentry of spacecraft, and also provid debris-protection capabilities (i.e., they serve as a multifunctional component). Recently, the need for additional Space Shuttle TPS has been recognized, and this could lead to unexpected opportunities for the use of cellular ceramics with intrinsic protection capabilities.

5.11.8.2
Armour Systems

Interest in composite, lightweight armour with enhanced performance for vehicular or blast protection led researchers at Ceramic Protection Corporation Inc. (Calgary, AB, Canada) and the University of Bologna to undertake the development and ballistic testing of 3D-reinforced composites based on polymer-infiltrated ceramic foams (see Fig. 9a). The armour system configuration comprised a thin (6 mm) ballistic-grade alumina face tile bonded to a backing system based on infiltrated SiC foams [20 mm thick, 10 or 20 ppi, samples courtesy of Foseco (Cleveland, OH)]. The alumina face tile, with optimized microcrystalline structure, high hardness, and mechanical properties, was used to fragment the bullet and to dissipate its impact energy. The functions of the polymer-infiltrated ceramic foams were to absorb the impact energy and stop the bullet. Two types of polyurethane (PU) – rigid cross-linked and elastomeric – were selected and vacuum infiltration was applied. Possible advantages of this design include the presence of a continuous polymeric phase with damping capability, the absorption of impact energy by crushing of the ceramic foams, and the presence of a continuous cellular ceramic structure which interferes with the propagation of the shock wave. Some polymers could also increase the strength and damage tolerance of the ceramic foams. This composite structure may be bonded to a metallic or plastic board that can be easily installed in mobile or stationary systems.

These composites were ballistically tested using 5.56 × 45 mm SS109 (steel tip ball), 7.62 × 51 mm NATO Ball FMJ (with a lead core), and 7.62 × 63 mm AP

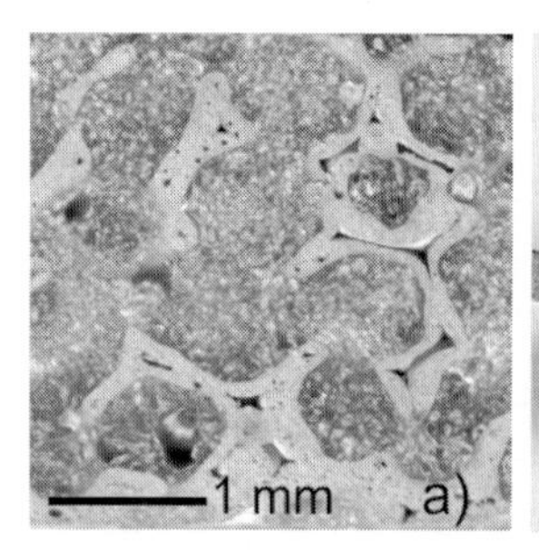

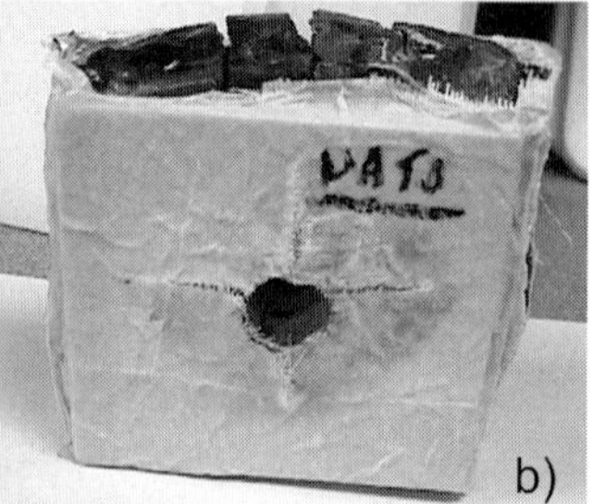

Fig. 9 a) Microstructure of the 3D composite (SiC-foam-reinforced cross-linked PU). b) Composite made from alumina ceramic tile bonded with SiC foam infiltrated by cross-linked PU after ballistic testing (7.62 × 51 mm NATO Ball FMJ). Images courtesy of P. Colombo (University of Bologna) and E. Medvedovski (Ceramic Protection Corporation).

M2 (with a tungsten carbide core) ammunition. The projectile velocities were 990–1000, 840–890, and 845–870 m s^{-1}, respectively. Strong influence of the design and mechanical properties of the composite components, especially the nature of the selected polymer, on the performance was observed. In the case of SiC foam infiltrated with cross-linked PU, the 5.56 × 45 mm SS109 and 7.62 × 51 mm NATO rounds were stopped in the backing (Fig. 9b), while the 7.62 × 63 mm AP round was stopped only in the trauma pack. Similar systems made with elastomeric PU could not defeat even the 7.62 × 51 NATO round. Preliminary testing of the system based on the SiC foam infiltrated by cross-linked PU but without face alumina tile resulted in unsatisfactory performance against the 7.62 × 51 NATO projectile.

Further experimental work dealing with optimization of the system design, assessment of the influence of cell size and ceramic foam strength, and selection of a preferred polymer for foam infiltration are currently underway.

5.11.9
Heat Exchangers

Open-cell ceramic foams fabricated by CVD have been proposed by Ultramet as cores of compact heat exchangers for actively cooled, high-power, SiC-based electronics. Due to its high thermal conductivity, excellent corrosion resistance, and the perfect match of its coefficient of thermal expansion (CTE) with that of the electronic components, SiC is used (see Fig. 10a). The foam structure functions as cooling fin of high thermal conductivity and high surface area, in which the struts act as extended surface for heat dissipation. By forcing cooling air or water through the foam, heat can be removed very rapidly. The turbulent mixing caused by the three-dimensional interconnected porosity of the foam structure continuously breaks down boundary layers within the fluid, causing more effective heat transfer. The completely open cell foam with high porosity offers low pressure drop for the cooling fluid.

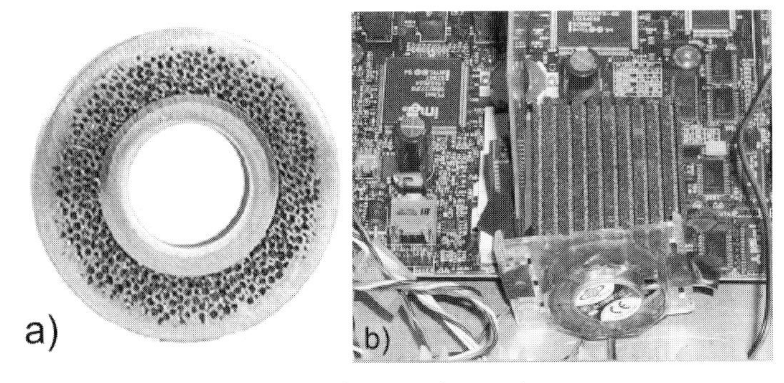

Fig. 10 a) Cross section of a SiC foam-core heat-exchanger tube. Image courtesy of E. Stankiewicz (Ultramet). b) Carbon-foam heat sink with a fin design, used to cool a computer chip. Image courtesy of J. Klett (ORNL).

Heat exchangers have also been built from metal foams (with the advantage of a more readily available joining technology, but a generally lower corrosion resistance in aggressive environments) and carbon foams. Researchers at Oak Ridge National Laboratories (Oak Ridge, TN) developed heat sinks based on special high-thermal-conductivity carbon foams (thermal conductivity of 150 $W\ m^{-1}\ K^{-1}$ for a density of 0.54 $g\ cm^{-3}$, see Chapter 2.7). Several applications have been proposed and tested. They range from computer chip cooling (see Fig. 10b) to power electronics cooling (a power density of 150 $W\ cm^{-2}$ was attained at die temperatures of only 70 °C; higher power densities allow faster processing speeds), transpiration/evaporative cooling (for electronics and leading edges), and personal cooling devices [33]. The use of these graphite foams significantly reduces temperatures compared to traditional heat sinks, such as Al foam or Al finned plate. However, in some applications the pressure drop can be exceedingly large when graphite foam block is used, so that pumping power rises. Thus, special engineering designs to reduce pressure penalties while maintaining heat transfer are being developed; they allow a reduction of the overall pressure drop of more than an order of magnitude.

In heavy vehicles, the radiator, located in the front, creates significant drag accounting for up to 12 % of fuel consumption at highway speeds. It has been shown that by utilizing a graphite foam as the fins of a radiator, the same heat can be dissipated in a significantly smaller package. For example, a radiator was designed for a racing car that can dissipate up to 33 kW in 60 % less volume than a typical radiator (see Fig. 11). However, the pressure drop across the radiator was higher than allowable due to fan considerations. Research is underway to minimize pressure drop while maintaining heat transfer. A smaller, lighter radiator will allow the redesign of the front cab and dramatically improve efficiency by allowing more heat to be dissipated in the same size and by improving aerodynamics, fuel efficiency, and load-carrying capacity of heavy vehicles.

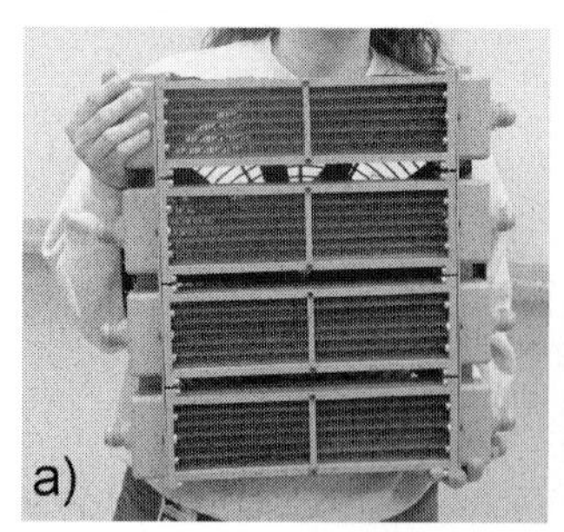

Fig. 11 a) Vehicle modular radiator based on a high-thermal-conductivity graphite foam. b) Closeup of one element. Blocks of graphite foam were machined and soldered to the cooling tubes on one side; the assembled foam and cooling tubes were then stacked together and soldered to the manifold. Images courtesy of J. Klett (ORNL).

5.11.10
Ceramic Foams for Semiconductor Applications

Carbon (C) or silicon carbide (SiC) foams developed at ERG Materials and Aerospace Corporation (Oakland, CA) are being used to perform a number of critical functions in the manufacturing of integrated circuit chips. The processes of dry etching and chemical vapor deposition (CVD) require precise control of gas flows and the ionized plasma field to enable the uniform deposition and removal of materials. Through such precise control, the yield of chips with submicron path widths can be greatly improved, and this can be achieved by using a specifically designed input-gas foam diffuser disk. This component consists of a 40 ppi ceramic foam that performs several functions. It allows uniform gas distribution, because the disk provides uniform pressure drop and distribution of gases to all areas of the shower head regardless of turbulence or flow asymmetry in the diffusion chamber. The foam also operates as a fluid flow straightener, allowing gasses to enter all shower head ports uniaxially with uniform velocity minimizing any chugging in the orifices. Moreover, the diffuser disk acts as a grounding plate and flame arrestor that prevents backfiring or arcing of the plasma back through the shower head and into the diffusion chamber and thus stabilizes the plasma. Furthermore, the foam can collect and remove particulate contamination that might otherwise enter the diffusion chamber. Finally, their open-cell structure allows compact condensation of by-products and process chemicals for disposal or recycling. High surface area foams doped with catalytic materials are also being examined as a means of artificially biasing the chemical population distribution in etching and CVD process chambers. Diffusers are commonly made of special-grade Al alloy, but C and SiC ceramic foams can be used in applications which involve higher temperatures or chemicals incompatible with Al.

5.11.11
Duplex filters

A Duplex filter combines two different porosities in a single filter, usually with a combination of porosities of 30/50 ppi or 40/60 ppi, but no particular restrictions exist. The alumina filters developed by Drache Umwelttechnik (Diez, Germany) allow a larger amount of particles to be filtered from a liquid-metal flow than conventional ceramic foam filters, are less expensive than any stacked filtration system, for which a special filterbox is needed, and have a lower release of particles, especially during longer casts, because any coarse particles released from the upper layer are caught by the finer bottom layer.

In Duplex filters constructed at Ultramet for gas-filtration applications, an oxide ceramic porous membrane filter is integrated into the first (surface) cells of a SiC foam on the inlet side of the foam filter structure (see Fig. 12a). The foam ligaments support and reinforce the submicrometer porous filtration membrane. The open-cell foam component of the duplex filter fulfils several functions, as a coarse filter,

Fig. 12 a) Cross section of a hot-gas duplex filter (SiC foam coated with a ceramic membrane). b) Hot-gas duplex filters (1 m long, based on SiC foam), covered with captured particulate, after high-temperature, high-pressure testing. Images courtesy of E. Stankiewicz (Ultramet).

as a structural element, and as a flow distributor/mixer for the process gas. Catalysts can also be coated onto the foam ligaments to add the function of catalytic converter to the duplex filter. These foams have been used in coal-combustion applications (see Fig. 12b). Other companies have also developed similar systems.

5.11.12
Lightweight Structures

Sandwich panels are valuable in applications where low weight and high stiffness are crucial [34]; consequently, lightweight structures have been fabricated with ceramic foam cores, in which the foam acts both as structural element and functional porous medium. Face sheets can be bonded to one or more sides of a ceramic foam to produce lightweight structures. Development of a sandwich panel made of a ceramic-fiber-reinforced SiC face shield and a CVI-SiC (CVI = chemical vapor infiltration) foam core for application in thermal protection systems (TPS) was reported as early as 1985 [35]. An example of a lightweight, high-strength, high-stiffness foam structure, developed at Ultramet, is an optical mirror for use in space applications (Fig. 13a). The open-cell foam serves as a stiff, light platform for the high-density CVD SiC mirror face sheet. The foam material can be chosen to match the CTE of the face-sheet mirror to minimize distortion.

Fig. 13 a) Lightweight foam mirrors (foam core with face sheet). Image courtesy of E. Stankiewicz (Ultramet). b) Pyrolyzed ceramic sandwich structure. Image courtesy of T. Hoefner, J. Zeschky, and P. Greil (Erlangen University).

Utilizing a novel process with a single processing step, researchers at Erlangen University fabricated ceramic sandwich panels from Si and SiC filler-loaded polymethylphenylsilsesquioxane [36]. Green tapes are bonded to a preceramic polymer-derived foam while it grows by self-blowing (see Chapter 2.1). Since the green tapes are made of filler-loaded preceramic polymer [37], they have the same composition as the foam material, and no interfacial adhesive is necessary to attach the face sheets to the foam core. Because of the fillers, shrinkage on pyrolysis at 1600 °C is low (ca. 3.5 % linear shrinkage). The resulting panel is shown in Fig. 13b. These panels are proposed as carrier substrates for silicon-based photovoltaic cells, but their impact behavior is also currently being investigated.

SiSiC sandwiched foams could also be produced by conventional powder technology (Schwartzwalder + tape casting) without any shrinkage and with the advantage of achieving a more uniform porosity in the core material [38].

5.11.13 Ceramic Foams as Substrates for Carbon Nanotube Growth

Catalytic chemical vapor deposition (CCVD) is the most promising route for the synthesis of carbon nanotubes (CNTs), which have very high specific surface area and exceptional mechanical, electrical, and thermal characteristics that make them of particular interest for a variety of applications [39]. CCVD consists of the catalytic decomposition or dismutation of a carbonaceous gas on nanometric metal particles at high temperature (ca. 700–1100 °C) [40]. Researchers at CIRIMAT (CNRS-Université Paul-Sabatier) recently proposed a method in which metal nanoparticles (< 5 nm) are formed in situ at the appropriate temperature by the reduction of a FeO/CoO-doped ceramic powder support (Al–Mg–O system) and are thus immediately active for the catalytic decomposition of CH_4 [41]. Besides its composition, many parameters of this solid solution, such as specific surface area and granulometry, are important to ultimately determine the formation of the desired kind and proportion of CNTs. On the one hand, because the active metal particles are those formed at the surface of the oxide grains, it is of importance that the solid solution offers a large surface area. On the other hand, because the formation of each carbon nanotube needs a great amount of carbon, easy access of a large supply of carbonaceous gas (CH_4) to the metal nanoparticles is of utmost importance. For these two reasons, the use of catalytic materials in the form of porous ceramic foam supports instead of powders becomes extremely advantageous. It was shown that an $Mg_{0.9}Co_{0.1}Al_2O_4$ solid solution foam prepared by gel casting (total porosity 98 %, cell window size < 300 μm), as compared to the corresponding powder, gave a fourfold increase in production of CNTs (> 95 % with only 1 or 2 walls, and ca. 70% single-wall nanotubes, SWNT, [42]), because it allowed for a higher quantity of surface metal nanoparticles, better dispersion of the metal particles (which hampers growth and therefore favors selectivity for CNT formation), and more space for the CNT to grow. In addition, a CNT–ceramic composite foam, with CNT in the open porosity of the matrix, would be of interest for many applications using the CNT as a catalyst substrate.

5.11.14 Metal Oxide Foams as Precursors for Metallic Foams

Researchers at Lehigh University developed a process for producing closed-cell metallic foams using a ceramic foam precursor constituted of Fe_2O_3 [43]. The oxide foam was obtained by reaction of an acidic solution with a metal blowing agent to generate hydrogen while cementitious materials simultaneously reacted to form a fast-setting hydrate. The mixture thus foams and sets simultaneously, and the evolved gas is enclosed to produce a ceramic foam with a predominantly closed cell structure [44]. The addition of carbon black increases the viscosity of the mixture, which in turn hinders the drainage of the foam and reduces cell coalescence. This ensures that a predominantly closed cell structure is developed. The ceramic precursor foam is then reduced by annealing at 1240 °C in a nonflammable hydrogen/inert gas mixture to obtain a metallic foam with variable density and cell size (typical values obtained so far are ca. 0.23 for relative density and 1.32 ± 0.32 mm for the average cell diameter). Linear shrinkage on conversion from ceramic to metallic foam is about 25 %. The normalized strengths of the metal foams obtained with this method compare favorably with those of steel foams produced by other techniques. Closed-cell foams are more advantageous for structural applications.

This process employs cheap raw materials, the reduction process requires only standard equipment which is already widely used in industry, and it allows easy scale-up and fabrication of complex shapes with inexpensive mold materials. Moreover, this novel approach circumvents the typical problem of the production of steel foams, in which foaming normally can only be achieved at temperatures close to the melting point, which makes control of the metal foam morphology very challenging. Experimentation is currently continuing towards the adaptation of this ceramic foam precursor process to a wide range of steel compositions (e.g., to obtain stainless steel foams by adding other oxides such as Cr_2O_3 and /or NiO).

5.11.15
Zeolite Cellular Structures

Zeolites are widely used in industry as active components in heterogeneous catalysts, adsorbents, and ion exchangers, and novel applications such as sensing/molecular recognition, biomolecule separation, and chromatography are being explored. They are usually produced in the form of micrometer-sized powders, and, if inserted as such in a reactor, they would give rise to very high pressure drop. Conventional processing for making shaped bodies in the millimeter range includes dispersing these components into an inorganic binder matrix and shaping by pressing, extrusion, or droplet formation by peptization, as well as depositing zeolites on a substrate (see also Chapter 5.4 and references therein). Both approaches have advantages, such as good mechanical strength, and disadvantages, such as pore blocking by inorganic binders, diffusion limitations, single-sided mass transport or easy detachment of the microporous layer from the substrate due to thermal stresses. There is widespread interest in developing macroporous bodies constituted of microporous zeolites. Researchers from Sogang University produced self-supporting pure zeolite open-cell foams with high thermal stability of the monolith (no dimensional variation on heating at 550 °C for 1 d), limited mechanical strength (but sufficient for some applications), and high specific surface area (445 $m^2\ g^{-1}$) [45]. This was achieved by dipping a polyurethane sponge in a sol–gel-based solution and placing the system in an autoclave. During zeolite synthesis the polyurethane template slowly decomposed, and the residues were eliminated by calcination at 550 °C in air. With this process very large monoliths (13 cm diameter, 27 cm height) can be fabricated.

An alternative approach, exploited by researchers at the University of Erlangen, is that of using biotemplates for the fabrication of zeolite-based micro/macroporous components. A natural sponge (loofah gourd) was used as a macrostructural template for the development of self-supporting biomimetic zeolite macrostructures with hierarchical micro-/meso-/macroporosity. This was accomplished by two-step hydrothermal synthesis (in situ zeolite seeding and secondary crystal growth) utilizing a dual-template technique (quaternary ammonium salts as template molecules to tailor the micropore network of the zeolite crystals and the loofah sponge for the

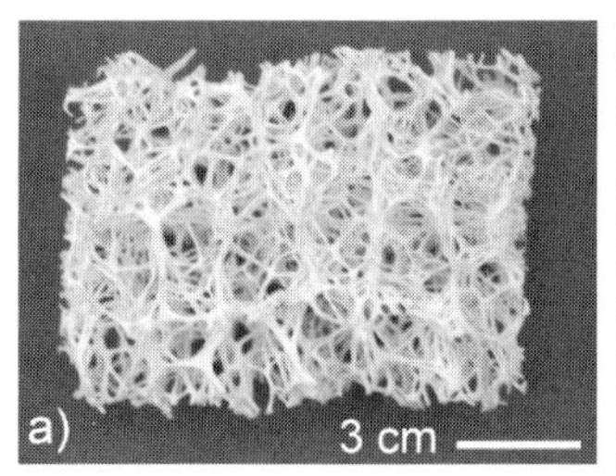

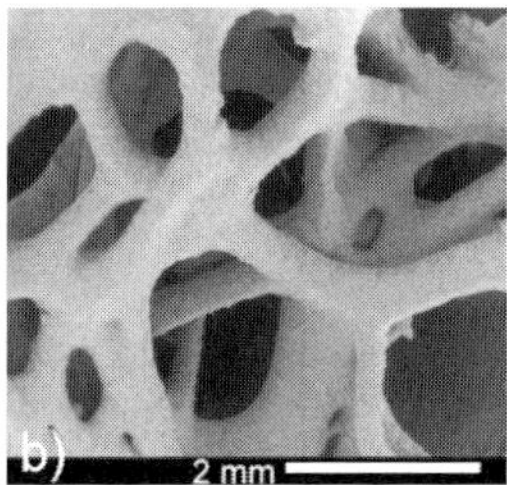

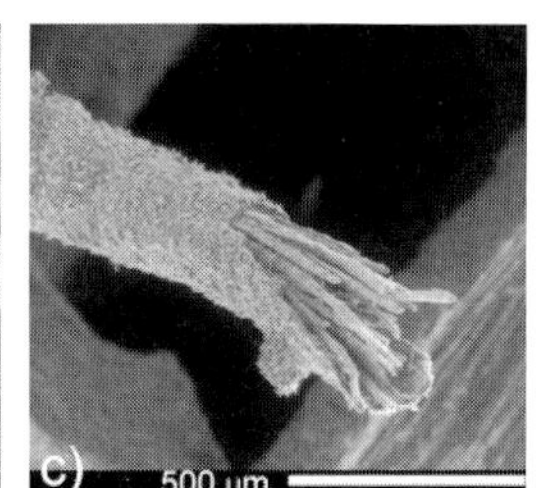

Fig. 14 Silicalite-1 (MFI-type) zeolite replica of the loofah sponge. Images courtesy of A. Zampieri and W. Schwieger (Erlangen University).

micro-/macrostructure development). After calcination, the pure zeolitic replica inherits the micro/macro-architecture of the biotemplate in detail (see Fig. 14b,c), and mechanical stability is maintained. An external scaffold of well-intergrown zeolite crystals supports the entire structure. The struts of this hierarchical porous foam are characterized by a vascular bundle of zeolitic hollow microchannels or microrods, made of crystals tightly packed together, running continuously along the network. Self-supporting structured pure zeolite monoliths with hierarchical porosity (e.g., see Refs. [46–50]) are a promising class of materials that are expected to find applications in adsorption/separation, catalysis, and the development of sensors.

5.11.16 Current Collectors in Solid Oxide Fuel Cells

A key problem in the successful use of solid oxide fuel cells (SOFCs) is the collection of current on the cathode side. Most metallic current collectors developed to date have a coarse design. The sizes of the current contacts are typically on the order of millimeters, and this leads to inhomogeneous current densities. $La_{0.84}Sr_{0.16}Co_{0.02}MnO_3$ perovskite open-cell ceramic foams, with porosities greater than 90 vol % and pore sizes smaller than 700 μm, have been proposed by researchers at ETH Zurich (Switzerland) as air distributors, current collectors, and load-bearing structural parts at the cathode or the anode side of an SOFC [51]. The foams are electrically conductive (100 S cm^{-1} at 634 °C, 60 ppi foam), and the electric current in this design is collected by many more and much smaller contact points. The foams, produced by the replica technique, successfully fulfil three important requirements, namely, high electrical conductivity, sufficient mechanical strength, and good creep resistance at 580–690 °C.

5.11.17 Sound Absorbers

The sound absorbing property of cellular structures has been thoroughly exploited with polymer and metal foams. Ceramic cellular materials expand this capability to environments where heat or corrosion resistance is required. Mufflers for hypersonic engines have been tested in Japan [52], and Volkswagen developed aeroplane mufflers based on reticulated alumina or SiC foams [53], while silicate ceramic foams obtained by direct foaming have been proposed by researchers at RWTH-IKKM (Aachen, Germany) [54]. Open-cell carbon foams have also been demonstrated by Ultramet as acoustic liners in mufflers for general aviation aircraft, but these must be employed below the temperature at which the carbon is oxidized. Open-cell carbon foam coated with CVD SiC has been used as an effective broadband sound absorber in a high-temperature, oxidizing gas environment [55]. Glass foams developed by Pittsburgh Corning Corp. (Pittsburgh, PA) as cellular insulation for building applications, can provide sound attenuation of up to 56 dB.

The mechanisms for sound absorption in a porous material are viscous and thermal dissipation (see Chapter 4.5), and to effectively dissipate acoustic power, the characteristic pore dimension should be on the order of the viscous and thermal penetration depths. Furthermore, to obtain good broadband attenuation, similar to that of existing liner materials, a comparatively low flow resistance is required [52].

5.11.18 Bacteria/Cell Immobilization

Kent Marine Inc. (Acworth, GA) and other suppliers sell open-cell ceramic foams or sintered glass products that are used as hosts for bacteria to purify water in fish tanks through wet/dry biofiltration [56]. Since inert ceramic foams combine a very high (geometric) surface area with a large and open pore structure, which allows water flow with limited resistance, they can be used as very compact filters for efficiently converting toxic fish wastes into nontoxic compounds and reducing the level of organics and algae. Decaying organic material (plant and animal waste) in a water tank is transformed to ammonia, and suitable bacteria grown on the surface of the foam metabolize the ammonia by converting it to nitrite. Another kind of bacteria grown on the foam consume nitrite producing nitrate, which in turn is consumed by plants, which give off oxygen.

Bacteria have been also immobilized on ceramic foams (12–55 ppi, reticulated Al_2O_3) for aerobic wastewater treatment (containing various organics, nontoxic solvents, and various salts), and results show that bacteria degrade organic carbon more efficiently when immobilized in a ceramic foam fixed bed in comparison with nonimmobilized bacteria, resulting in shorter residence times for wastewater and smaller reactors [57].

Immobilizing cell cultures on ceramic foams resulted in a high-density system for the growth of animal cells [58], with a five- to tenfold increase in volumetric cell density per unit surface area over standard culture systems. The foams used (30–100 ppi, reticulated Al_2O_3) were able to withstand repeated autoclaving, and this allowed sterilization and reuse. The best results were obtained for cellular substrates of larger cell size (ca. 550 μm). Less than 5 % of the total number of cells detached from the substrate during the experiment, and the productivity of the immobilized cells (BHK, which produce human transferrin) was similar to that observed in suspension cultures. Such compact systems offer the advantages of significant reduction in serum requirement and the potential for scale-up.

5.11.19 Light Diffusers

Ceramic foams have also been tested in applications where the light signature needs to be reduced or emitted light needs to be diffused. Besides these technical applications, ceramic foams can be used for other purposes. Architect Harry Allen (Harry

Allen and Associates, New York, NY) noticed how the material filters light, and created a line of lamps utilizing the material as the shade (see Fig. 15). After showing the lamps at the International Contemporary Furniture Fair, they were included in the "Mutant Materials in Contemporary Design" show at the Museum of Modern Art, New York, NY, in 1995. The lamps have since been added to the museum's permanent design collection. All the lamps were designed by Harry Allen in 1994; the diffuser is made of alumina foam (reticulated) and the base is made of painted steel.

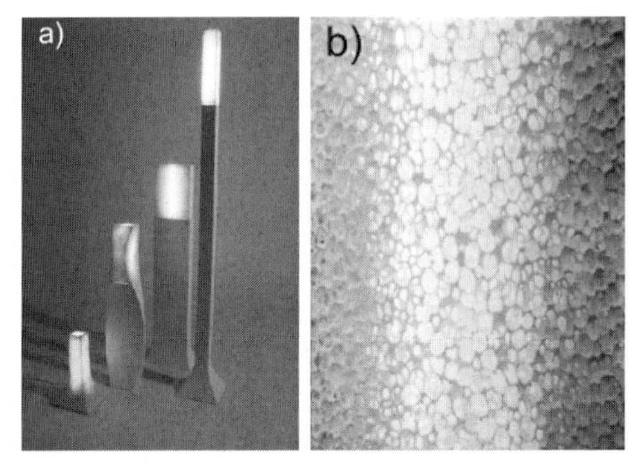

Fig. 15 a) The entire family of lamps (from left to right: Tower 1, Twist, Tower 2, and Tower 3). b) Detail of a ceramic foam lamp. Images courtesy of H. Allen (Harry Allen and Associates).

5.11.20
Summary

Cellular ceramics play a key role in a variety of most diverse innovative applications, besides well-established industrial roles such as filtering liquid metals or particules in gas streams. Novel fabrication processes and developments in established manufacturing technologies allow the production of components with improved properties, a wider range of compositions, and varied morphology, which can all be optimized for a targeted application. Because of this, there are increasing efforts to utilize cellular ceramics in very diverse fields, as engineers are coming to realize more and more that these materials are especially suited for fulfilling some of the unique requirements typical of advanced technological developments.

References

1 Brezny, R., Green, D.J., Mechanical Behavior of Cellular Ceramics in *Materials Science and Technology – A Comprehensive Treatment*, Vol. 11, Cahn, R.W., Haasen, P., Kramer, E.J. (Eds.), VCH, Weinheim, 1994, pp. 463–516.

2 Luyten, J., Mullens, S., Cooymans, J., De Wilde, A.-M., Thijs, I., *Adv. Eng. Mater.* **2003**, 5[10], 715–718.

3 Claussen, N., Le, T., Wu, S. *J. Eur. Ceram. Soc.* **1989**, *5*, 29–35.

4 Yoshihisa, K., Masashi, F., British Patent 2168337, 1986.

5 Adler, J., Apel, H., Ralf, D., Heymer, H., Launer, B., Standke, G., European Patent 1329229, 2003.

6 Zhu, X., Jiang, D., Tan, S., Zhang, Z., *J. Am. Ceram. Soc.* **2001**, *84*, 1654–1656.

7 Pu, X., Liu, X., Qiu, F., Huang, L., *J. Am. Ceram. Soc.* **2004**, *87*, 1392–1394.

8 Ortona, A., Molinari, M., Romelli, L., European Patent 1382590, 2004.

9 Celades, C.R., Velasco Diez, M.-A., European Patent 0369098, 1990.

10 P.R. Crooke, British Patent 2097777, 1982.

11 Adler, J., Teichgräber, M., Standke, G., Jaunich, H., Stöver, H., Stötzel, R., European Patent 0907621, 1999.

12 Fitzgerald, T.J., Michaud, V.J., Mortensen, A., *J. Mater. Sci.* **1995**, *30*, 1037–1045.

13 Tang, F., Fudouzi, H., Sakka, Y., *J. Am. Ceram. Soc.* **2003**, *86*, 2050–2054.

14 Colombo, P., Bernardo, E., Biasetto, L., *J. Am. Ceram. Soc.* **2004**, *87*,152–154.

15 Kim, Y.-W., Kim, S. H., Wang, C., Park C. B., *J. Am. Ceram. Soc.* **2003**, *86*, 2231–2233.

16 Fukasawa, T., Ando, M., Ohji, T., Kanzaki, S., *J. Am. Ceram. Soc.* **2001**, *84*, 230–232.

17 Fukasawa, T., Deng, Z.-Y., Ando, M., Ohji, T., Kanzaki, S., *J. Am. Ceram. Soc.* **2002**, *85*, 2151–2155.

18 Koch, D., Andresen, L., Schmedders, T., Grathwohl, G., *J. Sol-Gel Sci. Technol.* **2003**, *26*, 149–152.

19 Nakahira, A., Nishimura, F., Kato, S., Iwata, M., Takeda., S., *J. Am. Ceram. Soc.*, **2003**, *86*, 1230–1232.

20 Guo-Jun Zhang, G.-J., Yang, J.-F. Ohji, T., *J. Am. Ceram. Soc.*, **2001**, *84*, 1395–1397.

21 Dittrich, R., Tomandl, G., Mangler, M., *Adv. Eng. Mater.*, **2002**, *4*, 487–490.

22 Ding, X.-J., Zhang, J.-Z., Wang, R.-D., Feng, C.-D., *J. Eur. Ceram. Soc.*, **2002**, *22*, 411–414.

23 Salama, K., Lee, D.F., *Supercond. Sci. Technol.* **1994**, *7*, 177–193.

24 Sudhakar Reddy, E., Schmitz, G.J., *Am. Ceram. Soc. Bull.* **2002**, *81*, 35–37.

25 Sudhakar Reddy, E., Hari Babu, N., Shi, Y., Cardwell, D.A., Schmitz, G.J., *Supercond. Sci. Technol.* **2003**, *16*, L40–L43.

26 Sudhakar Reddy, E., Herweg, M., Schmitz, G.J., *Supercond. Sci. Technol.* **2003**, *16*, 608–612.

27 Couts, T.J., *Sol. Energy Mater. Sol. Cells* **2001**, *66*, 441–452.

28 Bitnar, B., Durisch, W., von Roth, F., Palfinger, G., Sigg, H., Grützmacher, D., Gobrecht, J., Meyer, E.-M., Vogt, U., Meyer, A., Heeb, A., Progress in TPV Converters in *Next Generation Photovoltaics: High Efficiency through Full Spectrum Utilization*, Marti, A., Luque, A. (Eds.), Institute of Physics, London, 2004, pp. 223–245.

29 Colombo, P., Bernardo, E., *Compos. Sci. Technol.* **2003**, *63*, 2353–2359.

30 Christiansen, E.L., Crews, J.L., Williamsen, J.E., Robinson, J.H., Nolen, A.M., *Int. J. Impact Eng.* **1995**, *17*, 217–228.

31 Angrilli, F., Pavarin, D., De Cecco, M., Francescani, A., *Acta Astronautica* **2003**, *53*, 185–189.

32 Colombo, P., Arcaro, A., Francescani, A., Pavarin, D., Rondini, D., Debei, S., *Adv. Eng. Mater.* **2003**, *5*, 802–805.

33 Klett, J., Trammell, M., *IEEE Transactions on Device and Materials Reliability*, **2004**, *4* (3), 626–637.

34 Van Voorhees, E.J., Green, D.J., *J. Am. Ceram. Soc.* **1991**, *74*, 2747–2752.

35 Fisher, R.E., Burkland, C.V., Bustamante, W.E., *Ceram. Eng. Sci. Proc.* **1985**, *6*, 806–819.

36 Höfner, T., Zeschky, J., Scheffler, M., Greil, P., *Ceram. Eng. Sci. Proc.*, **2004**, *25*, 559–564.

37 Cromme, P., Scheffler, M., Greil, P., *Adv. Eng. Mater.* **2002**, *4*, 873–877.

38 Adler, J., Standke, G., Si-SiC LigaFill Foams and Related Net-like Structures – New Lightweight and Low-cost Materials for Spaceborne Applications in EUROMAT 99, Vol. 1, *Materials for Transportation Technology*, Winkler, P.-J.

(Ed.), Wiley-VCH, Weinheim, 2000, pp. 270–276.
39 Monthioux, M., Serp, Ph., Flahaut, E., Razafinimanana, M., Laurent, Ch., Peigney, A., Bacsa, W., Broto, J.-M., Introduction to Carbon Nanotubes in *Springer Handbook of Nanotechnology*, Bhushan, B. (Ed.), Springer-Verlag, Berlin, 2004, pp. 39–98.
40 Colomer, J.-F., Stephan, C., Lefrant, S., Tendeloo, G.V., Willems I., Konya, Z., Fonseca, A., Laurent, Ch., Nagy, J.B., *Chem. Phys. Lett.* **2000**, *317*, 83–90.
41 Peigney, A., Laurent, Ch., Dobigeon, F., Rousset, A., *J. Mater. Res.* **1997**, *12*, 613–615.
42 Rul, S., Laurent, Ch., Peigney, A., Rousset, A. *J. Eur. Ceram. Soc.* **2003**, *23*, 1233–1241.
43 Verdooren, A., Chan, H.M., Grenestedt, J.L., Harmer, M.P., Caram, H.S., Production of Metallic Foams from Ceramic Foam Precursors in Cellular Metals: Manufacture, Properties, Applications, Proceedings International Conference on Cellular Metals and Metal Foaming, Banhart, J., Fleck, N., Mortensen, A. (Eds.), Verlag MIT Publishing, Berlin, 2003, pp. 243–248.
44 Motoki, H., US Patent 4084980, 1978.
45 Lee, Y.-J., Lee, J.S., Park, Y.S., Yoon, K.B., *Adv. Mater.* **2001**, *13*, 1259–1263.
46 Dong, A., Wang, Y., Tang, Y., Ren, N., Zhang, Y., Yue, Y., Gao, Z., *Adv. Mater.* **2002**, *14*, 926–929.
47 Valtchev, V., Smaihi, M., Faust, A.C., Vidal, L., *Chem. Mater.* **2004**, *16*, 1350–1355.
48 Anderson, M. W., Holmes, S. M., Hanif, N., Cundy, C., *Angew. Chem. Int. Ed.* **2000**, *39*, 2707–2710.
49 Zhang, B., Davis, S. A., Mann, S., *Chem. Mater.* **2002**, *14*, 1369–1375.
50 Tao, Y., Kanoh, H., Kaneko, K., *J. Am. Chem. Soc.* **2003**, *125*, 6044–6045.
51 Will, J., Gauckler, L.J., Ceramic Foams As Current Collectors in Solid Oxide Fuel Cells (SOFC): Electrical Conductivity And Mechanical Behaviour in Proceedings of the Fifth International Symposium on Solid Oxide Fuel Cells (SOFC-V), Stimming, U., Singhal, S.C., Tagawa, H., Lehnert, W. (Eds.), PV 97-40, The Electrochemical Society, Pennington, NJ, 1997, pp. 757–764.
52 Nakamura, Y., Beck, J., Oishi, T., Application of Porous Ceramics as Engine Ejector Liner in 3rd AIAA/CEAS Aeroacoustics Conference: a Collection of Technical Papers, American Institute of Aeronautics and Astronautics, Reston, VA, 1997, pp. 858–864.
53 Siebels, J, Weber, W., *VDI Ber.* **1995**, *1235*, 221–236.
54 Drobietz, R., *VDI Ber.* **1999**, *1472*, 155–172.
55 Stankiewicz, E.P., Heng, S., Noise Reduction System for General Aviation Aircraft, Phase II, Final Report (ULT/TR-98-6989), Contract NAS3-27633, Ultramet for NASA Lewis Research Center, Cleveland, OH, June 1998.
56 Schenck, R.C., US Patent 6245236, 2001.
57 Mizrah, T., Gabathuler, J.-P., Gauckler, L., Baiker, A., Padeste, L., Meyer, H.P., SELEE: Ceramic Foam for Pollution Reduction in Proceedings EnviCeram '88, Cologne, Germany, December 7–9, 1988, cfi/Beihefte, Köln, 1988, pp. 143–170.
58 Lee, D.W., Grace, J.R., Chow, B.K.C., MacGillivray, R.T.A., Kilburn, D.G., *Cytotechnology* **1991**, *5*, 233–241.

Concluding Remarks

Michael Scheffler and Paolo Colombo

We try here to briefly summarize and give an outlook concerning the future of this fascinating field of cellular ceramics, based on our knowledge of the published literature and several discussions that we had in the period between the conception of the book (summer 2003) and its release (spring 2005) with specialists at the forefront of research and development. As in many technical evolutions the need for new applications will affect processing technologies. Novel processing technologies lead to new structures and advanced properties and this in turn stimulates conceiving novel, broader applications. A deeper understanding of the structure–properties relationship with the aim of producing components with improved reliability can also be achieved by employing novel characterization and modeling tools. In this context the following developments can be expected in the coming years:

Processing and applications of cellular ceramics – foams, honeycombs, fibers, hollow-sphere assemblages, 3D periodic structures – will be a major topic. Main efforts will be:

- *Improvement of properties.* The main applications will be found in the automotive industry for exhaust gas cleaning applications, driven, for instance, by the EURO-V or the U.S. Federal Emission Standard effective by 2010. Suitable materials will allow the fabrication of lightweight components with integrated multifunctionality (e.g., high specific surface area for catalyst support, high relative strength and corrosion resistance to withstand the thermal cycles and severe environment, and electrical conductivity for soot elimination by direct heating).
- *Upscaling of monoliths.* The petrochemical industry is expected to invest a great amount of resources in the coming years to update old refineries. Environmental restrictions already require a more efficient use of fossil energy resources. To meet this goal, undesired side reactions in chemical processes must be minimized and efficiency in energy conversion increased. Many technical papers point out the superiority of cellular ceramic monoliths (especially ceramic foams) in some chemical processes compared to common solutions. To make them usable for this task, upscaling in size and improving reliability under service conditions are essential.

Cellular Ceramics: Structure, Manufacturing, Properties and Applications.
Michael Scheffler, Paolo Colombo (Eds.)

ISBN: 3-527-31320-6

- *Foams and honeycombs in environmental technology.* While honeycomb-supported catalysts have already become one of the most widespread technical solutions for exhaust gas cleaning (also in stationary applications), ceramic foams and fiber mats are being developed and tested as well for this purpose. In future there will be more applications for cellular ceramics in environmentally benign processes.
- *Bottom-up manufacturing.* Besides honeycombs and foams produced by well-established industrial processes, periodic structures and tailor-made structures fabricated by bottom-up processing technologies (inkjet printing, robocasting, fused deposition) will become a major topic in the ceramic and processing area, because they allow components to be fabricated with a very well controlled morphology and their dimensions to be reduced to the nanoscale. Main applications will be found in exhaust gas cleaning (catalyst supports), tissue engineering (artificial bone grafting), and porous preform manufacturing for single parts or small series.
- *Monoliths with hierarchical porosity.* In heterogeneously catalyzed processes macroporous monoliths with microporous modified surfaces promise to combine excellent fluid dynamics and reaction-adjusted pore diffusion. Modern technologies of surface modification like supported crystallization might allow the tailored production of a broad variety of hierarchically arranged porous materials of all classes of structure (honeycombs and related structures, foams, fiber mats, and periodic structures). The present combination of macro/microporosity will be extended to macro/mesoporosity, and meso/microporosity systems.
- *Lightweight technology.* An increasing demand for lightweight components for more efficient fuel consumption has been recognized in both the automotive and aerospace industries. The need for ultrahigh-temperature materials in aerospace applications will affect the development of novel lightweight/highly temperature stable materials and structure combinations. Additional properties, such as impact absorption and sound damping, will lead to more advanced applications for cellular ceramics in these fields.
- *UV-transparent glass foams and honeycombs.* Cellular ceramics made of UV-transparent materials (silica glass) will play an increasing role in environmental cleaning processes. They provide a greater surface area in comparison to glass tubing when used as support/substrate, and thus increase the efficiency of photocatalytic processes.

Morphology characterisation and modeling of cellular (ceramic) structures will help to understand the structure formation of random cellular materials and lead to novel manufacturing processes resulting in increased structural homogeneity. One of the most promising tools for morphology characterization is X-ray microcomputer tomography. The resolution of common equipment currently allows measurements down to the micrometer range, but the physical limit has not been reached yet. In fact, high-energy synchrotron radiation has already started to be applied to the characterization of microcellular foams. The development of advanced modeling tools will allow developers to confidently incorporate cellular ceramics into their design, taking full advantage of their positive characteristics, while at the same time minimizing their weaknesses.

We look forward to the coming years together with the readers and authors of this book, and hope that this work will contribute to developing further worldwide interest in the important field of cellular materials.

Index

Cellular Ceramics: Structure, Manufacturing, Properties and Applications.
Michael Scheffler, Paolo Colombo (Eds.)

ISBN: 3-527-31320-6

c

d

t